HANDBUCH DER PHYSIK

HERAUSGEGEBEN VON

S. FLÜGGE

BAND LIII

ASTROPHYSIK IV: STERNSYSTEME

MIT 189 FIGUREN

SPRINGER-VERLAG
BERLIN · GÖTTINGEN · HEIDELBERG
1959

ENCYCLOPEDIA OF PHYSICS

EDITED BY

S. FLÜGGE

VOLUME LIII

ASTROPHYSICS IV: STELLAR SYSTEMS

WITH 189 FIGURES

SPRINGER-VERLAG
BERLIN · GÖTTINGEN · HEIDELBERG
1959

ISBN-13: 978-3-642-45934-4 e-ISBN-13: 978-3-642-45932-0
DOI: 10.1007/978-3-642-45932-0

Softcover reprint of the hardcover 1st edition 1959

Inhaltsverzeichnis.

Kinematical Basis of Galactic Dynamics.

By

Frank K. Edmondson.

With 12 Figures.

1. Introduction and observational data. A correct kinematical description of the galaxy is a necessary first step before it is profitable to develop detailed dynamical theories. For example, if the motions in the galaxy are inward along the spiral arms the resulting dynamical theory will differ from one that explains circular motions.

The data of observation from which the kinematical description is obtained are, of course, well known to astronomers. However, scientists in other fields are frequently puzzled by the astronomical terminology and units. Hence, a brief survey of the data of observation will first be given. This will be followed by a description of the general features of observed stellar motions, and a discussion of basic kinematical considerations.

The astronomical data relevant to the present discussion are: α) Positions, β) Proper motions, γ) Radial velocities, δ) Parallaxes, ε) Magnitudes (and colors), ζ) Spectral types (and luminosity classes). A good general survey has been given by Irwin[1].

α) Positions. Astronomical positions are given in terms of spherical co-ordinates. Fundamental observations are made in the *equatorial system*, but it is more convenient to discuss stellar motions in *galactic co-ordinates*. The equatorial co-ordinates are *right ascension* (α) and *declination* (δ). The galactic co-ordinates are *galactic longitude* (l or λ) and *galactic latitude* (b or β). Tables for transforming from one system to the other have been published by Ohlsson[2]. Graphs[3] and nomograms[4] are also available when approximate values will suffice.

Fundamental positions must be measured with a meridian circle and an accurate clock (i.e., with a constant rate). After a sufficient number of fundamental positions have been established, it is possible to interpolate others either by meridian observations or by photography. The Earth's axis precesses in a period of about 26000 years. Hence, fundamental positions measured at different epochs will disagree owing to the motion of the co-ordinate system, the motions of the stars, and systematic errors in the observations.

β) Proper motions. Proper motions, denoted by the symbol μ, are angular displacements on the celestial sphere caused by the motions of the stars relative to the Sun. The largest known proper motion is 10″ per year, and only 329 stars have proper motions larger than 1″ per year. The average for the naked-eye stars is approximately 0″.1 per year.

Meridian observations at different epochs yield "absolute" proper motions only if the effects of precession and systematic errors can be completely removed. Photographic observations yield "relative" motions, unless they are calibrated by meridian observations. The Lick Observatory[5] is now conducting a photographic program to observe "absolute" proper motions by referring them to the "stationary" framework of very distant galaxies with nearly stellar images. Lack of a similar instrument and program in the southern hemisphere is serious and should be corrected as soon as possible.

γ) Radial velocities. Stellar radial velocities are usually measured using slit spectra, with a comparison spectrum to define the unshifted wavelength scale. Numerous attempts have

---

[1] J. B. Irwin: Vistas in Astronomy, vol. 3 (in press).
[2] J. Ohlsson: Lund Obs. Annals 1932, Nr. 3.
[3] J. A. Pearce and S. N. Hill: Dominion Astrophys. Obs. Publ. **4**, 49 (1927).
[4] H. Strassl: Veröff. Univ.-Sternw. Bonn 1949, Nr. 36.
[5] W. H. Wright: Proc. Amer. Phil. Soc. **94**, 1 (1950).

been made in the past to measure radial velocities on objective prism plates, the most successful being the recent work by FEHRENBACH[1]. Radial velocities of the clouds of neutral hydrogen can be measured from the shift of the 21 cm line[2].

δ) Parallaxes. Direct, or trigonometric parallaxes are obtained by observing the displacements of stars at 6 month intervals. Observations must extend over several years in order to separate the parallax from the proper motion.

Indirect determinations of parallax can be made using formula (1.4), below, if the absolute magnitude M can be estimated. For example, spectroscopic parallaxes are found by using spectroscopic features that are correlated with M[3,4,5], and parallaxes of cepheid variables are found using the period-luminosity relation[6]. Dynamical parallaxes of visual binary stars are found by forcing agreement with the mass-luminosity relation[7].

Statistical parallaxes may be found by combining proper motion and radial velocity data (Sect. 2).

The distance of a star in parsecs is the reciprocal of the parallax in seconds of arc:

$$d\,(\text{parsecs}) = \frac{1}{p''}\,. \tag{1.1}$$

ε) Magnitudes. The scale of stellar magnitudes goes back to HIPPARCHUS and PTOLEMY, who divided the naked-eye stars into six classes with the brightest stars being called first magnitude. HERSCHEL showed that the intensity ratio of a first magnitude star to a sixth magnitude star was nearly 100:1, and POGSON used this as a precise definition of the magnitude scale. From FECHNER's law, the ratio for one magnitude should be $\sqrt[5]{100} = 2.512\ldots$ Hence, the ratio of intensities of two stars is given by

$$\frac{I_1}{I_2} = (2.512\ldots)^{m_2 - m_1} \tag{1.2}$$

or in general

$$\operatorname{Log} I = \text{const} - 0.4\,m\,. \tag{1.3}$$

The numerical value of the constant is in principle defined by the stars of the *north polar sequence*[8] and other secondary standards.

Stellar magnitudes depend on the wavelength range of the receiver. Visual magnitudes were used first, followed by photographic and photovisual magnitudes derived from blue and yellow sensitive photographic plates. Multicolor photometry has been introduced following the development of sensitive photoelectric cells, and the U, B, V system[9,10] is replacing the older pg and pv magnitudes. The older definition of *color index* was pg-pv. Now one must specify which index is being used (e.g., B-V).

The *absolute magnitude* of a star is defined as the value the apparent magnitude would have if the star were placed at a distance of 10 parsecs. It is given by

$$M = m + 5 + 5 \operatorname{Log} p - A \tag{1.4}$$

where p is the parallax in seconds of arc, and A is the extinction caused by interstellar matter.

ζ) Spectral types[11]. The first classification of stellar spectra was by SECCHI[12] who studied several hundred stars down to the sixth magnitude and divided them into four types, denoted by I to IV. The Harvard system[13] has the following types

$$O\,B\,A\,F\,G\,K\,M\,|\,R\,N\,|\,S$$

---

[1] C. FEHRENBACH: Ann. d'Astrophysique **10**, 306 (1947).

[2] H.C. VAN DE HULST, C.A. MULLER and J.H. OORT: Bull. astronom. Inst. Netherl. **12**, 117 (1954).

[3] W.S. ADAMS and A. KOHLSCHÜTTER: Astrophys. Journ. **40**, 385 (1914).

[4] B. LINDBLAD: Upsala Obs. Medd. Nr. 28 (1925).

[5] A.N. VYSSOTSKY: Astrophys. Journ. **104**, 239 (1946).

[6] H.C. ARP: Astronom. J. **63**, 45 (1958).

[7] H.N. RUSSELL and C.E. MOORE: The Masses of the Stars. Chicago 1940.

[8] Trans. Int. Astronom. Un. **1**, 71–75 (1922).

[9] H.L. JOHNSON and W.W. MORGAN: Astrophys. Journ. **117**, 313 (1953).

[10] H.L. JOHNSON: Ann. d'Astrophysique **18**, 292 (1955).

[11] For details see Vol. L of this Encyclopedia.

[12] A. SECCHI: Astronom. Nachr. **59**, 193 (1863).

[13] A.C. MAURY: Harvard Ann. **28** (1897).

with decimal subdivisions between classes. The classes *O B A F G K M* are arranged in a temperature sequence. The stars of classes *R*, *N*, and *S* are cool stars and are quite rare. More than 400000 stars have been classified at Harvard[1].

The MK system[2,3] is a two-dimensional classification with coordinates, spectral type and luminosity (classes I to V). Spectroscopic parallaxes may be obtained from the luminosity classes after these have been calibrated by direct parallax measurements, statistical parallaxes, and studies of relative magnitudes in star clusters.

I. General features of observed stellar motions.

2. Solar motion. It is evident that observed proper motions and radial velocities are caused by changes of position relative to the Sun, and that part of these changes may be due to the motion of the Sun. Thus, Sir WILLIAM HERSCHEL[4] was the first to demonstrate that the Sun is moving toward a point in the sky not far from the bright star Vega. HERSCHEL and those who followed him defined the standard of rest as the "average of the stars", but this refers to the geometrical center and not to the center of mass since masses are not known for individual stars.

During the next century and a quarter the determination of the solar motion from proper motions was one of the major problems of stellar astronomy, and many eminent astronomers (GAUSS, ARGELANDER, AIRY *et al.*) made contributions. A partial list of solar motion determinations[5] gives 43 investigations by 16 investigators during the period 1783 to 1908.

Trustworthy radial velocities were not available in sufficient numbers for solar motion determinations until the beginning of the present century. W.W. CAMPBELL inaugurated the great Lick program in 1896, and had measured radial velocities of more than 300 stars by the end of 1900. His first determination of the solar apex[6] was based on 280 stars, known spectroscopic binaries having been excluded. His second determination in 1910 was based on 1047 stars[7]. Finally, in 1928 CAMPBELL and MOORE published their analysis of the radial velocities of 2148 stars[8], giving the results:

$$\begin{aligned} \alpha_0 &= 270^\circ\!.6, \\ \delta_0 &= +29^\circ\!.2, \\ V_0 &= 19.7 \text{ km/sec}, \\ K &= +1.3 \text{ km/sec}. \end{aligned}$$

Later determinations based on more stars are in substantial agreement with these values. The *K*-term is the average radial velocity after correction for solar motion. It was introduced by CAMPBELL to allow for a possible expansion of the stellar system or errors in the adopted wavelengths of the lines that were measured.

If different groups of stars have different mean motions, the solar motion will depend on the stars or objects to which it is referred. CAMPBELL and MOORE

---

[1] A.J. CANNON: Harvard Ann. **91**–**99** (1918–1924); **100** (1925–1936); **105**, 1 (1937); **112** (1949) with M.W. MAYALL.

[2] W.W. MORGAN, P.C. KEENAN and E. KELLMAN: An Atlas of Stellar Spectra. Chicago 1943.

[3] P.C. KEENAN and W.W. MORGAN: In: Astrophysics, edit. J.A. HYNEK, pp. 12–28. New York 1951.

[4] W. HERSCHEL: Phil. Trans. **73**, 247 (1783).

[5] W.W. CAMPBELL: Stellar Motions, p. 142–143. New Haven and London 1913.

[6] W.W. CAMPBELL: Astrophys. Journ. **13**, 80 (1901).

[7] W.W. CAMPBELL: Stellar Motions, p. 189. New Haven and London 1913.

[8] W.W. CAMPBELL and J.H. MOORE: Lick Obs. Publ. **16** (1928).

investigated the dependence on spectral type, and this has been done more recently by Nordström[1]. The solar motion relative to the globular clusters[2], and high velocity stars[3,4] differs considerably from the value referred to bright stars.

Proper motions have played an increasing role in recent years, following publication of the Boss Catalogue[5] and the McCormick proper motion surveys[6,7].

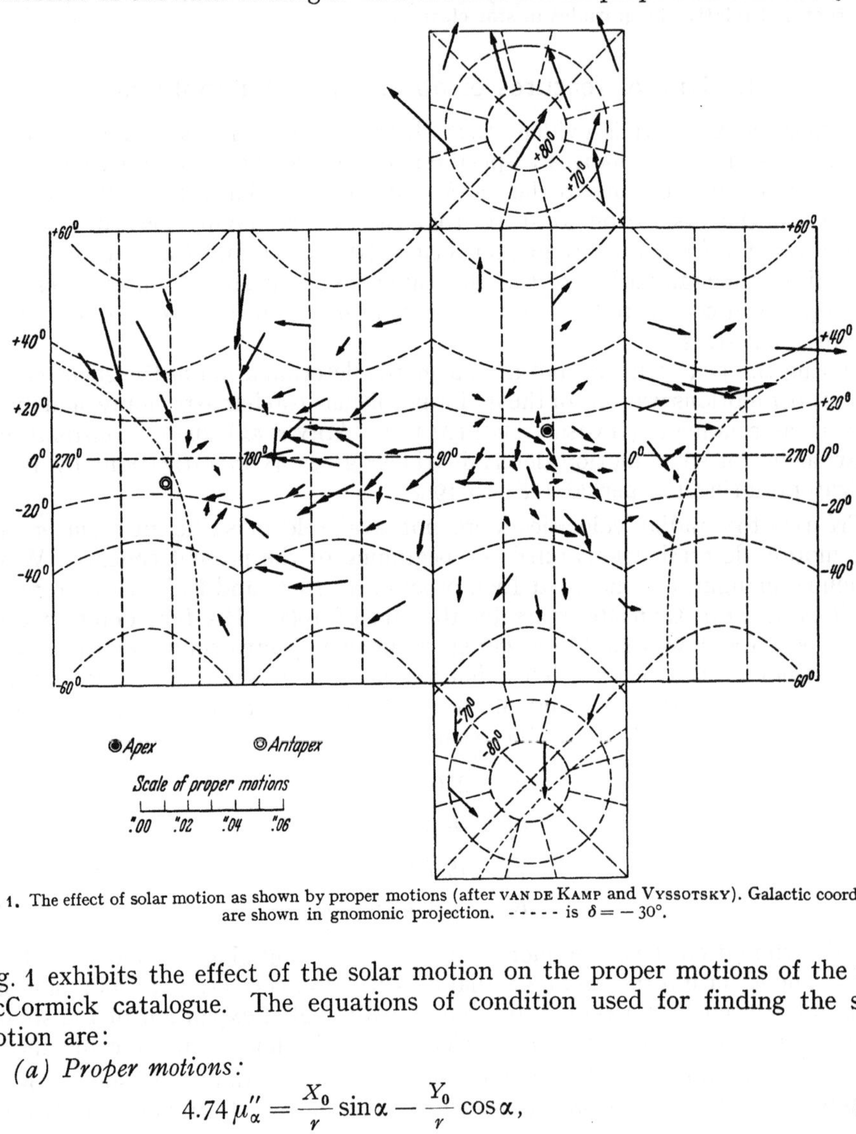

Fig. 1. The effect of solar motion as shown by proper motions (after van de Kamp and Vyssotsky). Galactic coordinates are shown in gnomonic projection. ----- is $\delta = -30°$.

Fig. 1 exhibits the effect of the solar motion on the proper motions of the first McCormick catalogue. The equations of condition used for finding the solar motion are:

(a) Proper motions:

$$4.74\,\mu''_\alpha = \frac{X_0}{r}\sin\alpha - \frac{Y_0}{r}\cos\alpha, \tag{2.1}$$

$$4.74\,\mu''_\delta = \frac{X_0}{r}\cos\alpha\sin\delta + \frac{Y_0}{r}\sin\alpha\sin\delta - \frac{Z_0}{r}\cos\delta. \tag{2.2}$$

---

[1] H. Nordström: Lund Obs. Medd., Ser. II 1936, Nr. 79.

[2] N.U. Mayall: Astrophys. Journ. **104**, 290 (1946).

[3] G. Miczaika: Astronom. Nachr. **270**, 249 (1940).

[4] W. Fricke: Astronom. Nachr. **277**, 241 (1949).

[5] B. Boss *et al.*: General Catalogue of 33,342 Stars for the epoch 1950, Carnegie Inst. Publ. 468 (1937).

[6] P. van de Kamp and A.N. Vyssotsky: McCormick Obs. Publ. **7** (1937).

[7] A.N. Vyssotsky and E.T.R. Williams: McCormick Obs. Publ. **10** (1948).

(*b*) *Radial velocities:*

$$\varrho = -X_0 \cos\alpha \cos\delta - Y_0 \sin\alpha \cos\delta - Z_0 \sin\delta + K, \tag{2.3}$$

where X_0, Y_0, Z_0, are the components of the solar motion in km/sec, K is CAMPBELL's K-term, and r is the distance in parsecs. 4.74 is the conversion factor required if μ is measured in seconds of arc per year, X_0 etc., in km/sec, and r in parsecs.

The "standard" solar apex which has been widely used has co-ordinates

$$\alpha_0 = 18^{\mathrm{h}} = 270^\circ,$$
$$\delta_0 = +30^\circ,$$
$$V_0 = 20 \text{ km/sec}.$$

Its significance will be discussed further in Sect. 8.

Statistical mean parallaxes can be derived from secular parallaxes. The mean secular parallax is computed from

$$\overline{\left(\frac{h}{r}\right)} = \frac{\overline{(v \sin\varDelta)}}{\overline{(\sin^2\varDelta)}} \tag{2.4}$$

where $\varDelta$ is the distance from the solar apex and v is the component of the proper motion directed away from the apex. The mean secular parallax is related to the mean annual parallax by the equation

$$\overline{\left(\frac{h}{r}\right)} = \overline{\left(\frac{V_0}{4.74\, r}\right)} = \frac{V_0}{4.74}\,\bar{\pi}_v \tag{2.5}$$

V_0 is known from radial velocities, Eq. (2.3).

3. Velocity dispersion. The peculiar motions of the stars were assumed to be at random in all of the early investigations of the solar motion. However, KOBOLD[1] found evidence that the motions of the stars were not at random, and ten years later KAPTEYN[2] announced his discovery of the *preferential motion* of the stars which he explained as being caused by "two star streams". KARL SCHWARZSCHILD[3] introduced an alternative "ellipsoidal hypothesis" to explain this preferential motion. The ellipsoidal hypothesis is more meaningful from the physical point of view, and represents observations almost as well as the two-streams hypothesis even though it has one less adjustable constant. A detailed treatment of the formulae used in these hypotheses has been given by EDDINGTON[4] and SMART[5].

SCHWARZSCHILD's frequency function is of the form

$$\exp\left[-K^2 U^2 - H^2 (V^2 + W^2)\right], \tag{3.1}$$

where U, V, and W are the components of the linear velocity of a star, the U-axis being the axis of preferential motion. Symmetry around this axis is assumed.

---

[1] H. KOBOLD: Astronom. Nachr. **137**, 393 (1895).
[2] J.C. KAPTEYN: Congress of Arts and Sciences, St. Louis **4**, 413 (1904).
[3] K. SCHWARZSCHILD: Göttinger Nachr. **1908**, 191.
[4] A. S. EDDINGTON: Stellar Movements and the Structure of the Universe. London 1914.
[5] W.M. SMART: Stellar Dynamics. Cambridge 1938.

The U- and V-axes lie in the galactic plane, and the U-axis is observed to point approximately toward the galactic center.

The *asymmetry* of stellar motions was discovered by BOSS[1], ADAMS and JOY[2], and discussed in detail by STROMBERG[3,4] and OORT[5]. The asymmetry is entirely in the V-axis, the shorter of the axes in the galactic plane. It shows up as an avoidance of motions toward the hemisphere centered on Cygnus ($l=57°$) by the high velocity stars (i.e., with peculiar radial velocities larger than 63 km/sec), as shown in Fig. 2.

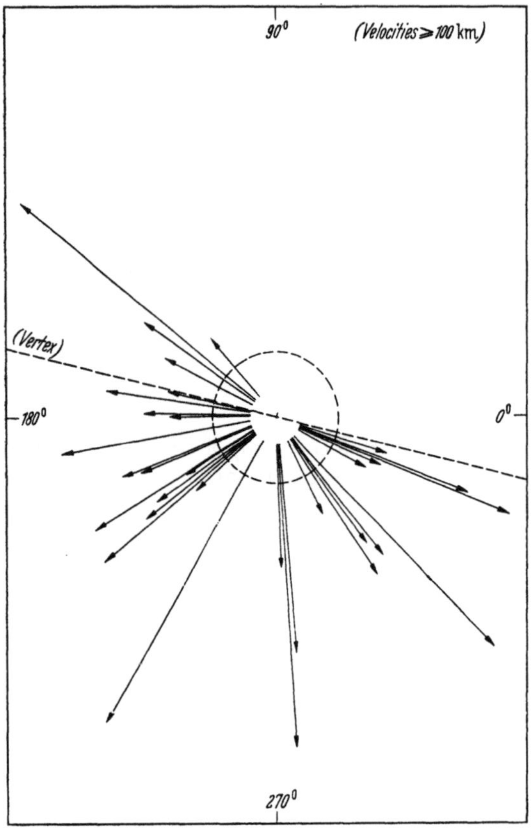

Fig. 2. The asymmetry of the motions of high velocity stars (after OORT). The dotted circle has a radius of 63 km/sec.

The asymmetry leads to a correlation between solar motion and velocity dispersion[6,3,4] as shown in Fig. 3.

4. Galactic rotation. CHARLIER[7,8] was the first to suggest that proper motions indicate a rotation of the galactic system around a distant center. FOTHERINGHAM[9] and SCHILT[10] came to a similar conclusion. The mean value of the four determinations is

$$-0''.0037 \pm 0''.0013 \text{ (m.e.)/yr}.$$

This corresponds to a rotation period of about 300 million years.

LINDBLAD[11] suggested galactic rotation as a dynamical explanation of the observed asymmetry of stellar motions. He postulated that the galaxy was composed of a number of subsystems with different degrees of flattening, different velocity dispersions, and different rotational velocities.

OORT[12–14] identified LINDBLAD's most highly flattened subsystems with stars moving in circular orbits in the galactic plane. The less flattened systems were

[1] B. BOSS: Pop. Astronom. **26**, 686 (1918).
[2] W.S. ADAMS and A.H. JOY: Astrophys. Journ. **49**, 179 (1919).
[3] G. STROMBERG: Astrophys. Journ. **59**, 228 (1924).
[4] G. STROMBERG: Astrophys. Journ. **61**, 363 (1925).
[5] J.H. OORT: Groningen Publ. 1926, No. 40.
[6] See footnote 3, p. 4.
[7] C.V.L. CHARLIER: Lund Obs. Medd., Ser. II 1913, Nr. 9, 78.
[8] C.V.L. CHARLIER: Mem. Univ. Calif. **7**, 32 (1926).
[9] J.K. FOTHERINGHAM: Monthly Notices Roy. Astronom. Soc. London **86**, 414 (1926).
[10] J. SCHILT: Astrophys. Journ. **64**, 161 (1926).
[11] B. LINDBLAD: Upsala Obs. Medd. **1925**, Nr. 3.
[12] J.H. OORT: Bull. astronom. Inst. Netherl. **3**, 275 (1927).
[13] J.H. OORT: Bull. astronom. Inst. Netherl. **4**, 79 (1927).
[14] J.H. OORT: Bull. astronom. Inst. Netherl. **4**, 91 (1927).

supposed to be stars moving in elliptical orbits inclined to the galactic plane. The absence of velocities larger than 63 km/sec directed toward galactic longitude 57° is explained by saying that 63 km/sec is the difference between the circular velocity and the velocity of escape. The solar motion is explained as the deviation of the Sun's motion from circular motion, and star streaming (the velocity ellipsoid) is explained as the result of small deviations from circular orbits.

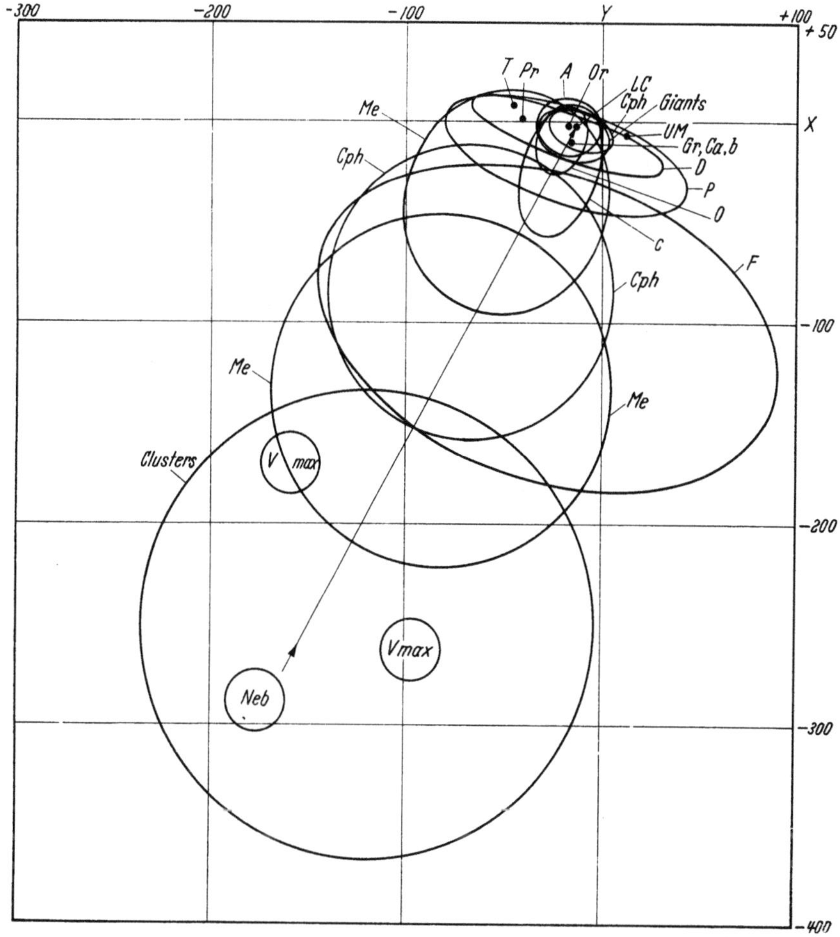

Fig. 3. The correlation between solar motion and velocity dispersion (after STROMBERG).

Fig. 4 is based on work by HAAS[1] and BOTTLINGER[2], and Fig. 5 shows observed space velocities according to OORT[3]. The relationships between relative motions and orbit eccentricities and dimensions may be seen by comparing these two diagrams.

OORT[4] also derived simple first-order expressions for the effect of galactic rotation on radial velocities and proper motions. Fig. 6 shows the geometrical

---

[1] J. HAAS: Astronom. Nachr. **239**, 97 (1930).
[2] K.F. BOTTLINGER: Veröff. Sternw. Babelsberg **10**, Nr. 2 (1937).
[3] J.H. OORT: Bull. astronom. Inst. Netherl. **4**, 269 (1928).
[4] See footnote 12, p. 6.

relationships. The radial velocity due to galactic rotation is

$$V_G = V \sin(\vartheta + l - l_0) - V_0 \sin(l - l_0), \tag{4.1}$$

$$= R_0 \left(\frac{V}{R} - \frac{V_0}{R_0}\right) \sin(l - l_0), \tag{4.2}$$

$$\sim -2A\,(R - R_0) \sin(l - l_0), \tag{4.3}$$

$$\sim r\,A \sin 2\,(l - l_0) \tag{4.4}$$

where

$$A = \frac{1}{2}\left(\frac{V}{R} - \frac{dV}{dR}\right)_0 = -\frac{1}{2} R_0 \left(\frac{d\omega}{dR}\right)_0 .$$

The proper motion in galactic longitude is

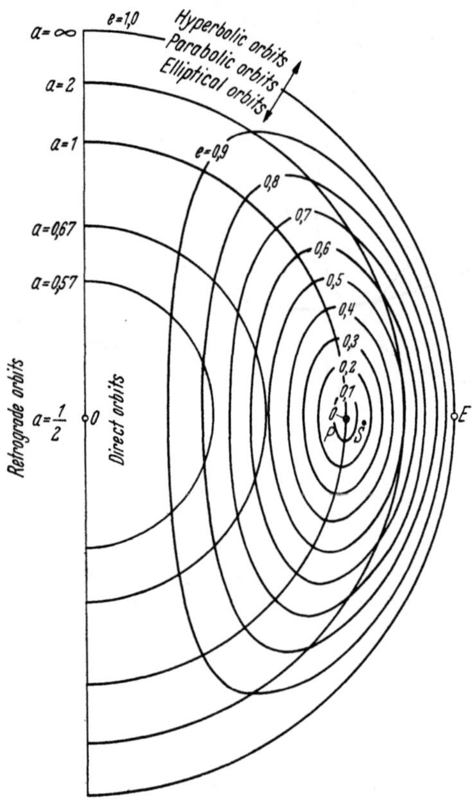

Fig. 4. Relationship between velocity, semi-major axis, and eccentricity of orbit in an inverse square field (after BOTTLINGER).

$$\mu_l'' \sim \frac{A}{4.74} \cos 2\,(l - l_0) + \frac{B}{4.74} \tag{4.5}$$

where

$$B = A - \frac{V_0}{R_0} .$$

One should note that A and B derived from proper motions are independent of the distances of the stars; however, they may be seriously influenced by systematic errors in the proper motions[1]. Derivation of A from radial velocities requires an accurate knowledge of the distance, since one obtains $\bar{r}\,A$ from the observations.

HAYFORD[2] carried the expansion for these formulae to the 5th order, obtaining:

$$\left.\begin{aligned} V_G = {} & A_1 \sin(l - l_0) + A_2 \sin 2\,(l - l_0) + \\ & + A_3 \sin 3\,(l - l_0) + A_4 \sin 4\,(l - l_0) + \\ & + A_5 \sin 5\,(l - l_0) + A_6 \sin 6\,(l - l_0) \end{aligned}\right\} \tag{4.6}$$

and

$$\left.\begin{aligned} \mu_l'' = {} & \frac{1}{4.74\,R_0} \left[B_0 + B_1 \cos(l - l_0) + \right. \\ & \left. + B_2 \cos 2\,(l - l_0) + \cdots\right]. \end{aligned}\right\} \tag{4.7}$$

The coefficients A_i and B_i are given as functions of the angular velocity and its derivatives.

Following OORT's discovery of the "double wave", Eqs. (4.4) and (4.5), PLASKETT and PEARCE[3] undertook an extensive observing program on distant O and B stars, in order to increase the amplitude of the observed double wave. Their result was

$$A = (15.5 \pm 0.9)\ \text{km/sec/kpc},$$

$$l_0 = 324.^\circ4 \pm 2.^\circ4$$

---

[1] See footnote 7, p. 4.

[2] P. HAYFORD: Lick Obs. Bull. **16**, 53 (1932).

[3] J.S. PLASKETT and J.A. PEARCE: Dominion Astrophys. Obs. Publ. **5**, 294 (1936).

based on radial velocities of 849 stars, and

$$B = (-12.0 \pm 2.7) \text{ km/sec/kpc},$$

based on proper motions of 717 stars of spectral types *O* to *B*7. The best values of these constants based on a combination of numerous determinations according

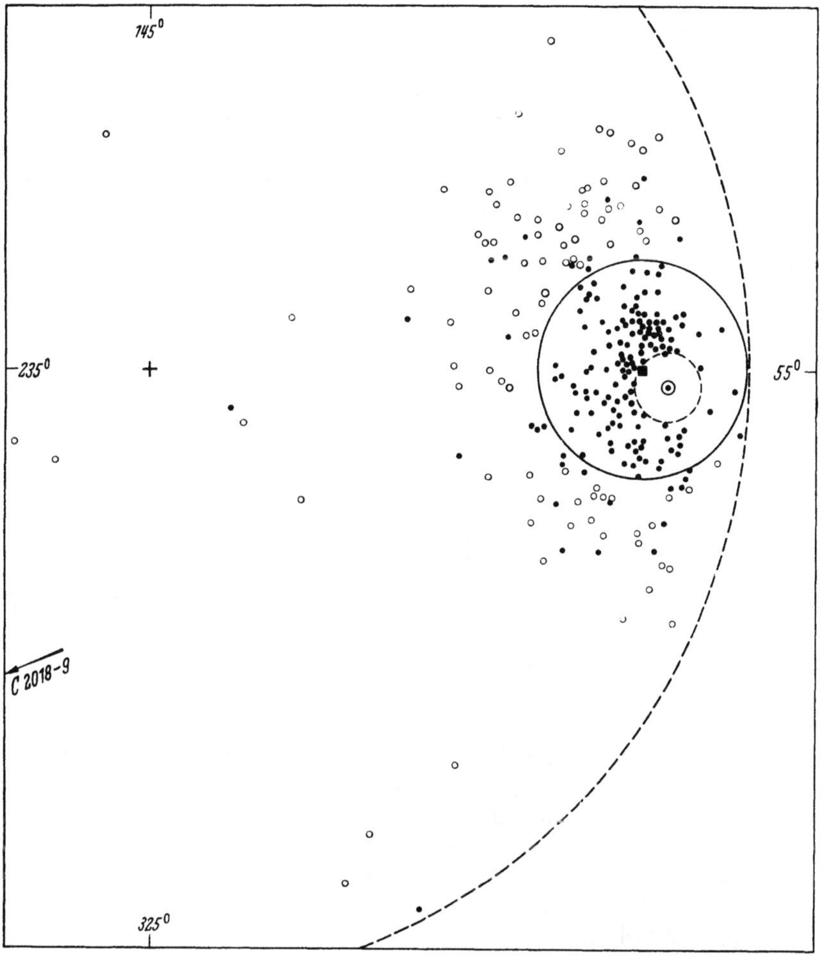

Fig. 5. Observed space velocities of stars (after OORT). This should be compared with Fig. 4. The small dotted circle centered on the solar velocity (⊙) has a radius of 19.5 km/sec. The solid circle whit the square as center has a radius of 65 km/sec. The large broken circle corresponds to the velocity of escape from the galaxy.

to OORT[1] is

$$A = (20.6 \pm 1.4) \text{ km/sec/kpc},$$
$$B = (-6.9 \pm 0.7) \text{ km/sec/kpc},$$
$$l_0 = 325^\circ \pm 1^\circ.$$

Several recent investigations give a slightly smaller value of *A*. PETRIE, CUTTLE and ANDREWS[2] derive $A = 17.7 \pm 1.1$ from *B* stars; STIBBS[3] derives $A = 19.5 \pm 1.9$

[1] J.H. OORT: Astrophys. Journ. **116**, 233 (1952).
[2] R.M. PETRIE, P.M. CUTTLE and D.H. ANDREWS: Astronom. J. **61**, 289 (1956).
[3] D.W.N. STIBBS: Monthly Notices Roy. Astronom. Soc. London **116**, 453 (1956).

and GASCOIGNE and EGGEN[1] derive 17.5 ± 1.9 from cepheids; THACKERAY[2] derives $A = 17.5 \pm 1.5$ from B stars; and EDMONDSON[3] finds $A \gtrsim 17.5$ from 11th magnitude K giants. However, WEAVER[4,5] and BAHNG, CODE and WHITFORD[6] find a much smaller value, approximately 10 km per sec/kpc.

The disagreement between PETRIE *et al.* and BAHNG *et al.* is important, because essentially the same stars were used in the two discussions. This shows that we are dealing with a distance-scale problem. The distances employed in the two investigations differ systematically by a factor of 1.5 owing to different methods of distance calibration.

WEAVER[4] has urged use of (4.2), a formula originally employed by CAMM[7,8] in place of (4.4). He[9] has criticized Eqs. (4.4) and (4.5) as being poor approximations which introduce mathematical bias leading to systematic errors in A and B. This question will be discussed in Sect. 7.

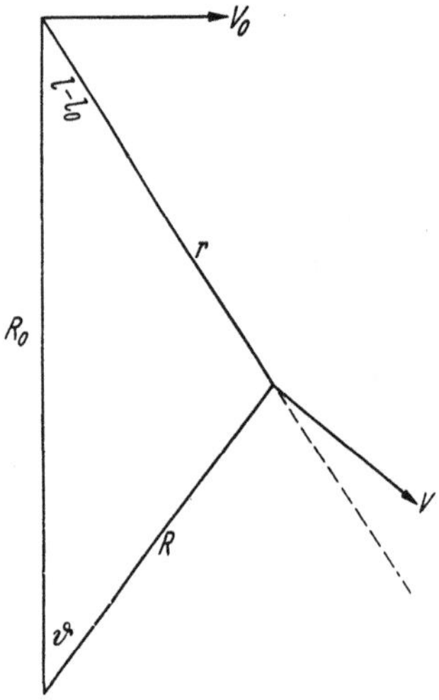

Fig. 6. Geometrical relationships for relative radial velocity and proper motion due to galactic rotation (circular orbits).

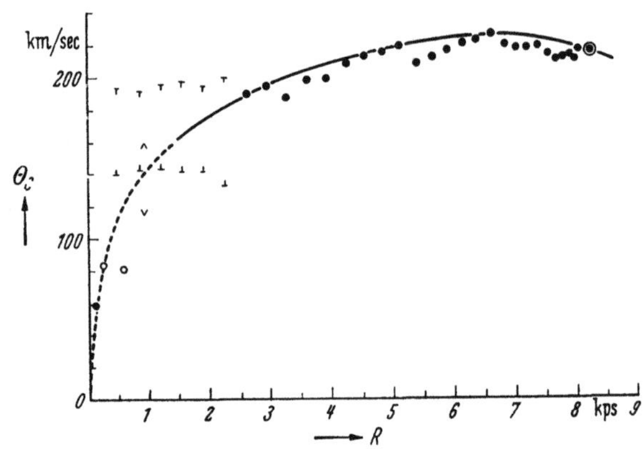

Fig. 7. Relationship between circular velocity and distance from the galactic center (after KWEE, MULLER, and WESTERHOUT) derived from 21 cm observations.

Observations of the 21 cm line of neutral hydrogen[10,11] have extended our knowledge of galactic kinematics to much larger distances. The relationship between circular velocity and distance from the galactic center for $R < R_0$ can be derived from "maximum velocities" for $|l - l_0| < 90°$. The maximum velocity $V_{\max}$ occurs at $r = R_0 \cos(l - l_0)$, and zero velocity occurs at $r = 0$ and $r = 2R_0 \times \cos(l - l_0)$. Substituting $R = R_0 \sin(l - l_0)$, corresponding to the point of maximum velocity, in (4.3) we obtain

$$V_{\max} = 2A\, R_0 \left[1 - \sin(l - l_0)\right] \sin(l - l_0). \tag{4.8}$$

[1] S.C.B. GASCOIGNE and O.J. EGGEN: Monthly Notices Roy. Astronom. Soc. London **117**, 430 (1957).
[2] A.D. THACKERAY: Observatory **78**, 47 (1958).
[3] F.K. EDMONDSON: Astronom. J. **61**, 3 (1956).
[4] H.F. WEAVER: Astronom. J. **60**, 202 (1955).
[5] H.F. WEAVER: Astronom. J. **60**, 208 (1955).
[6] J. BAHNG, A.D. CODE and A.E. WHITFORD: Unpublished.
[7] G.L. CAMM: Monthly Notices Roy. Astronom. Soc. London **99**, 71 (1938).
[8] G.L. CAMM: Monthly Notices Roy. Astronom. Soc. London **104**, 163 (1944).
[9] H.F. WEAVER: Astronom. J. **60**, 211 (1955).
[10] H.C. VAN DE HULST, C.A. MULLER and J.H. OORT: Bull. astronom. Inst. Netherl. **12**, 117 (1954).
[11] K.K. KWEE, C.A. MULLER and G. WESTERHOUT: Bull. astronom. Inst. Netherl. **12**, 211 (1954).

From measures at $(l-l_0)=15°$ to $30°$, the Leiden observers have found $A R_0 = 161$ km/sec [1]. Table 1 shows some corresponding values of A and R_0 based on this determination.

BAADE's[2] value of R_0 from RR Lyrae variables in the Sagittarius cloud is 8.16 kpc, corresponding to $A=19.8$.

The relationship between V_{max} and the circular velocity at R may be obtained by substituting $R=R_0 \sin(l-l_0)$ in (4.2), giving

$$V_{max} = V - V_0 \sin(l - l_0). \qquad (4.9)$$

Table 1.

A	R_0
10	16.1
12.5	12.9
15	10.7
17.5	9.2
20	8.0

The result obtained by the Leiden observers[3] is shown in Fig. 7. WEAVER's discussion of the cepheids[4] leads to much smaller values of V in the range $R=5.0$ to 7.5 kpc. The curves by BAHNG, CODE and WHITFORD[5] indicate a solid-body type of rotation extending almost all the way from the center to the Sun. These inconsistencies seem to be primarily a result of uncertainties in the distances, and the distance-scale problem thus becomes the central problem in galactic structure studies.

II. Kinematical considerations.

5. First order effects. The following treatment is based on MILNE's discussion[6].

Let

$$f(X, Y, Z, t, U, V, W)\, dX\, dY\, dZ\, dU\, dV\, dW \qquad (5.1)$$

be the number of stars at time t in the volume element $dX\,dY\,dZ$ centered on X, Y, Z with velocities in the velocity element $dU\,dV\,dW$ centered on U, V, W. The density of stars at X, Y, Z at time t is given by the integral of f over all U, V, W, and the mean velocity of these stars has components $\overline{U}, \overline{V}, \overline{W}$, where

$$\overline{U} = \overline{U}(X, Y, Z, t) = \frac{1}{n} \iiint f(X, Y, Z, t, U, V, W)\, U\, dU\, dV\, dW \qquad (5.2)$$

etc.

Let the co-ordinates of the Sun be X_0, Y_0, Z_0, and its velocity be U_0, V_0, W_0. Then the density of stars at X_0, Y_0, Z_0, at time t is $n_0 = n(X_0, Y_0, Z_0, t)$, and the mean velocity is $\overline{U}_0 = \overline{U}(X_0, Y_0, Z_0, t)$, etc.

The "local solar motion" is defined as the velocity of the Sun relative to its *immediate* surroundings, and its components are denoted by

$$S_1^0 = U_0 - \overline{U}_0, \quad S_2^0 = V_0 - \overline{V}_0, \quad S_3^0 = W_0 - \overline{W}_0.$$

The "solar motion" with respect to more distant stars at X, Y, Z at time t is denoted by

$$S_1 = U_0 - \overline{U}, \quad S_2 = V_0 - \overline{V}, \quad S_3 = W_0 - \overline{W}.$$

In general, $S_1^0 \neq S_1$, etc. Hence, a "solar motion" derived from distant stars will not provide a physically meaningful reduction to the "local standard of rest".

[1] See footnote 9, p. 10.
[2] W. BAADE: Mimeographed notes, University of Michigan Symposium 1953.
[3] See footnote 11, p. 10.
[4] See footnote 4, p. 10.
[5] See footnote 6, p. 10.
[6] E. A. MILNE: Monthly Notices Roy. Astronom. Soc. London **95**, 560 (1935). See this paper for references to earlier work.

Let x, y, z be the co-ordinates of a star relative to the Sun, so that

$$x = X - X_0, \quad y = Y - Y_0, \quad z = Z - Z_0$$

and

$$r^2 = x^2 + y^2 + z^2.$$

Let u, v, w be the components of velocity of the star relative to the Sun, so that

$$u = U - U_0, \quad v = V - V_0, \quad w = W - W_0.$$

The observed radial velocity of the star relative to the Sun is

$$\varrho(x, y, z, t) = u\frac{x}{r} + v\frac{y}{r} + w\frac{z}{r} \tag{5.3}$$

$$= (U - U_0)\frac{x}{r} + (V - V_0)\frac{y}{r} + (W - W_0)\frac{z}{r}. \tag{5.4}$$

The mean radial velocity of all stars at x, y, z at time t is

$$\left.\begin{aligned}\bar{\varrho}(x, y, z, t) = \frac{1}{n}\iiint\left[(U - U_0)\frac{x}{r} + (V - V_0)\frac{y}{r} + (W - W_0)\frac{z}{r}\right]\times\\ \times f(X, Y, Z, t, U, V, W)\, dU\, dV\, dW,\end{aligned}\right\} \tag{5.5}$$

$$= (\bar{U} - U_0)\frac{x}{r} + (\bar{V} - V_0)\frac{y}{r} + (\bar{W} - W_0)\frac{z}{r}. \tag{5.6}$$

But

$$\bar{U} - U_0 = (\bar{U} - \bar{U}_0) - (U_0 - \bar{U}_0), \tag{5.7}$$

$$= \bar{U} - \bar{U}_0 - S_1^0 \tag{5.8}$$

etc. If r is small this gives as a first approximation

$$\bar{U} - U_0 = x\left(\frac{\partial \bar{U}}{\partial X}\right)_0 + y\left(\frac{\partial \bar{U}}{\partial Y}\right)_0 + z\left(\frac{\partial \bar{U}}{\partial Z}\right)_0 - S_1^0. \tag{5.9}$$

The suffix 0 means that the differential coefficients are to be evaluated for $X = X_0$, etc., at time t. Substituting in (5.6) we find

$$\left.\begin{aligned}\bar{\varrho}(x, y, z) + S_1^0\frac{x}{r} + S_2^0\frac{y}{r} + S_3^0\frac{z}{r}\\ = \frac{1}{r}[A x^2 + B y^2 + C z^2 + 2F y z + 2G z x + 2H x y]\end{aligned}\right\} \tag{5.10}$$

where the coefficients $A, B, \ldots$ are functions of the differential coefficients in (5.9) and its analogues. The left-hand side of (5.10) is the mean radial velocity of stars at x, y, z, corrected for local solar motion, and will be denoted by $\bar{\varrho}'(x, y, z)$.

Let the xy-plane be the fundamental plane, and introduce longitude and latitude variables λ and β. Then

$$x = r\cos\lambda\cos\beta, \quad y = r\sin\lambda\cos\beta, \quad z = r\sin\beta$$

and

$$\bar{\varrho}'(r, \lambda, \beta) = r(a_0 + a_1\cos\lambda + a_2\cos 2\lambda + b_1\sin\lambda + b_2\sin 2\lambda) \tag{5.11}$$

where

$$\left.\begin{aligned}
a_0 &= \frac{1}{2}\left(\frac{\partial \overline{U}}{\partial X} + \frac{\partial \overline{V}}{\partial Y}\right)_0 \cos^2\beta + \left(\frac{\partial \overline{W}}{\partial Z}\right)_0 \sin^2\beta,\\
a_1 &= \left(\frac{\partial \overline{U}}{\partial Z} + \frac{\partial \overline{W}}{\partial X}\right)_0 \sin\beta\cos\beta,\\
a_2 &= \frac{1}{2}\left(\frac{\partial \overline{U}}{\partial X} - \frac{\partial \overline{V}}{\partial Y}\right)_0 \cos^2\beta,\\
b_1 &= \left(\frac{\partial \overline{V}}{\partial Z} + \frac{\partial \overline{W}}{\partial Y}\right)_0 \sin\beta\cos\beta,\\
b_2 &= \frac{1}{2}\left(\frac{\partial \overline{U}}{\partial Y} + \frac{\partial \overline{V}}{\partial X}\right)_0 \cos^2\beta.
\end{aligned}\right\} \tag{5.12}$$

There are similar expressions for tangential velocity in longitude and latitude.

Eq. (5.11) can be rewritten in the form

$$\bar{\varrho}'(r, \lambda, \beta) = r\left[a_0 + b_1' \sin(\lambda + \lambda') + b_2' \sin 2(\lambda + \lambda'')\right]. \tag{5.13}$$

This shows that, whatever the velocity distribution, there will in general be a non-zero K-effect, a first harmonic term that imitates the solar motion, and a second harmonic (or Oort term) in radial velocities as a function of longitude in any arbitrary plane. These first order terms are all proportional to the distance.

If we introduce the specialization for galactic symmetry, $\overline{U}$ and $\overline{V}$ independent of Z and $\overline{W} = 0$, and transfer to cylindrical co-ordinates R, ϑ, Z, with respect to the galactic center, the coefficients (5.12) become

$$\left.\begin{aligned}
a_0 &= \frac{1}{2}\left(\frac{1}{R}\frac{\partial \overline{M}}{\partial \vartheta} + \frac{\partial \overline{L}}{\partial R} + \frac{\overline{L}}{R}\right)_0 \cos^2\beta,\\
a_2 &= \frac{1}{2}\left(-\frac{1}{R}\frac{\partial \overline{M}}{\partial \vartheta} + \frac{\partial \overline{L}}{\partial R} - \frac{\overline{L}}{R}\right)_0 \cos^2\beta,\\
b_2 &= \frac{1}{2}\left(-\frac{1}{R}\frac{\partial \overline{L}}{\partial \vartheta} - \frac{\partial \overline{M}}{\partial R} + \frac{\overline{M}}{R}\right)_0 \cos^2\beta,
\end{aligned}\right\} \tag{5.14}$$

where L is the velocity along the radius, and M the velocity of rotation taken to be positive in the clockwise direction. Eq. (5.13) reduces to

$$\bar{\varrho}'(r, \lambda, \beta) = r\left\{a_0 + (a_2^2 + b_2^2)^{\frac{1}{2}} \sin\left(2\lambda + \arctan\frac{a_2}{b_2}\right)\right\}. \tag{5.15}$$

There are similar formulae for tangential velocity in longitude and latitude.

We consider two special cases:

(a) For the case of pure galactic expansion parallel to the galactic plane, $\overline{M} = 0$, $\overline{L}$ independent of ϑ, Eq. (5.15) becomes

$$\bar{\varrho}'(r, \lambda, \beta) = r\cos^2\beta\left[\frac{1}{2}\left(\frac{\partial \overline{L}}{\partial R} + \frac{\overline{L}}{R}\right)_0 + \frac{1}{2}\left(\frac{\partial \overline{L}}{\partial R} - \frac{\overline{L}}{R}\right)_0 \cos 2\lambda\right]. \tag{5.16}$$

(b) For the case of pure galactic rotation, $\overline{L} = 0$, $\overline{M}$ independent of ϑ, Eq. (5.15) becomes

$$\bar{\varrho}'(r, \lambda, \beta) = r\cos^2\beta\left[\frac{1}{2}\left(-\frac{\partial \overline{M}}{\partial R} + \frac{\overline{M}}{R}\right)_0 \sin 2\lambda\right]. \tag{5.17}$$

This is identical with (4.4) for stars in the galactic plane.

The foregoing discussion shows that an observed second harmonic in radial velocities and proper motions is *by itself* not sufficient evidence for galactic rotation. Additional independent information is required, such as the longitude of the galactic center from radio astronomy data, or the solar motion relative to the globular clusters.

6. Second order effects. Extension of MILNE's discussion to second order terms leads to the following form for the expressions for the radial velocity and the components of tangential velocity in longitude and latitude:

$$a_0 + a_1 \cos\lambda + a_2 \cos 2\lambda + a_3 \cos 3\lambda + b_1 \sin\lambda + b_2 \sin 2\lambda + b_3 \sin 3\lambda. \tag{6.1}$$

The coefficients a_i, b_i are functions of r, β, and the first and second partial derivatives of $\overline{U}$, $\overline{V}$, and $\overline{W}$ with respect to X, Y, and Z at the Sun. The general expressions have been published by EDMONDSON[1].

The solar motion as ordinarily defined is obtained by using (2.1) to (2.3) as equations of condition for a least squares solution. The first harmonic terms of (6.1) for distant stars will combine with the corresponding terms of (2.1) to (2.3) to give a "solar motion" which differs from the "local solar motion". The revised equations (in galactic co-ordinates) are

(*a*) *Proper motions:*

$$4.74\,\mu_\lambda'' = \frac{S_1^0}{r}\sin\lambda - \left(\frac{S_2^0}{r} - 3r\,Q_1\cos^2\beta\right)\cos\lambda, \tag{6.2}$$

$$4.74\,\mu_\beta'' = \frac{S_1^0}{r}\cos\lambda\sin\beta + \left(\frac{S_2^0}{r} - r\,Q_1\cos^2\beta\right)\sin\lambda\sin\beta - \frac{S_3^0}{r}\cos\beta. \tag{6.3}$$

(*b*) *Radial velocities:*

$$\varrho = -S_1^0\cos\lambda\cos\beta - (S_2^0 - r^2 Q_1\cos^2\beta)\sin\lambda\cos\beta - S_3^0\sin\beta. \tag{6.4}$$

It should be noted that only the Y-component is affected.

Q_1 may be expressed in terms of the linear velocity or the angular velocity of galactic rotation and their derivatives, or in terms of the Oort constant A_1 and an associated second order quantity A_2:

$$Q_1 = \frac{1}{8}\left[\left(\frac{d^2V}{dR^2}\right)_0 + \frac{1}{R_0}\left(\frac{dV}{dR}\right)_0 - \frac{V_0}{R_0^2}\right], \tag{6.5}$$

$$= \frac{1}{8}\left[3\omega_0' + R_0\,\omega_0''\right], \tag{6.6}$$

$$= \frac{1}{4}\left[-3\frac{A_1}{R_0} - A_2\right], \tag{6.7}$$

where

$$A_1 = -\tfrac{1}{2}R_0\,\omega_0', \quad\text{and}\quad A_2 = -\tfrac{1}{2}R_0\,\omega_0''.$$

Even if ω is a linear function of R, Q_1 will be -1.5 km/sec/kpc^2 if $A_1 = 17$ and $R_0 = 8.2$. This will change the Y-component of the solar motion by 3.4 km/sec for stars at a distance of 1.5 kpc, and by 6.0 km/sec for stars at a distance of 2.0 kpc.

7. Accuracy of the Oort approximation. WEAVER's criticism[2] is based on a numerical comparison of Eqs. (4.3) and (4.2). The comparison is incomplete because the galactic rotation first harmonic was not taken into account, although the expected amplitude is one-fourth the amplitude of the Oort term. The para-

---

[1] F. K. EDMONDSON: Monthly Notices Roy. Astronom. Soc. London **97**, 473 (1937).
[2] H. F. WEAVER: Astronom. J. **60**, 211 (1955).

meters of the model were

$$r = 1.25 \text{ kpc}, \qquad R_0 = 8.2 \text{ kpc},$$
$$A_1 = 10.8 \text{ km/sec/kpc}, \qquad A_2 = 5.1 \text{ km/sec/kpc}^2.$$

A_2 was estimated from the diagram in WEAVER's paper[1]. Using (6.7), the amplitude of the first harmonic is

$$(1.25)^2 \frac{1}{4}\left[-3\,\frac{10.8}{8.2} - 5.1\right] = -3.6 \text{ km/sec}.$$

The amplitude of the second harmonic is

$$1.25 \times 10.8 = 13.5 \text{ km/sec}.$$

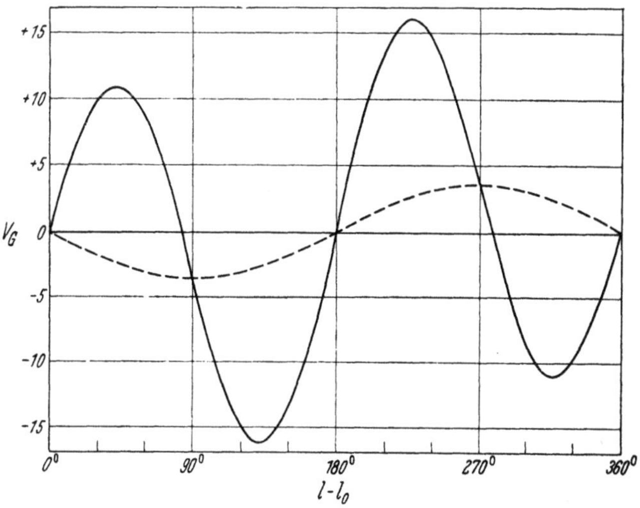

Fig. 8. Harmonic components of WEAVER's model. Dashed curve is first harmonic. Solid curve is sum of first and second harmonics.

Adding ordinates at $\lambda = 135°$ and 225°, and subtracting at $\lambda = 45°$ and 315°, as illustrated in Fig. 8, we find

$$\text{for } \lambda \text{ between } 90° \text{ to } 270°: \quad A = \frac{13.5 + 0.7 \times 3.6}{1.25} = 12.8 \text{ km/sec},$$

$$\text{for } \lambda \text{ between } 270° \text{ to } 90°: \quad A = \frac{13.5 - 0.7 \times 3.6}{1.25} = 8.8 \text{ km/sec}.$$

WEAVER's detailed calculations gave 13.9 km/sec and 9.3 km/sec.

In practice one solves for both solar motion and galactic rotation as unknowns (e.g. PLASKETT and PEARCE[2]). This avoids the incompleteness of WEAVER's model analysis and yields a correct value of A. However, the so-called solar motion will not be the local solar motion, for it will include the galactic rotation term in the Y-component as shown in (6.4).

8. The local standard of rest. VYSSOTSKY and JANSSEN[3] have derived the following elements for the solar motion:

$$A = 265°.0 \pm 1°.2, \qquad D = +20°.7 \pm 1°.4,$$

[1] H.F. WEAVER: Astronom. J. **60**, 202 (1955).
[2] J.H. PLASKETT and J.A. PEARCE: Dominion Astrophys. Obs. Publ. **5**, 294 (1936).
[3] A.N. VYSSOTSKY and E. JANSSEN: Astronom. J. **56**, 58 (1951).

or

$$L = 12\overset{\circ}{.}4 \pm 1\overset{\circ}{.}6, \qquad B = +22\overset{\circ}{.}3 \pm 0\overset{\circ}{.}9,$$

and

$$V_0 = 15.5 \pm 0.4 \text{ km/sec}.$$

They call this the "basic solar motion". Table 2 compares it with several other values of the solar motion.

Table 2.

$l=327\overset{\circ}{.}8$ X_0	$l=57\overset{\circ}{.}8$ Y_0	$b=+90°$ Z_0	Source
+10.2	+10.1	+5.9	Basic solar motion
11.6	14.4	7.6	Ca^+. BLAAUW: Bull. astronom. Inst. Netherl. No. 436, 1952
12.5	14.6	(6.8)	Cepheids. WEAVER: Astronom. J. **60**, 202 (1955)
10.5	15.4	7.3	Standard solar motion
10.6	17.3	(6.6)	$O5-B7$. BLAAUW: Bull. astronom. Inst. Netherl. No. 363, 1944
8.4	18.9	(7.3)	Cepheids. RAIMOND: Bull. astronom. Inst. Netherl. No. 450, 1954

Inspection of this table shows that only the Y-component varies in a systematic manner, and that it increases with increasing distance of the material to which it is referred. This indicates that the "basic solar motion" is the correct reduction to the local standard of rest in the sense of MILNE's "local solar motion", and that the other values in Table 2 differ from it owing to the effect of the galactic rotation first harmonic which increases as the square of the distance from the Sun.

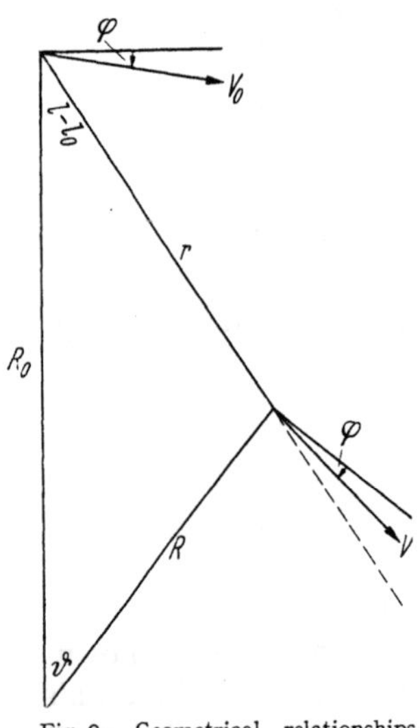

Fig. 9. Geometrical relationships for galactic rotation, including a constant deviation from circular motion.

9. Non-circular motion. Star streaming and asymmetry are, of course, caused by individual deviations from circular motion. The effect of systematic deviations from circular motion has been discussed by EDMONDSON[1-3] and RUBIN[4]. RUBIN postulated motions along the spiral arms, whereas EDMONDSON postulated a constant deviation from the normal to the radius. More general expressions can be derived from the coefficients published by EDMONDSON[5].

Fig. 9 is similar to Fig. 6 except for the constant deviation from circular motion denoted by φ, which is taken to be positive when the velocity vector deviates toward the center of the rotation. The galactic rotation equations are

(a) Radial velocities:

$$V_G = V \cos\left[90° - (l - l_0) - \vartheta - \varphi\right] - V_0 \cos\left[90° - (l - l_0) - \varphi\right] \tag{9.1}$$

---

[1] F.K. EDMONDSON: Publ. Astronom. Soc. Pacific **67**, 10 (1955).
[2] F.K. EDMONDSON: Astronom. J. **60**, 160 (1955).
[3] F.K. EDMONDSON: Int. Astronom. Union Symposium No. 4, *Radio Astronomy*, 19, Cambridge 1957.
[4] V.C. RUBIN: Astronom. J. **60**, 177 (1955).
[5] See footnote 1, p. 14.

for stars in the galactic plane. After elimination of ϑ this becomes

$$V_G = \left[V \frac{R_0 - r\cos(l-l_0)}{R} - V_0\right] \sin(l - l_0 + \varphi) + V \frac{r \sin(l-l_0)}{R} \cos(l - l_0 + \varphi) \quad (9.2)$$

$$= R_0(\omega - \omega_0) \sin(l - l_0 + \varphi) - \omega r \sin\varphi. \quad (9.3)$$

Eq. (9.3) reduces to (4.2) when $\varphi = 0°$.

If $(R - R_0)$ is not too large, we may put

$$\omega = \omega_0 + (R - R_0)\,\omega_0'.$$

Defining $A = -\frac{1}{2} R_0\, \omega_0'$, we have

$$V_G = -2A\,(R - R_0)\left[\sin(l - l_0 + \varphi) - \frac{r}{R_0}\sin\varphi\right] - r\,\omega_0 \sin\varphi. \quad (9.4)$$

For very small distances, we may put $R - R_0 = -r\cos(l - l_0)$. Thus we finally obtain

$$V_G = r\,A \sin 2\left(l - l_0 + \frac{\varphi}{2}\right) + r\,B\sin\varphi \quad (9.5)$$

where $B = A - \omega_0$.

(b) Proper motions:

For stars in the galactic plane, the tangential velocity in longitude is

$$T_G = 4.74\, r\, \mu_G''$$

$$= V\sin[90° - (l - l_0) - \vartheta - \varphi] - V_0 \sin[90° - (l - l_0) - \varphi] \quad (9.6)$$

$$= R_0(\omega - \omega_0)\cos(l - l_0 + \varphi) - \omega\, r\cos\varphi. \quad (9.7)$$

Hence, the proper motion in longitude is

$$\mu_G'' = \frac{(\omega - \omega_0)}{4.74}\left[\frac{R_0\cos(l - l_0 + \varphi)}{r} - \cos\varphi\right] - \frac{\omega_0\cos\varphi}{4.74}. \quad (9.8)$$

If $(R - R_0)$ is not too large, we may make the same approximation that was used to derive Eq. (9.4) and obtain

$$\mu_G'' = -2A\,\frac{(R - R_0)}{4.74\, r}\left[\cos(l - l_0 + \varphi) - \frac{r}{R_0}\cos\varphi\right] - \frac{\omega_0\cos\varphi}{4.74}. \quad (9.9)$$

Finally, for *very* small distances we obtain

$$\mu_G'' = \frac{A}{4.74}\cos 2\left(l - l_0 + \frac{\varphi}{2}\right) + \frac{B}{4.74}\cos\varphi. \quad (9.10)$$

Eqs. (9.5) and (9.10) differ very little from (4.4) and (4.5), and the differences are too small to be detected with certainty from observations. The longitude of the center derived both from radial velocities and proper motions will be in error by $\varphi/2$, which is smaller than the present errors of observation for φ as large as 4°. The constant term in the proper motions is multiplied by $\cos\varphi$, and therefore unchanged for small values of φ. There is a K-term in the radial velocities ($= r\,B\sin\varphi$) which is less than 1 km/sec at 1 kpc for $\varphi = 4°$, and therefore not detectable.

The exact formula (9.3) differs considerably from the exact formula (4.2) at large distances from the Sun, such as can be observed using the 21 cm hydrogen line. Fig. 10 shows the relationship between V_G and r for $l - l_0 = 45°$ and 135° and $\varphi = 0°$ and 4°.

Fig. 11 shows the locus of zero radial velocity (solid curves), $+50$ km/sec radial velocity (dashed curves), and -50 km/sec radial velocity (dotted curves) for (a) circular motion, $\varphi=0°$, and (b) non-circular motion, $\varphi=4°$. This type

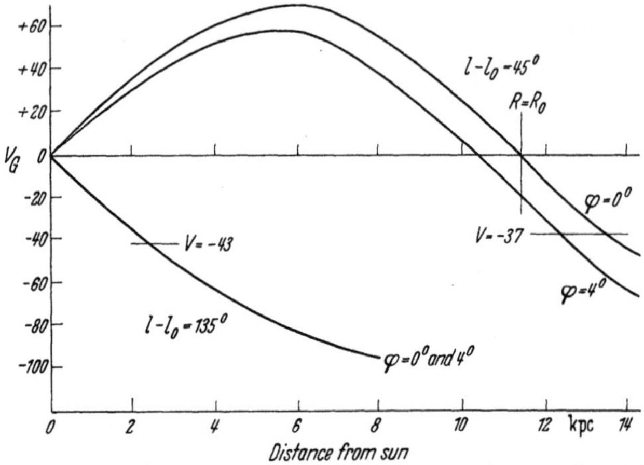

Fig. 10. Relationship between V_G and r showing effect of a 4° deviation from circular motion at $l-l_0=45°$ and 135°. Observed 21 cm velocities at these longitudes are shown.

of non-circular motion decreases the values of $V_{\max}$ in the northern hemisphere and increases them in the southern hemisphere. If $\varphi=4°$ and $V_0=216$ km/sec the difference between the two hemispheres should be $[30 \cos(l-l_0)]$ km/sec

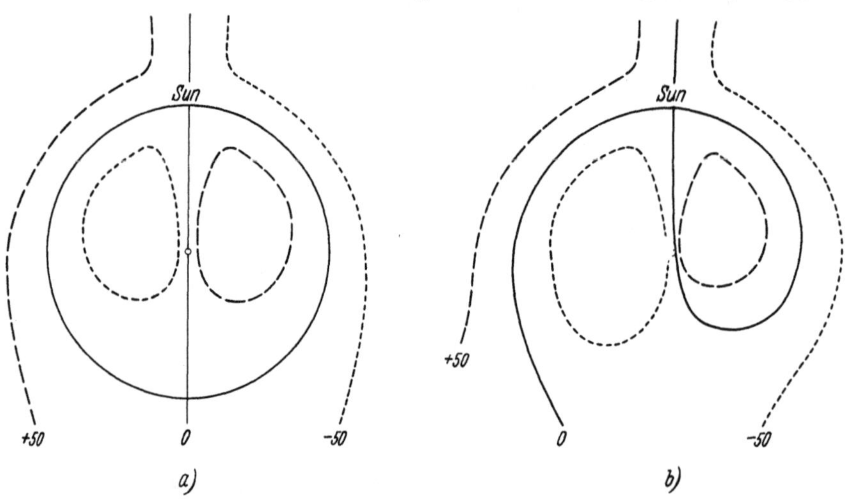

Fig. 11 a and b. Loci of equal radial velocity (+50, 0, −50 km/sec) for (a) Circular motion, and (b) a 4° constant deviation from circular motion. o indicates the galactic center.

for corresponding longitudes $(l-l_0)$ and (l_0-l). There is no evidence for a difference of this size in the comparison between the Australian and Dutch observations[1].

Fig. 12 shows the loci of equal radial velocities from $+100$ km/sec to -100 km per sec in the direction of the galactic center, with the longitude scale expanded by a factor of 10. An antenna with a beam width of 1° will accept a large range

[1] M. S. Carpenter: Int. Astronom. Union Symposium No. 4, *Radio Astronomy*, 14. Cambridge 1957.

of velocities, a circumstance which should be taken into consideration in the interpretation of broad "wings" and "tails" of observed 21 cm lines. The systematically negative velocities for distances beyond the galactic center if $\varphi = 4°$ have not been detected.

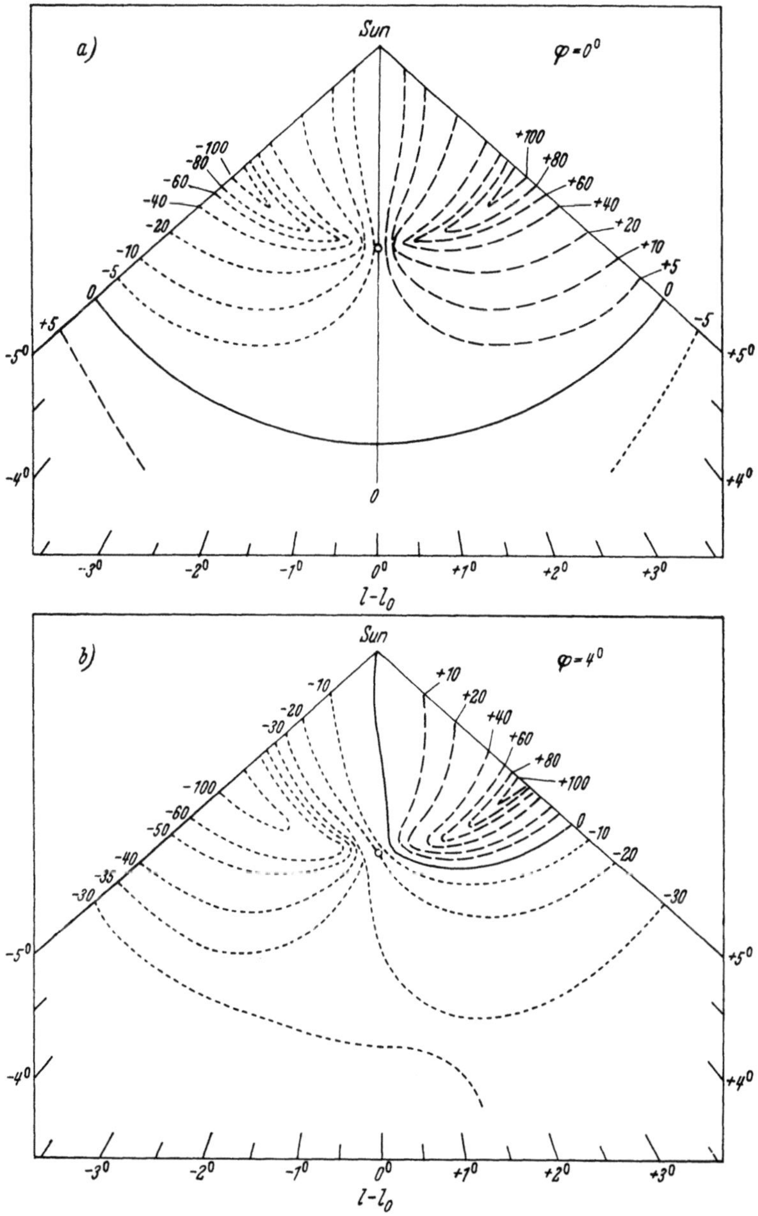

Fig. 12a and b. Loci of equal radial velocities in the direction of the galactic center (denoted by o).

10. Concluding remarks. A complete dynamical theory of the galaxy cannot be worked out until we have a correct kinematical description. The basic kinematical principles were stated by MILNE[1], but have largely been overlooked. The

[1] See footnote 6, p. 11.

discussion of Sect. 8 leads to certain precepts which should be followed in all galactic rotation studies:

(a) One should not derive the solar motion from a group of stars when using the exact formula (4.2) for galactic rotation to analyze their residual velocities. If the stars are so far away as to require use of the exact formula, they cannot possibly define the local solar motion. The observed radial velocities should be corrected using the Vyssotsky-Janssen "basic solar motion".

(b) The "basic solar motion" should also be employed when the intermediate approximation (4.3) is used.

(c) A first harmonic in the Y-component should always be included as an unknown when using the Oort approximation (4.4). This should be done even though the "basic solar motion" has already been taken out.

The distance-scale problem still remains to be solved before a consistent set of galactic rotation parameters can be obtained.

Galactic Dynamics.

By

Bertil Lindblad.

With 20 Figures.

I. Introduction.

1. General review of galactic dynamics before 1930. The empirical foundations for theories concerning galactic dynamics are the results of observation concerning stellar motions and the state of motion of interstellar matter, as well as on the distribution in space of stars and of interstellar gas and dust. Lately the occurrence of interstellar magnetic fields and the possible importance of electromagnetic forces have come into the picture.

During the 19th century the data on "proper motions", i.e. the apparent motions of the stars on the celestial sphere, had accumulated to such an extent that statistical investigations on stellar motions in the surroundings of the Sun could be undertaken with success. The apparent effect in the proper motions due to the motion of the Sun relative to the "centroid" of the surrounding stars had been known since the days of William Herschel. The question most near at hand was whether, after eliminating this effect, the motions of the stars occur at random in analogy with the motions of the molecules in a gas, or whether there exists some kind of preferential motion. The answer was given with Kapteyn's discovery (1904) of a preferential motion which he described as two star-streams pervading our surroundings of space. An alternative description of the phenomenon was proposed by K. Schwarzschild (1907), who introduced an ellipsoidal distribution of stellar motions in the velocity space. The mathematical analysis of the two-drifts theory was developed by Eddington (1906). A very complete description of the star-drift theory and the ellipsoidal theory has been given by Smart [*10*]. Galactic dynamics may be said to begin with certain fundamental works of Eddington and Jeans. Turner[1] had advanced the hypothesis that the two star-streams could be due to in- and out-going motion in orbits which pass nearly through the centre of the stellar system. Eddington[2] found that in globular stellar systems a radial, everywhere ellipsoidal star-streaming can exist. In a later paper Eddington[3] investigated the possible forms of stellar systems in which Schwarzschild's ellipsoidal law of velocities is rigorously obeyed. The ellipsoidal law is taken here with homogeneous expressions of second order in the velocities, though a superposed rotation is separately considered. Eddington found that (if a certain case of spheroidal velocity ellipsoid is excluded) ellipsoidal velocity distributions are possible only under a certain form of the governing potential, which includes the globular symmetry as a special case. It has been pointed out by Chandrasekhar ([*1*], p. 43) that the fundamental assumption made by Eddington that the axes of the velocity ellipsoids at various

[1] H. H. Turner: Monthly Notices Roy. Astronom. Soc. London **72**, 387, 474 (1912).

[2] A. S. Eddington: Monthly Notices Roy. Astronom. Soc. London **74**, 5 (1913); **75**, 366 (1915).

[3] A. S. Eddington: Monthly Notices Roy. Astronom. Soc. London **76**, 37 (1915).

points generate an orthogonal system of "principal velocity surfaces" limits the generality of EDDINGTON's analysis. In the case of general rotational symmetry, however, EDDINGTON's postulate applies. JEANS[1] developed a more general analysis, introducing the frequency function as a function of the integrals of the motion. In the application to ellipsoidal distributions he is still limited to homogeneous expressions of second order in the velocities, not taking into account mean differential motions in the system. KAPTEYN[2] formulated a dynamical theory for the "typical stellar system" derived in his statistical work on stellar distribution. The analysis of the "typical system" was very much advanced by JEANS[3] in an important investigation which gave much of the mathematical tools for the investigations of galactic dynamics.

A general explanation of the evidently very great flattening of the galactic system would be that it has a motion of rotation. POINCARÉ[4] has shown that for a continuous medium of density ϱ the upper limit of the angular speed of rotation ω is given by

$$\frac{\omega^2}{2\pi G\varrho} < 1,$$

where G is the constant of gravitation. From the density ϱ in our surroundings the minimum of the time of rotation T is of the order 10^7 years. As the astronomical coordinate system should in principle be an inertial system, and the determinations of stellar positions, as CHARLIER points out, can be reduced on the invariable plane of the planetary system, it should be possible to find a rotation term in stellar proper motions. Rotation terms were introduced by L. STRUVE[5] in the equations for determination of the constant of precession. CHARLIER[6] found in his last determination a mean motion in galactic longitude of $-0''.0024$ per year. It should be mentioned here that the mean motion in galactic longitude is identical with the constant B of the differential galactic rotation defined below.

GYLDÉN[7] tried to trace, by Fourier developments of stellar proper motions, analogies between stellar motions and the geocentric motions of asteroids. OPPENHEIM[8] has tried to follow up GYLDÉN's idea and for the B stars extended the harmonic analysis to the radial velocities.

The fundamental researches by H. SHAPLEY[9] on the distribution in space of the globular clusters extended the limits of the stellar system very widely over those of KAPTEYN's "typical system", and gave the direction of the galactic centre at great distance in the constellation Sagittarius, in galactic longitude about 327°. It is of great importance that in the field of stellar velocities also a phenomenon appears which cannot be placed within the "typical system". This is the asymmetrical drift of the mean motion with increasing velocity dispersion. The peculiar one-sided distribution of large velocity vectors relative

---

[1] J. H. JEANS: Monthly Notices Roy. Astronom. Soc. London **76**, 70 (1915).

[2] J. C. KAPTEYN: Mt. Wilson Contr. No. 230 = Astrophys. Journ. **55**, 302 (1922).

[3] J. H. JEANS: Monthly Notices Roy. Astronom. Soc. London **82**, 122 (1922).

[4] H. POINCARÉ: Bull. Astronomique **2**, 109 (1885).

[5] L. STRUVE: Mém. Acad. Imp. St. Pétersbourg (VII) **35**, No. 3 (1887).

[6] C. V. L. CHARLIER: The Motion and the Distribution of the Stars. Mem. Univ. Calif. **7**, 32 (1926).

[7] H. GYLDÉN: Övers. K. Vetenskapsakad. Förhandl. **28**, 956 (1871).

[8] S. OPPENHEIM: Astronom. Nachr. **188**, 137 (1911); **201**, 241, 417 (1915); **202**, 89 (1916); **204**, 417 (1917). — Wien Denkschr., Math.-Nat. Kl. **87** (1912); **92** (1915); **93** (1916). — Seeliger-Festschr. **1924**, 131.

[9] H. SHAPLEY: Mt. Wilson Contr. 151—157, 160, 161, 175, 176, 190, 195. — Astrophys. Journ. **48—52** (1918—1920).

to the Sun seems first to have been pointed out by B. Boss[1]. It was further studied by ADAMS and JOY[2], and especially by STRÖMBERG[3] and OORT[4].

STRÖMBERG in his first paper remarks that the motions of the brighter stars have well been explained by JEANS through the gravitational action of the stars in the local or "KAPTEYN" universe, but that "in order to account for the existence of stars of very high velocity through gravitational action of the system as a whole, we must include in our study a system of much larger dimensions". He further remarks that the direction of asymmetry is nearly in the galactic plane and nearly perpendicular to the direction toward the centre of the globular clusters as determined by SHAPLEY, that the velocity of translation is about 300 km/sec, and that it is possible that the local system moves around the centre of the large system with this velocity. Similar thoughts had been advanced by LUNDMARK[5] who was the first to point out the high velocity of the Sun relative to the system of globular clusters. Later STRÖMBERG favoured the possibility that the connection between the larger and the smaller systems is not of gravitational or material nature, but represents a velocity-restriction in a "world-frame" that coincides with the large system of clusters and spirals.

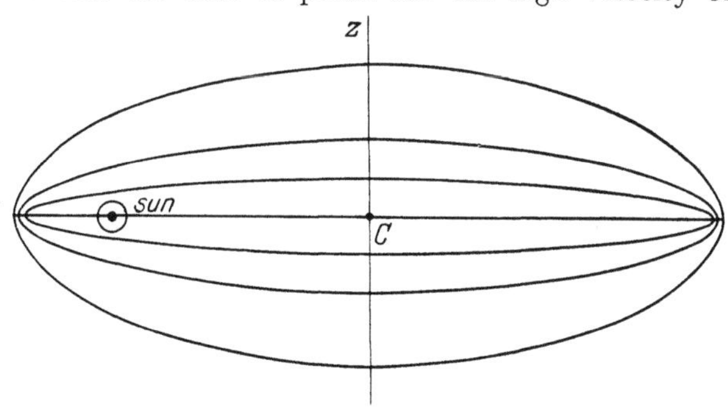

Fig. 1. Schematic representation of rotating sub-systems in the Galaxy.

OORT tried to define the velocity relative to the centroid at which the asymmetry sets in with an entirely one-sided distribution of the velocity vectors, and found this limit to be close to 62 km/sec. He tried further to define the physical properties of the high velocity stars compared to the "ordinary" stars (cf. Sect. 33).

A general explanation of the phenomenon of the asymmetrical drift based on the idea of "sub-systems" of nearly the same diameter in the galactic plane rotating about a distant centre situated in the direction of SHAPLEY'S centre of the globular clusters was given by LINDBLAD[6] (Fig. 1). This theory abandoned the idea of a local system, and assumed that in the most flattened sub-system the dispersion of the velocities is everywhere small compared with the systematic motion relative to the centre of gravity of the system as a whole, which can be identified with a general motion of rotation about the distant centre. The asymmetrical drift was interpreted as due to a decrease of the velocity of rotation with increasing dispersion of the relative velocities in the sub-system, and was explained quantitatively by the existence of a general "effective" limit of the

[1] B. Boss: Pop. Astr. **26**, 686 (1918).
[2] W. S. ADAMS and A. H. JOY: Mt. Wilson Contr. No. 163 = Astrophys. Journ. **49**, 179 (1919).
[3] G. STRÖMBERG: Mt. Wilson Contr. No. 275 and No. 292 = Astrophys. Journ. **59**, 228 (1924); **61**, 363 (1925).
[4] J. H. OORT: Publ. Astronom. Lab. Groningen **1926**, No. 40.
[5] K. LUNDMARK: K. Svenska Vetenskaps. Handl. **60**, 22–24, No. 8 (1920). — Publ. Astronom. Soc. Pacific **35**, 318 (1923). — Monthly Notices Roy. Astronom. Soc. London **84**, 747 (1924).
[6] B. LINDBLAD: Ark. Mat., Astronom. Fys., Ser. A **19**, No. 3 (1925); A **19**, No. 27 (1926); A **19**, No. 35 (1926); B **19**, No. 7 (1926). = Upsala Medd. No. 3, 4, 6, 13.

system in the galactic plane. The variation of the true galactic concentration from one sub-system to another is a direct consequence of the variation in the velocity dispersion, because for a greater general dispersion of the velocities relative to the centroid the binding of the objects to the galactic plane becomes less strong, the amplitudes in the motion at right angles to the plane on the average greater, and the true galactic concentration less pronounced.

This theory was soon confirmed very strongly by OORT's[1] discovery of the differential galactic rotation, which is due to the fact that the angular velocity of rotation in the central galactic layer decreases with the distance from the centre (Fig. 2). If $\omega(R)$ is the angular speed of rotation about the centre, and ω_0 the value of ω valid for the radius vector R_0 of the Sun, the differential effect W in the radial velocity is

$$W = R_0(\omega - \omega_0)\sin l', \tag{1.1}$$

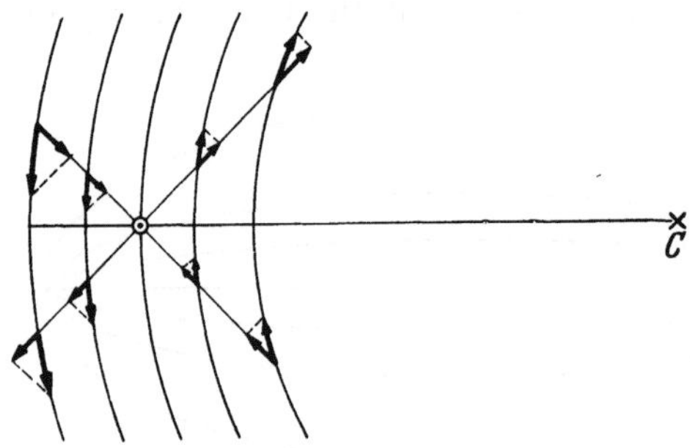

Fig. 2. The differential effects of galactic rotation.

where l' is the galactic longitude reckoned from the centre. The effect T in the linear tangential velocities is

$$T = R_0(\omega - \omega_0)\cos l' - \omega r, \tag{1.2}$$

where r is the distance from the Sun. For small relative distances r, the expressions are

$$\left.\begin{aligned} W &= rA\sin 2l', \\ T &= rA\cos 2l' + rB. \end{aligned}\right\} \tag{1.3}$$

Here

$$A = \frac{1}{2}\left(\frac{\Theta_0}{R} - \frac{\partial\Theta_0}{\partial R}\right), \quad B = \frac{1}{2}\left(-\frac{\Theta_0}{R} - \frac{\partial\Theta_0}{\partial R}\right), \tag{1.4}$$

where $\Theta_0 = \omega R$. A and B are OORT's constants of differential galactic rotation. Obviously we have

$$\omega = A - B. \tag{1.5}$$

Modern values of A, B and ω will be briefly discussed in Sect. 16.

The differential galactic rotation immediately gave a quantitative dynamical explanation of the ellipsoidal distribution of stellar velocities[2].

II. Mass motions and velocity distribution in the gravitational field of the Galaxy.

2. The equations of mass motion. The matter contained in our stellar system is a mixtures of stars, gas, dust and meteoric particles including dark bodies of considerable size. The stars are of a great variety in mass, and can be divided into natural groups which differ considerably in their dynamical properties. The interstellar gas is mainly made up of hydrogen, with the other elements entering in the "cosmical" abundance ratio. The dust particles will be of widely different size, but it seems from the optical properties of the dust that the distribution function is remarkably uniform from one place in the system to another. The number of particles per unit volume must decrease rapidly with increasing size. The amount of meteoric particles of different size in interstellar space is not very

[1] J. H. OORT: Bull. Astr. Inst. Netherl. **3**, 275 (1927); **4**, 79, 91 (1927).

[2] B. LINDBLAD: Monthly Notices Roy. Astronom. Soc. London **87**, 553 (1927). = Upsala Medd. No. 24. — Ark. Mat., Astronom. Fys. Ser. A **20**, No. 17 (1927). = Upsala Medd. No. 26. — J. H. OORT: Bull. Astr. Inst. Netherl. **4**, 269 (1928).

well known. Especially the part played by dark bodies of considerable size is an open question. For any physically defined group of particles we may define a density function and a function describing the distribution of velocities, which may vary from one point to another. In the absence of exterior disturbances there will be a certain plane passing through the centre of gravity, "the invariable plane", for which the sum of the area velocities about the centre of gravity is a maximum. This plane may be taken as fundamental plane of reference. We introduce a coordinate system x, y, z with this plane as xy-plane and with the origin in the centre of gravity of the system. The coordinate system is allowed to rctate with the angular velocity ω about the z-axis. At a given point x, y, z, let the mean motion along the axes be u, v, w. The velocities of individual particles may be $U+u$, $V+v$, $W+w$, so that U, V, W is the peculiar velocity of a particle. Denoting the potential function with $\varphi(x, y, z)$, and forming the tensor components $\overline{U^2}$, $\overline{UV}$, etc., we have the equations of mass motion and the equation of continuity, which are derived directly by considering the exchange of matter and momentum for an element of volume. These equations, as well as equations for higher moments of the velocities, can also be derived from the continuity Eq. (4.6) given below for the sixdimensional space as a consequence of LIOUVILLE's theorem. We give these equations here in a general form which is related to the equations of motion of a rotating liquid, using the notation

$$\frac{D}{Dt} = \frac{\partial}{\partial t} + u\frac{\partial}{\partial x} + v\frac{\partial}{\partial y} + w\frac{\partial}{\partial z}$$

as symbol for differentiation following the mean motion (u, v, w).

$$\left.\begin{aligned}
\frac{Du}{Dt} - 2\omega v + \frac{1}{\varrho}\left[\frac{\partial}{\partial x}(\varrho\overline{U^2}) + \frac{\partial}{\partial y}(\varrho\overline{UV}) + \frac{\partial}{\partial z}(\varrho\overline{UW})\right] &= \frac{\partial\varphi}{\partial x} + \omega^2 x,\\
\frac{Dv}{Dt} + 2\omega u + \frac{1}{\varrho}\left[\frac{\partial}{\partial x}(\varrho\overline{UV}) + \frac{\partial}{\partial y}(\varrho\overline{V^2}) + \frac{\partial}{\partial z}(\varrho\overline{VW})\right] &= \frac{\partial\varphi}{\partial y} + \omega^2 y,\\
\frac{Dw}{Dt} + \frac{1}{\varrho}\left[\frac{\partial}{\partial x}(\varrho\overline{UW}) + \frac{\partial}{\partial y}(\varrho\overline{VW}) + \frac{\partial}{\partial z}(\varrho\overline{W^2})\right] &= \frac{\partial\varphi}{\partial z},\\
\frac{D\varrho}{Dt} + \varrho\left(\frac{\partial u}{\partial x} + \frac{\partial v}{\partial y} + \frac{\partial w}{\partial z}\right) &= 0.
\end{aligned}\right\} \tag{2.1}$$

In these equations ϱ may denote the total density or the density in a certain sub-system, if such a system is considered separately. The potential φ is determined by the combined mass of all sub-systems which may be defined in the system. Summing over all such sub-systems the potential function $\varphi(x, y, z, t)$ is then given by POISSON's equation

$$\nabla^2\varphi = -4\pi G\sum_s \varrho_s. \tag{2.2}$$

3. Energy and area velocity of a particle. Let R, ϑ be polar coordinates in the invariable plane about the centre of gravity, which together with the z-coordinate perpendicular to the plane define a system of cylindrical coordinates. The total linear velocity components of a particle in the R, ϑ, and z directions may be Π, Θ, Z, respectively. We can define I_1 and I_2 as twice the energy per unit mass and twice the area velocity in the fundamental plane, so that

$$I_1 = \Pi^2 + \Theta^2 + Z^2 - 2\varphi, \quad I_2 = R\Theta. \tag{3.1}$$

We put $\varphi = 0$ for infinite distance from the system. I_1 and I_2 will be variables, unless the potential function φ is stationary, and has symmetry of rotation. We have in fact, following a particle in its motion,

$$\frac{D I_1}{D t} = -2 \frac{\partial \varphi}{\partial t}, \qquad \frac{D I_2}{D t} = \frac{\partial \varphi}{\partial \vartheta}. \tag{3.2}$$

In a coordinate system with I_2 and I_1 as axes (Fig. 3) the limit of possible combinations of I_1 and I_2 is defined by the space above the parabola ("characteristic parabola")

$$I_2^2 = R^2 (I_1 + 2\varphi). \tag{3.3}$$

In the case of particles (like the stars) which seldom suffer individual encounters, and if the potential φ approaches a steady state, we may further put the condition

$$I_1 < 0, \tag{3.4}$$

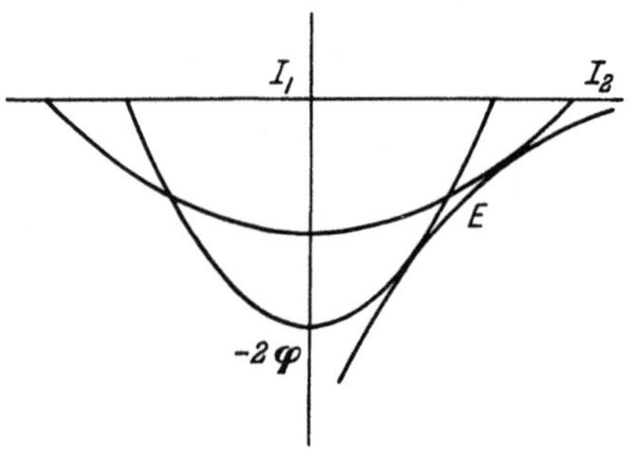

Fig. 3. The characteristic diagram between area velocity and energy for a system of rotational symmetry.

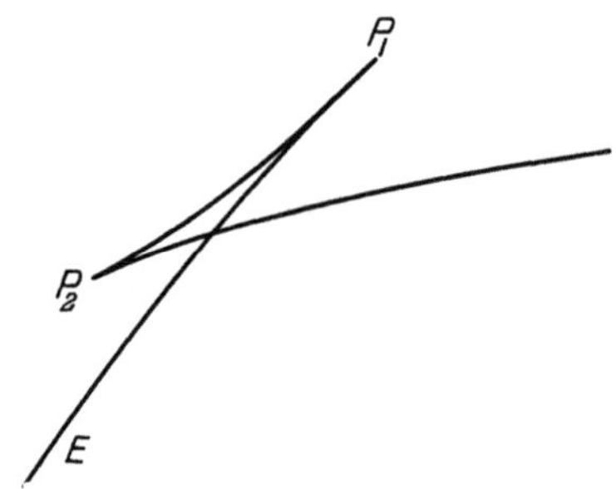

Fig. 4. The characteristic envelope curve E in the case of unstable circular motions.

as particles for which $I_1 > 0$ during any considerable time will leave the system. For a point on the parabola it is easily shown that

$$\frac{d I_1}{d I_2} = 2 \frac{\Theta}{R}, \tag{3.5}$$

i.e. twice the angular velocity of the particle about the centre. For points situated along a radius vector of given angle ϑ in the fundamental plane the characteristic parabolae will have an envelope E.

If φ is stationary, and has rotational symmetry about the z-axis, as well as symmetry with respect to the plane $z = 0$, the curve E represents circular motions in the plane of symmetry. The equation of E is

$$I_2^2 = -R^3 \frac{\partial \varphi}{\partial R}, \qquad I_1 = -R \frac{\partial \varphi}{\partial R} - 2\varphi. \tag{3.6}$$

The curve E is regular for

$$\frac{\partial^2 \varphi}{\partial R^2} + \frac{3}{R} \frac{\partial \varphi}{\partial R} < 0, \tag{3.7}$$

i.e. for stable circular orbits. For unstable orbits in an interval between R_1 and R_2, the curve has two arrow points P_1 and P_2 corresponding to R_1 and R_2, between which E runs "backwards" with $\frac{d I_2}{d R} < 0$ (Fig. 4). It follows that stable circular orbits represent a minimum of I_1 for given I_2. Therefore, if for any reason particles lose energy I_1 while keeping I_2 intact, the particles will ultimately sink to the curve E in the characteristic diagram, and will describe circular orbits in the fundamental plane[1]. The inclination of the tangent of E at the point where

[1] B. LINDBLAD: Monthly Notices Roy. Astronom. Soc. London **94**, No. 3 (1934). = Stockholm Medd. No. 13.

it has contact with the parabola of the radius vector R is given by

$$\frac{dI_1}{dI_2} = 2\,\omega_c \tag{3.8}$$

where ω_c is the angular velocity of circular orbits. If $\varphi = C_1 - C_2 R^2$, the envelope E is a straight line and ω_c is constant. For the "Newtonian" case $\varphi = C/R$, the equation of E is

$$I_2^2 I_1 = -C^2. \tag{3.9}$$

It is evident that the "characteristic diagram" defines very essential properties of the system. Consider the six-dimensional space defined by the coordinates R, ϑ, z and the velocity components Π, Θ, Z. If φ tends to be stationary with rotational symmetry about the z-axis, I_1 and I_2 will be first integrals of the motion and the particles will move on the surfaces $I_1 = \text{const}$, $I_2 = \text{const}$ in a quite complicated way, so that there will be a mixing of particles on these surfaces. This does not mean that every particle will reach every point on its I_1, I_2 surface. For instance, for particles moving in the equatorial plane, in the absence of local encounters, the motion will be confined to this plane and the particles will stay on the intersection of the I_1, I_2-surfaces with the planes $z=0$, $Z=0$. The exchange of particles between certain regions of the I_1, I_2-surfaces may therefore be very slow. In the motion along these surfaces a six-dimensional element $R\,dR\,d\vartheta\,dz\,d\Pi\,d\Theta\,dZ$ containing a certain number of particles will remain constant. This follows from LIOUVILLE's theorem, which has a far wider application because it is valid for a conservative system even when φ varies with time.

4. LIOUVILLE's theorem. In a conservative dynamical system of n degrees of freedom we let $q_1, q_2, \ldots, q_n$ be the Hamiltonian coordinates and $p_1, p_2, \ldots, p_n$ the conjugate variables. We then have

$$\frac{dq_i}{dt} = \frac{\partial H}{\partial p_i}, \qquad \frac{dp_i}{dt} = -\frac{\partial H}{\partial q_i}, \tag{4.1}$$

where H is the Hamiltonian function. If f is the frequency function in the $2n$-dimensional phase space, $q_1, q_2, \ldots, p_n$, it is readily shown that we have the equation of continuity

$$\frac{\partial f}{\partial t} + \sum_{i=1}^{n}\left(\frac{\partial f}{\partial q_i}\frac{dq_i}{dt} + \frac{\partial f}{\partial p_i}\frac{dp_i}{dt}\right) = 0. \tag{4.2}$$

If we define the symbol of differentiation when following the motion of an element in the phase space

$$\frac{D}{Dt} = \frac{\partial}{\partial t} + \sum_{i=1}^{n}\left(\frac{dq_i}{dt}\frac{\partial}{\partial q_i} + \frac{dp_i}{dt}\frac{\partial}{\partial p_i}\right), \tag{4.3}$$

we may write

$$\frac{Df}{Dt} = 0. \tag{4.4}$$

This means that the density of any element in phase space remains constant during its motion. In a non-rotating system x, y, z, we may write, if U, V, W are the total velocity-components of a particle,

$$H = \tfrac{1}{2}(U^2 + V^2 + W^2) - \varphi(x, y, z, t) \tag{4.5}$$

and have from LIOUVILLE's theorem

$$\frac{\partial f}{\partial t} + U\frac{\partial f}{\partial x} + V\frac{\partial f}{\partial y} + W\frac{\partial f}{\partial z} + \frac{\partial \varphi}{\partial x}\frac{\partial f}{\partial U} + \frac{\partial \varphi}{\partial y}\frac{\partial f}{\partial V} + \frac{\partial \varphi}{\partial z}\frac{\partial f}{\partial W} = 0. \quad (4.6)$$

This partial differential equation is equivalent with

$$dt = \frac{dx}{U} = \frac{dy}{V} = \frac{dz}{W} = \frac{dU}{\frac{\partial \varphi}{\partial x}} = \frac{dV}{\frac{\partial \varphi}{\partial y}} = \frac{dW}{\frac{\partial \varphi}{\partial z}}. \quad (4.7)$$

These are the equations of motion. If $I_1 = \text{const}$, $I_2 = \text{const}$, ... $I_6 = \text{const}$, are six independent integrals of the motion, we shall then have

$$f(x, y, z, U, V, W, t) = F(I_1, I_2, \ldots, I_6), \quad (4.8)$$

where F is an arbitrary function. This consequence of LIOUVILLE's theorem was first proved by POINCARÉ[1]. It was established by JEANS as a basis for stellar dynamics in 1915 (Sect. 1). A specification of all the six integrals would mean a detailed description of the actual motions in the system. The importance of the theorem as to the general state of motion in a stellar system depends on the fact that, if the system settles down to a state in which φ is stationary (in a rotating or non-rotating system), certain simple integrals can be found. Moreover, the general complication of the individual motions along the surfaces defined by these integrals in phase space will mean a general process of mixing of matter, so that we can disregard a certain number of the six integrals in the frequency function f to be expected as a result of the mixing process.

5. Quasi-stationary system of rotational symmetry. The most important case is the one which has already been mentioned above, when φ is stationary and has rotational symmetry. In this case I_1 and I_2 in (3.1) are two of the integrals. Except in special cases, and in limited regions of the system, we cannot expect other integrals which can be written down explicitly in analytical form. The equations of mass motion in cylindrical coordinates analogous to (2.1) may be written

$$\left.\begin{aligned}
&\frac{\partial}{\partial t}(\varrho\overline{\Pi}) + \frac{\partial}{\partial R}(\varrho\overline{\Pi^2}) + \frac{1}{R}\frac{\partial}{\partial \vartheta}(\varrho\overline{\Pi\Theta}) + \frac{\partial}{\partial z}(\varrho\overline{\Pi Z}) + \frac{1}{R}\varrho(\overline{\Pi^2} - \overline{\Theta^2}) = \varrho\frac{\partial \varphi}{\partial R},\\
&\frac{\partial}{\partial t}(\varrho\overline{\Theta}) + \frac{\partial}{\partial R}(\varrho\overline{\Theta\Pi}) + \frac{1}{R}\frac{\partial}{\partial \vartheta}(\varrho\overline{\Theta^2}) + \frac{\partial}{\partial z}(\varrho\overline{\Theta Z}) + \frac{2}{R}\varrho\overline{\Theta\Pi} = \frac{\varrho}{R}\frac{\partial \varphi}{\partial \vartheta},\\
&\frac{\partial}{\partial t}(\varrho\overline{Z}) + \frac{\partial}{\partial R}(\varrho\overline{Z\Pi}) + \frac{1}{R}\frac{\partial}{\partial \vartheta}(\varrho\overline{Z\Theta}) + \frac{\partial}{\partial z}(\varrho\overline{Z^2}) + \frac{1}{R}\varrho\overline{Z\Pi} = \varrho\frac{\partial \varphi}{\partial z},\\
&\frac{\partial \varrho}{\partial t} + \frac{\partial}{\partial R}(\varrho\overline{\Pi}) + \frac{1}{R}\frac{\partial}{\partial \vartheta}(\varrho\overline{\Theta}) + \frac{\partial}{\partial z}(\varrho\overline{Z}) + \frac{1}{R}\varrho\overline{\Pi} = 0.
\end{aligned}\right\} \quad (5.1)$$

Still admitting that I_1 and I_2 may vary with time for an individual star by local deviations from dynamical equilibrium, we shall assume that the matter in the system is "well mixed". In order to get a definite meaning to these words, we shall assume that the velocity distribution at a given point resembles a frequency function $F(I_1, I_2)$ at least so far that it is symmetrical with respect to the planes $Z = 0$ and $\Pi = 0$ in the velocity space. We need not assume, however, that there is rotational symmetry about the velocity component Θ. We shall then have at an arbitrary point (R, ϑ, z) of the system

$$\overline{\Pi} = 0, \quad \overline{Z} = 0, \quad \overline{\Pi Z} = 0, \quad \overline{\Theta Z} = 0, \quad \overline{\Pi\Theta} = 0. \quad (5.2)$$

[1] H. POINCARÉ: Leçons sur les hypothèses cosmogoniques, p. 100. Paris: A. Hermann et Fils 1911.

We define at the same point $\Theta_0 = \overline{\Theta}$, and write for the relative velocity dispersions α, β, γ,

$$\alpha^2 = \overline{\Pi^2}, \quad \beta^2 = (\overline{\Theta - \Theta_0})^2, \quad \gamma^2 = \overline{Z^2}. \tag{5.3}$$

If the frequency function were in reality $f = F(I_1, I_2)$ we should have $\alpha = \gamma$, but in consequence of what has been said about the slow exchange of matter between I_1, I_2-surfaces in the neighbourhood of $z = 0$, $Z = 0$, we admit that α and γ may actually differ. The linear velocity Θ_c of the motion in a circular orbit is given by

$$\Theta_c^2 = -R \frac{\partial \varphi}{\partial R}. \tag{5.4}$$

The Eqs. (5.1) are then reduced in the present case to the following two equations

$$\left.\begin{aligned} \frac{\partial (\varrho \alpha^2)}{\partial R} - \frac{\varrho}{R} (\Theta_0^2 + \beta^2 - \alpha^2) &= \varrho \frac{\partial \varphi}{\partial R}, \\ \frac{\partial (\varrho \gamma^2)}{\partial z} &= \varrho \frac{\partial \varphi}{\partial z}. \end{aligned}\right\} \tag{5.5}$$

If we know $\varrho(R, z)$ for a certain sub-system and the potential function $\varphi(R, z)$ these equations can serve to determine Θ_0 and the general velocity dispersion, if we assume that $\beta^2 - \alpha^2$ may be ignored compared with Θ_0, and that $\alpha = \mu \gamma$, where μ is known. The procedure is analogous to that used by JEANS[1] in his analysis of KAPTEYN'S model of the stellar system.

6. Theory of the asymmetrical drift. Limiting velocity dispersions. JEANS'S method was applied by LINDBLAD[2] to the series of sub-systems (Fig. 1), in order to explain the relation between asymmetrical drift and velocity dispersion found by STRÖMBERG[3]. We assume that in the region about the Sun the surfaces of constant density within a sub-system can be approximated by flattened spheroidal surfaces with the small axis coinciding with the axis of rotation of the system, and with nearly one and the same radius a in the galactic plane. This means that the variation of density in the sub-systems with R in the surroundings of the Sun is small. Moving from (R, z) to (R', z') on the same spheroidal surface the component of attraction $\partial \varphi / \partial z$ can be assumed to change approximately in the proportion z'/z. For, if we approximate the entire attracting mass by an infinite series of spheroids, each of density $\delta \varrho$, this relation will hold true for the attraction from an elementary spheroid enveloping the piece of density surface of the sub-system considered, and will be approximately true for the attraction from an inner elementary spheroid, if the piece of density surface considered coincides nearly with a surface confocal with the inner spheroid, as well may be the case if the inner spheroids are very much flattened with not too widely different a. It follows by integrating the second Eq. (5.5) from $z = 0$ to z_0, and from $z' = 0$ to z_0', respectively, where z_0 and z_0' mark the "effective" limit of the sub-system in z for R and R', that we have for points in the galactic plane $(\varrho \gamma^2)' = (z_0'/z_0)^2 (\varrho \gamma^2)$, and thus

$$\varrho \gamma^2 = C c^2 \left(1 - \frac{R^2}{a^2}\right), \tag{6.1}$$

where C is a constant for the sub-system in question, and c and a are the axes of the limiting effective surface of the sub-system. The first Eq. (5.5) gives, if

[1] J. H. JEANS: Monthly Notices Roy. Astronom. Soc. London **82**, 122 (1922).

[2] B. LINDBLAD: Ark. Mat., Astronom. Fys., Ser. A **19**, No. 21 (1925). = Upsala Medd. No. 3.

[3] G. STRÖMBERG: Mt. Wilson Contr. 292 = Astrophys. Journ. **61**, 353 (1925).

we ignore $\beta^2-\alpha^2$,

$$\Theta_0^2 = -R\frac{\partial\varphi}{\partial R} + \frac{R}{\varrho}\frac{\partial(\varrho\alpha^2)}{\partial R}. \tag{6.2}$$

If γ and α are proportional to each other in the region of the Sun, we can write by (6.1)

$$\varrho\alpha^2 = C_1 c^2\left(1-\frac{R^2}{a^2}\right), \tag{6.3}$$

and

$$\frac{\partial(\varrho\alpha^2)}{\partial R} = -2C_1\frac{c^2}{a^2}R = -2\varrho\alpha^2\frac{R}{a^2-R^2}. \tag{6.4}$$

This gives

$$\Theta_0^2 + 2\alpha^2\frac{R^2}{a^2-R^2} = -R\frac{\partial\varphi}{\partial R}. \tag{6.5}$$

We have $\Theta_0=\Theta_c$ for $\alpha=0$, and identify the asymmetrical drift S as

$$S=\Theta_c-\Theta_0. \tag{6.6}$$

If S is small compared with Θ_c, we have by (6.5)

$$S=\frac{\alpha^2}{\Theta_c}\frac{R^2}{a^2-R^2}=\frac{\alpha^2}{\omega_c}\frac{R}{a^2-R^2}. \tag{6.7}$$

The relation derived here corresponds to the nearly parabolic relation between S and α found by STRÖMBERG, if $\frac{a-R}{R}$ is nearly the same for different sub-systems. In the mean we derive[1] for 27 stellar groups treated by STRÖMBERG

$$\frac{a-R}{R}=0.26\pm 0.03 \text{ (m.e.)}.$$

The relation (6.7) has been derived here without any specification concerning the frequency distribution, except that in our neighbourhood $\beta^2-\alpha^2$ can be ignored compared with Θ_0^2 and that α and γ are proportional to each other in this region. Instead, certain assumptions have been made concerning the spatial configuration of the system, which implies a very slow variation of ϱ with R in the neighbourhood of the Sun for the sub-system considered.

An alternative treatment was given by OORT[2]. We may derive directly from the second Eq. (5.5), if S is small compared to Θ_c,

$$S=\frac{\alpha^2}{2\omega_c}\left[-\frac{\partial\log(\varrho\alpha^2)}{\partial R}-\frac{1}{R}\left(1-\frac{\beta^2}{\alpha^2}\right)\right]. \tag{6.8}$$

Assuming the velocity distribution to be ellipsoidal OORT derives (Sect. 8) $\alpha^2=$ const, so that $\frac{\partial\log(\varrho\alpha^2)}{\partial R}=\frac{\partial\log\varrho}{\partial R}$. OORT applies this equation to a number of groups of stars with different internal dispersion. For several groups the data given by STRÖMBERG were adopted. In spite of considerable individual variations in the computed density gradient, there is a general concordance and the mean for $F-M$ stars gives

$$\frac{\partial\log_{10}\varrho}{\partial R}=-0.19\pm 0.05 \text{ (m.e.) per kpc}.$$

---

[1] Cf. B. LINDBLAD [6], p. 1051. In this work the theoretical relation has been derived in a different way, whereas the present development reproduces more nearly the original treatment. In the present case a modern value of $\Theta_c=217$ km/sec has been used.

[2] J. H. OORT: Bull. Astr. Inst. Netverl. 4, 269 (1928).

In this case we have made no explicit assumption concerning the shape of the system, but have assumed the velocity distribution to be ellipsoidal. As we shall see this implies a certain law for the dependence of Θ_0 on R, and thus indirectly a condition on the large scale distribution of matter in the system. As both methods give nearly equivalent results concerning the character of the phenomenon, it is evident that the increase of the asymmetrical drift with increasing velocity dispersion is due to the general limitation of the system in space, and that it is largely independent of local conditions in our neighbourhood.

Integrating the second Eq. (5.5) from z to infinity, where $\varrho = 0$, we may write

$$\left.\begin{aligned} \varrho\gamma^2 &= -\int_z^\infty \varrho \frac{\partial\varphi}{\partial z}\, dz, \\ \text{or}\qquad \varrho\gamma^2 &= \varrho\varphi + \int_z^\infty \varphi \frac{\partial\varrho}{\partial z}\, dz. \end{aligned}\right\} \tag{6.9}$$

As $\partial\varrho/\partial z$ may be assumed to be of the same sign from z to ∞, we may write

$$\gamma^2 = \varphi - \overline{\varphi}, \qquad \overline{\varphi}\int_z^\infty \frac{\partial\varrho}{\partial z}\, dz = \int_z^\infty \varphi\frac{\partial\varrho}{\partial z}\, dz. \tag{6.10}$$

When we approach the limits of the system, $\varphi - \overline{\varphi}$ must decrease to zero, because φ is a continuous, monotonous function of z. If we let z_s denote the practical limit in z beyond which the density of matter may be ignored, we may take the integrals in (6.10) between z and z_s, so that we must have

$$\gamma^2 \to 0 \quad \text{for} \quad z \to z_s. \tag{6.11}$$

This shows that the velocity dispersion will vanish in the outer regions of a limited system. It may be shown in a similar way, that at an effective limit R_s

$$\alpha^2 \to 0 \quad \text{for} \quad R \to R_s. \tag{6.12}$$

7. Ellipsoidal frequency function, based on certain integrals of the motion, as approximation in a limited region. Our experience of the velocity dispersion in our neighbourhood suggests that at least in a limited region about the Sun the function f may be a function of second order in the velocities. The most general function of I_1 and I_2 of the ellipsoidal type will be

$$F(I_1 - 2k_1 I_2 + k_2 I_2^2). \tag{7.1}$$

This gives, however, a velocity ellipsoid where the axis in the radial direction is equal to that in the z-direction. If we ascribe the observed fact, that the axis which lies nearly in the radial direction is the largest one, to disturbances which are very slowly mixed out on the I_1, I_2-surfaces in the neighbourhood of $z=0$, $Z=0$, we can account for this by introducing a third integral which is approximately valid of the form

$$Z_0^2 = Z^2 - 2G(R, z), \tag{7.2}$$

where

$$G(R, z) = \varphi(R, z) - \varphi(R, 0). \tag{7.3}$$

We then find

$$\left.\begin{aligned} &F(I_1 - 2k_1 I_2 + k_2 I_2^2 + k_3 Z_0^2) \\ &= F\{\Pi^2 + (1+k_2R^2)(\Theta - \Theta_0)^2 + (1+k_3)Z^2 - [2\varphi + 2k_3 G + (1+k_2R^2)\,\Theta_0^2]\} \end{aligned}\right\} \tag{7.4}$$

where

$$\Theta_0 = \frac{k_1 R}{1 + k_2 R^2}. \tag{7.5}$$

If α, β, γ are the axes of the velocity ellipsoid in the radial, transverse, and vertical directions, we have

$$\alpha^2 = \text{const}, \qquad \frac{\beta^2}{\alpha^2} = \frac{1}{1 + k_2 R^2}, \qquad \frac{\gamma^2}{\alpha^2} = \frac{1}{1 + k_3}. \tag{7.6}$$

Eqs. (5.5) may then be written

$$\left.\begin{aligned} \frac{\partial \log \varrho}{\partial R} &= \frac{1}{\alpha^2} \frac{\partial \varphi}{\partial R} + \frac{\left(\frac{k_1^2}{\alpha^2} - k_2\right) R - k_2^2 R^3}{(1 + k_2 R^2)^2}, \\ \frac{\partial \log \varrho}{\partial z} &= \frac{1}{\gamma^2} \frac{\partial \varphi}{\partial z}. \end{aligned}\right\} \tag{7.7}$$

If the function F is exponential, we may write explicitly

$$\left.\begin{aligned} \log_{10} \varrho = C + \log_{10} \left[(2\pi)^{\frac{3}{2}} \alpha^2 \gamma\right]^{\frac{3}{2}} - \frac{1}{2} \log_{10} (1 + k_2 R^2) + \\ + \left[\frac{1}{\gamma^2} \varphi(R, z) + \left(\frac{1}{\alpha^2} - \frac{1}{\gamma^2}\right) \varphi(R, 0) + \frac{1}{2\alpha^2} \frac{k_1^2 R^2}{1 + k_2 R^2}\right] \log_{10} e. \end{aligned}\right\} \tag{7.8}$$

The constancy of α and γ, however, for a finite system must be subject to the limitations (6.11) and (6.12).

OORT's constants A and B for differential rotation are [cf. Eq. (1.4)]

$$A = \frac{1}{2}\left(\frac{\Theta_0}{R} - \frac{\partial \Theta_0}{\partial R}\right), \qquad B = -\frac{1}{2}\left(\frac{\Theta_0}{R} + \frac{\partial \Theta_0}{\partial R}\right).$$

This gives

$$A = \frac{k_1 k_2 R^2}{(1 + k_2 R^2)^2}, \qquad B = -\frac{k_1}{(1 + k_2 R^2)^2}. \tag{7.9}$$

We then have

$$\omega = \frac{\Theta_0}{R} = A - B = \frac{k_1}{1 + k_2 R^2}, \tag{7.10}$$

and

$$\frac{\beta^2}{\alpha^2} = 1 - \frac{A}{\omega}. \tag{7.11}$$

8. OORT's analysis of the velocity distribution in the neighbourhood of the Sun. The results just obtained are in several respects equivalent with those derived by OORT[1] in an important analysis of the velocity distribution in our neighbourhood. OORT assumes for the frequency function an ellipsoidal distribution, homogeneous of the second order in a coordinate system following the mean rotational motion Θ_0 at the point considered,

$$f = f_0 \, e^{-h^2 \Pi^2 - k^2 (\Theta - \Theta_0)^2 - l^2 Z^2 - m \Pi (\Theta - \Theta_0) - n \Pi Z - p (\Theta - \Theta_0) Z} \tag{8.1}$$

and inserts this function in the equation of continuity, which in this case may be written, assuming that φ has rotational symmetry.

$$\Pi \frac{\partial f}{\partial R} + \frac{\Theta}{R}\left(\Theta \frac{\partial f}{\partial \Pi} - \Pi \frac{\partial f}{\partial \Theta}\right) + Z \frac{\partial f}{\partial z} + \frac{\partial \varphi}{\partial R} \frac{\partial f}{\partial \Pi} + \frac{\partial \varphi}{\partial z} \frac{\partial f}{\partial z} = 0. \tag{8.2}$$

[1] J. H. OORT: Bull. Astr. Inst. Netherl. 4, 269 (1928).

The general solution (valid also if f is not exponential) is found to be

$$\left.\begin{aligned} h^2 &= c_1 + \tfrac{1}{2} c_5 z^2 + c_6 z, & k^2 &= c_1 + c_2 R^2 + \tfrac{1}{2} c_5 z^2 + c_6 z, \\ l^2 &= c_4 + \tfrac{1}{2} c_5 R^2, & n &= -c_5 R z - c_6 R, \\ m &= p = 0, & \Theta_0 &= \frac{c_3}{k^2} R. \end{aligned}\right\} \quad (8.3)$$

If we take $z=0$ to be a plane of symmetry, we must put $c_6=0$. In addition there is an integrability condition for f_0,

$$c_5\left\{3\left(z\frac{\partial\varphi}{\partial R} - R\frac{\partial\varphi}{\partial z}\right) + (z^2-R^2)\frac{\partial^2\varphi}{\partial R\,\partial z} + Rz\left(\frac{\partial^2\varphi}{\partial R^2} - \frac{\partial^2\varphi}{\partial z^2}\right)\right\} + 2(c_1-c_4)\frac{\partial^2\varphi}{\partial R\,\partial z} = 0. \quad (8.4)$$

In case we put no other restriction on φ than that it shall have a rotational symmetry, we must infer that $c_5=0$, $c_1=c_4$. In the case of a very flattened system of stars, however, $\frac{\partial^2\varphi}{\partial R\,\partial z}$ is likely to be small, so that we may admit $c_1 \neq c_4$ in (8.4). This is equivalent to assuming, as we have done in Sect. 7, that the motion in the z-coordinate is very nearly independent of the motion in the galactic plane. Results identical with Eqs. (7.5) to (7.11) are readily derived from (8.3) with $c_5=c_6=0$. — OORT's treatment of the asymmetrical drift in the same work has been described in Sect. 6.

9. CHANDRASEKHAR's analysis. In a profound analysis CHANDRASEKHAR [*1*] to [*3*] has investigated the conditions under which a general field of differential motions and a generalized ellipsoidal distribution of the residual velocities can occur. The assumptions are:

I. At any point (x, y, z) we can uniquely define a local standard of rest which is a continuous function of position and time.

II. The distribution function[1] $\Psi(x, y, z; U, V, W; t)$ is of the generalized Schwarzschild type, i.e.,

$$\Psi(x, y, z; U, V, W; t) \equiv \Psi(Q+\sigma), \quad (9.1)$$

where Q stands for

$$\left.\begin{aligned} Q = {} & a(U-U_0)^2 + b(V-V_0)^2 + c(W-W_0)^2 + \\ & + 2f(V-V_0)(W-W_0) + 2g(W-W_0)(U-U_0) + \\ & + 2h(U-U_0)(V-V_0); \end{aligned}\right\} \quad (9.2)$$

further, the coefficients of the velocity ellipsoid a, b, c, f, g, and h and the function σ all are continuous functions of position and time.

U_0, V_0, W_0 denote the components of the motion of the local centroid at (x,y,z) and are continuous functions auf x, y, z and of the time t.

III. The motions of the individual stars are governed by a potential function $\mathfrak{V}(x, y, z, t)$ per unit mass.

Substituting the assumed form for Ψ in the equation of continuity following from LIOUVILLE's theorem, we have

$$\frac{D(Q+\sigma)}{Dt} = 0. \quad (9.3)$$

[1] In order to facilitate consultation of the original works, the symbols used by CHANDRASEKHAR are preserved here.

We have

$$Q+\sigma = a\,U^2 + b\,V^2 + c\,W^2 + 2f\,V\,W + 2g\,W\,U + 2h\,U\,V - 2\Delta_1 U - 2\Delta_2 V - 2\Delta_3 W - \chi, \tag{9.4}$$

where

$$\left.\begin{aligned}\Delta_1 &= a\,U_0 + h\,V_0 + g\,W_0,\\ \Delta_2 &= h\,U_0 + b\,V_0 + f\,W_0,\\ \Delta_3 &= g\,U_0 + f\,V_0 + c\,W_0,\end{aligned}\right\} \tag{9.5}$$

and

$$-\chi = Q_0 + \sigma = a\,U_0^2 + b\,V_0^2 + c\,W_0^2 + 2f\,V_0\,W_0 + 2g\,W_0\,U_0 + 2h\,U_0\,V_0 + \sigma. \tag{9.6}$$

There result twenty partial differential equations which break up into three distinct sets of equations. The first group of ten equations involves only the coefficients a to h of the velocity ellipsoid. These equations do not involve any differentiations with respect to time. They are sufficient to determine the dependence of the coefficients of the velocity ellipsoid on the space-coordinates. The second group of six equations involves the Δ's and the time derivatives of the coefficients of the velocity ellipsoid. These equations are sufficient to determine the dependence of the Δ's on the space co-ordinates. The last group of four equations contain the potential function $\mathfrak{V}$ and lead to six other integrability conditions.

We shall consider here the steady state, when the time derivatives vanish. In this case the second group of equations are found to be

$$\left.\begin{aligned}\frac{\partial\Delta_1}{\partial x} &= 0, & \frac{\partial\Delta_2}{\partial x} + \frac{\partial\Delta_1}{\partial y} &= 0,\\ \frac{\partial\Delta_2}{\partial y} &= 0, & \frac{\partial\Delta_3}{\partial y} + \frac{\partial\Delta_2}{\partial z} &= 0,\\ \frac{\partial\Delta_3}{\partial z} &= 0, & \frac{\partial\Delta_1}{\partial z} + \frac{\partial\Delta_3}{\partial x} &= 0,\end{aligned}\right\} \tag{9.7}$$

and the equation in the third group, imposing the condition on $\mathfrak{V}$, runs

$$\Delta_1\frac{\partial\mathfrak{V}}{\partial x} + \Delta_2\frac{\partial\mathfrak{V}}{\partial y} + \Delta_3\frac{\partial\mathfrak{V}}{\partial z} = 0. \tag{9.8}$$

Let Δ be the vector with components $\Delta_1, \Delta_2, \Delta_3$. Then (9.8) may be written

$$\Delta\cdot\operatorname{grad}\mathfrak{V} = 0. \tag{9.9}$$

The solution will be

$$\mathfrak{V}(x, y, z) = \mathfrak{V}(I_1, I_2), \tag{9.10}$$

where I_1, I_2 are any two integrals of the equations

$$\frac{dx}{\Delta_1} = \frac{dy}{\Delta_2} = \frac{dz}{\Delta_3}. \tag{9.11}$$

Under the steady state conditions the Δ's reduce to

$$\left.\begin{aligned}\Delta_1 &= \beta_3\,y - \beta_2\,z + \delta_1,\\ \Delta_2 &= \beta_1\,z - \beta_3\,x + \delta_2,\\ \Delta_3 &= \beta_2\,x - \beta_1\,y + \delta_3\end{aligned}\right\} \tag{9.12}$$

where $\beta_1, \beta_2, \beta_3, \delta_1, \delta_2, \delta_3$ are constants or in vector form

$$\Delta = \boldsymbol{r} \times \boldsymbol{\beta} + \boldsymbol{\delta} \tag{9.13}$$

where $\boldsymbol{r}$, $\boldsymbol{\beta}$, $\boldsymbol{\delta}$ represent the vectors (x, y, z), $(\beta_1, \beta_2, \beta_3)$ and $(\delta_1, \delta_2, \delta_3)$. Choosing the z-axis in the direction of $\boldsymbol{\beta}$ we can write

$$\boldsymbol{\beta} = (0, 0, \beta). \tag{19.14}$$

If $|\beta| \neq 0$, a translation of the origin to $\left(-\frac{\delta_2}{\beta}, +\frac{\delta_1}{\beta}, 0\right)$ gives

$$\Delta_1 = \beta y, \quad \Delta_2 = -\beta x, \quad \Delta_3 = \delta_3. \tag{9.15}$$

Two independent integrals of (9.11) are now found to be

$$\left.\begin{aligned} x^2 + y^2 &= \text{const}, \\ z + \frac{\delta_3}{\beta} \arccos \frac{x}{\sqrt{x^2+y^2}} &= \text{const}. \end{aligned}\right\} \tag{9.16}$$

In cylindrical coordinates $(\tilde{\omega}, \vartheta, z)$ these may be written

$$\left.\begin{aligned} I_1 &= \tilde{\omega} = \text{const}, \\ I_2 &= z + \frac{\delta_3}{\beta} \vartheta = \text{const}. \end{aligned}\right\} \tag{9.17}$$

Finally

$$\mathfrak{B} \equiv \mathfrak{B}\left(\tilde{\omega}, z + \frac{\delta_3}{\beta} \vartheta\right). \tag{9.18}$$

This means that for stellar systems in steady states and with differential motions the potential $\mathfrak{B}$ must necessarily be characterized by *helical symmetry*. In the case of a stellar system of finite extent we must have $\delta_3 = 0$. Thus for stellar systems with differential motions, which are in steady states and are of finite extent, the potential $\mathfrak{B}$ must necessarily by characterized by *axial symmetry*.

Solving the twenty partial differential equations determining the coefficients of the velocity ellipsoid, the Δ quantities oi the mean motions, and the potential, CHANDRASEKHAR performed a complete survey of the general problem of ellipsoidal distributions with differential motions, including both the steady and non-steady states. In addition to the monograph [*3*] the preceding papers [*1*] and [*2*] contain a wealth of theoretical results. We only mention the discussion of the case

$$\mathfrak{B} = \mathfrak{B}_0(t) + \tfrac{1}{2}\left[\alpha_1(t)\, x^2 + \alpha_2(t)\, y^2 + \alpha_3(t)\, z^2\right],$$

which is of importance for the problem of the evolution and stability of ellipsoidal systems ([*2*], pp. 523—573), but also as the first terms in the expansion of the potential for a study of the conditions in a limited region of a system (cf. Sect. 11).

In the case of non-steady states of general ellipsoidal velocity distribution the rotational symmetry will remain, and the characteristics of the motions in the galactic plane may be treated as a two-dimensional case. The essential difference with the presentation in Sects. 7 and 8 ist that, in addition to differential rotations represented by Θ_0, we have motions in the radial directions specified by a velocity Π_0 proportional to R. A consequence is that we shall have a K-term proportional to the distance (positive or negative) in the radial velocities.

The case of spherical velocity distributions is developed in an approach to a general explanation of spiral structure.

SCHÜRER[1] has examined the principles of stellar dynamics and points out the equivalence of the two principal methods: (1) The insertion of a polynomial of second order in the velocities into the fundamental partial differential equation and the determination of the coefficients. This is the procedure applied by EDDINGTON, OORT, CHANDRASEKHAR, GRATTON[2] and others. (2) The solution of the partial differential equation according to LAGRANGE's method, limiting oneself to dynamical systems which allow integrals of first or second order in the velocities. This method has been followed by JEANS, LINDBLAD, HECKMANN and STRASSL (cf. Sect. 11), SHIVESHWARKAR[3], and others. The mathematical conditions for systems of linear or quadratic integrals have been generally formulated without reference to stellar dynamics. SCHÜRER introduces transformations involving changes of scale of space and time, as well as transformations to moving coordinate systems, which he applies to various types of potential functions independent of time occurring in stellar dynamics. The transformations produce non-steady states which are of the same generality as those derived in CHANDRASEKHAR's analysis of non-steady states.

10. Special types of potential functions. Within the general type of potential functions of rotational symmetry there occur certain special cases of interest. These are all contained in the form of potential function derived by EDDINGTON[4] (Sect. 1).

(1) We shall first consider the case $\frac{\partial^2 \varphi}{\partial R\, \partial z} = 0$. It has been shown in Sects. 7 and 8 that this case can be of importance as an approximation in a limited region. Extended to a system as a whole the relation gives $\varphi = F(R) + G(z)$, which cannot be striclty fulfilled in a limited system, because both F and G must become zero, when either R or z go to infinity separately.

(2) In the general case of rotational symmetry we must have $c_5 = 0$ in the formulae of Sect. 8. CAMM[5] has shown that more specialized potential fonctions of rotational symmetry may be obtained, if $c_5 \neq 0$.

If $c_5 \neq 0$, $c_1 = c_4$, the solution is

$$\varphi = \frac{1}{r^2} [F(r) - G(\chi)], \tag{10.1}$$

where r is the distance from the centre, χ the latitude above the central plane, F and G arbitrary functions. In this model the velocity ellipsoid has three unequal axes, except at the centre where the ellipsoid reduces to a sphere. Two of the axes always lie in the plane through the axis of symmetry. If the expression (10.1) shall be valid to $r = 0$, $G(\chi)$ must be a constant, and we have here the spherical case.

If $c_5 \neq 0$, $c_1 \neq c_4$, there is the solution

$$\varphi = \frac{F(\xi) - G(\eta)}{\xi - \eta}, \tag{10.2}$$

where ξ and η are found from the relations

$$R^2 = \frac{1}{2\beta} (\xi + \beta)(\eta + \beta), \quad z^2 = -\frac{1}{2\beta} (\xi - \beta)(\eta - \beta), \quad \beta = \frac{1}{c_5} (c_1 - c_4). \tag{10.3}$$

[1] M. SCHÜRER: Astronom. Nachr. **273**, 230 (1940).
[2] L. GRATTON: Atti R. Acad. Italia, Ser. VII **2**, 1 (1940).
[3] S. W. SHIVESHWARKAR: Monthly Notices Roy. Astronom. Soc. London **95**, 655 (1925).
[4] A. S. EDDINGTON: Monthly Notices Roy. Astronom. Soc. London **76**, 46 (1915).
[5] G. L. CAMM: Monthly Notices Roy. Astronom. Soc. London **101**, 195 (1941).

In this case the velocity ellipsoid has three un-equal axes even at the centre of the system, but also here two of the axes lie in the plane through the axis of symmetry.

Kuzmin[1] has developed the case (10.2) in a somewhat different form, and has applied it for deriving a model of the Galaxy.

Though the special cases (2) are of theoretical interest, it seems very difficult to see how such highly specialized potential functions (except in the spherical case) may arise by an irregular process of mixing, such as described in Sects. 3, 4 and 5, and therefore in the actual application to the galactic structure we shall assume a potential function of general rotational symmetry, and with an equatorial plane of symmetry, without further specifications a priori.

11. Star-streaming in a local region. The theory of a well mixed state of motion, as it has been presented in previous sections, aims at an explanation of the main features of the observed velocity distribution, and does not take into consideration the finer details of the state of motion in the surroundings of the Sun. We may assume that the system fluctuates about a certain mean state of motion, taken during a long interval of time, and that the theory thus far developed refers to such a mean state. Heckmann and Strassl[2] have developed an analysis which aims at a general representation of the detailed properties of the velocity distribution in a local region. The fluctuations in the potential function due to the fluctuations in the density ϱ will be very small, as the potential is derived by integration over the whole system, so that the influence of the density variations on the potential will be largely compensated. Introducing galactocentric coordinates x, y, z, it is therefore assumed that the potential Φ has axial symmetry about the z-axis and that it has the x, y-plane as plane of symmetry. In the neighbourhood of the point $(x_0, 0, z_0)$, where z_0 is small, we define $x - x_0 = \xi$, $y = \eta$, $z - z_0 = \zeta$, and develop the potential function to the second order in ξ, η, ζ, thus

$$\Phi = \Phi_0 + A_1 \xi + C_1 \zeta + \tfrac{1}{2} (A_2 \xi^2 + B_2 \eta^2 + C_2 \zeta^2) \tag{11.1}$$

where we must have

$$A_1 = x_0 B_2, \qquad C_1 = z_0 C_2. \tag{11.2}$$

The development to the second order in $x - x_0$, $y - y_0$, $z - z_0$, is appropriate, if we prescribe that the integrals of the six-dimensional equation of continuity

$$\frac{\partial f}{\partial t} + u \frac{\partial f}{\partial x} + v \frac{\partial f}{\partial y} + w \frac{\partial f}{\partial z} - \Phi_x \frac{\partial f}{\partial u} - \Phi_y \frac{\partial f}{\partial v} - \Phi_z \frac{\partial f}{\partial w} = 0, \tag{11.3}$$

where f is the frequency function $f(x, y, z, u, v, w)$, shall be linear or quadratic in the velocity components u, v, w. The function f is positive, limited, and becomes arbitrarily small for sufficiently large velocities. These conditions may be fulfilled most simply by a function of Gaussian form, or more generally by a sum of such functions,

$$f = \sum A_i \, \mathrm{e}^{-\frac{1}{2} Q_i(u, v, w)} \tag{11.4}$$

where Q_i only contains quadratic or linear terms.

Solving the Hamiltonian equations in the coordinates ξ, η, ζ, and the velocities p_ξ, p_η, p_ζ we obtain six integrals of motion, which may be so arranged

[1] G. G. Kuzmin: Publ. Tartu Obs. **32**, No. 5 (1953). — Publ. Acad. Sci. Est. SSR. **2**, No. 3 (1953).

[2] O. Heckmann and H. Strassl: Veröff. Univ.-Sternw. Göttingen **3**, Nr. 41, 43 (1934).

that ξ, η, ζ remain small as long as t is small. Accepting a Gaussian form $f = e^{-\frac{1}{2}Q}$, the final expression for Q becomes

$$Q = 2\alpha E + 2\alpha_\zeta E_\zeta + \sum_0^3 G_{ik}\gamma_i\gamma_k + \sum_0^3 h_i\gamma_i \quad (G_{ik} = G_{ki}), \tag{11.5}$$

where E, E_ζ, G_{ik}, h_i are arbitrary constants, and

$$\left.\begin{aligned}
2\alpha &= p_\xi^2 + p_\eta^2 + 2A_1\xi + A_2\xi^2 + B_2\eta^2,\\
2\alpha_\zeta &= p_\zeta^2 + 2C_1\zeta + C_2\zeta^2,\\
\gamma_0 &= p_\xi\eta - p_\eta(x_0 + \xi),\\
\gamma_1 &= (A_1 + A_2\xi)\cos\sqrt{A_2}\,t - \sqrt{A_2}\cdot p_\xi \sin\sqrt{A_2}\cdot t = a_0 + a_1\xi - a_2 p_\xi,\\
\gamma_2 &= B_2\eta\cos\sqrt{B_2}\cdot t - \sqrt{B_2}\cdot p_\eta \sin\sqrt{B_2}\cdot t = b_0 + b_1\eta - b_2 p_\eta,\\
\gamma_3 &= (C_1 + C_2\eta)\cos\sqrt{C_2}\cdot t - \sqrt{C_2}\cdot p_\zeta \sin\sqrt{C_2}\cdot t = c_0 + c_1\zeta - c_2\zeta^2.
\end{aligned}\right\} \tag{11.6}$$

Here $b_0 = 0$, and a_2, b_2, c_2 are small quantities. If $U(\xi,\eta,\zeta;t)$, $V(\xi,\eta,\zeta;t)$, $W(\xi,\eta,\zeta;t)$ denote the first order moments of f, which are the *stream components* in the point ξ,η,ζ at the time t, we can define relative velocities u, v, w in a coordinate system following the stream motion by putting

$$p_\xi = u + U, \quad p_\eta = v + V, \quad p_\zeta = w + W. \tag{11.7}$$

The stream motions U, V, W may be expressed to the second order by

$$\left.\begin{aligned}
U &= \sum U_{ijk}\xi^i\eta^j\zeta^k,\\
V &= \sum V_{ijk}\xi^i\eta^j\zeta^k, \qquad (i + j + k = 0, 1, 2),\\
W &= \sum W_{ijk}\xi^i\eta^j\zeta^k.
\end{aligned}\right\} \tag{11.8}$$

The coefficients U_{ijk}, V_{ijk}, W_{ijk} depend on $x_0, a_1, b_1, c_1, a_2, b_2, c_2, E, E_\zeta$ and

$$\left.\begin{aligned}
&K_{00} = G_{00}x_0^2, && K_{01} = G_{01}x_0a_2, && K_{02} = G_{02}x_0b_2, && K_{03} = G_{03}x_0c_2,\\
& && K_{11} = G_{11}a_2^2, && K_{12} = G_{12}a_2b_2, && K_{13} = G_{13}a_2c_2,\\
& && && K_{22} = G_{22}b_2^3, && K_{23} = G_{23}b_2c_2,\\
& && && && K_{33} = G_{33}c_2^2,
\end{aligned}\right\} \tag{11.9}$$

$$\left.\begin{aligned}
L_0 &= x_0\left(G_{01}a_0 + G_{02}b_0 + G_{03}c_0 + \frac{h_0}{2}\right),\\
L_1 &= a_2\left(G_{11}a_0 + G_{12}b_0 + G_{13}c_0 + \frac{h_1}{2}\right),\\
L_2 &= b_2\left(G_{21}a_0 + G_{22}b_0 + G_{23}c_0 + \frac{h_2}{2}\right),\\
L_3 &= c_2\left(G_{31}a_0 + G_{32}b_0 + G_{33}c_0 + \frac{h_3}{2}\right).
\end{aligned}\right\} \tag{11.10}$$

The half-axes of the velocity ellipsoid are

$$\sigma_1 = \frac{1}{\sqrt{E}}, \quad \sigma_2 = \frac{1}{\sqrt{E + K_{00} + 2\left(\frac{\xi}{x_0}K_{00} + K_{02}\right)}}, \quad \sigma_3 = \frac{1}{\sqrt{E_\zeta}}. \tag{11.11}$$

If (l_1, l_2, l_3), (m_1, m_2, m_3), (n_1, n_2, n_3), denote the direction cosines of the axes $\sigma_1, \sigma_2, \sigma_3$, respectively, we have

$$\left.\begin{array}{lll} l_1 = 1, & l_2 = \dfrac{\eta}{x_0} - \dfrac{K_{01}}{K_{00}}, & l_3 = 0, \\ m_1 = -\dfrac{\eta}{x_0} + \dfrac{K_{01}}{K_{00}}, & m_2 = 1, & m_3 = \dfrac{K_{03}}{E - E_\zeta + K_{00}}, \\ n_1 = 0, & n_2 = -\dfrac{K_{03}}{E - E_\zeta + K_{00}}, & n_3 = 1. \end{array}\right\} \tag{11.12}$$

The formulae (11.11) and (11.12) become identical with the results of Sects. 7 and 8, if we put $h_i = 0$ and $G_{ik} = 0$, when one of the indices $i, k \neq 0$.

Among the stream coefficients

$$\begin{gathered} U_{200}, U_{002}, U_{101}, \\ V_{002}, V_{011}, \\ W_{200}, W_{020}, W_{002}, W_{110}, W_{101}, W_{011}, \end{gathered}$$

vanish. The remaining 16 coefficients of first and second order, together with $\sigma_1, \sigma_2, \sigma_3$, and two direction cosines, e.g. l_2 and m_3, are 21 in principle observable quantities whithin the local region which may serve for the determination of the 20 parameters

$$x_0, \frac{a_1}{a_2}, \frac{b_1}{b_2}, \frac{c_1}{c_2}, E, E_\zeta, L_k \text{ and } K_{ik}.$$

HECKMANN and STRASSL apply an analysis in accordance with their stream formula to the results obtained by MINEUR and GUINTINI[1] for the two-dimensional streaming parallel to the galactic plane of the B stars in our surroundings. In addition to the effect of solar motion the differential rotation is clearly shown. The K-effect is small on the average.

HECKMANN and STRASSL have assumed a potential function of such symmetry as is reconcilable with a stationary state. Their analysis describes in detail the fluctuations of the distribution function in a local region which are possible with the prescribed form of the potential function and under a Gaussian form for the velocity distribution. This specification as to the velocity distribution is essential, because without any specification concerning the form of the frequency function the problem is indeterminate, as it is obvious that the distribution of density and velocities at a certain time in a local region is then entirely arbirary. The same problem will be studies in Chap. III of this paper on the basis of a study of differential orbital motions.

12. Existence problems connected with POISSON's equation. When we consider only a certain sub-system, which is defined by certain physical characteristics of the members, we may treat the potential function φ as prescribed and determined by the main mass of the system as a whole. When considering the dynamics of the Galaxy as a whole, including matter in the form of gas and dust, the system must be self-supporting, i.e. the potential must be determined by the density ϱ according to POISSON's equation (2.2) which we write in this case

$$\nabla^2 \varphi = -4\pi G \varrho. \tag{12.1}$$

EDDINGTON showed that the forms of potential function found by him for ellipsoidal velocity distributions (Sect. 10) do *not* admit a self-supporting system (except

[1] H. MINEUR and G. GUINTINI: Bull. Astr., Sér. II, **8**, 227 (1932).

in the spherical case). CAMM[1] has shown, in addition, in the case of an ellipsoidal velocity distribution with equal stellar masses and with a potential function of general rotational symmetry, expanding φ in descending integral powers of the distance from the centre, that there is no solution of (12.1).

KURTH[2] has deduced an important result concerning the non-steady stellar systems introduced by CHANDRASEKHAR and SCHÜRER, which can be derived from steady solutions of LIOUVILLE's equation by transformation of the scales of space and time (Sect. 9). It is shown that a non-steady state demands the density of the system to be constant in space, and that in the case of a varying space-density the solutions are reduced to the conditions of a steady state. CAMM[3] has shown that steady solutions of LIOUVILLE's and POISSON's equations exist in the one-dimensional case of stratification in infinite parallel layers, which is of interest as an approximation to the distribution of stars at right angles to the galactic plane (Sect. 15). Spherical systems have been studied by many authors. A frequency function of exponential ellipsoidal velocity distribution gives infinite radius and mass of the system. CAMM[4] defines an integrated frequency function $F = \int m f \, dm$, where f is the frequency function of particles of mass m, and finds that a frequency function of the type

$$F = \begin{cases} A(-I_1)^{\mu}, & \text{for } I_1 < 0, \\ 0, & \text{for } I_1 \geqq 0, \end{cases}$$

where μ is positive, gives a density distribution corresponding to polytropic gas spheres of index $n = \mu + \frac{3}{2}$. These have finite radius and finite mass for $n < 5$, and infinite radius but finite mass for $n = 5$. For $n > 5$ both radius and mass are infinite[5].

It is important to emphasize that, though the ellipsoidal distribution is a useful approximation in the statistical description of the frequency function of stellar velocities, it is by no means strictly obeyed. It is often found appropriate to use more general frequency functions, e.g. the A-series introduced by CHARLIER. Also an approximation by superposition of sub-systems of different internal velocity dispersions will mean a departure from a unique ellipsoidal distribution. In dynamical theory a strictly ellipsoidal distribution may introduce serious limitations, which are not inherent in the physical problem.

LINDBLAD[6] has approached the problem in the case of a frequency function of type $F(I_1, I_2)$, considered as a sum of similar functions for particles of different mass. The density may be written, after integration over the velocity components in F

$$\varrho = \Psi(R, \varphi), \tag{12.2}$$

and POISSON's equation is equivalent to

$$\varphi(R, z) = G \int_T \varrho' \frac{d\tau'}{r'} \tag{12.3}$$

where r' is the distance from the element of volume $d\tau'$ with density ϱ' to the point (R, z), and T is the entire volume of the system, supposed finite. In the

[1] G. L. CAMM: Monthly Notices Roy. Astronom. Soc. London **101**, 195 (1941).
[2] R. KURTH: Z. Astrophys. **26**, 100 (1949).
[3] G. L. CAMM: Monthly Notices Roy. Astronom. Soc. London **110**, 305 (1950).
[4] G. L. CAMM: Monthly Notices Roy. Astronom. Soc. London **112**, 155 (1952).
[5] For further reference on related subjects the reader may consult the monograph by R. KURTH, Introduction to the Mechanics of Stellar Systems, pp. 124—146. London-New York-Paris: Pergamon Press 1957.
[6] B. LINDBLAD: Stockholms Obs. Ann. **13**, No. 5 (1940).

case of an infinitely thin circular disk it is possible to achieve formal equilibrium by choosing a certain distribution in circular orbits with appropriate angular velocities; this corresponds to the existence of an envelope curve E in the I_1, I_2-diagram (Fig. 3). Along E there will then be a formally infinite frequency F, though the total mass is finite. To this frequency may be added a more general frequency $\Delta F(I_1, I_2)$, while keeping the circular disk intact in space by adjusting the angular velocities of the circular orbits represented by E.

The problem has been taken up in an elaborate discussion by W. FRICKE[1]. He assumes for $F(I_1, I_2)$ an expansion

$$F(I_1, I_2) = \sum_{i,k} \alpha_{ik} I_1^i I_2^k \tag{12.4}$$

where i and k are positive integers or fractions, including zero, which leads to a density function

$$\varrho(R, z) = 2\pi \sum_{\substack{i \\ k=0,1,\ldots n}} \frac{\alpha_{i,2k}}{2k+1} B\left(k + \tfrac{3}{2}, i + 1\right) R^{2k} (2\varphi)^{i+k+\frac{3}{2}} \tag{12.5}$$

where $i > 0$. B is the beta-function, $B = \dfrac{\Gamma(x)\,\Gamma(y)}{\Gamma(x+y)}$. In order that, according to POISSON's equation, $\varphi \to \dfrac{M}{R}$ for $R \to \infty$, and in order that φ shall be finite for $R = 0$ with $\left(\dfrac{\partial \varphi}{\partial R}\right)_0 = 0$, further limitations on i and k appear, so that finally

$$\left.\begin{aligned} F &= f_1 + f_2, \\ f_1 &= I_1^{\frac{5}{2}} \sum_{\substack{k=0,1,\ldots n \\ l=0,1,\ldots m}} \alpha_{k+l,2k} I_1^{k+l} I_2^{2k}, \\ f_2 &= \sum_{\substack{i \\ k=0,1,\ldots n'}} \alpha_{i,2k+1} I_1^i I_2^{2k+1}. \end{aligned}\right\} \tag{12.6}$$

The exponents i in f_2 are arbitrary positive broken or integral numbers. FRICKE shows that for an appropriate choice of the coefficients α several important characteristics of the actually observed velocity distribution are reproduced. An example is

$$\left.\begin{aligned} f = (2\varphi - V^2)^{\frac{5}{2}} \{ & 0.5 + 100\,(2\varphi - V^2)^{11} (R v)^{20} + \\ & + 49.5\,(2\varphi - V^2)^{10} (R v)^{19} [1 + (2\varphi - V^2)^2 (R v)^2] \}, \end{aligned}\right\} \tag{12.7}$$

where V is the total velocity, v the velocity component in the direction of rotation, expressed in the circular velocity as unit. The velocity distribution in the galactic plane is shown in Figs. 5 and 6.

The flattening of the velocity distribution towards the galactic plane is not reproduced in this theory. It can be ascribed to the action of the interstellar matter close to this plane and to the star-clouds coupled to this matter.

13. Deviations from a stationary state studied by means of the mass motions. In many ways the properties of a stellar system in equilibrium may be compared with a figure of equilibrium of a continuous medium. The deviations from the form of the stationary state may be defined by a deformation of the effective boundary in the equatorial plane and by general variations of the internal distribution of density, and we have to take into account the variations of the gravitation potential due to such variations. As the formation of stars and dust out of the gas in a primitive system will mean a reduction of the pressure in the extended medium, and a greater concentration of matter towards the equatorial

[1] W. FRICKE: Astronom. Nachr. **280**, 193 (1951).

plane, it is important to consider how an increase in the general degree of flattening of a system will effect such variations. For instance certain waves of

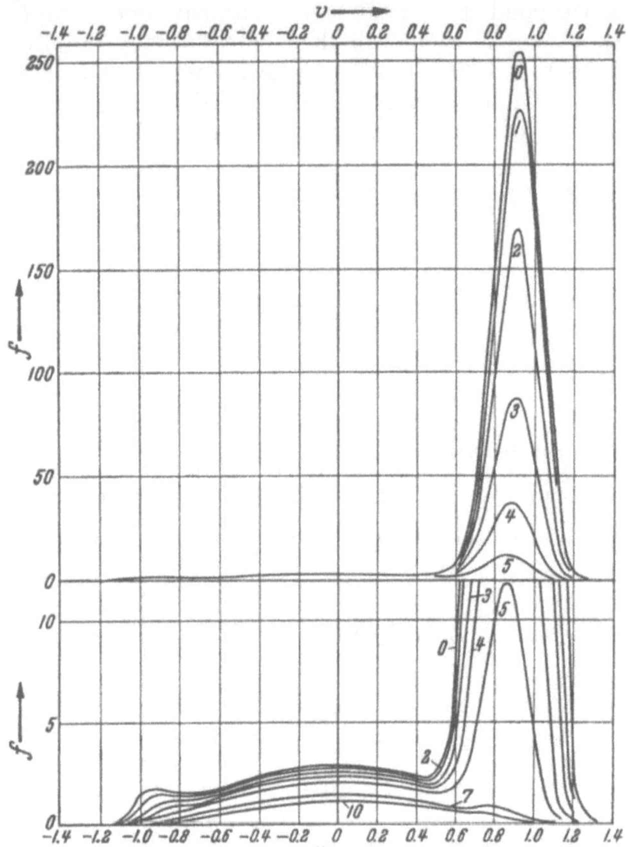

Fig. 5. Equidensity curves in the velocity space (Fricke), ΠZ-plane.

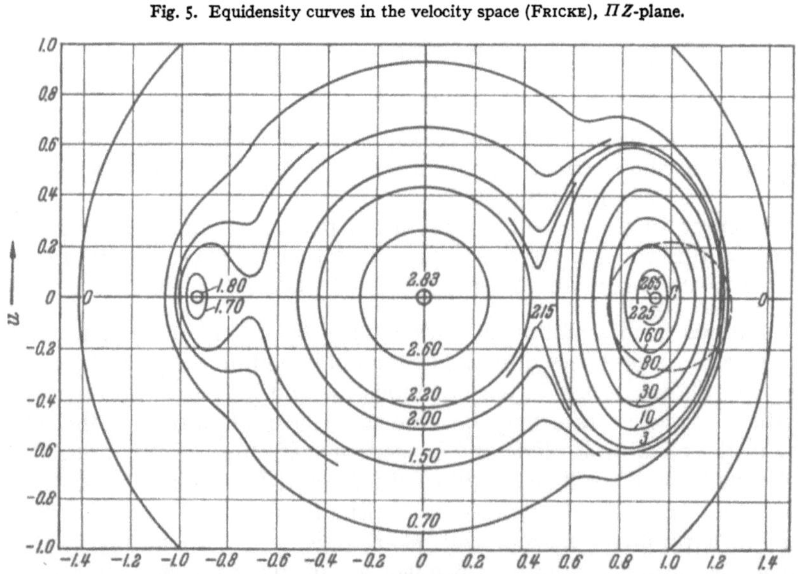

Fig. 6. Equidensity curves in the velocity space (Fricke), $\Pi \Theta$-plane.

deformation which render the spheroidal equilibrium form "ordinarily" unstable at a certain flattening (sectorial harmonic waves) will exist also in the case of "quasi-spheroidal" stellar systems of nearly uniform angular speed. The internal density variations, however, are probably of greater importance in an actual system like our Galaxy. The difficulty is that, considering conditions in a small region of the system, we have no immediate equation of state between density and pressure, which in our case we may define as $\frac{1}{3}\varrho(\overline{U^2}+\overline{V^2}+\overline{W^2})$, using the notations of Eq. (2.1). It has been shown by LINDBLAD[1], however, that, considering not too small a region in the neighbourhood of the equatorial plane, there will be an approximately adiabatic relation between the internal velocity dispersion and the density variation under the corresponding variation of the gravitation potential. Reference may be given here, in addition to the paper just mentioned, to works by LINDBLAD[2], COUTREZ[3], LANGEBARTEL[4], LINDBLAD and LANGEBARTEL[5]. We can obtain a first order theory by neclecting terms of higher order than the first in the variation from the stationary state. With the help of the "adiabatic" relation just mentioned, and integrating along the z-axis, it is possible to write down the partial differential equations of second order for the total variation $\mathfrak{D}_1$ of the mass $\mathfrak{D}$ in cylinders of unit cross-section and with axes parallel to the z-axis.

The most important modes of density variation in a region where the angular velocity of rotation does not change rapidly with the distance from the centre, appear to be of the type

$$\mathfrak{D}_1 = A_s\,\mathfrak{D}_0\left(\frac{r}{a}\right)^s \cos(\sigma t + s\vartheta)\,;\quad s = 1, 2, 3, \ldots \tag{13.1}$$

where $\mathfrak{D}_0$ is the value of $\mathfrak{D}$ for the stationary state. r and ϑ are polar coordinates in the equatorial plane, and a is the "effective" radius in the equatorial plane, in a coordinate system following the angular speed of rotation, A_s a constant. It is of particularly great importance if σ may become complex, so that there will be a wave increasing indefinitely in amplitude with the time. The conditions for instability have been investigated in detail for the waves $s=1, 2, 3$, and the results may be summarized briefly as follows.

The case $s=1$ represents a simple asymmetry. This wave will possibly be unstable for a small interval of flattening about $c/a=0.23$. The wave $s=2$ becomes unstable with increasing flattening at $c/a=0.21$. The wave $s=3$ is unstable for $c/a<0.18$.

The greatest interest attaches to the wave $s=2$, because it tends to explain the formation of "barred" spiral structures. The density variation is accompanied by the development of four whorl motions. Two of these whorls occur in the direction of rotation. They are situated on maxima of $\mathfrak{D}_1/\mathfrak{D}_0$ on one and the same diameter, symmetrically with respect to the centre at the points $r=0.775\,a$. The two other whorls occur in the other direction and are situated also at $r=0.775\,a$, on a diameter at right angles to that of the direct whorls and on minima of $\mathfrak{D}_1/\mathfrak{D}_0$. The two maxima of $\mathfrak{D}_1/\mathfrak{D}_0$, together with the central condensation, give an immediate explanation of the phenomenon of "the bar".

---

[1] B. LINDBLAD: Stockholms Obs. Ann. **16**, No. 1 (1950).

[2] B. LINDBLAD: Stockholms Obs. Ann. **12**, No. 4 (1936); **13**, No. 10 (1941); **14**, No. 1 (1942); **15**, No. 4 (1948).

[3] R. COUTREZ: Stockholms Obs. Ann. **15**, No. 3 (1948), and [*4*].

[4] R. LANGEBARTEL: Stockholms Obs. Ann. **17**, No. 3 (1951).

[5] B. LINDBLAD and R. LANGEBARTEL: Stockholms Obs. Ann. **17**, No. 6 (1953).

14. Detailed analysis of sub-systems in the Galaxy. The sub-systems introduced in the dynamical description of the Galaxy may be taken from a formal point of view as an artificial means of dissecting a complicated state of distribution and motion into simple units. In this way the number of sub-systems might be infinite, together representing a continuous, though very complicated, distribution function. It is clear, however, that in many cases sub-systems of different kinematical properties can be defined by isolating groups of different physical properties of the individuals. This is the case already when grouping the stars according to the ordinary spectral types, for instance by considering separately *O* and *B* stars, *A* stars, late type giants, late type dwarfs. Such sub-systems become even more pronounced when we consider the planetary nebulae, the variables of spectral type *Me*, the cluster type variables, and the globular clusters. BOTTLINGER[1], who discussed further the properties of various groups, recognised by STRÖMBERG and OORT (Sects. 1 and 6), wanted to distinguish between three main sub-systems, as follows.

System I of the strongest degree of flattening formed by the *W, O, B, c* stars, the long period δ Cephei stars, diffuse nebulae and calcium clouds, most of the open clusters and part of the *A* stars; system II of about 4 to 5 times the thickness of system I, containing the great mass of the stars, *G, K, M* giants, a great many dwarf stars, part of the *A* stars, further *N* stars, planetary nebulae, most of the novae, probably part of the *Me* stars of long period, and part of the stars of spectral type *R* and *S*; system III containing most of the *Me* stars, RR Lyrae stars and stars of high velocity relative to the Sun, also the globular clusters.

BOTTLINGER studied approximations to the law of central force, especially the Newtonian law $F=\frac{C}{R^2}$ and the law $F=\frac{aR}{1+bR^3}$. He concludes that the orbits of the objects belonging to system III no doubt pass partly through the nuclear regions of the Galaxy, and represent perhaps part of the physical properties of these regions.

We have here ideas which get their clear formulation in BAADE's theory of the two stellar populations, and which will be further discussed in Sect. 33.

Detailed studies of the distribution in space and of the kinematical properties of sub-systems have been performed especially by KUKARKIN[2] and PARENAGO[3]. Various sub-systems correspond to different sequences on the spectrum-luminosity diagram (Hertzsprung-Russell diagram). Other sub-systems of importance are defined by the Cepheid variables and the long-period variables.

Adopting the expression given in Sect. 7 and 8 for the velocity of rotation Θ_0, PARENAGO writes [cf. Eq. (8.3)].

$$\frac{R}{\Theta_0}=\frac{c_1}{c_3}+\frac{c_2}{c_3}R^2. \tag{14.1}$$

For classical Cepheids ranging from $R=5$ to $R=11$ kpc he derives $\omega(R)$ from the observed radial velocities W relative to the centroid, using the rigorous formula of differential galactic rotation (1.1) in the form,

$$\omega(R)=\frac{1}{R_0}[f(R,R_0)+\Theta_0(R_0)], \tag{14.2}$$

where

$$f(R,R_0)=W\operatorname{cosec} l'\sec b, \tag{14.3}$$

[1] K. F. BOTTLINGER: Veröff. Univ.-Sternwarte Berlin-Babelsberg **10**, H. 2 (1933).

[2] B. V. KUKARKIN: Investigation of the structure and evolution of stellar systems on the basis of variable star studies. Publ. House of Techn. and Theor. Lit., Moscow-Leningrad 1949. German ed. Akademie-Verlag, Berlin 1954.

[3] P. P. PARENAGO: Course of stellar astronomy, 3rd ed. Moscow 1954.

l' being the galactic longitude counted from the centre and b the galactic latitude. Assuming $R_0 = 7.2$ kpc and Θ_0 at the Sun 234 km/sec, he finds the values of $\Theta_0(R)$ in Table 1.

The corresponding relation between R/Θ_0 and R^2 is illustrated in Fig. 7. The straight line corresponds to

$$\frac{c_1}{c_3} = 0.014 \sec \cdot \text{kpc km}^{-1}; \qquad \frac{c_2}{c_3} = 0.00063 \sec \cdot \text{kpc}^{-1} \text{km}^{-1}.$$

Table 1.

R kpc	$\Theta_0(R)$ km/sec
5	219
6	238
7	237
7.2	234
8	226
9	220
10	215
11	212

For a very flattened sub-system Θ_0 coincides nearly with the velocity Θ_c of the circular orbit, for which $\Theta_c^2 = -R\frac{\partial \varphi}{\partial R}$. PARENAGO then derives as an approximation for the potential in the equatorial plane of the galactic system

$$\varphi(R, 0) = \frac{\varphi_c}{1 + \varkappa R^2}, \tag{14.4}$$

where

$$\varphi_c = \frac{c_3^2}{2 c_1 c_2}, \qquad \varkappa = \frac{c_2}{c_1}. \tag{14.5}$$

The values of φ_c and $\varkappa$ in this approximation may be derived from accurately determined values of the coefficients of differential rotation A and B. We have by (14.1)

$$\frac{c_1}{c_3} + \frac{c_2}{c_3} R_0^2 = \frac{1}{A - B}. \tag{14.6}$$

Eqs. (7.9) correspond to

$$A = \frac{c_2 c_3 R^2}{(c_1 + c_2 R^2)^2}, \qquad B = \frac{-c_1 c_3}{(c_1 + c_2 R^2)^2}. \tag{14.7}$$

Applied to $R = R_0$ these relations give

$$\frac{c_2}{c_1} = -\frac{A}{B R_0^2}, \tag{14.8}$$

and finally

$$\left.\begin{aligned} \frac{c_1}{c_3} &= -\frac{B}{(A - B)^2}, \\ \frac{c_2}{c_3} &= \frac{A}{(A - B)^2} \frac{1}{R_0^2}. \end{aligned}\right\} \tag{14.9}$$

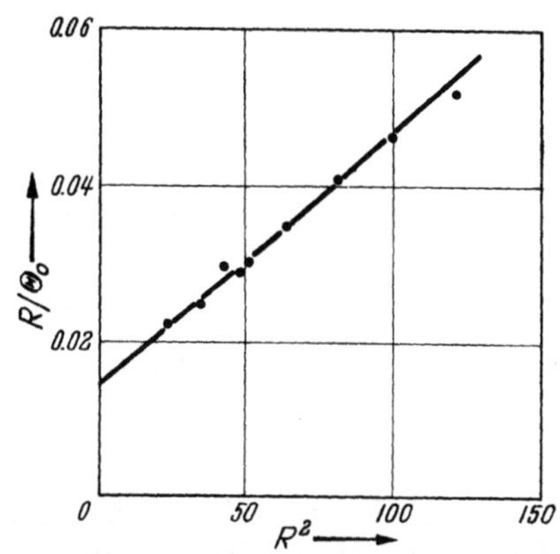

Fig. 7. Relation between velocity of rotation and distance from the centre for Cepheids in the surroundings of the Sun (PARENAGO).

PARENAGO assumes as best values $A = 19.4$, $B = -13$ (km/sec)/kpc and finally derives

$$\left.\begin{aligned} \varphi_c &= -\frac{1}{2} R_0^2 \frac{(A - B)^4}{A B} = 11.3 \cdot 10^{14}\,\text{cm}^2/\text{sec}^2, \\ \varkappa &= -\frac{A}{B R_0^2} = 0.0288\,\text{kpc}^{-2}. \end{aligned}\right\} \tag{14.10}$$

The velocity of escape $\Theta_\infty = \sqrt{2\varphi}$ may readily be calculated. For the neighbourhood of the Sun, PARENAGO obtains $\Theta_\infty - \Theta_c = 68$ km/sec, which corresponds well to the value of 65 km/sec estimated by OORT from the distribution of the velocity vectors of the high velocity stars (Sect. 1).

15. The distribution of stars and the component of force at right angles to the galactic plane. Of considerable interest from the dynamical point of view is the distribution of the component of force per unit mass

$$K(z) = \frac{\partial \varphi}{\partial z},$$

close to the perpendicular to the galactic plane passing through the Sun. Under the very general conditions of "mixing" of orbital motions leading to Eqs. (5.5) we have

$$K(z) = \frac{1}{\varrho}\frac{\partial}{\partial z}(\varrho\gamma^2). \tag{15.1}$$

This shows that $K(z)$ may be determined, if we know the distribution of ϱ with z and the change of γ^2 with z.

Following Oort[1] we may approach somewhat more in detail the process of mixing of the motions in z. Following a star in its orbital motion we call Z_0 the value of the velocity component Z when the star crosses the galactic plane $z=0$. Let a full period of oscillation, until the star again crosses the galactic plane in the same direction, be T. The fraction of T spent in the layer between z and $z+dz$ will be $\frac{2dz}{ZT}$. Most of the stars surrounding the Sun will move in orbits of a fairly limited range of variation in R so that the variation of $K(z)$ for a given z along the orbit will be small. Taking the means of T and Z_0 for several oscillations of one and the same star, T will be a function of Z_0. If the present population about the Sun is "well mixed up", we may therefore set $T=T(Z_0)$, and further write

$$Z_0^2 = Z^2 - 2\int_0^z K(z)\,dz, \tag{15.2}$$

where $K(z)$ is valid for the perpendicular through the Sun. If $p(Z_0)$ is the frequency function of Z_0, we may write for all stars in a cylinder of unit cross section with axis perpendicular to the galactic plane,

$$f(z,Z)\,dz\,dZ = p(Z_0)\,dZ_0\frac{2dz}{ZT(Z_0)}.$$

As $Z\,dZ = Z_0\,dZ_0$, we have

$$f(z,Z) = \frac{2p(Z_0)}{Z_0T(Z_0)} = f(Z_0), \tag{15.3}$$

i.e. $f(z,Z)$ is a pure function of Z_0. It can be shown that in this case (15.1) is valid for an arbitrary velocity function $f(Z_0)=f(0,Z)$. If $f(Z_0)$ is a Gaussian function

$$f(Z_0) = \frac{l}{\sqrt{\pi}}\,e^{-l^2Z_0^2} \tag{15.4}$$

it follows by integration in Z

$$\varrho(z) = \varrho(0)\,e^{2l^2\int_0^z K(z)\,dz}. \tag{15.5}$$

The density in this case follows the same law as that of a gas in isothermal equilibrium. This equation was first used for determination of $K(z)$ by Kapteyn[2]. If the function $f(0,Z)$ differs somewhat from a Gaussian distribution, it can most often be approximated by a sum of Gaussian components of different moduli, which gives a corresponding sum for $\varrho(z)$ in (15.5).

Oort finds the distribution in Z from radial velocites in the polar zones. His main results for various spectral types are given in Tables 2 and 3.

The variation of density with height above and below the galactic plane was taken from van Rhijn's results[3]. Results for A stars and $G0-K2$ giant stars

[1] J. H. Oort: Bull. Astr. Inst. Netherl. **6**, 249 (1932).
[2] J. C. Kapteyn: Mt. Wilson Contr. No. 230 = Astrophys. Journ. **55**, 302 (1922).
[3] P. J. van Rhijn: Publ. Astronom. Lab. Groningen No. 38.

Table 2.

Spectrum	M	l_1	l_2	ϑ_1
$B0-B9$		0.200		
$A0-A9$		0.084	0.020	0.957
$F0-F9$		0.049	0.020	0.957
$G0-G9$	$\leqq +0.4$	0.059		
$G0-G9$	$\leqq +0.5$	0.042	0.020	0.910
$K0-Mc$	$\leqq +3.4$	0.039	0.020	0.915
$K0-Mc$	$\geqq +3.5$	0.050	0.020	0.775

Table 3.

M_{vis}	l_1	l_2	l_3	ϑ_1	ϑ_2	ϑ_3
$< +1.5$	0.084	0.043	0.020	0.33	0.62	0.05
$+1.5$ to $+3.5$	0.084	0.046	0.020	0.22	0.73	0.05
$+3.5$ to $+5.5$		0.046	0.020		0.92	0.08
$+5.5$ to $+7.5$		0.046	0.020		0.85	0.15
$\geqq +7.5$		0.050	0.020		0.78	0.22

derived by LINDBLAD[1] and PETERSSON[2] were also considered. If $K(z)$ is expressed in cm/sec², z in parsecs, and if $\overline{Z^2}$ denotes the average value of the square of the velocities in km/sec, we find from (15.1)

$$K(z) = 7.48 \cdot 10^{-9}\, \overline{Z^2} \frac{\partial}{\partial z} \log_{10} (\overline{Z^2} \varrho) . \qquad (15.6)$$

Table 4.

z	$K(z)$ 10^{-9} cm/sec²
50	-0.77
100	-1.55
150	-2.59
200	-3.52
250	-3.78
300	-3.86
400	-3.68
600	-4.44

The results adopted are the means of four groups of different absolute magnitudes, and are given in Table 4.

The absolute value of $K(z)$ increases proportionally with z from $z=0$ to $z=200$, between $z=200$ and $z=500$ it remains practically constant and equal to 3.8×10^{-9} cm/sec².

Knowing $K(z)$ and the parameters of the velocity distribution the density ϱ may be calculated. If the luminosity function of the stars in a space element is known, the number of stars $A(m)$ with apparent magnitudes $m-0.5$ to $m+0.5$ may be computed and compared with the observed $A(m)$. The results show good agreement, and differences often indicate that corrections are needed to the luminosity function. For the faintest magnitudes the computation of $A(m)$ needs an extrapolation of $K(z)$ to heights about 500 pc. Two such extrapolations have been given based on different assumptions concerning the shape of the main bulk of the central part of the system.

The decrease of $K(z)$ with z near the galactic plane may be used for computing the total density of matter ϱ_t in the neighbourhood of the Sun. OORT found

$$\frac{\partial K(z)}{\partial z} = -5.62 \cdot 10^{-3} \sec^{-2}.$$

We may write

$$\frac{\partial K(z)}{\partial z} = -4\pi G x \cdot \varrho_t . \qquad (15.7)$$

---

[1] B. LINDBLAD: Ark. Mat., Astronom. Fys., Ser. B **19**, No. 15 (1926). = Upsala Medd. No. 14.

[2] H. PETERSSON: The distribution of distances and velocities of stars in the Carrington zone on the basis of spectrophotometric analysis. Inaug.-Diss. 1926. = Upsala Medd. No. 29.

Oort computes x on various assumptions concerning the shape and mass of the galactic system as a whole. The resulting values of ϱ_t vary only slightly and Oort derives as most probable value $\varrho_t = 0.092$ solar masses per cubic parsec or 6.3×10^{-24} g/cm³.

It may be remarked here that the factor x may be found directly from Oort's constants A and B for circular motions in the neighbourhood of the Sun. Writing

$$\frac{\partial^2 \varphi}{\partial R^2} + \frac{1}{R}\frac{\partial \varphi}{\partial R} + \frac{\partial^2 \varphi}{\partial z^2} = -4\pi G \varrho_t, \tag{15.8}$$

and

$$A = \frac{1}{4\omega_c}\left(\frac{\partial^2 \varphi}{\partial R^2} - \frac{1}{R}\frac{\partial \varphi}{\partial R}\right), \quad B = \frac{1}{4\omega_c}\left(\frac{\partial^2 \varphi}{\partial R^2} + \frac{3}{R}\frac{\partial \varphi}{\partial R}\right), \quad \omega_c = A - B \tag{15.9}$$

we easily find

$$\frac{\partial^2 \varphi}{\partial z^2} = -4\pi G \varrho_t - 2(A^2 - B^2). \tag{15.10}$$

This relation has been derived independently by Lindblad[1] and Parenago[2]. If $A = 0.021$, $B = -0.007$ (km/sec)/pc, we have $2(A^2 - B^2) = 0.83 \cdot 10^{-30}$ sec^{-2}. This gives, in Oort's case

$$4\pi G \varrho_t = 4.79 \times 10^{-30} \text{ sec}^{-2},$$

and $\varrho_t = 0.085$ solar masses per cubic parsec, or 5.8×10^{-24} g/cm³.

In a recent work Oort[3] has revised his results on the basis of new material, especially as to the distribution in z of giant stars of type K. Matter assumed to belong to population type I, including gas and dust, is treated separately. The force $K(z)$ has been determined which is consistent with the observations available, and which at the same time fulfills the requirement that $K(z)$ must be due to the attraction by stars and dark matter. The quantity $\frac{\partial^2 \varphi}{\partial R^2} + \frac{1}{R}\frac{\partial \varphi}{\partial R}$, which is equal to $2(A^2 - B^2)$ near the galactic plane, can be estimated outside of this plane by means of Schmidt's model of the galactic system (Sect. 16). Poisson's equation (15.8) then gives $\frac{\partial K(z)}{\partial z}$, if $\varrho_t(z)$ is known. Our knowledge of the distribution of this attracting mass, though incomplete, is sufficient to put rather stringent conditions on $K(z)$, so that it can be determined much more reliably than if the requirements of Poisson's law are left out of consideration. Knowing the form of $K(z)$ the density $\varrho(z)$ and the star numbers $A(m)$ may be computed for any group of stars of known velocity distribution and luminosity function. Numerically undetermined parameters may be adjusted by comparison with observation. On the supposition that the z-distribution of the mass of general stellar origin is the same as that of the K giants Oort derives the solution given in Table 5.

The total mass density $\varrho_t(0)$ at $z = 0$ was found to be 10.0×10^{-24} g/cm³, or 0.15 solar masses per cubic parsec, with an estimated uncertainty of about 10%. The ratio of mass-to light-density, M/L, in solar units is 2.4. For the total contents of a cylinder perpendicular to the galactic plane $M/L = 4.2$. The mass-density corresponding to the stars within 20 parsec from the Sun has been estimated by Gliese[4] to be 0.057 solar masses per cubic parsec. This would mean, in comparison to tha total density at $z = 0$ just mentioned, that an amount of

[1] B. Lindblad: Festskrift för Östen Bergstrand. Upsala 1938.
[2] P. P. Parenago: Course of stellar astronomy, 3rd ed. Moscow 1954.
[3] J. H. Oort: Bull. Astr. Inst. Netherl. (in press).
[4] W. Gliese: Z. Astrophys. **39**, 1 (1956).

about 62% of the mass in our neighbourhood consists of gas, dust, and matter of unknown nature.

A very much higher density in the immediate neighbourhood of the galactic plane has been proposed by SCHILT[1], who maintains that in our neighbourhood there is no clear evidence of a Schwarzschild distribution in the Z velocities, but that the stars appear to show strong perference for definite Z-values in the galactic plane. Trying to find from radial velocities and proper motions the gradient of the mean square of the Z-velocities towards the plane of symmetry (which is estimated by him to lie 6 parsec to the north of the position of the Sun) he derives the very high density 5.3×10^{-22} g/cm³.

As a guide to solutions of $K(z)$ we may consider the abstract case of a stellar system which is stratified in plane parallel layers. Assume that all the stars are of the same mass m. Let $\nu(z)$ be the number of stars per unit volume, $\Omega(z)$ the potential energy per unit mass, with $\Omega(0)=0$, $(z, Z)\, dz\, dZ$ as before the number of stars in a region of unit cross section and thickness dz with Z-velocities in the range Z to $Z+dZ$. In a steady state we have in analogy with (4.6)

$$Z \frac{\partial f}{\partial z} - \frac{d\Omega}{dz}\frac{\partial f}{\partial Z} = 0 \qquad (15.11)$$

and POISSON's equation

$$\frac{d^2\Omega}{dz^2} = 4\pi G m \nu(z). \qquad (15.12)$$

Table 5.

z parsec	$K(z)$ 10^{-9} cm/sec²
50	−1.37
100	−2.50
200	−4.28
300	−5.40
400	−6.17
600	−7.10
800	−7.65
1000	−8.05

CAMM[2] has shown that these conditions lead to a non-linear partial differential equation in two independent variables. He derives three special solutions. One is a Gaussian distribution in Z. The two others correspond mathematically to the equilibrium conditions in an adiabatic gas, one to the case when the ratio γ between the specific heats is less than unity and the other to $1<\gamma<3$, which gives systems of finite thickness. The general solution of (15.11) is

$$f(z, Z) = F[\tfrac{1}{2} Z^2 + \Omega(z)], \qquad (15.13)$$

analogous to (15.3), where F is arbitrary. PRENDERGAST[3] has shown that a general solution of (15.11) and (15.12), in which the force $K(z)$ vanishes in the plane $z=0$ dividing the system in equal masses, can be obtained by means of two quadratures in the following way

$$\frac{1}{2}\left(\frac{d\Omega}{dz}\right)^2 = 4\pi G m \left\{\sigma^2 \nu(0) - \int_{-\infty}^{+\infty} Z^2 F\left[\frac{1}{2} Z^2 + \Omega\right] dZ\right\}, \qquad (15.14)$$

where σ is the velocity dispersion. In this case $f(z, Z)$ is an even function of z and the system is symmetrical with respect to the plane $z=0$. CAMM and PRENDERGAST have derived the following expression for the total mass M in a column of unit cross section

$$M^2 = \frac{2\sigma^2}{\pi G} \varrho_0, \qquad (15.15)$$

where ϱ_0 is the density in the plane $z=0$.

[1] J. SCHILT: Astronom. J. 55, 97 (1950).

[2] G. L. CAMM: Monthly Notices Roy. Astronom. Soc. London, **110**, 305 (1950).

[3] K. H. PRENDERGAST: Astronom. J. **59**, 260 (1954).

A variation of the procedure of obtaining the force $K(z)$ has been given by NAHON[1]. Suppose that the analytical form of $K(z)$ is known, with certain unknown parameters to be determined. We define

$$\Omega(z) = -\int_0^z K(z)\,dz. \tag{15.16}$$

If the density function $\varrho(z)$ is known, we may then form the function $\varrho(\Omega)$. In terms of the frequency function f we have on the other hand

$$\varrho(\Omega) = 2\int_0^\infty f(Z^2 + 2\Omega)\,dZ. \tag{15.17}$$

Let m_{2n} be the moment of order $2n$ in the velocity distribution at a certain value of Ω, i.e.

$$\varrho\, m_{2n} = 2\int_0^\infty f(Z^2 + 2\Omega)\,Z^{2n}\,dZ. \tag{15.18}$$

We also have by (15.17)

$$\int_0^\infty \varrho(\Omega + y)\,y^{n-1}\,dy = 2\int_0^\infty y^{n-1}\,dy\int_0^\infty f[Z^2 + 2(\Omega + y)]\,dZ.$$

By putting

$$Z = R\cos\vartheta, \qquad \sqrt{2y} = R\sin\vartheta$$

it is found that

$$\int_0^\infty \varrho(\Omega + y)\,y^{n-1}\,dy = \frac{(n-1)!}{1\cdot 3\ldots(2n-1)}\,2\int_0^\infty f(R^2 + 2\Omega)\,R^{2n}\,dR,$$

so that by (15.18)

$$\varrho\, m_{2n} = \frac{1\cdot 3\ldots(2n-1)}{(n-1)!}\int_0^\infty \varrho(\Omega + y)\,y^{n-1}\,dy. \tag{15.19}$$

We may then write, integrating from the level z_0 to another level z_1 with densities ϱ_0 and ϱ_1,

$$\varrho_0\, m_2(z_0) - \varrho_1\, m_2(z_1) = \int_{\Omega_0}^{\Omega_1} \varrho(\Omega)\,d\Omega, \tag{15.20}$$

and for the moments of fourth order

$$\tfrac{1}{3}\left[\varrho_0\, m_4(z_0) - \varrho_1\, m_4(z_1)\right] = \int_{\Omega_0}^{\Omega_1} \varrho(\Omega)\,(\Omega - \Omega_0)\,d\Omega + (\Omega_1 - \Omega_0)\,\varrho_1\, m_2(z_1). \tag{15.21}$$

If z_1 is a high level, the terms multiplied by ϱ_1 are small.

According to OORT's results we may put with good approximation

$$\left.\begin{aligned} K(z) &= -k^2 z, & \Omega &= k^2\frac{z^2}{2}, & &\text{for } z<a,\\ K(z) &= -k^2 a, & \Omega &= k^2 a\left(z - \frac{2}{a}\right), & &\text{for } z>a. \end{aligned}\right\} \tag{15.22}$$

[1] F. NAHON: Bull. Astronomique 21, 56 (1957).

Now we can accept this form of $K(z)$, but treat k^2 and a as unknown parameters to be determined. Then we have

$$\frac{1}{k^2}\left[\varrho_0 m_2(z_0) - \varrho_1 m_2(z_1)\right] = \int_{z_0}^{z_1} \varrho\, \varkappa\, dz, \quad \text{where} \quad \begin{cases} \varkappa = z & \text{for } z < a, \\ \varkappa = a & \text{for } z > a \end{cases} \tag{15.23}$$

and

$$\left.\begin{aligned} &\frac{1}{3k^4}\left[\varrho_0 m_4(z_0) - \varrho_1 m_4(z_1)\right] - (\omega_1 - \omega_0)\frac{1}{k^2}\varrho_1 m_2(z_1) = \int_{z_0}^{z_1} \varrho\,\varkappa\,(\omega - \omega_0)\, dz, \\ &\text{where} \qquad \omega = \frac{z^2}{2} \qquad \text{for } z < a, \\ &\qquad\qquad\;\; \omega = a\left(z - \frac{a}{2}\right) \quad \text{for } z > a. \end{aligned}\right\} \tag{15.24}$$

The integrals in the right-hand members of (15.23) and (15.24) can be computed from $\varrho(z)$ for different values of a. The dependence of m_2/k^2 and m_4/k^4 on z_0 derived in this way has to be compared with observed values of m_2 and m_4 in order to determine a and k^2. Instead of m_4 we may use the "excess" of the distribution

$$\varepsilon = \frac{m_4 - 3m_2^2}{3m_2^2} \tag{15.25}$$

which may easily be formed according to (15.23) and (15.24). The small terms containing ϱ_1 may be computed in accordance with one of CAMM's types of systems. If we disregard the terms multiplied by ϱ_1, ε is independent of k^2. If ε is found to vary with a according to Eqs. (15.23) and (15.24), a single observed value of ε will theoretically determine a. An observed value of m_2 then determines k^2. If we do not use ε, but assume that the run of m_2 according to (15.23) varies with a, we shall need the values of m_2 for at least two values of z_0 in order to determine both a and k^2.

NAHON has used the determinations of $\varrho(z)$ by T. ELVIUS[1] for the spectral types $B8$—$A3$, $F2$—$F5$, $F8$—dG, gG—K. Information about the velocity distribution has been obtained from R. E. WILSON's General Catalogue[2] and from an investigation by VYSSOTSKY and JANSSEN[3]. NAHON finds a pronounced increase of m_2 with distance, especially for the late type giants. The data derived from the density distributions agree well with the observed values of m_2 when taking $a = 200$ pc, in good agreement with OORT, and taking $a k^2 = 14 \times 10^{-30}\,\text{sec}^{-2}$. The density in the central galactic stratum would then be about 0.23 solar masses per cubic parsec. It should be pointed out, however, that the value of k^2 will depend on the scale of ϱ, which is very sensitive to the scale of the distances of the stars. This again depends on the mean absolute magnitudes of the stars, and even a small systematic error in these will have a perceptible influence on the scale of ϱ. The agreement found between the A stars and the late type giants, however, and the agreement of the constant level of K for $z > 200$ pc with that found by OORT and VAN WOERKOM[4] for the RR Lyrae stars, no doubt speak in favour of the larger value of k^2.

WOOLLEY[5] determines the form of the potential $\varphi(z)$ by POISSON's equation for the stratified case assuming single or composite Gaussian functions for the

[1] T. ELVIUS: Stockholms Obs. Ann. **16**, No. 5 (1951).

[2] R. E. WILSON: Papers of the Mount Wilson Obs. **8**, 601. Publ. Carnegie Inst. Washington 1953.

[3] A. N. VYSSOTSKY and EDITH M. JANSSEN: Astronom. J. **56**, 58 (1951).

[4] J. H. OORT and A. J. J. VAN WOERKOM: Bull. Astr. Inst. Netherl. **9**, 185 (1941).

[5] R. v. D. R. WOOLLEY: Monthly Notices Roy. Astronom. Soc. London **117**, 198 (1957).

velocity distribution of the stars in general. Application to the observed distribution $\varrho(z)$ of the A stars and with a new determination of the dispersion in Z for this type (median velocity 6.7 km/sec with an assumed observational dispersion of 3.0 km/sec) gives for the total density in the galactic plane 0.18 solar masses per cubic parsec. WOOLLEY treats also the case of an infinitely long cylinder of uniform density and square section, roughly corresponding to a spiral arm. The solution corresponds well to the actual distribution of the A stars for a single Gaussian distribution, but leads to an increase of the general density.

16. The potential field and the general distribution of mass in the Galactic system. The rotation theory of the Galaxy immediately gave certain general information regarding the potential field of the system. The phenomenon of the asymmetrical drift interpreted as a differential velocity of rotation of sub-systems in the Galaxy indicated the rotational velocity at the distance R_0 of the Sun from the centre to be of the order 200 to 300 km/sec. The geometrical centre of the system of globular clusters, and the distribution of short period Cepheids in a region near the centre[1], in both cases with reasonable correction for interstellar absorption, gave a value of the order 10 kpc for the distance R_0. This corresponds to a period of revolution of about $2 \cdot 10^8$ years for circular orbits in the neighbourhood of the Sun. The corresponding value of the central force $\partial \varphi / \partial R$ gives a mass of the system inside the orbit of the Sun of the order of 10^{11} solar masses.

Plotting the apices of high velocity stars as a function of galactic longitude, no apices of higher velocity than 65 km/sec are found in the longitudes 20 to 85°[2]. The limiting velocity may correspond to the velocity required to reach the boundary of the system and is likely to be close to the velocity of escape[3] for velocity vectors near the Sun. In this case the velocity of escape near the Sun should be close to the sum of the circular velocity and 65 km/sec.

Further data bearing on the central field close to the Sun were obtained from the constants A and B of differential galactic rotation. A and B may differ from one sub-system to another, but for sub-systems of small internal velocity dispersion we may assume that A and B coincide with the values for circular motion. We may write

$$A = -\frac{1}{2} R_0 \frac{\partial \omega}{\partial R}, \qquad B = -\omega - \frac{1}{2} R_0 \frac{\partial \omega}{\partial R}, \tag{16.1}$$

where ω is the angular velocity of the mean motion in the sub-system at the distance R_0 from the centre. We may add here the expression for the ratio of the axes α and β of the "regular" velocity ellipsoid (Sect. 7, 8 and 18)

$$\frac{\beta^2}{\alpha^2} = 1 - \frac{A}{\omega} = \frac{-B}{A-B},$$

which is of value in the determination of B.

At last the observations of the 21-cm line of hydrogen in regions closer to the centre than the Sun give an important estimate of the quantity $A R_0$.

A determination of R_0 which carries a high weight has been made by BAADE[4] from the RR Lyrae type variables in a field near the galactic centre. He finds

$$R_0 = 8.2 \text{ kpc}.$$

[1] Cf. B. LINDBLAD: [6], p. 1069 (1933).

[2] J. H. OORT: Publ. Astronom. Lab. Groningen **1926**, No. 40.

[3] B. LINDBLAD: Vjschr. Astr. Ges. **61**, 265 (1926). — J. H. OORT: Bull. Astr. Inst. Netherl. **4**, 269 (1928).

[4] W. BAADE: Report of Commission 33 of the I.A. U., Dublin Meeting 1955. Trans. Internat. Astronom. Union **9** (1957).

Recent values of A and B have been derived by OORT[1] and by H. R. MORGAN and OORT[2]. The results are in close agreement with the values of A and B accepted by the Leiden observers in the reductions of the observations of the 21-cm line. The values of R_0, A, and B adopted are.

$$R_0 = 8.2\,\text{kpc}, \quad A = 19.5\,(\text{km/sec})/\text{kpc}, \quad B = -6.9\,(\text{km/sec})/\text{kpc},$$

which give for the angular velocity close to the Sun $\omega_0 = 26.4$ (km/sec)/kpc and for the circular velocity $\Theta_c = 216.5$ km/sec.

Considerably lower values of A have been derived by WEAVER[3] for cepheid variables and B stars, $A = 10.8$ and 13.2 (km/sec)/kpc, respectively, by determining from radial velocities and distances the slope of the $\Delta\omega(R)$ curve near the Sun. PARENAGO[4] finds for B a value $B = -13$ (km/sec)/kpc. On the other hand BLAAUW[5] found from distant $B2-B5$ stars $A = (20.0 \pm 1.8)$ (km/sec)/kpc, and HINS and BLAAUW[6] derived from the axial ratio of the velocity distribution of faint stars $\frac{-B}{A-B} = 0.24 \pm 0.04$; these results agree well with the values adopted by the Leiden observers.

In a certain galactic longitude l' the radial velocity of the 21-cm line will be a maximum at the circular orbit which touches the line of sight, if we assume that the motions are essentially circular motions about the centre of the system, and that the angular velocity $\omega(R)$ increases with decreasing R. SCHMIDT[7] derived for the maximum radial velocity, differentiating the formula $W = R_0(\omega - \omega_0)\sin l'$, the approximation

$$W_{\max} = 2A R_0 \sin l' (1 - \sin l')\left[1 - \frac{B}{\omega}(1 - \sin l')\right], \qquad (16.2)$$

From observations in $l' = 15$ to $30°$ he found

$$A R_0 = 156\,\text{km/sec},$$

which is again in good agreement with the values of A and R_0 accepted for the reduction of the 21-cm observations.

With values of R_0 and ω_0 accepted the maxima of the radial velocities for different l' give immediately the function $\omega(R)$. The results of the observations of the 21-cm line by the Leiden observers and the parallel work by Australian radio astronomers in the southern part of the Galaxy open a new era in the studies of the Galactic structure and dynamics. In the present connection we may remark that the knowledge of the function $\omega(R)$ has laid a new foundation for a more detailed analysis of the dynamics of the stellar system. The run of $\omega(R)$ for regions interior to the orbit of the Sun has been derived by KWEE, MULLER and WESTERHOUT[8]. A brief summary of their results is given in Table 6.

Table 6.

R kpc	ω (km/sec)/kpc
0.84	163
1.90	91.6
2.93	66.9
3.90	53.6
5.09	43.0
6.13	36.7
7.16	31.4
8.20	26.4

A study of the general distribution of mass in the Galactic system on the basis of the determinations of $\omega(R)$ has been made by SCHMIDT[9]. Various models

[1] J. H. OORT: Coll. Int. Centre Nat. Rech. Sci. **25**, 60 (1950).

[2] H. R. MORGAN and J. H. OORT: Bull. Astr. Inst. Netherl. **11**, 379 (1951).

[3] H. WEAVER: Astronom. J. **60**, 202, 208, 211 (1955).

[4] P. P. PARENAGO: Course of stellar astronomy, 3rd ed. Moscow (1954). — Astronom. J. USSR **32**, No. 3 (1955).

[5] A. BLAAUW: Report of Comm. 33 of the I. A. U. Tans. Internat. Astronom. Union **8**, 505 (1954).

[6] C. H. HINS and A. BLAAUW: Bull. Astr. Inst. Netherl. **10**, 365 (1948).

[7] M. SCHMIDT: Bull. Astr. Inst. Netherl. **13**, 15 (1956).

[8] K. K. KWEE, C. A. MULLER and G. WESTERHOUT: Bull. Astr. Inst. Netherl. **12**, 211 (1954).

[9] M. SCHMIDT: Bull. Astr. Inst. Netherl. **13**, 15 (1956). — SCHMIDT uses the symbol $\tilde{\omega}$ for R.

have earlier been used to represent the general mass distribution in the Galaxy. LINDBLAD introduced spheroidal sub-systems (Fig. 1) of nearly one and the same extension in the galactic plane, later adding a mass at the centre to account for the variation of ω with R. WYSE and MAYALL[1], in a study of the rotational velocities in extragalactic nebulae, introduced a plane circular disk in which the density changes gradually outwards. PEREK[2] introduced a model consisting of a non-homogeneous spheroid. PARENAGO[3] used a model which gives a variation of potential in the galactic plane admitting an ellipsoidal distribution everywhere in the system (cf. Sect. 14). OORT has constructed several models. In the most recent one[4] he introduced seven homogeneous spheroids of different dimensions. SCHMIDT gets a starting point for his approximation by the observation that the central force in the galactic plane $K(R, 0)$ follows nearly the law

$$K(R, 0) = PR + Q, \quad \text{where} \quad P < 0. \tag{16.3}$$

As a first approximation of the mass distribution he introduces a non-homogeneous spheroid, built up by an infinite number of homogeneous spheroids of density $d\varrho$ with varying great axis a but with the same eccentricity e. The central component of force parallel to the galactic plane K_R at a point (R, z) inside the spheroid may be written

$$K_R(R, z) = 4\pi e^{-3} R \sqrt{1 - e^2} \int_0^{\arcsin e} \varrho \sin^2\beta \, d\beta, \tag{16.4}$$

where ϱ is a function of a, and the relation between β and a for fixed (R, z) is determined by

$$R^2 \sin^2\beta + z^2 \tan^2\beta = a^2 e^2. \tag{16.5}$$

The value of a for the spheroid passing through (R, z) is given by

$$a^2 = R^2 + \frac{z^2}{1 - e^2}. \tag{16.6}$$

$K(R, 0)$ will be linear in R for a density law

$$\varrho = p + \frac{q}{a}, \tag{16.7}$$

where p and q are constants, and we obtain

$$\left.\begin{aligned} P &= 2\pi e^{-3} \sqrt{1 - e^2} \,(\arcsin e - e\sqrt{1 - e^2})\, p, \\ Q &= 4\pi e^{-2} \sqrt{1 - e^2} \,(1 - \sqrt{1 - e^2})\, q. \end{aligned}\right\} \tag{16.8}$$

P and Q may be derived from the coefficients A and B of differential rotation in the surroundings of the Sun, and are given by

$$P = \omega(\omega - 4A), \qquad Q = 4R_0 \omega A. \tag{16.9}$$

From the assumed values of A and B follows

$$P = -1362 \text{ (km/sec)}^2/\text{kpc}^2, \qquad Q = +16885 \text{ (km/sec)}^2/\text{kpc}.$$

A third equation for determining the three unknowns p, q, e is given by the condition that the density of matter about the Sun shall agree with the observed

[1] A. B. WYSE and N. U. MAYALL: Astrophys. Journ. **95**, 5 (1942).
[2] L. PEREK: Contr. Astr. Inst. Masaryk Univ. **1**, No. 6 (1948).
[3] P. P. PARENAGO: Astronom. J. USSR. **27**, 329 (1950); **29**, 245 (1952).
[4] J. H. OORT: Astrophys. Journ. **116**, 233 (1952).

value ϱ_0. Assuming the Sun to be inside the spheroid this gives

$$p + \frac{q}{R_0} = \varrho_0. \tag{16.10}$$

SCHMIDT adopts a value of ϱ_0 corresponding to 0.085 solar masses per cubic parsec derived from OORT's determination of the force $K(z)$ (Sect. 15). p, q, and e are derived from Eqs. (16.8) and (16.10). For the "shell" made up of spheroids of $a > R_0$ a density law $\varrho = s/a^4$ is assumed, where s is determined to give the best representation of the density distribution of faint stars at high galactic latitudes according to the results obtained by OORT[1].

In following approximations SCHMIDT superimposes several non-homogeneous spheroids of different degree of flattening in an attempt to represent various

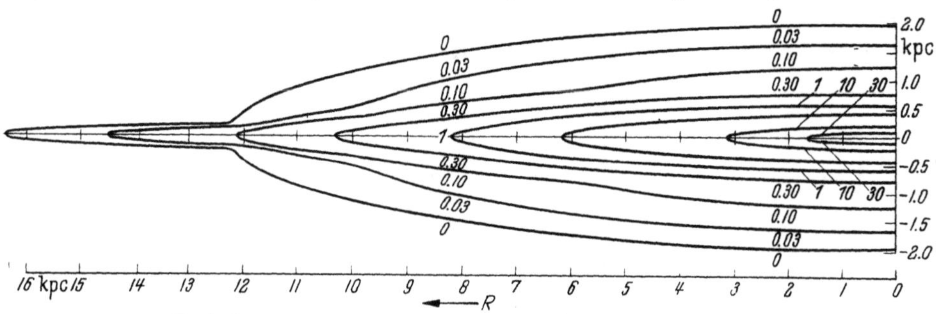

Fig. 8. Schematic model of the mass distribution in the galactic system (SCHMIDT).

stellar populations. In the third approximation the system is composed of four spheroids. The first one is aimed to represent population group I, composed of the hydrogen gas and the O, B and A stars, the second one the ordinary F—M stars, the third one high-velocity F—M stars, the fourth one unknown objects. In the derivation of the second spheroid SCHMIDT has taken into account the density surfaces of faint stars. The third spheroid has a larger ratio of the axes c and a, and the fourth spheroid has been introduced to give, together with the three others, the total observed $K(R, 0)$. The main data of the four spheroids

Table 7.

	Spheroid			
	1	2	3	4
a_r	16.4	12.3	12.3	8.68
c_r	0.33	0.86	1.97	0.61
ϱ	$-100 + \frac{1640}{a}$	$-300 + \frac{3690}{a}$	$-70 + \frac{861}{a}$	$-1670 + \frac{14499}{a}$
ϱ_0	100	150	35	98
M	18500	81800	43600	160300

are given in Table 7. The density ϱ is given in the unit $1/G$ (km/sec)²/kpc² $= 1.578 \times 10^{-26}$ g/cm³ $= 2.32 \times 10^{-4}$ ⊙/pc³ and the total mass M in the unit $1/G$ kpc (km/sec)² $= 4.63 \times 10^{38}$ g $= 2.32 \cdot 10^5$ ⊙, G being the constant of gravitation. a_r and c_r are the limiting axes of the non-homogeneous spheroids expressed in kpc.

[1] J. H. OORT: Bull. Astr. Inst. Netherl. **8**, 233 (1938).

A slight retouch to the system in order to produce the observed $\omega(R)$ is made by the homogeneous spheroids of Table 8, which have all the axial ratio $c/a=0.07$.

Table 8.

a	0.84	1.90	2.93	3.90	5.09	6.13	7.16	8.10	8.20
ϱ	+27000	+608	+99	−113	−199	−69	+1	+9	+20

The total mass is here negative, $M=-3000$. In order that the four components 1 to 4 in Table 7 shall correspond to true sub-systems in the Galaxy in the sense defined in earlier sections it is necessary that the distribution oı density and velocity dispersion obeys Eqs. (5.5). The value of γ is obtained by direct integration of the second equation. Assuming $\alpha=\mu\gamma$, where the factor μ is supposed to be known, we obtain from the first equation $\overline{\Theta^2}=\Theta_0^2-\beta^2$. This is an analysis first applied by JEANS[1] to KAPTEYN's local system. SCHMIDT in this way examines the third and fourth components and gets at least a partial verification that they are possible sub-systems.

Table 9.

R kpc	Θ_c km/sec	ω (km/sec)/kpc
8.2	216	26.4
9	205	22.8
10	195	19.5
12	175	14.6
15	150	10.0
20	127	6.3
30	102	3.4

SCHMIDT extensively tabulates the potential φ and the components of force $K_R=\partial\varphi/\partial R$ and $K_z=\partial\varphi/\partial z$ as functions of R and z. It is of great importance to get a reliable extension of the function $\omega(R)$ outside of the Sun's orbit. For every model which has been derived from $\omega(R)$ in the interior this will mean an extrapolation which is uncertain as long as strong independent evidence concerning the density distribution in the exterior regions is not at hand. SCHMIDT's model yields the values of Θ_c and ω for $R>R_0$ given in Table 9.

SCHMIDT finds that the velocity of escape at the Sun according to his model should exceed Θ_c by 70 km/sec, in good agreement with the results from observation.

The total mass of the system is found to be $0.702\cdot 10^{11}$ solar masses. The value of the total mass is not very sensitive to the exact model used. If we assume the system to be composed of a spherical central mass M_1 entirely inside the Sun and a strongly flattened homogeneous spheroid of radius R_1 and mass M_2 we have the formulae[2]

$$M_1=\frac{4R_0^3}{3G}\,\omega A\,,\quad M_2=\frac{4R_1^3}{3\pi G}\,\omega\left(\omega-\frac{4}{3}A\right). \tag{16.11}$$

If R_0 and R_1 are expressed in parsec and Θ_0 in km/sec, we have

$$M_1=0.617\cdot 10^{36}\,\frac{A}{\omega}\,R_0\Theta_0^2\,g\,,\qquad \frac{M_2}{M_1}=\frac{1}{\pi}\left(\frac{R_1}{R_0}\right)^3\left(\frac{\omega}{A}-\frac{4}{3}\right). \tag{16.12}$$

With the values of A, ω and R_0 accepted above, and with R_1 slightly exceeding R_0, we then obtain $M=M_1+M_2=0.9\cdot 10^{11}$ solar masses, with only a small fraction in the mass M_2.

Among the physically defined sub-systems of the Galaxy the system made up of the globular clusters is of great interest. An analysis based on JEANS's

[1] J. H. JEANS: Monthly Notices Roy. Astronom. Soc. London **82**, 122 (1922).
[2] B. LINDBLAD, [6], p. 1068 (1933).

principles was tried by LINDBLAD[1] with an attempt to determine the rotation of the system. The distribution of the clusters was studies by DE KORT[2] and OORT[3] trying to eliminate the effects of absorption. OORT compared the distribution with the one to be expected from different mass models. MAYALL[4], in his publication of an extensive list of new radial velocities determined by him, states that "no satisfactory interpretation of the individual peculiar velocities in terms of a differential or general rotation has been found". PARENAGO[5] has tried to determine the run of the rotational velocities with the distance from the centre. PEREK[6] found a general retrograde rotational motion of the system. Von HOERNER[7] concluded that the majority of clusters have highly eccentric orbits. SCHMIDT[8] has analysed the distribution of clusters of SHAPLEY'S concentration classes I to VIII in the regions of space where we may expect that all clusters of these classes have been discovered. He finds the axial ratio of surfaces of constant space density intersecting the galactic plane at distances from the centre $R=4$, 9, 16 kpc to be 0.24, 0.60, 0.84, respectively. There is thus with certainty a general flattening, which appears to decrease considerably when we go outwards from the centre. The total number of clusters of concentration classes I to VIII is estimated to be about 160, of which 62 have been observed. As to the motion of rotation SCHMIDT finds that a general direct rotation of about 80 km/sec is indicated.

III. Velocity distribution from statistics of differential orbital motions.

17. Epicyclic orbits. We may assume that in the neighbourhood of the Sun the centroid of sub-systems with small velocity dispersion coincides nearly with the motion in a circular orbit about the centre of the system. It is therefore natural, when analyzing the individual motions in the system, to introduce a coordinate system in which the origin follows the circular motions at a certain distance R_0 from the centre. We introduce therefore a coordinate system ξ, η, ζ, where

$$\xi = R - R_0, \qquad \eta = R_0 \vartheta. \tag{17.1}$$

R is the radius vector and ϑ the angular coordinate at the centre C. ζ is the coordinate at right angles to the galactic plane. Let $F(R)$ denote the force per unit mass directed towards the centre. It is of advantage to introduce provisionally as reference circle the circular orbit for which the area velocity of the motion is equal to that of the particle. This is always possible, except for certain cases of unstable circular motions. We denote the radius of this orbit with R_1 and write $\xi_1 = R - R_1$, $\eta_1 = R_1\vartheta$. For the motion in the galactic plane, the general theory of central motion gives for ξ_1 and η_1 the differential equations

$$\left.\begin{aligned} \frac{d^2\xi_1}{dt^2} + \varkappa_1^2\xi_1 &= a_1\xi_1^2 + b_1\xi_1^3 + \cdots \\ \frac{d\eta_1}{dt} &= R_1\omega_1\left(\frac{R_1^2}{\xi_1^2} - 1\right). \end{aligned}\right\} \tag{17.2}$$

[1] B. LINDBLAD: Ark. Mat., Astronom. Fys., Ser. A **19**, No. 27 (1926). = Upsala Medd. No. 4.
[2] J. DE KORT: Bull. Astr. Inst. Netherl. **9**, 189 (1941).
[3] J. H. OORT: Bull. Astr. Inst. Netherl. **9**, 193 (1941).
[4] N. U. MAYALL: Astrophys. Journ. **104**, 290 (1946).
[5] P. P. PARENAGO: Astronom. J. USSR. **24**, 167 (1947).
[6] L. PEREK: Ann. d'Astrophys. **11**, 185 (1948).
[7] S. VON HOERNER: Z. Astrophys. **35**, 255 (1955).
[8] M. SCHMIDT: Bull. Astr. Inst. Netherl. **13**, 15 (1956).

We define for an arbitrary R

$$\left.\begin{aligned} a &= \frac{6\omega^2}{R} - \frac{1}{2}\frac{d^2F}{dR^2}, & b &= -\left(10\frac{\omega^2}{R^2} + \frac{1}{6}\frac{d^3F}{dR^3}\right), \\ \varkappa^2 &= 4\omega(\omega - A), & A &= -\frac{1}{2}R\frac{d\omega}{dR}. \end{aligned}\right\} \tag{17.3}$$

Let the notations a_1, b_1, etc. indicate the values of a, b, etc. taken for $R = R_1$. Ignoring higher powers of ξ_1 than the first, the approximate solution becomes

$$\left.\begin{aligned} \xi_1 &= c\cos\varkappa_1(t - t_0), \\ \eta_1 &= -c\sqrt{\frac{\omega_1}{\omega_1 - A_1}}\sin\varkappa_1(t - t_0). \end{aligned}\right\} \tag{17.4}$$

Defining

$$R_1 = R_0 + c_1,$$

we have for the reference circle R_0

$$\left.\begin{aligned} \xi &= c_1 + c\cos\varkappa_1(t - t_0), \\ \eta &= -R_0(\omega_0 - \omega_1)(t - t_1) - \frac{R_1}{R_0 + c_1}c\sqrt{\frac{\omega_1}{\omega_1 - A_1}}\sin\varkappa_1(t - t_0). \end{aligned}\right\} \tag{17.5}$$

We may write $\varkappa$ for $\varkappa_1$, only remembering that $\varkappa$ is now valid for $R_0 + c_1$. Ignoring second order quantities in the coefficients of the periodic functions we may then write

$$\left.\begin{aligned} \xi &= c_1 + c\cos\varkappa(t - t_0), \\ \eta &= -2\bar{A}c_1(t - t_1) - c\sqrt{\frac{\omega_0}{\omega_0 - A_0}}\sin\varkappa(t - t_0). \end{aligned}\right\} \tag{17.6}$$

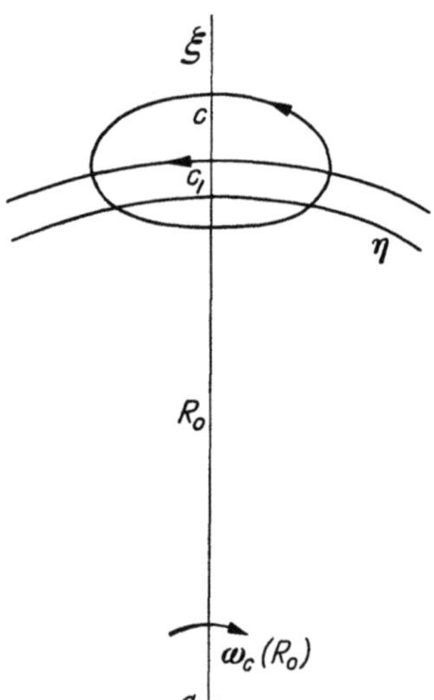

Fig. 9. Epicyclic orbit in a coordinate system following the rotational motion at a certain distance from the centre.

This is the approximation introduced by LINDBLAD[1]. In these equations c, c_1, t_0, t_1 are arbitrary parameters for a given R_0. Here ω_0 and A_0 are the values for $R = R_0$, c_1 being assumed to be a quantity of the first order. $\bar{A}$ is a mean value of A between R_0 and $R_0 + c_1$.

For $\xi, \frac{1}{R_0}\frac{d\eta}{dt}, \zeta$ small, the linearized differential equations may be written

$$\left.\begin{aligned} &\frac{d^2\xi}{dt^2} - 2\omega\frac{d\eta}{dt} = 4\omega A\xi, \\ &\frac{d^2\eta}{dt^2} + 2\omega\frac{d\xi}{dt} = 0, \\ &\frac{d^2\zeta}{dt^2} + \zeta\left(\frac{\partial^2\varphi}{d\zeta^2}\right)_0 = 0. \end{aligned}\right\} \tag{17.7}$$

They are sometimes convenient to use. Choosing first the reference circle $R_0 = R_1$ the formulae (17.4) to (17.6) may be derived from these equations as well.

It may be remarked here that there is an important consequence of the linearity of these equations, if we disregard any variation of the coefficients ω and A in the field considered. Let the orbit of a certain

[1] B. LINDBLAD: Stockholms Obs. Ann. **18**, No. 6 (1955).

reference particle be $\bar{\xi}(t), \bar{\eta}(t)$, where

$$\left.\begin{aligned} \bar{\xi} &= \gamma_1 + \gamma \cos \varkappa (t - \tau_0), \\ \bar{\eta} &= -2A\gamma_1 (t - \tau_1) - \gamma \sqrt{\frac{\omega}{\omega - A}} \sin \varkappa (t - \tau_0). \end{aligned}\right\} \tag{17.8}$$

We then may introduce differential coordinates relative to this motion by

$$x = \xi - \bar{\xi}(t), \quad y = \eta - \bar{\eta}(t). \tag{17.9}$$

It is evident that x and y will obey the differential equations (17.7) so that we may write the solution

$$\left.\begin{aligned} x &= c_1' + c' \cos \varkappa (t - t_0'), \\ y &= -2A c_1' (t - t_1') - c' \sqrt{\frac{\omega}{\omega - A}} \sin \varkappa (t - t_0'). \end{aligned}\right\} \tag{17.10}$$

Including terms of higher order in ξ or ξ_1 leads to expressions which contain elliptic and hyper-elliptic functions. A development of ξ in Fourier series, when the second order term is included, has been given by CONTOPOULOS[1]. In this development, which is of the form

$$\xi = c_1 + c \cos \varkappa (t - t_0) + d \cos 2\varkappa (t - t_0) + \cdots \tag{17.11}$$

the coefficients of the terms with arguments $2\varkappa(t-t_0)$, $3\varkappa(t-t_0)$, etc., are found to be small compared with c for moderate velocities relative to the centroid. Moreover, for given c_1 the change of $\varkappa$ with the amplitude factor c is found to be slow. We may make this dependence explicit by introducing

$$\xi_1 = \varepsilon + c \cos \varkappa (t - t_0) + d \cos 2\varkappa (t - t_0)$$

in the differential equations (17.2), and preserving terms of second order in c. We define a slightly different reference circle with radius $R_1' = R_1 + \delta$, such that *the resulting epicyclic orbit is closed.* We then define c_1 by $R_1' = R_0 + c_1$ for an arbitrary R_0. This means that for $c_1 = 0$ there is no secular term in η. (The three circles, the reference circle (R_1') for a closed epicycle, the circular orbit (R_1) of the same area velocity as that of the particle, and the circle $(R_1 + \varepsilon)$ defining the oscillation centre of the periodic terms in ξ, which all coincide in the first approximation, are here as a rule separate.) The result can be written

$$\left.\begin{aligned} \xi &= c_1 + \frac{c^2}{16\omega_1 A_1} F_1'' + c \cos \varkappa (t - t_0) - c^2 \frac{a_1}{6\varkappa^2} \cos 2\varkappa (t - t_0), \\ \eta &= -R_0(\omega_0 - \omega_1)(t - t_1) - \frac{R_0}{R_1} c \sqrt{\frac{\omega_1}{\omega_1 - A_1}} \sin \varkappa (t - t_0) + \\ &\quad + c^2 \frac{\omega_1 R_0}{6 R_1 \varkappa^3} \left(a_1 + \frac{9}{2} \frac{\varkappa^2}{R_1}\right) \sin 2\varkappa (t - t_0), \\ \varkappa &= 2\sqrt{\omega_1(\omega_1 - A_1)} + \frac{c^2}{2\varkappa} \left[\frac{6\omega_1^3 A_1}{\varkappa^2 R_1^2} - \frac{\omega_1^2}{\varkappa^2 R_1} F_1'' + \right. \\ &\quad \left. + \frac{6\omega_1 - 5A_1}{24\varkappa^2 A_1} (F_1'')^2 + \frac{1}{8} F_1'''\right], \\ F_1'' &= \left(\frac{d^2 F}{d R^2}\right)_1, \quad F_1''' = \left(\frac{d^3 F}{d R^3}\right)_1. \end{aligned}\right\} \tag{17.12}$$

[1] G. CONTOPOULOS: Stockholms Obs. Ann. **19**, No. 10 (1957).

Here ω_1, A_1, etc., denote the values of ω, A, etc., for $R=R_1'=R_0+c_1$, and ω_0 the value of ω for $R=R_0$. The difference δ between the value of R_0+c_1 used here, which defines the reference circle of the closed epicyclic orbit, and the radius R_1 first introduced, which defines the circular motion of the same area constant as the particle, may be written

$$\delta = c^2 \frac{\omega_1}{2A_1\varkappa_1^2}\left(\frac{6\omega_1 A_1}{R_1} - \frac{1}{2}F_1''\right). \tag{17.13}$$

The degree of approximation reached by the simple formulae (17.6) may be judged by considering the second order approximation in two special cases, the Newtonian field $F=C/R^2$, and the case of uniform angular velocity $F=CR$. In the Newtonian case we can write

$$\left.\begin{aligned}
\xi &= c_1 + \frac{c^2}{2R_1} + c\cos\varkappa(t-t_0) - \frac{c^2}{2R_1}\cos 2\varkappa(t-t_0),\\
\eta &= -R_0(\omega-\omega_0)(t-t_1) - \frac{R_0}{R_1}2c\sin\varkappa(t-t_0) +\\
&\quad + \frac{5}{4}\frac{R_0}{R_1^2}c^2\sin 2\varkappa(t-t_0),\\
\varkappa &= \omega_1.
\end{aligned}\right\} \tag{17.14}$$

In the case of uniform ω we have

$$\left.\begin{aligned}
\xi &= c_1 + c\cos\varkappa(t-t_0) - \frac{c^2}{4R_1}\cos 2\varkappa(t-t_0),\\
\eta &= -\frac{R_0}{R_1}c\sin\varkappa(t-t_0) + \frac{1}{2}\frac{R_0}{R_1^2}c^2\sin 2\varkappa(t-t_0),\\
\varkappa &= 2\omega.
\end{aligned}\right\} \tag{17.15}$$

In these cases the second order terms are evidently still insignificant for c/R and c_1/R_0 of the order $^1/_{10}$. Moreover, the second order term in $\varkappa$ vanishes.

In the general case we may write for $\varkappa$, if we ignore the term second order in c,

$$\varkappa^2 = 3\omega^2 + \frac{dF}{dR}, \tag{17.16}$$

which gives

$$\frac{d^2F}{dR^2} = \frac{4A}{R}\left(3\omega - \varkappa\frac{d\varkappa}{d\omega}\right). \tag{17.17}$$

This relation is of interest when $\varkappa$ is given as a function of ω. We may further write

$$\frac{d^3F}{dR^3} = \frac{4\omega}{R^2}A + \frac{4}{R}(6A-\omega)\frac{dA}{dR} - 4\omega\frac{d^2A}{dR^2}. \tag{17.18}$$

For the motion in ζ we have in the first approximation

$$\zeta = c_2\cos\varkappa_2(t-t_2), \quad \varkappa_2^2 = \left(\frac{\partial^2\varphi}{\partial\zeta^2}\right)_0. \tag{17.19}$$

The three-dimensional motion has been studied by CHANDRASEKHAR[1], who show the importance of terms containing $\left(\frac{\partial\varphi}{\partial z}\right)_0$ and $\left(\frac{\partial^2\varphi}{\partial R\,\partial z}\right)_0$ when the Eqs. (17.7) are generalized to describe the motions at points which are situated outside of the galactic plane. The first order theory still admits the axes of the velocity

[1] S. CHANDRASEKHAR: Monthly Notices Roy. Astronom Soc. London **98**, 710 (1938).

ellipsoid in the $\xi\zeta$-plane to be different (cf. Sect. 18), but there will be a tilt of the axes against the ξ- and ζ-directions, if $\left(\frac{\partial^2\varphi}{\partial R\,\partial z}\right)_0 \neq 0$.

COUTREZ [4] has studied individual three-dimensional trajectories. CONTOPOULOS[1] discusses the motion in R, z in relation to the section of the torus surface inside which the three-dimensional orbit must lie, and the recurrence of the orbit to an initial course. INGRID TORGÅRD[2] has studied extensively particle orbits in the equatorial planes of stellar systems.

We have treated the theory of the epicyclic orbits in the galactic plane with some completeness, because they give an approach to the problems of star-streaming and to the velocity-ellipsoid which is in several respects more powerful than those described in preceding sections. It is in general sufficient here to use the approximation (17.6). Differentiating these equations with respect to t we derive the velocity components $d\xi/dt$ and $d\eta/dt$. We may define, however, the *velocity components u, v relative to the circular motion at the point ξ, η where the particle is situated at the moment t*, thus

$$u = \frac{d\xi}{dt}, \qquad v = \frac{d\eta}{dt} + 2A\xi. \tag{17.20}$$

This gives

$$\left.\begin{aligned} u &= -c\varkappa\sin\varkappa(t-t_0), \\ v &= 2cB\cos\varkappa(t-t_0), \end{aligned}\right\} \tag{17.21}$$

where

$$B = A - \omega.$$

By means of (17.6) and (17.20) we derive the following expressions of the arbitrary parameters c, c_1, t_0, t_1

$$\left.\begin{aligned} c^2 &= \frac{u^2}{\varkappa^2} + \frac{v^2}{4B^2}, \\ c_1 &= \xi - \frac{v}{2B}, \\ c\sin\varkappa t_0 &= \frac{u}{\varkappa}\cos\varkappa t + \frac{v}{2B}\sin\varkappa t, \\ c\cos\varkappa t_0 &= -\frac{u}{\varkappa}\sin\varkappa t + \frac{v}{2B}\cos\varkappa t, \\ 2A\,c_1 t_1 &= \eta + \frac{u}{2B} + 2A\,c_1 t, \end{aligned}\right\} \tag{17.22}$$

and for the motion in ζ by (17.19), defining $w = d\zeta/dt$,

$$\left.\begin{aligned} c_2^2 &= \zeta^2 + \frac{w^2}{\varkappa_2^2}, \\ c_2\sin\varkappa_2 t_2 &= \zeta\sin\varkappa_2 t + \frac{w}{\varkappa_2}\cos\varkappa_2 t, \\ c_2\cos\varkappa_2 t_2 &= \zeta\cos\varkappa_2 t - \frac{w}{\varkappa_2}\sin\varkappa_2 t. \end{aligned}\right\} \tag{17.23}$$

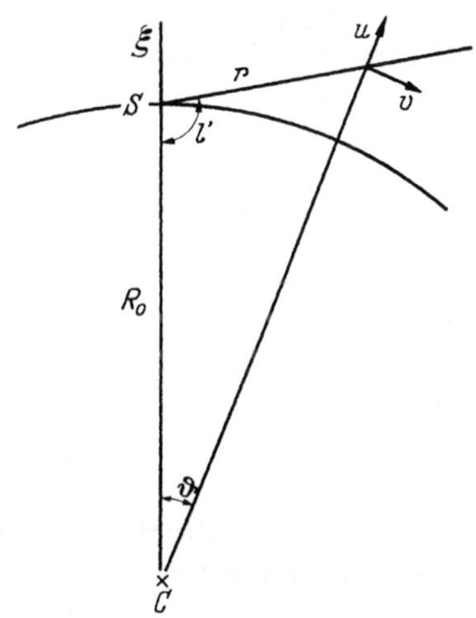

Fig. 10. The velocity components u, v relative to the local circular motion.

18. Theories of the velocity ellipsoid based on epicyclic motions. We shall assume that the variation with R of the component of force at right angles to the galactic plane for given ζ will be small, i.e. $\frac{\partial^2\varphi}{\partial R\,\partial z}$ to be numerically small (cf.

[1] G. CONTOPOULOS: Stockholms Obs. Ann. **19**, No. 10 (1957).
[2] I. TORGÅRD: Medd. Lunds. Astron. Obs. Ser. II **14**, No. 133 (1956).

Sect. 7 and 8). The motion in ζ may then be considered independent of the motion in ξ.

The frequency of stars of a given physical class to be expected in the six-dimensional space element $d\xi\, d\eta\, d\zeta\, du\, dv\, dw$ may be written $f(\xi, \eta, \zeta, u, v, w)\, d\xi\, d\eta\, d\zeta\, du\, dv\, dw$, where

$$f(\xi, \eta, \zeta, u, v, w) = f_1(\xi, \eta, u, v) \cdot f_2(\zeta, w). \tag{18.1}$$

We may write

$$f_1(\xi, \eta, u, v) = F_1(c, c_1, t_0, t_1); \quad f_2(\zeta, w) = F_2(c_2, t_2), \tag{18.2}$$

$$d\xi\, d\eta\, du\, dv = J_1\, dc\, dc_1\, dt_0\, dt_1; \quad d\zeta\, dw = J_2\, dc_2\, dt_2. \tag{18.3}$$

Here

$$J_1 = -16\omega A B^2 c c_1, \quad J_2 = c_2 \varkappa_2^2, \tag{18.4}$$

so that $d\xi\, d\eta\, du\, dw$ and $d\zeta\, dw$ are constant during the motion. We shall consider for the moment only the function f_1. The simplest assumption we can make is that in the region considered the frequency f_1 depends only on the size of the epicyclic orbits, thus only on the parameter c. If the distribution in c is Gaussian, we may put

$$f_1 = C\, e^{-h^2 c^2}. \tag{18.5}$$

If we write

$$f_1 = C\, e^{-\frac{1}{2}\left(\frac{u^2}{\alpha^2} + \frac{v^2}{\beta^2}\right)}, \tag{18.6}$$

we have

$$\alpha^2 = \frac{\varkappa^2}{2h^2}, \quad \beta^2 = \frac{2B_c^2}{h^2}, \tag{18.7}$$

and

$$\frac{\beta^2}{\alpha^2} = 1 - \frac{A_c}{\omega_c}, \tag{18.8}$$

where we have introduced the notations A_c, B_c, ω_c, to indicate that these quantities are valid for the circular motions. We may call this the "natural" velocity ellipsoid. If the frequency function depends also on c_1, we may put

$$f_1 = F(c, c_1), \tag{18.9}$$

and in the case when f is *ellipsoidal* in u, v

$$f_1 = C\, e^{-(h^2 c^2 + k_1 c_1 + k_2 c_1^2)}. \tag{18.10}$$

This gives

$$f = C\, e^{-\frac{h^2}{h^2 + k_2}(k_1 \xi + k_2 \xi^2)}\, e^{-\left[\frac{h^2}{\varkappa^2} u^2 + \frac{h^2 + k_2}{4B_c^2}(v - v_m)^2\right]} \tag{18.11}$$

where

$$v_m = \frac{B_c}{h^2 + k_2}(k_1 + 2k_2 \xi). \tag{18.12}$$

We may call this distribution the "regular" velocity ellipsoid. The axes of this ellipsoid are given by

$$\alpha^2 = \frac{\varkappa^2}{2h^2}, \quad \beta^2 = \frac{2B_c^2}{h^2 + k_2}, \tag{18.13}$$

and thus

$$\frac{\beta^2}{\alpha^2} = \frac{h^2}{h^2 + k_2}\left(1 - \frac{A_c}{\omega_c}\right). \tag{18.14}$$

Defining $\Theta_0 = \Theta_c + v_m$, where Θ_c is the linear velocity of the circular motion, and introducing the Oort coefficients

$$A = \frac{1}{2}\left(\frac{\Theta_0}{R} - \frac{\partial \Theta_0}{\partial R}\right), \quad B = -\frac{1}{2}\left(\frac{\Theta_0}{R} + \frac{\partial \Theta_0}{\partial R}\right),$$

we have to the first order

$$1 - \frac{A}{\omega} = \frac{h^2}{h^2 + k_2}\left(1 - \frac{A_c}{\omega_c}\right),$$

so that

$$\frac{\beta^2}{\alpha^2} = 1 - \frac{A}{\omega}. \tag{18.15}$$

By the analysis given in Sect. 7 we know that this relation is, in fact, rigorous. If $\varrho(\xi, \eta)$ is the density distribution in the galactic plane we may show that

$$v_m = \frac{\alpha^2}{2\omega}\frac{\partial \log \varrho}{\partial \xi}, \tag{18.16}$$

which is an approximation to the rigorous expression (6.8) when R is large.

19. Dissolution of a local distribution. In the general case, when all four parameters enter into the frequency function, we may write

$$f_1 = F(c_1,\ c \sin \varkappa t_0,\ c \cos \varkappa t_0,\ 2A\, c_1 t_1). \tag{19.1}$$

If we have a cloud of particles defined at the time $t=0$ by

$$f_0 = \Phi(\xi_0, \eta_0, u_0, v_0), \tag{19.2}$$

ξ_0, η_0, u_0, v_0 may be expressed in the parameters c, c_1, t_0, t_1, by putting $t=0$ in (17.6) and then by Eqs. (17.22) in ξ, η, u, v, t, as follows[1]

$$\left.\begin{aligned}
\xi_0 &= \xi - \frac{u}{\varkappa}\sin \varkappa t - \frac{v}{2B}(1 - \cos \varkappa t),\\
\eta_0 &= \eta + 2A\left(\xi - \frac{v}{2B}\right)t + \frac{u}{2B}(1 - \cos \varkappa t) - \frac{\varkappa}{2B^2}\, v \sin \varkappa t,\\
u_0 &= u \cos \varkappa t + \frac{\varkappa}{2B}\, v \sin \varkappa t,\\
v_0 &= -\frac{2B}{\varkappa}\, u \sin \varkappa t + v \cos \varkappa t.
\end{aligned}\right\} \tag{19.3}$$

In this formula we may interchange ξ_0, η_0, u_0, v_0 and ξ, η, u, v, changing at the same time the sign of the time interval t. In the case of an ellipsoidal distribution at $t=0$ we write

$$f_0 = C\, \mathrm{e}^{-Q_0}, \tag{19.4}$$

with

$$\left.\begin{aligned}
Q_0 = {} & a_{10}\xi_0 + a_{01}\eta_0 + a_{20}\xi_0^2 + a_{02}\eta_0^2 + a_{11}\xi_0\eta_0 +\\
& + b_{10}u_0 + b_{01}v_0 + b_{20}u_0^2 + b_{02}v_0^2 + b_{11}u_0 v_0 +\\
& + c_{11}\xi_0 u_0 + d_{11}\eta_0 u_0 + e_{11}\xi_0 v_0 + f_{11}\eta_0 v_0
\end{aligned}\right\} \tag{19.5}$$

and obtain the frequency function f_1 at the time t by inserting ξ_0, η_0, u_0, v_0 according to (19.3).

20. Dispersion orbits. We can define a curve in space along which the process of dissolution of a local cloud moving in the galactic plane takes place, and we may try to define a coordinate system such that the orbit of an individual particle in the cloud coincides with the curve along which the mass of matter dissolves in the long run of time[2]. In order to analyse this point we shall assume ω

---

[1] B. LINDBLAD and F. NAHON: Stockholms Obs. Ann. **18**, No. 2 (1954).

[2] B. LINDBLAD: Stockholms Obs. Ann. **19**, No. (1956); **20**, No. 4 (1958).

and $\varkappa$ to be known functions of R. We may then form $\varkappa$ as a function of ω, and define

$$\frac{\overline{d\varkappa}}{d\omega} = n \tag{20.1}$$

in a certain region about $R = R_0$. Introduce a coordinate system with angular rotation ω', where

$$\omega' = \omega_0 - \frac{1}{n}\varkappa_0 . \tag{20.2}$$

The angular velocity relative to the ξ, η-system, defined for $R = R_0$, will then be $-\frac{1}{n}\varkappa_0$. As convenient coordinates in this system we define

$$\xi = R - R_0, \quad \eta' = R_0 \vartheta', \tag{20.3}$$

where ϑ' is the angular coordinate in the coordinate system with angular speed ω'. Let further $\eta = \eta'$ for $t = 0$. Then

$$\eta' = R_0 \frac{\varkappa_0}{n} t - 2A\, c_1 (t - t_1) - c \sqrt{\frac{\omega}{\omega - A}} \sin \varkappa (t - t_0) , \tag{20.4}$$

where $\varkappa$ and ω are valid for the circular orbit of radius $R_0 + c_1$. Writing $\Delta\varkappa = \varkappa - \varkappa_0$, $\Delta\omega = \omega - \omega_0$ we have with sufficient accuracy $\Delta\varkappa = n\,\Delta\omega$, and further $\Delta\omega = -\frac{2A}{R_0} c_1$. This gives, if $\eta' = \eta_0'$ for $t = t_0$,

$$\eta' - \eta_0' = R_0 \frac{\varkappa}{n} (t - t_0) - c \sqrt{\frac{\omega}{\omega - A}} \sin \varkappa (t - t_0) \tag{20.5}$$

and

$$\varkappa (t - t_0) = n (\vartheta' - \vartheta_0') + n \frac{c}{R_0} \sqrt{\frac{\omega}{\omega - A}} \sin n (\vartheta' - \vartheta_0') + \cdots . \tag{20.6}$$

Thus we may write

$$\left.\begin{aligned} \xi &= c_1 + c \cos n (\vartheta' - \vartheta_0') . \\ \vartheta' - \vartheta_0' &= \frac{\varkappa}{n} (t - t_0) - \frac{c}{R_0} \sqrt{\frac{\omega}{\omega - A}} \sin \varkappa (t - t_0) , \\ \varkappa &= \varkappa_0 - n \frac{2A}{R_0} c_1 . \end{aligned}\right\} \tag{20.7}$$

Considering an adjacent orbit with the parameters $c + \delta c_1\; c_1 + \delta c_1$, $\vartheta_0' + \delta\vartheta_0'$, we have for a given ϑ'

$$\delta\xi = \delta c_1 + \delta c \cdot \cos n (\vartheta' - \vartheta_0') + \delta\vartheta_0' \cdot n\, c \sin n (\vartheta' - \vartheta_0') . \tag{20.8}$$

The orbits will therefore lie close together, but there will be a differential motion along the orbit by

$$\delta\vartheta' = -\frac{2A}{R_0} \delta c_1 \cdot t . \tag{20.9}$$

An orbit of this kind may be called a "dispersion orbit", indicating that in a cloud which dissolves in the zone in question the particles will remain close to such an orbit. There will be a smoothing process in the distribution of matter along such orbits, which will tend to smooth the frequency f_1 along the orbit. This property of the dispersion orbit is of great importance for the frequency function $f_1(\xi, \eta, u, v)$ in the galactic plane. We have by (20.4)

$$\vartheta_0' = \frac{\varkappa_0}{n} t_0 - 2A \frac{c_1}{R_0} (t_0 - t_1) , \tag{20.10}$$

We may then write quite generally in (18.2)

$$f_1(\xi, \eta, u, v) = F_1(c, c_1, \vartheta_0', t_1). \tag{20.11}$$

Table 10.

R kpc	ω km/sec/kpc	$\varkappa$ km/sec/kpc	$\omega-\frac{1}{2}\varkappa$	R kpc	ω km/sec/kpc	$\varkappa$ km/sec/kpc	$\omega-\frac{1}{2}\varkappa$
2.08	86.2	137.1	17.7	6.47	34.9	49.1	10.4
2.42	77.2	124.8	14.8	6.68	33.8	49.2	9.2
2.76	70.2	110.8	14.7	6.88	32.8	46.6	9.6
3.10	64.3	101.3	13.6	7.16	31.5	44.6	9.2
3.42	59.5	92.5	13.2	7.48	29.9	37.6	11.1
3.74	55.4	86.6	12.1	7.74	28.6	33.8	11.7
4.06	52.0	80.4	11.8	7.94	27.6	29.4	13.0
4.36	49.1	77.0	10.6	8.10	26.8	28.6	12.5
4.66	46.4	68.8	12.0	8.60	24.6	21.7	13.7
4.95	44.0	67.0	10.5	9.50	21.2	21.5	10.4
5.23	42.1	65.3	9.5	11.00	17.0	15.6	9.2
5.50	40.4	61.6	9.6	13.50	12.3	9.8	7.4
5.76	38.8	57.2	10.2	17.50	8.2	7.4	4.5
6.01	37.4	56.1	9.3	25.00	4.8	4.9	2.4
6.24	36.1	53.5	9.4				

We can expect t_1 to be "mixed out" of the frequency function in the course of time, especially if n is an integer, so that the orbit is closed. In this case, moreover, ϑ_0' can be chosen in a unambiguous way to represent the longitude of one of the maxima of ξ in the orbit. The frequency function may thus be written

$$f_1(\xi, \eta, u, v) = F_1(c, c_1, \vartheta_0'). \tag{20.12}$$

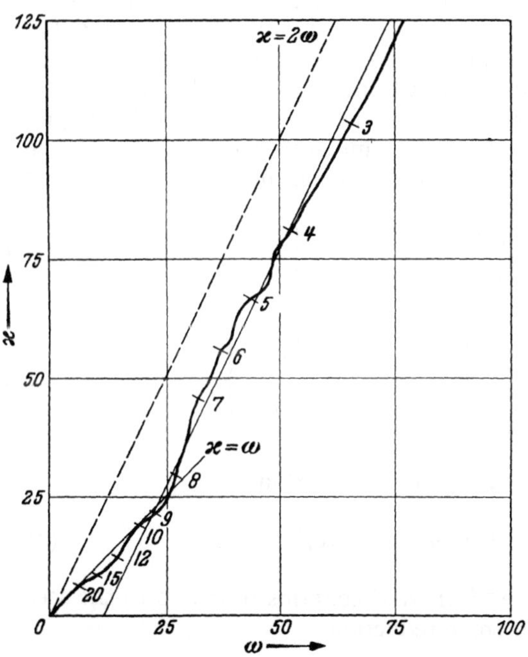

Fig. 11. The frequency $\varkappa$ of radial motion as function of the angular velocity ω of circular motions.

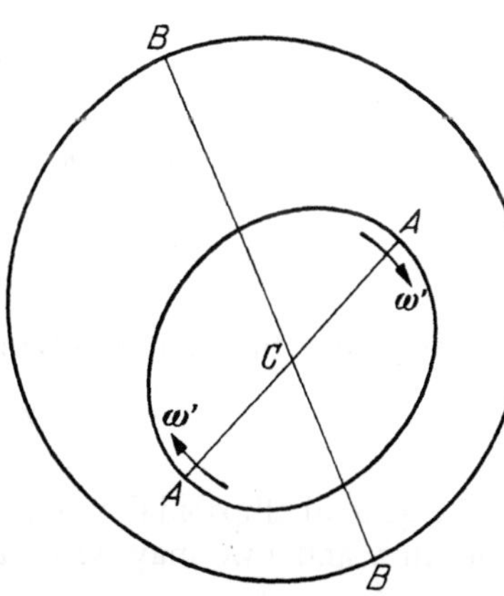

Fig. 12. The two types A and B of dispersion orbits.

Table 10 gives the values of ω and $\varkappa$ for different R according to the results of Kwee, Muller and Westerhout for $R < R_\odot$ and according to Schmidt's model of the system for $R > R_\odot$. The results are illustrated in Fig. 11. It is evident that in parts of the system interior to $R_\odot$ the points adhere closely to the line $\omega - \frac{1}{2}\varkappa = 11.5$, and in the exterior regions to the line $\varkappa - \omega = 0$. There are, therefore, two most important cases, namely in the *inner regions*

$$n = 2, \quad \text{with } \omega' = \omega - \tfrac{1}{2}\varkappa \text{ approximately constant}, \tag{20.13}$$

and in the *outer regions*

$$n = 1, \quad \text{with } \omega' = \omega - \varkappa \to 0 \text{ with increasing } R. \tag{20.14}$$

The corresponding orbits are illustrated in Fig. 12. It is of great importance that these orbits are simple closed curves, in the first case (AA) ellipses with their centres coinciding with the centre of the system, in the second case (BB) ellipses with focus in the centre of the system. In intermediate regions there will be other values of n, and when n is not an integer there will be more complicated "dispersion orbits" which are not closed curves. According to what has been developed above such an orbit is valid for a coordinate system of angular speed $\omega - \frac{1}{n}\varkappa$. In a coordinate system of angular speed $\omega - \frac{1}{2}\varkappa$, on the other hand, the orbit of a particle is always a closed ellipse with its centre in the centre of the system. If n differs from 2, the apsidal lines of such orbits of different c_1 will move relatively to each other.

21. Velocity ellipsoid on the basis of dispersion orbits. It is evident from Fig. 11 that in the region about the Sun, for R about 8.2 kpc, the variation of $\omega - \frac{1}{2}\varkappa$ with R is very small. It follows then that in a coordinate system of angular speed $\omega' = \omega_0 - \frac{1}{2}\varkappa_0$, valid for $R = R_0$, the orbits in the zone surrounding $R = R_0$ are closed ellipses with fixed apsidal lines. In this zone the longitude ϑ_0' of the apsidal line may therefore be considered as a third integral of the motion besides c and c_1. Through the smoothing process along the dispersion orbit the frequency will approach to be a function of the three first integrals c, c_1, and ϑ_0'. The components of velocity u, v relative to the local circular motion may be written

$$\left.\begin{aligned} u &= -c\varkappa \sin 2(\vartheta' - \vartheta_0'), \\ v &= 2cB\cos 2(\vartheta' - \vartheta_0') \end{aligned}\right\} \tag{21.1}$$

which gives

$$\left.\begin{aligned} c\sin 2\vartheta_0' &= \frac{u}{\varkappa}\cos 2\vartheta' + \frac{v}{2B}\sin 2\vartheta', \\ c\cos 2\vartheta_0' &= -\frac{u}{\varkappa}\sin 2\vartheta' + \frac{v}{2B}\cos 2\vartheta'. \end{aligned}\right\} \tag{21.2}$$

The frequency function involving c, c_1, ϑ_0' may be written

$$f_1 = F_1(c_1, c\sin 2\vartheta_0', c\cos 2\vartheta_0'). \tag{21.3}$$

In the general ellipsoidal case $f = C\,e^{-Q}$ where Q contains terms up to the second order in u and v we may write after some reductions

$$\left.\begin{aligned} Q = h^2c^2 + k_1c_1 + k_2c_1^2 + l_1\left[\frac{u}{\varkappa}\cos 2(\vartheta' + \psi_1) + \frac{v}{2B}\sin 2(\vartheta' + \psi_1)\right] + \\ + l_2\left[\left(\frac{u^2}{\varkappa^2} - \frac{v^2}{4B^2}\right)\cos 4(\vartheta' + \psi_2) + \frac{uv}{B\varkappa}\sin 4(\vartheta' + \psi_2)\right] + \\ + l_3\left(\xi - \frac{v}{2B}\right)\left[\frac{u}{\varkappa}\cos 2(\vartheta' + \psi_3) + \frac{v}{2B}\sin 2(\vartheta' + \psi_3)\right]. \end{aligned}\right\} \tag{21.4}$$

Here h^2 $k_1, k_2, l_1, l_2, l_3, \psi_1, \psi_2, \psi_3$ are arbitrary constants. We reduce this expression to the coordinate system ξ, η following the rotation ω by substituting

$$\vartheta' = \frac{\eta}{R_0} + \frac{\varkappa_0}{2} t. \tag{21.5}$$

For a given ξ, η the velocity distribution varies periodically with time.

The presence of terms in $u\,v$ means that there are deviations of the vertex line from the direction towards the centre. It may be shown that strict preservation of rotational symmetry in the density ϱ demands that $l_1 = l_3 = 0$. We assume for simplicity also that k_1, k_2, vanish, which means that we ignore for the moment a variation of the velocity distribution with ξ and a deviation of the centre of the velocity ellipsoid from the circular motion. The remaining arbitrary constants are then h^2, l_2, ψ_2, and the velocity ellipse will be

$$L\,u^2 + 2M\,u\,v + N\,v^2 = \text{const}, \tag{21.6}$$

where

$$\left.\begin{aligned} L &= \frac{1}{\varkappa^2}\left[h^2 + l_2 \cos 4(\vartheta' + \psi_2)\right], \\ 2M &= \frac{1}{B\varkappa}\, l_2 \sin 4(\vartheta' + \psi_2), \\ N &= \frac{1}{4B^2}\left[h^2 - l_2 \cos 4(\vartheta' + \psi_2)\right]. \end{aligned}\right\} \tag{21.7}$$

The vertex deviation χ will be given by

$$\tan 2\chi = \frac{4B\varkappa\, l_2 \sin 4(\vartheta' + \psi_2)}{h^2(4B^2 - \varkappa^2) + l_2(4B^2 + \varkappa^2)\cos 4(\vartheta' + \psi_2)}. \tag{21.8}$$

It may be shown that in this case Q may be written

$$Q = h^2 c^2 - l_2 c^2 \cos 4(\vartheta_0' + \psi_2). \tag{21.9}$$

For a given c^2 there will therefore be a maximum of frequency for orbits of apsidal lines

$$(\vartheta_0')_{\max} = -\psi_2, \quad \text{and} \quad -90^\circ - \psi_2. \tag{21.10}$$

The half-axes α and β of the velocity ellipsoid are found by solving the equation

$$t^2 - (L + N)\,t + L\,N - M^2 = 0. \tag{21.11}$$

If the roots are in order of size t_1 and t_2, we have

$$\frac{1}{2\alpha^2} = t_1, \qquad \frac{1}{2\beta^2} = t_2. \tag{21.12}$$

Evidently

$$t_1 t_2 = L\,N - M^2 = \frac{1}{4\varkappa^2 B^2}(h^4 - l_2^2). \tag{21.13}$$

Thus

$$\alpha^2 \beta^2 = \frac{\varkappa^2 B^2}{h^4 - l_2^2}. \tag{21.14}$$

Integrating f with respect to the velocities from zero to infinity gives the density proportional to $\alpha\beta$. This shows that the *space density is independent of* ϑ'. The rotational symmetry of density and potential is thus preserved. If the half-axes α, β, and the vertex deviation χ are known by observation, we may write

$$\left.\begin{aligned} \mu &= B\varkappa\left(\frac{1}{\alpha^2} - \frac{1}{\beta^2}\right)\sin 2\chi, \\ \nu &= \frac{1}{\alpha^2}\left(\frac{\varkappa^2}{2}\cos^2\chi - 2B^2\sin^2\chi\right) + \frac{1}{\beta^2}\left(\frac{\varkappa^2}{2}\sin^2\chi - 2B^2\cos^2\chi\right) \end{aligned}\right\} \tag{21.15}$$

and have by the expressions of L, M, N in terms of the observed quantities

$$\left.\begin{aligned} 2h^2 &= \frac{1}{\alpha^2}\left(\frac{\varkappa^2}{2}\cos^2\chi + 2B^2\sin^2\chi\right) + \frac{1}{\beta^2}\left(\frac{\varkappa^2}{2}\sin^2\chi + 2B^2\cos^2\chi\right), \\ \tan 4(\vartheta' + \psi_2) &= \frac{\mu}{\nu}, \\ 4l_2^2 &= \mu^2 + \nu^2, \quad l_2 > 0. \end{aligned}\right\} \tag{21.16}$$

By (21.14) we have also

$$l_2^2 = h^4 - \frac{\varkappa^2 B^2}{\alpha^2\beta^2}.$$

We may take as an example the velocity ellipse for the stars of types $B-M$ brighter than $6^{\mathrm{m}}.0$ as determined by H. NORDSTRÖM[1]. He gives $\alpha = 20.5$, $\beta = 14.9$ km/sec and the direction of the large axis towards $L = 340°.1$, $B = -2°.4$. With a direction towards the centre $L = 328°$ this gives $\chi = -12°$, the vertex deviation being counted in the positive direction. With $B = -0.007$, $\varkappa = 0.027$ km per sec/pc this gives

$$\mu = -0.1633 \times 10^{-6}, \qquad \nu = +0.4677 \times 10^{-6},$$

and if we choose $\vartheta' = 0$ for the present position of the Sun, we have for $l_2 > 0$, $\psi_2 = -4°.8$, which gives the two values $(\vartheta_0')_{\max} = +4°.8$, and $-85°.2$ for the orbits of maximum frequency for given c^2. Further we get

$$h^2 = 0.666 \times 10^{-6}, \qquad l_2 = 0.248 \times 10^{-6}.$$

The result is sensitive to a change of B, when we preserve the value of ω. For $B = -0.010$, $\varkappa = 0.0325$ km/sec/pc we obtain $\psi_2 = -8°.4$, and thus

$$(\vartheta_0')_{\max} = +8°.4, \quad \text{and} \quad -81°.6,$$

$$h^2 = 1.094\; 10^{-6}, \qquad l_2 = 0.232\; 10^{-6}.$$

In addition to the increase of the first value of $(\vartheta_0')_{\max}$, the ratio l_2/h^2 has decreased considerably. This means in principle a less pronounced deviation from the "natural" velocity ellipsoid.

We may show by Eqs. (21.14) and (21.16) that if we let B vary for given α, β, χ a minimum of l_2/h^2 will be obtained for

$$|B| = \omega\frac{\beta^2}{\alpha^2}\,\frac{1 + \dfrac{\alpha^2}{\beta^2}\tan^2\chi}{1 + \dfrac{\beta^2}{\alpha^2}\tan^2\chi} \geqq \omega\frac{\beta^2}{\alpha^2}, \qquad \text{if } \alpha > \beta.$$

We would then get for $\omega = 0.026$ km/sec/pc and for the ratio of the axes $\beta/\alpha = 0.727$ the value $B = -0.014$.

[1] H. NORDSTRÖM: Medd. Lunds Obs., Ser. II 8, No. 79 (1936).

The orbits corresponding to the two apsidal lines of maximum frequency may be associated with KAPTEYN's early ideas concerning the two star-streams. In this way the present theory combines the ellipsoidal theory with the theory of the two star-streams, and at the same time explains the vertex deviation of the ellipsoid. The "dualistic" character of the law of distribution, which has two preferential directions $(\vartheta_0')_{\max}$ of the apsidal lines differing by 90° from each other, is only apparent. (From a formal point of view it depends on the fact that we count c only with positive sign. If we would admit negative values of c as well, in which case the diameter of minimum length is taken as the apsidal line, the two values of ϑ_0' may be reduced to one.) From a physical point of view we may try to outline a schematic process by which a distribution of the kind considered here may have originated. We have seen that the observed axes α, β and the vertex deviation χ determine the constants h^2, l_2, ψ_2 of the general law of distribution. The change of sign of χ in the galactic plane in consequence of this distribution is shown in Fig. 13. The four points P denote points where χ is zero. The same occurs at points situated 45° from P, but in P the ratio α/β is smaller than for the intermediate points. The condition that the velocity distribution in P shall have circular symmetry, i.e. that $\alpha=\beta$, is easily seen to lead to the relation

$$\frac{l_2}{h^2}=\frac{\varkappa^2-4B^2}{\varkappa^2+4B^2}.$$

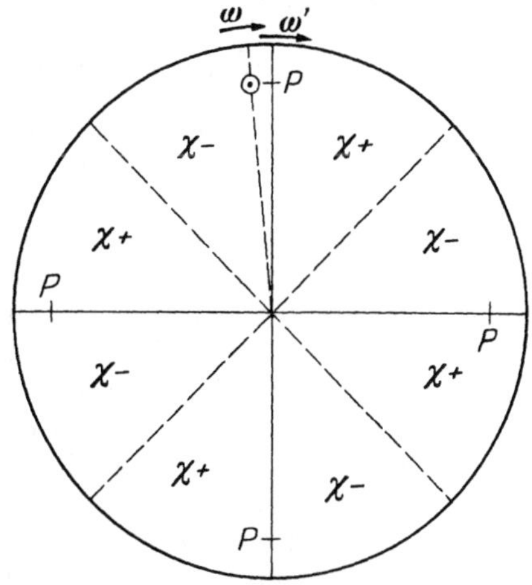

Fig. 13. The variation of the vertex deviation in the case of a dissolving cloud of stars.

For the values of $\varkappa$ and B found in our surroundings this would demand $l_2/h^2=0.58$, whereas we have found from the velocity ellipsoid near the Sun $l_2/h^2=0.37$. The points P move with the angular velocity ω', which is about half the velocity ω of circular motions close to the Sun. The present angular distance between one of the points P and the Sun is equal to ψ_2. We can imagine that around one of the points P there has once existed a vast condensation, perhaps an enormous turbulence element, which may have held together under a certain individual motion of rotation. The stars born out of this cloud can be supposed to have had originally a relative velocity distribution of circular symmetry in the galactic plane. In the process of dissolution the cloud has obtained differential galactic rotation. The mean motion of the cloud follows the angular speed ω, whereas the points P follow the smaller angular velocity ω'. The cloud has been successively drawn out in dispersion orbits about a certain circle of radius R_0, and the distribution will approach to a stationary distribution of the type considered here. The stars of population type I (cf. Sect. 33) which are responsible for the vertex deviation of the velocity ellipsoid about the Sun may thus have originated in one and the same vast cloud, which by spreading its stars has caused the characteristic velocity distribution in our surroundings.

22. The dispersion of stellar associations. We shall return at this point to our treatment in Sect. 19 of the dissolution of loose clusters or associations of stars in the general field of the Galaxy. It has immediate application to the capital discovery by AMBARZUMIAN[1] that massive stars are born in loose clouds

[1] V. A. AMBARZUMIAN: Astronom. J. USSR. **26**, 3 (1949). — Izv. Akad. Nauk USSR., Phys. Ser. **14**, 15 (1950).

to form expanding associations. The problem of expansion has been treated extensively by A. BLAAUW[1] in connection with a study of the Scorpio-Centaurus cluster. He has investigated the shape and motion of this vast cluster in the southern sky using modern material of proper motions and radial velocities, and has then tried to answer the question how these can be explained dynamically. The shape of the cluster is found to be considerably elongated, the distance between the extreme points in the direction of elongation being about three times its transverse section. The angle between the direction of elongation and the radius vector through the Sun is 52°. The mean velocity of the cluster relative to the circular motion at the position of the cluster expressed by the component of velocity in the direction outward from the galactic centre, Π_{rel}, and in the direction of galactic rotation, Θ_{rel}, is

$$\Pi_{\mathrm{rel}} = -6.9\ \mathrm{km/sec}, \qquad \Theta_{\mathrm{rel}} = -9.5\ \mathrm{km/sec}.$$

In the first one of the papers cited[1], BLAAUW gives tentatively the following explanation of the shape and motion of the cluster. Between 0.4×10^8 and 1.5×10^8 years ago the cluster, which up to that time may have been spherical, was liable to a disturbing force which gave it a velocity between 9 and 18 km/sec relative to an imaginary star, moving in a circle around the galactic centre and coinciding with the centre of the cluster at the time of disturbance. Differences occurred between the values of this relative velocity for different members of the cluster, the directions of these relative velocities being the same. The present shape and motion of the cluster are explained by the evolution since this disturbance. The explanation is based mainly on the property that stars which move with different velocities but in the same direction relative to the imaginary star just mentioned, are situated permanently on a straight line through the imaginary star. In the second paper quoted, BLAAUW adopts the hypothesis of expanding motions in groups of stars of recent formation introduced by AMBARZUMIAN in order to explain the forms and dimensions of the O-associations. This hypothesis has been confirmed by BLAAUW[2] for the ζ Persei group of O and B stars, which includes, among others, the bright stars ζ Per, ξ Per and o Per. In the case of the Scorpius-Centaurus cluster the initial expanding motions with respect to the centre of the group are assumed to be uniform, i.e. not slowing down by the mutual attractions of the stars in the group. If the initial velocity of a star at the origin $\xi_0 = 0$, $\eta_0 = 0$ be s_0 in the direction defined by the angle φ, so that

$$u_0 = s_0 \sin\varphi, \qquad v_0 = s_0 \cos\varphi, \tag{22.1}$$

we easily get by Eqs. (19.3) for a later epoch t (interchanging ξ_0, η_0, u_0, v_0 against ξ, η, u, v and changing the sign of t)

$$\left.\begin{aligned} \xi &= \frac{s_0}{\varkappa}\left[\sin\varkappa t \sin\varphi + \beta\,(1 - \cos\varkappa t)\cos\varphi\right], \\ \eta &= \frac{s_0}{2B}\left[(1 - \cos\varkappa t)\sin\varphi + (2At - \beta\sin\varkappa t)\cos\varphi\right]. \end{aligned}\right\} \tag{22.2}$$

Here

$$\beta = \sqrt{\frac{\omega}{\omega - A}}, \qquad \varkappa = 2\sqrt{\omega(\omega - A)}, \qquad \omega = A - B.$$

Instead of studying the orbits of individual stars for a certain value of φ, as was done in the earlier paper, BLAAUW examines the configuration of a group of stars

[1] A. BLAAUW: Publ. Kapteyn Astronom. Lab. **1946**, No. 52. — Bull. Astr. Inst. Netherl. **11**, 414 (1952).

[2] A. BLAAUW: Bull. Astr. Inst. Netherl. **11**, 405 (1952).

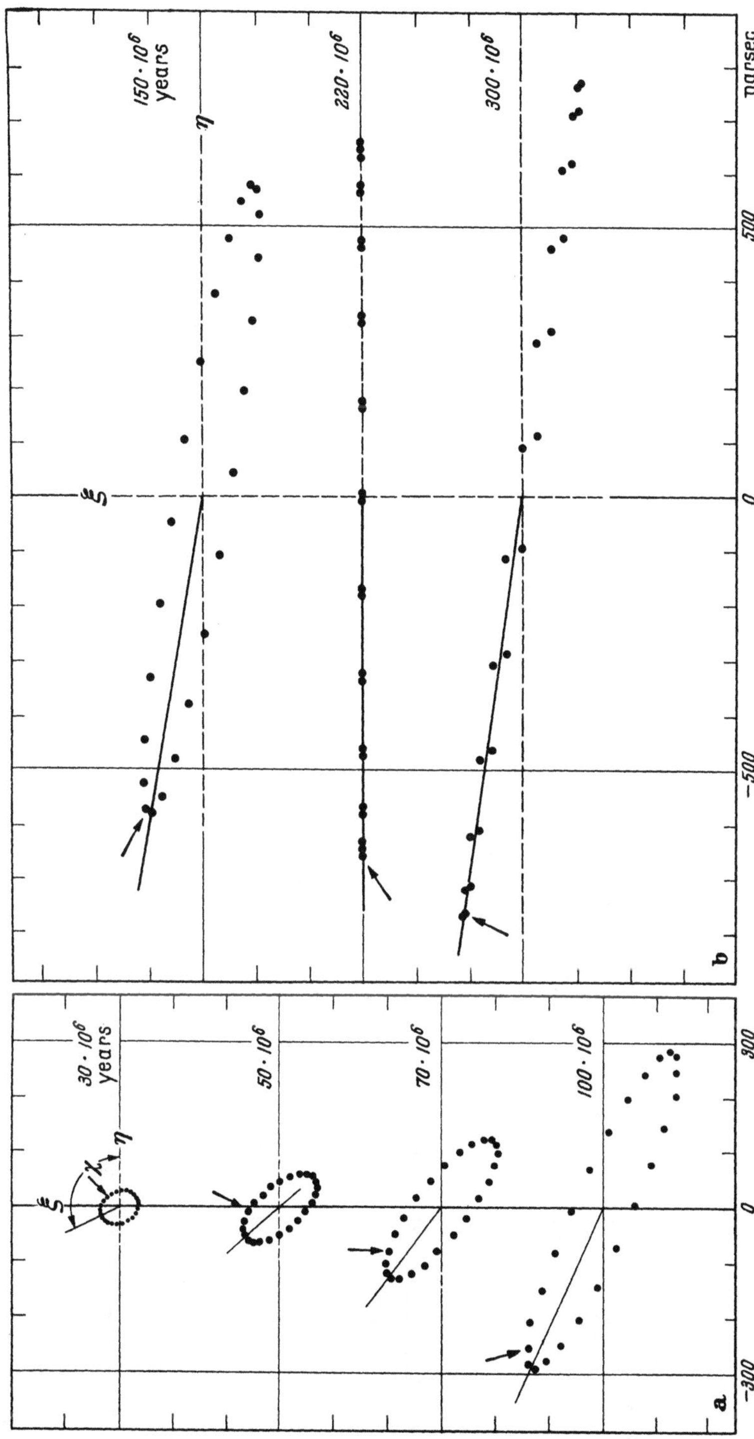

Fig. 14 a and b. The variation with time of the vertex direction and the extension of a stellar association (Blaauw).

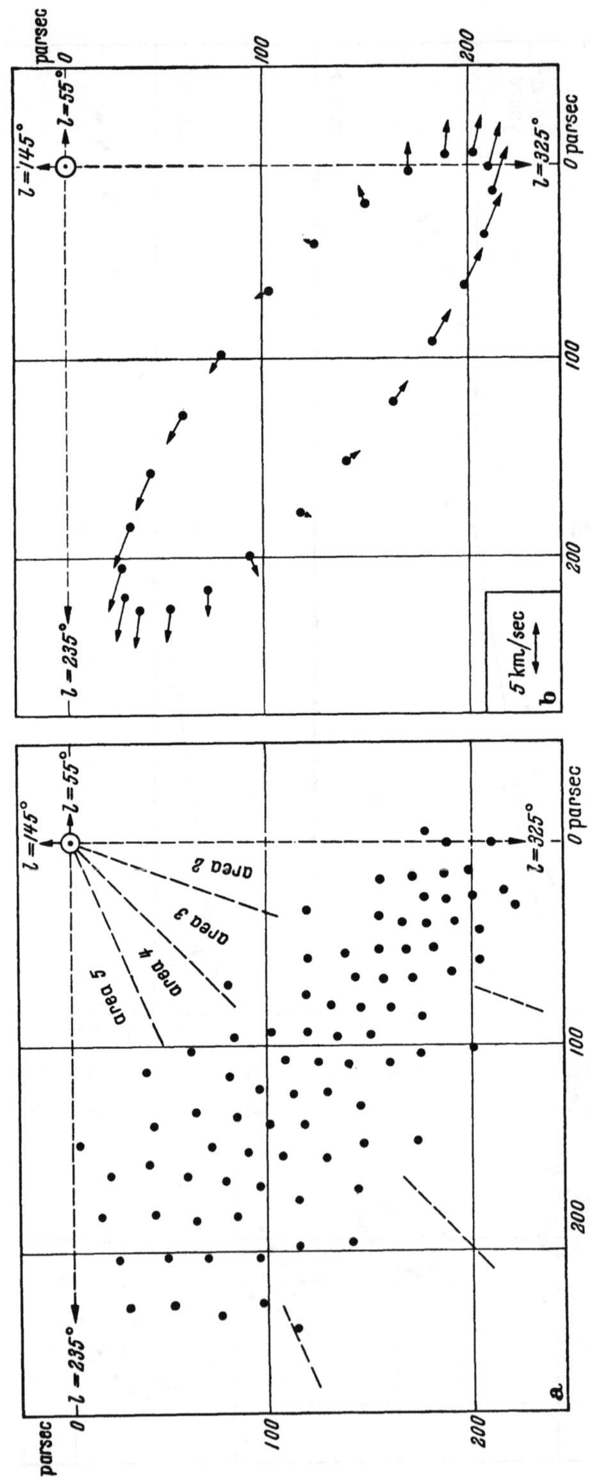

Fig. 15 a and b. The orientation and dimensions of the Scorpius-Centaurus cluster (BLAAUW).

with equal initial velocities s_0 in different directions φ. The configurations are always ellipses with the centre at the origin O. The direction of the major axis of the ellipse is given by

$$\tan 2\chi = -\frac{1-\cos\varkappa t}{A t - \frac{A-2B}{\varkappa}\sin\varkappa t}, \tag{22.3}$$

where χ is reckoned from the η-axis in the same way as φ. χ is independent of s_0. It is important to notice, in agreement with our Eqs. (17.10), that the relative orbits in the present approximation remain the same if the origin O follows another orbit than the circular motion. If the mean motion of the group differs from the circular motion, and if O follows the mean motion, the expressions for ξ and η referred to this origin and the expression of χ remain the same. The age of the group may thus be obtained from its orientation. This age, which may be called the "expansion age" is independent of the initial velocity s_0. (χ is unambiguous only if χ is below 170°.) If t is known from χ, the velocity s_0 can be derived from the dimensions of the group. Fig. 14 gives χ and the dimensions of the ellipse as functions of t for $s_0=1$ km/sec, and Fig. 15 the observed shape and orientation of the cluster. In the construction of Fig. 14 the numerical values

$$A = +0.020\ (\text{km/sec})/\text{pc} = +2.05\ 10^{-8}/\text{year},$$
$$B = -0.007\ (\text{km/sec})/\text{pc} = -0.72\ 10^{-8}/\text{year}$$

have been assumed. The observed shape gives $\chi=142°$ and an expansion age of 72 million years. We may define the begin of expansion as the initial stage of the cluster. The present mean motion of the cluster is found to indicate an initial velocity of 19 km/sec in the direction forming an angle of 87° with the direction of the circular motion, that is approximately in the direction outward from the galactic centre. The position of the cluster was at that time at the distance 2200 parsec from the present position of the Sun in the direction of galactic longitude 26° (Vulpecula-Sagitta).

IV. The dispersion of stellar velocities as function of the time.

23. Effects of encounters between individual stars. Already in the pioneer work of JEANS[1] it was shown that the influence of individual encounters between stars on the velocity dispersion must be exceedingly slow under the conditions of star density in our surroundings. The "time of relaxation" was found to be of the order of 10^{14} years. This result was confirmed by CHARLIER[2] in an elaborate theory of statistical mechanics based on the law of NEWTON. He found a time of relaxation of the same order. CHANDRASEKHAR [3] has made a rigorous examination of the problem. Considering the theoretical orbit for a star in the potential field $\varphi(x, y, z)$ of the smoothed-out distribution governed by the equation

$$\frac{d^2\boldsymbol{r}}{dt^2} = \operatorname{grad}\varphi, \tag{23.1}$$

which has the integral

$$\frac{1}{2}(\dot{x}^2+\dot{y}^2+\dot{z}^2) - \varphi(x,y,z,t) = -\int\frac{\partial\varphi}{\partial t}\,dt + \text{const}, \tag{23.2}$$

the question is: How long will it take for the cumulative effect of stellar encounters to deviate the star so much from the orbit derived from (23.1) that it can

[1] J. H. JEANS: Monthly Notices Roy. Astronom. Soc., London **74**, 109 (1913).
[2] C. V. L. CHARLIER: Medd. Lunds Obs. Ser. II **16** (1917).

no longer be described even approximately by this theoretical orbit? It is the estimation of this time that specifies the *time of relaxation* of the stellar system. In order to evaluate quantitatively the time of relaxation we may suppose to begin with that the different encounters can each be treated independently of the others and that they can be regarded individually as two-body encounters. Each encounter will result in a deflection, $\pi - 2\Psi$, of the star from its original direction of motion and an exchange of eneryg, ΔE, between the two stars taking part in the encounter. CHANDRASEKHAR considers the effect of encounters on a star of mass m_2 and initial velocity $\boldsymbol{v}_2$ during its motion through other stars. Let m_1 and $\boldsymbol{v}_1$ denote the mass and velocity of a typical *field star*. The parameters defining such an encounter are five, namely the magnitude v_1 of the vector $\boldsymbol{v}_1$, the angle ϑ between the vectors $\boldsymbol{v}_1$ and $\boldsymbol{v}_2$ when drawn from the same point, the

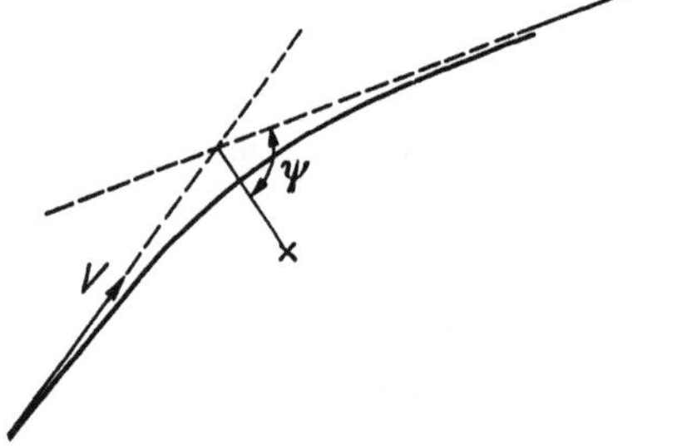

Fig. 16. The relative orbit of a close encounter.

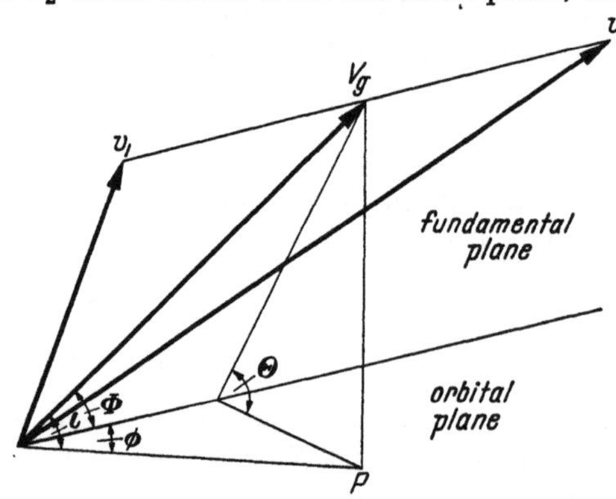

Fig. 17. The relative orientation of "fundamental plane" and "orbital plane".

azimuthal angle ω of the *fundamental plane* containing the vectors $\boldsymbol{v}_1$ and $\boldsymbol{v}_2$ about the axis defined by the direction of $\boldsymbol{v}_2$, the impact parameter D, and finally the angle Θ between the *orbital plane* and the fundamental plane just defined.

The velocity of the centre of gravity $\boldsymbol{V}_g$ remains constant during the encounter. Each star describes a hyperbola about the other (Fig. 16), and at the end of the encounter the direction of the relative velocity $\boldsymbol{V}$ is deflected by an amount $\pi - 2\psi$ in the orbital plane, where

$$\cos\psi = \frac{1}{\sqrt{1 + \frac{D^2 V^4}{G^2 (m_1 + m_2)^2}}}. \tag{23.3}$$

If i is the angle between $\boldsymbol{V}_g$ and its projection on the orbital plane, and φ the angle between $\boldsymbol{V}$ and the projection of $\boldsymbol{V}_g$ (Fig. 17), the change of energy of the mass m_2 is found to be

$$\Delta E = -2 \frac{m_1 m_2}{m_1 + m_2} V_g V \cos(\varphi - \psi) \cos\psi \cos i. \tag{23.4}$$

Let $N(v_1, \vartheta, \omega)\, dv_1\, d\vartheta\, d\omega$ be the number of field stars per unit volume with velocities in the range $(v_1, v_1 + dv_1)$ and in directions confined to the element of solid angle $\sin\vartheta\, d\vartheta\, d\omega$. The number of encounters in time dt in an infinitesimal range of the parameters $v_1, \vartheta, \omega, D, \Theta$ will be

$$N(v_1, \vartheta, \omega)\, dv_1\, d\vartheta\, d\omega\, d\Theta\, D\, dD\, V\, dt. \tag{23.5}$$

The contribution of these encounters to the sum $\sum(\Delta E)^2$ will be

$$\left.\begin{aligned}\sum \Delta E^2 = 4N(v_1, \vartheta, \varphi)\, V_g^2\, V^3 \left(\frac{m_1 m_2}{m_1+m_2}\right)^2 \times \\ \times \cos^2 i \cos^2(\varphi-\psi)\cos^2\psi\, D\, dD\, d\Theta\, dv_1\, d\vartheta\, d\omega\, dt.\end{aligned}\right\} \tag{23.6}$$

We may write

$$D\,dD = \frac{G^2(m_1+m_2)^2}{V^4}\,\frac{\sin\psi}{\cos^3\psi}\,d\psi \tag{23.7}$$

The integration with respect to ψ from 0 to $\pi/2$ diverges logarithmically. This divergence arises essentially from the improper use of the two-body encounters to describe distant encounters. By appropriately choosing an upper limit $D_{\max}$ for the impact parameter D we can compensate for the errors introduced by overestimating ΔE for encounters $D \leqq D_{\max}$ and, by ignoring ΔE for the encounters with $D \geqq D_{\max}$ we may assume for $D_{\max}$ a value of the same order as the average distance D_0 between the stars. Introducing

$$q = \frac{D_0}{G(m_1+m_2)}, \tag{23.8}$$

and integrating with respect to D, Θ, ϑ, ω, and v_1, assuming the frequency function of v_1 to be

$$N(v_1)\,dv_1 = \frac{4j^3}{\sqrt{\pi}} N\,e^{-j^2 v_1^2} v_1^2\, dv_1 \tag{23.9}$$

and retaining only dominant terms, gives finally, if $x_0 = j v_2$,

$$\sum \Delta E^2 = 8\pi N G^2 m_1^2 m_2^2 v_2 G(x_0) \log q v_2^2 \cdot dt \tag{23.10}$$

where $G(x_0)$ is expressed in terms of the error function $\Phi(x_0)$ thus

$$G(x_0) = \frac{1}{2x_0^2}\left[\Phi(x_0) - x_0 \Phi'(x_0)\right]. \tag{23.11}$$

In the range $0.6 \leqq x_0 \leqq 1.8$ we have $\overline{G(x_0)} \approx 0.18$.

We may define the time of relaxation T_E by

$$\frac{\Sigma \Delta E^2}{E^2} = \frac{dt}{T_E}, \tag{23.12}$$

from which it follows that T_E gives the time scale which would be required for stellar encounters to invalidate seriously the energy integral (23.2) in the orbit described in the general gravitational field of the smoothed-out distribution of matter in the system. It we express the mass, D_0, v_2, and N in the units of solar mass, parsec, 20 km/sec, and number per cubic parsec, respectively, we have, to a sufficient accuracy,

$$T_E = 1.0 \cdot 10^{13} \frac{v_2^3}{N m_1^2 \log_{10}[9.3 \cdot 10^4 D_0 v_2^2/(m_1+m_2)]} \text{ years.} \tag{23.13}$$

In the neighbourhood of the Sun we have approximately

$N = 0.1$ parsec^{-3}, $D_0 = 2.7$ parsecs, $m_1 = 0.50$, $j = 0.032$ sec/km.

For an incident star with $v_2 = 20$ km/sec we have then

$$T_E = 7 \cdot 10^{13} \text{ years}.$$

Since the period of galactic rotation is about $2 \cdot 10^8$ years, it follows that a star can describe at least a hundred revolutions before suffering appreciable perturbations by encounters with individual stars. The mean time of relaxation for a group of stars, $\overline{T_E}$, may be defined as the time required for $\sum \overline{\Delta E^2}$ to become equal to E^2, where $E = \frac{1}{2} m_2 v_2^2$, and is obtained by integration in v_2 to be, with the same units as above,

$$\overline{T_E} = 1.12 \cdot 10^{13} \frac{(\overline{v^2})^{\frac{3}{2}}}{N m_1^2 \log_{10} [9.3 \cdot 10^4 D_0 \overline{v^2}/(m_1 + m_2)]} \text{ years.} \tag{23.14}$$

The time of relaxation may also be defined with respect to the accumulation of deflections, $\pi - 2\Psi$, suffered at an encounter. We get

$$\sum \sin^2 2\Psi = 8\pi N G^2 m_1^2 \frac{1}{v_2^3} H(x_0) \log q v_2^2 \cdot dt \tag{23.15}$$

where

$$H(x_0) = \frac{1}{2 x_0^2} [x_0 \Phi'(x_0) + (2 x_0^2 - 1) \Phi(x_0)].$$

This time of relaxation, called T_D in CHANDRASEKHAR's work, will be of the same order as T_E.

24. Dynamical friction. In later works[1] CHANDRASEKHAR has developed a theory of *dynamical friction*, which describes how a star must suffer from a systematic tendency to be decelerated in the direction of its motion. This dynamical friction which stars experience is one of the direct consequences of the fluctuating force acting on a star due to the varying complexion of the near neighbours. It is directly connected with the demand that the stochastic variations in the velocity which a star suffers must be such as to leave an initial Maxwellian distribution of the velocities invariant. We may suppose that the acceleration, $\Delta \boldsymbol{u}$, which a star suffers in a time Δt, which is short compared to the time intervals during which $\boldsymbol{u}$ may change appreciably, but long compared to the periods of the elementary fluctuations in the field of force, can be expressed as the sum of two terms in the form

$$\Delta \boldsymbol{u} = \delta \boldsymbol{u} - \eta \boldsymbol{u} \Delta t. \tag{24.1}$$

Here the first term on the right-hand side is governed by the probability distribution

$$\Psi(\delta \boldsymbol{u}) = \frac{1}{(4\pi k \Delta t)^{\frac{3}{2}}} e^{-|\delta \boldsymbol{u} + \mathrm{grad}_{\boldsymbol{u}} k \Delta t|^2/(4 k \Delta t)}. \tag{24.2}$$

k has the character of a diffusion coefficient. The second term represents a deceleration of the star in the direction of its motion by an amount proportional to $|\boldsymbol{u}|$. The constant of proportionality, η, can therefore be properly defined as the coefficient of dynamical friction. The distribution function $W(\boldsymbol{u}, t + \Delta t)$ can be derived from the distribution $W(\boldsymbol{u}, t)$ by means of the integral equation

$$W(\boldsymbol{u}, t + \Delta t) = \int_{-\infty}^{+\infty} W(\boldsymbol{u} - \Delta \boldsymbol{u}, t) \Psi(\boldsymbol{u} - \Delta \boldsymbol{u}; \Delta \boldsymbol{u}) d(\Delta \boldsymbol{u}) \tag{24.3}$$

where $\Psi(\boldsymbol{u}; \Delta \boldsymbol{u})$ denotes the transition probability

$$\Psi(\boldsymbol{u}; \Delta \boldsymbol{u}) = \frac{1}{(4\pi k \Delta t)^{\frac{3}{2}}} e^{-|\Delta \boldsymbol{u} + \mathrm{grad}_{\boldsymbol{u}} k \Delta t + \eta \boldsymbol{u} \Delta t|^2/(4 k \Delta t)}. \tag{24.4}$$

[1] S. CHANDRASEKHAR: Astrophys. Journ. **97**, 255, 263 (1943); **98**, 54 (1943).

Developing and passing to the limit $\Delta t = 0$, we find

$$\frac{\partial W}{\partial t} = \mathrm{div}_{\boldsymbol{u}}(k\,\mathrm{grad}_{\boldsymbol{u}} W) + \mathrm{div}_{\boldsymbol{u}}(\eta W \boldsymbol{u}). \tag{24.5}$$

The condition that the Maxwellian distribution

$$\left(\frac{3}{2\pi\overline{|\boldsymbol{u}|^2}}\right)^{\frac{3}{2}} \mathrm{e}^{-3|\boldsymbol{u}|^2/2\overline{|\boldsymbol{u}|^2}} \tag{24.6}$$

satisfy this differential equation identically, requires

$$\frac{k}{\eta} = \frac{1}{3}\overline{|\boldsymbol{u}|^2} = \mathrm{const}. \tag{24.7}$$

In the case of the two-body approximation for stellar encounters which has been developed above we may easily derive the increments in velocity, $\Delta v_{\parallel}$ and $\Delta v_{\perp}$, which a star of velocity $v_2 = |\boldsymbol{v}_2|$ and mass m_2 suffers as the result of an encounter in directions which are respectively parallel and perpendicular to the direction of motion. These increments are

$$\left.\begin{aligned} \Delta v_{\parallel} &= -\frac{2m_1}{m_1+m_2}\left[(v_2 - v_1\cos\vartheta)\cos\psi + v_1\sin\vartheta\cos\Theta\sin\psi\right]\cos\psi,\\ \Delta v_{\perp} &= \pm\frac{2m_1}{m_1+m_2}\left[v_1^2 + v_2^2 - 2v_1v_2\cos\vartheta - \right.\\ &\quad \left. - \{(v_2 - v_1\cos\vartheta)\cos\psi + v_1\sin\vartheta\cos\Theta\sin\psi\}^2\right]^{\frac{1}{2}}\cos\psi. \end{aligned}\right\} \tag{24.8}$$

As CHANDRASEKHAR has shown, we find

$$\sum \Delta v_{\parallel} = -4\pi N m_1(m_1+m_2)\frac{G^2}{v_2^2}\log(q\overline{|\boldsymbol{u}|^2})\,\Delta t\,[\Phi(x_0) - x_0\Phi'(x_0)]. \tag{24.9}$$

This means that the star experiences dynamical friction with the coefficient

$$\eta = 4\pi N m_1(m_1+m_2)\frac{G^2}{v_2^3}\log(q\overline{|\boldsymbol{u}|^2})\,[\Phi(x_0) - x_0\Phi'(x_0)]. \tag{24.10}$$

We have, moreover,

$$\sum(\Delta v_{\parallel})^2 = \frac{8}{3}\pi N m_1^2\frac{G^2}{v_2^3}\overline{|\boldsymbol{u}|^2}\log(q\overline{|\boldsymbol{u}|^2})\,\Delta t\,[\Phi(x_0) - x_0\Phi'(x_0)]. \tag{24.11}$$

This gives

$$\frac{\sum(\Delta v_{\parallel})^2}{\eta\,\Delta t} = \frac{2}{3}\,\frac{m_1}{m_1+m_2}\,\overline{|\boldsymbol{u}|^2}, \tag{24.12}$$

which is to be compared with (24.7). It is thus seen that a detailed analysis of the fluctuating field of the near-by stats in terms of individual stellar encounters idealized as two-body problems fully confirms the conclusions reached on the basis of certain general principles. We shall return later to this kind of dynamical friction in connection with the dynamics of stellar clusters.

25. Influence of cloud formations. A far more active perturbation process which affects the general velocity dispersion may be found in the interaction between stars and clouds of interstellar matter. In recent years evidence has been accumulating which shows that the interstellar matter is far from evenly distributed but is rather concentrated into clouds. The uneven distribution of the interstellar matter will produce perturbations in the stellar orbits. This possibility was suggested by EDDINGTON[1] and has been subject to a detailed

[1] A. S. EDDINGTON: Nature, Lond. **151**, 91 (1943).

analysis by SPITZER and M. SCHWARZSCHILD[1] and by OOSTERBROCK[2]. When particles attract one another according to inverse-square forces, the cumulative effect of many small collisions, produced by relatively distant encounters, is more important than occasional large deflections by close encounters. The analogy with the Brownian motion has been pointed out by CHANDRASEKHAR (Sect. 23) and ROSSELAND[3]. In their first paper on the subject SPITZER and SCHWARZSCHILD applied a diffusion equation in velocity space analogous to the one derived by COHEN, SPITZER, and ROUTLY[4] for the interaction between electrically charged particles. If f is the velocity distribution function for some particular type of star, and f_c the corresponding function for the clouds, the Boltzmann equation is of the type

$$\frac{\partial f}{\partial t} = -K(f f_c), \tag{25.1}$$

and in the present case obtains the ultimate form

$$\frac{\partial f}{\partial \vartheta} = \frac{1}{x^2}\frac{\partial}{\partial x}\left(\frac{\varphi - x\varphi'}{x}\frac{\partial f}{\partial x}\right), \tag{25.2}$$

where

$$\vartheta = t\cdot\frac{\pi}{\sqrt{2}}\,G^2\frac{m_c^2\,n_c}{V_c^3}\log\alpha, \tag{25.3}$$

and $\varphi(x)$ the error function $\frac{2}{\sqrt{\pi}}\int\limits_0^x e^{-u^2}\,du$. m_c and n_c are the mass and number per unit volume of the clouds, V_c the root mean square of one component of the cloud velocities. α is essentially the ratio of the largest impact parameter to the radius of the cloud, and is estimated to be of the order 10.

Eq. (25.2) was integrated numerically starting from $f = e^{-x^2}$ at $\vartheta = 0$. If we assume that most of the interstellar matter is gathered into cloud complexes with masses of the order of 10^6 solar masses and if the velocity dispersion of these massive clouds is about 10 km/sec, it is possible to account on this basis for the fact that the root-mean-square random velocity for each spectral type along the main sequence increases with advancing spectral type from 10 km/sec for the O and B stars to some 20 km/sec for the F stars. The O and B stars have an average short age compared with the galaxy because of their high rate of energy output, whereas the average age of early F stars (older stars of population type I), if they have been continuously formed from interstellar clouds, must be half the age of the galaxy, approximately 1.5×10^9 years.

In their second paper SPITZER and SCHWARZSCHILD take into account the presence of differential galactic rotation. As a result of such non-uniform rotation, the clouds will have velocities relative to one another even if their random velocities are zero. The gravitational potential of these clouds will be a fluctuating function of time, and the stars will gain kinetic energy from the clouds. We accept here as basis the analysis of individual stellar motions in the preceding chapter and keep the notations employed there. If we consider a star in the neighbourhood of the Sun whose velocity component u in the direction of the radius vector is 10 km/sec, the axes of the epicyclic ellipse about its centre P will be 1400 and 700 parsecs. If the clouds move with the local circular velocity,

---

[1] L. SPITZER and M. SCHWARZSCHILD: Astrophys. Journ. **114**, 385 (1951); **118**, 106 (1953).
[2] D. E. OOSTERBROCK: Astrophys. Journ. **116**, 164 (1952).
[3] S. ROSSELAND: Monthly Notices Roy. Astronom. Soc. London **88**, 208 (1928).
[4] R. S. COHEN, L. SPITZER and P. M. ROUTLY: Phys. Rev. **80**, 230 (1950).

the effective velocity spread of the clouds encountered by the star along its path will be ± 15 km/sec. The effect of each encounter is to deflect the star by some angle $\Delta\vartheta$ in the ξ, η-plane without changing its local velocity $\sqrt{u^2+v^2}$. Thus the changes of u and v are given, to the first order, by

$$\Delta u = -v\,\Delta\vartheta, \qquad \Delta v = u\,\Delta\vartheta. \tag{25.4}$$

If we write $K = c\varkappa$, when c is the radial half-axis of the epicycle, and $\varkappa = 2\sqrt{\omega(\omega - A)}$, we have from Eqs. (17.22)

$$K^2 = u^2 - \frac{\omega}{B} v^2. \tag{25.5}$$

The change of K resulting from a deflection $\Delta\vartheta$ will be

$$K\,\Delta K = -\frac{A}{B} u\,v\,\Delta\vartheta. \tag{25.6}$$

Taking half the mean square V^2 of the velocity around the star's orbit,

$$V^2 = \frac{1}{2}\left(\overline{u^2} + \overline{v^2}\right) = \frac{1}{4} K^2 \left(1 - \frac{B}{\omega}\right). \tag{25.7}$$

Hence

$$V\,\Delta V = \frac{A}{4B}\left(1 - \frac{B}{\omega}\right) u\,v\,\Delta\vartheta. \tag{25.8}$$

If V_n be the value of V after n deflections

$$\overline{V_n^2} = V_0^2 + \sum_{i=1}^{n} \overline{(\Delta V_i)^2}. \tag{25.9}$$

We may write

$$\frac{d\overline{V^2}}{dt} = \frac{1}{16 V^2}\left[\frac{A}{B}\left(1 - \frac{B}{\omega}\right)\right]^2 \overline{u^2 v^2 \langle(\Delta\vartheta)^2\rangle}, \tag{25.10}$$

where $\langle(\Delta\vartheta)^2\rangle$ is the sum of all the mean square deflections per second. This may be written, following CHANDRASEKHAR [3],

$$\langle(\Delta\vartheta)^2\rangle = \frac{4\pi G^2 n_c m_c^2 \log\alpha}{(u^2+v^2)^{\frac{3}{2}}}. \tag{25.11}$$

This gives, inserting the expressions for u, v, V,

$$\frac{d\overline{V^2}}{dt} = \frac{\pi^{\frac{3}{2}} G^2 n_c m_c^2 \log\alpha}{2 V_m} F\left(\frac{B}{\omega}\right), \tag{25.12}$$

$$\left.\begin{aligned} F\left(\frac{B}{\omega}\right) &= -\frac{\omega}{B}\left(1 - \frac{B^2}{\omega^2}\right)^2 \left(1 - \frac{B}{\omega}\right)^{-\frac{1}{2}} \frac{2}{\pi} \int_0^{\pi/2} \frac{\sin^2\vartheta\cos^2\vartheta}{\left(\sin^2\vartheta - \frac{B}{\omega}\cos^2\vartheta\right)^{\frac{3}{2}}}\, d\vartheta \\ &= -\frac{\omega}{8B}\left(1 - \frac{B^2}{\omega^2}\right)^2 \left(1 - \frac{B}{\omega}\right)^{-\frac{1}{2}} G\left(1 + \frac{B}{\omega}\right), \end{aligned}\right\} \tag{25.13}$$

where

$$G(x) = 1 + \frac{3}{4}x + \frac{75}{128}x^2 + \frac{245}{512}x^3 + \frac{6615}{16384}x^4 + \cdots. \tag{25.14}$$

If $V_m(0)$ denotes the value of V_m at time zero, we have

$$V_m = V_m(0)\left(1 + \frac{t}{t_E}\right)^{\frac{1}{3}}, \tag{25.15}$$

where

$$t_E = \frac{4\,V_m^3(0)}{3\pi^{\frac{3}{2}}\,G^2\,n_c\,m_c^2\,\log\alpha\,F\left(\frac{B}{\omega}\right)}. \tag{25.16}$$

In the case of an inverse-square field, $1+\frac{B}{\omega}$ is 0.75. Numerical integration in this case gives $G(0.75)=2.48$. $\log\alpha$ may be assumed equal to 3.

If the velocity V_m for the F stars is to increase from 10 km/sec to 20 km/sec in 1.5×10^9 years, t_E must be equal to $2\cdot10^8$ years. If this value is inserted into (25.16), and $n_c\,m_c$, the average density of the interstellar matter, is taken to be $3\cdot10^{-24}\,\mathrm{g/cm^3}$, we find

$$m_c = 4.9\times10^5 \text{ solar masses}.$$

If only about half of the interstellar medium is concentrated in clouds, m_c must be about 10^6 solar masses. If the density in a cloud corresponds to 15 H atoms/cm³, the radius of such a cloud will be about 100 parsecs, and the mean distance between centres of neigh-bouring clouds about 350 parsecs. The spacing between the clouds is then about equal to the thickness of the galaxy. If the stars move in a plane, encountering clouds which are situated in the same plane, and if n_c' is the density of such clouds per unit area and a is the radius of a cloud, we find, ignoring encounters in which the star passes through a cloud,

$$\langle(\Delta\vartheta)^2\rangle = \frac{8\,G^2\,n_c'\,m_c^2}{a\,(u^2+v^2)^{\frac{3}{2}}}. \tag{25.17}$$

The ratio of this equation to the corresponding Eq. (25.11) above is not far from unity, so that the difference for encounters in the galactic plane is small. Though there will also be deflections in the z direction in the actual case and some increase of the Z velocities, this increase will certainly be less than is found in the simple three-dimensional picture. The theory therefore yields a simple qualitative explanation of the persistence of low Z velocities in the older stars. Another complication is the large size of the cloud complexes relative to the stellar epicyclic orbits, so that the curvature of the orbit is not negligible during an encounter. This might give a somewhat more rapid rate of deflection of a star.

26. Dynamics of stellar clusters. In this connection we shall devote some attention to the dynamics of stellar clusters.

If we denote the coordinates of a star i relative to the centre of gravity by ξ_i, η_i, ζ_i, we may write the internal kinetic energy

$$T = \frac{1}{2}\sum_i m_i\left[\left(\frac{d\xi_i}{dt}\right)^2 + \left(\frac{d\eta_i}{dt}\right)^2 + \left(\frac{d\zeta_i}{dt}\right)^2\right], \tag{26.1}$$

and the total potential energy

$$\Omega = -\sum_{ij} G\,\frac{m_i\,m_j}{r_{ij}}, \tag{26.2}$$

where r_{ij} is the distance between the stars i and j, and where the summation extends to all pairs i and j, $i \neq j$, each pair occurring only once. If M is the total mass of the cluster, the *virial theorem* states that

$$\frac{1}{2M}\,\frac{d^2}{dt^2}\left(\sum_{ij} m_i\,m_j\,r_{ij}^2\right) = 2T + \Omega, \tag{26.3}$$

where the sum in the left-hand member is extended over all pairs i, j, each pair occurring only once. This relation is due to LAGRANGE and JACOBI. In a quasi-stationary case the left-hand member may be assumed to vanish, and we then have the important relations

$$2T + \Omega = 0, \quad \text{or} \quad 2E - \Omega = 0, \tag{26.4}$$

if E is the total energy of the cluster. As Ω is negative, a necessary condition for the stability of a cluster is that E is negative. As an example we may consider the case when all members of the cluster are assumed to have the same mass m. If n is the number of stars, we can then write

$$2T = n m \overline{v^2} = M \overline{v^2}. \tag{26.5}$$

Further

$$\Omega = -\frac{1}{2} G M^2 \frac{n(n-1)}{\overline{R}}, \tag{26.6}$$

if $\overline{R}$ is the average distance between the individuals. To the order of magnitude we may write

$$\Omega = -\frac{1}{2} \frac{G M^2}{R}, \tag{26.7}$$

where R is the mean radius of the cluster. Expressing the mass and the radius in units of solar mass and parsec, respectively, CHANDRASEKHAR [3] finds

$$\sqrt{\overline{v^2}} = 4.63 \cdot 10^{-2} \sqrt{\frac{M}{R}} \text{ km/sec}. \tag{26.8}$$

For the Pleiades we may assume approximetely $M = 300 \odot$, $R = 3.5$ parsecs, which gives $\sqrt{\overline{v^2}} = 0.43$ km/sec. For the time of relaxation, T_E, in the Pleiades CHANDRASEKHAR finds 2.9×10^7 years.

For the question of disintegration of clusters by the escape of stars, CHANDRASEKHAR'S theory of dynamical friction is of great importance. We may suppose that, in order that a star may escape from a cluster, it is only necessary that it acquire a velocity greater than the velocity of escape. On this assumption the probability that a star will have acquired the necessary velocity for escape during a certain time can be determined in terms of the probability, $p(v_0, t)\, dt$, that a star having initially a velocity $|\boldsymbol{u}| = v_0$ at time $t = 0$ will acquire for the first time the velocity $|\boldsymbol{u}| = v_\infty$ between t and $t + dt$. This probability function can be derived from the spherically symmetric function of Eq. (24.5) which satisfies the boundary conditions

$$\left.\begin{aligned} &W(|\boldsymbol{u}|, t) = 0 \quad \text{for} \quad |\boldsymbol{u}| = v_\infty \quad \text{for all} \quad t > 0 \\ &W(|\boldsymbol{u}|, t) \to \frac{1}{4\pi v_0^2} \delta(|\boldsymbol{u}| - v_0) \quad \text{as} \quad t \to 0 \end{aligned}\right\} \tag{26.9}$$

where δ stands for the δ-function of DIRAC. The second Eq. (26.9) means that we start from a thin shell in the velocity space with radius v_0. The probability that a star with original velocity $|\boldsymbol{u}| = v_0$ during the time between t and $t + dt$ exceeds the limit v_∞ is

$$p(v_0, t) = -\left(4\pi k |\boldsymbol{u}|^2 \frac{\partial W(|\boldsymbol{u}|, t)}{\partial |\boldsymbol{u}|}\right)_{|\boldsymbol{u}| = v_\infty}. \tag{26.10}$$

The probability during the time 0 to t will then be

$$Q(v_0, t) = \int_0^t p(v_0, t)\, dt \tag{26.11}$$

and finally the probability for all v_0 during the same time

$$Q(t) = \int_0^{v_\infty} Q(v_0, t)\, f(v_0)\, d v_0, \tag{26.12}$$

where $f(v_0)$ is the frequency function of v_0. For η we have, according to (24.10), for $m_1 = m_2 = m$,

$$\eta = 8\pi N m^2 G^2 \left(\log \frac{D_0 \overline{|\boldsymbol{u}|^2}}{2Gm}\right) \frac{1}{|\boldsymbol{u}|^3} \left[\Phi(j|\boldsymbol{u}|) - j(|\boldsymbol{u}|)\, \Phi'(j(|\boldsymbol{u}|)\right], \tag{26.13}$$

where Φ is the error function and j is the parameter which occurs in the assumed Maxwellian distribution function

$$\frac{j^3}{\pi^{\frac{3}{2}}} e^{-j^2|\boldsymbol{u}|^2} d\boldsymbol{u}; \quad j = \left(\frac{3}{2\overline{|\boldsymbol{u}|^2}}\right)^{\frac{1}{2}}. \tag{26.14}$$

We may introduce the notations

$$\left.\begin{aligned} &\eta = \eta_0 \nu(\varrho), \\ &\varrho = j|\boldsymbol{u}|, \quad \nu(\varrho) = \frac{3\pi^{\frac{1}{2}}}{4} \varrho^{-3} \left[\Phi(\varrho) - \varrho\, \Phi'(\varrho)\right]. \end{aligned}\right\} \tag{26.15}$$

Here

$$\left.\begin{aligned} &\nu(\varrho) \to 1 \quad \text{as} \quad \varrho \to 0, \\ &\nu(\varrho) \sim \frac{3\pi^{\frac{1}{2}}}{4} \varrho^{-3} \quad \text{as} \quad \varrho \to \infty. \end{aligned}\right\} \tag{26.16}$$

The expression for η_0 becomes by (26.13)

$$\eta_0 = 8\pi N m^2 G^2 \left(\log \frac{D_0 \overline{\boldsymbol{u}^2}}{2Gm}\right) \left(\frac{3}{2\overline{|\boldsymbol{u}|^2}}\right)^{\frac{3}{2}} \frac{4}{3\pi^{\frac{1}{2}}}. \tag{26.17}$$

Again, since $k = \frac{1}{3} \overline{|\boldsymbol{u}|^2}\, \eta$, we have

$$k = \tfrac{1}{3} \overline{|\boldsymbol{u}|^2}\, \eta_0 \nu(\varrho). \tag{26.18}$$

We may introduce new variables τ and ϱ instead of t and u, thus

$$\tau = \eta_0 t, \quad \boldsymbol{u} = \left(\tfrac{2}{3} \overline{|\boldsymbol{u}|^2}\right)^{\frac{1}{2}} \boldsymbol{\varrho}. \tag{26.19}$$

Eq. (24.5) now takes the dimensionless form

$$\frac{\partial W}{\partial t} = \operatorname{div}_{\boldsymbol{\varrho}} \left[\frac{1}{2} \nu(|\boldsymbol{\varrho}|) \operatorname{grad}_{\boldsymbol{\varrho}} W + \nu(|\boldsymbol{\varrho}|)\, W \boldsymbol{\varrho}\right]. \tag{26.20}$$

For a spherically symmetric solution, Eq. (26.20) reduces to

$$\varrho \frac{\partial w}{\partial \tau} = \frac{\partial}{\partial \varrho} \left[\nu(\varrho) \left\{\frac{1}{2} \varrho \frac{\partial w}{\partial \varrho} + \left(\varrho^2 - \frac{1}{2}\right) w\right\}\right] \tag{26.21}$$

where

$$\varrho = |\boldsymbol{\varrho}|, \quad w = W\varrho. \tag{26.22}$$

We require a solution of (26.21) which satisfies

$$\left.\begin{aligned} &w(\varrho, \tau) = 0 \quad \text{for} \quad \varrho = 0 \quad \text{and} \quad \varrho = \varrho_\infty \quad \text{for all} \quad \tau > 0, \\ &w(\varrho, \tau) \to \frac{1}{4\pi \varrho_0} \delta(\varrho - \varrho_0) \quad \text{as} \quad \tau \to 0. \end{aligned}\right\} \tag{26.23}$$

Eq. (26.21) is separable ϱ and τ. We may write

$$w = e^{-\lambda\tau - \frac{\varrho^2}{2}} \Psi, \tag{26.24}$$

which gives

$$\frac{d^2\Psi}{d\varrho^2} + \frac{d\log v}{d\varrho}\frac{d\Psi}{d\varrho} + \left[2\frac{\lambda}{v(\varrho)} + 3 - \varrho^2 - \frac{d\log v}{d\varrho}\left(\frac{1}{\varrho} - \varrho\right)\right]\Psi = 0. \tag{26.25}$$

In order to satisfy the boundary conditions (26.23) it is necessary that λ take one of an infinite enumerable set of discrete values

$$\lambda_1, \lambda_2, \ldots, \lambda_n, \ldots$$

which may properly be called the "characteristic values" of the problem. Further, if

$$\Psi_1, \Psi_2, \ldots, \Psi_n, \ldots$$

denote the solutions of (26.25) which satisfy the boundary conditions (26.23) and belong, respectively, to the values $\lambda_1, \lambda_2, \ldots, \lambda_n, \ldots$ then it can be shown that these solutions form a complete set of orthogonal functions. We may suppose that these functions are all properly normalized. In terms of fundamental solutions $w_n = e^{-\lambda_n \tau} e^{-\frac{\varrho^2}{2}} \Psi_n(\varrho)$ which satisfy the first condition (26.23) we may then construct a solution which satisfies the second one, and may then write

$$W = \frac{1}{4\pi\varrho\varrho_0} e^{-(\varrho^2-\varrho_0^2)/2} \sum_{n=1}^{\infty} e^{-\lambda_n \tau} \Psi_n(\varrho)\Psi_n(\varrho_0). \tag{26.26}$$

CHANDRASEKHAR finally finds for the function $Q(\tau)$

$$Q(\tau) = \sum_{n=1}^{\infty} Q_n(\tau), \tag{26.27}$$

$$Q_n(\tau) = A_n(1 - e^{-\lambda_n \tau}), \tag{26.28}$$

$$A_n = \frac{1}{2\lambda_n} \varrho_\infty \nu(\varrho_\infty) e^{-\varrho_\infty^2/2} \left(-\frac{\partial \Psi_n}{\partial \varrho}\right)_{\varrho=\varrho_\infty} \left[\frac{e^{\varrho_0^2/2}}{\sigma_0}\Psi_n(\varrho_0)\right]. \tag{26.29}$$

Since in a star cluster the root-mean-square velocity of escape is twice the root-mean-square velocity of the stars in the system, it is clear that the values of ϱ_∞ which come under discussion are in the general neighbourhood of

$$\varrho_\infty = \sqrt{6} \sim 2.45.$$

For such values of ϱ_∞, $Q(\tau)$ can be represented with ample accuracy by the first term on the right-hand side of (26.27). By numerical integration CHANDRASEKHAR finds the values of λ, ϱ_∞ and Q_1 given in Table 11. As half-life of a cluster we may define $1/\lambda_1\eta_0$, which for $\lambda_1 = 0.0075$ becomes

Table 11.

λ	ϱ_∞	$Q_1(\tau)$
0.0025	2.6642	$1.0000\,(1 - e^{-0.0025\,\tau})$
0.0050	2.5320	$0.9978\,(1 - e^{-0.0050\,\tau})$
0.0075	2.4518	$0.9966\,(1 - e^{-0.0075\,\tau})$
0.0100	2.3936	$0.9941\,(1 - e^{-0.0100\,\tau})$
0.0125	2.3476	$0.9921\,(1 - e^{-0.0125\,\tau})$

$$133\,\eta_0^{-1}. \tag{26.30}$$

For the Pleiades η_0^{-1} is of the order of $2 \cdot 10^7$ years so that its half-life is of the order of $3 \cdot 10^9$ years. When dynamical friction is ignored, CHANDRASEKHAR finds that $Q(\tau)$ can be approximately represented by $Q(\tau) \approx 1.3\,(1 - e^{-0.82\,\tau})$, in contrast to the expression $Q(\tau) \approx (1 - e^{-0.0075\,\tau})$ when dynamical friction is included. CHANDRASEKHAR therefore states that there can thus be hardly any doubt that dynamical friction provides the principal mechanism for the continued existence of the galactic clusters like the Pleiades for times of the order $3 \cdot 10^9$ years. But, even with dynamical friction properly allowed for, it will be hard to account for such clusters' half-lives of the order of 10^{10} years. This, in turn, provides another strong argument in favour of the "short-time scale".

For stars with masses appreciably different from the average value, ϱ_∞ may be expected to change according to

$$\varrho_\infty(m) = \left(6\frac{m}{\bar{m}}\right)^{\frac{1}{2}}. \tag{26.31}$$

From Table 11 we now see that even a 10% increase of ϱ_∞ prolongs the half-lives by a factor of the order 3, while a similar decrease in ϱ_∞ shortens the half-life by a factor of the order 2. Therefore, a cluster loses its less massive members rather more rapidly than the average ones, while the more massive members continue to remain, on the average, for longer times.

There are reasons to expect that the equipartition of energy between stars of different masses is well advanced in clusters. This result is in accordance with VON ZEIPEL'S[1] method of determining the relative masses of stars of different colour in clusters by their degree of concentration toward the centre.

27. Stability of clusters. When analyzing the dynamics of a star cluster we must take into consideration that the cluster forms part of the galactic system, moving in this system under the general galactic field of potential φ. The variation of φ over the spatial extent of the cluster will give rise to what is essentially a tidal field. The simplest case of the problem occurs when the centre of gravity of the cluster describes circular orbit in the galactic plane. This problem was first considered by BOK[2] and MINEUR[3]. In analogy with Sect. 17, we introduce a coordinate system ξ, η, ζ with origin coinciding with the centre of gravity of the cluster at the distance $R = R_0$ from the centre and rotating uniformly with the angular velocity ω_c about the centre of the system. ξ, η, ζ are the distances from the origin along the radial, the transverse, and the z-directions, respectively. Let Ω be the potential due to the cluster

$$\Omega = -G\sum \frac{m_i m_j}{r_{ij}} \; (i \neq j). \tag{27.1}$$

If Ω vanishes, to the first order the Eqs. (17.7) will hold. It is readily seen that with Ω present the corresponding equations of first order for a star of index i will be

$$\left.\begin{aligned} \frac{d^2\xi_i}{dt^2} - 2\omega_c \frac{d\eta_i}{dt} + \alpha_1 \xi_i &= -\frac{1}{m_i}\frac{\partial \Omega}{\partial \xi_i}, \\ \frac{d^2\eta_i}{dt^2} + 2\omega_c \frac{d\xi_i}{dt} &= -\frac{1}{m_i}\frac{\partial \Omega}{\partial \eta_i}, \\ \frac{d^2\zeta_i}{dt^2} + \alpha_3 \zeta_i &= -\frac{1}{m_i}\frac{\partial \Omega}{\partial \zeta_i}. \end{aligned}\right\} \tag{27.2}$$

Here

$$\alpha_1 = -\left(\frac{\partial^2\varphi}{\partial R^2} - \frac{1}{R}\frac{\partial\varphi}{\partial R}\right), \qquad \alpha_3 = \frac{\partial^2\varphi}{\partial z^2}. \tag{27.3}$$

Following CHANDRASEKHAR [3] we may derive the *energy integral* as follows

$$\frac{1}{2}\sum m_i\left[\left(\frac{d\xi_i}{dt}\right)^2 + \left(\frac{d\eta_i}{dt}\right)^2 + \left(\frac{d\zeta_i}{dt}\right)^2\right] + \Omega + \frac{1}{2}\alpha_1 \sum m_i \xi_i^2 + \frac{1}{2}\alpha_3 \sum m_i \zeta_i^2 = \text{const}. \tag{27.4}$$

[1] H. VON ZEIPEL and J. LINDGREN: K. Svenska Vetenskaps. Akad. Handl. **61**, No. 15 (1921).
[2] B. J. BOK: Harvard Obs. Circ. **1934**, No. 384.
[3] H. MINEUR: Ann. d'Astrophys. **2**, No. 1 (1939).

If $\sum m_i \xi_i \eta_i = 0$, we have the *angular-momentum integral*

$$\sum m_i\left(\xi_i \frac{d\eta_i}{dt} - \eta_i \frac{d\xi_i}{dt}\right) + \omega_c \sum m_i (\xi_i^2 + \eta_i^2) = A_\zeta = \text{const}. \tag{27.5}$$

If we further suppose that

$$\sum m_i \eta_i \zeta_i = \sum m_i \zeta_i \xi_i = \sum m_i \zeta_i \frac{d\xi_i}{dt} = \sum m_i \zeta_i \frac{d\eta_i}{dt} = 0, \tag{27.6}$$

we have the additional integrals

$$\sum m_i\left(\eta_i \frac{d\zeta_i}{dt} - \zeta_i \frac{d\eta_i}{dt}\right) = A_\xi = \text{const}, \tag{27.7}$$

$$\sum m_i\left(\zeta_i \frac{d\xi_i}{dt} - \xi_i \frac{d\zeta_i}{dt}\right) = A_\eta = \text{const}. \tag{27.8}$$

The generalized Lagrangian identity ("virial theorem") will be

$$\left.\begin{aligned} \frac{1}{2\sum m_i} \frac{d^2}{dt^2}\left(\sum m_i m_j r_{ij}^2\right) &= \sum m_i \left[\left(\frac{d\xi_i}{dt}\right)^2 + \left(\frac{d\eta_i}{dt}\right)^2 + \left(\frac{d\zeta_i}{dt}\right)^2\right] + \\ &+ \Omega + 2\omega_c A_\zeta - 2\omega_c^2 \sum m_i (\xi_i^2 + \eta_i^2) - \alpha_1 \sum m_i \xi_i^2 - \alpha_3 \sum m_i \zeta_i^2. \end{aligned}\right\} \tag{27.9}$$

For a cluster in a statistically steady state we therefore have

$$2T + \Omega + 2\omega_c A_\zeta = \alpha_1 \sum m_i \xi_i^2 + \alpha_3 \sum m_i \zeta_i^2 + 2\omega_c^2 \sum m_i (\xi_i^2 + \eta_i^2). \tag{27.10}$$

In order to find the stability of a cluster against the shearing effects of differential galactic rotation we may suppose as a schematic case that the smoothed-out distribution of matter in the cluster can be represented by a homogeneous ellipsoid. Then Ω takes the form

$$\Omega_i = -\tfrac{1}{2} m_i \beta_0 + \tfrac{1}{2} m_i (\beta_1 \xi_i^2 + \beta_2 \eta_i^2 + \beta_3 \zeta_i^2). \tag{27.11}$$

We may write

$$\left.\begin{aligned} \beta_1 &= \pi G \varrho \beta_1'(a:b:c), \\ \beta_2 &= \pi G \varrho \beta_2'(a:b:c), \\ \beta_3 &= \pi G \varrho \beta_3'(a:b:c), \end{aligned}\right\} \tag{27.12}$$

where ϱ is the density of the ellipsoid and $\beta_1', \beta_2', \beta_3'$ are pure numbers depending on the ratio of the axes a, b, c.

The differential equations take the form

$$\left.\begin{aligned} \frac{d^2\xi}{dt^2} - 2\omega_c \frac{d\eta}{dt} + (\alpha_1 + \beta_1)\xi &= 0, \\ \frac{d^2\eta}{dt^2} + 2\omega_c \frac{d\xi}{dt} + \beta_2 \eta &= 0, \end{aligned}\right\} \tag{27.13}$$

and

$$\frac{d^2\zeta}{dt^2} + (\alpha_3 + \beta_3)\zeta = 0. \tag{27.14}$$

Eq. (27.14) gives immediately

$$\zeta = \zeta_0 \cos q_3 (t + t_3) \tag{27.15}$$

where ζ_0 and t_3 are constants of integration and

$$q_3^2 = \alpha_3 + \beta_3. \tag{27.16}$$

In the case of (27.13) we may set

$$\xi = \xi_0 e^{iqt}, \qquad \eta = \eta_0 e^{iqt}. \tag{27.17}$$

This gives

$$q^2 = \tfrac{1}{2}(\alpha_1 + \beta_1 + \beta_2 + 4\omega_c^2) \pm \tfrac{1}{2}\sqrt{(\alpha_1 + \beta_1 + \beta_2 + 4\omega_c)^2 - 4\beta_2(\alpha_1 + \beta_1)}\,. \tag{27.18}$$

Both roots are positive if $\alpha_1 + \beta_1 > 0$, but if $\alpha_1 + \beta_1 < 0$ one of the roots for q^2 becomes negative. The condition for dynamical stability then is

$$\alpha_1 + \beta_1 > 0\,. \tag{27.19}$$

This is readily shown to be equivalent to

$$\left.\begin{aligned} &\varrho > \varrho^* \\ \text{where}\qquad &\varrho^* = \frac{4}{\pi G \beta_1'}\,\omega_c A\,. \end{aligned}\right\} \tag{27.20}$$

Expressing ϱ in terms of solar masses per cubic parsec, CHANDRASEKHAR derives the data in Table 12. The critical densities given in Table 12 are considerably smaller than those which we ordinarily encounter in galactic clusters. The factor ϱ/ϱ^* for the Pleiades and the Praesepe is of the order of 20.

Table 12.

c/a	β_1'	ϱ^*	c/a	β_1'	ϱ^*
0.1	0.278	0.59	0.5	0.946	0.17
0.2	0.499	0.33	0.6	1.048	0.16
0.3	0.677	0.24	0.8	1.211	0.14
0.4	0.824	0.20	1.0	1.333	0.12

Another possible cause of instability of clusters would be encounters between stars belonging to the cluster and stars in the general galactic field penetrating the cluster. CHANDRASEKHAR has shown that these encounters are far less effective than the encounters between cluster members. This follows because the density of the stars in the cluster is several times the density of the field stars whereas the velocity dispersion of the cluster members is much smaller than that of the field stars.

V. The problem of spiral structure and problems concerning the evolution of the system.

28. Evidence concerning spiral structure. Our present knowledge about the spiral structure in the Galaxy is mainly due to observations of the radial velocities of hydrogen clouds by the 21 cm line and the location of these clouds in space by using our knowledge concerning the differential circular motions in the Galaxy. The main results so far obtained are found in a publication by VAN DE HULST, MULLER and OORT[1], and in papers by MULLER and WESTERHOUT, OLLONGREN and VAN DE HULST, WESTERHOUT, SCHMIDT and RAIMOND[2]. Preliminary results from the Southern sky have been published by CHRISTIANSEN and HINDMAN[3], briefly reviewed by PAWSEY[4], and further results by KERR, HINDMAN and MARTHA STAHE CARPENTER[5]. Important results on the spiral structure in our surroundings from the distribution in space of clusters of blue giant stars, which are often associated with regions of ionized hydrogen, have been derived by MORGAN, SHARPLESS and OOSTERBROCK[6] and by MORGAN, WHITFORD and CODE[7]

[1] H. C. VAN DE HULST, C. A. MULLER and J. H. OORT: Bull. Astr. Inst. Netherl. **12**, 117 (1954).

[2] C. A. MULLER and G. WESTERHOUT, A. OLLONGREN, H. C. VAN DE HULST, G. WESTERHOUT, M. SCHMIDT and E. RAIMOND: Bull. Astr. Inst. Netherl. **13**, No. 475 (1957).

[3] W. N. CHRISTIANSEN and J. V. HINDMAN: Austral. J. Sci. Res. A **5**, 437 (1952).

[4] J. L. PAWSEY: Austral. J. Sci. **18**, 3A (1956).

[5] F. J. KERR, J. V. HINDMAN and MARTHA STAHE CARPENTER: Nature, Lond. **180**, 677 (1957).

[6] W. W. MORGAN, S. SHARPLESS and D. OOSTERBROCK: Astronom. J. **57**, 3 (1952).

[7] W. W. MORGAN, A. E. WHITFORD and A. D. CODE: Astrophys. Journ. **118**, 318 (1953).

(Fig. 18). A general discussion of the distribution of the early type stars and of the spiral structure in our neighbourhood of the system has been made by WEAVER[1]. The distribution of Cepheids has been investigated by INGRID TORGÅRD[2]. That a spiral arm also contains a very considerable amount of stars of the "ordinary" types $A-K$ has been clearly demonstrated by RAMBERG[3]. Contributions to our knowledge of the structure in the Southern sky have been given for

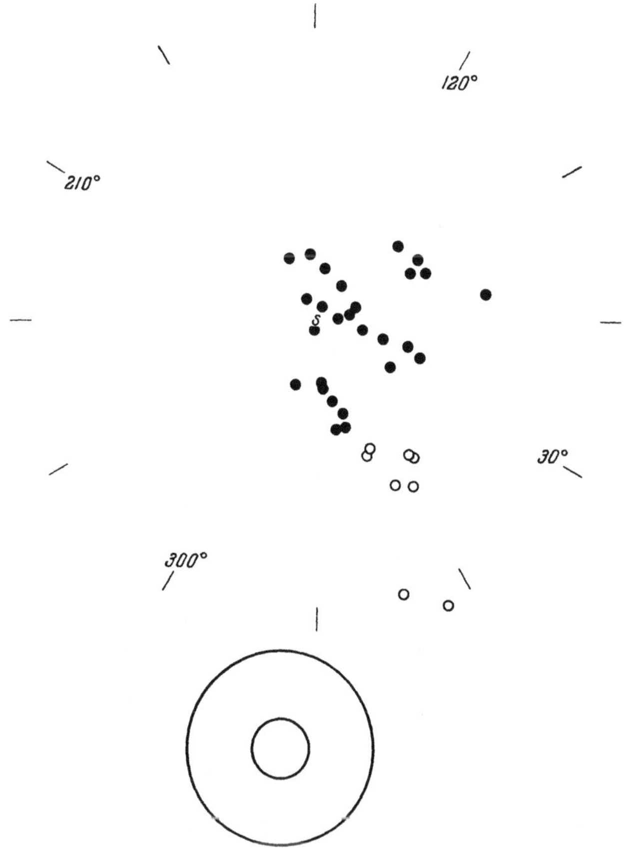

Fig. 18. Spiral arms derived by spectrographic and photometric analysis of stellar associations (MORGAN, WHITFORD, and CODE). S marks the position of the Sun.

early type stars especially by BOK, BESTER, and WADE[4] and by DORRIT HOFFLEIT[5]. The distribution of Southern Cepheids has been determined by WALRAVEN, MULLER, and OOSTERHOFF[6].

The spiral arms, which clearly belong to population type I, thus contain largely giant stars of types $O-B$ often connected with HII regions, a vast amount of neutral hydrogen, stars of ordinary spectral types, and in addition obscuring clouds of small particles.

[1] H. F. WEAVER: Astronom. J. **58**, 177 (1953).
[2] INGRID TORGÅRD: Astrophys. Journ. **120**, 170 (1954).
[3] J. M. RAMBERG: Stockholm Obs. Ann. **20**, No. 1 (1957).
[4] B. J. BOK, M. J. BESTER and C. M. WADE: Phys. Blätter **9**, 245 (1945). — Daedalus **86**, 9 (1955). — Harvard Reprint, Ser. II, No. 45 (1955).
[5] D. HOFFLEIT: Astrophys. Journ. **124**, 61 (1956).
[6] TH. WALRAVEN, C. A. MULLER and P. TH. OOSTERHOFF: Bull. Astr. Inst. Netherl. **14**, 125 (1958).

29. Possible importance of dispersion orbits. An inspection of bright and dark clouds in the spiral arms of external galaxies shows that there is in general a very pronounced elongation of such clouds along the arms. If we consider a condensation of matter, like a stellar association or a cloud of gas, which dissolves by the differential motion between individual parts of the condensation, we can define a curve in space along which the process of dissolution takes place. This follows because the dissolution is essentially one-dimensional, as it is dominated by the differential motion in the angular coordinate about the centre. We may now compare this curve with the orbital motion of an individual particle of the mass, and in particular we try to find a coordinate system such that the orbit adheres to the dispersion curve in the long run of time. It is evident that the "dispersion orbits" defined by Eqs. (20.7) fulfill the conditions of such orbits in the case of a central field of circular symmetry. We must therefore expect that there will be a close osculation between the spiral arms and the "dispersion orbits" thus defined. There is one point which seems to be of fundamental importance here. Considering the relation between ω and $\varkappa$ in Fig. 11, it is evident that $n = d\varkappa/d\omega$ in the inner regions is not everywhere equal to 2, and in the outer regions not everywhere equal to 1. In the zones where n differs from these values, the dispersion orbits are in general fairly complicated and not closed curves. This adds another dimension to the mixing process compared with the closed orbits in the case of $n=2$ or 1, and the matter will be spread here over a zone covering a considerable interval of the radius vector. In the zones where n is nearly equal to 2 or 1, on the other hand, the circumstance that the dispersion orbits are closed curves implies that we must expect permanent, slightly elongated rings to occur in these zones. It seems, therefore, that distinct rings of matter are likely to occur preferably in those zones where n is equal to 2 or 1. Considering the gravitational action of a slightly elongated ring in a zone where $n=2$, it seems that the ring may take up matter from surrounding zones where $n \neq 2$, due to the difference in the value of $\omega - \frac{1}{2}\varkappa$ in the two zones. We remember that $\omega - \frac{1}{2}\varkappa$ is in either case the angular speed of the apsidal line of the closed elliptical orbit. The production of rings of considerable ellipticity is likely to be enhanced by a "partial instability" in the central layer of the Galaxy. In a region, for instance, where n is nearly equal to 2 a growing production of rings along certain dispersion orbits, and the interference of such rings with each other is likely to lead to a somewhat uneven distribution of the apsidal lines which varies slowly with the time. This distribution will give a disturbing potential $\Delta\varphi$, which when expanded into a Fourier series along a nearly circular orbit in the region has the principal term

$$\Delta\varphi = N \cos 2\vartheta'$$

in a coordinate system of a certain mean angular speed $\omega_0' = \omega_0 - \frac{1}{2}\varkappa_0$. N is supposed to be >0 and not to vary very rapidly with R. If the value of ω' at the circular orbit is only slightly smaller than ω_0', the relative angular motion in this orbit, $\omega - \omega_0'$, will fall only slightly below $\frac{1}{2}\varkappa$, which is half the frequency of the radial oscillation in the orbit. There will then be resonance between the frequency of the disturbing force $2(\omega - \omega_0')$ and the frequency of oscillation $\varkappa$. This will give a considerable elongation of the orbit along the diameter $\vartheta' = 0$. In the case when $\omega' > \omega_0'$ the elongation will be along the diameter $\vartheta' = 90°$.

Analogous conditions may occur in the outer regions for $n=1$ in which case the apsidal lines are fixed in space.

In the actual case, inspecting the curve $\varkappa(\omega)$, we find values of R for which $n=2$ as follows

$$R = 4.3,\ 4.7,\ 5.3,\ 6.5,\ 8.3 \text{ kpc}$$

and values of R for which $n=1$,

$$R = 10.5,\ 13\ \text{kpc}.$$

It may be remarked here that the Leiden observers find from the 21 cm line, in the quadrant preceding the Sun in the rotation, rings of matter (spiral arms) at about the distances

$$R = 4.3,\ 5.0,\ 6.5,\ 8.0,\ 10.5\ \text{kpc}.$$

There is an important point to be considered here from the observational point of view concerning the location of the run of the spiral arms in space. This

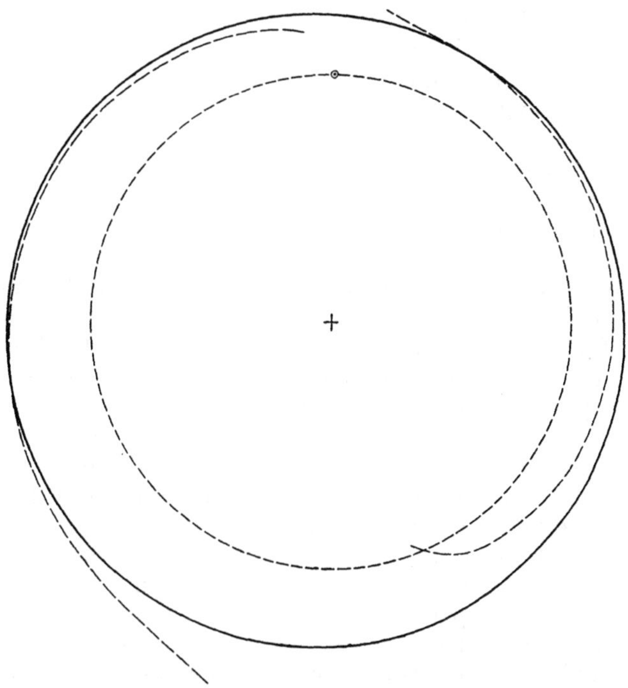

Fig. 19. Dispersion orbit of type B (fulldrawn curve) and the apparent run of the orbit (dotted curve) derived from the differential radial velocity W on the assumption of purely circular motions.

has been done from the observed radial velocity, assuming that the differential motion corresponds everywhere to purely circular motions about the galactic centre. If the motion of the gas follows instead a dispersion orbit of some ellipticity this method will lead to considerable displacements from the true location[1]. An example is given here in Fig. 19, where the full-drawn curve is a dispersion orbit of type $n=1$, and the dotted curve gives the inferred location in space on the assumption that the differential radial velocity corresponds to purely circular motions. Other systematic deviations from circular motion have been suggested. EDMONDSON[2] and VERA RUBIN[3] introduce inwards spiralling motions along the arms, and PARENAGO[4] a general negative K-effect proportional to the distance from the centre.

---

[1] B. LINDBLAD: Stockholms Obs. Ann. **19**, No. 9 (1957); **20**, No. 4 (1958).

[2] F. K. EDMONDSON: Publ. Astronom. Soc. Pacific **67**, 10 (1955). — Astronom. J. **60**, 160 (1955).

[3] V. RUBIN: Astronom. J. **60**, 177 (1955).

[4] P. P. PARENAGO: Astronom. J. USSR. **32**, 226 (1955).

In Fig. 12 the inner ellipse with apsidal line AA indicates a dispersion orbit of $n=2$, whereas the outer ellipse with apsidal line BB indicates a dispersion orbit of $n=1$. It seems very likely that the spiral structure depends in the first instance on a coupling between dispersion orbits of types A and B, whereby in a region of transition a pair of symmetrically situated orbits of type B become populated by matter originally belonging to orbits of type A [1]. This process may then be continued by the tidal action of the extended arms of the orbits of type B on the interior part of the system, as is strongly suggested for instance in the case of the spiral M 81. As to the initiation of the process there are various possibilities. Thanks to the law of areas in the motion there will be an accumulation of matter towards the vertices of a ring of type A. The corresponding disturbing field will move with the angular speed $\omega'=\omega-\frac{1}{2}\varkappa$. For greater distances from the centre, at $R=R'$, say, there will be a circular motion which has the angular speed ω'. Matter distributed along this orbit will suffer considerable secular disturbances, and will be broken up into two opposite whorls of matter. There is another very important circumstance concerning the stability of a ring of type A. For the outermost rings of this type the angular speed ω' will approach to $\frac{1}{2}\omega$, as $\varkappa$ approaches to ω. If we take into account the gravitational field of the ring on its own motion, the angular speed of the apsidal line will be enhanced, but for a massive ring somewhat closer to the centre the angular speed of the apsidal line may be equal to $\frac{1}{2}\omega$ at the place of the ring. Such a massive ring will have a considerable influence also on the surrounding matter, and will show a certain general similarity to the sectorial harmonic wave in a homogeneous spheroid which has two opposite maxima (cf. Sect. 13). We know that the spheroid becomes "ordinarily" unstable for this kind of wave, when the eccentricity of the meridian section of the spheroid exceeds $e=0.9529$ (ratio of the exes $c/a=0.30$). We may conclude that a heavy system of rings in the outer parts of the stellar system would be in a corresponding way unstable. The resulting picture is likely to be a heavy main ring A which is successively breaking up, with a spiral structure on its outside where the spiral arms start near the ring and an interior spiral structure where the outer parts of the spiral arms approach towards the main ring A. Such a picture corresponds to the actual shape of a great number of spiral nebulae, and it is well possible that it also applies to the galactic system.

It should be pointed out here that the importance of the occurrence of rings of matter for the spiral phenomenon has been pointed out strongly by HUBBLE and by RANDERS [2], who showed the importance of viscous forces in the interstellar gas for the creation of circular rings.

The stability of annular formations in a stellar system may be investigated by numerical computations on suitable models, using electronic computers of high capacity. Such investigations are at present pursued at the Stockholm Observatory. Preliminary results have been given by B. LINDBLAD and P.O. LINDBLAD [3].

30. Possibility of a barred structure in the central region. Though there are no indications of a "barred" spiral structure in the outer regions of the galactic system, it is well possible that an asymmetry of the distribution of density of the kind which is characteristic for the barred spirals is present in the central regions of our Galaxy (cf. Sect. 13). A suggestion in this direction has been given

[1] B. LINDBLAD: Stockholms Obs. Ann. **19**, No. 7 (1956).

[2] G. RANDERS: Mt. Wilson Contr. No. 634 = Astrophys. Journ. **92**, 255 (1940).

[3] B. LINDBLAD and P. O. LINDBLAD: Symposium on Large Scale Structure of the Galactic System held at I. A. U. General Assembly in Dublin 1955. I. A. U. Symposium, No. 5, University Press, Cambridge 1958.

also by HUGH M. JOHNSON[1]. It is of interest to see what the effect of such a formation on the structural details of the system would be.

We may suppose that in the original state we have in the system a series of circular rings of type I matter in the equatorial plane, as described in the previous section. The disturbing effect of a growing bar, which follows a certain mean angular speed of the system, on the series of rings will be an expansion of the rings in the quadrants following the bar in the rotation and a contraction of the rings in the two quadrants preceding the bar in the rotation. The effect will be a breaking up and recombination of rings leading to a spiral shape. The further change of shape of the spiral structure will follow from a differential rotation of outer and inner parts which will tend to straighten out and shorten the arms, which can, however, still remain concave towards the centre. Of special importance for conditions in the galactic system is that there will be, in the interior, matter moving along the bar on either side of the bar. The presence of such matter is evident from dark lanes parallel to the bar in external galaxies, and it may be expected to reach very high outward velocities in the course of time. This may be seen as follows.

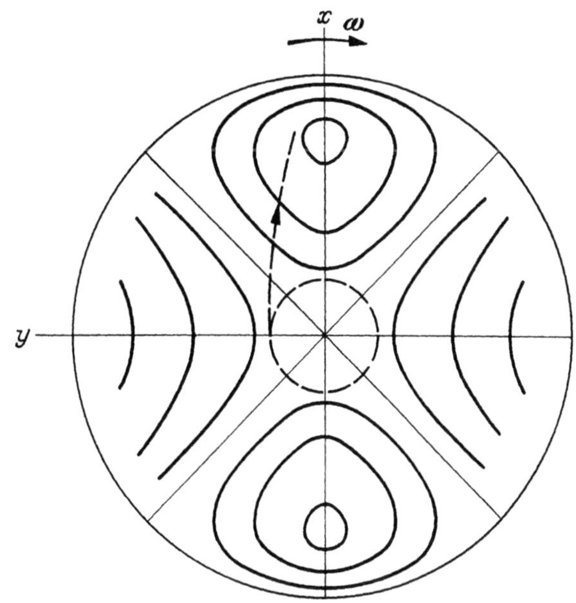

Fig. 20. The motion in a rotating "barred" structure.

We introduce a coordinate system (x, y) with the origin in the centre of the system and rotating with angular speed ω in the sense indicated in Fig. 20, where the full drawn contours schematically indicate the curves of constant $\mathfrak{D}_1 + \mathfrak{D}_0$. The x-axis is the central line of the "bar". We may suppose that the picture represents an eventually barred structure in the central region of the Galaxy, within, say, the distance 3 kpc from the centre and that ω does not change perceptibly with r in the region of the bar. We assume, therefore, $a < 3$ kpc, and ignore the mass of the system outside this limit. For the motion of a particle in x, y we have the differential equations

$$\left.\begin{aligned} \frac{d^2 x}{dt^2} + 2\omega \frac{dy}{dt} + \left(\frac{1}{r} F - \omega^2\right) x = K_x, \\ \frac{d^2 y}{dt^2} - 2\omega \frac{dx}{dt} + \left(\frac{1}{r} F - \omega^2\right) y = K_y, \end{aligned}\right\} \qquad (30.1)$$

where F is the central force, and K_x, K_y are the components of disturbing force due to the "bar". We readily obtain the integral

$$\frac{1}{2}\left[\left(\frac{dx}{dt}\right)^2 + \left(\frac{dy}{dt}\right)^2\right] = \varphi_1 + \int_{r_0}^{r} \left(\omega^2 - \frac{1}{r} F\right) r\, dr + C \qquad (30.2)$$

[1] H. M. JOHNSON: Astronom. J. **62**, 19 (1957).

where φ_1 is the disturbing potential and C is a constant. We may fix ω for the interval of r considered such that

$$\int_{r_0}^{r} \left(\omega^2 - \frac{1}{r} F\right) r\, dr = 0. \tag{30.3}$$

If we write

$$\mathfrak{D}_0 = \sqrt{1 - \frac{r^2}{a^2}},$$

and

$$\mathfrak{D}_1 = \lambda \sqrt{1 - \frac{r^2}{a^2}} \left(\frac{r}{a}\right)^2 \cos 2\vartheta, \tag{30.4}$$

the corresponding potential φ_1 then is[1], choosing units such that $G=1$,

$$\varphi_1 = \lambda \frac{a\pi^2}{2^4} \frac{\Gamma(4)}{\Gamma(2)\,\Gamma(3)} \left[\left(\frac{r}{a}\right)^2 + \frac{5}{6}\left(\frac{r}{a}\right)^4\right] \cos 2\vartheta. \tag{30.5}$$

Assuming the flattening of the system to be great, we have

$$\omega_c^2 = \frac{\pi^2}{2a}, \tag{30.6}$$

so that

$$\varphi_1 = \frac{3}{8} \lambda (\omega_c r)^2 \left(1 - \frac{5}{6} \frac{r^2}{a^2}\right) \cos 2\vartheta. \tag{30.7}$$

For an orbit starting with the velocity v very small and close to the centre, we have

$$v \approx \sqrt{2\varphi_1}. \tag{30.8}$$

For a well developed bar we may assume $\lambda \approx 1$, and then have for small ϑ and $r < a$,

$$v \approx \omega_c r. \tag{30.9}$$

If y remains fairly constant, i.e. if the Coriolis force nearly balances the attraction at right angles to the bar, the motion will be mainly in x and we shall have radial motions of the order of the velocity of the circular orbits $\omega_c r$. It may be possible to explain in this way very large velocities observed by means of the 21 cm line close to the nucleus[2]. As the disturbing force K increases with time, matter belonging to the central nucleus will be successively deflected outwards. There is likely to be a small spiral structure at the centre from which long strings of matter extend, in analogy with the case of the barred spirals.

31. Turbulence theory of HEISENBERG and VON WEIZSÄCKER. Concerning the formation of spiral structure a great number of important theoretical arguments have been raised in recent times. Reference should be made here to CHANDRASEKHAR's work mentioned in Sect. 9. Special account should also be taken of the "turbulence theory" of spiral structure developed by HEISENBERG and VON WEIZSÄCKER[3]. In the exceedingly thin slowly rotating interstellar gas there will appear a hierarchy of turbulence elements and in the outer regions a general expansion. Some such turbulence elements may form clouds which disperse by differential rotation and form the spiral arms.

[1] R. LANGEBARTEL: Stockholm Obs. Ann. **17**, No. 3 (1951).
[2] H. VAN WOERDEN, W. ROUGOOR and J. H. OORT: C. R. Acad. Sci. Paris **244**, 1691 (1957).
[3] C. F. VON WEIZSÄCKER: Z. Naturforsch. **3**a, 524 (1948). — W. HEISENBERG and C. F. VON WEIZSÄCKER: Z. Physik **125**, 290 (1949).

An important point when deciding between the merits of different theories is whether the arms "trail" or "open up" in the direction of rotation. In the case of the purely gravitational theory exposed above the arms should in general be expected to open up in the direction of rotation by out-going motion, and eventually "trail" by in-going motion, whereas in the theory of HEISENBERG and VON WEIZSÄCKER the arms "trail" in out-going motion.

From the observational point of view this question has not yet been definitely settled (cf. Sects. 28, 29).

32. Electromagnetic forces in spiral arms. About the size of electromagnetic forces in interstellar space very little is known with certainty. From the interstellar polarisation of light it seems evident that magnetic fields exist which cause the necessary alignment of the scattering particles. Moreover, it seems likely that the magnetic lines of force follow the spiral arms. This is in good agreement with our interpretation of the arms as having narrow osculation with "dispersion orbits" along which clouds of matter extend by differential motion, as the magnetic lines of force will tend to follow the matter in its motion. Estimates of the strength of the magnetic field in a spiral arm have been made by DAVIS[1] and by CHANDRASEKHAR and FERMI[2], who followed two procedures, the first of which had been used by DAVIS. As we observe distant stars in a direction approximately perpendicular to the spiral arm, it appears that the direction of polarization is only approximately parallel to the arm, and the mean angular deviation of the plane of polarization from the direction of the spiral arm appears to be about $\alpha = 0.2$ radians[3]. There must be a relation between this angle α and the strength of the magnetic field H. The velocity of the transverse magneto-hydrodynamic wave is given by[4]

$$V = \frac{H}{\sqrt{4\pi\varrho}}, \tag{32.1}$$

where ϱ is the density of diffuse matter in the form of gas and dust, which has a sufficiently high electric conductivity to be effectively attached to the magnetic lines of force in such a way that only longitudinal relative displacements are possible. The transverse oscillations of particular line of force can be written

$$y = a \cos k(x - Vt). \tag{32.2}$$

It follows that

$$V^2 \overline{\left(\frac{dy}{dx}\right)^2} = \overline{\left(\frac{dy}{dt}\right)^2}. \tag{32.3}$$

If v denotes the root-mean-square velocity of the turbulent motion of the diffuse matter, we should have

$$\overline{\left(\frac{\partial y}{\partial t}\right)^2} = \frac{1}{3} v^2, \tag{32.4}$$

as only one component of the velocity is active in shifting the lines of force in the y-direction. $\partial y/\partial x$, on the other hand, represents the apparent deviation of the line of force from a straight line, so that

$$\overline{\left(\frac{\partial y}{\partial x}\right)^2} = \alpha^2. \tag{32.5}$$

---

[1] L. DAVIS: Phys. Rev. **81**, 890 (1951).
[2] S. CHANDRASEKHAR and E. FERMI: Astrophys. Journ. **118**, 113 (1953).
[3] W. A. HILTNER: Astrophys. Journ. **114**, 241 (1951).
[4] H. ALFVÉN: Cosmical Electrodynamics. Oxford: Clarendon Press 1950.

This gives, introducing the value of V,

$$H = \left(\frac{4}{3}\pi\varrho\right)^{\frac{1}{2}} \frac{v}{\alpha}. \tag{32.6}$$

CHANDRASEKHAR and FERMI take as typical values $\varrho = 2 \cdot 10^{-24}$ g/cm^3, $v = 5 \cdot 10^5$ cm/sec, $\alpha = 0.2$ radians, which give

$$H = 7.2 \times 10^{-6} \text{ gauss}.$$

The second procedure for estimating H consists in equating the gravitational pressure in the spiral arm with the sum of the material pressure and the pressure due to the magnetic field. Assuming for simplicity that the spiral arm is a cylinder of radius R with uniform density one finds for the gravitational pressure $p_{\text{grav}} = \pi G \varrho \varrho_t R^2$, where ϱ is the density of the diffused matter only, and ϱ_t is the total mean density, including the contribution of the stars. The kinetic and magnetic pressures are $p_{\text{kin}} = \frac{1}{3}\varrho v^2$ and $p_{\text{mag}} = \frac{H^2}{8\pi}$, respectively. For the equilibrium we must have $p_{\text{grav}} = p_{\text{kin}} + p_{\text{mag}}$. Assuming the radius R of the spiral arm to be 250 parsecs, or 7.7×10^{20} cm, and $\varrho_t = 6 \cdot 10^{-24}$ g/cm^3, in accordance with OORT's[1] estimate of the density in our surroundings, one gets $\frac{H^2}{8\pi} = 1.3 \times 10^{-12}$, or

$$H = 6 \cdot 10^{-6} \text{ gauss}.$$

The value of H obtained, of the order of 10^{-5} gauss, is ten times smaller than that which DAVIS and GREENSTEIN[2] have estimated for producing an adequate orientation of the dust particles to account for the interstellar polarization, which may indicate that the mechanism of orientation is somewhat more effective than has been assumed by DAVIS and GREENSTEIN.

Problems of gravitational stability in the presence of a gravitational field have been treated by CHANDRASEKHAR and FERMI[3]. We are specially interested here in the problem about the stability of a spiral arm.

With the usual assumptions of magneto-hydrodynamics, the equations of motion governing an inviscid fluid can be written

$$\varrho \frac{du_i}{dt} = -\frac{\partial}{\partial x_i}\left(p + \frac{H^2}{8\pi}\right) + \varrho \frac{\partial V}{\partial x_i} + \frac{1}{4\pi} \frac{\partial}{\partial x_j} H_i H_j, \tag{32.7}$$

where ϱ denotes the density, u_i the velocity components of mass motion, p the pressure, V the gravitational potential, and H the intensity of the magnetic field. (Summation over repeated indices is to be understood.) CHANDRASEKHAR and FERMI derive from this equation the virial theorem, putting $I = \int_0^M r^2\, dm$,

$$\frac{1}{2}\frac{d^2 I}{dt^2} = 2T + 3(\gamma - 1)\,\mathfrak{U} + \mathfrak{M} + \Omega \tag{32.8}$$

where T is the kinetic energy of mass motion, $\mathfrak{U}$ the heat energy of molecular motion, $\mathfrak{M}$ the magnetic energy of the prevailing field, and Ω the gravitational potential energy. γ denotes the ratio of the specific heats. In equilibrium we have $T = 0$, and then

$$3(\gamma - 1)\,\mathfrak{U} + \mathfrak{M} + \Omega = 0. \tag{32.9}$$

[1] J. H. OORT: Astrophys. Journ. **116**, 233 (1952).
[2] L. DAVIS and J. L. GREENSTEIN: Astrophys. Journ. **114**, 206 (1951).
[3] S. CHANDRASEKHAR and E. FERMI: Astrophys. Journ. **118**, 116 (1953).

It follows, if we introduce the total energy $E=\mathfrak{U}+\mathfrak{M}+\Omega$,

$$E=-\frac{3\gamma-4}{3(\gamma-1)}(|\Omega|-\mathfrak{M}). \tag{32.10}$$

From this equation for the total energy it follows that a necessary condition for the dynamical stability of an equilibrium configuration is

$$(3\gamma-4)(|\Omega|-\mathfrak{M})>0. \tag{32.11}$$

Thus, even when $\gamma>\frac{4}{3}$ (the condition for dynamical stability in the absence of a magnetic field) a sufficiently strong magnetic field can induce dynamical instability in the configuration.

In the case of an infinite cylinder in which the magnetic field is in the direction of the axis of the cylinder, and where all the variables are functions only of the distance r from the axis of the cylinder, the equations of motion reduce to a single one

$$\varrho\frac{du_r}{dt}=-\frac{\partial}{\partial r}\left(p+\frac{H^2}{8\pi}\right)-\frac{2Gm(r)}{r}\varrho, \tag{32.12}$$

where $m(r)$ is the mass per unit length at distances less than r. Choosing the time t and the mass per unit length $m(r)$ as the independent variables, we can write the equations of continuity and motion in the forms

$$\left.\begin{aligned}\frac{\partial}{\partial m}(\pi r^2)&=\frac{1}{\varrho},\\ \frac{\partial^2 r}{\partial t^2}&=-2\pi r\frac{\partial P}{\partial m}-\frac{2Gm(r)}{r},\\ P&=p+\frac{H^2}{8\pi}.\end{aligned}\right\} \tag{32.13}$$

Introducing variations of small amplitude $\delta r, \delta P, \delta\varrho$, etc., from equilibrium values r_0, P_0, ϱ_0, etc., thus

$$r=r_0+\delta r,\qquad P=P_0+\delta P,\qquad \varrho=\varrho_0+\delta\varrho,\qquad \text{etc.,} \tag{32.14}$$

we find

$$\frac{\partial^2}{\partial t^2}\delta r=-2\pi r_0\frac{\partial}{\partial m}\delta P+\frac{4Gm}{r_0^2}\delta r. \tag{32.15}$$

For adiabatic pulsations in a medium of infinite conductivity the expression for δP can be found in the case under consideration, using the relation for $\delta\boldsymbol{H}$, as we follow the motion,

$$\delta\boldsymbol{H}=\operatorname{curl}(\delta r\times\boldsymbol{H}_0)+(\delta r\cdot\operatorname{grad})\boldsymbol{H}_0, \tag{32.16}$$

which gives

$$\frac{\boldsymbol{H}_0\cdot\delta\boldsymbol{H}}{4\pi}=-\frac{H_0^2\varrho_0}{4\pi}\frac{\partial}{\partial m}(2\pi r_0\delta r). \tag{32.17}$$

The equation for the variation δr then becomes

$$\frac{\partial^2}{\partial t^2}\delta r=4\pi^2 r\frac{\partial}{\partial m}\left\{\left(\gamma p+\frac{H^2}{4\pi}\right)\varrho\frac{\partial}{\partial m}(r\delta r)\right\}+\frac{4Gm}{r^2}\delta r. \tag{32.18}$$

When all the physical variables vary with time like $e^{i\sigma t}$, the equation reduces to

$$\left(\sigma^2+\frac{4Gm}{r^2}\right)\delta r=-4\pi^2 r\frac{d}{dm}\left\{\left(\gamma p+\frac{H^2}{4\pi}\right)\varrho\frac{d}{dm}(r\delta r)\right\}, \tag{32.19}$$

where δr has now the meaning of an amplitude. The boundary conditions $\delta r = 0$ at $m=0$ and $\delta P = 0$ at $m = M$, in conjunction with this equation, will determine for σ^2 a sequence of possible characteristic values σ_k^2. The solutions δr_k are orthogonal, so that $\int_0^M \delta r_k \, \delta r_l \, dm = 0$. We may derive the relation

$$\left.\begin{aligned} \sigma^2 \int_0^M (\delta r)^2 \, dm = \gamma \int_0^M \frac{P}{\varrho} \left[\frac{1}{r} \frac{d}{dr} (r \, \delta r)\right]^2 dm - 4G \int_0^M \left(\frac{\delta r}{r}\right)^2 m \, dm + \\ + \frac{1}{8\pi} (2-\gamma) \int_0^M \frac{H^2}{\varrho} \left[\frac{1}{r} \frac{d}{dr} (r \, \delta r)^2\right]^2 dm. \end{aligned}\right\} \quad (32.20)$$

It can be shown that this equation gives a minimum value of σ^2 when the true solution δr belonging to the lowest characteristic value of the pulsation equation is substituted; and any other function δr (satisfying the boundary equations) will give a larger value for σ^2. These facts can be made the basis of a variational procedure for determining σ^2. We may expect to get a good estimate of σ^2 for the fundamental mode by setting $\delta r \propto r$, which is shown by CHANDRASEKHAR and FERMI to lead to

$$\sigma^2 \int_0^M r^2 \, dm = 4\left[(\gamma - 1)^2 \mathfrak{U} + \mathfrak{M}\right]. \quad (32.21)$$

The frequency of the radial pulsations has been studied further by LYTTKENS[1].

CHANDRASEKHAR and FERMI have proceeded to study the stability for transverse oscillations of an infinite cylinder of incompressible fluid along the axis of which a magnetic field is acting. It is shown that the cylinder is unstable for all periodic deformations of the boundary with wavelengths exceeding a oertain critical value, depending on the strength of the field. The wavelength of maximum instability is also determined. This wavelength gives the length cf the "pieces" into which the cylinder will ultimately break up, because the corresponding component in the Fourier analysis of an arbitrary pertubation is the one whose amplitude will increase most rapidly with time. It is found that the magnetic field has a stabilizing effect both in increasing the wavelength of maximum instability and in prolonging the time needed for the instability to manifest itself. For a cylinder of radius $R = 250$ parsecs and $\varrho = 2 \cdot 10^{-24}$ g/cm^3 a magnetic field in excess oi $7 \cdot 10^{-6}$ gauss effectively removes the instability.

33. Evolutionary aspects in stellar dynamics. The dynamics of sub-systems in the Galaxy becomes of special importance in conjunction with the types of stellar population discovered by BAADE[2]. The idea that the matter of a stellar system may be divided into two different populations of entirely different space distributions originated in BAADE's study of the Andromeda nebula, M 31, and its two satellites M 32 and NGC 205. He concluded that, in dealing with galaxies, we have to distinguish two types of stellar population, one which is represented by the ordinary Hertzsprung-Russell diagram (type I), the other by the H-R diagram of the globular clusters (type II). Characteristic of the first type are highly luminous O and B stars and open clusters; of the second globular clusters and short-period Cepheids. Early-type nebulae (E—Sa) seem to have populations of pure type II. Both types coexist, although differentiated by their spatial arrangement, in the intermediate spirals like the Andromeda nebula and our

[1] E. LYTTKENS: Astrophys. Journ. **119**, 413 (1954). = Medd Uppsala Obs. No. 110.
[2] W. BAADE: Mt. Wilson Contr. No. 696 = Astrophys. Journ. **100**, 137 (1944).

own Galaxy. In the late type spirals and in most of the irregular nebulae the highly luminous stars of type I are the most conspicuous feature. The matter of type I also contains clouds of gas and dust, whereas the "amorphous" matter of population II, which is concentrated towards the centre, contains little, if any, of gas and dust. In galaxies of intermediate type, however, the population type II extends to considerable distances from the centre and the system may be roughly described as a huge disk of population II in which the matter of the spiral structure of population type I is embedded.

Similar conditions prevail in our own system, the Galaxy. This is inherent already in STRÖMBERG's[1] analysis of the asymmetrical drift, where various physically defined groups of stars have different characteristics as to mean motion and internal velocity dispersion. OORT[2] showed that the high-velocity stars in our neighbourhood have on the average different physical characteristics compared with the slow-moving stars. Some classes of objects with peculiar spectra contain an unusually large proportion of high velocities. These are long period variables of spectral types $M\,1\,e$–$M\,5\,e$, planetary nebulae, stars of spectral type R, further variables of RR Lyrae type and globular clusters. The stars belonging to the upper main branch of the ordinary H-R diagram, are very rare among the high velocity stars. In the B–A types the percentage is low. The mean absolute magnitude of dwarfs of a given spectral type seems to be the same for high- and low-velocity stars. The relative proportion of dwarfs to giants is much higher among the high-velocity stars than among the ordinary stars. The percentage of double stars is two or three times lower among the high-velocity stars. Compared with BAADE's results the high velocity stars belong to population type II, the slow-moving stars mainly to population type I.

The physical difference between the population types is firstly due to their different H-R diagrams between spectral type (or colour) and absolute magnitude. In our neighbourhood of the galactic system the kinematical difference between the two population types is the different internal velocity dispersion. The stars of type I have comparatively small (for the B stars very small) internal velocity dispersion, whereas the type II stars belong to the high velocity groups with large internal dispersion. In these respects the population types have the characteristics of sub-systems differing in physical characteristics as well as in kinematic properties, and the distinction between population types I and II has much in common with BOTTLINGER's[3] identification of three main sub-systems in the Galaxy (cf. Sect. 14). A very important property of the population types introduced by BAADE is, however, that there appear physical differences between the population types which may be said to add a new dimension to the Hertzsprung-Russell diagram. In the spectral types F and G the stars of type II are characterized by considerably weakened metallic lines compared with ordinary stars, and often the distinction of population type cuts through other physical groupings. This is the case with the Cepheid variables, where the short period variables with period less than $1^{\mathrm{d}}.8$ are the most pronounced type II objects, and are very frequent in globular clusters (cluster type variables). For periods between 2^{d} and 10^{d} it is more difficult to separate the two population types, but they differ as to the period-luminosity relation. Even among the globular clusters differences in the integral spectral type may be noticed (A. DEUTSCH, W. W. MORGAN), the most pronounced type II characteristics (weak metallic lines) being found for the clusters in the galactic halo with very large distances from the

[1] Mt Wilson Contr. No. 275 and No. 292 = Astrophys. Journ. **59**, 228 (1924); **61**, 363 (1925).
[2] J. H. OORT: Publ. Astronom. Lab. Groningen **1926**, No. 40.
[3] K. F. BOTTLINGER: Veröff. Univ.-Sternwarte Berlin-Babelsberg **10**, H. 2 (1933).

galactic plane. Typical type II objects are found in the red long-period variables of comparatively small amplitude and not too late spectral type ($<M4e$).

Modern theories of stellar evolution have made possible to correlate the population types with the age of the stars in the population groups. The tracks of evolution in the colour—absolute magnitude diagram are corroborated in a very interesting way by the colour—magnitude diagrams of clusters, which make possible to establish a scale of evolutionary age for the clusters. In the computation of an evolutionary track the most important parameters are the mass and the chemical composition, defined by the "metal content" Z, i.e. the fraction of mass composed of other elements than H and He (M. SCHWARZSCHILD). The evolutionary track has several pronounced stages, a "Kelvin contraction" with increasing temperature, until in the H-R diagram the star reaches the main sequence, where the nuclear processes start in the stellar interior, beginning with conversion of hydrogen the helium in the core, whereafter the star moves from the main sequence towards higher luminosity until it reaches a stage where loss of mass may occur in the case of stars of high initial mass. After the nuclear energy begins to be exhausted, the luminosity of the star must decrease. Whereas the rising stage is already well covered by computations on stellar models, the track of decreasing luminosity is not yet well defined. It is generally accepted that the series of "white dwarfs" represent a final stage in the evolution.

Table 13.

	Age (Unit 10^9 years)	Z	σ km/sec
Young galactic clusters	0 to 1	0.04	10
Strong-line stars . . .	1 to 3	0.03	20
Weak-line stars . . .	3 to 5.5	0.02	30
High velocity stars . .	5.5 to 6	0.01	50
Globular clusters . . .	6 to 6.5	0.003	130

It is of great interest that the extreme type II characteristics are connected with high age and low metal content Z. The population types are therefore essentially groups of different age and at the same time groups of different degree of metal contents. SCHWARZSCHILD[1] has given the age sequence with corresponding metal content Z and velocity dispersion in Table 13.

The initial "metal content" Z thus appears to be a function of the present age of the star. This indicates that in the interstellar matter from which the stars have been formed the metal content has been continually increasing. The reason for this phenomenon may be the contribution to the interstellar matter from stars shedding mass at their highest peak of expansion. The super-novae represent a phenomenon of wide distribution of originally stellar matter which may be very effective in this direction.

The correlation of age with the velocity dispersion σ, and thus with the distribution in space, expresses the property of the population types as sub-systems of different age in the Galaxy. The transition between the population types is a gradual one. From globular cluster of types similar to M 3 and M 92 and subdwarfs as extreme population II we may proceed over long period variables (spectral type $<M4e$) and other high velocity stars to a "disk population" composed of planetary nebulae, novae and cluster type variables of very short period, to an older population I composed of A stars and later spectral types with strong lines, and finally to an extreme population I represented by gas and dust, super-giants and galactic clusters[2].

[1] M. SCHWARZSCHILD: Stellar Populations, Specola Astron. Vaticana. Ric. Astron. **5**, 207 (1958).

[2] Cf. J.H. OORT: Stellar Populations, Specola Astron. Vaticana. Ric. Astron. **5**, 415 (1958).

The correlation between age and velocity dispersion indicates that stars began to be formed in the gaseous medium far from the galactic plane, and that the process of star formation has gradually approached this plane. In the neighbourhood of the galactic plane the production of stars from interstellar gas is still going on at a certain rate.

Several attempts have been made to approach the physical processess by which stars may have been produced out of an "original" gaseous mass. We may mention here the hydrodynamical scheme of evolution developed by WEIZSÄCKER[1], and the evolution process of star clouds into galaxies and stars suggested by HOYLE[2]. VAN WIJK[3] has considered the origin of the high velocity stars, which may have been formed originally with an isotropic Maxwellian distribution. LINDBLAD[4] has considered the continuous formation of sub-systems of decreasing thickness, following a successive cooling and contraction of the "effective surface" of the gas in the system. The details of such theories, however, fall largely outside of the scope of the present work.

General references.

[1] CHANDRASEKHAR, S.: The Dynamics of Stellar Systems. I—VIII. Astrophys. Journ. **90**, No. 1 (1939).

[2] CHANDRASEKHAR, S.: The Dynamics of Stellar Systems. IX—XIV. Astrophys. Journ. **92**, No. 3 (1940).

[3] CHANDRASEKHAR, S.: Principles of Stellar Dynamics (Astrophysical Monographs sponsored by the Astrophysical Journal). Chicago, Ill.: Chicago University Press 1942.

[4] COUTREZ, R.: Contribution a l'Étude de la Dynamique des Systèmes Stellaires. Annales de l'Observatoire R. de Belgique, Troisième Sér., Tome IV, Fasc. 3. 1949.

[5] JEANS, J. H.: Problems of Cosmogony and Stellar Dynamics. Cambridge: Cambridge University Press 1919.

[6] JEANS, J. H.: Astronomy and Cosmogony. Cambridge: Cambridge University Press 1928.

[7] LINDBLAD, B.: Die Milchstraße, Handbuch der Astrophysik, Das Sternsystem. Teil 1/II. Berlin: Springer 1933.

[8] OGORODNIKOV, K.F.: The Dynamics of the Stellar System. [In Russian.] State Publications of Physical and Mathematical Literature, Moscow 1958.

[9] PAHLEN, E. VON DER: Einführung in die Dynamik von Sternsystemen (Lehrbücher und Monographien aus dem Gebiete der exakten Wissenschaften 10, Astronomisch-geophysikalische Klasse, Bd. I). Basel: Birkhäuser 1947.

[10] SMART, W. M.: Stellar Dynamics. Cambridge: Cambridge University Press 1938.

[1] C. F. VON WEIZSÄCKER: Astrophys. Journ. **114**, 165 (1951).

[2] F. HOYLE: Astrophys. Journ. **118**, 513 (1953).

[3] U. VAN WIJK: Astronom. J. **61**, 277 (1956).

[4] B. LINDBLAD: Vistas in Astronomy, II, p. 1711. London and New York: Pergamon Press 1956.

Radio-frequency Studies of Galactic Structure.

By

J. H. Oort.

With 15 Figures.

Introduction.

The discovery of radio-frequency radiation from the Galactic System is due to Jansky[1]. In 1931, while engaged in a study of receiver noise in the region of short wavelengths ($\lambda = 14.6$ m) he observed that part of this noise came preferentially from a certain azimuth. From the variation of this azimuth with the time of the day, and with the time of the year, he concluded that the source was related to the nuclear region of the Galactic System.

During a surprisingly long time Jansky's discovery remained almost unnoticed by astronomers. It was in the first place the work by Grote Reber in 1940[2], and especially his more extensive investigation published in 1944[2], that drew general attention to the importance of this radiation for the investigation of the Galactic System. Reber worked at a wavelength of 1.87 m, using a parabolic reflector of 9.5 m aperture. With this equipment he produced the first map of the distribution of the radio-frequency radiation from the Galactic System.

Since then, a considerable number of surveys have been made at various wavelengths, between 10 cm ($\nu = 3000$ Mc/s) and 32.8 m ($\nu = 9.15$ Mc/s).

The atmosphere is transparent from a wavelength of about 1 mm to several dozens of metres. The upper limit is set by reflection and absorption of the ionosphere. Near the magnetic poles of the earth this limit appears to lie well above 100 m. The domain of radiowaves from outer space which are accessible to observation is thus extremely large compared to the domain of optical radiation. It extends over a range of approximately 100000:1. In the centimetre and millimetre range the radiation is generally weak, so that with present techniques only regions of exceptional density, and the brightest radio sources, can be adequately observed at these wavelengths.

1. General surveys. A number of the principal surveys are listed in Table 1. The list contains only the surveys of *continuous* radiation. Part of the data were taken from a publication by Hanbury Brown and Hazard [*1*]. Surveys of the hydrogen-line emission at 21 cm will be discussed in Sect. 7.

At the time of writing complete surveys at λ 21 and 75 cm over the sky north of declination $-35°$ are being carried out with a paraboloid of 25 m aperture, giving beamwidths of 0°.5 and 2°.1 respectively. The survey at 3.5 m by Mills is being extended over the larger part of the sky, covering the entire Milky Way.

The beamwidth given in Table 1 is defined as the average distance between points where the intensity received from a point source would be half that received

[1] K. G. Jansky: Proc. Inst. Radio Engrs. **20**, 1920 (1932); **21**, 1387 (1933).

[2] Grote Reber: Astrophys. Journ. **91**, 621 (1940); **100**, 279 (1944).

Table 1. *Surveys of continuous radiation.*

λ (metres)	ν (Mc/s)	Instrument*	Aperture (metres)	Beamwidth (degrees)	Region	T_c (degrees K)	Reference
0.10	3000	paraboloid	2.2	3.4	δ −33° to −20°	2.6	1
0.22	1400	paraboloid	7.5	2.3	l 323° to 353°	20	2
0.25	1200	paraboloid	5	2.8	δ −33° to −20°	17	1
0.33	910	paraboloid	7.5	3.6	δ −32° to +52°	23	3
0.50	600	merid. parab.	11	3.3	δ −90° to +51°	200	4
0.625	480	paraboloid	9.5	4.5	δ −35° to +50°	107	5
0.75	400	fixed parab.	24	2	δ −49° to −17°	260	6
0.75	400	paraboloid	25	2.1	l 324° to 338°	320	2
1.20	250	helix array		1.2×8	δ −40° to +50°		7
1.50	200	Yagi array	1.6	25	δ −90° to +45°	447	8
1.50	200	dipole array	5	16.5	δ −20° to +90°		9
1.87	160	paraboloid	9.5	12	δ −35° to +90°	1370	10
1.89	158	fixed parab.	66	2	δ +38° to +68°		11
3.0	100	Yagi array	6	17	δ −90° to +56°	3860	12
3.5	85	MILLS' cross	450	0.8	l 323° to 333°	35000	13
3.7	81.5	cyl. parab.	98×12	2×15	δ −28° to +82°		14
15.2	19.7	MILLS' cross	1000	1.4	δ −33° to −25°	500000	15
16.4	18.3	dipole array	26×17	40	δ −34°	140000	16
32.8	9.15	dipole array	40	29	δ −32°	1000000	17

* Unless otherwise indicated "paraboloid" in this column refers to a paraboloidal reflector movable in two co-ordinates. A "merid. parab." is movable only in the meridian.

[1] J. H. PIDDINGTON and H. C. MINNETT: Austral. J. Sci. Res. **4**, 459 (1951).
[2] G. WESTERHOUT: Bull. Astr. Inst. Netherl. **13**, 105 (1956).
[3] J. F. DENISSE, E. LEROUX and J. L. STEINBERG: C. R. Acad. Sci. Paris **240**, 278 (1955).
[4] J. H. PIDDINGTON and G. H. TRENT: Austral. J. Phys. **9**, 481 (1956).
[5] G. REBER: Proc. Inst. Radio Engrs. **36**, 1215 (1948).
[6] R. X. MCGEE, O. B. SLEE and G. J. STANLEY: Austral. J. Phys. **8**, 347 (1955).
[7] H. C. KO and J. D. KRAUS: Astronom. J. **61**, 182 (1956) (abstract).
[8] C. W. ALLEN and C. S. GUM: Austral. J. Sci. Res. **3**, 224 (1950).
[9] F. DRÖGE and W. PRIESTER: Z. Astrophys. **40**, 236 (1956).
[10] G. REBER: Astrophys. Journ. **100**, 279 (1944).
[11] R. HANBURY BROWN and C. HAZARD: Monthly Notices Roy. Astronom. Soc. London **113**, 109 (1953).
[12] J. G. BOLTON and K. C. WESTFOLD: Austral. J. Sci. Res. **3**, 19 (1950).
[13] B. Y. MILLS: Observatory **76**, 65 (1956). The temperatures were corrected in accordance with a private communication.
[14] J. E. BALDWIN: Monthly Notices Roy. Astronom. Soc. London **115**, 684 (1955).
[15] C. A. SHAIN: Austral. J. Phys. **10**, 195 (1957).
[16] C. A. SHAIN: Austral. J. Sci. Res. **4**, 258 (1951).
[17] C. S. HIGGINS and C. A. SHAIN: Austral. J. Phys. **7**, 460 (1954).

when the source is in the centre of the beam. Most surveys suffer from low resolution, their beamwidths being generally larger, in many cases even much larger, than the structural features investigated. For this reason our knowledge of the true distribution of radio brightness is still very incomplete. The surveys with a beamwidth of 0°.5 at 21 cm, of 2° at 75 cm and of 0°.8 at 3.5 m which are now in progress will radically improve this situation.

The column T_c gives the beam temperatures observed when the radio telescope points to the galactic centre (for the derivation of T_c see HANBURY BROWN and HAZARD, Ref. [*1*]). The temperature T_c is meant to refer to the background of galactic radiation with the point source near the centre removed. But in several cases there were no data available to subtract the radiation of the point source.

2. Units. The radiation received from a source of small angular size is expressed in watts per square metre per cycle per second $\{\mathrm{W\,m^{-2}(c/s)^{-1}}\}$. The intensity per solid angle ("brightness") may be expressed in $\{\mathrm{W\,m^{-2}\,sterad^{-1}}$ $(\mathrm{c/s})^{-1}\}$, but is more usually given in a temperature scale. This so-called "brightness temperature" is defined as the temperature of a black body that, at the frequency considered, would give the observed surface brightness.

At the frequencies concerned PLANCK's law can be represented by

$$B_\nu = \frac{2kT}{c^2}\,\nu^2, \tag{2.1}$$

where B_ν is the energy emitted by unit area per second per steradian in a band of 1 c/s at frequency ν in a direction perpendicular to the area. T is the temperature of the black body, k is BOLTZMANN's constant, and c the velocity of light. If there is no absorption the radiation *received* per unit area from a steradian covered by a black body of temperature T will also be B_ν. The brightness temperature T_b can thus be defined by (2.1). Inserting numerical values we have, in m.k.s. units,

$$B_\nu = 2.77\times 10^{-23}\, T_b\, \lambda^{-2}\ \mathrm{watt\ metre^{-2}\,(c/s)^{-1}\,sterad^{-1}}, \tag{2.2}$$

or, in cgs units,

$$B_\nu = 2.77\times 10^{-16}\, T_b\,\lambda^{-2}\ \mathrm{erg\ sec^{-1}\,cm^{-2}\,(c/s)^{-1}\,sterad^{-1}}. \tag{2.3}$$

For details see the recent textbook by PAWSEY and BRACEWELL[1].

3. Origin of radiation. The radiation observed in the radio domain is mostly continuous. Line emission and line absorption has, so far, been found only in one case, namely the hyperfine-structure line of atomic hydrogen at 21.1 cm wavelength ($\nu = 1420$ Mc/s). This one line has, however, become of the greatest importance for the exploration of the Galactic System. It was first suggested by VAN DE HULST[2] that the interstellar hydrogen might show an observable emission corresponding to this transition. The first observations of the emission were made by EWEN and PURCELL[3].

The continuous radiation may be caused by at least three mechanisms: (a) free-free transitions in ionized gas ("thermal" radiation), (b) deceleration of high-energy electrons in large-scale magnetic fields ("synchrotron" radiation), (c) large-scale oscillations in an ionized gas (plasma oscillations). The last of these processess is probably of no importance for the galactic radiation, and will not be considered in the present chapter.

The observed continuous radiation comes partly from a thin galactic layer and partly from an almost spherical, and nearly homogeneous atmosphere, or "corona" enveloping the Galactic System. In addition, a probably quite appreciable amount comes from other galaxies or from intergalactic space. Part of the radiation from the galactic layer comes from discrete regions, called "radio sources". Some of these are clouds of ionized hydrogen emitting thermal radiation. The very intense galactic sources, however, have been identified with peculiar nebulosities, the radio-frequency radiation of which may be due to the synchrotron process (see below). It is possible that the whole of the non-thermal galactic radiation concentrated near the Milky Way is due to the integrated effect of such sources, but it is more likely that most if it is due to radiation of relativistic electrons in interstellar magnetic fields.

4. Thermal emission. We shall first consider the thermal emission. At radio frequencies this comes wholly from the interstellar gas; the thermal radiation

[1] J. L. PAWSEY and R. N. BRACEWELL: Radio Astronomy. Oxford 1955.
[2] H. C. VAN DE HULST: Nederl. Tijdschr. Natuurk. **11**, 201 (1945).
[3] H. I. EWEN and E. M. PURCELL: Nature, Lond. **168**, 356 (1951).

of the stars is entirely negligible. Even the Sun's radiation at metre waves is not so very much larger than that received from an equal area in the Milky Way.

Along an outward normal the emission of a spherical mass of gas of diameter l containing N_e electrons and N_e protons per cm³, at an electron temperature T_e is

$$I_\nu = 5.3 \times 10^{-39} g_{III}\, l\, N_e^2\, T_e^{-\frac{1}{2}}\, (\text{c/s})^{-1}\, \text{sterad}^{-1}, \tag{4.1}$$

if the optical depth is small. For the actual interstellar gas, in which there is some admixture of heavier ions, the coefficient in (4.1) will be about 10% higher.

The formula is based on KRAMERS' theory[1] of the emission by electron-ion encounters in an ionized gas. In the particular form given above it was taken from an article by J. L. GREENSTEIN and R. MINKOWSKI[2]. The authors define a quantity d as follows

$$d = 8.1 \times 10^{-6} \nu\, T_e^{-\frac{1}{2}} N_e^{-\frac{1}{3}}. \tag{4.2}$$

If d is larger than unity—as it will usually be in interstellar space—the factor g_{III} is given by

$$g_{III} = 1.27\,(8.19 + \tfrac{3}{2}\log_{10} T_e - \log_{10}\nu). \tag{4.3}$$

With $N_e = 10$, $T_e = 10000°$ and $Z = 1$ (which may be more or less representative for an average region of ionized hydrogen) we find, for $\nu = 10^8$ (100 Mc/s), $d = 3.8$, while for 1000 Mc/s $d = 38$. For these frequencies g_{III} is 7.9 and 6.6, respectively.

The absorption coefficient for an ionized region of the interstellar mixture will be

$$\varkappa(\nu) = 0.019 g_{III}\, T_e^{-\frac{3}{2}} \nu^{-2} N_e^2. \tag{4.4}$$

The optical depth is then given by

$$\tau(\nu) = \int \varkappa(\nu)\, ds' = 0.019\, g_{III}\, T_e^{-\frac{3}{2}} \nu^{-2} \int N_e^2\, ds', \tag{4.5}$$

in which ds' is measured in cm. If we measure ds in parsec (pc), we get, inserting $g_{III} = 7$ and $T_e = 10000°$,

$$\tau(\nu) = 4.1 \times 10^{11} \nu^{-2} \int N_e^2\, ds. \tag{4.6}$$

$\int N_e^2\, ds$, as occurring in this expression, is called the "emission measure"[3]. The brightness temperature is, evidently,

$$T_b = T_e\{1 - e^{-\tau(\nu)}\}. \tag{4.7}$$

As $\varkappa(\nu)$ is always small, we obtain for 1 cm

$$T_b = T_e \varkappa(\nu).$$

Replacing T_b by B_ν by means of (2.3) and substituting $\varkappa(\nu)$ from (4.4), we again get (4.1), except that the coefficient comes out 10% higher because of the inclusion of heavier ions.

From optical data we may make a very rough estimate of the intensity of thermal radiation which might be expected. SHARPLESS and OSTERBROCK[4] list 9 H_α emission regions within 1 kpc. The region surveyed by these astronomers extends from about 320 to 220° galactic longitude. The average distance from the galactic plane is 49 pc for these 9 emission nebulae. This agrees with the average distance from the galactic plane found for O and B stars in general.

---

[1] H. A. KRAMERS: Phil. Mag. **46**, 836 (1923).
[2] J. L. GREENSTEIN and R. MINKOWSKI: Astrophys. Journ. **118**, 1 (1953).
[3] B. STRÖMGREN: Astrophys. Journ. **108**, 261 (1948).
[4] S. SHARPLESS and D. OSTERBROCK: Astrophys. Journ. **115**, 89 (1952).

Adopting, accordingly, an equivalent thickness of 0.10 kpc for the layer of emission regions, the equivalent volume within 1 kpc would be 0.23 kpc³. The average number of emission regions per kpc³ near the galactic plane would thus be 39. The average diameter of the 9 regions is 34 pc; the average density may very roughly be estimated at 10 protons and 10 electrons per cm³. These numbers correspond to an emission measure of 3400. It should be kept in mind that these data are extremely uncertain and incomplete. In the galactic plane a line of sight of 1 kpc length would, on the average, meet 0.035 of such emission regions.

Putting g_{III} again equal to 7, and $T_e = 10000°$, we find from (4.6) that the optical depth of such a schematic cloud is

$$\tau(\nu) = 1.4 \times 10^{15} \nu^{-2}. \tag{4.8}$$

At 100 Mc/s ($\lambda = 3$ m) this becomes 0.14, corresponding to a brightness temperature of 1400°. If we would suppose the frequency and other characteristics of ionized clouds to remain the same all over the galactic disk, which has a diameter of about 20 kpc, the optical depth of the entire system would be of the order of 0.1 at this frequency. Although the data are uncertain it appears probable that for wavelengths less than about 5 m the optical depth, and therefore also any absorption effects, would in general be negligible.

5. Synchrotron radiation. That radiation by high-energy electrons in magnetic fields might play a role in radio astronomy, was first suggested by ALFVÉN and HERLOFSON[1] to explain the radiation of radio sources. We shall call it "synchrotron radiation", as this type of radiation was first observed in a synchrotron[2]. Recently, strong evidence of the presence of synchrotron radiation in radio sources has been found (see the chapter on galactic radio sources). The possible significance of this mechanism for explaining the "radio corona" of the Galactic System will be discussed in Sect. 11. According to SCHWINGER[3] the power in ergs emitted per second by a high-energy electron per c/s is given by

$$P(\nu) = \frac{3^{\frac{3}{2}}}{2} \frac{e^2}{R} \left(\frac{E}{m c^2}\right)^4 \frac{\nu_0 \nu}{\nu_c^2} \int\limits_{\nu/\nu_c}^{\infty} K_{\frac{5}{3}}(\eta)\, d\eta. \tag{5.1}$$

In this expression e is the charge of the electron in e.s.u., m its rest mass, c the velocity of light, R the radius of curvature of the orbit, all in cgs units. The integral can be computed from tables of Bessel funtions. R is related to the energy E (in ergs) and the magnetic field H (in gauss) by

$$R = \frac{E}{e H_\perp}. \tag{5.2}$$

$H_\perp$ stands for $H \sin\vartheta$, where ϑ is the angle between the velocity and H. Further, ν_0 is the Larmor frequency

$$\nu_0 = \frac{c}{2\pi R}. \tag{5.3}$$

The critical frequency ν_c is

$$\nu_c = \frac{3c}{4\pi R} \left(\frac{E}{m c^2}\right)^3. \tag{5.4}$$

[1] H. ALFVÉN and N. HERLOFSON: Phys. Rev. **78**, 616 (1950).

[2] F. R. ELDER, A. M. GUREWITSCH, R. V. LANGMUIR and H. C. POLLOCK: Phys. Rev. **71**, 829 (1947).

[3] J. SCHWINGER: Phys. Rev. **75**, 1912 (1949). These phenomena have also been discussed elsewhere, cf. VLADIMIRSKY in J. Exp. a. Techn. Phys. **18**, 392 (1948) (in Russian), who also gives numerical values. Essentially the same problem has also been treated by G. A. SCHOTT in his classical text book on "Electromagnetic Radiation", p. 109 a.f. Cambridge: Cambridge University Press 1912.

The total energy radiated per second is given by

$$\frac{dE}{dt} = -\frac{2e^2c}{3R^2}\left(\frac{E}{mc^2}\right)^4. \tag{5.5}$$

Eliminating R with the aid of (5.2), we can write

$$P(\nu) = \frac{3^{\frac{1}{2}}e^3}{mc^2} H_\perp \frac{\nu}{\nu_c}\int\limits_{\nu/\nu_c}^{\infty} K_{\frac{5}{3}}(\eta)\,d\eta = C H_\perp \frac{\nu}{\nu_c}\int\limits_{\nu/\nu_c}^{\infty} K_{\frac{5}{3}}(\eta)\,d\eta, \tag{5.6}$$

in which $C = 2.343\times 10^{-22}$;

$$\nu_c = \frac{3e}{4\pi mc} H_\perp\left(\frac{E}{mc^2}\right)^2 = 6.268\times 10^{18} H_\perp E^2 = L H_\perp E^2_{\text{Gev}}, \tag{5.7}$$

where $L = 1.608\times 10^{13}$.

Further,

$$\left.\begin{aligned} \frac{dE}{dt} &= -\frac{2e^4}{3m^2c^3} H_\perp^2\left(\frac{E}{mc^2}\right)^2 = -2.368\times 10^{-3} H_\perp^2 E^2,\\ \text{or}\qquad \frac{dE_{\text{Gev}}}{dt} &= -A H_\perp^2 E^2_{\text{Gev}}, \end{aligned}\right\} \tag{5.8}$$

with $A = 3.793\times 10^{-6}$.

E_{Gev} is the energy of the electron in units of 1 Gev $= 10^9$ ev. The solution of (5.8) may be written

$$E_{\text{Gev}} = \frac{1}{AH_\perp^2}\left(t + \frac{1}{AH_\perp^2 E_a}\right)^{-1}, \tag{5.9}$$

where E_a is the electron's initial energy in units of 10^9 ev, and t is the time in seconds elapsed since. The energy will diminish to half of its original value in a time

$$T_{\frac{1}{2}} = \frac{1}{3.156\times 10^7 A H_\perp^2 E_a} = \frac{0.00835}{H_\perp^2 E_a}\ \text{years}. \tag{5.10}$$

The radiation is totally polarized, with the electrical vector parallel to the radius of curvature of the orbit. The radiation is concentrated in a very sharp cone, the axis of which co-incides with the velocity of the electron. The average angle in radians, between the light-rays and the velocity is according to SCHWINGER

$$\bar{\vartheta} = mc^2/E. \tag{5.11}$$

$P(\nu)$ has a maximum near $\nu/\nu_c = 0.30$. It varies relatively little over a large range of frequencies, decreasing to half of its maximum value at $\nu/\nu_c = 0.011$ and $\nu/\nu_c = 1.47$. Numerical values for

$$\frac{\nu}{\nu_c}\int\limits_{\nu/\nu_c}^{\infty} K_{\frac{5}{3}}(\eta)\,d\eta = F(\nu/\nu_c)$$

are shown in Table 2 and Fig. 1.

In the following we shall write

$$\frac{\nu}{\nu_c} = \alpha.$$

For small values of α we have approximately

$$P(\nu) = 2\times 3^{\frac{1}{6}}\Gamma\left(\frac{2}{3}\right)\frac{e^2}{R}\left(\frac{\nu}{\nu_0}\right)^{\frac{1}{3}} = 5.04\times 10^{-22} H_\perp \alpha^{\frac{1}{3}}, \tag{5.12}$$

and

$$F(\alpha) = 2.15\,\alpha^{\frac{1}{3}}. \tag{5.13}$$

Table 2. $F(\alpha) = \alpha \int_{\alpha}^{\infty} K_{\frac{5}{3}}(\eta)\, d\eta \quad (\alpha = \nu/\nu_c)$.

α	$F(\alpha)$	α	$F(\alpha)$	α	$F(\alpha)$	α	$F(\alpha)$
0.001	0.213	0.25	0.917	1.20	0.566	4.00	0.0522
0.005	0.358	0.30	0.919	1.40	0.486	4.50	0.0339
0.010	0.445	0.40	0.901	1.60	0.414	5.00	0.0214
0.025	0.583	0.50	0.872	1.80	0.354	5.59	0.0124
0.050	0.702	0.60	0.832	2.00	0.301	6.25	0.0068
0.075	0.772	0.70	0.788	2.50	0.200	6.93	0.0037
0.10	0.818	0.80	0.742	3.00	0.130	8.73	0.00064
0.15	0.874	0.90	0.694	3.50	0.0845	10.67	0.00010
0.20	0.904	1.00	0.655				

These expressions give an approximation which is sufficient for our purpose when α is less than about 0.01; at $\alpha = 0.01$ they are 4% higher than the rigorous formulae. For $\alpha > 10$, $P(\nu)$ and $F(\alpha)$ may be written

$$\left.\begin{aligned} P(\nu) &= \left(\frac{3\pi}{2}\right)^{\frac{1}{2}} \frac{e^3}{m c^2} H_{\perp}\, \alpha^{\frac{1}{2}} e^{-\alpha} \\ &= 2.94 \times 10^{-22} H_{\perp}\, \alpha^{\frac{1}{2}} e^{-\alpha}, \end{aligned}\right\} \quad (5.14)$$

$$F(\alpha) = 1.26\, \alpha^{\frac{1}{2}} e^{-\alpha}. \quad (5.15)$$

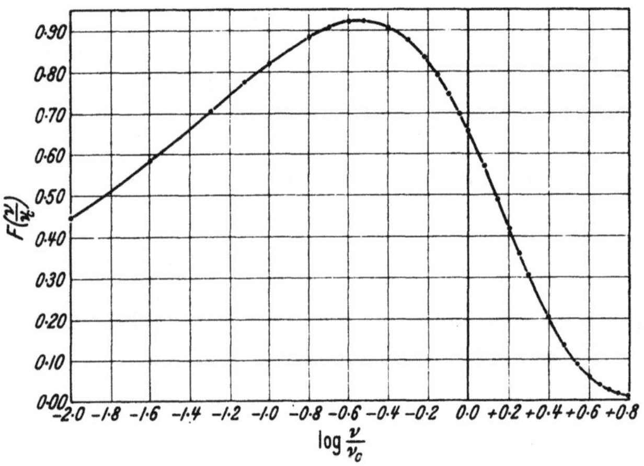

Fig. 1. Radiation of electron in magnetic field as a function of log (ν/ν_c).

6. Line emission. Let $n_m(V, r)\, dV$ be the number of hydrogen atoms per cm³ in the upper levels of the hyperfine structure at a distance r and with radial velocities between V and $V + dV$. Let $N_m(V)\, dV$ be the number of atoms in the same levels and in the same velocity range in a column of cross-section 1 cm² along the entire line of sight. Then

$$N_m(V) = \int_0^{\infty} n_m(V, r)\, dr.$$

The same notation with index n will refer to the lower level.

The radiation received per cm² per sec per unit solid angle in a frequency interval $d\nu$ from an optically thin layer is

$$I(\nu)\, d\nu = \frac{h\nu}{4\pi} A_{mn} N_m(V)\, dV. \quad (6.1)$$

Here A_{mn} is the transition probability for spontaneous emission and $dV = \frac{c}{\nu}\, d\nu$. The 21-cm line has[1]

$$A_{mn} = \frac{64\pi^4 \nu^3}{3 h c^3} \beta_0^2 = 2.84 \times 10^{-15}\ \text{sec}^{-1},$$

where β_0 is the Bohr magneton. The atomic absorption coefficient $\varkappa(\nu)$ (cm²) is related to the transition probability for absorption, B_{nm}, by the well known relation

$$\int \varkappa(\nu)\, d\nu = B_{nm} \frac{h\nu}{4\pi}.$$

[1] P. Wild: Astrophys. Journ. **115**, 206 (1952).

Upon the assumption that Doppler effect is the dominant effect of line broadening, this relation leads to the following expression for the absorption coefficient per unit length

$$k(\nu) = n_n(V, r)\, B_{nm} \frac{hc}{4\pi}. \tag{6.2}$$

In the radio (or classical) domain the true absorption is largely compensated by the stimulated emission. The difference is the effective absorption coefficient

$$k'(\nu) = \frac{h\nu}{kT} k(\nu). \tag{6.3}$$

Here T is the excitation temperature defined by

$$\frac{n_m}{n_n} = \frac{g_m}{g_n} e^{-h\nu/kT} \approx \frac{g_m}{g_n}\left(1 - \frac{h\nu}{kT}\right).$$

PURCELL and others[1] have shown that this excitation temperature is made equal to the kinetic temperature in the neutral hydrogen clouds by means of the highly effective process of electron exchange during atom-atom encounters. The temperature found for the neutral clouds is 125°, so that $h\nu/kT = 0.0007$, which is small enough for the usual approximations for the radio domain to be permitted.

Combining the preceding formulae with the relation

$$B_{nm} = \frac{g_m c^2}{g_n 2 h \nu^3} A_{mn},$$

we obtain the optical depth of the entire column in a certain direction

$$\tau(\nu) = \int_0^\infty k'(\nu)\, dr = N_n(V)\, A_{mn} \frac{g_m c^3 h\nu}{g_n 8\pi\nu^3 kT}. \tag{6.4}$$

Inserting the following numerical values: $\nu = 1.4204056 \times 10^9$, $c/\nu = \lambda = 21.105$ cm, $h\nu/k = 6.813 \times 10^{-2}$ degrees, $g_m/g_n = 3$ (the upper level being triple), $A_{mn} = 2.84 \times 10^{-15}$, and introducing the total number of hydrogen atoms in the ground state, N, instead of N_n ($N = 4N_n$), we obtain

$$N(V) = 1.835 \times 10^{13}\, T\, \tau(\nu). \tag{6.5}$$

The intensity received from this direction is

$$I(\nu) = B_\nu(1 - e^{-\tau(\nu)}), \tag{6.6}$$

where B_ν is the intensity of a black body with temperature equal to the electron temperature of the neutral hydrogen clouds. B_ν is given by (2.1).

7. Distribution of neutral hydrogen. Spiral structure[2]. If B_ν is known, (6.5) and (6.6) permit us to derive $N(V)$ from measures of the line profile $I(\nu)$ for a given galactic longitude and latitude, at least if τ is not much larger than unity. This latter condition is sufficiently fulfilled in nearly all longitudes differing more than 15° from the direction towards the centre of the Galactic System and from the opposite direction. The value of B_ν can be determined from measures in the few directions where the optical thickness is large. In this way a temperature of 125° K has been derived[3] (in the following the intensities will always

[1] H. I. EWEN and E. M. PURCELL: Nature, Lond. **168**, 356 (1951). — E. M. PURCELL and G. B. FIELD: Astrophys. Journ. **124**, 542 (1956).

[2] For a general article on this subject see Ref. [6].

[3] H. C. VAN DE HULST, C. A. MULLER and J. H. OORT: Bull. Astr. Inst. Netherl. **12**, 117 (1954).

be expressed in a temperature scale). The temperature thus determined is the harmonic mean temperature of the clouds in the region investigated[1].

The Doppler broadening of the line is due partly to the rotation of the Galactic System, and partly to irregular motions in the interstellar medium. Let $\omega(R)$ be the angular velocity of rotation at a distance R from the galactic centre. If in Fig. 2 S represents the interstellar material in the vicinity of the Sun, which is supposed to be in the galactic plane at a distance R_0 from the centre and is rotating with an angular velocity ω_0, the radial velocity of the average medium at P relative to that at S will be

$$V_g = R_0 \{\omega(R) - \omega_0\} \sin l'. \tag{7.1}$$

l' is the difference between the galactic longitude in which the observation is made and that of the centre. The galactic longitude of the centre has been determined from radio as well as optical data; it is 327°.7, with an uncertainty of less than 0°.1. Let $n_H(r)$ be the density of hydrogen atoms at a distance r from the observer. If there were no random motions the number of hydrogen atoms in a column of 1 cm² cross-section and in a unit interval of V_g would be equal to

$$\mathfrak{N}(V_g) = \frac{n_H(r)}{dV_g/dr}. \tag{7.2}$$

The actual velocity distribution will be smeared out by random radial velocities v. If the frequency function of v is represented by $\varphi(v)$, we have

$$N(V) = \int_{-\infty}^{+\infty} \mathfrak{N}(V_g)\, \varphi(V - V_g)\, dV_g. \tag{7.3}$$

The integral may formally be extended from $-\infty$ to $+\infty$, as $\mathfrak{N}(V_g)$ is zero below a minimum value and beyond a maximum value of V_g. In principle we can find $\mathfrak{N}(V_g)$ from the observed function $N(V)$ by solving the integral equation by numerical methods, provided we know $\varphi(v)$. If we know also V_g as a function of r, we can then determine the hydrogen density $n_H(r)$ from (7.2).

The distribution of the random motions $\varphi(v)$ can be found from the observations of interstellar absorption lines in the optical region of the spectrum, and partly also from the 21-cm observations. In the general region around the Sun the average random velocity in one co-ordinate is about 6 km/sec. It appears to increase somewhat with decreasing R, to a value of around 10 km/sec at $R = 3$ kpc. Inside 3 kpc the random motions become much higher, averaging around 50 km/sec at $R = 1.5$ kpc, while still closer to the centre they may even increase to 80 or 100 km/sec. For details concerning the determination of $\varphi(v)$ (see Refs. [2] to [4].

The angular rotation $\omega(R)$ can partly be determined from 21-cm observations see Sect. 8), and partly from stellar data. For $R > 10$ kpc one has to rely largely on rotational velocities computed from the general distribution of mass in the Galactic System, as inferred from all radio and optical data available. As these data are quite uncertain, the rotations for $R > R_0 + 2$ are likewise uncertain. Details concerning the determination of $\omega(R)$ for this region may be found in Ref. [5]. Fig. 8 shows the variation of ω with R. The linear circular velocity Θ_c is given in Fig. 9. These curves have been used to compute V_g as a function of distance from the Sun, r, for various directions in the galactic plane. R_0, the distance of the Sun from the centre, was assumed to be 8.2 kpc, as found from optical data.

[1] Gas Dynamics of Cosmic Clouds, Int. Astr. Union Symposium No. 2, p. 67. Amsterdam 1955.

We have, then, all the data we need to compute hydrogen densities from the observed profiles of the 21-cm line. However, the actual determination of these densities still meets with several difficulties. In the first place an ambiguity arises in the calculation of $n_{\mathrm{H}}(r)$ for the part of the system in which R is less than R_0. In Fig. 2 let D be the point where the line SP comes closest to the centre, and let A and A' be situated symmetrically with respect to D. In the case of a simple rotation the average radial velocity of the inter-

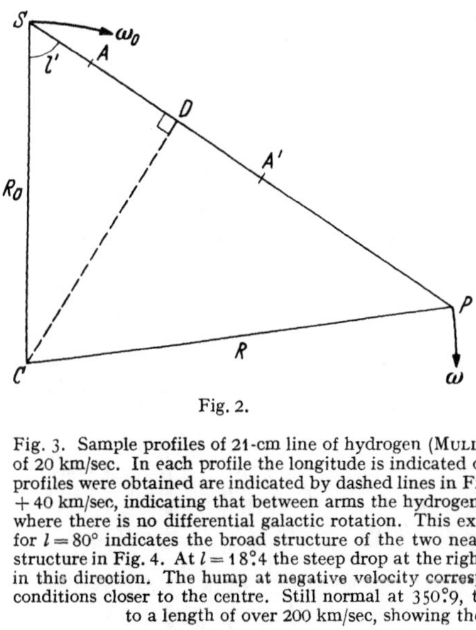

Fig. 2.

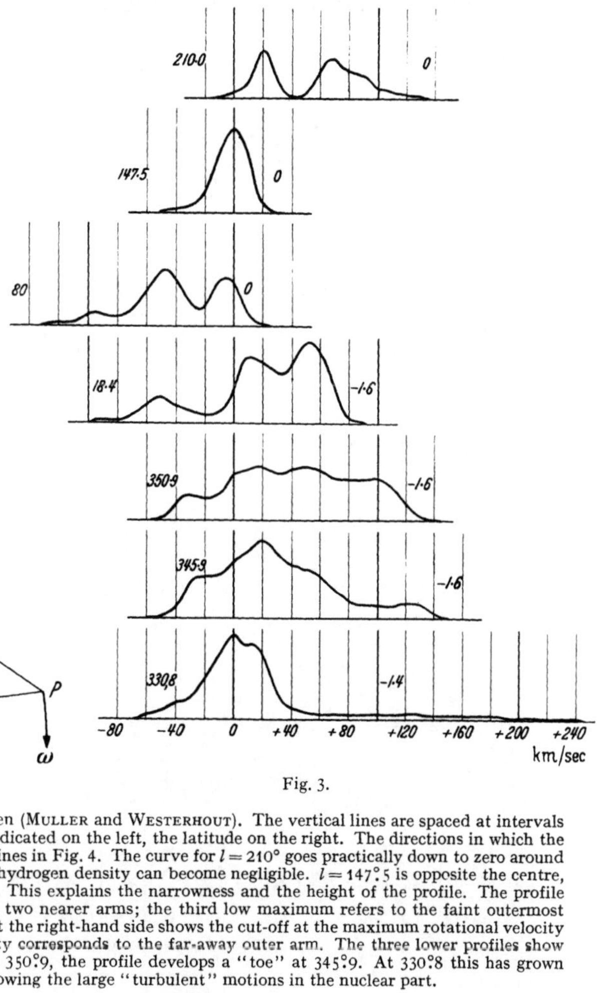

Fig. 3.

Fig. 3. Sample profiles of 21-cm line of hydrogen (MULLER and WESTERHOUT). The vertical lines are spaced at intervals of 20 km/sec. In each profile the longitude is indicated on the left, the latitude on the right. The directions in which the profiles were obtained are indicated by dashed lines in Fig. 4. The curve for $l = 210°$ goes practically down to zero around +40 km/sec, indicating that between arms the hydrogen density can become negligible. $l = 147°.5$ is opposite the centre, where there is no differential galactic rotation. This explains the narrowness and the height of the profile. The profile for $l = 80°$ indicates the broad structure of the two nearer arms; the third low maximum refers to the faint outermost structure in Fig. 4. At $l = 18°.4$ the steep drop at the right-hand side shows the cut-off at the maximum rotational velocity in this direction. The hump at negative velocity corresponds to the far-away outer arm. The three lower profiles show conditions closer to the centre. Still normal at 350°.9, the profile develops a "toe" at 345°.9. At 330°.8 this has grown to a length of over 200 km/sec, showing the large "turbulent" motions in the nuclear part.

stellar gas at A must evidently be equal to that at A'. We cannot, therefore, distinguish between atoms at A and A' by their velocities. A rough separation can, nevertheless, be made by studying the distribution perpendicular to the galactic plane; the angular width of this distribution will evidently be much smaller for the hydrogen near A' than for that near A. In this way a general picture of the hydrogen distribution in the inner parts of the system could be obtained, but the uncertainty is greater than that for the outer regions.

In longitudes differing less than about 15° from those of the galactic centre and anticentre the radial components of the differential galactic rotation become so small that they are drowned by the superimposed irregular motions, so that a separation according to distance becomes practically impossible. In these sectors the hydrogen density remains unknown. For this reason the corresponding sectors in Fig. 4 have been left blank. The same holds for part of the region around 50° longitude, where the rotation effects are small up to about 4 kpc from the Sun.

In the analysis of the line profiles it is assumed that, on the average, the interstellar medium moves in circular orbits. Although there is evidence that this assumption is correct to a considerable extent, there is also evidence that

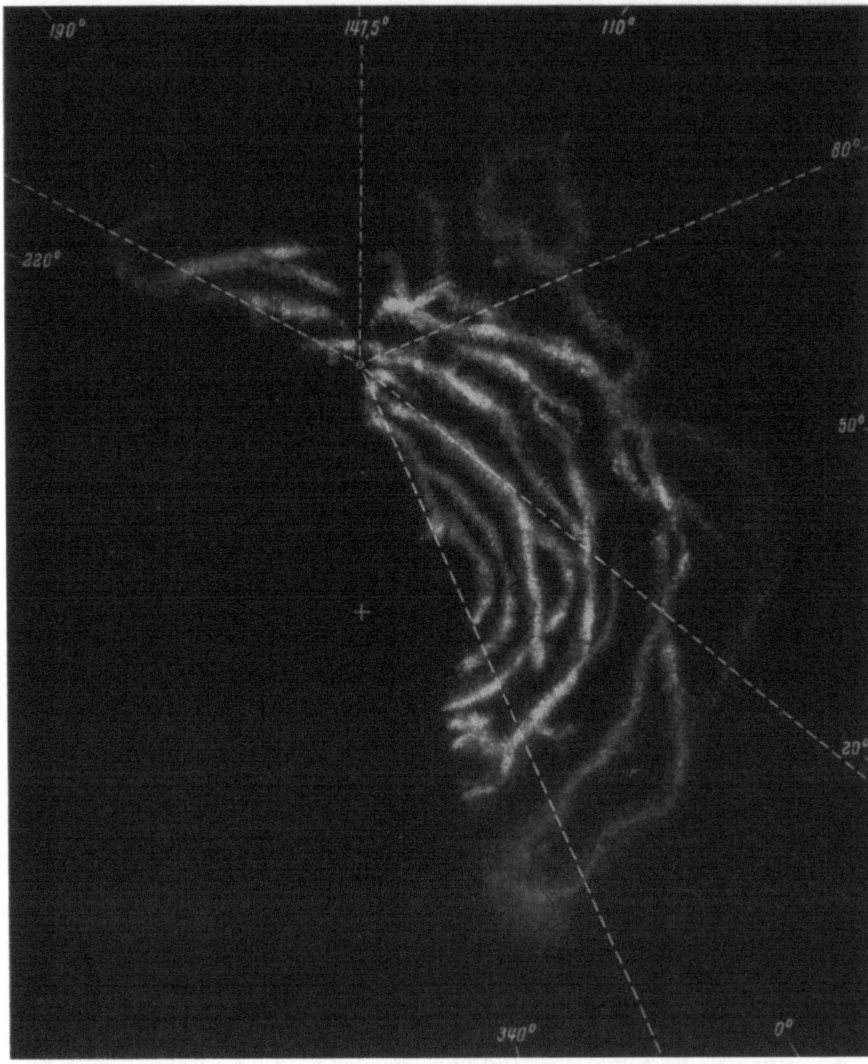

Fig. 4 a and b. Distribution of hydrogen in galactic plane, according to Westerhout and Schmidt [4]. The cross marks the position of the galactic centre, while the small circle indicates the Sun. The numbers near the edges are galactic longitudes; the centre is in 327°.7 longitude. The dotted lines correspond to the profiles in Fig. 3.

in some places notable systematic deviations from circular motions occur. These will have caused local errors in the determination of the hydrogen distribution.

Profiles of the 21-cm line have been determined from galactic longitude 320° through 0° to 220° in a belt extending to about 10° on either side of the galactic equator. At the time of writing no detailed data had been published on the more southern part of the galactic belt, between 220 and 320° longitude. Some typical line profiles are shown in Fig. 3. They were obtained with a paraboloid of 7.5 m aperture and a beamwidth of 2°.3[1]. The line profiles were corrected for the

[1] C. A. Muller and G. Westerhout: Bull. Astr. Inst. Netherl. 13, 151 (1957).

smoothing-out effect of this relatively large beam width before they were used for the computation of n_H.

Fig. 4a and b, taken from investigations by WESTERHOUT and SCHMIDT [4], shows the hydrogen densities in the galactic plane derived from such line pro-

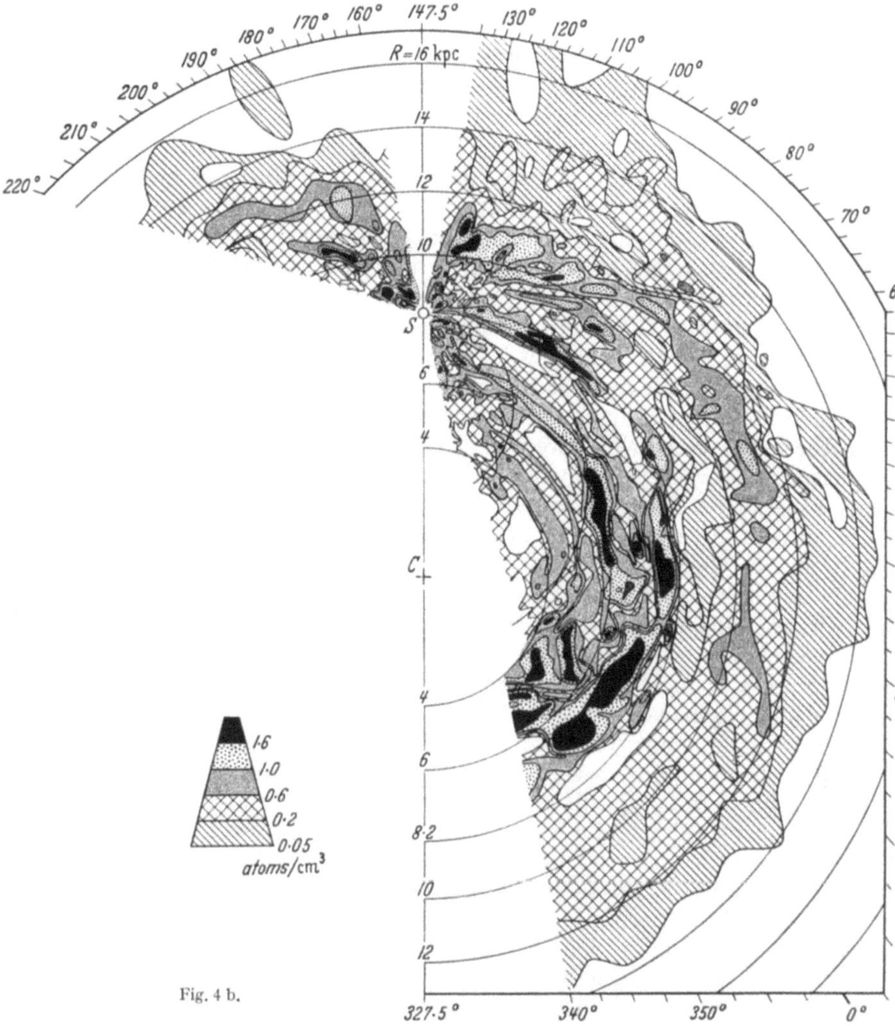

Fig. 4b.

files. The rotation of the system is clockwise. The directions of the seven profiles of Fig. 3 have been indicated by dotted lines.

It should be noted that there are only few optical data with which the radio picture can be compared. They refer mostly to associations of early-type stars, which appear all to be situated in regions of high hydrogen density. However, due to the strong absorption of light in the galactic layer, these data do not extend further than 2 or 3 kpc from the Sun. The great advantage of the radio investigations is that this radiation passes unhindered through the obscuring interstellar clouds. Fig. 4 shows clearly that the hydrogen is concentrated in long irregular lanes, which must doubtlessly be of the same nature as the spiral arms observed in external galaxies. The maxima of the profiles in Fig. 3 correspond to the principal spiral arms. The depth of the minimum in the profile for $l=210°$

illustrates that the space between the arms may be quite empty. The long extension to high negative velocities in the profile for $l=80°$ corresponds to a thin, irregular outer arm. The profile for $l=147°$ refers to the direction opposite the centre. In this direction there is no differential rotation in radial velocity. The profile, therefore, has only one peak, and is narrow, the width being determined mostly by random motions. The profile for $l=18\overset{\circ}{.}4$ shows the sharp cut-off at the positive side, which corresponds to the rotational velocity at the point where the corresponding line in Fig. 4 comes closest to the centre. The hump at the negative velocities corresponds to the long outer arm shown in Fig. 4.

The Sun is situated near the inner side of an arm, which passes through the Orion association between 170 and 180° longitude, and stretches away from the Sun in the general direction of Cygnus. This arm, which has been called the Orion arm, can be followed to 340° longitude, gradually approaching the centre, so that in the latter longitude its distance from the centre is about 7 kpc. In the quadrant from 55 to 145° longitude we find another dense arm about 2 kpc outside this. It contains the large association around h and χ Persei and is called the Perseus arm; in several directions it is double. Still further out is an extended and irregular arm of rather smaller density. Two other, fairly heavy arms can be seen inside the Orion arm, the outer one of these has been named Sagittarius arm. On the average the arms are spaced at about 2 kpc. Their widths in the plane, measured between points where the density is half the maximum density, may very roughly be estimated as 0.8 kpc. The average density between these points is perhaps about 1.5 atoms per cm^3. In the space between the arms it may be from 5 to 10 times lower. All these numbers give only orders of magnitude, and vary a great deal from one region to another. The picture presented by the arms is, indeed, far from a smooth continuous structure. They are cut up and have several dead branches. And even where they are not cut up, their internal structure is most irregular, the gas being often concentrated in large associations and complexes of clouds.

For R less than about 3 kpc the character of the spiral structure changes considerably. This may be seen from the sample profiles of Fig. 3. At 351° longitude the line profile is still normal; the large extent to positive velocities is due to the increased angular velocity of the inner parts of the system. At $l=346°$, however, the character of the profile has changed: a tail extends to the larger velocities, which is considerably lower than the main part of the profile. The same feature, but in a much more extreme way, is shown at $l=331°$. From a discussion of all the line profiles around the centre it has appeared that these tails are due partly to the increase of ω towards the centre, and partly to an increase of deviations from circular motion. The low intensity of the tail is caused by the large spread in velocity coupled with a somewhat lower density of atomic hydrogen in the nuclear part.

A considerable part of the hydrogen in the central region appears to be concentrated in arm-like structures. The strongest of these shows as a striking maximum on the negative-velocity side in the line profiles between $l=319°$ and $l=326°$ (cf. Bull. Astr. Inst. Netherl. No. 475, pp. 161—163). Recent observations in the nuclear region have indicated the existence of other similar structures, with large deviations from circular motion. The first-mentioned "arm" moves away from the centre with a velocity of 50 km/sec[1]. There is evidence that *all* the gas within 3 kpc from the centre is streaming away from it, with radial components between about 40 and 150 km/sec. These motions must be accurately in the galactic plane.

[1] H. van Woerden, G. W. Rougoor and J. H. Oort: C. R. Acad. Sci. Paris **244**, 1691 (1957).

Sample cross-sections perpendicular to the galactic plane are shown in Fig. 5. The contours again indicate curves of equal hydrogen density. It is evident

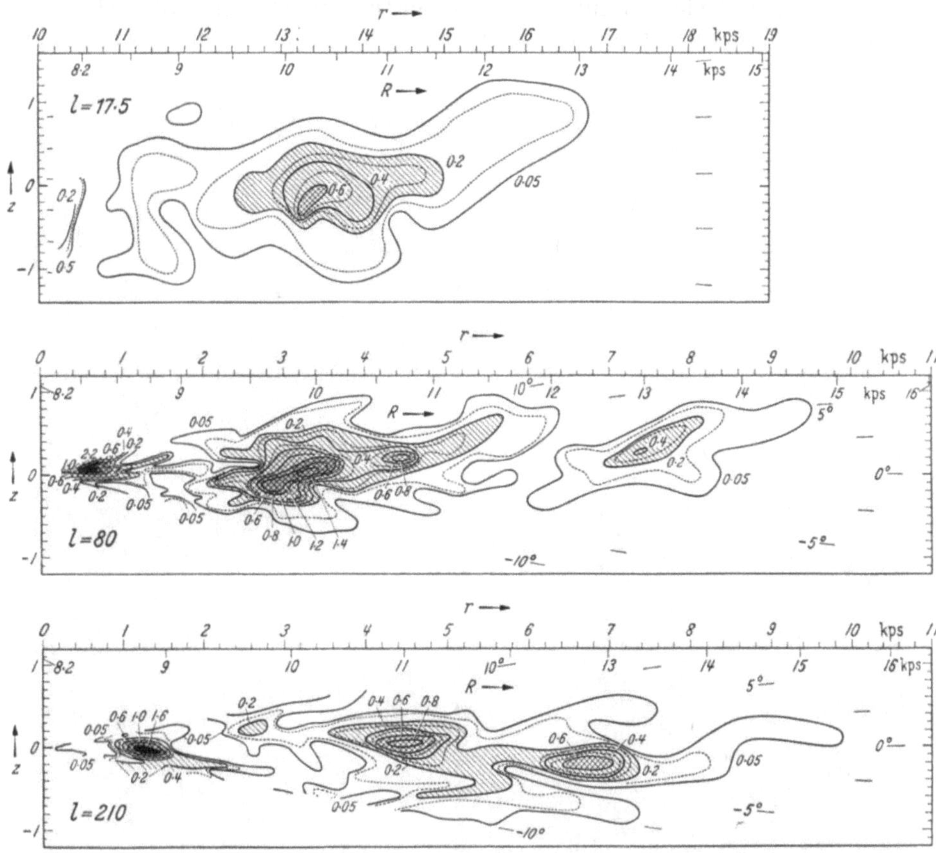

Fig. 5. Sample cross sections of hydrogen distribution in planes perpendicular to the galactic equator for three of the longitudes indicated by dashed lines in Fig. 4 (WESTERHOUT). Heavy numerals indicate scales of distance, r, from the Sun, and from the conventional galactic plane (z), in kpc. Small numerals indicate distances, R, from the galactic centre. The numbers on the equidensity contours represent numbers of hydrogen atoms per cm³.

that the gas is strongly concentrated to the galactic plane. From similar sections observed in all longitudes covered by Fig. 4 it was found that the thickness of the gas layer varies very little over the Galactic System. The distance between the surfaces where the density is half the average density in the plane is everywhere about 220 pc. Only in the region within 3 kpc from the centre the thickness is markedly smaller.

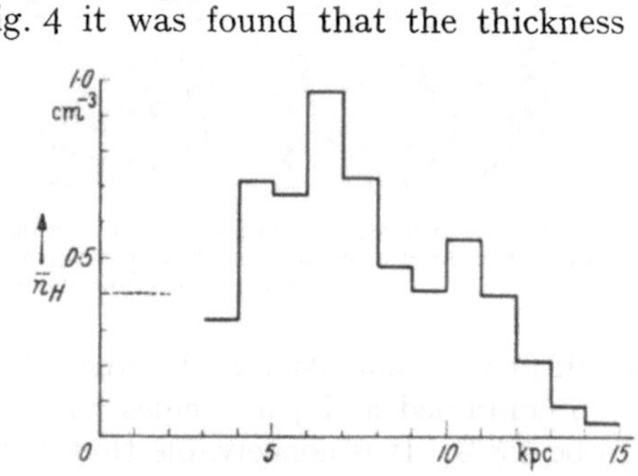

Fig. 6. Average hydrogen density in the galactic plane as a function of the distance from the centre (from Bull. Astr. Inst. Netherl. No. 475).

The smoothed-out overall density in this layer is 0.7 hydrogen atoms per cm³. Fig. 6 shows its variation with distance from the centre. Between 11 and 12 kpc the average density begins to drop sharply; at $R=15$ kpc it has decreased to about 0.05 H/cm³. In the region within $R=3$ kpc the overall density has been estimated to be 0.4; slightly lower, therefore, than in the main part of the system [4]. It should be noted that the density of hydrogen atoms is not necessarily representative of the total density of hydrogen. For part of it is ionized, while another part may be in molecular form. In our surroundings most of the hydrogen un-ionized[1] (observations of the continuous radiation at decimetre wavelengths

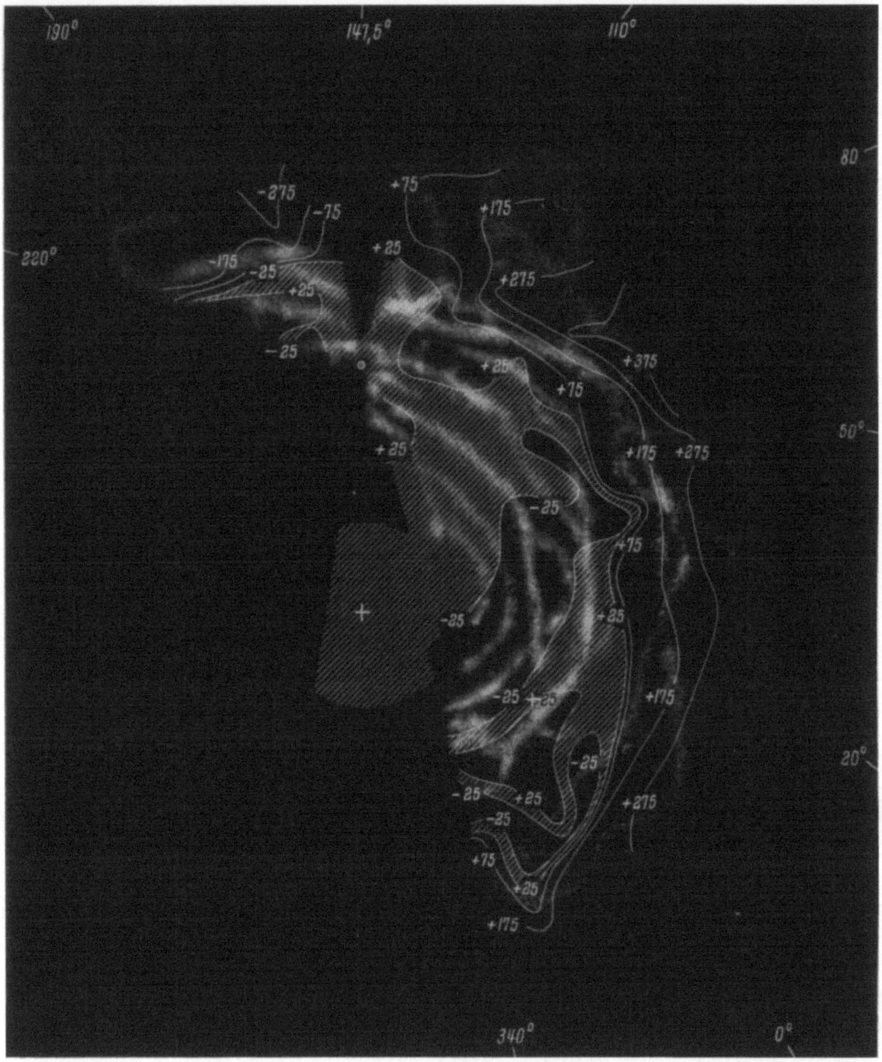

Fig. 7. Deviations of centre of hydrogen layer from the average galactic plane. The contours are marked by number indicating the corresponding distances in pc north (positive) or south (negative) of the average plane. In the hatched areas the centre of the hydrogen layer lies within 25 pc of this plane.

confirm this conclusion and extend it to the whole Galactic System). The fraction contained in H_2 molecules may likewise be small; Kahn[2] has estimated it to be 0.8%. It is conceivable that in the nuclear part conditions are sufficiently different to make the molecular part considerably greater, so that the total hydrogen density would not necessarily be lower than in the rest of the system.

[1] B. Strömgren: Astrophys. Journ. **108**, 242 (1948).

[2] Gas Dynamics of Cosmic Clouds, Int. Astr. Union Symposium No. 2, p. 60. Amsterdam 1955.

One of the most remarkable features of the hydrogen distribution is that the various arms lie so very nearly in one plane. This is particularly striking for the part within the Sun's distance from the centre, where the maximum deviations from the mean plane for stretches of 0.5 to 1 kpc length amount to about 50 pc; the *average* deviation is much smaller still. It is only in the outer parts, where the density in the arms becomes noticeably less, that arms are locally found to deviate up to 500 pc from the average plane. In these outer regions, beyond about $R=11$ kpc, there are likewise *systematic* deviations. The character of these deviations may be seen in Fig. 7, which gives the deviation of the hydrogen layer from the mean plane defined by these hydrogen observations. This plane makes an angle of 1°.4 with the conventional galactic plane with pole at $\alpha=12^{h}40^{m}.0$, $\delta=+28°0'$ (1900), on which the galactic co-ordinates in the present article are based. Recent Australian observations show that in longitudes from 180 to 280° there are, beyond $R=11$ kpc, systematic deviations from the mean plane comparable with those shown in Fig. 7, but in the opposite direction[1].

Not all of the gas is concentrated in the galactic layer. Optical observations as well as measures at 21 cm indicate that outside this layer there may be a thin "corona" of high-velocity clouds extending to considerable distances from the galactic plane.

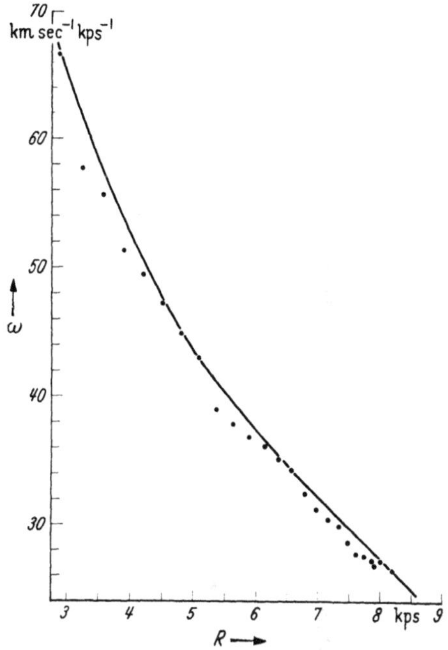

Fig. 8. Variation of angular rotation of the Galactic System with distance from the centre.

8. Galactic rotation from radio observations. The profiles of the 21-cm line may be used to find the velocity of rotation of the Galactic System at various distances from the centre up to $R=R_0$. The angular velocity of rotation increases with decreasing distance from the centre. Observing in the direction SP (Fig. 2) we see from (7.1) that the maximum radial velocity will occur at D. If there were no irregular motions, and if the galactic layer would everywhere contain atomic hydrogen, the line profile would be cut-off sharply at a velocity $\Theta_c(R_D)-\Theta_0\sin l'$; Θ_0 is the circular velocity near the Sun. If Θ_0 is known, we could then at once find $\Theta_c(R_D)$, the circular velocity at D. By observing in different directions we would obtain the variation of Θ_c with R, up to $R=R_0$. In reality random motions are superimposed on the rotation, so that the edge of the line is smoothed out. Knowing the average random velocity we can, however, correct for this effect, and determine the velocity of rotation at D, provided that this point lies in a spiral arm [3]. If, however, the hydrogen density near D happens to be small, the limit of the line will be reached at a smaller velocity, corresponding to the angular rotation at a point on the line SP farther away from the centre.

Maximum velocities have been determined in this manner with the same instrument as used for measuring the hydrogen distribution discussed in the

[1] F. J. Kerr, J. V. Hindman and Martha Stahr Carpenter: Nature, Lond. **180**, 677 (1957), Fig. 2.

preceding section. They are shown as filled circles in Fig. 9. The curve gives the estimated run of the circular velocity Θ_c with R. For the reasons just mentioned this curve has been drawn through the highest points, which are supposed to correspond to directions where the point D lies in a spiral arm. The sign ⊙ indicates the circular velocity near the Sun. The curve beyond R_0 is based on computations from a model of the mass distribution in the Galactic System by M. Schmidt [5].

For R below 3 kpc we get into the nuclear region, where the deviations from circular motion are so high that the derivation of the rotational velocity becomes highly uncertain. The rotation curve within 2 kpc from the centre is, therefore, nothing but a rough estimate.

The *angular* velocities corresponding to these rotational velocities are shown in Fig. 8. Details concerning these determinations are given in Ref. [3]. It should be noted that, except for the region near the centre, the available data are obtained from only one half of the Galactic System. Provisional results obtained in Australia for some points in the southern Milky Way confirm the rotations found in the north. A satisfactory derivation of the rotation curve will only be possible after the southern data are complete.

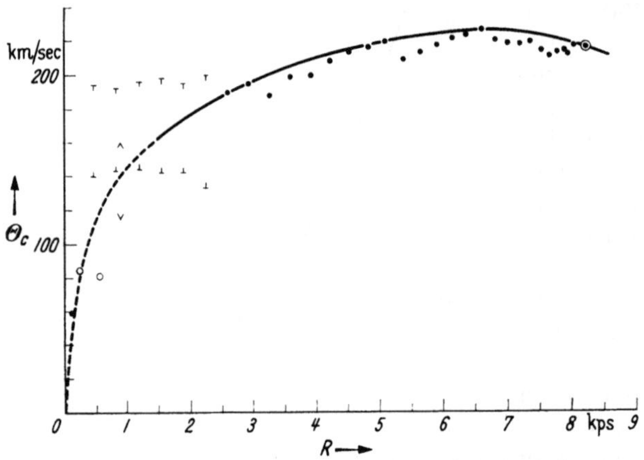

Fig. 9. Variation of linear velocity of rotation in the Galactic System with distance from the centre.

The data on rotation are of importance for the derivation of the hydrogen distribution through the Galactic System. They can also be used to derive the gravitational force in the galactic plane. This gives us valuable information on the distribution of mass in the inner regions of the system [5].

9. Distribution of ionized hydrogen. The separation of the various components contributing to the continuous radiation at radio frequencies is still rather uncertain. On theoretical grounds it has been found that the electron temperature in regions of ionized hydrogen will generally be about 10000°. At long wavelengths vastly higher brightness temperatures have been observed. At a wavelength of 15.2 m Shain, for instance, has found a temperature of 600000° for a region near the centre[1]. The beamwidth used was 1°.4. This indicates that for such long wavelengths non-thermal radiation is predominant. That there is considerable non-thermal radiation at shorter wavelengths also, is apparent from the manner in which the background radiation varies with ν. For the so-called isotropic component, or the galactic corona, this variation has been found to be [1]

$$T_b \propto \nu^{-2.5}. \tag{9.1}$$

Similarly, for a region of roughly 6° diameter around the galactic centre, Piddington[2] found:

$$T_b \propto \nu^{-2.6} \tag{9.2}$$

[1] C. A. Shain: Austral. J. Phys. **10**, 195 (1957).
[2] J. H. Piddington: Monthly Notices Roy. Astronom. Soc. London **111**, 45 (1951).

between $\nu = 30$ and $\nu = 100$ Mc/s. Recently ADGIE and SMITH[1] have derived an exponent -2.50 for the galactic corona as well as for the radiation closer to the galactic plane.

The spectra of the non-thermal radio sources vary from case to case. As an average for six bright sources we may adopt

$$I(\nu) \propto \nu^{-0.65}. \tag{9.3}$$

This simple law holds approximately from $\nu = 20$ to $\nu = 10000$ Mc/s. The individual exponents range for the six sources from -0.3 to -0.8[2]. Converted into temperature by means of (2.1) this becomes $T \propto \nu^{-2.65}$. The agreement of the exponent with that in (9.2) suggests that the non-thermal radiation from the Galactic System may originate in a similar way. But all data are so uncertain and the phenomena so complex that an inference like this should be considered with great reserve.

For thermal radiation we have, according to (4.5) and (4.7)

$$T_b \propto \nu^{-2}, \tag{9.4}$$

if the optical depth is small. For appreciable optical depth the absolute value of the exponent is smaller; it evidently approaches zero for very large τ.

It is clear from the above that the thermal radiation will become relatively more important as we go to shorter wavelengths. The exact ratio between the two types of radiation cannot, however, be determined from the data so far available.

As we have seen in Sect. 4, optical data permit us to make a rough estimate of the amount of thermal radiation produced per unit of volume in the general vicinity of the Sun. This gave us an optical depth of about $5 \times 10^{13} \nu^{-2}$ per kpc in the galactic plane. The estimate may well be in error by a factor of two. But it can serve at least to get a first estimate of the amount of thermal radiation to be expected from the Galactic System.

We have found in Sect. 7 (Fig. 6) that, after smoothing out the spiral arms, the overall density of atomic hydrogen does not vary very greatly up to a distance of 11 kpc from the centre, while it drops rapidly to negligible values beyond this distance. As a provisional working hypothesis we assume that the same distribution holds for ionized hydrogen. Very roughly we assume the density of HII to be constant up to $R = 11$ kpc, and zero beyond this distance. The total optical depths found on this hypothesis for different galactic longitudes are shown in Table 3 for a wavelength of 100 cm. For a wavelength of λ cm the numbers should be multiplied by $(\lambda^2/10000)$[3].

Table 3. *Calculated optical depths of ionized hydrogen at* $\lambda = 100$ cm *in the galactic plane, for the case of a homogeneous distribution* (l' *is the galactic longitude counted from the longitude of the centre*).

l'	$\tau(100)$	l'	$\tau(100)$	l'	$\tau(100)$	l'	$\tau(100)$
0°	0.0134	30°	0.0121	80°	0.0062	140°	0.0024
10	0.0133	40	0.0122	100	0.0042	160	0.0020
20	0.0129	60	0.0087	120	0.0030	180	0.0019

[1] R. ADGIE and F. G. SMITH: Observatory **76**, 181 (1956).

[2] CH. L. SEEGER: Report URSI Commission Vd at meeting in Boulder, Col., 1957. Cf. also ADGIE and SMITH[1].

[3] The optical depth per kpc used for computing the numbers in Tables 3 and 4 is 25% higher than the value just quoted. The increase is meant to take account of the fainter hydrogen emission regions which were not included in the surveys.

In Table 4 the brightness temperatures corresponding to this schematic model are compared to those found in surveys by WESTERHOUT[1] at $\lambda=22$ cm and by PIDDINGTON and TRENT[2] at $\lambda=50$ cm. In both surveys the irregularities have been somewhat smoothed out, and the bright sources like Sagittarius A, Cygnus A, Cassiopeia A, Taurus A, were left out. The temperatures given are counted from a zero level corresponding to the average temperature in high latitudes.

The observed temperatures are higher than those calculated on the supposition of a homogeneous distribution, in particular for smaller values of l'. It is possible that the true density of ionized hydrogen in the inner parts of the system is larger than has been assumed. But at least part of the difference must be due to superimposed non-thermal radiation. The importance of the latter is clearly shown—at least for $\lambda=50$ cm—when we consider the radiation at some distance from this plane. In the sector of 60° longitude around the centre the brightness temperatures given by PIDDINGTON and TRENT for $\lambda=50$ cm average about 55° at $b=5°$, and 32° at $b=10°$. If the ionized hydrogen is confined to a layer comparable in thickness to that of the atomic hydrogen, the thermal emission at $b=10°$ would have to come practically entirely from distances less than 1.5 kpc, for which, according to Sect. 4, the brightness temperature at $b=10°$ may be estimated at 0°5. We are thus led to conclude that the major part of the radiation at $\lambda=50$ cm is non-thermal. A similar conclusion was reached by SHAIN[3] in his discussion of a high-resolution survey at $\lambda=15.2$ m. He states that for the region between −12 and −42° declination (which, on the galactic circle, corresponds to the longitude interval 312 to 347°) there is down to about 8° galactic latitude a rise in temperature which is similar for the surveys at 15.2 m, 3.5 m and 50 cm. This may be considered as an indication that for latitudes 8° and higher the absorption by ionized hydrogen is small, even at 15.2 m wavelength. The effect of this absorption becomes evident only at lower latitudes, where the 15.2 m intensity rises more slowly than that at the lower wavelengths. Within about 4° from the galactic plane the measures at 15.2 m even show a trough, which must be due to absorption by ionized hydrogen. As SHAIN points out, this absorption should have been practically complete already at 8° latitude if a large fraction of the 50 cm radiation observed at that latitude were of thermal origin and if the 15.2 m radiation comes mainly from beyond the hydrogen layer (which appears plausible).

Table 4. *Comparison of observed temperatures in the galactic plane with those computed from a model.*

l'	$\lambda=22$ cm		$\lambda=50$ cm	
	obs.	comp.	obs.	comp.
5°	23°	6°5	170°	33°
8— 12	23	6°4	150	33
15— 25	23	6°2	140	32
30— 50	11	5°4	80	28
60—100	4:	3°0	20	16
100—140		1°5	6:	8
140—180		1°0	6:	5

Although the present evidence points to an important non-thermal component in the radiation at 50 cm wavelength, the alternative possibility that the ionized gas in the central parts of the Galactic System extends much further from the galactic plane than the atomic hydrogen, cannot yet be entirely ruled out.

A considerable amount of valuable information on the distribution of ionized hydrogen, as well as of possible non-thermal radiation near the galactic plane will become available in WESTERHOUT'S extensive survey at 22 cm with a beamwidth of 0°6, which is now being prepared for publication.

[1] To appear in Bull. Astr. Inst. Netherl. No. 488.

[2] J. H. PIDDINGTON and G. H. TRENT: Austral. J. Phys. **9**, 481 (1956).

[3] C. A. SHAIN: Austral. J. Phys. **10**, 195 (1957).

Striking evidence for a concentration of ionized hydrogen within 150 pc of the galactic centre is furnished by MILLS' survey at 3.5 m, which shows the absorbing effect of this hydrogen (cf. p. 122).

10. General radiation from the region close to the galactic plane. The only general surveys available at present which have a sufficiently good resolution to show details of the radiation close to the galactic plane are those at short wavelengths[1].

Fig. 10 gives the distribution of the radiation at $\lambda = 50$ cm as observed by PIDDINGTON and TRENT. The picture suggests that the low-latitude radiation at this wavelength is largely concentrated in the spiral arms. Separate maxima and "shoulders", accompanied by sharp drops in intensity, can in many cases be correlated with the structure found from the 21-cm observations. The question whether these structural details are wholly caused by the radiation of ionized hydrogen, or partly also by non-thermal emission, cannot yet be answered decisively. We know that there exists a type of non-thermal sources which are strongly concentrated to the galactic plane (the so-called class I sources) but it is still quite uncertain how much they contribute to the total radiation.

Of the two strong concentrations at 40° northern declination the one at higher right-ascension (the so-called Cygnus X source) co-incides with the direction of the Orion arm, and might be the combined radiation of a large number of emission regions strung along this arm. The second source, Cygnus A, is an extra-galactic radio source. The secondary maximum at about 10^h40^m, $-60°$ is in the direction of the large association around the η Carinae nebula and probably also runs lengthwise through a long stretch of a spiral arm. The maxima at 13^h22^m, $-42\overset{\circ}{.}8$ and 5^h31^m, $+22\overset{\circ}{.}0$ are due, respectively, to the Centaurus source (the galaxy NGC 5128) and the Taurus A source (the Crab nebula).

Within 15° from the centre we observe, in the lowest latitudes, a stronger increase in intensity than would correspond to a more or less homogeneous distribution of the radiating sources. Super-imposed on this general increase there is a sharp rise right at the centre, which is the radio source Sagittarius A. Measures at a wavelength of 3 cm with the 15 m paraboloid of the Naval Research Laboratory in Washington[2] indicate that its diameter is about 15′. The fact that the position co-incides within $0\overset{\circ}{.}1$ with that of the galactic centre (as determined from the general distribution of the radiation at metre waves, as well as from planetary nebulae) suggests that it is actually in the centre of the Galactic System. This is confirmed by observations of the hydrogen line at 21 cm wavelength, which may be seen in absorption against the continuous radiation of the source. One of the absorption features is caused by a spiral arm which can be shown to lie within 3 kpc from the centre[3].

The spectrum of Sagittarius A resembles that of thermal radiation from an ionized gas. If it actually consists of ionized hydrogen, and if its radius is 18 pc (corresponding to the measured diameter of 15′), its density would have to be about a hundred protons and electrons per cm^3.

Many of the discrete sources identified in the surveys of decimetre waves represent large emission nebulae, like the well-known nebula in Orion. So far,

[1] *Note added in proof:* After the writing of this chapter two surveys have been published which have added greatly to our knowledge, one at 350 cm, with a beamwidth of 50′ [B. Y. MILLS, E. R. HILL and O. B. SLEE: Observatory **78**, 116 (1958)] and one at 22 cm, with a beamwidth of $0\overset{\circ}{.}6$ [G. WESTERHOUT: Bull. Astr. Inst. Netherl. **14**, 215 (1959)]. These permitted a fairly good separation between thermal and non-thermal radiation. Both kinds of radiation show concentration toward the spiral arms.

[2] F. T. HADDOCK and T. P. M. MCCULLOUGH: Astronom. J. **60**, 161 (1955).

[3] H. VAN WOERDEN, G. W. ROUGOOR and J. H. OORT: C. R. Acad. Sci. Paris **244**, 1691 (1957).

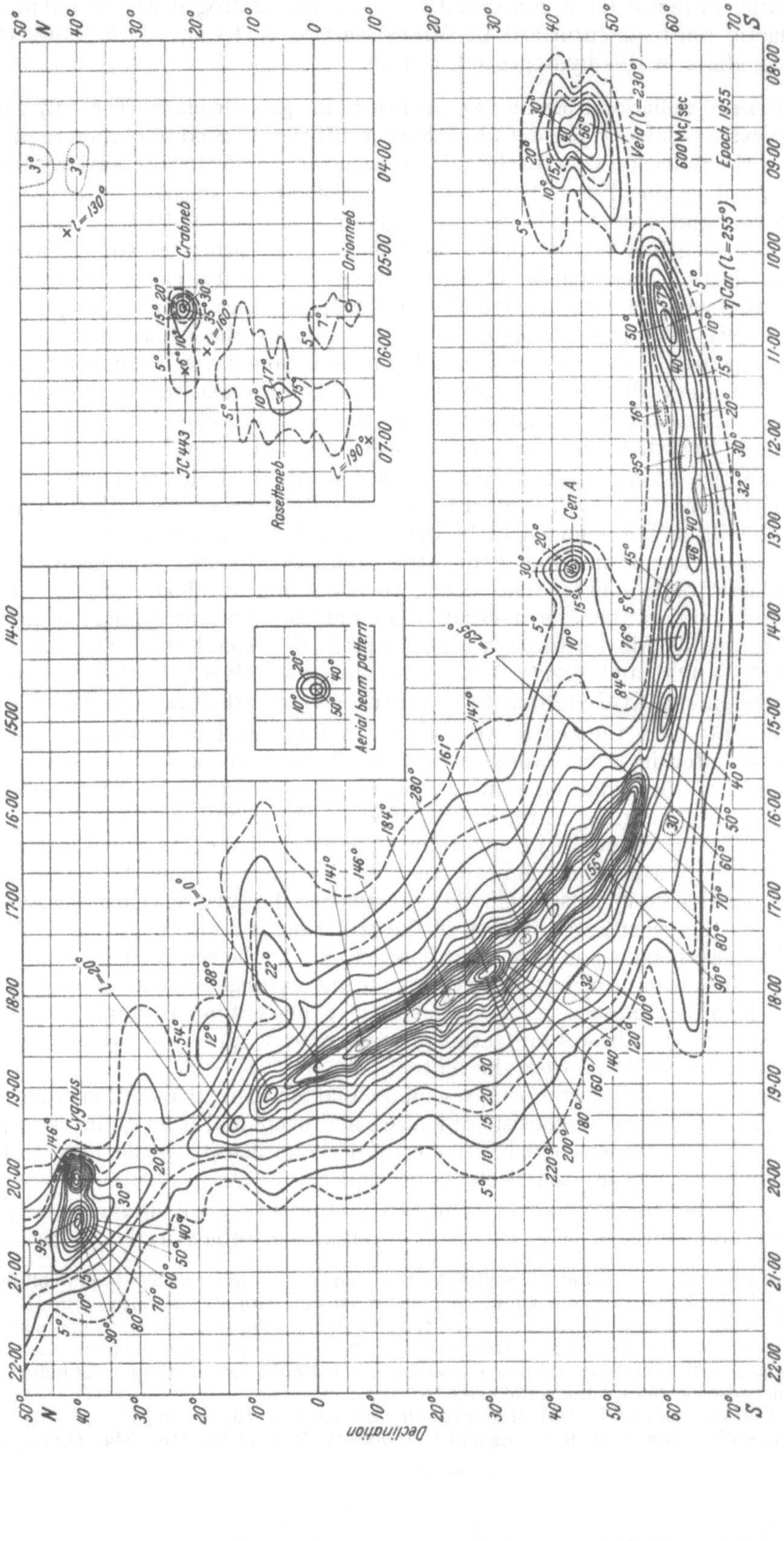

Fig. 10. Distribution of radiation at 50 cm wavelength (Piddington and Trent). Co-ordinates are right-ascension and declination for 1955; a few galactic longitudes have been indicated.

only the nearer nebulae have been found, all of which were already known from optical observations.

The general surveys at *metre* wavelengths have all been made with such large beamwidths that it is impossible to infer from them the true distribution of the component with strong galactic concentration. A high-resolution survey for $\lambda = 3.5$ m is at present being made by MILLS in Sydney. From the few results of this survey which have been published[1], it is evident that also at these wavelengths there is considerable radiation coming from a narrow band at the galactic equator. Fig. 11 shows MILLS' provisional results for three strips through the Milky Way at longitudes 327, 348 and 3°, respectively. We see a narrow maximum superimposed upon radiation which is less strongly concentrated towards the galactic circle. The width of the former is quite similar to that of the belt from which the thermal radiation comes. But the intensity of the narrow peaks is several times higher than the expected thermal radiation. The same phenomenon had been observed by SCHEUER and RYLE[2] in their interferometer measurements at 81.5 Mc/s ($\lambda = 3.68$ m). The temperatures given by these authors are somewhat lower, however.

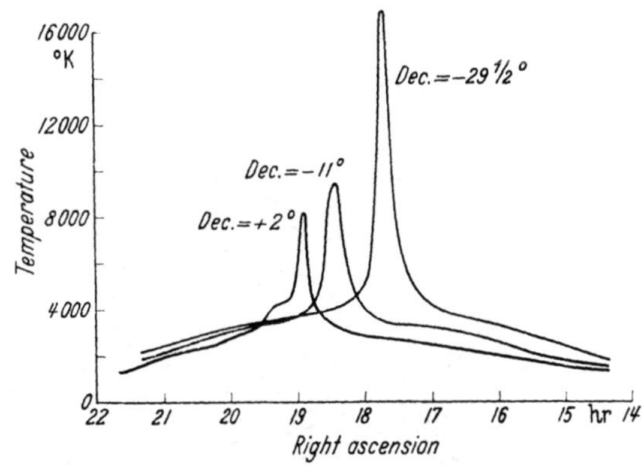

Fig. 11. Variation of radiation at $\lambda = 350$ cm with right-ascension, for three declinations crossing the galactic equator near longitudes 327° (highest peak), 348° and 3° (lowest peak) respectively; beamwidth 50′ (MILLS). The temperature scale should be revised according to recent information from Dr. MILLS: the sharp peaks must be increased by about 60% relative to the slowly varying component, which itself needs an increase of about 20%.

Detailed results of MILLS' pencil-beam survey at 3.5 m have been published for a region 10°×10° surrounding the centre[3]. These are represented in Fig. 12. We see two maxima roughly 1° on either side of the centre, which is marked by a cross. Thus, where the surveys for short wavelengths show a maximum, high-resolution longer-wavelength surveys indicate a minimum. This drop near the centre must be due to absorption by a mass of ionized hydrogen. If this is again supposed to be around the galactic nucleus it should extend to about 150 pc from the centre, with an average density between 5 and 10 electrons per cm³. Within this mass the hydrogen should be strongly concentrated toward the centre in order to explain the relatively sharp emission feature observed at short wavelengths. Fig. 12 shows that there is also a concentration of non-thermal emission within about 350 pc from the centre.

It is clear from Fig. 11 that the radiation received consists of at least two components having a radically different distribution. At present the distribution of the radiation coming from higher galactic latitudes can best be studied from surveys such as shown in Fig. 13, giving the general distribution at 3 m wavelength[4], and Fig. 14 which gives the distribution for 1.5 m wavelength[5]. In

[1] B. Y. MILLS: Austral. J. Phys. **8**, 368 (1955). — *Note added in proof:* A complete map has now been published by B. Y. MILLS, E. R. HILL and O. B. SLEE: Observatory **78**, 116 (1958).

[2] P. A. G. SCHEUER and M. RYLE: Monthly Notices Roy. Astronom. Soc. London **113**, 3 (1953).

[3] B. Y. MILLS: Observatory **76**, 65 (1956).

[4] J. G. BOLTON and K. C. WESTFOLD: Austral. J. Sci. Res. **3**, 19 (1950).

[5] F. DRÖGE and W. PRIESTER: Z. Astrophys. **40**, 236 (1956). The contours are partly based on data by ALLEN and GUM.

both distributions an attempt has been made to correct for the antenna pattern. In BOLTON and WESTFOLD's survey the beamwidth was 17°.

If we compare the distributions in Figs. 13 and 14 with those of known types of stars, the most striking difference is the relatively intense radiation from very high latitudes, and the fact that this is of nearly the same intensity as the radiation received from the Milky Way region opposite the galactic centre. There are no known members of the Galactic System which display a distribution resembling

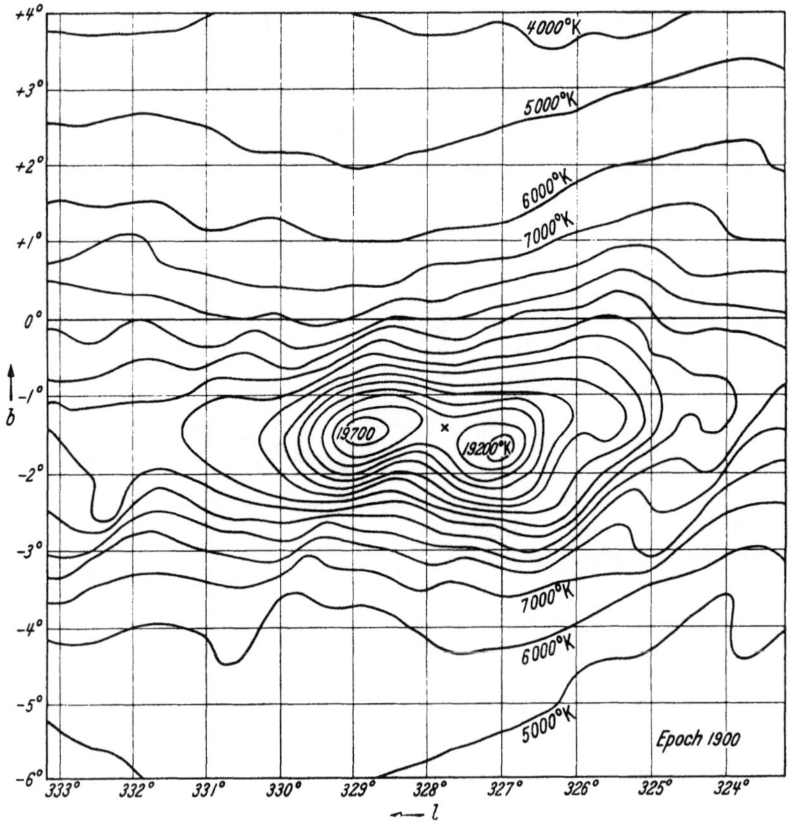

Fig. 12. Distribution of radiation at $\lambda = 350$ cm in region surrounding the galactic centre (MILLS). According to a private communication from Dr. MILLS the temperatures T indicated in this figure need modification, the true temperatures being approximately equal to $2T - 3500°$ K. The beam width used was 50′.

this. For this reason it has been thought that the radiation observed in high latitudes as well as opposite the centre would be mainly of extra-galactic origin. However, as pointed out by several authors[1], the minimum temperatures are not found at the galactic poles but at about 40° latitude in a longitude opposite the centre. This indicates that at least a considerable fraction of this radiation must be connected with the Galactic System.

On the other hand, there may well be *also* an important amount of *extra-*galactic radiation. It is still impossible to give a reasonable estimate of the relative importance of the two contributions. A possible interpretation of the galactic part of this "isotropic" component will be discussed in Sect. 11.

[1] I. S. SHKLOVSKY: Astronom. J. USSR. **29**, 418 (1952); **30**, 15, 577 (1953). — A. FOKKER: I. R. A. Rapport: No. 1. — S. P. WYATT: Astrophys. Journ. **118**, 304 (1953).

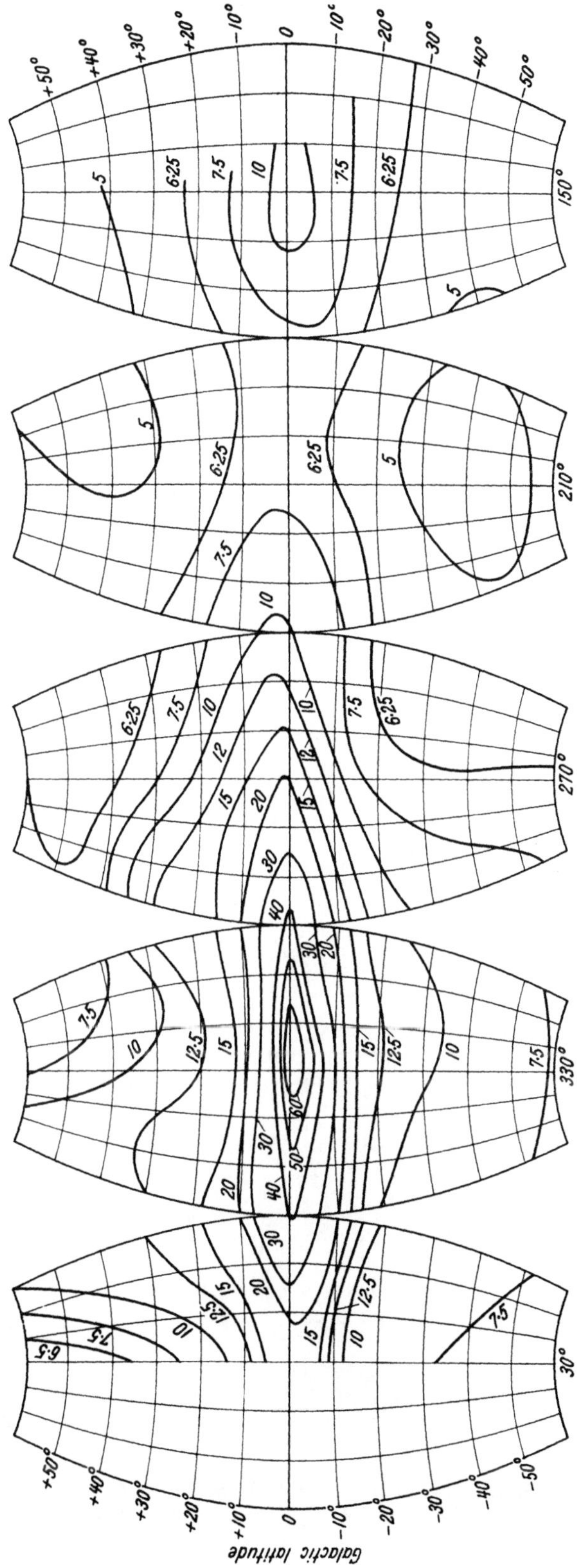

Fig. 13. Chart of brightness temperature at 3 metre wavelength (BOLTON and WESTFOLD). The co-ordinates are galactic longitude and latitude; the unit is 100° K. The sharp concentration to the galactic plane has been effaced to a certain extent by the large beamwidth.

Attempts to separate the galactic and isotropic components by means of existing surveys, and to evaluate the space distribution of the source of the galactic radiation have been made by several authors[1]. Fig. 15 shows the distribution as

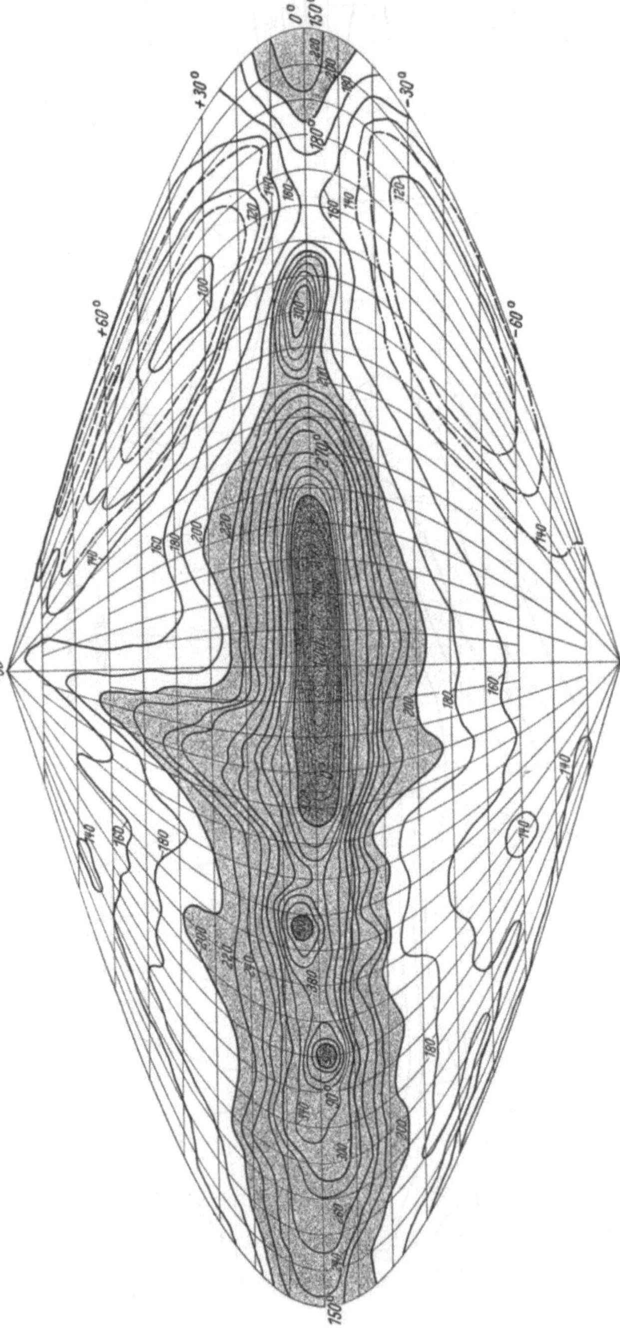

Fig. 14. Chart of brightness temperature at 1.5 metre wavelength in galactic co-ordinates. The numbers indicate temperatures in °K. Up to 300° the steps are 20°, between 300° and 500° they are 40°. The darkest parts are above 500°; the temperature near the centre is 1200°.

[1] J. G. Bolton and K. C. Westfold: Austral. J. Sci. Res. **4**, 476 (1951). — R. Hanbury Brown and C. Hazard: Phil. Mag. **44**, 939 (1953). — S. P. Wyatt: Astrophys. Journ. **118**, 304 (1953).

derived by HANBURY BROWN and HAZARD [1], the upper curve giving the variation of the space density in the galactic plane with distance from the centre, while the lower one gives the variation along a line through the galactic centre normal to the galactic plane. The curves were based on observations by REBER[1] at 480 Mc/s; they were also shown to agree closely with BOLTON and WESTFOLD's observations at 100 Mc/s after allowance was made for an isotropic component of 600°. The radiation by ionized hydrogen had been subtracted from the observed intensities; the estimates used for this thermal radiation were somewhat higher than those given in Sect. 9. In the central region, however, the actual thermal radiation may well be higher. HANBURY BROWN and HAZARD point out that the space distribution of sources is the same for different frequencies. The brightness temperature was found to vary as $\nu^{-2.8}$ over the range 18.3 to 1200 Mc/s.

Fig. 15 indicates that the sources responsible for this radiation are strongly concentrated to the galactic centre. The surfaces of equal density have an axial ratio of about 8:1. The authors suppose that the radiation is the integrated effect of localized sources. They point out that the so-called type I radio sources might suffice to explain the observed intensities. The distribution of these sources might then be like that of a disk population II, such as shown by planetary nebulae for example.

WESTERHOUT and OORT[2] had previously indicated that the distribution of radio sources is like that of the integrated mass of the stars. This again might indicate that they belong to an intermediate population II. UNSÖLD[3] and others have suggested that the radiation might be produced by high eruptive activity in faint dwarf stars. The failure to detect any near-by dwarf as radio emitter, as well as the difficulty to account for the galactic "corona" (cf. Sect. 11) have made this idea less attractive, and rather suggest that the radiation originates in interstellar space.

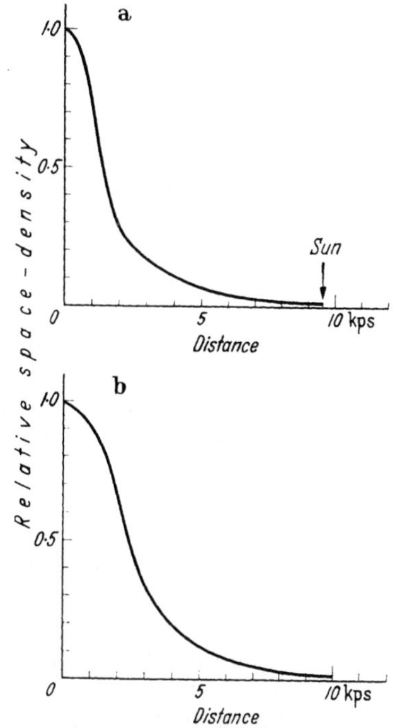

Fig. 15 a and b. Space distribution of the sources of the radiation near the galactic plane, according to HANBURY BROWN and HAZARD. (a) Gives the variation of space density in the galactic plane with distance from the galactic centre; (b) shows the variation along a line through the galactic centre normal to the galactic plane.

The evidence that the radiation discussed is due to type I radio sources is not convincing either, as it involves a considerable extrapolation from the region near the Sun where the source density is low. Moreover, observations of other galaxies give no evidence that radio emission is primarily related with population II; on the contrary, the galaxies in which population I dominates show relatively stronger radio emission. Another difficulty is in the striking irregularities of Figs. 13 and 14 at latitudes above 30°, especially the large protrusion of the northern isophotes around 0° longitude. These do not fit into the picture of a population II distribution.

[1] GROTE REBER: Proc. Inst. Radio Engrs. **36**, 1215 (1948).
[2] G. WESTERHOUT and J. H. OORT: Bull. Astr. Inst. Netherl. **11**, 323 (1951).
[3] A. UNSÖLD: Z. Physik **141**, 70 (1955).

The alternative possibility that the non-thermal radiation of strong galactic concentration as well as most of the radiation coming from higher latitudes originates largely in high-energy electrons in interstellar magnetic fields should be seriously considered. The curves in Fig. 15 might then reflect the distribution of the average field strength.

Leaving out the isotropic component, the total radiation from the Galactic System may be estimated at 3.4 watt/steradian per c/s at 100 Mc/s. The isotropic component gives roughly 1.5 times as much; it is uncertain which fraction of this is connected with the Galactic System[1].

Summarizing the somewhat confused picture given by the incomplete data available at the time of writing, we may state that, beside the thermal radiation described in Sect. 9, there is a considerable amount of non-thermal radiation, increasing in intensity towards the longer wavelengths. For wavelengths of 20 cm and less the latter component may be comparable to, or even smaller than the thermal emission; at metre waves it certainly preponderates. The non-thermal radiation consists of at least two components, one of which is strongly concentrated to the galactic plane, the other having a nearly isotropic distribution. The former shows a fairly strong concentration towards the galactic centre; it is still unknown whether its galactic concentration is quite as strong as that of the thermal radiation. It is also unknown whether there is, in addition, a source of radiation having a distribution intermediate between the thin-layer distribution of the interstellar gas and the nearly spherical distribution of the isotropic component.

11. Corona of radio emission around the Galactic System. The smallness of the variation of the 100- and 200-Mc/s radiation above 30° latitude with either latitude or longitude, indicates that the volume in which this radiation is generated must have an almost spherical shape, with a radius considerably larger than the distance from the centre to the Sun. The lack of variation with longitude suggests that, at least for $R > 4$ kpc, there is no important concentration towards the centre. For a first, tentative, investigation we shall, therefore, suppose that a fairly large ellipsoidal volume enveloping the Galactic System is homogeneously filled with sources of this radiation. Assuming a certain value for the extra-galactic radiation we can then derive the radius R_i in the galactic plane and the axial ratio. From his survey at 3.7 m wavelength BALDWIN[2] obtained the following values:

Table 5.

Adopted extra-galactic brightness temperature	R_i (kpc)	σ (watt ster^{-1} (c/s)$^{-1}$ pc^{-3})	min. axial ratio
0	16	1.8×10^8	0.7
250	13.5	1.8×10^8	0.6
500	11	1.8×10^8	0.5

He showed that the extra-galactic radiation cannot be larger than 500° and adduced some, quite uncertain, evidence that it would be at least 200°. σ gives the radiation which must be generated per pc^3.

---

[1] The numbers are from the article by HANBURY BROWN and HAZARD [*1*].
[2] J. E. BALDWIN: Monthly Notices Roy. Astronom. Soc. London **115**, 690 (1955).

Corroborating evidence for the existence of a large, more or less homogeneous corona has been found from the Andromeda nebula, where HANBURY, BROWN and HAZARD[1], BALDWIN[2], KRAUS[3] and SEEGER *et al.*[4] observed a similar aura at wavelengths of 1.89, 3.7, 1.2 and 0.75 m, respectively.

As there are no known stellar or nebulous objects in the Galactic System which show a distribution in the least like that found for the galactic radio corona, it is unlikely that it is made up of discrete sources. An attractive hypothesis has been advanced by SHKLOVSKY[5] and by PIKELNER[6]. They suggest that the corona is due to synchrotron radiation emitted by electrons moving in weak magnetic fields. The latter would be carried to large distances from the galactic plane by high-velocity gas clouds. The theory has recently been discussed also by BURBIDGE[7]. SPITZER[8] has suggested that the corona would consist mainly of gas at a temperature of the order of a million degrees. But the radiation would always have to be provided by a small number of relativistic electrons mixed with the ordinary high-temperature gas.

Using the formulae given in Sect. 5 we can make the following rough estimate of the number and energy of the relativistic electrons needed to yield the radiation given by BALDWIN. We shall make the calculations for two assumed values of $H_\perp$, as indicated in Table 6. From Table 5 we find $\sigma = 7 \times 10^{-40}$ erg/sec per c/s per cm^3. We suppose that for the electrons responsible for the radiation at 3.7 m ($\nu = 8 \times 10^7$) ν_c is equal to ν. Calculating E_{Gev} from (5.7) we obtain the values given in Table 6. From formula (5.6) and Table 2 we find, putting $\nu = \nu_c$, $P(\nu) = 1.6 \times 10^{-22} H_\perp$ for one electron. Equating this to the radiation found by BALDWIN we get the numbers of electrons per cm^3 as given in the column n. The following two columns give, respectively, the energy density of the magnetic field and that of the electrons required for the radiation.

Table 6.

$H_\perp$ (gauss)	E_{Gev}	n (cm^{-3})	E_H (erg/cm^3)	nE (erg/cm^3)	$T_{\frac{1}{2}}$ (years)
10^{-6}	2.2	4×10^{-12}	4×10^{-14}	1.4×10^{-14}	4×10^9
10^{-7}	7.0	40×10^{-12}	0.04×10^{-14}	44×10^{-14}	120×10^9

It should be noted that the true energy density of fast particles is likely to be rather higher than indicated. In the first place the particles considered can give only a small fraction of the radiation observed at higher frequencies, as will be evident from Table 2 and Fig. 1, so that we shall have to add comparable numbers of particles with higher energies. In the second place the electrons are probably accompanied by high-energy protons. For it is plausible to assume that there is a connection between the high-energy electrons, to which

[1] R. HANBURY, BROWN and C. HAZARD: Monthly Notices Roy. Astronom. Soc. London **111**, 357 (1951).
[2] J. E.. BALDWIN: Nature, Lond. **174**, 320 (1954).
[3] J. D.. KRAUS: Nature, Lond. **175**, 502 (1955).
[4] CH. L. SEEGER, G. WESTERHOUT and R. G. CONWAY: Astrophys. Journ. **126**, 585 (1957).
[5] I. S.. SHKLOVSKY: Cf. footnote 1 on p. 122.
[6] S. P. PIKELNER: Dokl. Akad. Nauk USSR. **88**, 229 (1953). See also V. GINSBERG: Uspekhi Phys. Nauk **51**, 343 (1953).
[7] G. R. BURBIDGE: Astrophys. Journ. **123**, 178 (1956).
[8] L. SPITZER jr.: Astrophys. Journ. **124**, 20 (1956).

the radiation of the galactic corona is ascribed, and cosmic rays. In the primary cosmic rays received on the Earth protons are known greatly to outnumber electrons of the same energies.

The last column of the table gives the times in which the energy of the electrons would be reduced by radiation to half the value given in the second column.

General references.

[1] Brown, R. Hanbury and C. Hazard: A Model of the Radio-Frequency Radiation from the Galaxy. Phil. Mag. **44**, 939 (1953).
[2] Hulst, H. C. van de, C. A. Muller and J. H. Oort: The Spiral Structure of the Outer Part of the Galactic System Derived from the Hydrogen Emission at 21 cm Wave Length. Bull. Astr. Inst. Netherl. **12**, 117 (1954).
[3] Kwee, K. K., C. A. Muller and G. Westerhout: The Rotation of the Inner Parts of the Galactic System. Bull. Astr. Inst. Netherl. **12**, 211 (1954).
[4] Hulst, H. C. van de, C. A. Muller, A. Ollongren, E. Raimond, M. Schmidt and G. Westerhout: Galactic Structure Determined from 21-cm Observations. Bull. Astr. Inst. Netherl. **13**, 151 (1957).
[5] Schmidt, M.: A Model of the Distribution of Mass in the Galactic System. Bull. Astr. Inst. Netherl. **13**, 15 (1956).
[6] Oort, J. H., F. G. Kerr and G. Westerhout: The Galactic System as a Spiral Nebula. Monthly Notices Roy. Astronom. Soc. London **118**, 379 (1958).

Star Clusters.

By

Helen Sawyer Hogg.

With 29 Figures.

A. Introduction.

1. Scope of article. Star clusters, studied for three centuries, have in the last decade assumed ever-increasing importance for several reasons. The emphasis on the two types of stellar populations puts them in the spotlight since galactic clusters represent population I and globular clusters population II. The increasing development of theories of stellar evolution rests on star clusters, which furnish a group of stars presumably created from the same material at about the same time and hence offer a chance to study evolutionary tracks of stars. Stellar associations, too, groups of stars spread loosely over a large area, are a new and important field of investigation.

Galactic clusters, groups of dozens or hundreds of stars, seem to form the backbone of the Milky Way, whereas the great globular clusters, systems of thousands to hundreds of thousands of stars, act as beacon lights to indicate the overall dimensions of our galaxy in all directions. Indeed, some globular clusters surrounding the galaxy are found at such great distances that they are actually intergalactic objects.

In this article, we shall deal with the observed characteristics of the galactic and globular clusters associated with our galaxy, and with some of the more important theoretical deductions from these. We may consider that the general subject of star clusters in our galaxy is bounded on one end by stellar associations and sparse groups of stars projected against such a rich Milky Way background that one can hardly be sure a definite cluster exists. It is bounded on the other end by globular clusters which belong to other galaxies or which are intergalactic, and by the problem of a possible sequence between the globular clusters and the elliptical galaxies.

Between these limits then, we will set down a few of the more important observations and ideas from the original papers in this field. Since these papers now total about three thousand, it is impossible within the limits of a short article to do justice to all the ideas and information, or to all the astronomers whose efforts have secured this knowledge.

The reader may refer to extensive bibliographies given as follows: for all clusters up to 1930 by H. Shapley [*1*]; for galactic clusters to 1931 by P. Collinder [*2*]; for globular clusters to 1947 by the writer [*3*]; and recently, the most complete bibliography of all for individual galactic and globular clusters and associations, from 1900 to 1957, by G. Alter, J. Ruprecht and V. Vanýsek [*4*].

2. Pre-twentieth-century work. For thousands of years keen-eyed observers who looked at the heavens have been conscious of the grouping of stars we call open or galactic clusters. References to the most conspicuous group, the Pleiades (Fig. 1) are very numerous, including the well-known one in the Book of Job.

The telescope started the study of open clusters when Galileo dissolved the cloud of Praesepe into what he described as "a multitude of more than forty small stars". And it led to the discovery of globular clusters when Hevelius noticed Messier 22 before 1665, Halley discovered ω Centauri in 1679 and Messier 13 in 1715, and KIRCH found Messier 5 in 1702.

Fig. 1. The Pleiades. Lick Observatory photograph, 2^h exposure with Carnegie astrograph by VASILEVSKIS, Feb. 9, 1951.

The first realization of the great distances of star clusters came in the middle of the 18th century when JOHN MICHELL of England pioneered in stellar statistics. By using the calculus of probability and the laws connecting distance and brightness of a light source, he deduced parallaxes for the stars. He derived a distance for the Pleiades by two different assumptions; first, that their mean absolute brightness is the same as the Sun's, which gives a distance of 170 light years; and second, that their mean mutual distance is the same as for stars in the Sun's neighbourhood, giving a distance of 320 light years, not far from the present value.

The observational astronomers of the latter part of the 18th and of the 19th centuries succeeded in cataloguing most of the brighter clusters we know today. These included MESSIER and MÉCHAIN, WILLIAM and JOHN HERSCHEL, Lord

ROSSE, and JAMES DUNLOP. Sir WILLIAM HERSCHEL made notable contributions to the knowledge of the form and extent of the galaxy from his studies of star clusters. The observational skill of some of these observers was phenomenal. They noted objects which even now are hardly more than visible in some of the world's largest telescopes. After many years, their efforts were coordinated by Sir JOHN DREYER in his New General Catalogue, with its supplements, the First and Second Index Catalogue [*5*]. There are still objects in these lists for which it has not yet been possible to assign a definite classification of globular or galactic cluster.

With the application of photography in the closing years of the last century, research on star clusters bounded forward. At Harvard, E. C. PICKERING[1] reported on the spectra of 1003 stars in galactic clusters, the Pleiades, Praesepe, Carina, Coma Berenices, and others, finding that *A*-type stars predominated. Simultaneously, astronomers were making star counts in the rich globular clusters, and at Harvard, S. I. BAILEY[2] found variable stars by the hundreds in some of the globular clusters.

3. Early 20th century work. The massive attack on globular clusters by HARLOW SHAPLEY [*1*], [*6*] with the powerful equipment on Mt. Wilson led to the recognition of the importance of these objects, both as individual systems and as outliers of our galactic system. By applying ingenious methods of deriving their distances from integrated magnitudes, apparent diameters, magnitudes of the 25 brightest stars or variables for 69 globular clusters, SHAPLEY was able to show the great overall extent of the galaxy, and that the distant center was in Sagittarius, around which a third of the globular clusters are congregated. SHAPLEY's researches on galactic clusters also helped to fill in the picture of galactic structure. His last catalogues [*6*] listed 93 globular clusters and 249 galactic, from which he derived a value of 70 kpc for the diameter of the galaxy in the galactic plane.

Almost simultaneously, the forefront of the attack on galactic clusters was being carried out by R. J. TRUMPLER, first at Allegheny[3] and later at Lick, where he produced his important work [*7*] on the distances of 100 open clusters from the magnitudes and spectral types of stars, with a catalogue of 334 clusters. From this study of the relation between diameters and distances of clusters, TRUMPLER demonstrated the important fact that there is absorption of light in our galaxy. While the absorption may be present in all galactic latitudes, it takes place mainly in a thin layer along the galactic plane, but the distribution of absorbing material is by no means as uniform as the blanket correction of 0.67 magnitude per 1000 psc derived by TRUMPLER would indicate.

The open clusters form a flattened disk-like system 1000 psc thick with a diameter about 10000 psc. The plane of symmetry is inclined $2\overset{\circ}{.}3$ to the galactic plane with a pole at R.A. $12^h 50\overset{m}{.}4$, Decl. $+27\overset{\circ}{.}7$ (1950). The system is strongly concentrated towards a point 350 psc away in galactic longitude 247°, near the rich open cluster NGC 3532. TRUMPLER gave a diagram pointing out that the space distribution of open clusters outlines spiral arms, supporting the hypothesis that our galaxy is a highly resolved spiral.

Other astronomers contributed to the progress of the study of star clusters. The catalogue of clusters by P. J. MELOTTE [*8*] from the Franklin-Adams charts yielded diameters used as distance indicators by SHAPLEY, and by C. V. L. CHARLIER for open clusters. S. RAAB [*9*] at Lund in 1922 first used the method of

[1] E.C. PICKERING: Harvard Obs. Ann. **26**, 260 (1897).
[2] S.I. BAILEY: Harvard Obs. Ann. **38** (1902).
[3] R.J. TRUMPLER: Allegheny Obs. Publ. **6**, 45 (1922).

fitting the H-R diagram of the cluster to that of the solar neighbourhood to derive distances, while A. WALLENQUIST[1] in a series of papers strengthened the spectroscopic parallaxes, especially for the southern stars, through the use of special criteria of absolute magnitude.

B. Galactic clusters.

I. Appearance and apparent distribution.

4. Description and distribution. In contrast to the symmetrical globular clusters, most galactic clusters are somewhat ragged and irregular groups of stars, often made very beautiful in a telescope by the color contrast presented

Table 1. *Galactic clusters in high latitudes (more than 20° from galactic equator).*

Name	R.A. 1950	Decl. 1950	λ	β
Blanco 1	00ʰ01ᵐ.7	−30°13′	341°9	−80°.7
NGC 189	35 9	+60 47	89 9	+22 8
Coll 21	01 47 3	+27 00	107 3	−33 1
NGC 752	54 8	+37 26	105 4	−22 7
Pleiades	03 44 5	+23 57	134 7	−22 2
Taurus	04 17	+15 30	147 0	−22 4
Coll 464	05 16 4	+73 14	107 0	+20 5
NGC 2314	07 03 8	+75 19	106 3	+28 1
NGC 2420	35 5	+21 41	165 8	+21 1
Praesepe	08 37 2	+20 10	173 3	+34 0
NGC 2682	47 8	+12 00	183 6	+33 3
Coma	12 22 5	+26 23	195 6	+85 2
Ursa Major	14 36 7	+69 47	76 5	+44 8
Mel 227	20 7 0	−79 24	281 4	−31 0
NGC 6994	56 2	−12 49	4 0	−35 6

by bright red and blue stars. Probably half the known galactic clusters contain fewer than a hundred stars, but the richest, like h and χ Persei may have a thousand. ALTER[2] has given an effective definition of a galactic cluster as a "star congregation with closer gravitational relation than that of the surrounding star field".

Table 2. *Appearance of Coma Berenices cluster at different distances.*

Distance (psc)	Δm (PG)	Angular diameter	Brightest cluster star	$m_{pg}<18$			$m_{pg}<16$			$m_{pg}<14$		
				Cluster stars	Back-ground stars	%	Cluster stars	Back-ground stars	%	Cluster stars	Back-ground stars	%
1000	6.4	18′	11.6	31	700	4	23	147	14	12	26	32
2000	8.7	9	13.9	21	180	10	11	37	23	3	6	....
3000	10.4	6	15.6	12	78	13	5	16	24	....	....	....
4000	11.8	4.5	17.0	5	44	10	....	....	....	....	....	....

It has long been noted that the galactic clusters show a strong concentration to the plane of the galaxy, except for those like the Coma Berenices cluster, which are situated so near us that they appear projected in high galactic latitudes and a few other unusual objects (Table 1). The distribution of the galactic clusters catalogued in Appendix A is shown in an Aitoff projection in Fig. 2. With increasing distance, galactic clusters shrink to a size where they cannot be detected

[1] A. WALLENQUIST: Upsala Medd. **32** (1927); **42** (1929); **56** (1932); Bosscha Ann. III, **2** (1929); III, **3** (1931).

[2] G. ALTER: Monthly Notices Roy. Astronom. Soc. London **102**, 205 (1942).

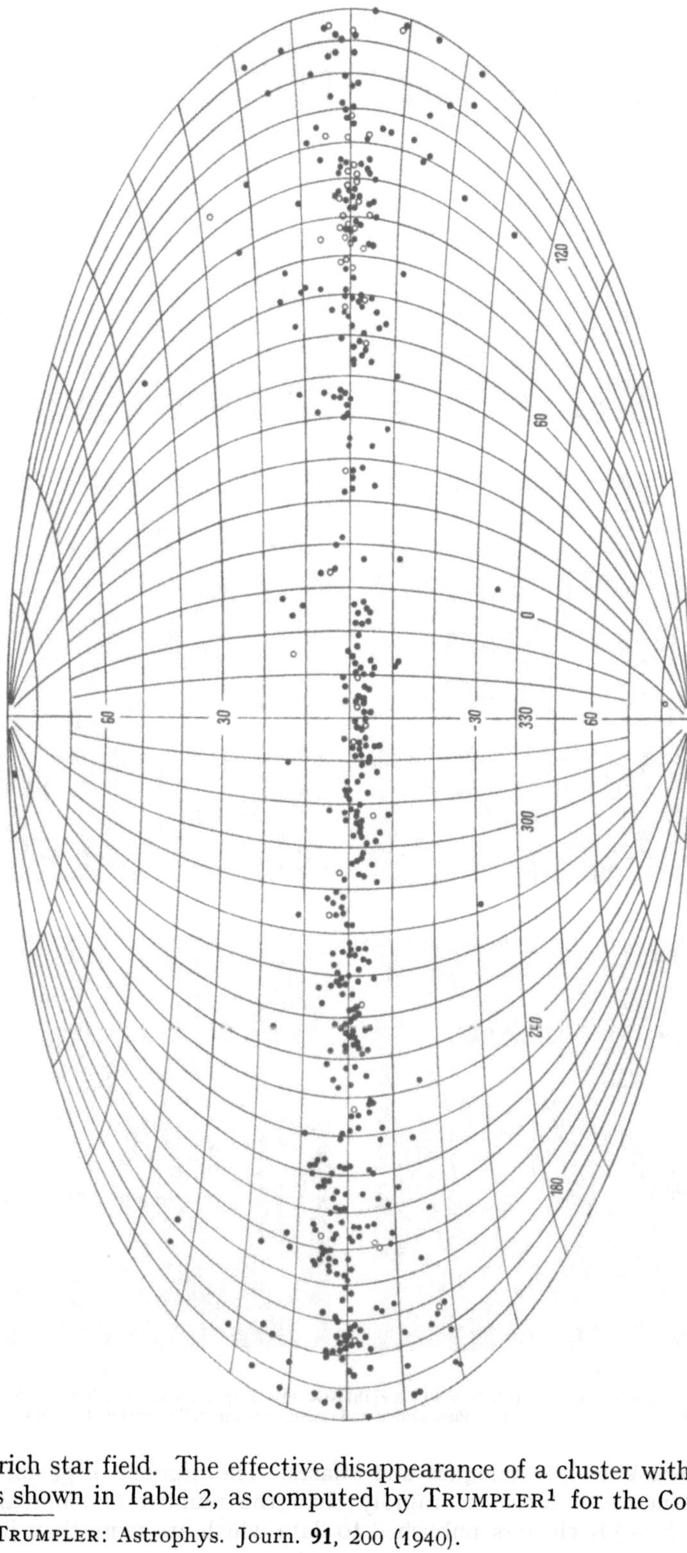

Fig. 2. Distributin of galactic clusters in galactic coordinates. Open circles are clusters added since COLLINDER's Catalogue.

against a rich star field. The effective disappearance of a cluster with increasing distance is shown in Table 2, as computed by TRUMPLER[1] for the Coma cluster.

[1] R. J. TRUMPLER: Astrophys. Journ. **91**, 200 (1940).

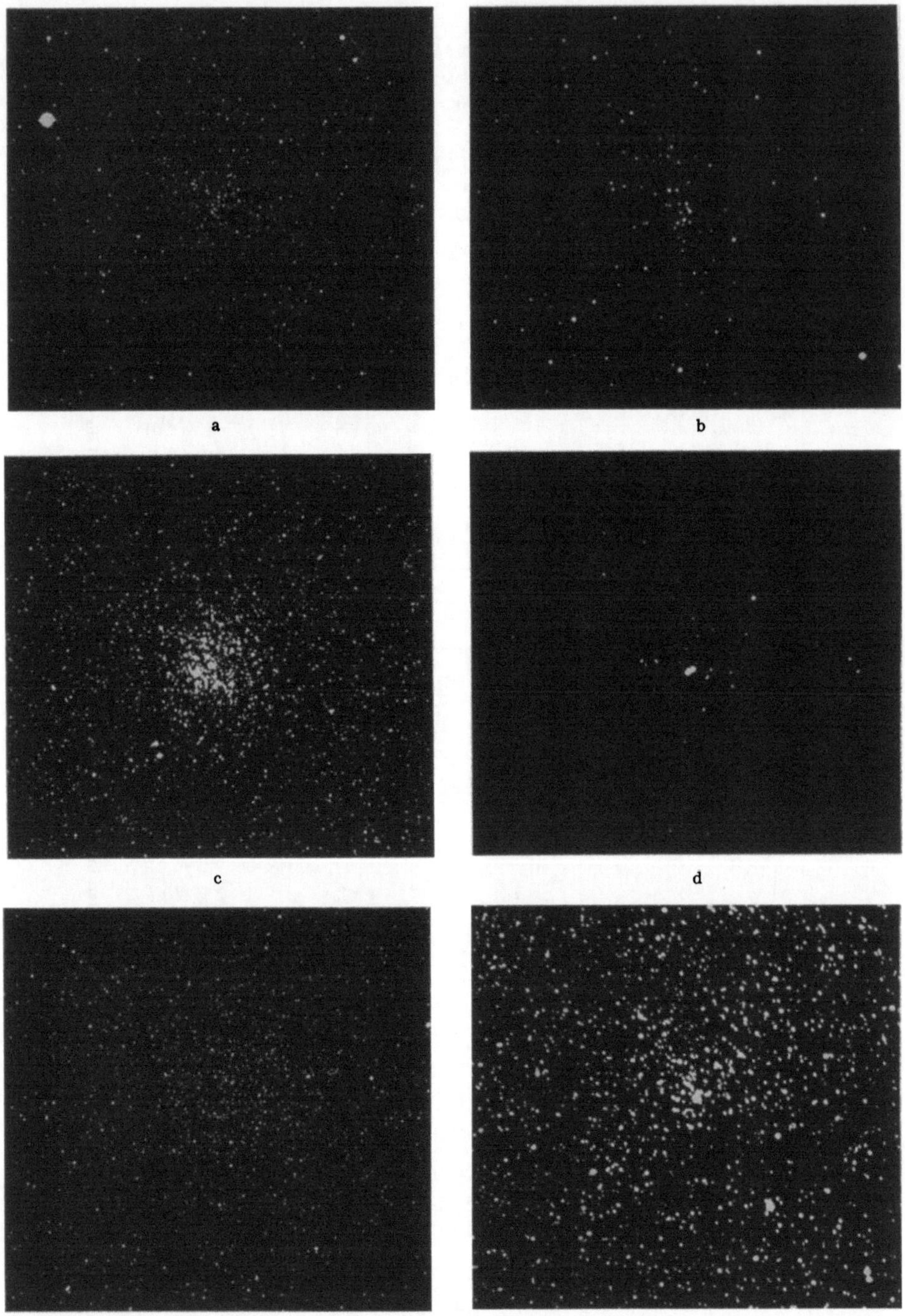

Fig. 3a–f. Typical galactic clusters. (a) NGC 6819, I 2 *r*; (b) NGC 2254, I 2*p*; (c) NGC 6705, II 2*r*; (d) NGC 1502, II 3*p*; (e) NGC 7789, III 1*r*; (f) NGC 1027, IV 3*m*. Photographs and classification from TRUMPLER, Lick Obs. Bull. 420 (1930).

5. Numbers known, catalogues and atlases. The number of galactic clusters known has increased far more rapidly than the number of globular clusters. Appendix A lists 514 clusters published to date which are more than just several stars together. It is based principally on COLLINDER'S catalogue [2], which was subsequent to those of SHAPLEY and TRUMPLER, with more recent items from various observers, some of which were kindly supplied by ALTER in advance of publication of his catalogue. For photographs of many representative clusters, the reader is referred to the numerous plates in RAAB'S paper [9], COLLINDER'S catalogue, and the excellent recent atlas by B. E. MARKARYAN [10].

6. Classification. It is difficult to look at many galactic clusters without trying to classify them by their appearance, and since the dimensions of a cluster are governed by its gravitational potential and are related to its constitution, a classification on the basis of appearance has significance. From a number of such classifications, that of TRUMPLER has been most widely used. He employed

Table 3. *Galactic clusters. Mean linear diameters of subgroups.*

Subgroup	No. of clusters	Mean diameter psc
I*p*	7	3.0
m	4	4.0
r	9	4.4
II*p*	13	3.4
m	11	4.4
r	9	4.8
III*p*	6	4.5
m	5	5.9
r	2	6.5
IV*p*		7.9
m	7	10.3
r	3	11.3

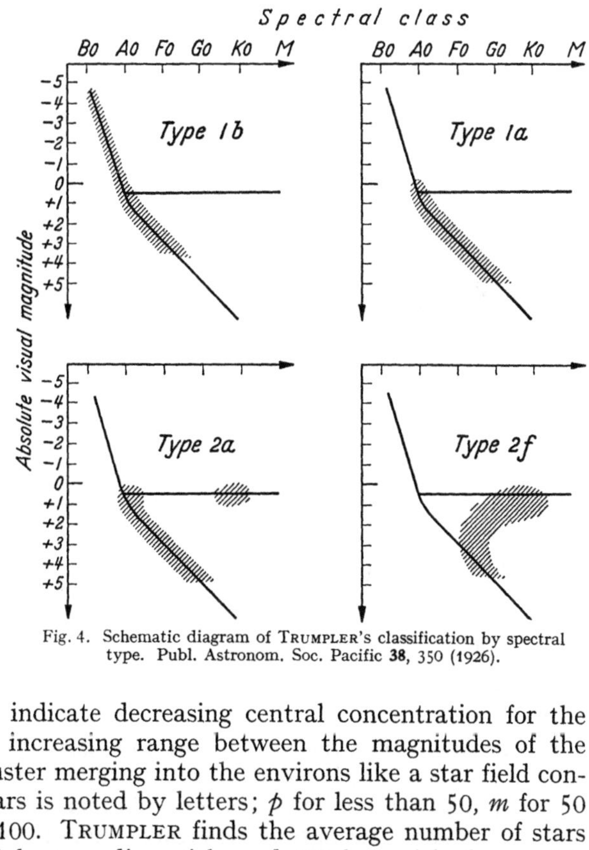

Fig. 4. Schematic diagram of TRUMPLER'S classification by spectral type. Publ. Astronom. Soc. Pacific **38**, 350 (1926).

Roman numerals I to IV to indicate decreasing central concentration for the cluster, numerals 1 to 3 for increasing range between the magnitudes of the cluster stars. Class IV is a cluster merging into the environs like a star field condensation. The number of stars is noted by letters; *p* for less than 50, *m* for 50 to 100, and *r* for more than 100. TRUMPLER finds the average number of stars to be 35 for a poor cluster, 75 for a medium rich, and 200 for a rich cluster. His determination of average linear diameter for each type of cluster is given in Table 3. Collinder noted four pronounced types of clusters. The Praesepe type is concentrated and regular with a continuous range of magnitudes, while the Pleiades type is relatively coarse and irregular with a considerable difference in magnitude. The third type is a cluster with nebulosity, while the fourth, μ Normae, is a bright star closely surrounded by faint stars with pronounced magnitude difference.

The most significant classification of galactic clusters, however, is one which depends upon the array of spectra of the individual stars. Classification on this basis was used by Mrs. FLEMING in early Harvard work for the seven conspicuous clusters, the Pleiades, Praesepe, Coma Berenices, IC 2602, NGC 3532, M 6 and M 7. Particularly noticeable is the difference between the Pleiades and Hyades. In the Pleiades, the brightest stars are type *B*, and no bright yellow stars are present. In the Hyades, there are no B-type stars; the bluest stars are of type A, and a few bright K stars occur. The spectral classification devised by TRUMPLER[1] is shown schematically in Fig. 4 and is defined as follows:

Type 1b. Brightest cluster stars are $B0$ to $B5$, including occasionally an O-type star. Fainter stars crowd around dwarf branch. Giants $A5$ to M are missing or rare. Examples, h and χ Persei, M 36, M 35, Pleiades.

Type 1a. Brightest cluster stars of class $B8$ to $A5$; fainter stars follow dwarf branch closely. No stars of type $B0$ to $B5$ present, and none (or very few) which are giants. Examples, M 34, M 39, NGC 1647.

Type 2a. Brightest cluster stars scattered along giant branch from $B8$ to K; fainter stars follow dwarf branch from $B8$ toward F and G. Giant branch not uniform but concentrated around class A. Often marked gap in giant branch between A and G makes $G5$ to $K0$ giants appear isolated. Dwarf branch continuous. Examples, Melotte 210, M 37, M 11, Praesepe, Taurus cluster, Coma Berenices.

Type 2f. No stars of classes B or A present. Brighter stars follow giant branch from K to F; fainter stars follow dwarf branch from $F0$ to G. NGC 752 is a representative of this rather scarce class.

Type 1o was later introduced for clusters with many O-type stars, and $3a$ for those with many red or yellow giants, but very few A-type.

Table 4. *Frequency of different types of magnitude-spectral class diagrams.*

Class	*o*	*b*	*b−a*	*a*	*a−f*	*f*	Total
1	7	24	5	3			39
1−2	3	15	10	3			31
2		1	5	18	1	1	26
2−3				3			3
3				1			1
No.	10	40	20	28	1	1	100

Obviously the age and evolution of the cluster determine the type of diagram. Trumpler classified 100 clusters for which spectral types were available [7] and the observed frequency of different types is given in Table 4.

Recently B. E. MARKARYAN[2] from extensive work on galactic clusters has devised a new scheme of classification, with the emphasis on the spectra of the brightest stars, whether O, B or A. The O group has a three-way division into clusters with nuclear condensations, star chains, and gaseous nebulosity.

It is obvious that a spectrum or color luminosity diagram is the most important tool for appraising a galactic cluster. For the more prominent clusters, these diagrams have been determined many times over, but for most clusters they are still completely lacking, despite the concentrated efforts of the observers mentioned, and many others, including especially G. ALTER[3], who studied a large number of clusters.

7. Integrated magnitudes. The total integrated magnitudes of galactic clusters range from those bright enough to be visible to the unaided eye down to those barely detectable on long-exposure photographs. Various investigators have made systematic attempts to determine them for as many clusters as possible[4]. Probably the most consistent set for the greatest number of clusters is that of LUNDMARK and COLLINDER [2]. These observers made independent estimates for all the clusters in their catalogue; the mean value of these estimates is given in Appendix A.

Although at first glance the total magnitude of a galactic cluster might not appear significant, because of the wide range in number of stars and the absolute magnitudes of the brightest stars, nevertheless recent work by I. M. KOPYLOW[5]

[1] R. J. TRUMPLER: Publ. Astronom. Soc. Pacific **38**, 350 (1926).

[2] B.E. MARKARYAN: Mitt. Obs. Bjurakan No. 5 (1950).

[3] G. ALTER: Monthly Notices Roy. Astronom. Soc. London **100, 101, 102, 103, 104** (1940—1944).

[4] J. HOLETSCHEK: Vienna, Ann. der K. K. Univ. Sternw. **20**, 40 (1907). — B. VORONTSOV-VELYAMINOV: Astronom. Nachr. **226**, 195 (1926); **228**, 325 (1926). — M. NABOKOV: Astronom. Nachr. **241**, 289 (1931).

[5] I.M. KOPYLOW: Mitt. Astrophys. Obs. Krim. **8**, 122 (1952).

shows an important correlation between the absolute magnitude of the cluster and the spectral type of its brightest stars. Using mainly the magnitudes of LUNDMARK and COLLINDER, spectral types of SANFORD, TRUMPLER and ZUG, with the best parallaxes available, KOPYLOW derived the absolute integrated magnitudes for 300 clusters. The maximum of the distribution function is around absolute magnitude -3.5 (Fig. 5). This is about four magnitudes fainter than the maximum frequency for globular clusters. Among the most luminous clusters in KOPYLOW's table are h and χ Persei with M equal to -8.0 and -8.1 respectively.

8. Angular diameters. The angular diameters of galactic clusters have been measured by a number of observers, and serve as distance indicators. There is much greater uncertainty in them than in the diameters of globular clusters. From a central nucleus, the stars of galactic clusters merge into the rich background of the MILKY WAY. Nevertheless, the diameters have proved useful. CHARLIER [*11*] in 1918, using MELOTTE's diameters as distance indicators, determined the scale of the stellar system by identifying the positions in space of the centers of the two systems of open clusters on one side and the extended cluster of B stars on the other. SHAPLEY in 1919[1], from the parallaxes of 14 clusters on the basis of observed colors or spectra, established a parallax-diameter relation from which parallaxes for 56 other clusters were obtained. RAAB [*9*], by using material of the Henry Draper catalogue derived photometric parallaxes by the Kapteyn integral formula for 46 open clusters, from which he derived a diameter parallax correlation. RAAB determined the diameter by a star-count method of CHARLIER, using the radial dispersion in density taken five to six times as an equivalent of the diameter. TRUMPLER's correlation of linear diameter with class of cluster has been noted (Table 4). The diameters determined by LUNDMARK and COLLINDER [*2*] from FRANKLIN ADAMS charts seem to be the most consistent set available and hence have been used for most of the clusters in Appendix A. They differ systematically from TRUMPLER's in that for apparent diameters between 10′ and 4′ they are larger, and below 4′ they are smaller.

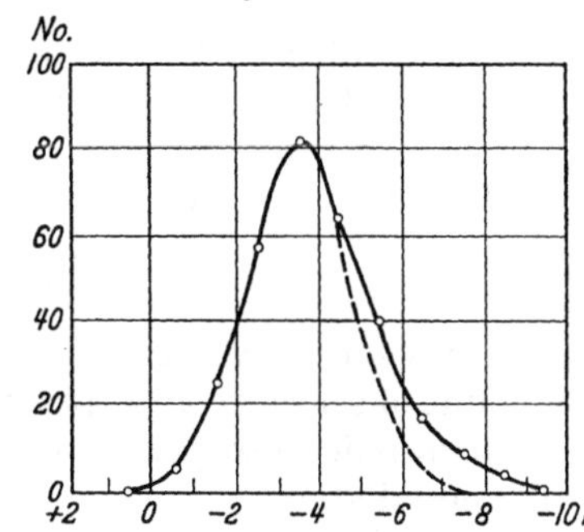

Fig. 5. Frequency distribution of absolute magnitudes of galactic clusters by KOPYLOW. Mitt. Astrophys. Obs. Krim. **8**, 122 (1952).

The largest open clusters have diameters of several degrees, Taurus 400′, Perseus 240′, Pleiades 120′. In linear diameters the largest are not over 15 psc, the smallest about 1.5 psc, while the majority range from 2 to 6 psc with the maximum at 2.5 psc. Most clusters have a flattening within narrow limits.

9. Absorption of light in space. The observations of TRUMPLER [*7*], showing that the linear diameters of galactic clusters appeared to increase with distance, led him to the important conclusion that the greater distances were estimated systematically too large because of absorption of light in the MILKY WAY. On the assumption that clusters of the same constitution have the same linear diameter, the distance r (in psc) is obtained by

$$r = 3438 \frac{D_c}{d} \tag{9.1}$$

where D_c is the linear diameter in psc and d is the angular diameter in minutes of arc.

[1] H. SHAPLEY: Mt. Wilson Obs. Comm. No. 62 (1919).

Shortly thereafter, R. S. ZUG[1] showed that the selective absorption had a pronounced dependence on galactic longitude, while the color work of STEBBINS and WHITFORD[2] on globular clusters showed the spottiness of the absorption. A blanket correction cannot be made; rather the selective absorption must be determined for each cluster or region individually, the ratio for the total to selective absorption must be known, and a corresponding correction applied to the distance.

A recent discussion of various determinations with new data by W. A. HILTNER and H. L. JOHNSON[3] gives the ratio of total visual absorption (A_V) to the selective (color-excess on B-V standard), as

$$R = \frac{A_V}{E_{(B-V)}} = 3.0 \pm 0.2\,(\text{p.e.})\,. \tag{9.2}$$

The ratio is higher for the total photographic absorption (A_{pg}), at least 4.0 ± 0.2, with some determinations as high as 5 or 6.

The same workers give the equation for the mean reddening line on the standard UBV system as

$$\frac{E_{(U-B)}}{E_{(B-V)}} = 0.72 + 0.05\,E_{(B-V)}\,. \tag{9.3}$$

It is obvious how greatly the true distances of objects may be affected by the amount of selective absorption, and by the factor used to convert this to the total photographic absorption to be applied to the measured apparent magnitudes of individual stars.

II. Methods of distance determination.

10. Geometric methods. This section considers in a general way some of the various methods of attacking the problem of distance determination. The distance measures fall into two general groups, geometric measures involving precise positions, and photometric and spectroscopic measures involving absolute magnitude of stars. Most of these methods are considered in greater detail, with references, under the sections in which they are applicable.

The geometric methods include:

Direct trigonometric parallaxes. Unfortunately, practically all the galactic clusters are so distant that these cannot be used. Several of the nearest clusters can be studied by this method, however.

The proper motions of individual cluster stars, as in the Hyades, provide a reliable distance indicator.

Some galactic clusters, like the Pleiades, contain binary stars whose orbits are well observed. These offer a chance for a dynamical parallax determination.

The angular diameters may be used to determine distances, on the assumption that clusters of the same constitution have the same linear diameter (Sect. 8).

For clusters with radial velocity measures available, the differential galactic rotation effect may be applied to determine a distance scale.

A variation of the foregoing is the use of the velocity of the interstellar lines in the spectra of cluster stars, early applied by C. S. BEALS[4] to the Perseus and Cygnus clusters.

---

[1] R. S. ZUG: Lick Obs. Bull. No. 454 (1932).
[2] J. E. STEBBINS and A. E. WHITFORD: Contrib. Mt. Wilson Obs. No. 547 (1936).
[3] W. A. HILTNER and H. L. JOHNSON: Astrophys. Journ. **124**, 367 (1956).
[4] C. S. BEALS: Monthly Notices Roy. Astronom. Soc. London **96**, 669 (1936).

11. Photometric or spectroscopic methods. For these to be reliable, some way of knowing or eliminating the effect of interstellar absorption must be found. W. BECKER has derived a method of three-color photometry which overcomes the effects of interstellar absorption and derives a distance independent of it (Sect. 19).

These methods include:

The total integrated magnitude. This is a very rough distance indicator because of the wide range in absolute magnitude of clusters, and the influence of absorption.

The apparent magnitudes of the brightest stars are a rough distance indicator, but with the same limitations.

The color-luminosity or spectrum-luminosity diagrams for clusters provide the distance criterion most commonly used at present. The distance modulus of the cluster, y, is given by

$$y = \overline{m}_s - \overline{M}_s \tag{11.1}$$

where m_s and M_s are apparent and absolute magnitudes of the cluster stars, and the mean value for all cluster stars is adopted.

The distance in psc, r, is then given by

$$y = 5 \log r - 5 + A(r) \tag{11.2}$$

where $A(r)$ is the amount of interstellar absorption affecting the light of a star at distance r.

The use of such a diagram depends upon a standard main sequence from which to derive the absolute magnitudes (Sect. 20).

Table 5. *Distances of some galactic clusters.*

Cluster	True modulus (mag.)		Adopted modulus (mag.)	True distance (psc)
	from B−V	from U−B		
M 39	7.0	....	7.0	250
III Cep[1]	9.4	9.3	9.3	730
h and χ Per	11.8	11.7	11.7	2250
M 34	8.2	....	8.2	440
I Per[1]	5.8	5.9	5.9	160
II Per[1]	7.6	7.8	7.7	350
Pleiades	5.5	....	5.5	126
M 36	10.6	10.4	10.5	1260
Hyades	....	....	3.0	40
NGC 2264	9.9	9.5	9.7	870
Praesepe	6.0	....	6.0	158
NGC 2244	10.9	11.2	11.1	1660
I Ori[1]	8.0	8.0	8.0	400
M 67	9.6	....	9.6	830
Coma Cluster	4.5	....	4.5	80
NGC 2362	10.8	10.8	10.8	1450
NGC 6530	11.2	10.9	11.0	1580
M 25	8.7	8.7	8.7	550
M 11	11.2	....	11.2	1740
IC 4665	7.6	7.6	7.6	330

[1] Association.

Any method of spectroscopic parallax determination may be applied to a cluster, provided it contains stars of the proper type bright enough to be studied with sufficient dispersion. Spectroscopic parallaxes from hydrogen line intensities by C.H. PAYNE and C. ANGER RIEKE showed up the clustering of supergiants around NGC 6231 in Scorpius[1].

The equivalent widths of the interstellar calcium lines have been used successfully by R.F. SANFORD[2] to derive distances in good agreement with other methods.

With the proof that some galactic clusters contain classical cepheid variables (Sect. 17), the period-luminosity relation may be a further tool for distance determination in these clusters. Because interstellar absorption is particularly heavy in the low latitudes frequented by galactic clusters, geometric methods appear to have the greatest reliability. For some clusters, the geometric and photometric values are in only fair agreement. Table 5 gives the best determined distances for twenty clusters from the work and compilation of H.L. JOHNSON[3].

III. Stellar content.

12. The main sequence. The luminosity of the brightest stars, or the starting point of the upper end of the main sequence, varies greatly from cluster to cluster, from about -8 to -2 visual magnitude. Stars of all spectral classes from *O* to *M* are found in galactic clusters. The frequency of type differs from cluster to cluster, as does the concentration to the center of different types. The red stars are concentrated to the center in M 67 [*1*], the white stars in M 36, while other clusters show no dependence[4]. This is doubtless related to the mass distribution in the cluster, and probably depends on the age of the cluster.

The absolute magnitude of the starting point of the main sequence furnishes a criterion for classification of the spectrum-luminosity diagram. The clusters with fainter luminosity in the upper end of the main sequence are increasingly older on a time scale. NGC 2352 and 2264, and h and χ Persei are very young galactic clusters, while NGC 752, with the upper end of the main sequence at $+2.2$, is very old. M 67 is comparable in age to globular clusters (Sect. 32).

13. The brightest stars. Giants appear when the upper end of the main sequence is not higher than about -0.5 absolute magnitude, around spectral type *A* 0. The existence of giants is possible only if all main-sequence stars with larger masses are absent, either because they have disappeared or because they never formed. TRUMPLER'S exclusion rule holds, that no giants are present with bright *B*-stars in a cluster, though a few late-type supergiants may occur. MICZAIKA[5] states that it takes about 5×10^8 years before ordinary giants and subdwarfs appear in an open cluster.

Supergiants are not uncommon. In the brilliant group around NGC 6231, O. STRUVE[6] determined radial velocities for 16 normal stars (six of which are spectroscopic binaries) of type *cO* 9 to *cB* 2, along with two Wolf-Rayet and two P Cygni stars, all of which belong to one physical group. He states that the tendency of galactic clusters to be surrounded by extended groups of supergiant stars is probably one of their most important structural characteristics. W.P. BIDELMAN[7] finds that later-type supergiants do not in general co-exist with early *O*-type stars. He has investigated *B*, *A* and *M* supergiants in h and χ Persei,

---

[1] C.H. PAYNE and C. J. ANGER: Harvard Obs. Circ. No. 363, 364, 365 (1931). — C. ANGER RIEKE: Harvard Obs. Circ. No. 397 (1935).

[2] R.F. SANFORD: Astrophys. Journ. **110**, 117 (1949).

[3] H.L. JOHNSON: Astrophys. Journ. **126**, 121 (1957).

[4] See footnote 1, p. 132.

[5] G. MICZAIKA: Les Processus Nucléaires dans les Astres, Louvain, 1954, p. 275.

[6] O. STRUVE: Astrophys. Journ. **100**, 189 (1944).

[7] W.P. BIDELMAN: Astronom. J. **59**, 315 (1954).

and early B supergiants in $\varkappa$ Crucis[1], while M supergiants have been noted in NGC 7419[2].

14. Some unusual types. *α) Metallic-line stars.* The presence of these in galactic clusters has been noted by H. L. WEAVER[3], who totalled the numbers in five representative clusters as follows: Pleiades, 9; Hyades, 9; Praesepe, 9; Coma Berenices, 6; M39, 1. Within these clusters, the metallic-line stars total 35% of the whole stellar population in the absolute magnitude interval from $+1.7$ to $+3.3$. This is a significantly higher percentage than SLETTEBAK found among the general population of nearby stars. Of the objects examined, only NGC 752, a cluster containing almost no stars earlier than F0, was lacking these objects. The metallic-line stars occur only in certain critical regions of the H-R diagram where the physical parameters of a star exist in a delicately balanced state. A study of 8 Comae[4] shows that high atmospheric turbulence is linked with metallicism.

β) Emission line stars. Be stars are unusually frequent in several clusters. In the Pleiades, STRUVE finds that 4 out of 13 stars of types B8 and earlier have bright lines, a much higher ratio than normal in the galaxy. TRUMPLER noted a similar tendency in the double cluster in Perseus.

15. White dwarfs. A class of stars of increasing importance in clusters of both types is the white dwarf. While at present it is observationally tedious or impossible to detect the fainter white dwarfs, the brighter ones can be found. LUYTEN'S[5] observational work indicates some ten white dwarfs in the Hyades, while theoretical work by A. R. SANDAGE [*12*] predicts the following numbers of stars which have disappeared from the main sequence by exhaustion of their energy sources, in other words, the white dwarfs: h Persei, 0; Pleiades, 2; Coma Berenices, 9; Hyades, 23; Praesepe, 20; M 67, 146. To make a complete observational search for these theoretically-present stars would require working to the following apparent magnitudes: Pleiades, 25.6; Coma, 24.5; Hyades, 23.0; Praesepe, 25.9; and M 67, 29.5. The numbers of white dwarfs already found in these clusters are in line with those predicted except that none is known in the Pleiades. BAADE has found a large number of faint blue stars in M 67. These numbers may be compared with the staggering estimate of 48500 white dwarfs in the globular cluster M 3 (Sect. 58).

16. Binary stars. Visual binaries are common in galactic clusters. From a few first noted in BURNHAM'S early list[6], the numbers have increased to include even stars which are presumed to be double because they stand off in magnitude from a precise color-magnitude diagram. O. HECKMANN[7] considers that a very high percentage of the stars in the Hyades may be binaries, even up to 70% of the brighter cluster members, while the percentage in Pleiades and Praesepe may be about 20. Data on six binaries in the Hyades[8] are given in Table 6 to show the types of binaries in galactic clusters.

---

[1] W. P. BIDELMAN: Astrophys. Journ. **105**, 492 (1947). — Publ. Astronom. Soc. Pacific **66**, 249 (1954).

[2] V. M. BLANCO, J. J. NASSAU, J. STOCK and W. WEHLAU: Astrophys. Journ. **121**, 637 (1955).

[3] H. L. WEAVER: Astronom. J. **55**, 82 (1950).

[4] G. R. MICZAIKA, F. A. FRANKLIN, A. J. DEUTSCH and J. L. GREENSTEIN: Astrophys. Journ. **124**, 134 (1956).

[5] W. J. LUYTEN: Astronom. J. **59**, 224 (1954).

[6] S. W. BURNHAM: Lick Obs. Publ. **2**, 211 (1895).

[7] O. HECKMANN: Mitt. astronom. Ges. **1952**, 20 (1953).

[8] G. P. KUIPER: Astrophys. Journ. **86**, 189 (1937).

Table 6. *Data on six binaries in the Hyades.*

ADS	m_{pv}	Sp.	Δm	P years	a
3135	6.99	*F* 6	1.12	92.2	0″.526
3169[1]	7.09	*F* 8, *G* 1	1.37	487	1 72
3210	7.70	*G* 5	0.6	30.4	0 236
3264[1]	5.73	*A* 6*n*	2.28	170	1 029
3475	6.68	*F* 7	0.0	16.34	0 1895
3483	6.34	*F* 5	2.0	116.2	0 790

[1] The primary is a spectroscopic binary with one spectrum visible.

Somewhat puzzling is the distribution of spectroscopic binaries in these clusters. With large numbers of visual binaries present, one would expect spectroscopic binaries in good proportion. This holds true for some clusters like NGC 6231, where almost half the normal stars investigated (seven out of sixteen) are spectroscopic binaries, but other clusters are curiously deficient in them. Only recently has one been found in the Pleiades and investigated by J. A. PEARCE[1]. Where spectroscopic binaries exist, there are sure to be eclipsing stars, such as HAFFNER's variable TX Cancri in Praesepe[2].

17. Variable stars. From the very nature of galactic clusters it has been much harder to get a true picture of their variable star content than in the case of globular clusters. From a superficial study, it is not possible to tell if a variable in a knot of a few dozen stars, seen against a rich star background, is a cluster member. For years, the galactic clusters have stood in marked contrast to certain of the globular clusters in which dozens of variables were found readily. It must be remembered, however, that in proportion to the total number of stars in the cluster, one cannot expect many variables in galactic clusters.

For some years it was thought that these clusters contained no variables, even though some were noted by E. E. BARNARD[3] near M 11, by E. LINDEMANN[4] near the double cluster in Perseus, and P. DOIG[5] placed T Serpentis in NGC 6633 and U Sagittarii in M 25. The final test of a variable as a cluster member must be its common motion with the cluster, and radial velocity is a better criterion than proper motion.

Recently the pattern for variable stars in galactic clusters is becoming clearer. These clusters do contain classical cepheid variables, as proved by the work of J. B. IRWIN[6], D. W. N. STIBBS[7], and M. W. FEAST[8], showing that S Normae is a member of NGC 6087, and U Sagittarii is a member of M 25. A hunt has been on to detect other cepheids in galactic clusters, by R. P. KRAFT[9] and S. VAN DEN BERGH[10] and Table 7, compiled from all these sources lists 11 cepheids within the boundaries of galactic clusters, which, except ER Carinae, are actual or probable members. Six others which possibly belong to galactic clusters are: GI Car, UX Car, R Cru, T Cru, AT Pup, and Y Sct.

[1] J. A. PEARCE: Roy. Soc. Canada, paper given in 1958.
[2] H. HAFFNER: Z. Astrophys. **14**, 285 (1937).
[3] E. E. BARNARD: Astronom. J. **32**, 102 (1919).
[4] E. LINDEMANN: Bull. St. Petersburg. Acad. Sci. **5**, 2, 55 (1895).
[5] P. DOIG: J. Brit. Astronom. Ass. **35**, 202 (1925).
[6] J. B. IRWIN: Monthly Notices Roy. Astronom. Soc. S. Africa **14**, 38 (1955).
[7] D. W. N. STIBBS: Monthly Notices Roy. Astronom. Soc. London **115**, 363 (1955).
[8] M. W. FEAST: Monthly Notices Roy. Astronom. Soc. London **117**, 193 (1957).
[9] R. P. KRAFT: Astrophys. Journ. **126**, 225 (1957).
[10] S. VAN DEN BERGH: Astrophys. Journ. **126**, 323 (1957).

Table 7. *Cepheids within boundaries of galactic clusters.*

Star	Cluster	Magnitudes		Period (days)
		max.	min.	
AO CMa	Uncatalogued	12.9	14.1	5.82
ER Car	NGC 3532	7.6	8.0	7.72
GH Car	Trumpler 18	9.8	10.1	5.73
CE_a Cas	NGC 7790	11.0	11.8	5.13
CE_b Cas	NGC 7790	11.1	11.8	4.45
CF Cas	NGC 7790	11.9	12.8	4.88
DL Cas	NGC 129	10.0	10.8	8.00
CV Mon	Uncatalogued	10.0	10.7	5.38
S Nor	NGC 6087	6.7	7.5	9.75
EV Sct	NGC 6664	10	10.2	3.09
U Sgr	M 25	6.9	8.0	6.74

Galactic clusters, particularly those associated with nebulosity, also contain T Tauri-type variables. In some clusters, large numbers of very faint variables are being found, which are apparently true cluster members as they do not occur in numbers far outside the cluster area. The variables have an absolute visual magnitude of +3 or fainter, and around forty percent of them show bright H_α. These stars are apparently recent formations from the cosmic complex. They have been noted in the Orion nebula cluster by G.H. HERBIG[1], totalling 115, and by M.F. WALKER[2] in NGC 2264, totalling 79, and in NGC 6530, 69. These are all very young clusters.

Eclipsing variables are present in galactic clusters (Sect. 16).

Another type of variable in these objects is the spectrum variable, exemplified by Pleione (28 Tauri), one of the Pleiads, which shows small and irregular light variation along with variation in the spectrum, in which emission lines appear at times. (See Sects. 24, 27 for the shell episode of Pleione.) Observations at Göttingen show other Pleiads slightly variable. The presence of spectral variability is probably not uncommon. O. J. EGGEN[3] found 21 Comae to be slightly variable, a typical strontium star of class *A 3 p*.

We conclude then that variable stars are definitely present in galactic clusters, and in substantial numbers, but that the types most frequent differ markedly from the types present in globular clusters. The variables most numerous in globular clusters, the RR Lyrae stars, appear to be completely absent in galactic clusters.

18. Neutral hydrogen and proto-stars. A successful search for neutral hydrogen in galactic clusters has been made with the Harvard 60-foot radio telescope by F. DRAKE[4]. He has found 21-cm radiation in h and χ Persei, the Pleiades, Coma Berenices and Praesepe, but none in M 67. This neutral hydrogen is in excess of that present in the general interstellar medium and it corresponds in position and radial velocity with the respective clusters. Since existing theories of stellar evolution predict that all stars of mass less than a certain limiting value will be in the proto-star stage, DRAKE considers this neutral hydrogen to be in proto-stars and not in an extended residue of cluster formation or material ejected from stars. The amount of neutral hydrogen in clusters is strongly correlated with the age of the cluster (Sect. 21). While the youngest clusters have about as much

[1] G.H. HERBIG: J. Roy. Astronom. Soc. Canada **46**, 222 (1952).
[2] M.F. WALKER: Astrophys. Journ. Suppl. **2**, 365, No. 23 (1956).
[3] O. J. EGGEN: Astrophys. Journ. **112**, 141 (1950).
[4] F. DRAKE: Astronom. J. **63**, 49 (1958).

neutral hydrogen as stellar mass, the ratio decreases with age and reaches zero for the old cluster M 67. We await other observations to confirm this important result.

IV. Color and spectrum luminosity diagrams, evolution and ages.

19. Color and spectrum arrays. The color or spectral class of individual stars correlated with their apparent or absolute magnitude forms one of the most essential pieces of information for galactic clusters. Not merely does such an array provide the most accurate distance of a cluster apart from trigonometric measures, it is also an important source of information for theories of stellar evolution.

The early work of SHAPLEY[1] at Mt. Wilson showed a marked difference in the color-magnitude array for globular and galactic clusters, and later P. TEN BRUGGENCATE [*13*] showed a division into two sorts of stellar populations. Meanwhile, B. LINDBLAD[2] emphasized the spectral differences in cluster stars and the application of spectroscopic parallax to galactic clusters, and TRUMPLER studied individual spectra in one hundred clusters.

When spectra are not available, colors may be determined and used to extend the arrays to additional clusters, or to fainter stars in a spectrum-luminosity diagram. The early color-luminosity diagrams were somewhat distorted because color measures at long wavelengths differ from those at short wave-lengths where the hydrogen absorption in the *A* stars (usually numerous in galactic clusters) can distort the color index. Furthermore, colors are sensitive to interstellar absorption in a way that spectral type is not. W. BECKER[3] has given an ingenious method whereby the distance modulus and interstellar absorption of open clusters may be determined without knowing the spectral types. He makes photometric observations at λ 638, 470, 373 μ from which the long-wave color index 470 μ/638 μ may be compared with the short-wave color index 373 μ/470 μ to give a color difference which is nearly independent of interstellar reddening. This determination is based on the fact that interstellar reddening and temperature reddening do not have the same effect at the three chosen wavelengths, especially for *B* and *A*-type stars.

The types of stars found in galactic clusters have been discussed in Division III, and actual luminosity diagrams are given in Figs. 6, 8, 12, 13, 14.

The more accurate the determination of the apparent magnitudes, the higher the precision obtained in the color-luminosity diagrams. Photoelectric work in recent years has reached a high standard of accuracy, and has led to studies of fine structure in the diagrams for near clusters like the Hyades and Praesepe.

20. Determination of standard main sequence. No longer is the standard main sequence derived from nearby stars with well-determined trigonometric parallaxes considered adequate as a comparison for the color-luminosity array of a galactic cluster. Such a main sequence is in error in that it is too bright and too red for purposes of comparison with cluster main sequences because of the evolutionary changes which have set in. The slope of the luminosity function for brighter stars of younger galactic clusters is noticeably less steep than the slope of the function for main sequence stars in the solar neighbourhood. The effects of evolution are apparent in the color-magnitude diagram to a point three magnitudes below the termination of the main sequence, but below this evolution may be neglected.

---

[1] H. SHAPLEY: Mt. Wilson Obs. Contrib. **116**, 51 (1915) and Commun. 34 (1916).
[2] B. LINDBLAD: Mt. Wilson Obs. Contrib. No. 228 (1922).
[3] W. BECKER: Z. Astrophys. **29**, 66 (1951).

A series of determinations of a standard main sequence appears in various recent papers by H. L. JOHNSON, W. A. HILTNER[1] and A. R. SANDAGE[2, 3], of which the latest values by JOHNSON[4] are given in Table 8. This sequence depends on

Table 8. *The standard main sequence.*

B-V	M_v	U-B	M_v	B-V	M_v	U-B	M_v
−0.25	−2.10	−0.90	−2.12	+0.50	+4.23	+0.08	+2.10
−0.20	−1.10	−0.80	−1.60	+0.60	+4.79	+0.07	+2.58
−0.15	−0.15	−0.70	−1.04	+0.70	+5.38	+0.02	+3.04
−0.10	+0.58	−0.60	−0.52	+0.80	+5.88	0.00	+3.56
−0.05	+1.21	−0.50	−0.03	+0.90	+6.32	0.00	+4.23
0.00	+1.62	−0.40	+0.34	+1.00	+6.78	+0.06	+4.79
+0.05	+1.84	−0.30	+0.75	+1.10	+7.20	+0.24	+5.38
+0.10	+2.10	−0.20	+1.02	+1.20	+7.66		
+0.20	+2.58	−0.10	+1.26	+1.30	+8.11		
+0.30	+3.04	0.00	+1.62	+1.40	+8.60		
+0.40	+3.56	+0.05	+1.84				

piecing together segments from different clusters whose parallaxes are known, and it rests on the well-determined distance of the Hyades from VAN BUEREN'S trigonometrical parallax and from new Bergedorf proper motions[5]. Two important assumptions have been made, that the Hyades are representative of main sequence stars and that any differences in chemical composition between the clusters do not affect the position of their main sequence. By use of this standard main sequence, JOHNSON has derived the photometric distances (Table 5) from B-V and U-B measures for twenty galactic clusters from data furnished by many observers. Recent work by M. SCHWARZSCHILD [*14*] shows that the slope of the luminosity function in young clusters compares favorably with the birth-rate function for the stars of the solar neighbourhood, though cluster data are still scanty. The birth-rate function is computed from the observed luminosity function and the theory of stellar interiors which gives the time a star of given luminosity remains on the main

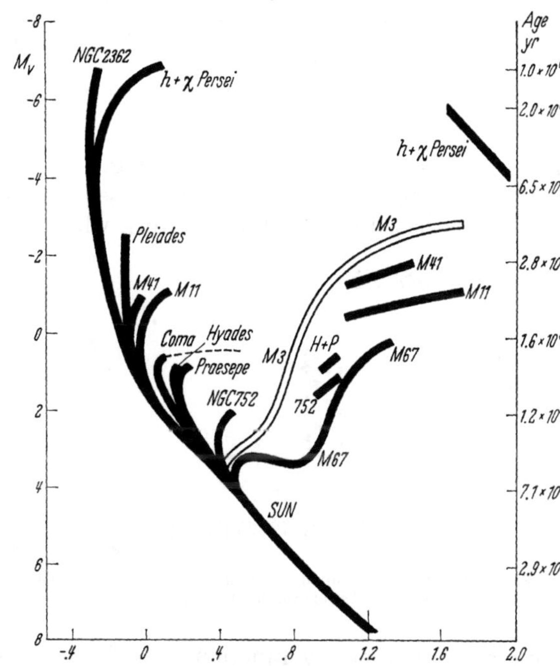

Fig. 6. Composite color-magnitude diagram by SANDAGE for 11 clusters. Astrophys. Journ. **125**, 435 (1957).

[1] H. L. JOHNSON and W. A. HILTNER: Astrophys. Journ. **123**, 267 (1956).
[2] H. L. JOHNSON and A. R. SANDAGE: Astrophys. Journ. **121**, 616 (1955).
[3] A. R. SANDAGE: Astrophys. Journ. **125**, 437 (1957).
[4] H. L. JOHNSON: Astrophys. Journ. **126**, 126 (1957).
[5] O. HECKMANN and H. L. JOHNSON: Astrophys. Journ. **124**, 477 (1956).

sequence. It is the time independent function. For a wide range of stellar masses it is the time required for about thirteen percent of the hydrogen to be transmuted to helium.

21. Evolution and ages. The presence of stars in groups makes a natural testing ground for theories of stellar formation and evolution. From work by B. STRÖMGREN that stars form a two-parameter family by mass and hydrogen content, G. P. KUIPER[1] postulated that galactic clusters are groups in which all members have the same hydrogen content, and form a one-parameter family. By the assumption that all the stars in a certain group originated on the main sequence at the same time and evolve in a similar way, J. B. OKE[2] derived H-R diagrams in terms of bolometric magnitudes and logarithms of effective temperatures for some galactic clusters, and theoretical evolutionary models for clusters of different ages.

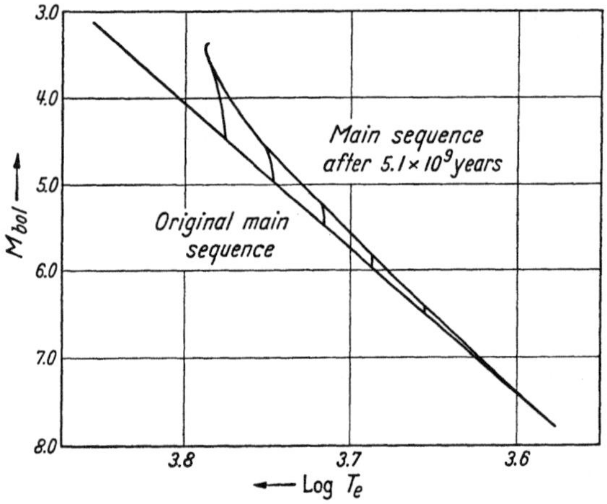

Fig. 7. The deviation of the main sequence from the original main sequence due to evolution. (Same as preceding.)

The composite color magnitude diagrams for ten galactic clusters and one globular cluster by SANDAGE[3] are shown in Fig. 6, while Fig. 7 gives the theoretical deviation of the main sequence from the original main sequence caused by evolution. This change of slope is based on SALPETER'S work which shows that the change in slope of

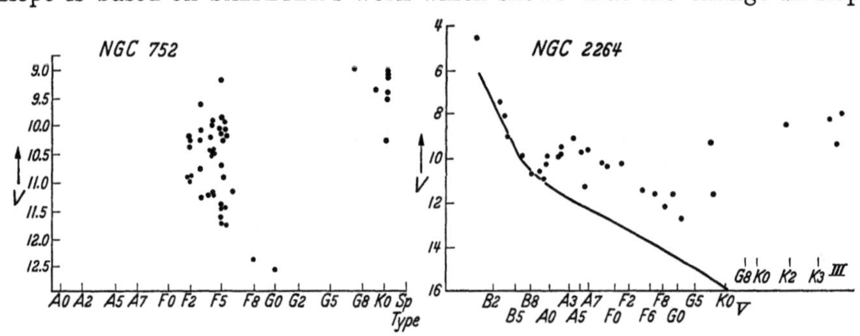

Fig. 8. Contrast in color-magnitude diagrams for the old cluster NGC 752 (ROMAN) and the very young cluster NGC 2264 (Walker). Astrophys. Journ. **121**, 454 (1955), and Astrophys. Journ. Suppl. No. 23 (1956).

the general luminosity function near $M_v = +3.5$ is probably a result of stellar evolution. The extremes of ages of galactic clusters from these diagrams are represented by the very young clusters NGC 2264, 2244, and 2362, with ages of the order of 10^6 years and the very old clusters like NGC 752 and M 67 with ages of 10^9 years. Fig. 8 shows the contrast in the luminosity diagrams determined by M. F. WALKER[4] for NGC 2264 and by Miss N. G. ROMAN[5] for NGC 752.

[1] G. P. KUIPER: Harvard Obs. Bull. No. 906 (1936).
[2] J. B. OKE: J. Roy. Astronom. Soc. Canada **49**, 8 (1955).
[3] A. R. SANDAGE: Astrophys. Journ. **125**, 437 (1957).
[4] M. F. WALKER: Astrophys. Journ. Suppl. **2**, 365, No. 23 (1956).
[5] N. G. ROMAN: Astrophys. Journ. **121**, 454 (1955).

V. Motions of stars.

22. Proper motions. Proper motions of cluster stars provide one of the most reliable means of identifying cluster members and, if well determined, may be used to find the mean internal motion of cluster stars and give a value of the time of relaxation. Unfortunately, except for the very nearest clusters, the time interval for accurate observations on this is still too short, but eventually, fundamental work like that of KÜSTNER of Bonn and SCHLESINGER of Yale should yield valuable results. The clusters with largest proper motion, the moving clusters, are discussed in Division VII (p. 157).

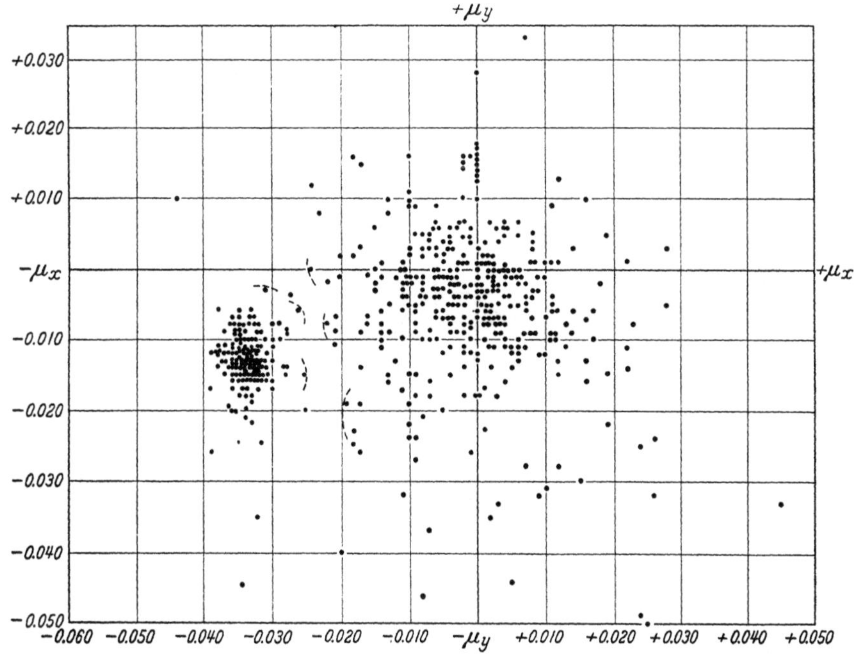

Fig. 9. The proper motions in Praesepe, by KLEIN-WASSINK. Dotted lines separate Praesepe stars from the background stars. Groningen Publ. No. 41 (1927).

We take the Pleiades as an example of proper motion work. E. HERTZSPRUNG [*15*] treated the stars magnitude by magnitude, with the mean errors of the proper motions increasing with faintness to a mean error of $7\frac{1}{2}''$ in 1000 years at magnitude 15 to 16. The apparent proper motion of the Pleiades is about 0''.048. HERTZSPRUNG[1] derived the internal motions of the stars as 0''.001 yearly, while J. TITUS[2] reached a value of 0''.00079. At a distance of 120 psc, the linear velocity is in approximate agreement with the value of 1 km/sec obtained from velocity measures of SMITH and STRUVE (Sect. 23).

For the two clusters Pleiades and Praesepe, measurements of the dispersion of the proper motions have been used by various investigators to determine the mass. From a thorough consideration of available observations U. VAN WIJK determined a mass for the Pleiades of about 500 solar masses, and for Praesepe of about 700, in excellent agreement with masses of these clusters determined by H. MINEUR in his classic paper (Sects. 40, 41).

For only three nearby galactic clusters, Pleiades, Hyades and Praesepe, have position measures given an observable motion of the center of gravity. The

[1] E. HERTZSPRUNG: Bull. Astronom. Inst. Netherl. **7**, 187 (1934).
[2] J. TITUS: Astronom. J. **47**, 25 (1938).

cluster members have motions far greater than the peculiar motions of the individual stars, all nearly equal toward the same spot on the sphere. The excellent diagram by W. J. KLEIN WASSINK[1] for Praesepe is shown in Fig. 9. The cluster stars are grouped around a central point which corresponds to the motion of the center of gravity, while the nearby field stars fall by chance around the origin of the coordinate system.

23. Radial velocities. These are also of great use in determining whether a star is a cluster member. In addition, for clusters far away, they may be used with galactic rotation formulae to derive the distance of the cluster. For clusters with early-type stars which show strong interstellar lines, interstellar velocities and equivalent widths provide an auxiliary study to the stellar velocities.

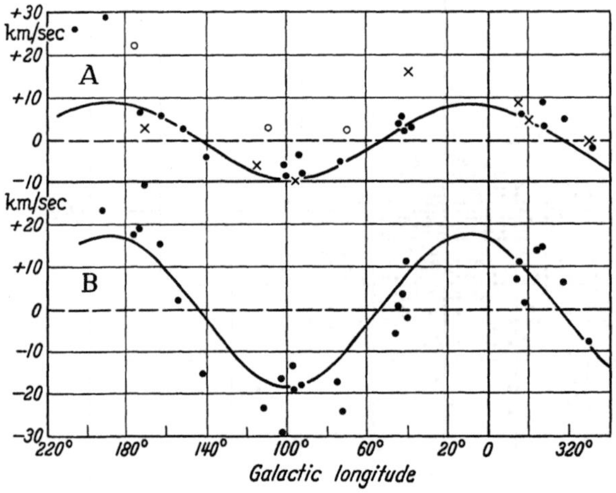

Fig. 10. SANFORD's correlation of velocities with galactic longitude (A) interstellar, (B) stellar. Open circles are velocities of low weight. Astrophys. Journ. **110**, 117 (1949).

Because the brightest stars in galactic clusters are within reach of many spectrographs, their velocities are to be found in many catalogues. The first large systematic attempt at velocities was by Miss HAYFORD[2], who obtained velocities of 174 stars in 29 open clusters, in 21 of which velocities of interstellar lines of Ca^+ were also determined. More recently, R.F. SANFORD[3] at Mt. Wilson has studied velocities of stellar and interstellar lines, as well as equivalent widths of the latter, in 31 galactic clusters. He showed that all three of these quantities provide useful means for distance determination in open clusters. A general comparison leads to a reduction of TRUMPLER'S distances beyond 1000 psc from 1600 to 1200 psc. SANFORD'S velocities are in good accord with galactic rotation as shown in Fig. 10, where *A* shows the measured interstellar velocities of clusters after correction for solar motion and to 1 kpc by distances derived from equivalent widths plotted against galactic longitude, with a curve for differential galactic rotation at $\frac{1}{2}$ kpc. *B* shows the measured stellar velocities similarly reduced and plotted, with a curve for galactic rotation at 1 kpc.

The Pleiades is an example of a cluster where concentrated effort has been made to derive velocities for many individual stars. B. SMITH and O. STRUVE[4] determined velocities for 71 stars of which 69 were cluster members. The mean median velocity of the cluster stars of low rotation was $+5.6$ km/sec, which was taken as the velocity of the center of gravity. The stars of low rotation gave a larger value of radial velocity than the stars of high rotation, for a reason not clear. These velocities are about 2 km/sec smaller than determinations by TRUMPLER at Lick and PEARCE at the Dominion Astrophysical Observatory. A difference between the external and internal mean errors of 1.2 km/sec suggests to SMITH

[1] W. J. KLEIN WASSINK: Groningen Publ. No. 41 (1927).
[2] P. HAYFORD: Lick Obs. Bull. **16**, 53 (1932).
[3] R.F. SANFORD: Astrophys. Journ. **110**, 117 (1949).
[4] B. SMITH and O. STRUVE: Astrophys. Journ. **100**, 360 (1944).

and STRUVE that the internal motions of the Pleiades may be about 1 km/sec, in accord with the values determined from proper motion work. The study of radial velocities in open clusters leads naturally into the study of stellar rotations.

24. Rotations of stars. The existence of large rotations in early-type stars is well established. However, STRUVE[1] has shown that the rotations of the B stars in the Pleiades are excessively large compared to similar stars in galactic fields. The bright-line stars in the Pleiades have very large rotations, 150 to 300 km/sec. The A stars in this cluster do not have an excess of rotational velocity compared with galactic A stars (Table 9). By comparison, the rotational velocity of our Sun,

Table 9.

Mean rotational velocities of Pleiades[1].

Type	Number	Mean $v_0 \sin i$ km/sec
B6 to B9	18	130
A0 to A5	37	53
A6 to F8	15	33

Mean rotational velocity in the galaxy.

Spectrum	Number	Mean $v_0 \sin i$ km/sec
B	275	73
A	413	87
F	112	23

a G-type star, is only 2 km/sec. There is a tendency in the Pleiades for the larger rotations to occur near the center of the cluster.

One of the largest of the rotational velocities, 300 km/sec, belongs to the shell star Pleione. STRUVE considers this star to be near the limit of equatorial break-up, of which the shell is the result (Sect. 27). In all stars, equatorial break-up will set an upper limit to the rotational velocity. The limiting equatorial velocity is

$$v = 4.4 \times 10^2 \sqrt{\frac{\mathfrak{M}}{R}} \tag{24.1}$$

where $\mathfrak{M}$, the mass, and R, radius are in solar units. For the Sun, $v = 440$ km/sec, but the effect of radiation pressure is to reduce the gravitational acceleration and extend this limit.

The only bright Pleiad with small rotational velocity is Maia, which may be exhibiting a line-of-sight effect. The observed rotational velocities are not the true, but are the projection on the line of sight, $v_0 \sin i$. When $i = 90°$, the axis of the star is at right angles to the line of sight.

By contrast with the Pleiades, the Hyades[1] do not show high rotational velocities. STRUVE interprets this to indicate that the environment determines the origin of the high rotation. Of two possible mechanisms, the building-up of rapid rotation from close stellar encounters or the building-up from the infall of nebular or interstellar matter, the latter mechanism is favored. The Pleiades are enveloped in a dense reflection nebula. Its density is 0.5 solar masses/psc^3 or 4×10^{-23} g/cm^3. If the entire cluster is in rotation, even slightly, then a small amount of mass accretion can cause high rotations of individual stars.

VI. Some well-known clusters.

a) Pleiades.

25. General. This cluster in Taurus is the best known of all galactic clusters. STRUVE[2] has stated that more photographs have been taken of the Pleiades than

[1] O. STRUVE: Popular Astronomy **53**, 201, 259, 265 (1945).
[2] O. STRUVE: Sky and Telescope **10**, 164 (1951).

of any other object in the stellar universe, and that more refined observations have been made of Pleiades than of any other distant object. Though the Pleiades were recorded in ancient times by the Chinese, Hindus, Chaldeans, and Greeks, and described in early literature, including the Bible, yet even with our powerful modern telescopes, we have not photographed all the members of the group, nor do we know the distance precisely.

In literature, the Pleiades are mentioned in terms of seven—the seven sisters, or seven pigeons—though only six are easily visible to the unaided eye. This raises the problem of the lost Pleiad; whether or not one star has faded in brilliance since the early records. Though this suspicion may be strengthened by the shell episode of Pleione (Sect. 24), keen observers under excellent conditions can see more than six Pleiads; the chart showing eleven stars, made by KEPLER's teacher, MAESTLIN, in 1579, before the invention of the telescope, is an example of this.

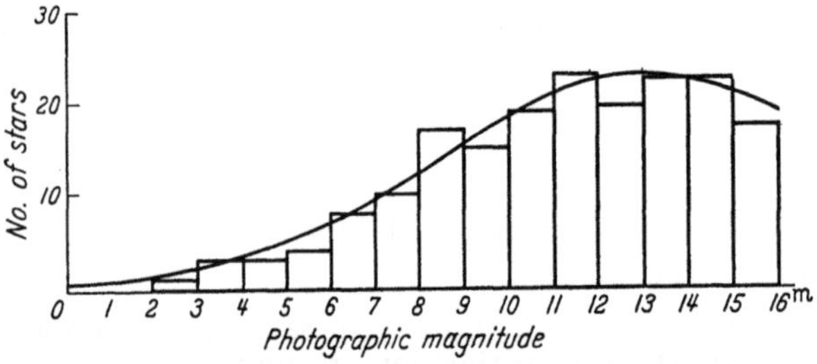

Fig. 11. Frequency of physical Pleiades of different photographic magnitudes, by HERTZSPRUNG. Monthly Notices Roy. Astronom. Soc. London **89**, 667 (1929).

According to TRUMPLER[1], the cluster is composed of three hundred to five hundred stars and has a diameter of 10 psc. The center of the Pleiades group is 6′ west of Alcyone, the lucida, at $3^h 41^m.1$, $+23° 48'$ (1900). Whereas most of the members seem to lie within an area of diameter 2°, TRUMPLER found stars with common motion as far as 3° from Alcyone. The Pleiades stars seem to be about 1 psc apart, but more densely distributed in the center. HERTZSPRUNG [*15*] found that the frequency curve of photographic magnitudes for known members (Fig. 11) probably reaches a maximum about apparent magnitude 13, but that explorations of fainter magnitudes could shift this to the fainter values. The faintest Pleiades found by VAN MAANEN[2] were about absolute magnitude +12. TRUMPLER noted many double stars among the Pleiades, eight doubles and one triple system with several other suspected pairs, though spectroscopic binaries are scarce. The most extensive catalogues of the Pleiades have been made at Leiden by HERTZSPRUNG[3] and collaborators, with a comprehensive study and full bibliography by L. BINNENDIJK[4].

The brightest stars in the Pleiades are late *B*-type. The H-R diagram on the MK system, as derived by E.E. MENDOZA[5] from accurate photoelectric colors of various observers, is shown in Fig. 12. The *F* stars in the Pleiades are relatively fainter in ultraviolet light than stars of similar type in the solar neighborhood. No white dwarfs have yet been found. The color-magnitude diagram has been studied by various investigators to derive an age of the Pleiades. MITCHELL and JOHNSON[6] find an age of 150×10^6 years from a study of the evolution of the brighter cluster stars. Also, the cluster main sequence has an observed gap at B-V = 0.90. If this is assumed to separate the stable main-sequence stars from those

---

[1] R. J. TRUMPLER: Lick Obs. Bull. No. 333 (1921).
[2] A. VAN MAANEN: Publ. Astronom. Soc. Pacific **53**, 193 (1941).
[3] E. HERTZSPRUNG: Leiden Obs. Ann. **19**, Part A (1947); Part B (1949).
[4] L. BINNENDIJK: Leiden Obs. Ann. **19**, Part 2 (1946).
[5] E.E. MENDOZA: Astrophys. Journ. **123**, 54 (1956).
[6] R.I. MITCHELL and H.L. JOHNSON: Astrophys. Journ. **125**, 414 (1957).

still in process of gravitational contraction, a similar age is derived. Mme. A. G. MASEVICH[1] finds that the early stars have evolved without mixing, off the main sequence, with loss of mass if rapidly rotating, in about 10^7 years.

A full discussion of the nebulosity of the Pleiades falls under the subject of gaseous nebulae. In a classical paper, E.C. SLIPHER[2] announced from a 21-hour exposure spectrogram at the Lowell Observatory that the Pleiades' nebulosity is a reflection type, whose spectrum is a true copy of that of the brighter Pleiades stars. It is a continuous spectrum crossed by strong hydrogen lines of the Balmer series and fainter helium lines. The brightest patches of the nebula reflect 1% of the light that a sheet of white paper in their place would reflect. Polarization measures by S. VAN DEN BERGH[3] of 217 Pleiades stars show that the polarization has a definite increase at a distance roughly that of the Pleiades, and deviations in it correlate with the filamentary structure of the nebulosity.

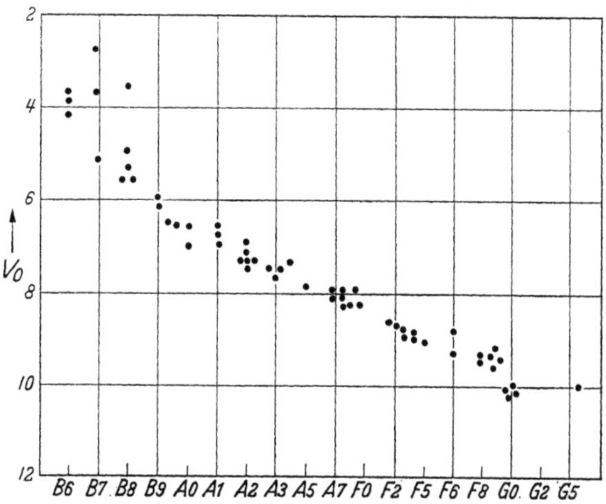

Fig. 12. *H-R* diagram for the Pleiades, by MENDOZA. V_0 is V magnitude corrected for absorption. Astrophys. Journ. **123**, 54 (1956).

26. Distance determination. This has been attacked in various ways, some of which do not give concordant results, as values of the distance modulus derived range from 6.0 to 7.6. In geometrical methods, the Pleiades are about at the limit of distance for reliable trigonometric parallax. H. L. ALDEN[4] at Leander McCormick obtained a parallax of 0.″009 ± 0.004 (p.e.) while J.H. PITMAN[5] at Sproul derived 0.″017 ± 0.006. The presence of numerous binary stars gives a chance to derive dynamical parallaxes, as done by TRUMPLER to obtain a value of 0.″010 ± 0.″006 and L. GRATTON[6], 0.″009 ± 0.″003 though a larger value of 0.″016 was obtained by HERTZSPRUNG[7] for AITKEN's binary No. 2755 which has had nearly a full revolution. The moving-cluster method fails for the Pleiades as the convergent point is not accurately determinable. But if they are assumed at rest with respect to the solar system, their motion indicates a parallax of 0.″014 ± 0.″006.

Amongst the photometric methods, the color-magnitude or spectrum-luminosity diagrams, compared with a standard main sequence, provide the best yardstick. S. GUENTHER[8] at Göttingen has shown that at two effective wavelengths, the main sequence for the Pleiades is 0.5 magnitude higher than the main sequence for Praesepe. If the spectral types are abscissae, the main sequence is only 0.1 magnitude above Praesepe. This shows a reddening of the Pleiades stars by interstellar or nebular particles, corresponding to a photographic dimming in

[1] A.G. MASEVICH: Astronom. J. UdSSR. **32**, 412 (1955).
[2] E.C. SLIPHER: Lowell Obs. Bull. No. 55 (1912).
[3] S. VAN DEN BERGH: Z. Astrophys. **40**, 437 (1956).
[4] H.L. ALDEN: Publ. Amer. Astronom. Soc. **4**, 349 (1922).
[5] J.H. PITMAN: Publ. Amer. Astronom. Soc. **5**, 401 (1926).
[6] L. GRATTON: Z. Astrophys. **15**, 48 (1938).
[7] E. HERTZSPRUNG: Bull. astronom. Inst. Netherl. **9**, 258 (1942).
[8] S. GUENTHER: Z. Astrophys. **26**, 31 (1949); **27**, 107 (1950).

the blue of 0.4 magnitude. GUENTHER adopts a distance of 140 psc. A modulus of 5.5 (distance 126 psc) was derived by MITCHELL and JOHNSON by fitting the main sequence stars of the Pleiades to the Hyades main sequence. At Heidelberg, G. MICZAIKA[1] used a spectroscopic parallax method of UNSÖLD and STRUVE, in counting the number of hydrogen absorption lines in each star to derive the intrinsic luminosity, and obtained a parallax of 0″.010. There has been no chance to try an unusual method of distance determination suggested by HERTZSPRUNG [*15*]. If an eclipsing star could be found in the Pleiades, sending out light in regular periods, it would intermittently illumine the surrounding nebulosity. The stripes observed in connection with a known period would give a lower limit of the parallax.

27. Spectroscopic studies of brightest stars. The bright individual members of the Pleiades, data on which are given in Table 10, have been extensively studied spectroscopically. Their radial velocities and high rotational velocities have been discussed in Sects. 23, 24.

Table 10. *Data table for brightest Pleiades.*

Name	Desig.	HD	R.A. 1900 Dec.	Sp.	Mag. *B*	Color *B-V*	Vel. km/sec	$v_0 \sin i$ km/sec	Tr. no.	Hertz-sprung
Celaeno	16 Tau	23288	$3^h 38^m 51^s$ +23° 58′	*B*8	5.22	−224	+ 4.3	219	139	447
Electra	17 Tau	23302	38 56 +23 48	B6*e*	3.40	−297	+14.9	158	148	468
—	18 Tau	23324	39 12 +24 32	*B*8	5.38	−255	− 1.2	178	166	541
Taygeta	19*q* Tau	23338	39 15 +24 09	*B*6	4.00	−289	+ 3.2	95	170	563
Maia	20 Tau	23408	39 52 +24 03	*B*7	3.61	−249	+ 5.1	≦30	231	785
Asterope	21 Tau	23432	39 57 +24 15	*B*8*e*	5.53	−215	− 0.2	119	240	817
Merope	23 Tau	23480	40 23 +23 38	*B*6	3.93	−242	+ 4.1	224	286	980
Alcyone	25η Tau	23630	41 32 +23 48	*B*7*e*	2.59	−274	+ 9.1	178	414	1432
—	—	23753	42 26 +23 07	*B*8	5.18	−251	−14.4	250	506	1823
Atlas	27 Tau	23850	43 13 +23 45	*B*8	3.35	−266	+ 1.7	198	594	2168
Pleione	28 Tau	23862	43 14 +23 50	*B*8*pe*	4.83	−257	+10	(300)	602	2181

Notes: Spectral types by W. W. MORGAN, in: Astronom. J. **112**, 361 (1950); magnitudes and colors on standard *U*, *B*, *V*, system by H. L. JOHNSON and W. W. MORGAN: Astrophys. Journ. **114**, 522 (1951); radial velocities from B. SMITH and O. STRUVE: Astrophys. Journ. **100**, 360 (1944); rotational velocities ($v_0 \sin i$) by E. VAN DIEN: J. Roy. Astronom. Soc. Canada **42**, 249 (1948); Trumpler numbers from Lick Obs. Bull. **10**, 110 (1921); Hertzsprung numbers from his later catalogue, Leiden Ann. **19**, pt. 1 (1947).

Pleione has been an object of special attention because of its recent shell episode. The shell formation began in 1938 when O.C. MOHLER at Cook and D.B. McLAUGHLIN at Michigan independently observed the reappearance of bright hydrogen lines which had first been noted at Harvard in December 1888, lasting until 1903. In the recent episode, hydrogen emission on either side of the dark core was strong at H_α, weak at H_β, and absent at H_γ, and the lines following. STRUVE and SWINGS[2] observed the sharp cores of the Balmer lines as far as H31, and noted oscillations in the radial velocity of Pleione with a range of 10 km/sec and period of four months, with a mean velocity of +5.5 km/sec. Numerous dark lines of metals reached their maximum intensity in 1945 or 1946, after which they became weaker until in 1951, they were nearly gone. In a summary of the shell episode, P.E. MERRILL[3] considers the shell was formed by atoms which moved outward from the photosphere. At first the motion was so slow that the observed radial velocity of the shell was the same as that of the

[1] G. MICZAIKA: Astrophys. Journ. **112**, 361 (1950).
[2] O. STRUVE and P. SWINGS: Astrophys. Journ. **93**, 446 (1941).
[3] P.E. MERRILL: Astrophys. Journ. **115**, 145 (1952).

star as a whole. In time, the outward velocity of the upper levels became appreciable. When the supply of atoms eventually failed, the shell blew away. STRUVE estimates the radius of the shell at about 1.5 times the radius of the star, or 7.5 times the Sun's radius. A bibliography of observations of Pleione has been given by L. BINNENDIJK[1].

Maia, the only bright Pleiad with low rotational velocity, is shown by observations in 1955[2] to be a spectrum variable. The helium lines vary in intensity and width in a non-periodic manner at short intervals of around two hours, but short-period variations in radial velocity or in light were not found.

b) Praesepe.

28. This cluster in Cancer, one of the earliest noted, has many names. Called by the Chinese, TSEIH SHE KEH, "the exhalation of piled-up corpses", it was later known as the Beehive, Manger, or Crib, and first resolved into many stars by Galileo. The Praesepe cluster is remarkably similar to the Hyades, though the brightest stars are fainter, of the 6th magnitude. The main sequences are identical, and both lack normal stars between $A\,1$ and $A\,5$[3]. Both contain four K-giant stars, some metallic line stars, and the earliest spectral type is the same in each, $A\,1$ (40 Cancri and 68 Tauri). However, there is a difference in the fine structure of the diagrams according to HECKMANN [*16*] who gives large-scale diagrams for comparison. An estimate that twenty percent of the stars are probably double or multiple accounts for some of the scatter in the H-R diagram, though this is a smaller proportion than in the Hyades. An interesting cluster member is HAFFNER'S eclipsing binary TX Cancri (Sect. 16) of the W Urs Maj type[4], with a spectroscopic orbit[5]. The Algol-type variable, S Cancri, however, is not a cluster member.

From proper motions of 600 stars, KLEIN WASSINK (Fig. 9) identified 200 cluster stars and found the brightest part of the cluster elongated parallel to the Milky Way. The proper motion of the group is $0\overset{''}{.}037$ in position angle 249°. The observed radial velocity of the cluster stars is $+33.0 \pm 0.6$ km/sec, by ADAMS and KOHLSCHÜTTER[6]. The idea was presented by K. SCHWARZSCHILD and E. HERTZSPRUNG[7], that the Hyades and Praesepe are moving through space in the same direction with the same velocity, a space velocity of 43.7 km/sec toward R.A. $92\overset{\circ}{.}0$, Dec. $+8\overset{\circ}{.}6$. The upper limit for the true inner motions in the cluster was found by VAN DE KAMP[8] to be 0''00073 in each coordinate.

Recent photoelectric work[9] seems to show that earlier observations indicating reddening and interstellar material in the Praesepe cluster are not confirmed, and that the amount of reddening material in front of or within the cluster is negligible.

c) Coma Berenices.

29. The noticeable clustering of naked-eye stars in the northwest part of the constellation probably helped to influence the name of the latter, after the hair

---

[1] L. BINNENDIJK: Astronom. J. **60**, 364 (1955).

[2] O. STRUVE, J. SAHADE, C.R. LYNDS and S.S. HUANG: Astrophys. Journ. **125**, 115 (1957).

[3] W.P. BIDELMAN: Publ. Astronom. Soc. Pacific **68**, 318 (1956).

[4] S.D. TSCHERNY: Astronom. J. **49**, 285 (1937).

[5] D.M. POPPER: Astrophys. Journ. **108**, 493 (1948).

[6] W.S. ADAMS and A. KOHLSCHÜTTER: Ann. Report Director Mt. Wilson Obs. Carnegie Yearbook **50**, 200 (1923).

[7] K. SCHWARZSCHILD and E. HERTZSPRUNG: Astronom. Nachr. **196**, 9 (1913).

[8] P. VAN DE KAMP: Astrophys. Journ. **81**, 297 (1935).

[9] H.L. JOHNSON: Astrophys. Journ. **116**, 640 (1952).

of the historical figure Berenice, wife of Ptolemy Euergetes of Egypt of the third century B.C. So sparse is the grouping that not until 1915 was it first catalogued as an open cluster by MELOTTE [8].

In an exhaustive investigation, TRUMPLER[1] gives an excellent summary of this sparse cluster. From a comparison of proper motion, radial velocity measures and magnitudes from a spectrum-luminosity diagram, 37 stars within a circle of 7° diameter brighter than magnitude 10.5 were identified as cluster members. The center is at R.A. $12^h\,18^m.0$, Dec. $+26°.5$ (1900). The annual proper motion, $\mu_\alpha \cos\delta$ is $-0''.013$, μ_δ, $-0''.017$, μ, $0''.021$ in position angle 218°, the radial velocity -0.4 km/sec, space motion relative to Sun 7.5 km/sec toward R.A. $8^h\,4^m$, Dec. $-47°$, corrected for solar motion, 13.5 km/sec toward R.A. $17^h\,12^m$, Dec. $+18°$. At a distance of 80 psc the cluster is heading in the general direction of the solar apex but at only two thirds the Sun's velocity.

About one-half the known galactic clusters are poor in stars, and the Coma cluster is typical of this large group. The earliest spectral type is *A* 0. WEAVER[2] finds that most of the fainter stars form an extraordinarily compact sequence. There are a few giant *A* and *F* stars and some metallic-line stars. Of the thirty-seven members, one is a visual binary, two are spectroscopic binaries, six have variable velocity. The brightest stars in Coma are much less luminous than in the Pleiades, and the cluster contains in a given volume only one-fourth as many stars as the Pleiades. Its total mass probably does not exceed seventy solar masses. With an average density of 0.12 solar masses/psc^3, it is close to the density of instability. Probably the less massive stars have already escaped, which might explain the high percentage of binaries that are less likely to reach the velocity of escape.

d) The double cluster in Perseus, h and χ Persei.

30. This pair of clusters, among the most conspicuous in the sky, presents problems to the investigator who attempts to determine the number of stars they contain, or their diameter. The apparent motion of the clusters is so small that available proper motions are not sufficiently accurate for a test of membership. Velocities of the brighter stars average -43 km/sec. That this velocity may be considered proof of membership is questioned by P. TH. OOSTERHOFF[3] in his comprehensive paper on these clusters, because galactic rotation is of the same order at the distance of the clusters. Brighter than 15.5, the number of cluster stars is estimated to be 340 for h and 300 for χ Persei. If both clusters have an angular diameter of 25′ at a distance of 2000 psc, the corresponding linear diameter is 15 psc and the distance between the centers of the clusters is 15 psc.

TRUMPLER showed that most members of both clusters brighter than magnitude 13 have *B*-type spectra. From the luminosity diagram, these are among the very young clusters. The bright stars are of high intrinsic luminosity, showing the *c*-characteristic in their spectra, while some of the stars of magnitudes 9 and 10 have bright hydrogen lines. The dark lanes across these clusters, readily seen in BARNARD'S Atlas of the Milky Way, lead to irregular absorption over the region, with a differential effect between the two clusters. AMBARTSUMIAN and GORDELADSE[4] noted a differencc of color excess between the two in that χ Persei has a positive color equivalent of $0^m.24$ with respect to h Persei. The velocities

[1] R. J. TRUMPLER: Lick Obs. Bull. **18**, 167 (1938).
[2] H. L. WEAVER: Astrophys. Journ. **116**, 612 (1952).
[3] P. TH. OOSTERHOFF: Leiden Obs. Ann. **17**, Part 1 (1937).
[4] V. A. AMBARTSUMIAN and SH. G. GORDELADSE: Bull. Abastumani Obs. **2**, 37 (1938).

and intensities of the interstellar lines determined by BEALS[1] indicated a distance of these clusters about double the earlier value of 1330 psc.

The bright red supergiants surrounding the clusters have been known for many years[2]. From work with the McDonald reflector, W. P. BIDELMAN[3] considered the double cluster unique in the great abundance of late- and early-type supergiants. He identified 13 M-type supergiants as cluster members, with visual absolute magnitudes to -5.0. The most luminous star in the cluster is the irregular variable S Persei, with a rough periodicity of 835 days, reaching an absolute visual magnitude of -7.0 at maximum. The cluster of supergiants extends over a larger region than the usual diameter of the cluster, thus introducing interesting dynamical implications.

e) Messier 11.

31. The fact that both BAILEY and MELOTTE classified this cluster in the Scutum cloud as globular, while SHAPLEY considered it a condensation of stars

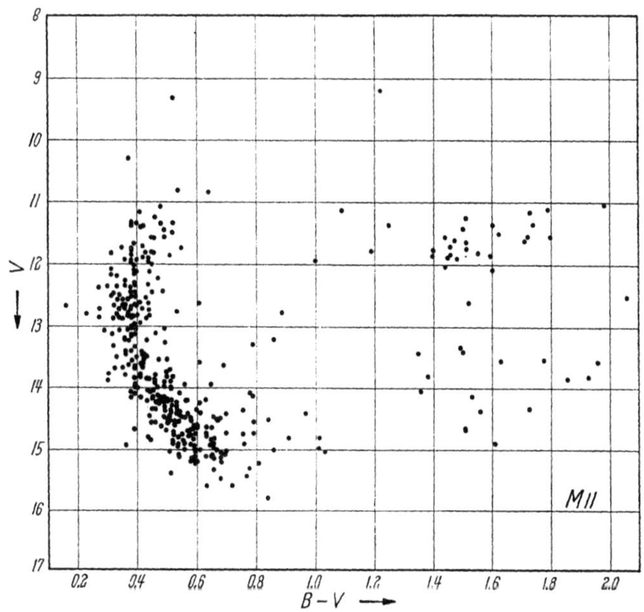

Fig. 13. The color magnitude diagram for M 11 by JOHNSON and SANDAGE. Astrophys. Journ. **124**, 81 (1956).

against a very rich field illustrates the uncertain position it has held for years in the astronomical picture. It has something of the symmetry and richness of a globular cluster. The advent of color-luminosity diagrams with evolutionary applications has clarified its situation. Magnitudes and colors by WALLENQUIST[4] for over 1000 stars were followed by proper motions by MEURERS[5] separating cluster and field stars. The color-magnitude diagram from high-precision, three-color photometry with large telescopes[6] is shown in Fig. 13. The well-defined main sequence terminates at the bright end at $M_v = -1.0$, between that of the Pleiades and Praesepe, and its age is considered intermediate between these two clusters.

[1] C. S. BEALS: Monthly Notices Roy. Astronom. Soc. London **96**, 669 (1936).
[2] T. ESPIN: Monthly Notices Roy. Astronom. Soc. London **52**, 154 (1892).
[3] W. P. BIDELMAN: Astrophys. Journ. **105**, 492 (1947).
[4] Å. WALLENQUIST: Upsala Medd. No. 65 (1936).
[5] J. MEURERS: Phys. Verh. **5**, 141 (1954).
[6] H. L. JOHNSON, A. R. SANDAGE and H. D. WAHLQUIST: Astrophys. Journ. **124**, 81 (1956).

It contains many more yellow giants than most galactic clusters and several of type later than $K0$. From 400 stars, the cluster reddening is found to be $E_{(B-V)} = 0.42$ mag.; the apparent distance modulus 12.3; and the true distance 1600 psc, somewhat greater than earlier determinations. The cluster is probably closer to the Sun than is the Scutum Cloud and is seen in projection against it.

f) Messier 67.

32. For some years, the rich cluster M67 in Cancer has been the center of argument as to whether it should be classified as globular or galactic. SHAP-

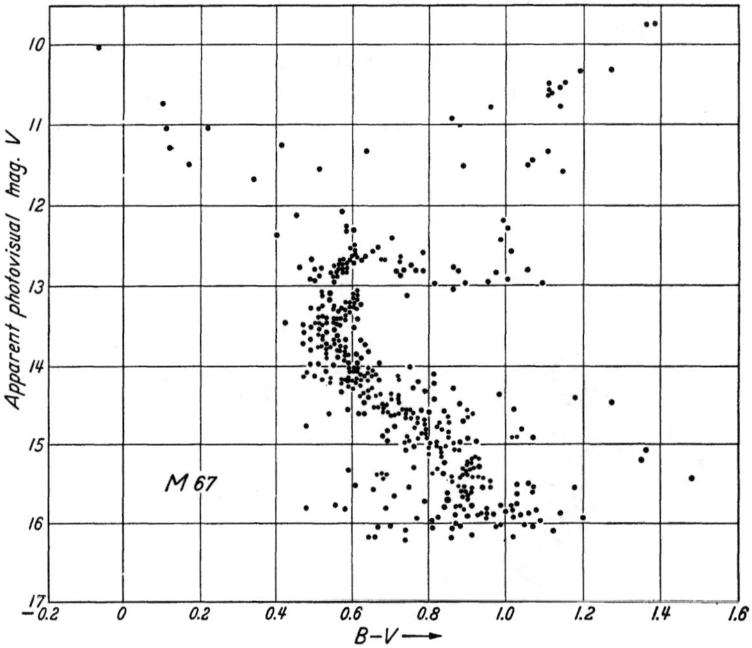

Fig. 14. The color-magnitude diagram for the very old cluster M67, by JOHNSON and SANDAGE. Astrophys. Journ. **121** 616 (1955).

LEY'S[1] early work at Mt. Wilson showed that the relation of magnitude to color in this cluster differed from that in globular clusters as well as from that in the Pleiades and Hyades. The total membership is about 500 stars between photovisual magnitude 10 and 15. SHAPLEY predicted that stars fainter than magnitude 15 may be totally absent and this result is corroborated by the recent luminosity function of VAN DEN BERGH[2] from Palomar Sky Survey prints which shows a sharp decrease in the number of stars fainter than $M = +5$, in an area of radius 4'.8 with counts to mag. 19.5 in the blue, and 18.5 in the red. The decrease may be due to the escape of the least massive cluster members, and suggests that the cluster is old for a galactic cluster. Assuming a present mass of 500 solar masses, and a radius $\overline{R} = 1.2$ psc, VAN DEN BERGH sets an age for M67 of several times 10^9 years.

The color-magnitude diagram is shown in Fig. 14, as determined for 500 stars by JOHNSON and SANDAGE[3], who consider M67 may be the oldest galactic cluster known. The great age of M67 compared with most galactic clusters seems to

[1] H. SHAPLEY: Mt. Wilson Contr. No. 117 (1916).
[2] S. VAN DEN BERGH: Astronom. J. **62**, 100 (1957).
[3] H.L. JOHNSON and A.R. SANDAGE: Astrophys. Journ. **121**, 616 (1955).

provide the logical explanation as to why it falls between globular and galactic clusters in some of its characteristics, and was classified as a globular cluster of type I (associated with stellar systems of type I population) by BECKER and STOCK[1]. POPPER'S[2] observations that the red stars are normal giants of luminosity class III and do not show the spectral peculiarities of the brightest red stars in globular clusters help to assign it to the galactic cluster category. Its distance is 1000 psc, its distance above the galactic plane is 500 psc. Its radial velocity is +33 km/sec, its space velocity 54 km/sec (reduced to the Sun, 35 km/sec), and its proper motion 0.″009.

g) ϰ Crucis, the Jewel Box.

33. This cluster (NGC 4755) seems to suffer observationally from its location in the far southern sky, at declination −59°. Whereas the early observers of this cluster, beginning with Sir JOHN HERSCHEL, describe it as one of the most beautiful objects to be seen with a telescope, detailed information about it from modern investigations is scanty. Miss MARY PROCTOR'S[3] description can bear repeating. "The star ϰ Crucis hovers on the edge of the Coal Sack, the dark, cavernous-looking space near the Southern Cross. A star of 7th magnitude surrounded by a cluster of 100 stars from the 7th magnitude down, 8 of the more conspicuous being coloured with various shades of green, red, and blue giving the cluster the appearance of a 'superb piece of jewellery' according to Sir JOHN HERSCHEL, who mapped 110 stars in 1830 from 7 to 16 mag." TRUMPLER classifies it as 1 to 2*b* in spectral classification (from the Henry Draper catalogue) and as I 3*r* with a distance of 1090 psc and an angular diameter of 12′. A fine photograph with the 60-inch reflector of the Boyden station is given by B. J. BOK and P.F. BOK [*17*].

VII. Moving clusters.

34. The Hyades. Certain galactic clusters for which the individual stars have a pronounced motion toward a convergent point are known as moving clusters. BECKER [*18*] has given an excellent table of data on these.

The Hyades is one of the earliest known and the nearest of the open clusters, as well as being the most prominent example of the moving cluster. It was early referred to as "the rainy Hyades" perhaps because it rises in the autumn when the fall rains come. The resemblance of this cluster to Praesepe, which is more distant, has long been noted in motions of stars and in spectral types. However, the parallelism of the motions might be an effect of the general Taurus stream.

On the basis of precise proper motions, VAN BUEREN[4] has catalogued some 159 members of the group down to magnitude 9.0 visual, and estimates that the total number of members is 350 and the total mass of the cluster about 320 solar masses. The cluster is strongly concentrated, half the mass being contained in a sphere of about 6 psc radius centered at R.A. $4^h\,24^m$, Dec. +15°45′ (1950) at a distance of 40 psc from the Sun. As first noted by RASMUSON[5], the cluster is flattened, with the equatorial plane roughly parallel to the galactic plane. The dispersion along the short axis is 2.2 psc, along the longer axis, 3.3 psc, corresponding to a flattening of 0.67. The motion of the Hyades as determined by VAN BUEREN is shown in Fig. 15. He points out that these motions really constitute what

[1] W. BECKER and J. STOCK: Z. Astrophys. **31**, 316 (1952).
[2] D.M. POPPER: Astronom. J. **59**, 445 (1954).
[3] M. PROCTOR: J. Brit. Astronom. Ass. **25**, 193 (1915).
[4] H.G. VAN BUEREN: Bull. Astronom. Inst. Netherl. **11**, 385 (1952).
[5] N. RASMUSON: Lund. Obs. Medd. **2**, No. 26 (1921.)

is commonly referred to as "the Taurus stream". VAN BUEREN derives 0.21 km per sec for the root-mean-square velocity of the cluster stars in one coordinate, assuming the radius of 6 psc and the total mass 320 solar masses. From this, the relaxation time is 1.1×10^8 years and the average period of revolution of a star in the cluster is 2×10^8 years. Irregularities in the shape of the cluster indicate that statistical equilibrium is not yet established.

Numerous determinations of the convergent point have been made, along with the cluster velocity. That of SMART[1] from Boss' proper motions at R.A. 91°50′ ± 0°.66, Dec. 8°9′ ± 0°.26 (1950) has been substantiated by a recent investigation by PEARCE[2] from 74 accurate proper motions on the N30 system, and radial velocities from Dominion Astrophysical Observatory spectra. PEARCE finds corresponding values of 91°.85 ± 0°.65, +9°.1 ± 0°.25 and adopts a cluster velocity of (43.2 ± 0.21) km/sec.

However, HECKMANN and LUBECK[3], with new proper motion material, have used R. E. WILSON's convergent point and EGGEN's space velocity, $A = 94°.0$, $D = +7°.6$, and $V = 45.3$ km/sec, to derive parallaxes for the individual stars according to the formula

$$m - M = -5 - 5 \log \frac{4.738 \mu}{V \sin \lambda} \quad (34.$$

where λ is the individual distance from the convergent point. The distance moduli so derived range from 2.56 to 3.69. Similar distance moduli result from work by JOHNSON and KNUCKLES[4] and HECKMANN and JOHNSON[5].

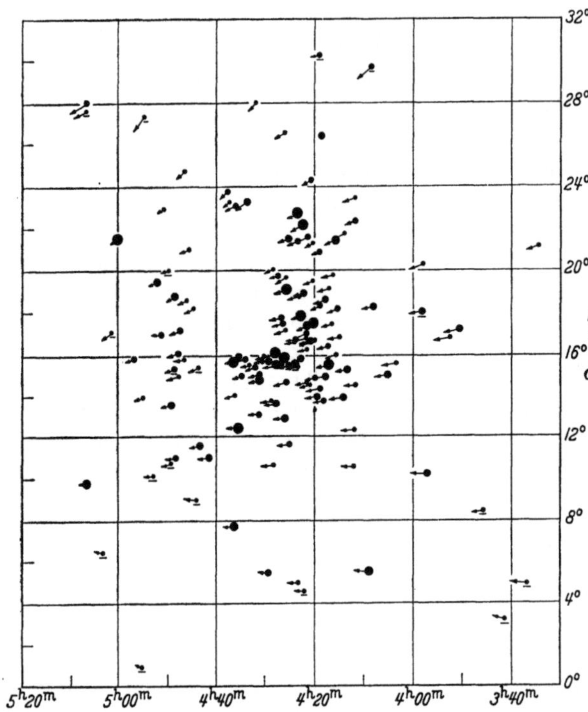

Fig. 15. The annual proper motion of the Hyades, brighter than 9m.0 visual, by VAN BUEREN. Bull. Astronom. Inst. Netherl. **11**, 385 (1952).

The spectral types of 72 of the brighter Hyades have been determined by TITUS and MORGAN[6] who state that the main sequence defined by the ordinary Hyades stars coincides with the main sequence from stars in the solar neighbourhood over the range *A* 5 to *G* 0. A group of six stars from *A* 1 to *A* 5 with very strong metallic lines appear of lower luminosity than the ordinary main sequence. The similarity with Praesepe is summarized by BIDELMAN (Sect. 28), but the use of accurate data on proper motions and magnitudes for 97 cluster members re-

---

[1] W.M. SMART: Monthly Notices Roy. Astronom. Soc. London **99**, 168 (1939).
[2] J.A. PEARCE: Publ. Astronom. Soc. Pacific **67**, 23 (1955).
[3] O. HECKMANN and K. LÜBECK: Z. Astrophys. **40**, 51 (1956).
[4] H.L. JOHNSON and C.F. KNUCKLES: Astrophys. Journ. **112**, 209 (1955).
[5] O. HECKMANN and H.L. JOHNSON: Astrophys. Journ. **124**, 477 (1956).
[6] J. TITUS and W.W. MORGAN: Astrophys. Journ. **92**, 256 (1940).

solves the fine structure in the color magnitude diagram and shows up differences between the two clusters, according to HECKMANN [*16*].

35. The Ursa Major group. This group of stars is of particular interest because it occupies a volume of space containing our Sun. The Ursa Major group is composed of two subgroups which consist of a moderately compact cluster of 14 stars sharing accurately the same space motion, and an extended stream of stars which has approximately the same space motion. A full investigation of the group by Miss ROMAN[1] with complete bibliography has been supplemented by MICZAIKA[2].

The five central stars in the Big Dipper and five fainter ones in the same region, in a volume of space less than 10 psc across, are moving with the same velocities. The nucleus is located about 23 psc from the Sun and occupies a roughly ellipsoidal region, $4\times 6\times 10$ psc in diameter. The shortest diameter is perpendicular to the galactic plane, the longest, in the direction of motion of the cluster. The nucleus consists of twelve stars which define the main sequence from $A\,0$ to $K\,3$, plus a peculiar A star and a metallic-line star. Solutions for the convergent point, utilizing SMART's work on proper motions, give R.A. $306{}^\circ\!.4\pm 1{}^\circ\!.9$, Dec. $-37{}^\circ\!.4\pm 4{}^\circ\!.6$, with a speed of 17.9 ± 0.5 km/sec. However, a combination of the Allegheny parallaxes and proper motions gives a value of (18.4 ± 1.3) km/sec for the speed of the cluster and $0''\!.040\pm 0''\!.004$ for the mean parallax of the nucleus. The motion in the local centroid is 29 km/sec.

The stars of the extended Ursa Major stream seem to fill a volume of space limited only by observational selection, about 100 psc in radius, which contains the Sun, the Hyades, the Pleiades, and the Coma cluster. The motion of the Ursa Major stream is toward a point 30° distant from the vertex of star-streaming, which makes it difficult to tell whether a particular star belongs to the Ursa Major stream or not. At least 70 members are known in the stream, which contains no B-type stars but differs from the nucleus in having some giants. Of the stars in the group later than $G\,0$, 75% show emission at H and K. The members of the stream show no concentration toward the nucleus and are probably not the result of disruption of a compact group.

The density of the nucleus, about 0.1 solar mass per cubic parsec is well below the critical density for a star cluster (Sect. 41) and is comparable to the average star field density. Miss ROMAN points out however (in contrast to ideas in the afore-mentioned section) that under certain conditions the Ursa Major cluster may have a very long life. The gravitational interactions between its cluster members and field stars are slight because the cluster has a high velocity with respect to the local standard of rest. Furthermore, the shearing effects of galactic rotation will play no part if the stars in the cluster have the same angular momentum rather than the same linear velocity, in their revolution about the galactic center. Applying CHANDRASEKHAR's method of analysis (Sect. 38), Miss ROMAN finds that a member of the cluster will probably change its kinetic energy with respect to the center of the nucleus by less than 20% in 3×10^9 years, and concludes that the lifetime of the cluster is of the same order as the galactic time scale.

36. The Perseus moving cluster. An hour of right ascension east, and 8 degrees of declination south of the double cluster in Perseus lies the group of B-type stars centred around the supergiant F-type star, α Persei, and known as the Perseus moving cluster. The existence of this group came to notice when EDDINGTON and BOSS independently pointed out that sixteen B-type stars in Perseus between

[1] N. G. ROMAN: Astrophys. Journ. **110**, 205 (1949).
[2] G. MICZAIKA: Astronom. J. **59**, 233 (1954).

$+40$ and $+50°$, and between $3^h\,14^m$ and $5^h\,18^m$ are moving in parallel motion, with a total annual proper motion of about 0ʺ.04 in position angle close to 140°. The investigation of ROMAN and MORGAN[1] determined that α Persei and 25 other stars of types $B3$ to $A2$ are members of a cluster 180 psc from the Sun moving with a velocity of 16 km/sec toward $l=215°$ and $b=+17°$. Recently HECKMANN[2] and his collaborators recognized 163 stars to 12.2 mag. pg., and spectral type $G5$, as possible members of this cluster on the basis of common proper motion. Different determinations of the modulus range from 4.8 to 6.8, but a recent one by EGGEN[3] at 5.85 is in close agreement with a spectroscopic parallax from five A-type stars by PETRIE[4], with a distance of 150 psc. VAINU BAPPU[5] notes an anomaly in the parallax of the supergiant α Persei which leads to too low an absolute magnitude for this star and points out that the proximity of the convergent point to the solar apex may introduce errors in the parallaxes derived from proper motions.

37. The Scorpio Centaurus cluster (The Southern Stream). Between galactic longitudes 210 and 330° and galactic latitudes $+30$ and $-30°$, a large proportion of the B-type stars have similar motions. The total annual proper motions range from 0ʺ.02 to 0ʺ.05 and the great circles along which the stars appear to move are approximately convergent. KAPTEYN[6] noted this and RASMUSON'S[7] comprehensive research on this southern stream was followed by that of J. S. PLASKETT and J. A. PEARCE[8].

From a study of the motions, SMART[9] concluded that the stars of the so-called Scorpio Centaurus cluster do not form a true moving cluster but are a representative sample of B stars in general. However, investigations by BLAAUW[10] show that the inclusion of faint field stars in SMART'S work masked the determination of the convergent point, and the presence of a true group of bright B stars. BLAAUW derives the stream motion with respect to the Sun, after elimination of differential galactic rotation as $V=25.9$ km/sec ± 0.8 (p.e.), toward $l=226°.9\pm 1°.6$ (p.e.), $b=-13°.8\pm 2°.1$ (p.e.). He concludes that from its form, dimensions and orientation as projected on the galactic plane the group must have been formed about 70 million years ago in a region 2200 psc distant in the direction of galactic longitude 26° with a mean initial velocity of expansion of 0.7 km/sec.

VIII. Disruption with time.

38. The virial theorem. A question which has long interested observers of star clusters is the length of time which will elapse before the cluster will disintegrate into a host of individual and unconnected stars. In his pioneer work, JEANS[11], investigating the disruptive forces of a star cluster, suggested that our whole galactic system might have been formed out of the debris of disintegrated clusters, born in the galactic plane, with stars packed more closely together than

[1] N. G. ROMAN and W. W. MORGAN: Astrophys. Journ. **111**, 426 (1950).
[2] O. HECKMANN, W. DIECKVOSS and H. KOX: Astronom. Nachr. **283**, 109 (1956).
[3] O. J. EGGEN: Astronom. J. **60**, 407 (1955).
[4] R. M. PETRIE: Publ. Dom. Astrophys. Obs. **9**, 251 (1953).
[5] M. K. VAINU BAPPU: Monthly Notices Roy. Astronom. Soc. London **114**, 687 (1954).
[6] J. C. KAPTEYN: Astrophys. Journ. **40**, 43 (1914).
[7] N. RASMUSON: Lund Obs. Medd. **2**, No. 26 (1912).
[8] J. S. PLASKETT and J. A. PEARCE: Monthly Notices Roy. Astronom. Soc. London **94**, 679 (1934).
[9] W. M. SMART: Monthly Notices Roy. Astronom. Soc. London **100**, 60 (1939).
[10] A. BLAAUW: Groningen Publ. No. 52 (1946). — Bull. Astronom. Inst. Netherl. **11**, 414 (1952).
[11] J. H. JEANS: Monthly Notices Roy. Astronom. Soc. London **82**, 132 (1922).

now. A detailed discussion of the dynamics of this problem has been given by CHANDRASEKHAR [19] to whom we refer the reader who wishes to follow all the equations.

After HECKMANN and SIEDENTOPF[1] showed that encounters between stars of a cluster establish an isothermal condition in times comparable with the short time scale, SPITZER[2] applied the Maxwell-Boltzmann distribution of velocities to get around the fact that an isothermal gaseous sphere has an infinite mass and radius. The fraction of stars with velocities greater than the escape velocity may be calculated for the ideal cluster from the virial theorem and the Maxwell-Boltzmann distribution.

From the virial theorem, for a cluster in a statistically steady state, the kinetic energy of the cluster stars is one-half the total gravitational energy measured in the negative sense:

$$2T + W = 0.$$

(Assuming $\sum_i m_i r_i^2 = \text{const}$, the cases of stellar systems which have positive, instead of negative energy are considered by AMBARTSUMIAN in his work on stellar associations given at the Liège Colloquium[3].)

Or the relations may be written $2T = m\overline{v^2}$, where M is the total mass and $\overline{v^2}$ the velocity dispersion squared is the mean value of squares of random velocities of the cluster stars. In terms of mass in solar units and R, the radius of the cluster in parsecs, an equation may be derived

$$\sqrt{\overline{v^2}} = 4.63 \times 10^{-2} \sqrt{\frac{M}{R}}\ \text{km/sec}. \tag{38.1}$$

The mean kinetic energy of the cluster stars is therefore one-half the potential energy of the cluster per star, and one-fourth the mean potential energy of the cluster stars. From this it follows that the mean square velocity of the cluster stars is one-fourth the mean square velocity of escape.

If the kinetic energy of the cluster is divided among the individual stars by the Maxwell-Boltzmann formula, some stars will obviously have enough energy to leave the cluster. Except in the richest clusters, the mean free path is of the order of the dimensions of the cluster. However, the stars will suffer appreciable changes in their energies and directions of motion even before traversing their mean free path. With time, the cluster will lose stars by "evaporation" and will gradually contract.

The mean square velocity of the stars sets a limit on the radius that a cluster of given mass may attain. Furthermore, the formula may be used in reverse, that is, the probable mass of a cluster may be obtained from information on the radius and random velocities of the cluster stars. For the Pleiades, assuming a radius of 8 light years, and velocity 0.6 km/sec, the total mass is 400 suns. For a globular cluster of radius 50 light years and velocity spreads of 15 km/sec, the mass is from 100000 to 1000000 suns.

39. The time of relaxation. An important quantity in the disintegration of a cluster is the time of relaxation, which is the time for an enclosed system of particles to establish a Maxwellian velocity distribution. A determination of the time of relaxation gives the rate at which stars leave the cluster by evaporation. The act of relaxation is continuous and occurs after every major disturbance

[1] O. HECKMANN and H. SIEDENTOPF: Z. Astrophys. **1**, 43 (1930).
[2] L. SPITZER jr.: Monthly Notices Roy. Astronom. Soc. London **100**, 396 (1940).
[3] V. A. AMBARTSUMIAN: Les Processus Nucléaires dans les Astres, Louvain, 1954. Mém. Soc. Roy. Sci. Liège, Ser. IV **14**, 293 (1954).

like the escape of a star from a cluster. SPITZER'S formula for the relaxation time is

$$T_{\text{rel}} = 8.0 \times 10^5 \frac{N^{\frac{1}{2}} R^{\frac{3}{2}}}{m^{\frac{1}{2}} (\log N - 0.5)} \text{ years} \tag{39.1}$$

where N is the total number of stars in the cluster, $\log N$ its common logarithm, R the radius of the cluster in parsecs, frequently taken as the radius containing half the mass of the cluster, and m the mass of a cluster star in units of solar mass. The time of relaxation for the Pleiades is 5×10^7 years, whereas for a typical globular cluster it is 10^{10}.

40. Rate of disintegration by escape of stars. AMBARTSUMIAN[1] showed that the escape of stars is an important factor in disintegration of a cluster. A short time of relaxation of 3×10^7 years brings on a relatively rapid disintegration of clusters. The rate of escape is given by

$$\frac{\Delta n}{n} \approx -0.0074 \frac{\Delta t}{T_{\text{rel}}}. \tag{40.1}$$

Though CHANDRASEKHAR considers that stars of 0.4 the average mass have the maximum rate of loss by escape, this conclusion is not confirmed by SPITZER[2] who finds equipartition of energy for the less massive stars to be set up in less than 10^8 years in most galactic clusters. This means that with higher velocities, these stars will evaporate more rapidly, and their distance from the cluster center will increase so that these stars form an aura around the cluster.

What happens as the clusters contract? The escape of stars and resultant contraction lead to a collapse of the remains in a period 100 times the time of relaxation. VAN WIJK[3] shows that the Pleiades are headed for collapse to 40 stars in 25 cosmic years*. For a cluster one-half the radius of the Pleiades and one-eighth the mass, collapse comes in 5 cosmic years. At the other end of the scale, globular clusters endure for several hundred cosmic years. The physical evolution of their individual stars is faster than the dynamical one of the cluster as a whole. KING[4] notes that the lifetime of a cluster, however, is substantially reduced by evolutionary contraction. If a cluster loses just under 1% of its stars in time equal to the relaxation time, then the time for half the stars to escape, $93\,T_0$, (where T_0 is the initial time of relaxation), is reduced by considerations of contraction to $39\,T_0$. For $\frac{9}{10}$ the stars to escape, the comparable time is $42\,T_0$.

41. Effects of stellar encounters and of galactic rotation. Whereas at one time it was thought that all clusters might eventually loosen up from energy transferred by passing field stars, only very loose clusters are affected by these outside stars. Encounters with other members of the cluster are much more important because the density in the cluster is several times the background density. For clusters like the Pleiades and Praesepe, VAN WIJK has shown that the effect of the escape of stars by the evaporation process is twenty times as large as the effects between encounters of cluster members and field stars. MINEUR[5] showed that a cluster cannot be stable unless its density is at least 0.6 solar masses/psc^3, the critical density in the neighbourhood of the Sun. The specific effects of galactic rotation are negligible for clusters whose densities are larger than the critical value.

* One cosmic year is a time of $2 \cdot 10^8$ years.

[1] V.A. AMBARTSUMIAN: Ann. Leningrad State Univ. No. 22, 19 (1938).

[2] L. SPITZER jr.: Astrophys. Journ. **127**, 17 (1958).

[3] U. VAN WIJK: Sky and Telescope **8**, 123 (1949). — Ann. d'Astrophys. **12**, 81 (1949).

[4] I. KING: Astronom. J. **62**, 144 (1957).

[5] H. MINEUR: Ann. d'Astrophys. **2**, 1 (1939).

The disruptive effect of passing interstellar clouds is very important however, as SPITZER[1] has recently found it to be 30 times more rapid in disruption rate than the values by BOK[2] for field star encounters, as applied to clusters with 5 psc radius. SPITZER's value of the disruption time t_d for values of central density (ϱ_c) up to 2.2 $\mathfrak{M}_\odot/\text{psc}^3$ is $t_d = 1.9\times 10^8\,\varrho_c\,(\mathfrak{M}_\odot/\text{psc}^3)$ years. The Hyades will be disrupted in 6×10^7 years, the Pleiades in around 10^9 years or less. For most galactic clusters, with mean densities between 0.5 and 5 $\mathfrak{M}_\odot/\text{psc}^3$, the disruption time is between 10^8 and 10^9 years. Only a very dense cluster like M67 with density 140 $\mathfrak{M}_\odot/\text{psc}^3$ can survive for 5×10^9 years, which probably explains the rarity of very old galactic clusters.

Table 11. *Disintegration time of clusters.*

Cluster	Mass (Sun=1)	Radius (light-years)	Velocity dispersion (km/sec)	Time of Relaxation (years)	Time of Relaxation (cosmic years)	Effect of galactic tidal forces	Future
Globular cluster	10^6	50	12−15	10^{10}	50	None	Not much change during next 50 cosmic years
Pleiades	400	8	0.6	3×10^7	0.15	Probably slight	In 10 cosmic years, mass of 200 suns and radius of 3 light years
Small, dense galactic cluster	50	4	0.3	8×10^6	0.04	Probably slight	Multiple star system; periodic orbits within 5 cosmic years
Nucleus of Hyades	150	16	0.2	10^8	0.5	Disintegration in 15 cosmic years	Disintegration in 15 cosmic years
Nucleus of Ursa Major cluster	100	75	0.1 ? ?	Irrelevant	Irrelevant	Disintegration in $2\frac{1}{2}$ cosmic years	Disintegration in $2\frac{1}{2}$ cosmic years

BOK: Sky and Telescope **10**, 240 (1951).

BOK's computations show that clusters below a critical density will disintegrate in a direction toward and away from the galactic center. The Taurus cluster with disintegration in 3×10^9 years will be highly flattened in a direction perpendicular to the galactic plane, but not elongated in the plane. The Ursa Major cluster, with a total mass in the nucleus of about 100 suns, is spread over a volume of space 150 light years in diameter. This cluster should double its radius in about 200 million years, and in 500 million it will be so broken up as to lose its identity. (But compare Sect. 35 for results from different assumptions.) The disintegration times of several well-known clusters, computed by BOK, are given in Table 11.

IX. Nebulous and very young clusters.

42. General. A great many of the galactic clusters, including some of the best-known like the Pleiades, the Orion cluster and the cluster around η Carinae are tangled in nebulosity. In some cases, like the Pleiades, this shines purely by reflected light from nearby stars; in other cases, such as Orion, it gives a bright-line glowing gas spectrum. Therefore, the vast subject of diffuse nebulosity and

[1] L. SPITZER, jr.: Astrophys. Journ. **127**, 17 (1958).
[2] B. J. BOK: Harvard Obs. Circ. 384 (1934). — Sky and Telescope **10**, 211, 239 (1951).

cosmic dust and gas is a close relative of the subject of galactic clusters. In recent years, studies show that some of these nebulous clusters, in particular the Orion cluster and NGC 2264, 2362 and 6530, are exceedingly young, with ages of only a few million years, and in the case of the Orion group perhaps only a few hundred thousand. The age determination may be made in several ways. One is from the point at which the color-luminosity diagram for the cluster stars departs from the main sequence (see Sect. 21), as used by P. P. PARENAGO[1] to show the group in Orion to be very young. Another method involves the age from the measured expansion rate of the cluster members, either by a proper motion study or a radial velocity study.

43. The Orion Nebula cluster. The Great Nebula in Orion so completely overshadows the nearby cluster of stars that no one bothered to catalogue the latter until TRUMPLER[2] called attention to the Trapezium Cluster. Recent work at Tonantzintla Observatory by P. PISMIS[3] has helped to untangle the various star groups in this region and give a clear picture. There are actually four groups of stars which lie on a line of constant right ascension, the well-known sword of Orion. The richest cluster, embedded in the Great Nebula, is at declination $-5°25'$. North of it are two open clusters. One, at $-4°28'$ is a loose cluster, NGC 1981; the other, at $-4°50'$ is a cluster embedded in the diffuse nebula NGC 1977. South of the Orion nebula cluster, at $-5°57'$, is the bright nebula NGC 1980 with another cluster. There may be a bridge of fainter stars connecting these clusters, which may be physically associated in a chain.

The Orion Nebula cluster is exceptional in the high concentration of its brightest stars to the center. It is estimated to contain 467 members. The brightest members, to magnitude 12.0 roughly, are contained within a radius of $3'.7$. This explains why earlier-determined diameters were only 4′ for this cluster. The radius of the cluster is increased to 15′ by star counts going to fainter magnitudes. The assumed distance of 500 psc gives a linear diameter of 4.3 psc for the cluster. The presence of the nebula renders difficult both star counts and the determination of selective absorption.

The motions of 196 of these stars brighter than magnitude 14.0 visual in an area of one-quarter square degree centered on the Trapezium have been recently investigated by K. AA. STRAND[4]. Based on ten pairs of plates with the 40-inch Yerkes refractor, with 50 years between the first and second epoch plates, STRAND'S conclusions were that the reference frame of the 50-year period showed an expansion of 1 in 6000. This gives an expansion age of the cluster of 300000 years ± 60000 (m.e.). Twenty stars in the vicinity of the Trapezium give a root-mean-square internal velocity in x and y of $0''.0012$ and $0''.0013$ per year, while the radial velocities give 3.6 km/sec. An assumption of the same velocity distribution in x, y and r direction gives a distance of 600 psc. The color-magnitude diagram shows a broad continuous sequence from $O6$ to $K5$ V, departing from the zero-age main sequence at $B3$ V. T Tauri-type objects are present at spectral type $A0$, whereas in the very young cluster NGC 2264, they do not appear earlier than $G0$. If the earliest spectral type of the T Tauri-stars is correlated with the age of the cluster in the sense that the younger clusters have earlier spectral types, then these observations lend support to the very young expansion age for this cluster.

[1] P. P. PARENAGO: Astronom. Zhur. **30**, 249 (1953).
[2] R. J. TRUMPLER: Publ. Astronom. Soc. Pacific **43**, 255 (1931).
[3] P. PISMIS: Tonantzintla a. Tacubaya Bol. No. 10, 26 (1954).
[4] K. AA. STRAND: Astronom. J. **62**, 247 (1957).

44. A very young cluster, NGC 2264. This cluster in Monoceros is at the other end of the age scale from Messier 67. Investigations of MERLE WALKER[1] in three colors to $V=17$, show a normal main sequence from *O 7* to *A 0*. Below this, the stars fall above the main sequence. The reality of this effect is confirmed by spectroscopic observation. WALKER states that the shape of the color-magnitude diagram agrees with the theory that the cluster is so young that the fainter stars are still contracting gravitationally from the pre-stellar medium and have not yet reached the main sequence. The age where the cluster stars depart is determined at 3×10^6 years. In spectra of stars later than *F 8* the rotational velocity appears to increase. This suggests that these stars have not had time to dissipate the original angular momentum of their material. WALKER notes the occurrence of T Tauri stars in the region, where 153 variable stars are known from the work of WOLF, HOFFMEISTER, HERBIG and WALKER. This concentration of T Tauri stars poses the question whether or not these stars represent a necessary stage in the gravitational contraction of all newly formed stars. The luminosity function, which is complete to $M_V=+8$, shows a maximum at $M_V=+6$.

A 21-cm investigation of the region of this cluster (as well as of NGC 2244, the cluster around the Rosette nebula) has been made by T.K. MENON[2] of Harvard. The profile on this cluster shows two clearly separated peaks. The cluster at a distance of 600 psc is on the distant edge of the first peak. A dark cloud, a region of low neutral hydrogen intensity, is immediately behind the cluster. Theories of VAN DE HULST, MULLER and OORT suggest that the formation of H_2 molecules might produce such an apparent decrease in neutral hydrogen density as is observed here. This decrease may indicate that processes of condensation and molecule formation may be well advanced in this dark nebulosity. The exhaustion of the neutral hydrogen may be related to the formation of the T Tauri stars.

X. Stellar associations.

45. Description and numbers. AMBARTSUMIAN[3] first drew attention to the importance of large stellar groupings, known as associations or aggregates, and later [*20*] defined them in this way: "*O*-associations are stellar systems where the partial density of *O* to *B* 2 stars is larger than the average field density of these stars, in such a way that this difference cannot be explained by chance fluctuations; moreover, *O* or *B* 0 stars are present". The radii of these associations range up to 200 psc, far exceeding that of a typical galactic cluster or even that of a globular cluster. The star densities are small. From a catalogue of 27 associations in 1953, MORGAN, WHITFORD and CODE[4] estimated their distances and found that they outlined the positions of three of the spiral arms of the galaxy. Simultaneously BLAAUW[5] made notable contributions to the study of the motions of expansion and estimates of age, as for the ζ Persei group. Studies of radial velocity by PETRIE[6] at the Dominion Astrophysical Observatory do not confirm the expansion idea in some groups, as that around the double cluster in Perseus.

Whereas some stellar associations appear too sparse to have been reckoned as galactic clusters, others, like the double cluster in Perseus may logically be

[1] M.F. WALKER: Astrophys. Journ. Suppl. **2**, No. 23 (1956).

[2] T.K. MENON: Astronom. J. **61**, 9 (1956).

[3] V.A. AMBARTSUMIAN: Astronom. J. UdSSR. **26**, 3 (1949).

[4] W.W. MORGAN, A.E. WHITFORD and A.D. CODE: Astrophys. Journ. **118**, 318 (1953).

[5] A. BLAAUW: Bull. astronom. Inst. Netherl. **11**, 405 (1952). — Smithson. Contrib. Astrophys. **1**, 159 (1956).

[6] R.M. PETRIE: J. Roy. Astronom. Soc. Canada **51**, 177 (1957).

counted as both star clusters and associations. The central double cluster of high density, so conspicuous on small-scale photography of the Milky Way, is surrounded by a large group of supergiants and giants of classes *B*, *A* and *M*, which is undoubtedly connected with the double cluster. They have the same radial velocities, the same interstellar calcium lines and the same distance moduli. BIDELMAN estimates the extent of this group as over 450 light years. A curious feature of a grouping such as this is that the greater the distance from which one views it, the more prominent does the large, loose aggregate of stars become. The intrinsically faint foreground stars would gradually fade out of the picture, leaving the intrinsically luminous stars of the large cluster increasingly conspicuous. Similar large groups of stars whose overall diameters run to several hundred light years have been noted in the Magellanic Clouds.

At present, 46 stellar associations have been catalogued by J. RUPRECHT [*4*]. AMBARTSUMIAN estimated that the total number in our galaxy might be a hundred, but STRUVE[1] has suggested that if we recognize groupings of T Tauri stars and other objects of low luminosity as real stellar associations, the number would then total several thousand. Associations are definitely perishable systems and will dissolve in the general mass of stars in 10 or 20 million years. This suggests that an association is created every thousand years. An association may contain about 100 stars. STRUVE estimates that 10 million stellar associations have passed through their evolutionary cycles during the lifetime of our galaxy, each ending in disintegration and the scattering of its members through the galaxy. However, ROBERTS[2] uses a basis of 1000 galactic clusters of 200 stars each, and 800 associations of 1000 stars each to compute the formation rates of early type stars in the galaxy.

The study of stellar associations is the newest branch of investigation of star clusters. Obviously of great importance, the study is still in its infancy and today's conclusions may be outdated tomorrow. Associations may more properly belong under a study of galactic structure, since an association appears to be the loosest kind of cluster.

C. Globular clusters.

I. Appearance and apparent distribution.

46. Numbers known and probable. As noted in the Introduction, most globular clusters recognized today were discovered in the 18th and 19th centuries. Since the publication of the New General Catalogue by DREYER, not many new objects have been found which are considered globular clusters. Recently the National Geographic-Palomar Sky survey with the 48-inch Schmidt telescope has added more than a dozen. Many of these are faint, rather non-descript objects, with brightest stars in the 19th magnitude. Certain brighter clusters like NGC 6838, previously considered galactic, have in recent years been moved into the globular cluster category on the basis of physical characteristics. Appendix B tabulates the 118 clusters now considered as globular.

The number actually belonging to our galaxy may be substantially higher. The work of F. K. EDMONDSON[3] on the motions indicated about 200 clusters might belong to our galaxy, while J. DE KORT[4] made computations which gave estimates

[1] O. STRUVE: Sky and Telescope **8**, 215 (1949).
[2] M. S. ROBERTS: Publ. Astronom. Soc. Pacific **69**, 59 (1957).
[3] F. K. EDMONDSON: Astronom. J. **45**, 1 (1935).
[4] J. DE KORT: Bull. astronom. Inst. Netherl. **9**, 189 (1941).

of even higher numbers, up to 2000, depending upon the assumption made for the concentration of mass to the galactic centre. From comparison with the one thousand globular clusters believed to surround the elliptical galaxy Messier 87 (Sect. 74), a large total may not be impossible.

47. Asymmetrical distribution. From the time of Sir WILLIAM HERSCHEL, the difference in distribution of globular and galactic clusters with respect to the Milky Way was recognized. While the globular clusters may be found in high galactic latitudes, they are strongly concentrated in the direction of Sagittarius in the southern Milky Way. HARLOW SHAPLEY [1] used this asymmetrical distribution as an indication that the center of our galaxy was in this direction. The system of globular clusters is not flattened like the galaxy, but has a roughly spherical arrangement around it.

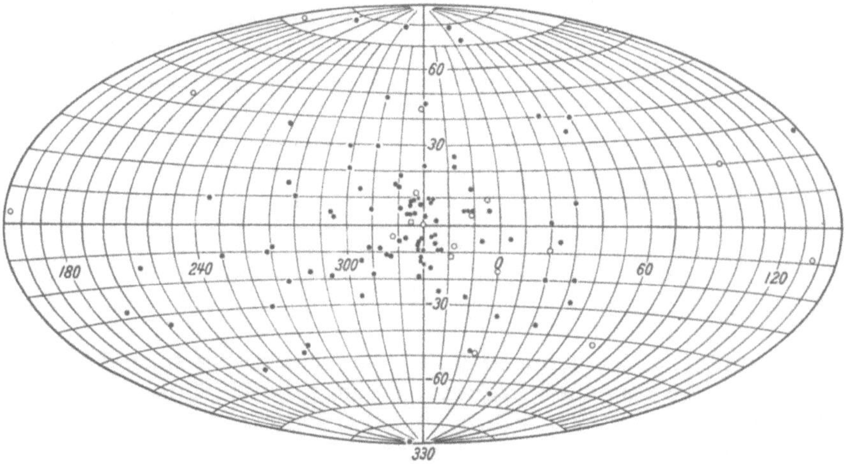

Fig. 16. Distribution of globular clusters in galactic co-ordinates. Open circles are clusters added to globular list since 1947.

On one of the long-exposure, small-scale Harvard plates of this region, almost one third of the total number of globular clusters can be seen. The globular clusters may then be considered as beacon lights, roughly outlining our galaxy and indicating the center of mass. From the concentration in Sagittarius and the dearth in the opposite part of the sky, SHAPLEY deduced that the Sun is a long way from the galactic center. It is now considered to be about 9000 psc or 30000 light years from it. The distribution in galactic coordinates of the 118 globular clusters catalogued at present is shown in Fig. 16.

48. Apparent and absolute magnitudes. Because the globular clusters are such concentrated systems, presenting a small area in ordinary-size telescopes, an estimate of integrated magnitude was early used as an indication of their relative distances, on the assumption that the scatter in their absolute magnitudes was not great. The apparent range is from naked-eye objects like ω Centauri and the great cluster in Hercules, Messier 13, down to clusters barely detectable with the most powerful instruments. SHAPLEY[1] at Mt. Wilson, using HOLETSCHEK'S[2] visual magnitudes, was able to derive an indication of the great extent of the globular cluster system. Other observers, VORONTSOV-VELYAMINOV, NABOKOV, and SAWYER and SHAPLEY, made systematic observations on as many clusters as possible.

[1] H. SHAPLEY: Mt. Wilson Contrib. 151, 152 (1918).
[2] HOLETSCHEK: Vienna, Ann. der K. K. Univ. Sternw. **20**, 40 (1907).

Further observations have shown that there is actually a great spread in the absolute magnitudes of globular clusters, which renders the apparent magnitudes of interest but reduces their reliability as distance indicators. The most uniform set of apparent magnitudes is probably that of CHRISTIE[1] for 68 clusters, determined with a schraffierkassette on a 10-inch refractor at Mt. Wilson. The integrated absolute magnitude can be determined free from the influence of interstellar absorption if the distance modulus $m - M$ has been determined photometrically by the formula

$$M_c = m_c - (m - M)$$

where the subscript c refers to integrated magnitude. CHRISTIE's frequency distribution of the absolute magnitudes is shown in Table 12. The brightest clusters, ω Centauri and 47 Tucanae (later added) have absolute magnitudes greater than -9. The faintest and feeblest clusters such as NGC 7492 and 4147 are about -5;

Table 12. *Frequency distribution of absolute magnitudes of globular clusters.*

Integrated Abs. Mag.	Number	Integrated Abs. Mag.	Number
−9.9 to −9.5	1	−6.9 to −6.5	8
−9.4 to −9.0	0	−6.4 to −6.0	0
−8.9 to −8.5	2	−5.9 to −5.5	3
−8.4 to −8.0	8	−5.4 to −5.0	4
−7.9 to −7.5	11	−4.9 to −4.5	1
−7.4 to −7.0	10		

the whole cluster has less luminosity than one very bright supergiant star. In CHRISTIE's list, the maximum frequency occurs close to $M = -7.7$, but SHAPLEY [*21*], with 31 clusters and slightly revised distances, found a median absolute photographic magnitude of -7.3. This average may drop in the years ahead as later discoveries will probably be feebler clusters.

49. Integrated colors and absorption of light. The selective reddening of many globular clusters escaped notice until STEBBINS[2] at Mt. Wilson observed it, following TRUMPLER's work on galactic clusters. The blueness of the stars in certain high-latitude clusters had previously been used as an argument against absorption in space. Measures of colors of 68 globular clusters (with effective wavelengths of cell and filters of 4340 and 4670 Å) made by STEBBINS and WHITFORD[3] showed uniform color for clusters in high latitudes, with maximum redness near the zone of avoidance in the heart of the Milky Way. The colors determined are strongly correlated with the number of extragalactic nebulae in a field; these are absent when the color excess is greater than $0^{m}.20$. The greatest observed color excess is for NGC 6539, $+0.82$, but NGC 6440, 6553, 6517 and 6760 also show very heavy coloration. NGC 2419 and 7492 are outstandingly blue. The actual amount of color excess is linked to the average value used for the color class and spectral type, and there is evidence (Sect. 59) that globular cluster stars show discrepancies, apart from interstellar absorption, in this relationship. Fig. 17 shows the colors of STEBBINS and WHITFORD, and especially the strongly reddened group around the galactic center.

Obviously a correction for interstellar absorption must be made in determining the distances of globular clusters. For high-latitude clusters, this correction

[1] W.H. CHRISTIE: Mt. Wilson Contr. No. 620 (1940).
[2] J. STEBBINS: Mt. Wilson Comm. 111 (1933).
[3] J. E. STEBBINS and A. E. WHITFORD: Mt. Wilson Obs. Contrib. No. 547 (1936).

is normally small, of the order of a tenth of a magnitude. For clusters in low latitudes it may be very large. No blanket formula can be used; it must be determined for each individual cluster. The application of a correction for absorption to the distances of globular clusters led STEBBINS and WHITFORD to a smaller estimate of the overall dimensions of our galaxy, of not more than 30000 psc. The correction factor depends on the ratio of total to selective absorption (Sect. 9). G.E. KRON and N.U. MAYALL at Lick, and J.B. IRWIN at Indiana, have under way extensive observations of colors of globular clusters.

50. Apparent and linear diameters. The range in apparent diameters of globular clusters is as obvious as the range in integrated magnitudes. At one end are diameters of a degree for the giant globular clusters, ω Centauri and 47 Tucanae; at the other end almost unresolved knots a minute in diameter for very distant or highly absorbed clusters. Diameters, as well as integrated magnitudes, were early used as distance indicators, and many attempts have been made to obtain as uniform a set as possible. Measures by MELOTTE [8] from Franklin-Adams plates were later supplemented by Harvard work of SHAPLEY and SAWYER[1], SHAPLEY and SAYER[2]. MOWBRAY[3] at Lick, with a homogeneous series of Crossley plates in blue and red light, determined surface-brightness profiles, finding the diameters at which the surface brightness falls to one-half or one-fifth that at the center, while VAN DEN BERGH[4] takes the distance of cluster-type variables as an indication of the diameter. For the estimated limiting diameter of clusters free of interstellar absorption, MOWBRAY found a loose linear relation between the absolute magnitude and the logarithm of the linear diameter, but no correlation between the absolute magnitude and log D for diameters determined from profiles of the clusters.

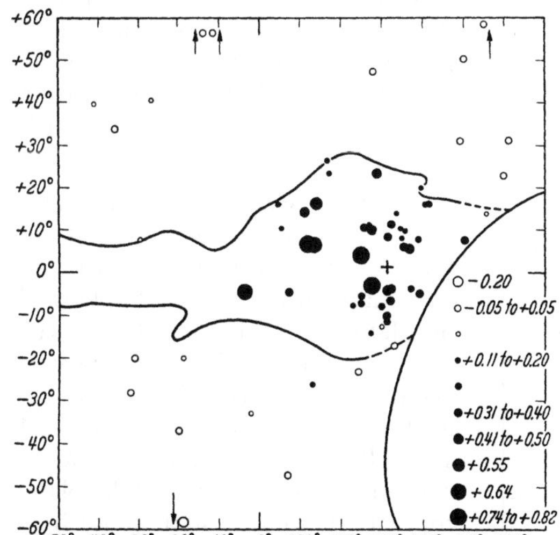

Fig. 17. Colors of globular clusters in galactic center region, by STEBBINS and WHITFORD. Astrophys. Journ. **84**, 132 (1936).

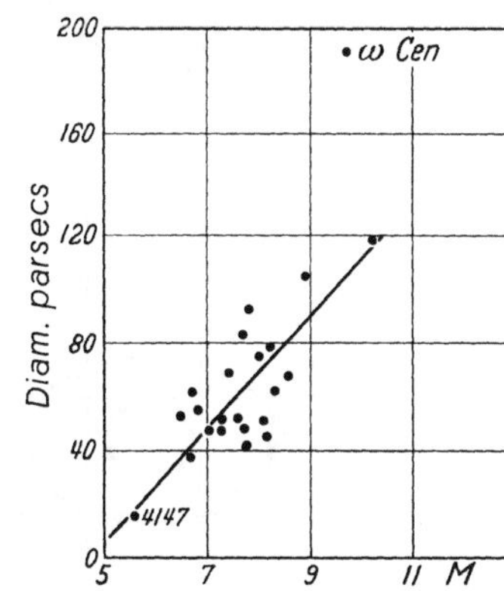

Fig. 18. Size and brightness of 23 globular clusters, by SHAPLEY. Popular Astronomy **57**, 212 (1949).

The linear diameters for 23 high latitude clusters correlated by SHAPLEY [*21*] with the absolute magnitudes are shown in Fig. 18. SHAPLEY'S point for ω Centauri has been replaced by the recent revision of LINDSAY [*16*] at absolute magnitude −10.0, distance 6.8 kpc, and linear diameter 190 psc. From long-exposure

[1] H. SHAPLEY and H.B. SAWYER: Harvard Obs. Bull. 852 (1927).
[2] H. SHAPLEY and A.R. SAYER: Proc. Nat. Acad. Sci. Wash. **21**, 593 (1935).
[3] A.G. MOWBRAY: Lick Obs. Contr., Ser. II, No. 14 (1946).
[4] S. VAN DEN BERGH: Z. Astrophys. **41**, 61 (1956).

ADH Schmidt plates from Bloemfontein, LINDSAY finds a diameter of 95′.7 in blue light, or 57′ in red, corresponding at a distance of 6.8 kpc to 620 and 370 light years respectively. At the lowest end of the scale is NGC 4147, with a linear diameter of only about 16 psc, recently shown by SANDAGE and WALKER[1] to be one of the feeblest globular clusters, with $M_{pg} = -5.84$. From the figure, it is obvious that the brighter and more massive clusters are larger. This factor, combined with the sensitivity of diameter measures to interstellar absorption, renders apparent diameters of little use as distance indicators.

The linear diameter of a globular cluster is also correlated with its distance from the galactic center, not its distance from the Sun, according to W. LOHMANN[2]. Clusters 25000 psc from the center are 30% larger than those at only 10000 psc.

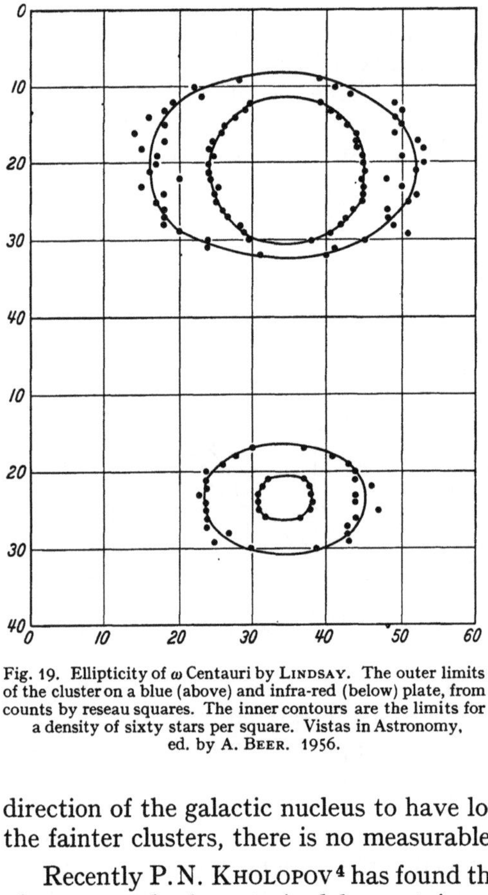

Fig. 19. Ellipticity of ω Centauri by LINDSAY. The outer limits of the cluster on a blue (above) and infra-red (below) plate, from counts by reseau squares. The inner contours are the limits for a density of sixty stars per square. Vistas in Astronomy, ed. by A. BEER. 1956.

51. Ellipticity. Anyone who measures the apparent diameters of globular clusters is struck with the fact that many of them are not spherical, but definitely elliptical. The greatest ellipticity observed is for the very asymmetrical cluster M 19, with 0.4 for the ratio of minor to major axis[3]. The ellipticity and position angle determined from small-scale plates compare favorably with those from star counts. SHAPLEY [*1*] found from 37 clusters that there is no strong evidence that the orientations are not at random, but there is a slight tendency for smaller values of the orientation to have small values of $R \sin \beta$, and for clusters in the direction of the galactic nucleus to have lower average inclinations. For many of the fainter clusters, there is no measurable ellipticity.

Recently P. N. KHOLOPOV[4] has found that globular clusters consist of a number of zones, each characterized by certain values of the average density gradient. The gradient of spatial density changes by one order of magnitude from one zone to the next. The orientation of the ellipsoid of equal apparent density in the inner regions may differ from the orientation in the outer regions. This is shown conspicuously by work of E. LINDSAY [*16*] on the giant globular cluster ω Centauri in Fig. 19, where the dots indicate the outer limits of the cluster and the inner contours are limits, for a density of sixty stars per reseau square, of 159′.5. There is high oblateness at the boundary of this cluster, but inside it is nearly circular. An infra-red plate count gives similar results, but with smaller diameters. This decreasing oblateness toward the center is a situation which is dynamically likely.

[1] A. R. SANDAGE and M. F. WALKER: Astronom. J. **60**, 230 (1955).
[2] W. LOHMANN: Z. Astrophys. **32**, 298 (1953).
[3] H. B. SAWYER: David Dunlap Obs. Publ. No. 14 (1943).
[4] P. N. KHOLOPOV: Publ. Astronom. Sternberg-Inst. **23**, 250 (1953).

52. Concentration class. Just as different types of extragalactic nebulae present a diversity of appearance, so do globular clusters. The early work of SHAPLEY

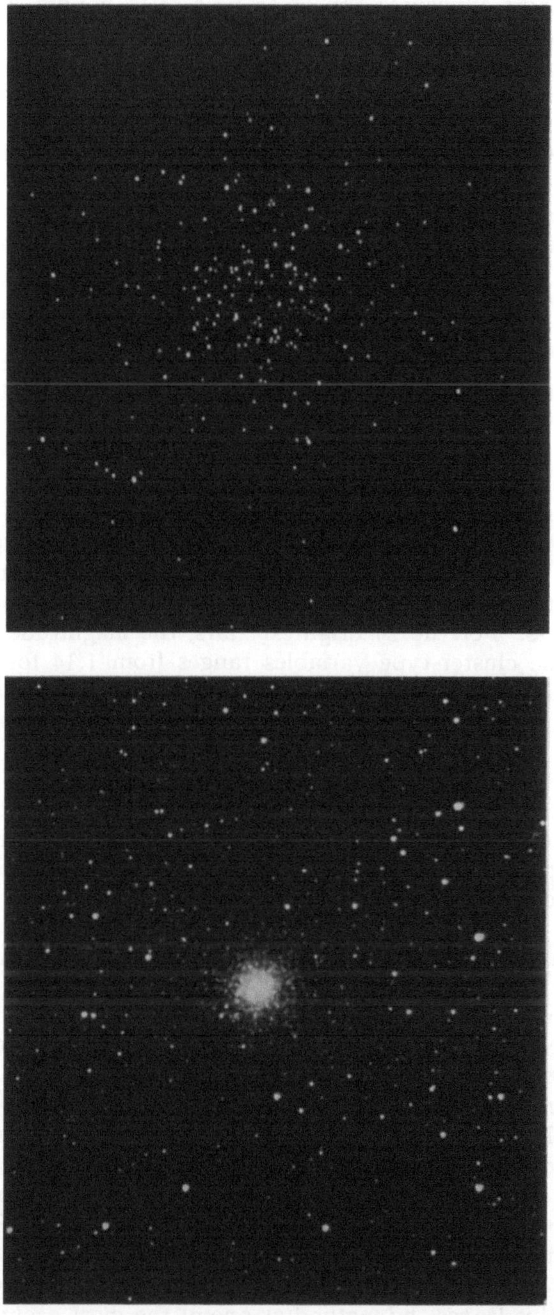

Fig. 20. Comparison of the cluster NGC 6356 class II (lower) with NGC 5466 class XII (upper). Photographs from 74-inch reflector, David Dunlap Observatory.

distinguished "giant poor" clusters, while later the clusters were assigned to twelve different classes, by SHAPLEY and SAWYER[1], on the basis of concentration

[1] H. SHAPLEY and H. B. SAWYER: Harvard Obs. Bull. No. 824 (1927).

to the center. Class I is represented by such compact objects as NGC 6864 and 7006, while the other extreme, class XII, contains clusters like NGC 5466 and 7492 which on short exposures appear to be a mere handful of scattered stars. An idea of the range in appearance of globular clusters of different classes may be gained from Fig. 20. Most of the better-known clusters are in the middle of this concentration sequence, with Messier 3 of class VI, and Messier 13 of class V. The concentration classes do not appear to be correlated with galactic positions or with quantities like the number of variable stars. They do bear a correlation with the absolute magnitudes of the brightest stars (Sect. 53), and also the most open clusters tend to average nearer and fainter. When more color-magnitude diagrams become available, the significance of the concentration classes may be more apparent.

II. Distance determinations.

53. Methods. From the telescopic power needed to resolve the individual stars, WILLIAM HERSCHEL[1] in the 19th century realized the great distances of the globular clusters. A century later, SHAPLEY at Mt. Wilson[2] extended these ideas to derive distances for the globular clusters from the magnitudes of their bright stars. Magnitudes of the 25 brightest stars were available in 28 globular clusters at that time, and variables had been studied in seven. This series was later expanded by SHAPLEY and SAWYER[3] to the brightest stars in 48 globular clusters, and to variables in 19. In the later work, the difference in magnitude between the brightest stars and the cluster-type variables was found to be correlated with concentration class. For the 25 brightest stars, the magnitude difference above the median of the cluster-type variables ranges from 1.34 for Class I down to about 0.9 for Class XII. The values of the 6th brightest star range accordingly from 1.77 to 1.3, and for the 30th from 1.04 to 0.7. For the giant poor clusters, the mean of the 25 brightest stars is only +0.44 above that of the cluster-type variables.

In order to derive an approximate distance for those globular clusters for which magnitudes of individual stars are unknown, Shapley used integrated magnitudes and angular diameters (Sects. 48, 50) standardized by correlation with clusters of known distances. These methods must be considered only a first approximation, because of the great range in absolute magnitude of globular clusters and the influence that interstellar absorption may have on the apparent angular diameters and integrated magnitudes.

Just as for galactic clusters, a well-determined color-magnitude array fitted against a main sequence provides an excellent criterion of distance, and distances for those clusters where arrays have been determined (Sect. 58) are the most accurately known.

An exceedingly handy method of distance determination in globular clusters, one not available in galactic clusters, is that of the median magnitudes of the RR Lyrae stars, or cluster-type variables. At present these have been found in 38 of the known globular clusters (Sect. 61). Evidence indicates that their mean absolute magnitude is not far from 0.0, though there is some suggestion of a spread of even up to 0.7 mag. in this value, probably related to the length of period. Still, this leaves the method as one of the most accurate. Difficulties arise in an attempt to determine the distance of these clusters from long-period cepheids. Only half a dozen clusters contain more than one of them, and there

[1] W. HERSCHEL: Phil. Trans. Roy. Soc. Lond. **108**, 429 (1818).
[2] H. SHAPLEY: Mt. Wilson Contr. No. 152 (1918).
[3] H. SHAPLEY and H.B. SAWYER: Harvard Obs. Bull. 869 (1929).

is some doubt as to the exact value of the period-luminosity relation for these. The series of curves obtained from a careful series of observations by ARP[1] is considered by REDDISH[2] to be unjustified as he postulates that for these type II cepheids, a radius-period relationship may be more fundamental than a period-luminosity relation.

54. Interstellar absorption. For the majority of the clusters, more difficult than determining the apparent magnitude of the brightest stars or variables is the problem of the correction to apply for stellar absorption. Although some globular clusters are in high latitudes, little affected by absorption, most of them are not. A glance at Fig. 16 will show that two thirds of the globular clusters are within 20° of the galactic equator and 40 are within 10°, where the absorption may be very heavy indeed. Many globular clusters therefore are affected by absorption just as much as the galactic clusters, and the corrections already discussed in Sect. 9 apply. For some of the low-latitude clusters the discrepancy in distance between published lists, such as those by LOHMANN[3] and ZAGAR[4], is rather appalling. For example, the distances in kpc given by these workers, in the order listed, (with galactic latitude in parenthesis) are:

NGC 5946 (+3°) 3.3 *vs.* 32.2; NGC 6121 (+15°) 2.2 *vs.* 7.2;
NGC 6171 (+22°) 2.6 *vs.* 21.2; NGC 6517 (+5°) 9.1 *vs.* 50.0;
NGC 6553 (−4°) 1.3 *vs.* 26.8.

Fortunately the distances of high-latitude clusters are more reliable, as shown by the revised distances of thirty of them by SHAPLEY[5], which are relatively little changed from earlier determinations. For these clusters, the agreement between published values of distance is much better, as for example for NGC 4147 (+78°), LOHMANN'S modulus of 16.7, distance 21.9 kpc, compares with ZAGAR'S distance of 20.0, SHAPLEY'S 20.0, and values from a recent intensive investigation by SANDAGE and WALKER[6] of a modulus of 16.85, distance 23.4 kpc without an absorption correction. Along with an excellent discussion of many aspects of globular clusters, W. A. BAUM[7] has recently given a careful compilation of the best-determined distances of globular clusters, which are incorporated in Appendix B.

55. Intergalactic globular clusters. One important point has recently become very obvious, namely that among the globular clusters listed as belonging to our galaxy are some at such enormous distances that they must be intergalactic objects. The first such object to be recognized, so far away that it raised the problem of whether it belonged to our galaxy, is NGC 2419, investigated by BAADE[8] in 1935. At a distance of 55.8 kpc, or 64.3 kpc from the galactic center, it was dubbed the "intergalactic tramp", as it seemed to be suspended in space. Now, as a result of the National Geographic-Palomar Sky Survey plates, other clusters of much fainter apparent magnitude, but at even greater distances are being uncovered. Of a catalogue of 14 globular clusters published by ABELL[9] (of which 12 are new objects), No. 3, a cluster in Sextans discovered by BAADE

[1] H. ARP: Astronom. J. **60**, 1 (1955).
[2] V. C. REDDISH: Monthly Notices Roy. Astronom. Soc. London **115**, 480 (1955).
[3] W. LOHMANN: Z. Astrophys. **30**, 234 (1952).
[4] F. ZAGAR: Contr. Oss. Astr. Milano-Merate, N.S., No. 50 (1954).
[5] H. SHAPLEY: Proc. Nat. Acad. Sci. Wash. **30**, 61 (1944).
[6] A. R. SANDAGE and M. F. WALKER: Astronom. J. **60**, 230 (1955).
[7] W. A. BAUM: Smithson. Contrib. Astrophys. **1**, 165 (1956).
[8] W. BAADE: Mt. Wilson Contrib. No. 529 (1935).
[9] G. O. ABELL: Publ. Astronom. Soc. Pacific **67**, 258 (1955).

and A. G. WILSON, has now been investigated by E. M. BURBIDGE and A. R. SANDAGE[1]. From the color-magnitude diagram of the brightest stars, a modulus of 20.6, and distance 130 kpc are determined. The same observers have obtained a modulus of 20.5 for Palomar No. 4, discovered by HUBBLE in Ursa Major, while ROSINO[2] at Asiago arrives at a modulus of 20.2, distance 98 kpc, from two red slow variables he has found in the cluster. In addition to ABELL'S clusters, from the charts of the Palomar Sky Atlas, Ambartsumian found a new globular cluster in Ursa Major, for which R. K. SHAKHBAZIAN[3] derived a modulus of 20.6, distance 130 kpc. It is a feeble cluster, with total visual absolute magnitude of only -5.6. Since we are much quicker to find absolutely bright objects than absolutely faint ones, it is quite possible that there is more of an intergalactic fringe of low-luminosity globular clusters than we are aware.

III. Content of globular clusters.

56. Star counts. In early work of PICKERING[4] and his associates, particularly S. I. BAILEY, at Harvard, stars by the thousands were counted in globular clusters, 6400 in ω Centauri and smaller numbers in 47 Tucanae and M 13. The number of stars per square minute of arc increased in arithmetical progression as the center of the cluster was approached. Pickering concluded that two constants are required to define the distribution of stars in the cluster, one dependent on the total number of stars, the other upon their distance. From extensive Mt. Wilson counts by PEASE and SHAPLEY on a larger number of clusters, the latter [*1*] concluded that the frequency is roughly proportional to the fourth power of the distance from the center. Star counts with magnitudes may be used not merely to derive the space density, but to determine the luminosity function of the cluster and estimate its total star content. The great power of the 200-inch telescope permits more extensive star counts than ever before obtained, as the count of 44500 stars by SANDAGE[5] in M 3 (Sect. 58).

57. Density distribution. *α) Observational.* The tedious star counts, only partly successful because of central congestion and faintness of stars, led to attempts to determine the energy distribution across the diameter of the cluster by a method less dependent on instrumental resolving power in congested areas. The methods involve the distribution of brightness as a function of distance from the center in integrated images, usually out of focus. The method developed by HERTZSPRUNG[6] of a grid and out-of-focus plate was later applied by F. S. HOGG[7] to six clusters, to show that for some clusters, star counts give a greater concentration to the center than do the intensity measures, while SCHILT[8] with a thermopile photometer obtained an opposite result for ω Centauri in which there was a much larger increase of stellar density toward the center than BAILEY'S star counts revealed. Since it is increasingly evident that the distributions of faint and bright stars are not necessarily the same, these accurate methods of measuring density distribution would appear to be important in getting the true picture of the cluster.

[1] E.M. BURBIDGE and A.R. SANDAGE: Amer. Astr. Soc., Dec. 1957 paper.
[2] L. ROSINO: Asiago Oss. Contrib. No. 85 (1957).
[3] R.K. SHAKHBAZIAN: Russ. Astr. Circ. No. 177 11, (1957).
[4] E.C. PICKERING: Harvard Obs. Ann. **26**, 213 (1897).
[5] A.R. SANDAGE: Astronom. J. **59**, 162 (1954).
[6] E. HERTZSPRUNG: Astronom. Nachr. **207**, 89 (1918).
[7] F.S. HOGG: Harvard Obs. Bull. No. 870 (1929). — Astronom. J. **42**, 77 (1932).
[8] J. SCHILT: Astronom. J. **38**, 109 (1928); Popular Astronomy **36**, 296 (1928).

β) Theoretical. The observations of globular clusters give the distribution of stars or of integrated light projected on a plane perpendicular to the line of sight. Many formulae have been given for deducing the spatial distribution from the real distribution. Foremost among them, VON ZEIPEL[1] showed that the density (of either stars or integrated light) in space at a distance ϱ from the center of the cluster is

$$\varphi(\varrho) = \frac{1}{\pi} \int\limits_{\varrho}^{R} \sqrt{r^2 - \varrho^2}\, \frac{d}{dr}\left(\frac{1}{r}\,\frac{df}{dr}\right) dr \tag{57.1}$$

in which $f(r)$ is the density in projection at a distance r from the projected center and R is the radius of the cluster, which may be infinite. PLUMMER[2] developed a simpler formula by dividing the cluster into a number of parallel strips of infinitesimal width dr. Let $F(r)\,dr$ be the total number of stars (or the sum of integrated light) in a strip with perpendicular distance r from the projected center of the cluster, then the space density is given by

$$\varphi(r) = -\frac{1}{2\pi r}\,\frac{dF}{dr}\,. \tag{57.2}$$

In the reduction of star counts, PLUMMER'S formula is simpler, but from measures of integrated light intensity, VON ZEIPEL'S formula needs to be applied. A practical simplification of von ZEIPEL'S formula has been given by I. S. KING[3], while CAMM[4] has shown that the density distribution from various models for four globular clusters does not correspond very closely to the observed distribution; the best fit is obtained from a modification of PLUMMER'S method in which angular momentum is taken into account. WOOLLEY[5] notes that a larger proportion of the massive stars will have low velocities and will be nearer the center of the cluster.

58. Luminosity functions and color-magnitude diagrams. *α) Luminosity function.* From counts of numbers of stars per unit area in clusters, the next obvious step was counts per magnitude interval and the derivation of frequency curves of apparent magnitude. SHAPLEY'S counts showed that a characteristic feature of the luminosity curve was a hump on the ascending branch. The counts did not go fainter than absolute photographic magnitude +3, and the curve was rising steeply at that point.

As an example of highly accurate and detailed counts, to absolute magnitudes fainter than our Sun, we cite the work of SANDAGE on M 3 (Fig. 21), already mentioned, with the Hale telescope. Ten plates were taken, with graduated exposure time from 5 sec to 50 min, Eastman 103-a-D plates behind a Schott GG 11 filter. A small region around the nucleus was so heavily exposed as to make counting impossible, and the number of stars in it was obtained by extrapolation, but the area is too small to affect the results markedly. The original counts of 44500 stars by SANDAGE in photographic and photovisual magnitudes have been revised to the standard B, V system now recommended, by photoelectric measures of JOHNSON[6]. The luminosity function as derived by SANDAGE[7]

[1] H. VON ZEIPEL: Ann. Obs. Paris Mém. **25** F (1908).
[2] H.C. PLUMMER: Monthly Notices Roy. Astronom. Soc. London **71**, 460 (1911).
[3] I. KING: Proc. Nat. Acad. Sci. Wash. **36**, 684 (1950).
[4] G.L. CAMM: Monthly Notices Roy. Astronom. Soc. London **110**, 305 (1950); **112**, 155 (1952).
[5] R. v. D. R. WOOLLEY: Observatory **74**, 96 (1954).
[6] H.L. JOHNSON and A.R. SANDAGE: Astrophys. Journ. **124**, 379 (1956).
[7] A.R. SANDAGE: Astrophys. Journ. **125**, 422 (1957).

is shown in Fig. 22. The apparent modulus is assumed to be 15.68. The function $\varphi(M_v)$ gives the total number of stars between $M_v+0.1$ and $M_v-0.1$ in a circular area with radius 8′ of arc, centered on the nucleus, for intervals of 0.2 mag. in V.

Fig. 21. M 3 with the 200-inch telescope. Mount Wilson and Palomar Observatories.

It was plotted from $M_v=-3.0$ to $M_v=+6.0$, at which point the function is still rising steeply. By certain approximations, SANDAGE estimates the total mass of M 3 to be $2.45\times 10^5\,\mathfrak{M}_\odot$. The amount of light contributed by stars of different M_v (Table 13) is in sharp contrast to the fractional mass of the cluster

Table 13. *Fractional light for stars at M_v and brighter in M 3.*

M_v	Fractional light	M_v	Fractional light	M_v	Fractional light
−3.0	0.004	+0.6	0.652	+4.2	0.922
−2.6	0.046	+1.0	0.684	+4.6	0.943
−2.2	0.123	+1.4	0.712	+5.0	0.959
−1.8	0.203	+1.8	0.739	+5.4	0.972
−1.4	0.282	+2.2	0.768	+5.8	0.981
−1.0	0.353	+2.6	0.800	+6.5	0.994
−0.6	0.416	+3.0	0.832	+7.5	0.998
−0.2	0.480	+3.4	0.865	+8.5	0.999
+0.2	0.615	+3.8	0.896	+9.5	0.9996

contributed by the brighter stars (Table 14). It is noteworthy that whereas 90% of the visual light of M 3 comes from stars brighter than $M_v = +4.0$, stars fainter

Table 14. *Fractional mass contributed by stars at M_v and brighter in M 3.*

M_v	Fractional mass	M_v	Fractional mass
−3.0	0.000	9.0	0.371
−2.0	0.000	10.0	0.425
−1.0	0.000	11.0	0.482
0.0	0.005	12.0	0.546
1.0	0.008	13.0	0.614
2.0	0.014	14.0	0.661
3.0	0.030	15.0	0.692
4.0	0.069	16.0	0.706
5.0	0.136	17.0	0.712
6.0	0.196	18.0	0.714
7.0	0.258	19.0	0.715
8.0	0.317	White dwarfs	0.285

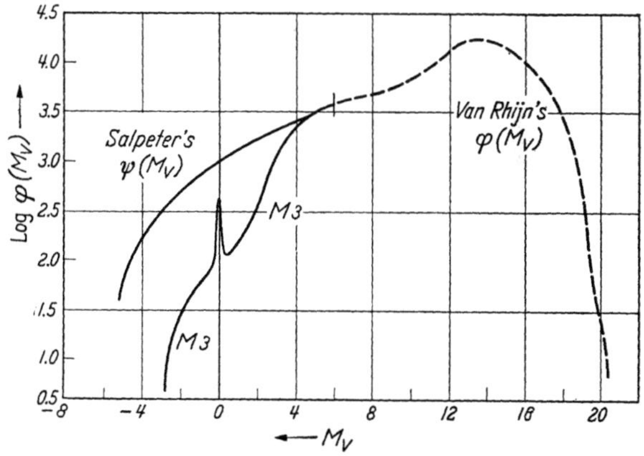

Fig. 22. The observed luminosity function for M 3 to $M_v = +6.0$ is compared by SANDAGE with the general luminosity function to $M_v = +20$. Astrophys. Journ. **125**, 422 (1957).

than this magnitude contribute more than 90% of the mass of the cluster. The number of white dwarfs present is estimated at 48500.

β) Color-magnitude diagrams. The unusual type which the globular clusters possess first showed up in work by SHAPLEY[1] when he published a color-luminosity array for M 13 showing that in this cluster the giant red stars were much brighter than the blue stars. Studies in other clusters, M 3, M 5, M 15 and M 68 confirmed the result. In all of these, the absolute luminosity brightened with increasing color, a distortion of the diagram for the solar neighbourhood. A few years later, TEN BRUGGENCATE [*13*] emphasized the two sorts of stellar population, but it was not until BAADE's dramatic resolution of the nucleus of the Andromeda galaxy and of certain elliptical galaxies that the significance of populations I and II was universally recognized.

Diagrams are now available to a considerable degree of accuracy for ten globular clusters, M 2, M 3, M 5, M 13, M 10, M 15, M 92, NGC 4147, NGC 6397, and NGC 6779. Obviously it is far more difficult to reach a significant sample of the main sequence than to reach the upper branches. Results for 2000 stars in 7 of these clusters have been intercompared by ARP[2]. The diagrams of globular clusters have features that are basically similar, but differences do exist. The

[1] H. SHAPLEY: Mt. Wilson Obs. Contr. No. 116, 51 (1915).
[2] H. ARP: Astronom. J. **60**, 317 (1955).

position of the main sequence in terms of absolute luminosity is not always easy to state, especially for clusters which have few or no cluster-type variables. Even for clusters with these variables, an uncertainty in their absolute magnitude results in uncertainty of the main sequence position. For example, the main sequence termination point for the galactic cluster M 67 is fainter than for M 3, unless M_v for the RR Lyrae stars in the latter is $+0.4$ mag. As representative of a cluster investigated to the highest degree of precision, we select M 3 (Fig. 23) where the color-magnitude diagram has been put on the standard V and B

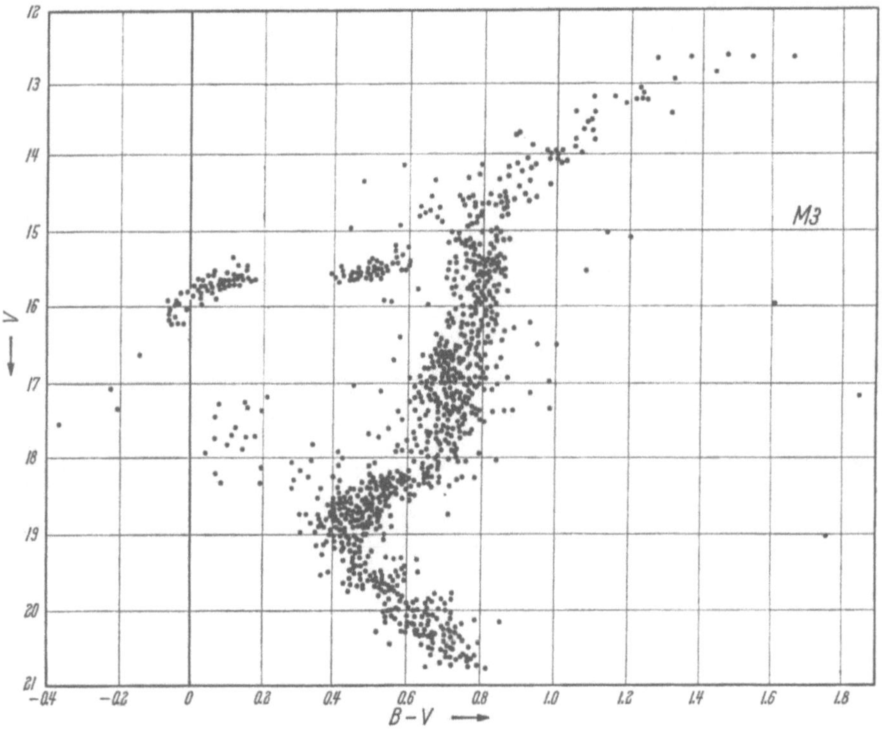

Fig. 23. Color-magnitude diagram for M 3 by JOHNSON and SANDAGE. Astrophys. Journ. **124**, 379 (1956).

system by work of JOHNSON and SANDAGE. This is the cluster richest in RR Lyrae stars, which stand out noticeably in the diagram in the portion of it they occupy (cf. Sect. 61). The bluest stars reach only to B-V $= -0.40$, while the color boundaries of the variable-star region stand out at B-V $= +0.17$ to $+0.39$. The break point from the main sequence, of real significance in determining an age for the cluster, occurs at $M_v = +3.3$, at V$=19.0$. This leads to a derived age of 6×10^9 years for this cluster, but the derived age is tied to the absolute magnitude of the RR Lyrae stars.

γ) Diagrams in three colors. The recent upsurge of interest in deriving magnitudes in three colors, U, B, and V has been applied to several globular clusters. The chief anomaly which appears from such an analysis is that certain clusters, M 3, M 92, and NGC 4147, have a noticeable ultra-violet excess similar to that noted for some weak-line *F*-type stars in the general field by Miss ROMAN[1]. The anomaly persists for the fainter stars in M 3 and may actually increase up to an excess of 0.55 magnitude for the faintest observed main sequence stars.

[1] N.G. ROMAN: Astronom. J. **59**, 307 (1954).

A suggestion made by STRÖMGREN to account for the excess is that a decrease in blanketing of metal lines (because of less metal abundance) will cause a change in U-B. Three-color photometry in M 92 which is nearly complete may test this hypothesis since the lines in M 92 are about 60% weaker than those in M 3.

59. Spectra, integrated. A clue to the type of stars in globular clusters is to be found from the integrated spectral class. The early classification by Miss CANNON[1] is chiefly of historical interest, as it was based on the distribution of light in the continuous spectrum, and made before the reddening due to interstellar absorption was recognized. It indicated later spectral types than actually exist. N. U. MAYALL[2], as a by-product of his radial velocity investigation (see Sect. 65), has given the integrated spectral types for fifty globular clusters with a precision not previously attained. The classification was based on a comparison of intensities of lines in cluster spectra with those in stellar standards established at Mt. Wilson.

For the fifty clusters, the spectral types are included in the interval from *A* 5 to *G* 5 with the majority (30) between *F* 8 and *G* 5, and an average type of *F* 7.6 ± 3.6 p.e. Table 15 shows the frequency distribution of observed spectral type.

Table 15. *Frequency distribution of observed spectral types of 50 globular clusters.*

Sp	No.	Sp.	No.	Sp.	No.	Sp.	No.
A 5	1	*F* 0	1	*F* 5	4	*G* 0	5
A 6	1	*F* 1	1	*F* 6	2	*G* 1	5
A 7	1	*F* 2	2	*F* 7	1	*G* 2	5
A 8	0	*F* 3	1	*F* 8	5	*G* 3	4
A 9	3	*F* 4	2	*F* 9	1	*G* 4	3
						G 5	2

Estimates of these spectral types are bound to be uncertain because of the composite and peculiar nature of the spectra. Clusters in the range from *A* 5 to *F* 8 were more difficult to classify than the later types. In many cases the *K*-line seemed to be "filled in" or weak. As a rule the *K*-line yielded the earliest type, H_δ and H_γ the average, and the *G* -band the latest. In only two cases, NGC 6356 and 6838, was the CN feature at λ 3883 strong enough to suggest giant characteristics. For clusters observed with the 36-inch refractor and higher dispersion (130 Å/mm at H_γ) the Sr II lines at λ 4077 and λ 4215 had a weakness which matched those in ordinary dwarfs. This is the same type of effect found in the high-velocity stars, and the RR Lyrae variables, population II objects.

MAYALL calls attention to a fairly compact group of about 10 clusters near the galactic center at longitude 330° and latitude −10°, which average nearly *G* 2 and contribute largely to the frequencies for the later classes. Recently,

Table 16. *Globular clusters of later type.*

Cluster NGC	Type (1955) λλ 4100–4350	*l*	*b*	*d*	Cluster NGC	Type (1955) λλ 4100–4350	*l*	*b*	*d*
6304	*G* 5	324°	+4°	3′.8	6637	*G* 5	329°	−12°	3′.8
6356	*G* 5	334	+9	3.5	6638	*G* 0 to *G* 2	336	−9	2.2
6440	*G* 5	335	+2	1.7	6652	*G* 2	329	−13	2.3
6441	*G* 2	321	−6	3.0	6712	*G* 0 to *G* 2	353	−6	4.2
6624	*G* 2	330	−9	2.7	6838	*G* 2	24	−6	6.1

[1] A. J. CANNON: Harvard Obs. Bull. No. 868 (1929).
[2] N. U. MAYALL: Astrophys. Journ. **104**, 290 (1946).

W. W. MORGAN[1] has applied his two-dimensional classification methods to the composite spectra obtained by MAYALL (Table 16), and has added some McDonald 82-inch Cassegrain spectra for 12 clusters, with dispersion 150 Å/mm from H_β

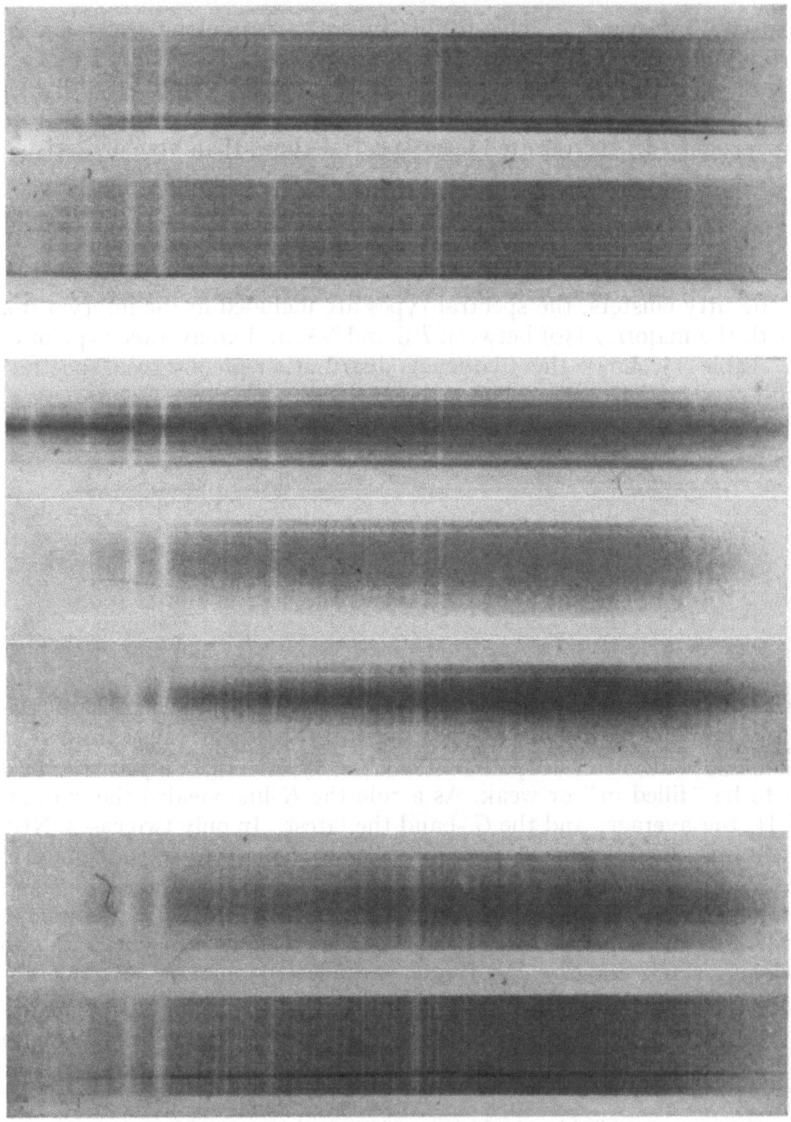

a b c d e f g

Fig. 24a—g. The spectra of three globular clusters obtained by MORGAN with the 82-inch McDonald telescope, in comparison with spectra of other objects. Publ. Astronom. Soc. Pacific **68**, 509 (1956). (a) HD 140283, an *F*-type weak-line star. (b) 45 Boo, *F* 5 V. (c), (d), (e) the globular clusters M 15, M 3 and NGC 6356. (f) the nuclear region of M 31. (g) the sun. W. W. MORGAN.

to H_ε. Spectral types were determined by direct comparison with a sequence of standard dwarfs on the MK system, in three ways, by a comparison of the *G*-band with H_γ; the absolute intensity of H_γ, and the intensity of the strong Fe I lines in the region λλ 4250—4400 to give a "metallic-line" type. MORGAN notes a remarkable variation of metallic line strength from cluster to cluster,

[1] W. W. MORGAN: Publ. Astronom. Soc. Pacific **68**, 509 (1956).

and a surprising absence of noticeable compositeness in some spectra. The spectra of most of these clusters match closely with subdwarf *F* stars, with metallic lines greatly weakened for the type indicated by H_γ and the *G* band.

For three clusters, NGC 6356, 6637 and 6440, the abundance of metals is more nearly normal: for 6356 even similar to the Sun. These three clusters are so different from the others in spectral appearance that MORGAN concludes it is hard to see how the stellar constitution can be similar. Fig. 24 shows the spectra of the globular clusters M 15, M 3 and NGC 6356 compared with other pertinent spectra.

60. Spectra, individual. Since very few globular cluster stars are brighter than apparent photographic magnitude 13, they are difficult objects to study spectroscopically. PEASE[1] pioneered in this field when he classified the spectra of 39 individual stars in Messier 13 from three plates exposed from 20 to 30 hours in a focal-plane spectrograph on the 60-inch Mt. Wilson reflector. He obtained an average of *F* 0 for the brightest stars. (This compares with MAYALL's *F* 2 estimate from integrated spectrum.) LINDBLAD[2], at Mt. Wilson, from a slitless spectrogram of M 13, first called attention to the weakness of the cyanogen absorption in three supergiant stars about spectral type *K* 0.

POPPER[3], in 1947, with the McDonald 82-inch reflector and the Cassegrain spectrograph and *f*/1 Schmidt camera with a dispersion of 150 Å/mm at H_γ, obtained spectra for 21 stars in M 13 and 14 stars in M 3. As PEASE picked his stars at random, the yellow-red brightest stars were missed, whereas Popper deliberately selected from among the 30 brightest. From POPPER's spectrograms the brightest stars, with absolute photovisual magnitudes between -2 and -3, and color-indices between $+0.9$ and $+1.6$ are of spectral type *G* 5 to *K* 0 and luminosity class Ib. They are peculiar in that the absorption shortward from λ 4215 is weak. Each cluster has at least one *O*- or *B*-type star among its brightest members but these are not of great luminosity.

Another attack on this problem has now been begun with the 8-inch coudé camera on the 200-inch telescope, by BOWEN, O. C. WILSON and DEUTSCH[4]. With this equipment at a dispersion of 38 Å/mm, photographic magnitude 14.0 can be reached in about three hours' exposure. BAUM[5] found that two red giants in M 3 and two in M 92 had spectral types far earlier than predicted from their color. For example, an *F* 5 type star in M 92 has a color class for *K* 3. KEENAN using ratios of lines of Fe and Cr classified the star as *G* 3.

For the velocities of individual stars, see Sect. 67.

61. Variable stars. α) *Numbers and types.* The discovery of large numbers of variable stars in globular clusters in the last decade of the 19th century gave an impetus to this study. Following the early work of E. C. PICKERING, DAVID PACKER and BARNARD, S. I. BAILEY of Harvard and his associates investigated several prominent globular clusters which proved very rich in variables, ω Centauri, M 3, M 5, and M 15. That not all clusters contain many variables was emphasized by the great Hercules cluster, M 13, which proved then, as now, to have relatively few.

In the present century, a systematic search of clusters, one by one, with large telescopes, is bringing us a good understanding of the over-all variable content of globular clusters. In particular, the systematic program of ROSINO[6] at Bologna

[1] F. G. PEASE: Publ. Astronom. Soc. Pacific **26**, 204 (1914).
[2] B. LINDBLAD: Mt. Wilson Obs. Contr. No. 228 (1922).
[3] D.M. POPPER: Astrophys. Journ. **105**, 204 (1947).
[4] A. J. DEUTSCH: Principes Fondamentaux de Classification Stellaire, p. 25. Paris 1955.
[5] W.A. BAUM: Astronom. J. **57**, 222 (1952).
[6] L. ROSINO: See series beginning 1944 in Bologna Obs. publications and continuing in Asiago.

Table 17. *Summary of variable stars in 75 globular clusters.*

NGC	No. Var's.	Per's Pub.	Types	NGC	No. Var's	Per's Pub.	Types
104	11	3	*M, I*	6287	3	0	
288	1 (2)	1	*SR*	6293	5	0	
362	14	10	*RR, C*	6333	13	11	*RR*
1851	2 (1)	0		6341	16	13	*RR, EW*
1904	5	3	*RR, I*	6356	5	0	
2298	(6)	0		6362	15	0	
2419	36	0	*RR, I*	6366	2	0	
2808	(4)	0		6397	3	3	*RR, M, SR*
3201	77	59	*RR, E, M*	6402	72	3	*RR, C*
Pal 4[1]	2	2	*M*	6426	11	0	
4147[2]	16	15	*RR*	6522	9	8	*RR, I*
4372	(3)	0		6528	0	0	
4590[3]	31	24	*RR, M*	6535	1 (1)	0	
4833	10	9	*RR, M, C*	6539	(1)	0	
5024	43	32	*RR*	6541	1	0	*M*
5053	10	10	*RR*	6553	6	3	*RR, C, N*
5139	164	153	*RR, C, I, RV, EW*	6558[5]	2	0	
5272	187	171	*RR, C, SR, EW*	6584	0	0	
5286	0	0		6626	16	1	*RR, C*
5466	18	18	*RR*	6656	24	22	*RR, C, M, SR*
5634	7	1	*RR*	6712	12	1	*SR*
5694	0	0		6715	28	0	*RR*
Pal 5[4]	5	3	*RR*	6723	19	19	*RR*
5897	4	0	*RR*	6752	1	0	
5904	97	92	*RR, C, I, UG*	6760	4 (2)	0	
5986	1	0		6779	12	4	*RR, C, RV, SR*
6093	7	1	*C, M, N*	6809	6	5	*RR*
6121	43	41	*RR, SI*	6838	4	0	*E*
6144	1	0		6864	11	0	
6171	24	0	*RR, M*	6934	51	0	*RR*
6205	10	6	*RR, C, SR*	6981	39	27	*RR*
6218	1	1	*C*	7006[6]	49	20	*RR, M*
6229	22	0	*RR, C*	7078[7]	97	62	*RR, C*
6235	2	0		7089	17	17	*RR, C, RV*
6254	3	2	*C, SR*	7099	4	3	*RR, UG*
6266	26	0		Pal 13[1]	4	0	*RR*
6273	4	0		7492	1 (8)	0	
6284	6	0					

Notes: Numbers of variables in parentheses are unpublished, and are included for clusters where they form a high percentage of the variables known.

Abbreviations (on KUKARKIN's system in the General Catalogue of Variable Stars): *RR* = RR Lyrae, *C* = cepheid, *M* = long period, *SR* = semiregular, *I* = irregular, *UG* = U Geminorum, *E* = eclipser, *EW* = eclipser, W Urs Maj type.

The sources of most of this data are to be found in David Dunlap Publ. **1**, No. 20, (1939), and **2**, No. 2 (1955). Footnotes refer to material which changes these earlier totals appreciably.

[1] L. ROSINO: Asiago Contr. No. 85 (1957).

[2] A. R. SANDAGE and M. F. WALKER: Astronom. J. **60**, 230 (1955). — R. L. NEWBURN jr.: Astronom. J. **60**, 197 (1957).

[3] L. ROSINO and S. PIETRA: Bologna Obs. Publ. **6**, No. 5 (1954).

[4] S. PIETRA: Bologna Obs. Publ. **6**, No. 16 (1956).

[5] L. ROSINO: Asiago Contr. No. 52 (1954).

[6] L. ROSINO and G. MANNINO: Asiago Contr. No. 59 (1955).

[7] I. IZSAK: Mitt. Sternw. Budapest No. 28 (1952). — G. MANNINO: Asiago Contr. No. 75 (1956).

and Asiago, and SAWYER[1] at the David Dunlap Observatory have filled in many gaps in the clusters examined, and determination of types of variables. At present, as shown in Table 17, 75 of the 118 clusters have been examined, and 1453 variables have been published in 67 of these. Periods are known for 876 stars in 33 clusters. A catalogue of the variable stars in globular clusters, first published by SAWYER[2] in 1939, appeared in a second edition in 1955. The difference from cluster to cluster in the number of variables is very great. Four clusters, NGC 5286, 5694, 6528, 6584 have yielded none, while NGC 5272 (M 3) has 187. Only 7 clusters contain more than 50 variables. Three-quarters of the clusters examined contain fewer than 20. Probably most of the clusters exceedingly rich in variables have now been examined.

The 10% of the variables which are not RR Lyrae types is made up of most other types of variables, but some of these are field stars, not cluster members. These variables are among the brightest stars in the region of the cluster and their light variations, usually greater than a magnitude, make them conspicuous. Most of the cepheids, either classical or W Virginis, the RV Tauri stars and the irregular variables seem to be actual cluster members. On the other hand, most of the long-period variables with periods greater than 200 days, except those in 47 Tucanae, appear to be field stars. The nova in NGC 6553 was probably not a member of it, while the membership of the nova in NGC 6093 is problematical. Most of the eclipsing stars, including W Ursae Majoris type, are not cluster members.

β) The RR Lyrae stars. About 90% of the variables found in globular clusters are RR Lyrae stars but, despite this large proportion, these stars are not present in all globular clusters. Nearly half the clusters so far examined give no positive indication that an RR Lyrae star is in their midst. The lack of these variables in a cluster was shown by M. SCHWARZSCHILD[3] to be caused by an absence of stars in the appropriate evolutionary stage, because stars in a certain position on the color-luminosity diagram are almost certainly RR Lyrae variables. ARP[4] has correlated the characteristics of the variables present with the position of the giant sequence in the color-magnitude diagram.

The RR Lyrae variables have been the center of many investigations. BAILEY[5] early showed that these stars fall into three main subclasses, *a*, *b* and *c* (Fig. 25), according to the period length and shape of light curves. For many years these stars have been assumed to have a mean absolute magnitude of 0.0, making them enormously useful as distance indicators. There is now evidence of a spread of some tenths of a magnitude for the absolute magnitude of some of these stars[6]. The constancy of period of these stars has also been studied, especially in several of the richer clusters such as M 3 and ω Centauri. Some RR Lyrae stars are remarkably constant in both period, and shape of light curve. Others exhibit a fluctuation in shape of curve, or time of maximum around a mean value, or a secular lengthening or shortening of the period. In ω Centauri, MARTIN[7] found 54 stars whose periods were lengthening so fast that in 10^6 years the period would increase by $0^{d}.2$. He noted that a variable of type *a* would become one of type *b* if the relation between the period and characteristics of light variation remained

[1] H. B. SAWYER: See series beginning 1934 in Dominion Astrophysical Publications and continuing in David Dunlap Observatory.
[2] H. B. SAWYER: David Dunlap Obs. Publ. **1**, 20 (1939); **2**, No. 2 (1955).
[3] M. SCHWARZSCHILD: Harvard Obs. Circ. No. 437 (1940).
[4] H. ARP: Astronom. J. **60**, 317 (1955).
[5] S. I. BAILEY: Harvard Obs. Ann. **38** (1902).
[6] P. NOTNI: Mitt. Univ. Sternw. Jena No. 26 (1957).
[7] W. CHR. MARTIN: Leiden Obs. Ann. **17**, pt. 2 (1938).

the same. There is great variation in the frequency of periods from cluster to cluster as noted by OOSTERHOFF[1] and SAWYER[2], probably of significance in the degree of evolutionary development of the cluster.

γ) The cepheids in globular clusters have been eagerly investigated because of the importance of the period-luminosity relationship. They have been shown, largely by the spectroscopic work of JOY[3] (Sect. 67) to be mostly of the W Virginis type, population II cepheids. Nevertheless, a few stars appear to be classical cepheids, numbers 1, 2 and 6 in M13, number 3 in M10, number 1 in M15. The scarcity of cepheids in any single cluster except ω Centauri renders the period-luminosity relation somewhat indeterminate. Only half a dozen clusters contain more than one long-period cepheid. A composite period-luminosity relation constructed[4] by adding together 25 cepheids from 13 clusters is shown in Fig. 26. ARP'S concentrated and accurate magnitude work in 6 clusters[5] did not erase the spread around the curve and was interpreted by him to indicate a series of relationships.

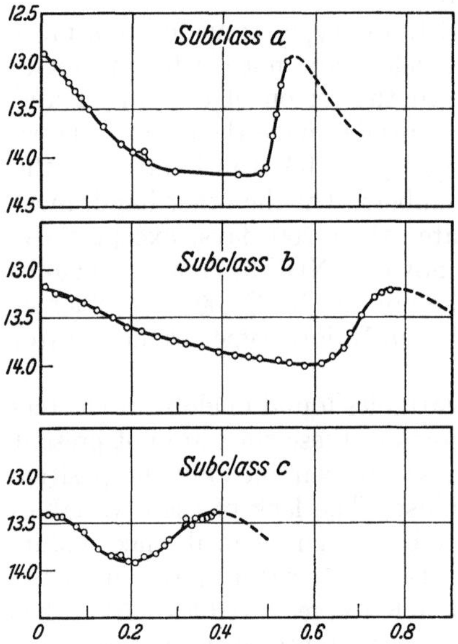

Fig. 25. BAILEY's subclasses of cluster-type variables. Harvard Ann. **38** (1902).

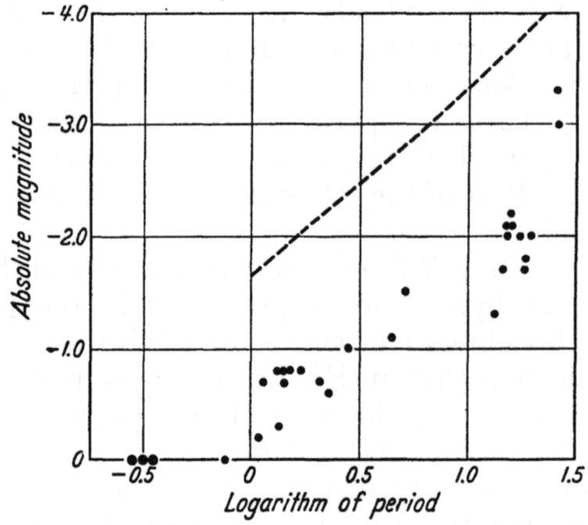

Fig. 26. A composite period-luminosity relation for cepheids in globular clusters. The dotted line suggests the relation for type I cepheids. Proc. Nat. Acad. Sci. Wash. **39**, 349 (1953).

62. Planetary nebula in M 15. This object is in a class by itself, being the only one of the type yet recognized in a globular cluster. Situated near the cluster center, it was discovered by F. G. PEASE[6] on a plate taken at the Cassegrain focus of the 100-inch telescope Aug. 30, 1927 through a Pulkova ultraviolet color filter. The star Küstner No. 648, pg. mag. 13.78 appeared very bright. Two spectrograms of 4 hours exposure each showed continuous *O*-type spectrum with strong bright lines. The radial velocity indicated it to be a cluster member.

More recently, JOY[3] has repeated the velocity measures and states that the nebula, with velocity from two plates, −122 km/sec compared with a cluster velocity of −114 km/sec, is without doubt a cluster member. JOY measured the bright lines of hydrogen and 4471 HeI, 3726 OII, 3728 OII, 3868 NeIII, 4363 OIII, 4958 OIII and 5006 OIII.

[1] P. TH. OOSTERHOFF: Observatory **62**, 104 (1939).
[2] H.B. SAWYER: David Dunlap Obs. Comm. No. 11 (1944).
[3] A.H. JOY: Astrophys. Journ. **110**, 105 (1949).
[4] H. SHAPLEY and V. MCKIBBEN NAIL: Proc. Nat. Acad. Sci. Wash. **39**, 349 (1953).
[5] H. ARP: Astronom. J. **60**, 1 (1955).
[6] F.G. PEASE: Publ. Astronom. Soc. Pacific **40**, 342 (1928).

63. Diffuse nebulosity in globular clusters. The problem of the presence of nebulous matter in these objects has been a very controversial problem for a hundred years, ever since the drawing by the Earl of ROSSE[1] of the dark lanes in the Hercules cluster, M 13. Many astronomers who have worked for years with big telescopes on globular clusters have been convinced that the dark lanes conspicuous on the photographs are not an illusion resulting from a chance grouping of stars, as some astronomers insist. M 13 is by no means the only cluster on which these are visible. SCHEINER[2], in his work on the Hercules cluster, commented that he had seen them both visually and on photographs.

Recent thought has tended to prescribe that nebulosity is impossible in globular clusters at present. They are admittedly old systems; and presumably most of them have already gone through the plane of the galaxy more than once, a process which would tend to sweep them clear of nebulous matter. A recent analysis by M. CIMINO[3] shows that remnants of gaseous material will no longer be retained in these systems because of the central temperature of 150° K, but will escape into interstellar space.

Is there, however, any reason why the wisps of nebulosity which appear so conspicuous on photographs of globular clusters cannot lie between us and the cluster? If we are to presume that small patches of nebulosity are scattered around our galaxy even in higher latitudes, there is no celestial object which would show them up more quickly than a globular cluster, which subtends a moderately large surface area with a moderately uniform pattern of stars. The stars in high latitudes do not provide a rich background against which dark nebulosity would be visible; the extragalactic nebulae have smaller surface areas, and in many cases have their own associated nebulosity "built in".

The argument that the nebulosity does not exist because longer exposures show stars behind it does not seem very valid since, if the nebula were small and shallow in depth so that it cut down the starlight only a magnitude or two, longer exposures would penetrate it.

When a sufficient time interval has elapsed it may be possible to show relative motion between the presumed dark streaks in the cluster and the stars belonging to it.

IV. Motions.

64. Proper motions. Although proper-motion work on globular clusters was begun half a century ago, because of the great distances of these objects the results to date have not been many. The early work of BARNARD, KÜSTNER, BALANOVSKY

Table 18. *Relative proper motions of eight globular clusters.*

Number	μ_x	$E\mu_x$	μ_y	$E\mu_y$
NCG 4147	+0.″0014	±0.″0002	0.″0000	±0.″0017
M 53	+0.0125	±0.0006	−0.0014	±0.0018
M 3	+0.0088	±0.0011	+0.0050	±0.0009
M 5	+0.0078	±0.0005	−0.0063	±0.0011
M 13	−0.0020	±0.0004	+0.0036	±0.0011
M 92	−0.0016	±0.0013	+0.0030	±0.0018
M 15	−0.0037	±0.0003	+0.0015	±0.0006
M 2	+0.0036	±0.0007	+0.0072	±0.0015

[1] Earl of ROSSE: Phil. Trans. Roy. Soc. Lond. **1861**, 732, plate 28.

[2] J. SCHEINER: Abh. kgl.-preuß. Akad. Wiss., Anhang 1, 1 (1892). — Himmel u. Erde **6**, 105 (1893).

[3] M. CIMINO: Oss. Astr. Roma Monte Mario Contr. Sci., N.S., No. 201 (1954).

and LUDENDORFF was followed by VAN MAANEN'S work with the 100-inch telescope. These observations have been more recently supplemented by N.V. GAMLEJ[1] for 8 globular clusters (Table 18). For all of these, the derived proper motions are larger than the probable errors. GAMALEJ uses these results to derive the solar motion of 196 km/sec toward the apex in galactic longitude 108° and latitude +21°, as well as the space motion of globular clusters. He says that the uncertainty of his material is too great to comment on the proper motions of the individual stars. The proper-motion work has concentrated on the big, bright clusters which are easily observed in the northern hemisphere, rather than on the nearest clusters like Messier 4 and Messier 22 in which motions would be expected to show up much more rapidly.

65. Radial velocities, integrated. The determination of the line of sight motion for globular clusters has proved more successful than the attempt at the proper

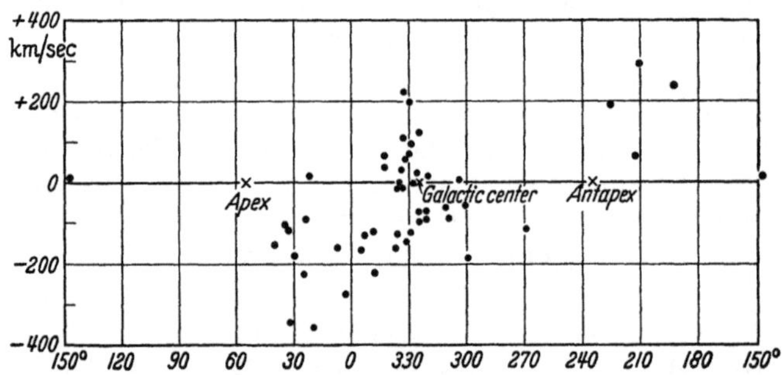

Fig. 27. Observed radial velocities of 50 globular clusters plotted against galactic longitude. Astrophys. Journ. **104**, 290 (1946).

or cross motion. The pioneer work by V.M. SLIPHER[2] between 1918 and 1924 yielded velocities for 17 clusters and various observers later added a few more. N.U. MAYALL[3] made a substantially complete study of clusters north of Dec. −40° and doubled the number of velocities known, bringing the total to 50. MAYALL'S velocities were determined with a low-dispersion spectrograph (430 Å per mm at H_γ) on the 36-inch Crossley reflector of the Lick Observatory.

The velocities range from −360 km/sec to +291 km/sec. There are 31 negative and 19 positive velocities; the center of gravity on the velocity axis falls at −58 km/sec. The excess of negative velocities is due to a selection effect in galactic longitudes, as shown by the distribution in Fig. 27. The gap in distribution of velocities between longitudes 40 and 190° is real, but the gap between 230 and 300° is due to the fact that this part of the sky cannot be observed from the north temperate zone. The radial velocities of the fifty globular clusters are included in Appendix B.

66. Galactic orbits of globular clusters and galactic rotation from globular clusters. A procedure for the analysis of globular-cluster motions was developed by F.K. EDMONDSON[4] for 26 observed clusters. Recently S. VON HOERNER[5] has considered the larger number of velocities available from MAYALL'S work to find whether globular-cluster motions are predominantly radial or circular. After subtracting the motion of the Sun (260 km/sec toward Cygnus and 20 km/sec

[1] N.V. GAMLEJ: Publ. Astr. Obs. Pulkova **17**, 6, No. 141 (1948).
[2] V.M. SLIPHER: Popular Astronomy **26**, 8 (1918); **30**, 11 (1922); **32**, 622 (1924).
[3] N.U. MAYALL: Astrophys. Journ. **104**, 183 (1946).
[4] See footnote 3, p. 166.
[5] S. VON HOERNER: Z. Astrophys. **35**, 255 (1955).

toward Hercules), VON HOERNER has computed the component in the line of sight for MAYALL'S 50 cluster velocities. The angle ϑ is taken as the angle between the Sun and the galactic center, whose distance is assumed to be 9000 psc. Clusters in circular orbits would be expected to show small line-of-sight motions for small values of ϑ (near zero), and larger values for ϑ around 90°. VON HOERNER finds the opposite true, in that 40 objects with ϑ between 0 and 45° have average line-of-sight motions of 106 km/sec, while those between 45 and 90° have only 74 km/sec. Almost one-half the clusters are moving too fast for circular orbits at their respective distances. The most distant clusters are about 50000 psc from the galactic center. This may be taken as the maximum radius for a circular orbit of a globular cluster. If the cluster has an almost rectilinear orbit of very high eccentricity, its semi-major axis would be about 25000 psc. By an application of KEPLER'S law, the Sun's orbit around the galactic center has a period of 200 million years and a semi-major axis of 9000 psc. For a distant globular cluster as cited, the corresponding periods are $2\frac{1}{2}$ and 0.9 billion years for a circular and rectilinear orbit, respectively. Even the most remote globular clusters therefore have galactic periods well under their lifetimes of 6 billion years. Nearer globular clusters have much shorter periods; Messier 13 has 200 million years for a circular orbit, or only 70 million for a radial.

From his radial velocities for 50 clusters, MAYALL made a number of solar motion solutions and found no appreciable deviation of the computed apex from the one generally assumed at $l=55°$ and $b=0°$. No K-term was indicated. The solar motion with respect to the clusters was (175 ± 25) km/sec, but observational selection probably makes this value too small. The most probable value deduced for the mean solar motion is (200 ± 25) km/sec. There now seems to be a real difference of 80 to 100 km/sec between the solar motion with respect to the galactic center and to the system of globular clusters. The obvious interpretation is that the cluster system partakes in the general galactic rotation. Attempts to find evidence of a general or differential rotation of the cluster system have been inconclusive. There is a possibility of a flattening of the system analogous with the projected distribution of clusters in the Andromeda galaxy. MAYALL considers that the radial velocities of these would provide the most direct means for further investigations of the dynamics of a system of globular clusters.

67. Radial velocities of individual stars. These are difficult to obtain because of the faintness and crowding of globular cluster stars. They are important for a study of the velocity dispersion of cluster stars, for the behaviour patterns of variable stars, and for proof of membership of a star in the cluster, as the planetary nebula in M 15 (Sect. 62).

The first attempt at velocities of globular cluster stars was that of SANFORD[1] who observed seven (without published identifications) in M 3. In recent years, A.H. JOY obtained an extensive series of spectra of slow variables in these clusters, which, because at maxima they are usually among the brighter stars, lend themselves to such an investigation (Sect. 61). With the 100-inch telescope, a dispersion of 110 Å/mm at H_γ, and exposures up to 4 hours, he obtained spectrograms of 35 variables in 14 globular clusters and was able to correlate their velocity curves with their light cycles (Fig. 28). JOY found the velocities of the individual stars to have insufficient precision to justify any conclusions about peculiar motions or rotation effects.

As a supplement to his spectral classification in Messier 13 and Messier 3, POPPER (Sect. 60) measured the radial velocities on 14 spectrograms of 13 Ib

[1] Ann. Rep. Dir. Mt. Wilson Obs. **1918**. — Popular Astronomy **27**, 99 (1919).

stars in M 13. His plates were taken at the Cassegrain spectrograph of the McDonald Observatory, with the $f/1$ Schmidt camera, dispersion 150 Å/mm at H_γ, with which equipment photographic magnitude 13.5 can be reached in one hour. Like JOY, POPPER states that his material is not well suited for velocity dispersion but he is able to conclude that the dispersion for the brightest stars is not more than a few kilometers per second. The powerful equipment of the 200-inch Hale telescope with the 8″.4 focus aplanatic sphere camera has been used by WILSON and COFFEEN[1] to obtain 23 spectrograms of 15 red giant stars in M 92. The general mean of all velocity measures is −121.0 km/sec, but some of the stars may have intrinsic velocity variations. The metallic lines are weak in this cluster, thereby rendering it unsuitable for extensive velocity work.

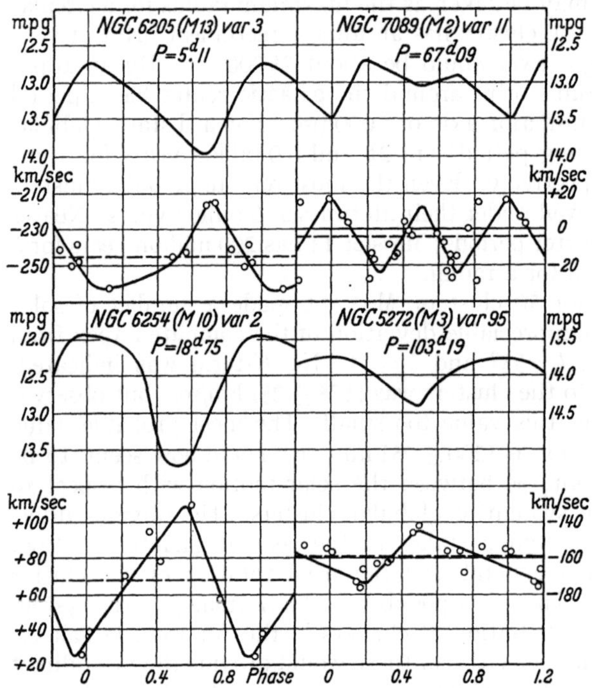

Fig. 28. Light and velocity curves of four variable stars in globular clusters, as given by JOY. Astrophys. Journ. **110**, 105 (1949).

Individual stars: theoretical orbits. The theoretical orbits of individual stars in globular clusters, first investigated by E. STRÖMGREN[2], have latterly been further studied by G. CONTOPOULOS[3] and R. KURTH[4], who showed that under certain reasonable assumptions about the motions and disregarding mutual interactions between stars, for a spherical cluster with finite radius in a steady state, the stars will describe circular orbits around the center, or oscillate along a straight line passing through the center, or in the most general case, show a rosette motion. The stars reaching the boundary of the cluster have rectilinear orbits; there are no circular orbits near the boundary.

V. Masses and densities.

68. Masses. The masses of globular clusters may be determined from the virial theorem which gives a simple relation between the total mass, the radius and the mean square velocity of particles in equilibrium in a spherical configuration. It has been shown by SCHWARZSCHILD and BERNSTEIN[5] that some of the earlier applications[6] of the virial theorem for this determination are unreliable because of difficulties in finding the true cluster radius and assumptions regarding the

[1] O. C. WILSON and M. COFFEEN: Astrophys. Journ. **119**, 197 (1954).
[2] E. STRÖMGREN: Astronom. Nachr. **203**, 17 (1916).
[3] G. CONTOPOULOS: Z. Astrophys. **35**, 67 (1954).
[4] R. KURTH: Astronom. Nachr. **282**, 241 (1955).
[5] M. SCHWARZSCHILD and S. BERNSTEIN: Astrophys. Journ. **122**, 200 (1955).
[6] R. KURTH: Z. Astrophys. **29**, 26 (1951).

distribution of potential energy. It is considered that the safest form of the virial theorem to use for the determination is:

$$\overline{V}^2 = \frac{G M}{\overline{r}} \quad \text{with} \quad \overline{r} = 2 \frac{\left(\int\limits_0^R S\, dq\right)^2}{\int\limits_0^R S^2\, dq} \tag{68.1}$$

where $\overline{r}$ is the effective distance directly determined from observed strip counts, S, which represent the number of stars in a strip of unit width as a function of the perpendicular distance, q, from the cluster center. It is assumed that the cluster is spherically symmetrical and that the density of the counted stars is proportional to the density of all stars.

Radial velocity measures of 23 spectrograms of 15 red giant stars in M 92 by WILSON and COFFEEN have been used by SCHWARZSCHILD and BERNSTEIN, along with star counts by TAYLER[1], to derive for M 92 a mass of $(1.4 \pm 0.7) \times 10^5\, \mathfrak{M}_\odot$, a value 2.4 times smaller than that derived from earlier applications of the virial theorem. With this new mass value, the mass-luminosity ratio for M 92 in solar units is reduced to 0.8 ± 0.4.

Another method of mass-determination is from the combination of extensive star counts supplemented by assumptions as to luminosity function. In this manner, SANDAGE (Sect. 58) has obtained a value of 2.45×10^5 solar masses for M 3 by evaluating the total mass from

$$\mathfrak{M}_t = \sum_{M=-3.0}^{+\infty} \mathfrak{M}\, \varphi(M). \tag{68.2}$$

Another approach to the problem is by application of an equilibrium tidal theory. HUGH M. JOHNSON[2], by assuming a galactic nucleus of point mass 10^{11} suns, derives a lower limit to the globular cluster masses of 4×10^5 suns.

69. Densities. The average density of globular clusters is about one star per two cubic parsecs, but star counts show that the density decreases about as the cube of the distance from the center. Near the center then the density is 100 times greater than average, or 50 stars per cubic parsec. This compares with a density in the solar neighbourhood of 1 star per 10 cubic parsecs. The average distance between globular cluster stars appears to be of the order of tens of thousands of astronomical units.

For the two most powerful globular clusters in our galaxy, namely 47 Tucanae ($M_p = -10.0$) and ω Centauri ($M_p = -10.45$), the star density in the center is enormous. By the current distance values, both these clusters are at least four times brighter intrinsically than any others. GASCOIGNE and BURR[3] conclude that whereas in the region of the Sun the light emission is equivalent to that from 0.0005 star of zero absolute magnitude per cubic parsec, the figure at the center of ω Centauri is 30000 times this, and at the center of 47 Tucanae a million times. These investigators state that at the center of 47 Tucanae the light of the stars would be equivalent to that from several thousand full Moons, and the resulting illumination would be comparable with terrestrial twilight.

---

[1] R. J. TAYLER: Astronom. J. **59**, 413 (1954).
[2] H. M. JOHNSON: Astrophys. Journ. **115**, 124 (1952).
[3] S. C. B. GASCOIGNE and E. J. BURR: Monthly Notices Roy. Astronom. Soc. London **116**, 570 (1956).

VI. Evolution, age and origin.

70. Evolution and age. Just as the color-magnitude diagrams for galactic clusters have been used for extensive calculations of evolutionary trends, with time intervals involved, so have those for globular clusters. HOYLE and SCHWARZSCHILD[1] have followed the evolution of the main sequence to the top of the red giant branch, from early stages of a partially degenerate isothermal helium core and radiative hydrogen envelopes. The models show the well-known turn-off from the main sequence in the H-R diagram. The helium core steadily increases in mass because of the burning of hydrogen. When the core has 20% of the stellar mass, the envelope has an appreciable extent and it is necessary to take into account photospheric boundary conditions and the hydrogen convective envelope. At the top of the red giant sequence the temperature in the hydrogen-burning shell reaches about 40 million degrees, while the helium core reaches about 50% of the stellar mass. The difference between the red giants of population I and population II may be explained on the basis of a lower metal abundance in population II. It is noteworthy, however, that the incidence of helium burning in the core determines the top of the giant sequence, independent of metal content, for either population I or population II. From their computations HOYLE and SCHWARZSCHILD estimate the age of stars in globular clusters at 6×10^9 years, but from the limits of physical approximations involved, the age might be as low as 4×10^9 or as high as 8×10^9 years.

A highly refined computation by HOYLE and HASELGROVE[2] on the Cambridge University electronic computer along the above lines leads to an age of 6.5×10^9 years for M3. Fifteen evolutionary steps have been computed for a star of mass 2.512×10^{33} grams from an initial composition of concentrations of hydrogen, 0.9309, helium 0.0666 and carbon-nitrogen 0.0025. The thirteenth step gives a color-magnitude diagram tallying closely with that of M3 with the above age.

71. Origin. In his theory of the evolution of the Milky Way, VON WEIZSÄCKER[3] postulates that the globular clusters have retained their spherical states since their formation. He proposes that about 6×10^9 years ago, our galaxy was a roughly spherical cloud of turbulent gas with an admixture of dust grains. The turbulence was reduced by encounters of eddies with one another, but gravitational effects gradually caused the cloud to break up into smaller cloudlets. In those cloudlets denser than 10^{-22} g/cm^3 protostars formed. The oldest stars, of population II, were formed at this very early stage and the globular clusters were produced in the dense original cloudlets. While the present population of stars in a typical cluster requires the above density, if a cluster has already lost much of its original mass in passing through the Milky Way, the density must have been much higher originally, probably a cloudlet density higher than anything observed in our Milky Way now. STRUVE[4] has computed how far from the galactic center the clusters were first formed, using VON HOERNER'S work (Sect. 66) and the finding of KURTH[5] that the linear diameters of globular clusters tend at first to decrease with distance from Sun, but then they increase again when the distance equals about 25000 psc. The true correlation of linear diameter is with the

---

[1] F. HOYLE and M. SCHWARZSCHILD: Astrophys. Journ. Suppl. **2**, No. 13 (1955).

[2] F. HOYLE and C.B. HASELGROVE: Monthly Notices Roy. Astronom. Soc. London **116**, 527 (1956).

[3] C. VON WEIZSÄCKER: Z. Astrophys. **35**, 252 (1955).

[4] O. STRUVE: Sky and Telescope **14**, 368 (1955).

[5] R. KURTH: Z. Astrophys. **30**, 200 (1952).

cluster's distance from the galactic center, not with its distance from the Sun, according to LOHMANN[1].

STRUVE, by applying the probability (from KEPLER'S laws) of finding any given cluster at $\frac{1}{10}$ its apogalactic distance from the center, at $\frac{2}{10}$, etc. and adopting a working hypothesis that all globular clusters were formed from dense knots of diffuse material in various shells at specific distances from the galactic center over a period of a billion years, finds that few if any clusters were born in the vicinity of the galactic center, while an appreciable number were formed at great distances up to 25000 psc. The observed and computed distribution in various shells on the above assumptions is shown in Table 19. The fact that clusters 25000 psc from the center are about 30% larger than those at only 10000 psc suggests that clusters formed near the outer boundaries of the galactic cloud were larger than those born near the center.

Table 19. *Distribution of globular clusters, present and past.*

1	2	3	4	5	6	7
Ring limits	Observed number in ring	Observed areal density	Computed number in ring	Computed areal density	Original distribution in shells	Original volume density
0 — 2½	12	61.2	3	14.4	0	0.0
2½ — 5	20	34.0	6	10.0	11	42.7
5 — 7½	27	27.5	7	7.5	25	35.8
7½ — 10	14	10.2	8	6.1	23	16.9
10 — 12½	7	4.0	10	5.3	16	7.1
12½ — 15	2	0.9	10	4.7	2	0.6
15 — 17½	2	0.8	11	4.3	0	0.0
17½ — 20	2	0.7	11	3.8	1	0.2
20 — 22½	2	0.6	12	3.5	4	0.5
22½ — 25	1	0.3	11	3.0	7	0.7

The ring limits are given in kpc, and areal densities are in number of clusters per 100 kpc² in the galactic plane. The volume density is the number of clusters per 1000 kpc³.

72. Contraction with age. From densitometric observations, FINLAY-FREUNDLICH[2] has shown that since globular clusters consist of a core of massive, bright stars with wide envelopes of faint stars surrounding them, they have a structure corresponding to an isothermal core and an adiabatic atmosphere. The contracting phase of a spherical cluster of stars leads to a state in which the structure corresponds to a solution of the differential equations of hydrostatic equilibrium. From the differential equation from which the virial theorem is derived

$$\frac{1}{4}\frac{d^2 I}{dt^2} = T + \frac{1}{2} V \tag{72.1}$$

in which I is the moment of inertia, T the kinetic energy, and V the potential energy of the cluster for a Newtonian gravitational field, FINLAY-FREUNDLICH concludes that the globular clusters have approached or reached the phase when d^2I/dt^2 vanishes or becomes very small. The contraction of the cluster will probably proceed at a cosmic time-scale. Near the state of greatest condensation, when d^2I/dt^2 becomes zero, the cluster will undergo only very small changes and for a long time its structure will correspond to the hydrostatic equilibrium

[1] W. LOHMANN: Z. Astrophys. **32**, 298 (1953).

[2] E. FINLAY-FREUNDLICH: Monthly Notices Roy. Astronom. Soc. London **105**, 237 (1945); **107**, 268 (1947).

equation. This probably explains the great similarity of globular clusters. If these clusters differ only slightly in total mass and energy, then there will be only moderate scatter in the values of their radii around a mean value.

VII. Relation to elliptical galaxies.

73. Observational comparison. Ever since LUNDMARK[1] raised the question of the relationship between the globular clusters and the elliptical extragalactic nebulae, the problem has been under consideration. Anyone who has studied

Table 20. *Central brightness of some globular clusters and galaxies, in numbers of tenth photographic magnitude stars per square minute.*

	Globular clusters				
Cluster	47 Tuc	M 15	M 92	M 5	ω Cen
Concentration class	III	IV	IV	V	VIII
Central brightness	30	8.5	6.55	4.4	3.4
	Elliptical galaxies				
NGC numbers	221	3115	3379	4494	4649
Type	*E*2	*E*7	*E*0	*E*0	*E*2
Central brightness	100	12	21	10	7

Note, inter alia, the tendency in the globular clusters for central brightness to vary regularly with concentration class.

original photographs of these objects is conscious of the similarity of their appearance, and wonders if an elliptical galaxy is a globular cluster on a larger, blown-up scale. The very luminous cluster 47 Tucanae, for example—from a greater distance would it appear as a small elliptical galaxy? The central brightness of globular clusters varies markedly with concentration class, and GASCOIGNE and BURR (Sect. 69) have shown that the central brightness of 47 Tucanae is considerably higher than that of other clusters and actually within the brightness range of elliptical galaxies (Table 20). However, if the central brightness is measured as a ratio with respect to the brightness at the effective radius, DE VAUCOULEURS[2] has shown this ratio to be 160 for elliptical galaxies, whereas for 47 Tucanae it is only 40.

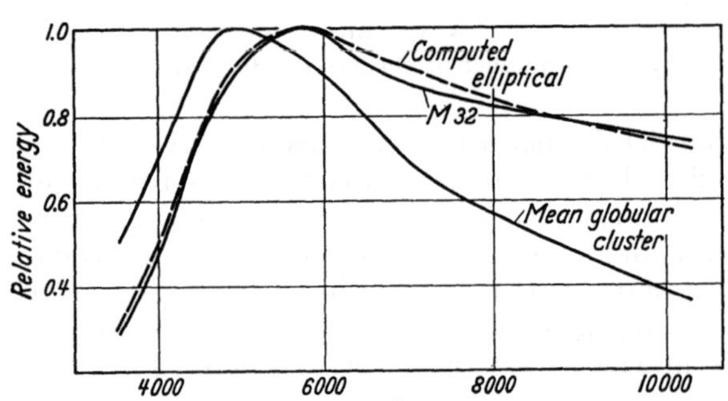

Fig. 29. Relative spectral-energy curves for a mean globular cluster and M 32 by ROBERTS. The dashed curve is for the computed elliptical nebula. Astronom. J. 61, 195 (1956).

The current evidence is against the assumption that globular clusters and elliptical galaxies represent two ends of a sequence in which other objects of intermediate mass and absolute luminosity will be found. M. S. ROBERTS[3] has

[1] K. LUNDMARK: Publ. Astronom. Soc. Pacific **42**, 23 (1929). — Lund Medd. 119 (1930).
[2] G. DE VAUCOULEURS: Monthly Notices Roy. Astronom. Soc. London **113**, 134 (1953).
[3] M. S. ROBERTS: Astronom. J. **61**, 195 (1956).

compared the two fundamental properties, the relative spectral-energy distribution and the mass-luminosity ratio. Although a color-magnitude diagram for an elliptical galaxy has not yet been determined, ROBERTS has constructed a spectral energy curve from the observational data available on the brightest stars in galaxies. The spectral energy curves for a mean globular cluster and computed elliptical galaxy, and Messier 32 (Fig. 29) show that elliptical galaxies are redder than globular clusters. They also have a much higher mass-luminosity ratio. SCHWARZSCHILD[1] has derived a mass-luminosity ratio of 200 for M32 and 100 for the elliptical galaxy NGC 3115. The mass-luminosity ratios of globular clusters are of a different order, as low as 1 (Sect. 68), more like those of galactic clusters, found by L. MAWRIDIS[2], to be 0.37 for the Hyades, 0.74 for Praesepe and 0.09 for the Pleiades. ROBERTS concludes that the luminosity curve for an elliptical galaxy predicts relatively more faint stars than in a globular cluster, and that a spectroscopic comparison in the red between globular clusters and an elliptical galaxy would be valuable.

VIII. Clusters associated with extragalactic systems.

74. Systems containing clusters. It is not difficult to catalogue numerous tight groupings of stars in some of the nearer external systems. The problem is to identify the type of objects which appear so compressed. G.E. KRON and N.U. MAYALL at the Lick Observatory are currently identifying the globular clusters in external systems by an intensive program of integrated magnitudes with photoelectric measures in two colors.

First clusters to attract attention were those in the Magellanic Clouds, where they were early catalogued by JAMES DUNLOP and Sir JOHN HERSCHEL, and in this century by SHAPLEY and collaborators at Harvard, and R. INNES at the Union Observatory. SHAPLEY [*22*] considers the Clouds have three types of clusters, globular, circular and open, on the basis of appearance. In the Small Cloud, LINDSAY[3] lists about 100 clusters, of which KRON[4] considers 10 as definitely globular, 9 as stellar associations, and about 50 as analogues of galactic open clusters. In two globular clusters belonging to the Small Cloud, NGC 121 and NGC 1978, a significant step has been the discovery of variable stars by A.D. THACKERAY and A.J. WESSELINK[5] and the derivation of periods of RR Lyrae stars and slow variables by THACKERAY[6] in NGC 121. In the Large Cloud, hundreds of clusters and nebulous objects are noted, of which at least seven appear to be globular clusters.

In Messier 31, the great spiral in Andromeda, HUBBLE[7] published a list of 140 nebulous objects or possible clusters, which was increased to 249 by BAADE in a study by SEYFERT and NASSAU[8]. It is much more difficult to identify these objects than those in the Clouds because Messier 31 is 10 times farther away. The resolution of the individual stars in the M31 clusters can be achieved only under superfine conditions with the largest telescopes. Besides, dust reddens the colors in M31. However, according to KRON[9] 71 objects are identified as globular,

[1] M. SCHWARZSCHILD: Astrophys. Journ. **59**, 273 (1954).
[2] L. MAWRIDIS: Z. Astrophys. **41**, 35 (1956).
[3] E. LINDSAY: Irish Astronom. J. **4**, 65 (1956).
[4] G.E. KRON: Publ. Astronom. Soc. Pacific **68**, 230 (1956).
[5] A.D. THACKERAY and A.J. WESSELINK: Nature, Lond. **171**, 693 (1953); Observatory **75**, 22 (1955).
[6] A.D. THACKERAY: In press.
[7] E. HUBBLE: Astrophys. Journ. **76**, 44 (1932).
[8] C.K. SEYFERT and J.J. NASSAU: Astrophys. Journ. **102**, 377 (1945).
[9] G.E. KRON: Astronom. Soc. Pacific Leaflet No. 339 (1957).

to an enormous spherical volume around M31, with open clusters close in. In the elliptical companion of M31, NGC 205, HUBBLE noted 8 clusters of which 2 are classed as globular from their colors. Two globular clusters were found by BAADE[1] in each of the similar elliptical galaxies, NGC 147 and 185.

In other spirals, HUBBLE noted 15 clusters in M33 (of which 4 are identified as globular), several in M81 and 6 in M101.

In the Fornax system, HUBBLE[2] found two globular clusters, for the brightest of which, NGC 1049, MAYALL[3] has actually obtained a radial velocity of −25 km per sec (with large uncertainty). No mention has been made of clusters in the similar Sculptor system. In the irregular system NGC 6822, HUBBLE[4] noted 10 clusters of which 3 may be globular.

The external system richest in globular clusters appears to be M87, a giant *E*0 galaxy in Virgo. According to BAUM[5] it has more than a thousand, whose distribution is roughly the same as the distribution of the surface brightness. These globular clusters have been compared photometrically with those in M31 to derive a new modulus of 30.2 for the Virgo cluster of galaxies. C.O.R. JASCHEK[6] has found that the numbers of globular clusters are correlated with the mass of the external galaxy with which they are associated.

75. Epilogue. So we conclude that, just as the subject of astronomy itself is one of receding horizons, so too is that subdivision of it which deals with star clusters. In no star cluster in our galaxy, whether galactic or globular, has the intrinsically faintest star yet been photographed. In virtually all clusters, such stars are beyond the reach of existing equipment. We do not yet know how many of each type of cluster exist in our galaxy, nor how great the distances at which they lie, but we are making progress in finding out. If the reader finds in this article statements to challenge, these may well provide a basis for future investigations. And if the future years should exhaust the possibilities for further profitable studies of the star clusters of our galaxy, then the intergalactic clusters and the clusters associated with external galaxies would offer scope for studies for centuries still to come.

Appendix A.

Catalogue of galactic clusters.

Description of the Catalogue. The catalogue contains 514 clusters published as galactic. Some of these have been studied very little, while the classification of others, like NGC 2682 and 7789 is controversial.

The first column gives the designation which is the NGC number unless otherwise indicated. The arrangement is in order of NGC number, or, for clusters without it, in order of right ascension. Abbreviations are: I = Index Catalogue, Tr = Trumpler, Cr = Collinder, Mel = Melotte, Harv = Harvard. The second and third columns give right ascension and declination for the equinox of 1950, while the fourth and fifth give galactic coordinates for 1900.

The sixth column, the apparent diameter, is by TRUMPLER [7] when possible, otherwise from COLLINDER [2] or miscellaneous sources. It is not corrected for absorption. The seventh column of integrated magnitude is from COLLINDER. The eighth column gives the class, by TRUMPLER when possible, otherwise by COLLINDER, or from miscellaneous sources. TRUMPLER's classification is discussed in Sect. 6, and *U* is unsymmetrical, *E* elongated, and *N* nebulosity. The ninth column gives some indication of the photographic magnitude of the brightest stars. For nearly half the clusters it is SHAPLEY's value [6] of the magnitude

[1] W. BAADE: Astrophys. Journ. **100**, 148 (1944).
[2] E. HUBBLE: Publ. Astronom. Soc. Pacific **51**, 42 (1939).
[3] N.U. MAYALL: Astrophys. Journ. **104**, 213 (1946).
[4] E. HUBBLE: Astrophys. Journ. **62**, 409 (1925); **76**, 69 (1932).
[5] W.A. BAUM: Publ. Astronom. Soc. Pacific **67**, 328 (1955).
[6] C.O.R. JASCHEK: Z. Astrophys. **44**, 23 (1957).

of the 5th brightest star, while for the others it is the approximate magnitude at which the cluster stars appear. The tenth column gives the distance in kpc, usually as given by Mme. K. A. Barkhatova[1] but also from occasional sources.

In the remarks column an asterisk means that the cluster has been noted as having a hundred or more stars. The Messier identifications are given as well as other familiar designations. For those clusters for which identification in other catalogues has been ambiguous, various designations are noted.

It should be emphasized that the observational part of this catalogue is compiled from heterogeneous sources, with a wide range in the accuracy of the data given. Many of the clusters discovered by Collinder have not yet been confirmed by other observers, and Collinder's distance estimates are too large by present standards. Reference to Alter's catalogue [4] will indicate the sources of data on various clusters.

Catalogue of galactic clusters.

Desig.	α 1950	δ	λ 1900	β	App. Diam.	Int. Mag.	Class	Mag. Br. St.	Dist. kpc	Remarks
Blanco 1	$00^h01^m.7$	−30° 13′	341°.9	−80°.7	90′		III *p*	]9	0.2	
King 13	07.5	+60 56	85.6	− 1.0	5		*p*	12		
King 1	19.2	+64 07	87.3	+ 2.0	9		*m*	13		
103	22.5	+61 04	87.5	− 1.1	4	10.8	I 2*p*	12.4	1.2	
129	27.1	+59 57	88.0	− 2.3	13	9.8	IV 2*p U*	11.0	1.5	
133	28.3	+63 05	88.3	+ 0.9	8	9.4	Plei	11.7	2	
136	28.7	+61 15	88.3	− 1.0	2.2	11.3	II 2*p*	13	2.5	
Alter	29.0	+62 53	88.4	+ 0.6	7		*p*	10	0.4	King 14
146	30.2	+63 00	88.5	+ 0.8	7	9.6	Plei	10.3	0.7	
188	39.5	+85 04	89.9	+22.8	14	9.3	II 1*r*	14.6	1.0	
189	35.9	+60 47	89.2	− 1.5	4	11.1	Plei	12	1.5	Cr α in error
225	40.5	+61 31	89.7	− 0.7	14	8.9	III 1*p*	8.5	0.9	
King 16	40.7	+63 55	89.8	+ 1.6	4		*p*	11		
King 2	48.1	+57 55	90.7	− 4.4	6		*r*	17		
281	49.9	+56 21	90.6	− 6.0	22	7.4	*O* type	9	2.1	I 1590
366	01 03.1	+62 04	92.4	0.0	3	11.8	Plei	10	2.5	
381	05.2	+61 19	92.7	− 0.8	7	9.3	III 2*p*	9.5	1.6	
433	12.1	+59 52	93.7	− 2.4	3.4			9	4.5	
436	12.5	+58 33	93.8	− 3.5	6	9.3	I 3*m*	9.8	1.7	
457	15.9	+58 04	94.4	− 3.9	12	8.0	I 3*r*	8.6	1.4	*
559	26.1	+63 03	94.9	+ 1.2	6	7.4	I 2*m*	11.7	1.3	
581	29.9	+60 27	95.8	− 1.3	6.5	6.9	II 3*m*	9.8	1.2	M 103
Tr 1	32.3	+61 01	96.0	− 0.6	4.5	8.9	I 3*p*	10	1.2	
King 3	32.9	+64 18	95.4	+ 2.5	3		*r*	16		
609	33.8	+64 17	95.5	+ 2.6	3	12.7	I 2*p*:	14.5	2.4	
637	39.5	+63 45	96.2	+ 2.2	3.5	7.3	II 3*p*:	10.2	1.6	
654	40.6	+61 38	96.8	+ 0.1	5.5	9.8	I 2*p*	9.8	1.2	
659	40.8	+60 27	97.1	− 1.0	5	9.9	II 2*p*	10.9	1.4	
663	42.6	+60 59	97.2	− 0.4	14	7.5	IV 2*m*	9.8	1.2	
Cr 463	44.3	+71 41	95.0	+10.5	45	5.8	Plei		0.5	
Cr 21	47.3	+27 00	107.3	−33.1	7	8.2	Plei		2	
I 166	49.1	+61 36	97.8	+ 0.3	4.5		II 1*r*	17		Tom 3
743	55.2	+59 56	98.9	− 1.0	1.6			9.5	1.1	
744	55.2	+55 14	100.2	− 5.6	14	9.0	I 2*p*	10	0.8	
752	54.8	+37 26	105.4	−22.7	45	6.6	III 1*m*	9.6	0.4	
Stock 2	02 11.4	+59 00	101.2	− 1.3	80		*OB*	9	0.3	*
869	15.5	+56 55	102.4	− 3.1	30	4.3	IV 3*r*	9.5	2.2	* h Per
884	18.9	+56 53	102.9	− 3.0	30	4.4	IV 3*r*	9.5	2.2	* χ Per
Tom 4	25.2	+61 33	101.8	+ 1.7	2.2		III 1*p*	16.5	9	
Mark 6	25.9	+60 26	101.9	+ 0.4	4.5		*O*-type		0.9	
I 1805	29.0	+61 13	102.4	+ 1.5	20	6.8	IV 3*m N*	]8.6	1.0	Mel 15
956	29.2	+44 25	109.3	−14.0	10	8.9	Plei	9	4	
957	30.0	+57 18	104.1	− 2.0	9	7.2	II 3*m U*	11.3	1.0	
King 4	32.0	+58 47	103.1	+ 0.5	3		*p*	13		

[1] K. A. Barkhatova: Russ. Astronom. J. **27**, 180 (1950).

Desig.	α 1950	δ	λ 1900	β	App. Diam.	Int. Mag.	Class	Mag. Br. St.	Dist. kpc	Remarks
Tr 2	$02^h 33^m.8$	+55°45′	105°.2	− 3°.2	18′	7.0	II 3*p*	]9	0.5	
1027	38.8	+61 20	103.5	+ 2.1	21	7.4	IV 3*m*	9.8	0.9	
1039	38.8	+42 34	111.7	−14.8	30	5.8	I 3*m*	]8.6	0.4	M 34
I 1848	47.3	+60 14	104.9	+ 1.6	22	7.0	IV 3*mN*		0.7	
Cr 33	55.4	+60 12	105.8	+ 1.8	50	5.9	Plei		0.4	Cluster?
Cr 34	57.1	+60 12	105.9	+ 2.1	25	6.8	Plei		0.7	Cluster?
1193	03 02.5	+44 11	114.8	−11.3	3.2	12.6	I 2*r*	14	5.1	
Tr 3	07.8	+63 04	105.6	+ 5.3	17	7.1	II 3*p*	10.8	0.5	Harv 1
1220	08.0	+53 09	110.8	− 3.2	2.0	11.8	I 2*p*	13	2.0	
King 5	11.0	+52 32	111.5	− 3.5	9		*m*	13		
1245	11.3	+47 03	114.5	− 8.0	7	8.9	III 2*r*	13:	2.6	
Pers. mov. cl.	18	+48 26	114.8	− 6.2	240	2.3	IV 3*m*	5	0.2	Mel 20
King 6	24.2	+56 17	111.0	+ 0.7	15		*m*	10		
1342	28.4	+37 10	123.0	−14.3	15	7.2	II 2*m*	9.1	0.9	
Gingrich	41.1	+32 08	128.3	−17.3				10	0.2	Cluster in Neb. I 348
Tom 5	43.7	+58 55	111.6	+ 4.4	17		III 2*m*	14	7:	
Pleiades	44.1	+23 57	134.7	−22.2	120	1.6	II 3*rN*	4.2	0.1	* M 45
1444	45.6	+52 30	115.8	− 0.4	3.5	6.4	II 3*p*		1.0	
King 7	55.2	+51 39	117.5	0.0	8		*m*	16		
1496	04 00.6	+52 29	117.8	+ 1.0	6.5	9.6	Plei	12	2.9	
1502	03.1	+62 11	111.3	+ 8.5	8	5.6	II 3*p*	9.8	0.9	
1513	06.3	+49 23	120.3	− 0.6	9	8.9	II 2*mU*	12.8	0.9	
1528	11.6	+51 07	119.7	+ 1.2	22	6.4	II 2*m*	9.2	0.5	
I 361	14.9	+58 11	115.1	+ 6.6	6	11.2	III 2*m*	14	1.6	
Taurus Cl.	17	+15 30	147.0	−22.4	400	0.8	II 3*m*	4	0.04	Hyades
1545	17.1	+50 08	121.1	+ 1.5	15	7.8	I 2*p*	9	0.7	
1582	28.6	+43 45	127.4	− 1.9	35	7.0	Plei	9	0.8	
1605	31.4	+45 09	126.3	− 0.5	4.5	11.0	III 2*m*		1.6	
1624	36.6	+50 21	123.2	+ 3.5	2.0	11.8	*O*-type	13	3.0	
1647	43.1	+18 59	148.2	−15.4	35	6.2	III 2*m*	9.7	0.5	
1662	45.7	+10 51	155.5	−19.7	14	8.0	II 2*p*	9	0.8	
1664	47.4	+43 37	129.4	+ 0.7	13	8.0	II 2*m*	10.9	0.7	
1746	05 00.6	+23 45	146.8	− 9.3	40	6.1	IV 2*m*	9.5	0.7	
1750	00.9	+23 35	147.0	− 9.3				8		
1778	04.7	+36 59	136.6	− 0.7	10	8.5	II 2*p*	8	0.9	
1807	07.8	+16 28	153.9	−12.1	14	8.3	I 2*p*	9.7	0.4	
1817	09.2	+16 38	153.9	−11.7	16	7.8	III 2*r*	11.1	1.2	
Cr 464	16.2	+73 14	107.0	+20.5	180	4.2	Plei		0.2	
1857	16.7	+39 18	136.1	+ 2.5	9	8.4	II 2*m*	10.6	1.1	
Cr 62	18.0	+40 56	135.7	+ 3.3	32	4.2	Plei		0.5	
I 410	19.3	+33 28	141.6	− 0.3	6		*O*-type	9	2.0	Mult. star
1883	22.2	+46 30	130.6	+ 7.4	3	12.0	I 2*m*	[12.6	2.9	
1893	19.4	+33 21	141.7	− 0.5	15	7.8	I 3*m*	9.0	0.9	
Cr 65	23	+16 03	156.2	− 9.3	240	3.0	Plei		0.1	
1907	24.7	+35 17	140.3	+ 1.6	5	10.3	I 2*m*	11.1	1.6	
1912	25.3	+35 48	140.0	+ 2.0	18	7.0	II 2*r*	9.7	0.8	* M 38
1931	28.1	+34 13	141.5	+ 1.6	1.2	12.8	neb. cl.			Cr δ in error
Cr. 69	32.4	+09 54	163.5	−10.8	65	2.8	neb. cl.		0.4	λ Ori cl.
Cr. 70	32.6	−01 08	172.5	−16.4	245	0.6	neb. cl.		0.3	* Orion belt
1960	32.8	+34 06	142.2	+ 2.4	16	6.5	I 3*m*	]8.7	1.3	M 36
1976	32.8	− 5 25	176.6	−17.9	30				0.5	* Orion neb. and Trapezium cluster
1977	32.8	− 4 50	176.1	−17.6						
1980	32.8	− 5 57	177.1	−18.2	14	2.5	Neb. cl.		0.3	
1981	32.8	− 4 28	175.7	−17.5					0.5	*
Cr 74	45.8	+07 23	167.0	− 9.0	3	14.4			10	
King 8	46.1	+33 37	144.1	+ 4.3	2		*p*	15		

Desig.	α 1950	δ	λ 1900	β	App. Diam.	Int. Mag.	Class	Mag. Br. St.	Dist. kpc	Remarks
Pismis 1	05h47m.3	+33° 37′	144°.2	+ 4°.7	4.8′		III m		4	
2099	49.1	+32 32	145.3	+ 4.5	24	6.2	I 1r	9.7	0.8	* M 37
2112	51.3	+00 23	173.6	−11.1	9	9.1	I 2m	10	1.7	
2126	59.1	+49 54	130.7	+14.4	5.5	10.2	III 2p	13	4.3	
2129	58.1	+23 18	154.3	+ 1.6	7	7.0	II 3p	11.2	1.0	
2141	06 00.3	+10 26	165.8	− 4.3	8	10.8	IV 2r	15	4.0	
I 2157	02.0	+24 00	154.2	+ 2.7	5.5	9.1	II 2p	11.2	1.1	Tr 4
2168	05.8	+24 21	154.3	+ 3.6	29	5.6	III 3r	9.0	0.7	* M 35
2169	05.6	+13 58	163.3	− 1.4	6	7.0	II 3p	9.0	0.7	
2175	06.8	+20 20	158.2	+ 1.9	16	6.8	neb. cl.	7.5	1.1	
2186	09.5	+05 28	171.2	− 4.7	4.5	9.2	I 3p	11.9	1.4	
2192	11.7	+39 52	141.0	+12.0	5	10.9	III 2m	13.7	3.3	
2194	11.0	+12 49	164.9	− 0.8	6	10.4	I 2r	[13	2.0	*
2204	13.5	−18 38	193.5	−14.7	9	9.4	I 2m	13	1.6	
9-12 Gem	15	+23 39	156.5	+ 5.2	40	5.7	Plei		0.3	
2215	18.4	−07 16	183.6	− 8.7	8.5	8.6	II 2p	11.2	1.3	
Cr 91	18.9	+02 23	175.4	− 4.2	25	6.0	Plei		0.9	
Cr 92	20.2	+05 10	173.0	− 2.8	12	8.6	Neb. cl.		1.9	
2232	24.3	−04 42	181.4	− 6.6	30	4.1	Plei		0.5	
2236	27.0	+06 52	172.1	0.0	6	11.6	I 2m	14.1	1.6	
Cr 96	27.7	+02 54	176.5	− 2.2	12	9.1	Plei		2	
Cr 95	27.8	+09 58	69.4	+ 1.2	23	6.8	Neb. cl.		0.9	
2239	28.3	+04 59	174.1	− 1.1	16					In neb 2237
2243	27.9	−31 15	206.9	−16.8	4	10.5	I 2m	14.3	4.0	
Cr 97	28.6	+05 57	173.1	− 0.7	25	5.4	Plei		0.6	
2244	29.7	+04 54	174.1	− 0.6	27	5.2	IV 3mN	]8.9	1.7	
2250	30.3	−05 00	182.9	− 5.9	10	8.9	Plei	12	2.2	
2251	32.0	+08 24	171.3	+ 1,6	7	8.5	II 2p		1.4	
2252	32.3	+05 25	174.3	− 0.1	14	7.7	Plei	9	1.6	
2254	33.3	+07 43	172.1	+ 1.6	3.8	10.9	I 2p	12	1.8	
Cr 104	33.8	+04 52	175.4	− 0.3	19	9.6	Chain		2.3	
Tr 5	34.0	+09 29	170.6	+ 2.5	9	10.9	III 1r	17	2.0	
Cr 106	34.6	+06 00	173.8	+ 1.1	50	4.6	μ Norm		0.3	
Cr 107	35.0	+04 47	175.5	+ 0.4	40	5.4	Plei		0.6	
2259	35.8	+10 56	169.5	+ 3.6	3.5	10.8	I 3m	15.5	3.3	
2262	35.8	+01 14	178.3	− 1.2	5	11.3	Plei		4.7	
Cr 110	35.8	+02 04	177.4	− 0.8	12	10.5	Praes		2.3	
Cr 111	36.0	+06 57	173.0	+ 1.5	3.2	7.0	μ Norm		2.8	
2264	38.3	+09 56	170.6	+ 3.7	30	4.1	II 3pN	]8.9	0.9	S Mon
2266	40.1	+27 01	155.4	+11.7	6	9.4	I 2r	11.3	2.5	
2269	41.3	+04 37	175.7	+ 1.9	3	9.8	II 2p	10	2.4	
Cr 115	43.9	+01 49	179.3	+ 1.8	9	9.1	Plei		2.4	
2281	45.8	+41 07	142.5	+18.4	15	7.2	I 3p	10.6	0.7	
2286	45.1	−03 07	183.0	− 0.9	9	8.3	III 2p	]8.9	1.2	
2287	44.9	−20 41	198.7	− 9.0	32	5.0	I 3r	8	0.4	* M 41
2301	49.2	+00 32	180.3	+ 1.7	15	6.3	I 3m	8.9	0.9	
2304	52.1	+18 05	164.9	+10.4	4	11.2	I 1m		2.9	
Cr 121	52.1	−24 34	202.5	− 9.6	55	5.9	Plei		0.5	
2309	53.6	−07 08	187.6	− 0.9	3	10.6	II 2p	13	1.9	
2311	55.3	−04 31	185.4	+ 0.7	4.5	9.7	II 2p	12	1.5	
Tom 1	58.2	−20 26	199.8	− 6.1	5		III 2p	14		
2314	07 03.8	+75 19	106.3	+28.1	6					Glob?
2323	00.6	−08 16	189.4	+ 0.1	16	7.2	I 2m	9:	0.7	* M 50
Tom 2	01.2	−20 47	200.1	− 5.4	2		III p	16		
2324	01.6	+01 08	181.2	+ 4.8	8	9.0	I 2m	13.7	1.2	
2331	04.1	+27 26	157.8	+16.6	20	8.5	Plei	9	1.3	
2335	04.2	−10 00	191.0	0.0	10	9.3	II 2p	10	0.9	
Cr 465	04.8	−10 33	192.0	+ 0.2	10	10.1	Praes		2.6	
Cr 466	04.9	−10 45	192.0	+ 0.2	5	11.1	Praes		4.6	
2343	05.9	−10 35	192.0	+ 0.2	6	7.8	II 3p	8	1.2	
2345	06.0	−13 06	194.2	− 1.0	11	8.1	II 2pU	9	1.1	

Desig.	α 1950	δ	λ 1900	β	App. Diam.	Int. Mag.	Class	Mag. Br. St.	Dist. kpc	Remarks
2353	$07^h12^m.2$	−10° 13′	192°.4	+ 1°.7	20′	5.2	I 3*mU*	9.4	0.8	
2354	12.1	−25 39	206.0	− 5.6	20	8.9	III 1*m*	11.2	0.7	
Cr 132	12.4	−31 05	210.9	− 8.6	115	3.9	Plei:		0.3	
2355	14.1	+13 52	171.1	+13.3	7	9.7	I 2*m*	13	3.2	
Cr 135	15.2	−36 45	215.6	−10.5	65	3.4	neb. cl.		0.5	
2360	15.5	−15 32	197.5	− 0.1	11	9.1	I 2*r*	9.4	0.8	
2362	16.7	−24 51	205.8	− 4.4	7	3.8	I 3*p*	9.4	1.4	τ Can Maj
2367	18.0	−21 50	203.3	− 2.6	4.5	7.9	II 3*pE*		2.1	
2368	18.6	−10 17	193.2	+ 3.1	4	11.8	II 2*p*		1.9	
2374	21.7	−13 09	196.1	+ 2.3	4.5	7.3	II 2*p*		0.9	
Cr 140	21.9	−32 06	212.7	− 6.8	62	4.3	Plei		3.6	
2383	22.6	−20 50	202.9	− 1.2	6	8.7	I 2*p*		1.3	
2384	22.9	−20 56	202.3	− 1.3	2.8	8.2		10.7	2.2	
Cr 143	23.0	−20 56	202.3	− 1.3	3.5	8.3			4.7	Forms double clus. with 2384
2395	24.3	+13 41	172.4	+15.4	14	9.4	III 2*p*:	9	1.1	
Tr 6	24.0	−24 12	206.0	− 2.6	5.5	10.0	I 2*p*		1.7	
Tr 7	25.2	−23 56	205.9	− 2.2	5.5	7.9	II 3*p*	10	1.8	
Mel 66	24.8	−47 38	227.0	−13.5	11	10.7	II 2*m*		1.2	
2396	26.7	−11 38	194.4	+ 3.1	20	7.4	Plei		1.1	
2401	27.1	−13 52	197.4	+ 3.1	2	12.6	I 2*p*:		2.7	
2414	31.0	−15 20	199.1	+ 3.2	8	8.2	I 3*p*		1.3	
2420	35.5	+21 41	165.8	+21.1	6	10.0	II 2*m*	12:	2.4	
2421	34.1	−20 30	204.0	+ 1.3	9	9.0	II 2*p*	9.4	1.0	
2422	34.3	−14 23	198.7	+ 4.4	30	4.3	II 3*m*	9.8	0.5	
2423	34.8	−13 45	198.2	+ 4.8	19	7.0	I 2*m*	10.8	0.6	
Mel 71	35.2	−11 57	196.7	+ 5.8	8	9.0	II 2*m*	10.8	1.7	
Mel 72	36.1	−10 34	195.6	+ 6.7	5	10.2	II 1*p*	12.8	1.9	
Cr 467	37.0	−10 26	195.3	+ 6.6	2.5	11.8	Praes		5.6	
2432	38.7	−18 58	203.2	+ 3.0	4.5	10.2	II 2*p*		1.8	
2437	39.5	−14 42	199.6	+ 5.3	27	6.6	II 2*r*	10.8	0.7	* M 46
2439	39.0	−31 32	214.0	− 3.4	9	7.1	II 3*mU*	10.6	1.0	
2447	42.5	−23 45	207.8	+ 1.3	18	6.5	I 3*r*	9.7	0.7	M 93
2451	43.6	−37 51	220.0	− 5.8	37	3.7	I 3*p*	6	0.3	
2453	45.7	−27 07	211.0	+ 0.1	4	9.0	I 3*p*		1.9	
2455	46.8	−21 10	206.1	+ 3.5	5.5	10.2	III 2*p*	12.7	2.8	
Pismis 2	49.1	−25 19	209.9	+ 1.8	2		Im	12	7	
2467	50.4	−26 16	210.8	+ 1.4	15	7.1	Neb. cl.		2.8	
2477	50.5	−38 25	221.2	− 4.9	25	5.7	I 2*r*	10.9	0.7	*
2479	52.8	−17 35	203.5	+ 6.6	8			11.6	2.6	
2482	52.8	−24 10	209.4	+ 3.1	11	8.7	II 2*p*	10.0	0.8	
Tr 8	53.1	−17 41	203.9	+ 6.6	9	9.6	II 1*p*		1.5	
Tr 9	53.2	−25 48	210.8	+ 2.3	5	9.1	II 3*p*	10.3	1.7	
2489	54.2	−29 56	214.4	+ 0.2	7	9.3	I 2*m*	11.1	1.8	
2506	57.6	−10 39	198.4	+11.2	11	8.4	I 2*r*	11.3	2.1	
2509	58.5	−18 56	205.7	+ 7.0	4.5	9.3	I 2*m*	10.2	1.8	
2516	57.4	−60 44	241.2	−15.5	50	3.3	I 3*r*	10.1	0.4	
Cr 173	08 01.6	−46 08	228.8	− 7.4	420	0.6	Plei		0.5	
2527	03.1	−28 01	213.9	+ 2.9	15	8.2	II 2*p*		0.6	
2533	05.0	−29 46	215.6	+ 2.3	4	9.9	I 2*p*		1.5	
2539	08.4	−12 41	201.6	+12.4	22	8.0	II 1*m*	10.8	1.1	*
2546	10.6	−37 29	222.6	− 1.1	45	5.2	III 2*p*	7	0.5	
2547	11.4	−48 49	232.1	− 7.9	17	5.0	II 3*p*	9.4	0.5	
2548	11.3	−05 39	195.8	+16.7	30	5.5	I 2*r*	9.4	0.5	
2567	16.6	−30 29	217.6	+ 3.9	11	8.4	II 2*m*	10.0	1.5	
2571	16.9	−29 35	216.5	+ 4.0	10	7.5	I 3*p*	9.4	1.1	
2579	19.2	−36 01	221.7	+ 0.2	10	9.2	Plei		2.4	
2580	19.6	−30 10	218.0	+ 4.5	7	9.7	Praes	10.9	3.0	
Cr 185	20.5	−36 00	221.9	+ 1.0	9	10.0	Plei		2.5	
2587	21.3	−29 20	217.1	+ 4.8	8.5	9.1	Plei	10.9	2.5	

Desig.	α 1950	δ	λ 1900	β	App. Diam.	Int. Mag.	Class	Mag. Br. St.	Dist. kpc	Remarks
2588	$08^h21^m.2$	−32° 49′	220°.0	+ 3°.3	2.5′	11.8	II 2*p*:		2.7	
Cr 187	22.1	−28 58	216.4	+ 5.8	9	9.6	Plei		2.6	
2627	35.2	−29 46	219.4	+ 7.6	8	8.4	II 2*m*	11.5	1.4	
2632	37.2	+20 10	173.3	+34.0	90	3.9	I 2*r*	7.5	0.2	* Praesepe, M 44
2635	36.5	−34 36	223.4	+ 4.8	3	10.2	II 2*m*:	12.7	3.1	
I 2391	38.8	−52 53	237.8	− 6.4	45	2.6	II 3*p*	3.5:	0.3	
I 2395	39.6	−48 01	236.0	− 5.0	20	4.6	II 3*p*	10.1	0.6	
2658	41.4	−32 29	222.4	+ 6.9	9	9.2	I 2*m*	12.2	1.3	
2659	40.9	−44 47	231.8	− 1.0	11	9.7	II 2*m*	12.5	1.4	
2660	40.6	−46 59	232.9	− 2.8	4	10.8	Glob.	12:	5.0	
Cr 196	43	−31 27	221.6	+ 7.0	5	10.5	Praes		4.1	
Cr 197	42.8	−41 11	229.3	+ 1.4	19	7.7	Plei		1.1	
Cr 198	43.3	−31 35	221.6	+ 7.6	6	11.2	Praes		4.2	
2669	43.4	−52 47	238.3	− 5.8	14	6.0	I 3*p*		0.6	
2670	43.9	−48 36	235.1	− 3.0	11	9.2	I 2*p*	11.2	1.0	
2671	44.4	−41 42	229.9	+ 1.5	4	11.6	II 2*p*:	13:	2.8	
Harv 3	44.8	−52 36	237.9	− 5.9	10	6.2	Plei	10.4	1.5	
Tr 10	46.0	−42 18	230.5	+ 1.3	30	5.0	II 3*p*		0.4	
2682	47.8	+12 00	183.6	+33.3	18	7.4	II 2*r*	10.8	0.7	* M 67
Mark 18	58.9	−48 47	236.7	− 1.3	3		*O* type		1.8	
Cr 205	59.0	−48 32	235.4	− 1.3	3	8.4	μ Norm		4.5	
2818	09 14.0	−36 24	229.9	+ 9.3	9	9.9	II 2*m*	13:	1.4	
2849	17.3	−40 19	223.0	+ 6.5	2.2	12.5	Praes		10	
I 2488	26.1	−56 43	245.4	− 4.1	18	7.4	IV 2*mU*	11.2	1.0	
2910	28.7	−52 41	243.0	− 0.8	6	8.3	II 2*p*	11.2	1.1	
2925	32.0	−53 13	243.7	− 0.9	16	8.3	II 2*p*	10.7	0.8	
2972	38.5	−50 06	242.4	+ 2.2	4.5	10.2	I 2*m*		2.6	
3033	47.1	−56 11	247.2	− 1.8	5	8.4	II 2*p*		1.7	
Cr 213	52.9	−50 29	244.4	+ 3.4	16	9.2	Plei		1.7	
3105	59.0	−54 32	247.6	+ 0.5	2.5	11.1	II 2*p*	[12.5	2.4	
3114	10 01.1	−59 53	250.9	− 3.6	37	4.5	II 3*r*	7:	0.5	
Tr 11	03.5	−61 21	252.0	− 4.7	4	10.8	II 2*p*		1.5	
Tr 12	04.8	−60 04	251.4	− 3.5	4	10.5	I 2*m*		2.0	
3228	19.8	−51 28	248.6	+ 4.8	20	6.4	I 2*p*	10.1	0.5	
Tr 13	22.1	−59 50	253.1	− 2.2	4	11.3	II 2*m*:		1.5	
3247	24.1	−57 41	252.1	+ 0.1	9.5	10.3	Plei		2.8	
3255	24.7	−60 25	253.7	− 2.5	3	12.6	I 2*p*:		1.9	
I 2581	25.6	−57 23	252.3	+ 0.1	10	5.3	II 3*p*	11.8	0.9	
Cr 223	28.6	−59 34	253.6	− 1.7	8.5	9.4	Plei.		2.6	
3293	33.9	−57 59	253.6	+ 0.2	8	7.5	II 3*r*	]7.8	0.9	
3324	35.4	−58 22	253.9	− 0.2	5				3.3	
3330	36.6	−53 53	251.9	+ 4.0	8	8.6	II 3*m*	10.2	1.3	Harv 4
Mel 101	40.4	−64 50	257.4	− 5.6	15	8.4	II 2*m*	11.4	0.5	
Cr 228	41.1	−59 45	255.0	− 1.2	24	4.9			0.6	
I 2602	41.2	−64 08	257.2	− 4.9	65	1.6	II 3*m*	6:	0.2	ϑ Car
Tr 14	42.0	−59 18	255.1	− 0.5	4	6.8	I 3*mN*		1.0	
Tr 15	42.7	−59 06	255.1	− 0.3	3.5	9.0	I 3*p*		1.0	
Cr 232	42.8	−59 18	255.2	− 0.5	3.2	6.9	Plei		1.9	
Tr 16	43.1	−59 27	255.3	− 0.6	10	6.7	IV 3*mNU*	10	1.3	η Car
Cr 234	43.3	−59 29	255.3	− 0.6	5	7.5	Plei		2.1	
Tr 17	54.2	−58 57	256.4	+ 0.5	5.5	9.4	I 3*p*		1.0	
Cr 236	55.0	−60 46	257.0	− 1.2	11	7.7			2.1	
3496	57.8	−60 04	257.2	− 0.4	7	9.2	III 2*m*		1.3	
3532	11 04.3	−58 24	257.4	+ 1.5	55	3.4	II 2*r*	8.1	0.4	
3572	08.3	−59 58	258.2	− 0.1	6	8.8	Plei	9.6	1.1	
Cr 240	09.1	−60 01	258.3	0.0	31	4.5	Plei		0.5	
Tr 18	09.3	−60 24	258.6	0.0	12	8.2	IV 3*m*		1.1	
3590	10.9	−60 25	258.8	− 0.2	4.5	8.0	*O* type	9.9	1.1	
Tr 19	12.2	−57 19	258.0	+ 2.8	13	9.6	IV 2*m*		1.3	

Desig.	α 1950	δ	λ 1900	β	App. Diam.	Int. Mag.	Class	Mag. Br. St.	Dist. kpc	Remarks
3603	$11^h12^m.9$	−60° 59′	259°.3	− 0°.6	2′	10.3	I 3*m*:		1.5	
I 2714	15.7	−62 26	260.1	− 1.8	11	8.2	II 2*r*	10.7	1.0	
Mel. 105	17.3	−63 14	260.5	− 2.5	4	9.4	I 3*m*	11.0	1.6	
3680	23.3	−42 57	254.9	+17.0	12	8.6	I 2*p*	10	0.9	
3766	33.8	−61 20	261.8	− 0.1	12	4.6	I 2*r*	8.1	0.9	
I 2944	34.3	−62 45	262.1	− 1.5	50	2.8	Neb. cl.	7	2.1	
I 2948	36.4	−63 15	262.5	− 1.9	15			8.6	0.7	
3960	48.4	−55 25	262.3	+ 6.1	7	8.8	I 2*m*	13.1	1.5	Mel. 108
4052	59.3	−62 55	265.1	− 1.0	10	8.8	II 2*mE*	11.6	0.9	
4103	12 04.0	−60 58	265.3	+ 1.0	9	7.4	I 3*m*	9.4	1.0	
4230	14.4	−55 02	265.9	+ 7.0	4.5	9.4	II 2*p*:		1.6	
4337	21.2	−57 51	267.1	+ 4.2	4	9.8	I 2*p*	13.1	1.5	
4349	21.7	−61 37	267.5	+ 0.6	17	8.0	II 2*r*	10.4	0.9	
Coma	22.5	+26 23	195.6	+85.2	300	2.9	II 3*p*	5	0.03	Mel. 111
Cr 257	23.2	−60 29	267.6	+ 1.7	6	9.1	Plei		2.9	
Harv. 5	25.2	−60 29	267.8	+ 1.9	7.2	8.4	Plei	9.6	2.3	
4439	25.7	−59 49	267.8	+ 2.4	3.5	8.7	II 2*p*	10.2	1.8	
4463	27.1	−64 31	268.3	− 2.3	5	8.2	I 3*p*	10.2	1.5	
Harv. 6	34.9	−68 12	269.2	− 5.9	7	10.7	Praes	13.6:	3.7	
Tr 20	36.8	−60 20	269.2	+ 1.9	10	10.1	III 2*r*	14:	1.7	
4609	39.4	−62 42	269.6	− 0.4	4.5	8.9	II 2*p*	9.3	1.4	
4755	50.7	−60 04	270.9	+ 2.2	12	5.2	I 3*r*	7:	1.2	ϰ Cru
4815	54.9	−64 41	271.3	− 2.4	4	10.4	I 3*m*	13.6:	1.8	
4852	57.1	−59 20	271.8	+ 2.9	11	8.9	II 2*m*	9.9	1.1	
Cr 268	13 15.4	−66 56	273.2	− 4.9	8	10.1	Plei	13.6	3.0	Harv 8
Cr 269	19.3	−65 51	273.8	− 3.8	15	9.2	Plei		1.8	
5138	24.1	−58 45	275.3	+ 3.1	8	9.8	II 2*p*		1.0	
Cr 271	26.4	−63 56	274.9	− 2.0	8.5	10.1	Plei		2.9	
Cr 272	27.3	−61 01	275.5	+ 1.0	13	9.2	Plei		2.0	
5168	27.9	−60 41	275.5	− 0.7	3.2	11.5	I 3*m*		1.9	
Tr 21	28.8	−62 33	275.3	− 0.7	5	9.6	I 3*p*		1.1	
Cr 275	31.3	−59 53	276.0	− 2.0	8	10.2	Plei		3.0	
5281	43.1	−62 39	276.8	− 1.2	4	8.2	I 3*m*	10.1	1.6	
Cr 277	45.1	−65 50	276.4	− 4.3	18	9.2	Praes		1.5	
5288	45.1	−64 26	276.7	− 3.0	3	11.8	II 2*p*		2.0	
5316	50.4	−61 37	277.9	− 0.4	11	8.8	II 2*p*	10.5	0.7	
5460	14 04.4	−48 05	283.8	+12.0	35	6.1	II 3*m*	9.1	0.5	
5593	22.4	−54 35	284.5	+ 4.9	8			11.5	2.5	
5606	24.1	−59 25	282.5	+ 0.4	2.2	10.0	II 2*p*		1.4	
5617	26.0	−60 30	282.4	− 0.7	14	8.5	I 2*r*	10.8	0.9	
Tr 22	27.5	−60 57	282.3	− 1.2	7	9.9	III 2*p*	12	1.8	
5662	31.6	−56 20	284.7	+ 2.8	15	7.7	II 3*pE*	9.7	0.6	
U Maj	36.7	+69 47	76.5	+44.8				1	0.02	
5715	39.7	−57 20	285.3	+ 1.4	9	9.8	II 2*m*	12.4	1.3	
5749	45.4	−54 19	287.4	+ 3.8	8	8.9	II 2*p*	10.5	0.9	
5764	50.0	−52 29	288.8	+ 5.1	2.2	12.6	I 2*p*:		3.4	
5822	15 01.5	−54 09	289.5	+ 2.8	40	6.4	III 1*m*	10.8	0.6	
5823	02.0	−55 14	288.9	+ 1.7	10	8.6	III 2*m*	10.8	1.3	
5925	23.9	−54 21	292.1	+ 0.8	20	8.4	III 2*m*		0.7	
Harv 9	30.0	−53 26	293.5	+ 1.1	3			13.6	6.6	
Cr 292	46.7	−57 31	286.6	+ 1.8	16	7.9	Plei		1.4	
5999	48.2	−56 19	293.7	− 2.8	6	9.0	I 2*r*	12.4	1.6	*
6005	51.8	−57 17	293.4	− 3.8	3.5	11.5	I 2*m*	13.6	2.0	
Tr 23	56.6	−53 23	296.5	− 1.4	5	11.2	III 2*p*		1.4	
6025	59.4	−60 22	292.1	− 6.8	11	6.0	II 3*p*	8.5	0.7	
6031	16 03.7	−53 55	296.9	− 2.5	2	12.2	II 2*p*:		1.7	
6067	09.3	−54 05	297.4	− 3.2	16	6.5	I 2*r*	10.9	0.8	*
Harv. 10	14.5	−55 00	298.6	− 3.4	24	6.8	Plei	10.6	0.5	Cr 299
6087	14.7	−57 47	295.8	− 6.0	18	6.0	II 3*m*	9.7	0.6	
6124	22.2	−40 33	308.6	+ 4.8	25	6.3	I 3*r*	9.1	0.6	*
Antares Cl	23.1	−26 07	319.5	+14.4	600	1.0	Neb. cl.		0.5	

Desig.	α 1950	δ	λ 1900	β	App. Diam.	Int. Mag.	Class	Mag. Br. St.	Dist. kpc	Remarks
6134	$16^h24^m.0$	−49° 02′	302°.6	− 1°.2	8′	8.8	II 2*r*	11.3	1.1	
6152	28.8	−52 31	300.6	− 4.2	20	8.1	I 2*m*	11.0	0.6	Called 6154 in Harv cat.
6167	30.5	−49 30	303.1	− 2.4	14	6.6	I 3*m*	10:	0.7	Includes Harv 11
6169	30.5	−43 57	307.8	+ 1.8	7	5.3	μ Norm		0.5	
Cr 307	31.4	−50 52	302.6	− 3.0	8.5	10.3	Plei		3.0	
6178	32.1	−45 32	306.1	+ 0.1	4	7.2	II 3*p*	10.3	0.5	
6192	36.8	−43 16	308.4	+ 1.0	7	8.5	I 3*m*	9.0	0.4	
6193	37.5	−48 40	304.8	− 2.4	20	5.4		10.4	0.5	
6200	40.5	−47 23	305.9	− 2.1	18	8.0	Plei		1.3	
6204	42.8	−46 56	306.3	− 2.2	6	8.4	II 2*m*	10.6	0.5	
6208	45.5	−53 44	301.3	− 6.8	20	9.2	IV 2*m U*	12.5	1.2	
6216	45.8	−44 39	308.4	− 1.1	3.8		II 2*m*		0.5	Mel 152?
6222	46.1	−44 39	308.7	− 0.9	3.9	10.2	Praes	12.8:	4.6	Mel 152?
6231	50.5	−41 43	311.2	0.0	16	8.5	I 3*r*	]7.4	0.6	*
Cr 316	52.0	−40 45	312.6	+ 0.5	120	3.4	Plei		0.2	
6242	52.2	−39 25	313.2	+ 1.2	11	8.2	I 3*m*	8.3	0.8	
Tr 24	53.5	−40 35	312.8	+ 0.4	60	8.5	IV 3*m N*	]7.4	0.9	* Harv 12
6249	54.0	−44 42	309.2	− 2.3	7	9.3	II 2*p*		0.4	
6250	54.3	−45 43	309.0	− 2.6	17	8.0	Plei		1.3	
6253	55.2	−52 38	303.0	− 7.3	5	10.2	I 2*m*	[13	1.9	
6259	57.1	−44 36	309.6	− 2.7	16	8.6	IV 2*r*	11.3	0.5	*
6268	58.7	−39 40	313.7	+ 0.1	7	9.5	II 2*p*	10.4	0.9	
6281	17 01.4	−37 50	315.6	+ 0.8	9	8.4	II 2*p*	7:	0.8	
Harv 13	01.6	−48 07	307.3	− 5.4	15			13	5.0	
6318	14.2	−39 23	315.7	− 2.2	5	11.9	II 2*m*	11.5	1.4	
6322	14.9	−42 53	312.9	− 4.3	9	6.5	II 3*p*	9.6	0.9	
I 4651	20.8	−49 53	307.6	− 9.0	14	8.0	II 2*r*	10.3	1.2	*
Tr 25	21.2	−38 57	316.9	− 3.1	5	11.7	I 2*p*	11.6	1.1	Harv 14
I 1257	24.5	− 7 03	345.5	+13.7						
Tr 26	25.3	−29 27	325.3	+ 1.5	8	8.9	Plei	8.7	0.8	Cr 331, Harv 15
Cr 332	27.4	−37 03	319.1	− 3.0	15	9.7	Plei		1.7	
Cr 333	28.0	−34 03	321.7	− 1.4	3	10.0	Plei		3.7	
Harv. 16	28.0	−36 49	328.5	− 3.0	15			8.4	0.6	
6374	29.0	−32 34	323.5	− 0.6	2.5	10.0	Plei		5.0	
6383	31.5	−32 32	323.4	− 1.3	5.5	5.4	II 3*p*	9.6	0.6	
Tr 27	32.9	−33 27	322.7	− 2.0	8	9.1	I 2*p*		0.8	
Tr 28	33.5	−32 27	323.7	− 1.6	7.5	9.4	II 2*p*		0.8	
Cr 338	34.9	−37 32	320.0	− 4.0	25	8.0	Plei		1.1	
6396	34.8	−34 58	321.7	− 3.2	4	10.9	I 2*p*		1.3	
6400	37.4	−36 54	321.0	− 4.3	7.5	8.9	I 2*p*	8.7	0.9	
6404	36.3	−33 13	323.3	− 2.5	5	10.6	II 3*m*	[9.6	1.4	
6405	36.8	−32 11	324.3	− 2.0	26	4.6	II 3*m*	8.3	0.4	M 6
Tr 29	38.2	−40 05	317.7	− 6.5	14	7.5	II 2*p*	9.6	0.6	Harv 17
6416	41.1	−32 20	324.6	− 2.9	22	8.7	III 2*p*	8.3	0.6	
Cr 345	41.3	−33 44	324.0	− 3.3	6	10.9	Plei		4.0	
Cr 347	43.2	−29 16	327.8	− 2.0	5	10.5	Plei		4.0	
6425	43.7	−31 30	325.6	− 3.0	10	9.2	II 1*p*		0.7	
I 4665	43.8	+05 44	358.2	+15.6	50	5.4	II 2*p*	7:	0.3	
Cr 350	45.5	+01 19	354.2	+13.8	50	6.2	Plei		0.6	
Cr 351	46.2	−28 43	328.9	− 1.4	8.8	9.3	Plei		2.5	
6451	47.5	−30 12	327.2	− 3.0	6	8.3	II 2*m*	10.3	1.2	
6469	49.9	−22 20	334.2	+ 0.5	15	8.2	IV 2*m*	11.8	1.0	
6475	50.6	−34 48	323.5	− 5.9	50	3.3	I 3*m*	7.4	0.2	M 7
Tr 30	53.0	−35 19	323.3	− 6.6	15	8.8	IV 3*m*		1.3	
6494	54.0	−19 01	337.6	+ 1.4	27	5.9	I 2*r*	10.2	0.5	* M 23
Tr 31	56.7	−28 10	329.9	− 3.7	4.5	9.8	II 2*m*		1.8	
6507	56.7	−17 23	339.6	+ 2.0	9	9.6	Plei	12	2.6	
Cr 359	58.5	+02 55	356.6	+11.1	310	3.0	Plei	]9	0.2	

Desig.	α 1950	δ	λ 1900	β	App. Diam.	Int. Mag.	Class	Mag. Br. St.	Dist. kpc	Remarks
6514	17h59m.3	−23°02′	334°.6	− 1°.6	28′	5.2	Neb. cl.	7	1.0	
6520	18 00.3	−27 54	330.5	− 4.3	6.5	7.6		8.7	1.7	
6523	00.1	−24 23	333.6	− 2.5	45	5.2			1.1	M 8 Lagoon Neb.
6530	01.7	−24 20	333.8	− 2.8	14	7.7	II 2*mN*	9.3	1.6	
6531	01.6	−22 30	335.4	− 1.9	12	7.2	I 3*p*	9.3	1.0	M 21
6540	03.2	−27 49	331.6	− 4.8	0.7	14.6	Praes	[13	31:	
Cr 468	03.6	−27 28	331.5	− 4.0	2	13.4	Plei		12.5	
6546	04.2	−23 19	335.0	− 2.8	13	8.6	IV 3*r*	11	1.3	
Cr 367	06.6	−24 00	334.6	− 2.9	48	6.4	Neb. cl.		1.7	
6568	09.8	−21 36	337.1	− 3.1	15	8.6	III 1*m*		1.0	
Mark 38	12.3	−19 01	339.7	− 2.5	2		0 type		1.0	
6583	12.8	−22 09	337.0	− 4.0	3.8	12.1	I 2*m*	13:	3.3	
6595	14.0	−19 54	339.9	− 2.6	15	7.2	Neb. cl.	9	0.2	
Tr 32	14.7	−13 22	344.9	− 0.2	5.5	12.2	I 2*p*		0.8	
6603	15.5	−18 26	340.6	− 2.8	4.5	11.4	I 2*r*	13.0	2.2	M 24
6604	15.3	−12 15	346.0	+ 0.2	2.8	8.2	I 3*p*	7	0.9	
6611	16.0	−13 48	344.7	− 0.7	8	6.6	II 3*mN*	10.6	1.2	M 16
6613	17.0	−17 09	341.8	− 2.5	7	8.0	II 3*p*	10.9	0.7	M 18
6618	17.9	−16 12	342.5	− 2.2	20	6.9	Neb. cl.	10.6	1.5	M 17, Horseshoe Neb.
Tr 33	21.8	−19 43	340.1	− 4.7	5.5	8.8	I 3*p*	11	1.8	
6631	24.4	−12 05	347.2	− 1.7	4.5	11.7	I 2*p*		0.9	
6633	25.1	+06 32	3.8	+ 6.8	25	5.6	I 2*pE*	8.0:	0.3	
I 4725	28.8	−19 17	341.3	− 6.0	35	6.2	IV 3*r*	9.3	0.6	M 25
6645	29.7	−16 56	343.5	− 5.1	13	8.4	I 2*r*	11.3	0.9	
6649	30.7	−10 26	349.4	− 2.3	7.5	10.3	I 2*m*	12.1	0.8	
Graff	32.5	+05 09	3.3	+ 4.6				9	0.3	*
6664	34.0	−08 16	351.7	− 2.0	20	8.9	IV 2*m*	11.3	0.9	
I 4756	36.5	+05 25	4.0	+ 3.8	50	5.4	III 1*m*	8:	0.5	*
Tr 34	37.1	−08 31	351.8	− 2.8	10	11.6	II 2*p*		1.0	
6683	39.5	−06 20	353.8	− 1.9	3	9.5				
Tr 35	40.4	−04 11	356.0	− 1.4	5	10.0	II 2*m*		0.8	
6694	42.6	−09 27	351.6	− 4.4	9	9.6	II 2*m*	11.9	1.2	M 26
6704	48.2	−05 16	355.9	− 3.7	5	9.4	I 2*p*	12	1.4	
6705	48.4	−06 20	355.0	− 4.3	12.5	6.8	II 2*r*	12	1.7	* M 11
6709	49.1	+10 17	9.8	+ 3.3	12	7.4	II 2*p*	9.0:	0.7	
6716	51.6	−19 57	343.1	−11.1	8	6.8	Plei		0.8	
Cr 394	50.5	−20 27	342.5	−11.0	25	6.6	Praes		0.9	
6738	59.0	+11 32	11.7	+ 1.9	17	8.2	Plei		1.2	
6755	19 05.3	+04 08	6.3	− 3.1	15	8.9	IV 2*m*	11.0	1.2	
6756	06.2	+04 35	6.8	− 3.1	4	10.6	I 2*p*	[13	1.3	
Cr 399	23.2	+20 06	22.3	+ 0.9	85	4.8	Plei		0.2	4-5 Vul
6802	28.4	+20 10	23.0	− 0.3	5	11.8	III 1*mE*	13.0:	1.7	
Stock 1	33.7	+25 06	27.7	+ 1.0	60			10	0.3	
Cr 401	35.8	+00 14	6.3	−11.4	1.2	7.0	μ Norm		2.6	
6811	36.7	+46 27	46.9	+11.2	13	9.2	III 1*p*	11.8	1.2	
6819	39.6	+40 04	41.5	+ 7.6	6	9.9	I 2*r*	13.0:	3.7	*
6820	40.5	+22 58	26.2	− 0.9	0.8	14.9	Neb. cl.		31:	
6823	41.0	+23 11	27.1	− 1.3	7	9.8	IV 3*p*	9.0:	1.8	
6830	48.9	+22 57	27.9	− 3.0	10	8.9	IV 2*m*	9.0	1.9	
6834	50.2	+29 17	33.4	+ 0.2	7	10.0	I 2*m*	13:	1.3	
Harv 20	50.9	+18 13	24.0	− 5.7	7.8	9.6	Plei	9.0:	0.8	
6846	54.5	+30 13	35.1	− 0.1	0.5	14.2	Plei	15	33:	
6866	20 02.1	+43 51	47.0	+ 6.0	10	8.9	I 2*p*	9.4	1.2	
6871	04.0	+35 38	40.3	+ 1.2	25	5.8	IV 3*p*	7	1.1	
Mel. 227	05	−79 28	281.4	−31.0	60	5.3	II 2*p*		0.2	
I 1311	08.7	+41 04	45.3	+ 3.5	8	13.1		13.0	3.2	
6882	09.6	+26 24	33.3	− 5.0	8	5.7	I 2*p*	11	1.1	
6883	09.4	+35 42	41.0	+ 0.3	15	8.0	IV 3*p*		1.0	
6885	09.9	+26 20	33.3	− 5.1	22	9.1	III 2*p*	9.8	1.0	20 Vul
I 4996	14.7	+37 29	43.0	+ 0.5	6	7.1	I 3*p*	9	1.0	

Desig.	α 1950	δ	λ 1900	β	App. Diam.	Int. Mag.	Class	Mag. Br. St.	Dist. kpc	Remarks
Cr 419	$20^h16^m.3$	+40° 34′	45°.4	+ 2°.4	4.8′	5.4	μ Norm		1.5	
6910	21.3	+40 37	46.3	+ 1.3	13	7.7	IV 3*p*	9.0	0.8	
Cr 421	21.5	+41 32	46.8	+ 2.4	6	10.0			3.6	
6913	22.1	+38 22	44.6	− 0.2	7	9.0	III 3*p*	9.4	0.7	M 29
6939	30.4	+60 28	63.3	+11.9	8	10.1	II 1*r*	13:	2.0	
6940	32.5	+28 08	37.8	− 8.1	26	6.7	III 1*m*	11.8	1.0	*
6994	56.2	−12 49	4.0	−35.6	2.8	8.9	Glob?	10	3.8	
6996	54.7	+44 26	53.0	− 1.0	8	10.0	Neb. cl.		2.9	
Cr 427	58.9	+67 58	71.2	+14.4	5	13.8	Plei		5.5	
7023	59.9	+67 58	71.3	+14.4	5	7.1	Neb. cl.		2.8	
Cr 428	21 01.4	+44 23	53.8	− 1.5	17	8.7	Neb. cl.		1.5	
7031	05.7	+50 38	59.0	+ 1.8	7	11.0	II 3*pU*		1.0	
7039	10.4	+45 27	55.7	− 2.4	25	6.9	Plei		0.9	
I 1369	10.5	+47 32	57.3	− 0.9	3	12.1	II 2*p*		1.6	
7044	11.1	+42 17	53.7	− 4.7	5	11.6	I 2*r*	15	2.2	
7062	21.4	+46 10	57.6	− 3.2	6	11.7	II 2*p*	11.4	1.3	
7063	22.4	+36 17	51.0	−10.6	8	8.9	II 2*p*		1.2	
7067	22.4	+47 48	58.8	− 2.3	2.5	13.0	II 2*p*		1.9	
7082	27.6	+46 52	58.9	− 3.4	25					
7086	28.8	+51 22	62.1	− 0.2	7.5	11.6	I 2*m*	11.8	1.6	
7092	30.4	+48 13	60.2	− 2.7	32	5.3	II 2*p*	6.5	0.2	M 39
I 1396	37.5	+57 16	66.9	+ 3.5	60	5.1	IV 3*rN*		0.3	Tr 37
7127	42.2	+54 23	65.6	+ 0.8	2.8			11.8	1.3	
7128	42.3	+53 29	65.0	+ 0.1	3.2	11.3	II 2*p*	11.8	1.6	
7129	40.1	+65 52	72.3	+10.0	2.5	11.4			7.7	
7142	44.7	+65 34	72.9	+ 9.3	11	10.0	II 1*m*	14	1.2	
I 5146	51.5	+47 02	62.8	− 6.3	12	8.3	Neb. cl.		4.2	
7160	52.3	+62 22	71.5	+ 6.3	10	6.4	I 3*p*	7	0.4	
7209	22 03.2	+46 15	63.4	− 7.7	20	8.0	II 2*p*	10.1	0.6	
Cr 471	05.9	+71 45	78.2	+13.4	180	3.3	Plei		0.2	
I 1434	08.6	+52 35	67.7	− 3.0	7	9.8	III 2*m*	12	2.9	
7226	08.7	+55 10	69.1	− 0.8	2.5	13.3	II 2*p*		2.8	
7235	10.8	+57 02	70.4	+ 0.6	4.5	9.2	II 3*p*		1.2	
7243	13.3	+49 38	66.7	− 5.8	21	6.7	III 2*p*	9.0	1.1	
7245	13.4	+54 05	69.1	− 2.1	4	11.5	II 2*p*	13	2.3	
King 9	13.6	+54 09	69.2	− 2.1	3		*r*	18		
I 1442	14.6	+53 48	69.1	− 2.5				12		
7261	18.6	+57 50	71.7	+ 0.7	7	9.8	II 2*p*		1.4	
7296	26.2	+52 02	69.7	− 4.8	4	9.7	II 2*p*	10	2.6	
7380	45.0	+57 50	74.8	− 0.9	9	8.8	III 2*p*	9.0	1.5	
7419	52.3	+60 34	76.8	+ 1.1	2.2	13.0	II 2*p*:	13	1.3	
King 10	52.9	+58 54	76.2	− 0.5	4		*m*	11		
King 19	23 06.2	+60 15	78.3	+ 0.2	4		*p*	12		
7510	09.4	+60 18	78.7	+ 0.1	3	9.3	II 2*mU*	11.7	1.1	
Mark 50	13.1	+60 12	79.5	0.0	1.5		*O* type		2.6	
7654	22.0	+61 19	80.4	+ 0.5	13	8.2	II 2*r*	11.0	0.7	* M 52
7686	27.8	+48 51	77.5	−11.6	13	8.2	I 3*p*		0.9	
King 20	31.0	+58 14	80.6	− 2.7	5		*p*	12		
King 11	45.4	+68 21	84.7	+ 6.6	4		*m*	17		
7762	47.4	+67 45	84.7	+ 6.0	11	10.0	I 1*mU*	12.8	0.9	
King 21	47.4	+62 26	83.6	+ 0.8	2		*p*	12		
King 12	50.5	+61 41	83.8	0.0	2		*p*	10		
Harv 21	51.7	+61 29	84.0	− 0.6	2	9.3	Plei	11.7	2.8	
7788	54.2	+61 07	84.1	− 0.6	11	9.4	IV 3*p*	11.0	1.6	
7789	54.5	+56 27	83.3	− 5.2	19	9.3	III 1*r*	12.5	1.2	*
7790	55.9	+60 56	84.3	− 0.8	4.5	7.2	II 2*p*	11.7	1.2	Correct position by KRAFT.

Appendix B.

Catalogue of globular clusters.

Description of the catalogue. This contains 118 clusters, all that are at present considered globular and not in external systems. It includes several which are intergalactic.

The first column gives usually the NGC number, or for very recent clusters, an abbreviation of the discovery source[1]. The order is that of NGC number which does not always correspond to the order of right ascension for 1950. The second column gives the Messier or Dunlop (Δ) number, or other identification. The next four columns give the right ascension and declination for the equinox of 1950, and the galactic longitude and latitude for 1900. The seventh column is the concentration class of SHAPLEY and SAWYER[2]. The eighth column gives the angular diameters, largely by MOWBRAY[3] or by SHAPLEY and SAYER[4]. In the ninth column most of the integrated photographic magnitudes are by CHRISTIE[5] or by SAWYER and SHAPLEY[6] reduced to CHRISTIE's system. The tenth column gives the spectral types by MAYALL[7] with a few recent revisions by MORGAN[8]. The colors in column eleven are those designated as *C* 2 by STEBBINS and WHITFORD[9], and the radial velocities in column twelve are by MAYALL[7].

The thirteenth column gives the magnitudes of the 25 brightest stars (excluding the five brightest), from such a wide variety of sources that they cannot be noted here. Some are quite recent, while others go back to SHAPLEY and SAWYER[10] in 1929. The apparent modulus in column fourteen is that derived only from magnitudes of individual stars, either the 25 brightest stars, or variable stars, or both. The absorption correction in column fifteen is the standard one of 0.25 cosec β for clusters 20° or more from the galactic plane. For clusters closer to the galactic plane it is usually the correction derived by LOHMANN[11] from measured colors. Certain workers have rejected any correction for high latitude clusters. To avoid confusion the distances, given in kiloparsecs in column sixteen, are usually those already published. The reader may standardize the absorption corrections for himself. For the high latitude clusters the distance is usually that given by BAUM[12], based on detail kindly supplied in a private communication. For most of the other clusters *I* have computed the distance from the best apparent modulus and absorption correction as indicated. The distances are given only for those clusters which have a modulus determined from magnitudes of individual stars. The angular diameters may be used to give a rough indication of the distance for other clusters[13], but this method introduces greater uncertainty, particularly for clusters without known colors which can give an indication of the reddening and absorption.

It should be emphasized that it is not possible to construct a truly homogeneous catalogue of such data as are given in Appendix B. For example, the magnitude of the 25 brightest stars and the apparent modulus for the same cluster may have to be drawn from different sources. Also, because there are different published values of the data given, those in the table may not always agree with other values noted in the text. Such a spread in values is useful in indicating the degree of present uncertainty.

---

1 Of clusters recently added to the globular list, the Palomar clusters as well as IC 1276 and NGC 6717 have been published by G. O. ABELL: Publ. Astronom. Soc. Pacific **67**, 258 (1955); the Russian cluster by R. K. SHAKHBAZIAN: Russ. Astronom. Circ. No. 177, 11, (1957); and that abbreviated as HP by J. DUFAY et al.: Publ. l'Obs. Haute-Provence **3**, No. 17 (1954). NGC 2158 is considered globular by L. ROSINO: Contr. Padova in Asiago No. 52 (1954), while NGC 6380 was found to be globular by A. D. THACKERAY on plates taken with the 74-inch telescope at Radcliffe Obs. (private communication).

2 H. SHAPLEY and H. B. SAWYER: Harvard Bull. No. 849, (1927).

3 A. G. MOWBRAY: Astrophys. Journ. **104**, 47 (1947)

4 H. SHAPLEY and A. R. SAYER: Harvard Obs. Repr. No. 116 (1935).

5 W. H. CHRISTIE: Astrophys. Journ. **91**, 8 (1940).

6 H. B. SAWYER and H. SHAPLEY: Harvard Bull. No. 848 (1927).

7 N. U. MAYALL: Astrophys. Journ. **104**, 290 (1946).

8 W. W. MORGAN: Publ. Astronom. Soc. Pacific **68**, 509 (1956).

9 J. STEBBINS and A. E. WHITFORD: Astrophys. Journ. **84**, 132 (1936).

10 H. SHAPLEY and H. B. SAWYER: Harvard Bull. 869 (1929).

11 W. LOHMANN: Z. Astrophys. **30**, 234 (1952).

12 W. A. BAUM: Smithson. Contrib. Astrophys. **1**, 165 (1956).

13 H. SHAPLEY: Popular Astronomy **57**, 218 (1949).

Catalogue of globular clusters.

NGC	Name	α 1950	δ	λ 1900	β	Conc.	App. Diam.	Int. Mag.	Sp. Type	Color	Rad. Vel.	Mag 25 Br.St.	App. Mod.	Abs. Corr.	D kpc
104	47 Tuc	$00^{h}21^{m}.9$	−72°21′	272°	−45′	III	44′	4.68				13.44	14.5	.35	6.6
288		50.2	−26 52	154	−88	X	12.4	8.96		−.08		14.80	15.7	.25	12.0
362	Δ 62	01 00.6	−71 07	268	−47	III	17.7	8.0				14.12	15.7	.34	12.0
1261		03 10.9	−55 25	236	−52	II	4.0	9.5							
Pal 1		25.7	+79 28	97	+20	XII	1.3					19			
Pal 2		04 43.1	+31 23	137	−9	IX	1.7								
1841		52.5	−84 05	264	−30		2.4	12.2							
1851	Δ 508	05 12.4	−40 05	211	−34	II	11.5	7.72	*dF* 5		+291				
1904	M 79	22.2	−24 34	195	−28	V	7.8	8.39	*dF* 3		+231	15.29	16.3	.00	17.3
2158		06 04.4	+24 06	154	+3		7.5	11.8				15.32	15.8		
2298		47.2	−35 57	213	−15	VI	4.2	10.48	*F* 8		+64				
2419		07 34.8	+39 00	148	+27	II	6.2	11.51	*F* 5	−.17	+14	17.84	18.8	.55	57.5
2808		09 10.9	−64 39	250	−11	I	18.8	7.8				14.9	16.2	1.8	7.6
Pal 3		10 03.0	+00 18	209	+43	XII	2.2								
3201	Δ 445	15.5	−46 09	245	+09	X	29.3	8.8					15.08	2.2	3.8
Russ		52.0	+40 44	143	+64		1.3	15.0				18	20.9	.33	130
Pal 4		11 26.6	+29 15	170	+73	XII	2.5	14.4				19.90	20.5	.00	125
4147		12 07.6	+18 49	224	+78	VI:	4.1	11.01	*A* 5	−.07	+191	16.58	17.06	.00	25.7
4372		23.0	−72 24	268	−10	XII	19.8	9.1							
4590	M 68	36.8	−26 29	268	+37	X	9.8	9.12	*A* 6		−116	14.80	16.1	.42	13.8
4833		56.0	−70 36	271	−08	VIII	12.7	8.5					15.65	2.4	4.4
5024	M 53	13 10.5	+18 26	306	+79	V	14.4	8.68	*A* 8*n*	−.08	−112	15.07	16.7	.26	19.1
5053		13.9	+17 57	308	+78	XI	8.9	10.9				15.6	16.4	.26	16.6
5139	ω Cen	23.8	−47 03	277	+15	VIII	65.4	4.25					14.65	1.3	4.6
5272	M 3	39.9	+28 38	07	+77	VI	18.6	7.21	*dF* 2	−.05	−150	14.23	15.7	.26	12.0
5286	Δ 388	43.0	−51 07	280	+10	V	13.6	9.5							
5466		14 03.2	+28 46	08	+72	XII	9.2	10.39		−.05		15.72	16.4	.26	16.6
5634		27.0	−05 45	311	+48	IV	3.7	10.8	*F* 4	−.07	−63	16.32	17.1	.34	22.9
5694		36.7	−26 19	299	+29	VII	2.2	10.87	*A* 9	−.07	−187	16.79	17.7	.52	27.5
I 4499		52.7	−82 02	275	−21	XI	6.2	11.6							
5824		15 00.9	−32 53	300	+21	I	3.7	10.08	*dF* 5	−.05	−58				
Pal 5		13.5	+00 05	329	+44	XII	10.3					16.90	17.22	.00	27.8
5897		14.5	−20 50	311	+29	XI	8.7	9.61		−.05		15.26	16.2	.52	13.8
5904	M 5	16.0	+02 16	332	+46	V	19.9	7.04	*dF* 7	−.05	+45	13.97	15.2	.35	9.1
5927		24.5	−50 29	294	+04	VIII	12.0	9.7							
5946		31.8	−50 30	295	+03	IX	2.6	11.0							
5986	Δ 552	42.8	−37 37	305	+13	VII	6.0	8.72	*G* 0	−.02	+2				
6093	M 80	16 14.1	−22 52	321	+18	II	5.1	8.39	*dF* 4	−.02	+18	14.88	16.05	0.9	10.5
6101		20.0	−72 06	285	−17	X	14.6	10.2							
6121	M 4	20.6	−26 24	319	+15	IX	22.8	7.41		+.05		13.11	13.97	1.9	2.5
6139		24.3	−38 44	310	+06	II	2.6	10.4		+.15					
6144		24.2	−25 56	320	+14	XI	6.2	10.85		+.01		15.76	16.0	1.3	8.7
6171		29.7	−12 57	331	+22	X	7.8	10.10	*G* 2	+.18	−147	15.5	16.2	.67	12.6
6205	M 13	39.9	+36 33	26	+40	V	23.2	6.78	*dF* 2	−.03	−228	13.75	14.8	.39	7.6
6218	M 12	44.6	−01 52	343	+25	IX	12.2	7.95	*F* 7	.00	+36	13.97	15.2	.59	8.3
6229		45.6	+47 37	40	+39	IV	3.8	10.26	*dF* 6	−.04	−150	16.18	17.7	.40	28.8
6235		50.4	−22 06	327	+12	X	1.9	10.8		.00		16.24	17.13	1.1	15.9
6254	M 10	54.5	−04 02	343	+22	VII	12.2	7.64	*G* 0	+.03	+73	14.06	15.2	.67	7.9
6266	M 62	58.1	−30 03	321	+06	IV	6.3	8.16	*F* 8	+.08	−81	15.87	16.74	2.4	7.2
6273	M 19	59.5	−26 11	325	+08	VIII	5.3	8.29	*F* 5	+.04	+102	14.80	15.95	2.0	6.2
6284		17 01.5	−24 41	326	+09	IX:	2.7	10.61	*F* 8	.00	+22	16.06	17.26	1.1	17.4
6287		02.1	−22 38	328	+10	VII	2.7	11.24		+.12		16.08	17.28	2.9	7.6
6293		07.1	−26 30	325	+07	IV	3.5	9.38	*F* 5	−.01	−73	15.39	16.69	1.3	12.0
6304		11.4	−29 24	324	+04	VI	3.8	9.82	*G* 5	+.18	−98				
6316		13.4	−28 05	325	+04	III	2.4	10.10		+.15					
6325		15.0	−23 42	329	+07	IV:	1.6	12.66		+.16					
6333	M 9	16.2	−18 28	333	+09	VIII	5.5	8.92	*F* 2	+.06	+224	15.50	16.39	2.4	6.3
6341	M 92	15.6	+43 12	35	+34	IV	12.2	7.30	*A* 5*n*	−.09	−118	13.86	15.2	.45	8.7
6342		18.2	−19 32	333	+08	IV	1.3	11.35		+.18					

NGC	Name	α 1950	δ	λ 1900	β	Conc.	App. Diam.	Int. Mag.	Sp. Type	Color	Rad. Vel.	Mag. 25 Br.St.	App. Mod.	Abs. Corr.	*D* kpc
6352		17h21m.6	−48°26′	309°	−08°	XI:	8.9′	9.1							
6355		20.9	−26 19	327	+04		1:	9.6							
6356		20.7	−17 46	334	+09	II	3.5	9.68	*G* 5	+.11	+31	16.38	17.2	2.8	7.6
HP		24.9	−29 57	325	+01	IX	1.3								
6362	Δ 225	26.6	−67 01	293	−18	X	8.5	8.3							
6366		25.1	−05 02	346	+15	XI	5.8	12.1		+.25		15.78	16.25	4.7	2.0
6380		31.9	−39 02	318	−05										
6388		32.6	−44 43	313	−08	III	6.8	8.7							
6397	Δ 366	36.8	−53 39	306	−13	IX	19:	7.3				12.61	13.0	1.5	2.0
6401		35.6	−23 53	331	+03		1								
6402	M 14	35.0	−03 15	349	+13	VIII	6.7	9.44	*G* 0	+.11	−131	15.44	16.72	2.9	5.8
Pal 6		40.6	−26 12	330	0	XI	1.8					20			
6426		42.4	+03 12	356	+15	IX	2.2	12.33		+.04					
6440		45.9	−20 21	335	+02	V	1.7	12.05	*G* 5	+.37	−133				
6441		46.8	−37 02	321	−06	III	3.0	8.93	*G* 2	+.14	−70				
6453		48.0	−34 37	323	−05	IV	3.6	11.4		+.10					
6496		55.5	−44 13	316	−11	XII	12.7	10.3							
6517		59.1	−08 57	347	+05	IV	1.0	12.90		+.30					
6522		18 00.4	−30 02	329	−05	VI	1.5	10.40	*F* 8	+.14			17.77	3.1	8.7
6528		01.6	−30 04	328	−06	V	1.2	11.04		+.22					
6535		01.3	−00 18	355	+09	XI:	1.3	11.9		.00		15.97	16.88	1.1	14.5
6539		02.1	−07 35	349	+05	X	3.5	12.39		+.41					
6541	Δ 473	04.4	−43 44	317	−12	III	23.2	7.9				13.35	14.67	1.6	4.2
6544		04.3	−25 01	334	−04		1:		*G* 1		−12				
6553		06.3	−25 56	333	−04	XI	3.2	10.20		+.36					
6558		07.0	−31 47	328	−07										
I 1276		08.0	−07 14	349	+04	XII	6.0					18			
6569		10.4	−31 50	328	−08	VIII	2.2	10.63		+.15					
6584	Δ 376	14.6	−52 14	310	−18	VIII	9.7	9.4							
6624		20.5	−30 23	330	−09	VI	2.7	9.53	*G* 2	+.07	+69				
6626	M 28	21.5	−24 54	335	−07	IV	15.0	8.48	*F* 9	+.09	+1	14.73	15.62	2.6	4.0
6637	M 69	28.1	−32 23	329	−12	V	3.8	8.94	*G* 5	+.12	+95				
6638		27.9	−25 32	336	−09	VI	2.2	10.24	*G* 2	+.06	−14	16.22			
6642		28.4	−23 30	338	−07		0.8	10.3							
6652		32.5	−33 02	329	−13	VI:	2.3	9.86	*G* 2	+.05	−124				
6656	M 22	33.3	−23 58	337	−09	VII	17.0	6.48	*F* 6	+.03	−148	12.93	14.15	1.8	3.0
Pal 8		38.5	−19 52	342	−08	X	1.6					19			
6681	M 70	40.0	−32 21	330	−14	V	4.1	8.95	*G* 2	−.04	+198				
6712		50.3	−08 47	353	−06	IX:	4.2	9.98	*G* 2	+.11	−131	15.35	16.48	2.7	5.8
6715	M 54	52.0	−30 32	333	−16	III	5.5	8.74	*F* 7	+.01	+107		17.4	1.4	15.9
6717		52.1	−22 47	341	−12	VIII	2.6					16			
6723	Δ 573	56.2	−36 42	328	−19	VII	7.5	7.75	*G* 3	−.06	−3	14.20	15.34	0.2	10.5
6752	Δ 295	19 06.4	−60 04	304	−27	VI	41.9	7.2				13.26	14.2	0.55	5.2
6760		08.6	+00 57	04	−05	IX:	2.4	11.25		+.30		16.77	17.45	5.4	2.6
6779	M 56	14.6	+30 05	30	+08	X	5.0	9.55	*F* 5	−.04	−154	15.31	16.3	0.7	13.2
Pal 10		16.0	+18 28	20	+01	XII	3.1					20			
6809	M 55	36.9	−31 03	337	−25	XI	14.8	7.08		−.09		13.58	14.5	0.59	6.0
Pal 11		42.6	−08 09	0	−17	XI	2.8					17			
6838	M 71	51.5	+18 39	24	−06		6.1	8.3	*G* 2		−80	14.82	14.6	1.6	4.0
6864	M 75	20 03.2	−22 04	348	−27	I	4.6	9.50	*G* 1	+.01	−222	17.06	18.1	.55	31.6
6934		31.7	+07 14	20	−20	VIII	6.2	10.01	*F* 9	−.04	−360	15.78	16.9	.73	17.4
6981	M 72	50.7	−12 44	03	−34	IX	5.1	10.24	*G* 2	−.04	−255	15.86	17.0	.45	20.0
7006		59.1	+16 00	32	−20	I	2.2	11.45	*F* 1	−.07	−348	17.1	18.8	.73	41.7
7078	M 15	21 27.6	+11 57	33	−28	IV	12.3	7.33	*dF* 0	−.08	−114	14.31	15.9	.53	12.0
7089	M 2	30.9	−01 03	22	−37	II	11.7	7.30	*dF* 0	−.06	−3	14.61	16.2	.42	14.5
7099	M 30	37.5	−23 25	355	−48	V	8.9	8.58	*A* 7 *n*	−.11	−164	14.63	15.54	.34	11.0
Pal 12		43.7	−21 28	359	−49	XII	2.1					17			
Pal 13		23 04.2	+12 28	57	−43	XII	4.0					19	18.0	.37	33.6
7492		05.7	−15 54	22	−65	XII	4.3	12.33		−.20		16.82	17.9	.28	33.3

General references.

[1] SHAPLEY, H.: Star Clusters. Cambridge, Mass. 1930 (Catalogues and bibliography).
[2] COLLINDER, P.: On Structural Properties of Open Galactic Clusters and their Spatial Distribution. Lund Obs. Ann. No. 2 (1931).
[3] SAWYER, H. B.: A Bibliography of Individual Globular Clusters, David Dunlap Obs. Publ. **1**, No. 20 (1947).
[4] ALTER, G., J. RUPRECHT and V. VANYSEK: Catalogue of Star Clusters and Associations. Prague: Czechoslovak Academy of Sciences 1958 (on cards with bibliography for each cluster).
[5] DREYER, J. L.: A New General Catalogue of Nebulae and Clusters of Stars. Mem. Roy. Astronom. Soc. London **49**, Part. 1, 1 (1888). — Index Catalogue of Nebulae found in the Years 1888—1894. Mem. Roy. Astronom. Soc. London **51** (1895). — Second Index Catalogue of Nebulae and Clusters of Stars. Mem. Roy. Astronom. Soc. London **59**, Part 2 (1908); republished, Roy. Astronom. Soc. London **1953**.
[6] SHAPLEY, H.: Stellar Clusters. Handbuch der Astrophysik, Vol. 5, Part 2, pp. 698—773. 1933.
[7] TRUMPLER, R. J.: Preliminary Results on the Distances, Dimensions and Space Distribution of Open Star Clusters. Lick Obs. Bull. No. 420 (1930).
[8] MELOTTE, P. J.: A Catalogue of Star Clusters shown on Franklin-Adams Chart Plates. Mem. Roy. Astronom. Soc. London **60**, Part, V, 175 (1915).
[9] RAAB, S.: A Research on Open Clusters. Lund Obs. Medd., Ser. II, No. 28 (1922).
[10] MARKARYAN, B. E.: Atlas der offenen Sternhaufen der verschiedenen Typen. Astronom. Rat Akad. Wiss. UdSSR, Moskau [Russian] 1952.
[11] CHARLIER, C. V. L.: Stellar Clusters and Related Celestial Phenomena. Lund Obs. Medd., Ser. II, 19 (1918).
[12] SANDAGE, A. R.: Observational Approach to Evolution. Astrophys. Journ. **125**, 422, 435 (1957).
[13] TEN BRUGGENCATE, P.: Sternhaufen. Berlin 1927.
[14] SCHWARZSCHILD, M.: Structure and Evolution of the Stars. Princeton, N. J. 1958.
[15] HERTZSPRUNG, E.: The Pleiades. Monthly Notices Roy. Astronom. Soc. London **89**, 660 (1929).
[16] Vistas in Astronomy. ed. by A. BEER. London and New York 1956.
[17] BOK, B. J., and P. F. BOK: The Milky Way, 3rd. ed., p. 12. Cambridge, Mass.: Harvard Univ. Press 1957.
[18] BECKER, W.: Sterne und Sternsysteme, p. 163. Dresden and Leipzig 1950.
[19] CHANDRASEKHAR, S.: Principles of Stellar Dynamics. Chicago, Ill.: Chicago University Press 1942.
[20] AMBARTSUMIAN, V. A.: Voprosy Kosmogonii, Moscow **1**, 198 (1952).
[21] SHAPLEY, H.: A Half Century of Globular Clusters. Popular Astronomy **57**, 212 (1949).
[22] SHAPLEY, H.: The Inner Metagalaxy, p. 55. New Haven: Yale Univ. Press 1957.

Discrete Sources of Cosmic Radio Waves.

By

R. Hanbury Brown.

With 15 Figures.

Introduction.

One of the principal features of the distribution over the sky of cosmic radio-frequency radiation is the presence of localized sources of relatively high intensity which are described by the term *discrete sources*. At the present time our knowledge of the nature of these sources is very limited and the description *discrete* is correspondingly vague. No attempt will be made here to define precisely what is classed as a discrete source, except to state that the term has been applied to sources with an angular extent small compared with the general features of the distribution of radiation over the sky; it has been applied indiscriminately to objects which differ considerably in nature and which may have apparent angular diameters varying from about one minute of arc to about one degree. The more popular term "radio star" has been avoided because the observational evidence of the past decade points to the conclusion that the discrete sources are not associated with bodies of stellar dimensions, but arise in objects, or in regions, of vastly greater size. Radio sources which are associated with the solar system, for example the Sun or the planets Venus and Jupiter, are not included in this article[1].

An early survey by Reber[2] in 1944 showed the presence of certain regions of high intensity, for example in Cassiopeia, which we now know to be associated with powerful discrete sources. However this survey was conducted with a 12 degree beam and was therefore not capable of showing the highly localized nature of these sources. Another clue to the existence of the sources was provided by Hey, Parsons and Phillips[3] who in the course of a survey in 1946 noted that, whereas the radiation from the rest of the sky appeared to be constant, a region in Cygnus exhibited irregular fluctuations over periods of the order of one minute. They concluded that, in this particular region, there must be a high intensity variable source subtending a small angle.

This clue was followed up by Bolton and Stanley[4] who investigated the region with an interferometer of comparatively high resolving power. In June 1947 they detected radiation from a source in Cygnus and showed that it had an angular diameter of less than 8′; they concluded that the source consisted of two components, one constant and the other variable. A further search disclosed a number of other sources and established the existence of a class of such objects. The results obtained on the source in Cygnus were confirmed in 1948 by Ryle

[1] Cf. the contribution of F. J. Kerr to Vol. LII of this Encyclopedia.

[2] G. Reber: Astrophys. Journ. **100**, 279 (1944).

[3] J. S. Hey, S. J. Parsons and J. W. Phillips: Proc. Roy. Soc. Lond., Ser. A **192**, 425 (1948).

[4] J. G. Bolton and G. J. Stanley: Nature, Lond. **161**, 312 (1948).

and SMITH[1] who also discovered a number of sources including the intense source in Cassiopeia.

Later work showed that the variability, which had originally drawn attention to the source in Cygnus, was a property of the terrestrial ionosphere and not of the sources themselves.

Following this early work several surveys of the sky have been made to map the positions of the sources. As increasingly powerful equipment has been brought into use the number of sources reported in the literature has increased from a total of about 10 in 1948 to about 2000 in 1956. At the same time attempts have been made to measure precise celestial coordinates, parallax, intensities, spectra, polarization etc., and also to determine the origin of the scintillations. A considerable effort has also been made to identify the sources with objects which can be observed photographically. The period covered by this work is barely ten years, and our present knowledge is still inadequate to decide most of the major questions about the sources.

An attempt will be made in the following sections to summarize the principal facts which are known with reasonable certainty and to give a brief account of some of the theoretical speculations about the origin of the radiation. Before discussing the experimental data a short account of some definitions and units will be given followed by a brief discussion of the principal techniques of observation.

A. Definitions and units.

1. Flux-density, brightness and brightness temperature. There are three principal quantities which are used to describe the intensity of the radiation from radio sources, *flux-density, brightness* and *brightness temperature.*

Flux-density is a measure of the flow of energy through unit area normal to the direction of energy flow, usually at the observing point. Thus if energy ΔE joules, in a frequency[2] range $\Delta\nu$ (c/s), is observed to flow through an area ΔA (m²) in a time Δt (sec), then the *flux-density* S from the source is,

$$S = \frac{\Delta E}{\Delta A\cdot\Delta t\cdot\Delta\nu} \quad \text{watt metre}^{-2}\,(\text{c/s})^{-1}. \tag{1.1}$$

It is usual, unless otherwise stated, to express flux-density as the total flux in both planes of polarization. Since this definition refers only to a plane wave, the unit of flux-density can only be used when the source has a small angular diameter and can be regarded as a point source.

Brightness is a measure of flux-density which is used when it is necessary to take into account the finite angular extent of a source, for example when the source is comparable with, or larger than, the aerial beam. It may also be necessary to take into account, or to describe, the variations of brightness over the disc of the source. Under these conditions it is convenient to define the *brightness* of the disc in a direction ϑ, φ as $B(\vartheta, \varphi)$, where

$$B = \lim_{\Delta\Omega\to 0}\frac{\Delta S}{\Delta\Omega} \quad \text{watt metre}^{-2}\,(\text{c/s})^{-1},\ \text{steradian}^{-1} \tag{1.2}$$

where ΔS is the flux-density (as defined above) due to an element with coordinates ϑ, φ, and $\Delta\Omega$ is the solid angle in steradians subtended by this element at

[1] M. RYLE and F. G. SMITH: Nature, Lond. **162**, 462 (1948).

[2] In the present article the symbol ν will be used to denote frequency when discussing the radiation field, and the symbol f will be used to denote the frequency of voltages or currents in electrical circuits e.g. in aerials, receivers etc.

the observer. As in the case of flux-density it is usual to express brightness in terms of the total in both polarizations.

Brightness temperature is a useful alternative method of expressing brightness, and is commonly used in the description of the background radiation from the sky; it is often referred to in the literature as effective temperature or apparent temperature. The *brightness temperature* of any source, or element of a source, is the temperature of a black-body of the same angular size which would give the same radiation as that actually observed. In the radio domain, where the Rayleigh-Jeans law can be used, the brightness temperature T in degrees Kelvin and the brightness B of a surface are related by,

$$T = B\frac{\lambda^2}{2k} \tag{1.3}$$

where B is defined above, λ is the wavelength of observation, and k is BOLTZMANN's constant in joules per degree Kelvin.

2. The principal characteristics of an aerial. Observations of the discrete sources are usually carried out with directional aerials of the largest possible size, to obtain the maximum angular resolving power and power-gathering capacity. The principal parameters which describe the performance of an aerial in these respects are its *beamwidth* and *effective area*. These quantities depend upon the wavelength, the physical dimensions of the aerial and the detailed way in which the aperture of the aerial is excited; they are discussed exhaustively in many of the standard texts on radio engineering.

A useful guide to the *beamwidth*, or angular resolving power of an aerial is given by the relation,

$$\vartheta \approx \lambda/l \quad \text{radian} \tag{2.1}$$

where ϑ is the width of the beam between half-power points and l is the physical width of the aerial in the plane in which ϑ is measured. This expression yields a reasonably accurate value for aerial systems which have physical apertures large compared with the wavelength.

The *effective area* of an aerial is a particularly valuable concept in radio astronomy, and is equal to the area over which the aerial can be considered to absorb all the incident power. It cannot be directly equated to the physical area of an aerial, but in most large aerial systems, it usually corresponds to between 0.5 and 0.9 of the physical area. The effective area varies with angle and is usually specified in the direction in which the aerial system has the maximum effective area.

If an aerial with an effective area A is directed at a source with a flux-density S, and the source subtends an angle small compared with the aerial beam, then the total power P_A which the aerial will receive and deliver to a matched receiver is,

$$P_A = \tfrac{1}{2} S A \quad \text{watt (c/s)}^{-1} \tag{2.2}$$

where the factor $\frac{1}{2}$ is included because any aerial system can only absorb one half of the energy in a randomly polarized wave.

If the source has an angular diameter which is comparable with, or larger than, the aerial beam, it is necessary to take account of the variation of the effective area of the aerial system with angle. Thus if the source has a brightness distribution over its disc given by $B(\vartheta, \varphi)$ and the effective area in a direction ϑ, φ is $A(\vartheta, \varphi)$, then the total power received by the aerial is,

$$P_A = \tfrac{1}{2} \iint\limits_{4\pi} B(\vartheta, \varphi)\, A(\vartheta, \varphi)\, d\vartheta \cdot d\varphi \quad \text{watt (c/s)}^{-1} \tag{2.3}$$

and, by means of Eq. (1.3), this can be written in terms of the brightness temperature of the disc as,

$$P_A = \frac{k}{\lambda^2} \iint_{4\pi} T(\vartheta, \varphi)\, A(\vartheta, \varphi)\, d\vartheta \cdot d\varphi \quad \text{watt (c/s)}^{-1}. \tag{2.4}$$

B. Techniques of observation.

3. General remarks. The principal difficulties in observing the discrete sources are the detection of extremely low levels of flux-density and the provision of adequate resolving power. In practice the detection of large numbers of sources calls for very large aerials and sensitive and stable receivers. At centimetre wavelengths the main difficulty is to obtain sufficient sensitivity to detect many sources; however, at metre wavelengths, experience has shown that, even for aerials of moderate size, the detection of sources is limited by lack of resolving power rather than sensitivity.

In the following discussion of techniques a division will be made between *single beam systems* in which the aerial is operated as a single unit, and *multiple beam systems* in which the aerial is distributed in separate units over an extended baseline or area.

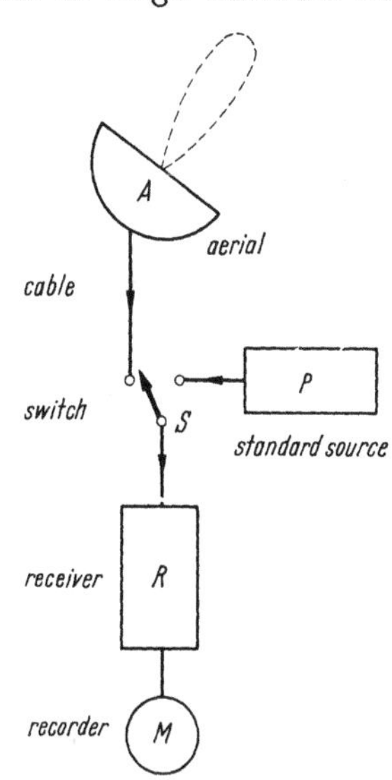

Fig. 1. The basic components of a simple single beam radio telescope.

4. Single beam systems. The basic components of a simple *single beam* radiotelescope are shown in Fig. 1. A directional aerial system A is connected by a cable to a receiver R. The output power from the receiver is recorded by the meter M, which is usually arranged to draw a line on a moving chart. By means of the switch S the power received from the aerial may be compared with that from a standard source P which therefore serves to calibrate the gain of the receiver.

If the receiver has a gain g, a bandwidth Δf and a noise factor F, it can be shown that its power output N_0 is given by,

$$N_0 = g\,k\,\Delta f\{(F-1)\,T_0 + T_A\} \tag{4.1}$$

where T_0 is the temperature at which the noise factor is measured (usually 290° K), T_A is the apparent temperature of the aerial, and it is assumed that there is no loss in the cable. The aerial temperature T_A is defined by the equation,

$$P_A = k T_A \tag{4.2}$$

where P_A is the power per unit bandwidth received by the aerial.

The sky is surveyed by scanning the aerial beam over the desired region, either by moving the aerial mechanically or by fixing it in a chosen azimuth and elevation and making use of the rotation of the earth. When a discrete source is in the beam the power received by the aerial increases by an amount P_A, given by Eq. (2.2), and there is a corresponding deflection on the record. The height of this deflection is proportional to the power received from the source, and its width depends upon the beamwidth and rate of scan. Fig. 2 shows a record of the transit of the intense source Cygnus A(19N4A) through the beam of the fixed 218 ft diameter paraboloid at Jodrell Bank. The record shows three different degrees of scintillation (see Sect. 14).

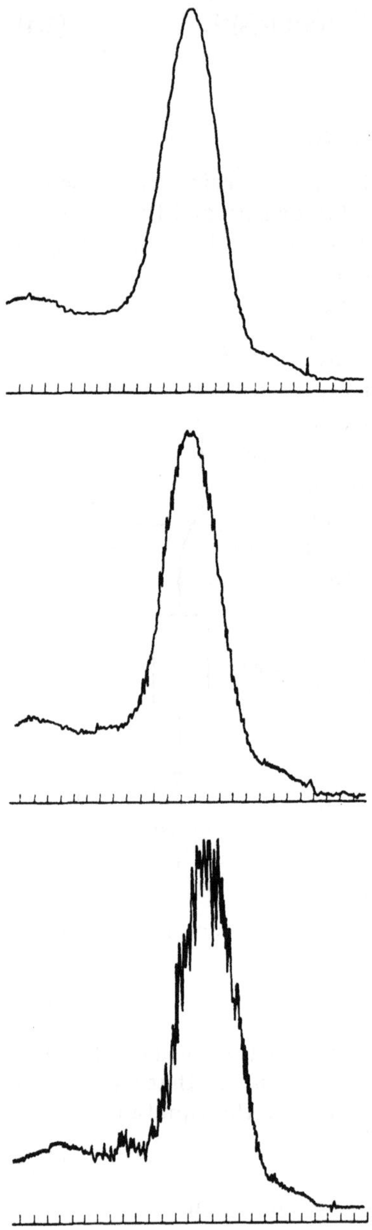

Fig. 2.

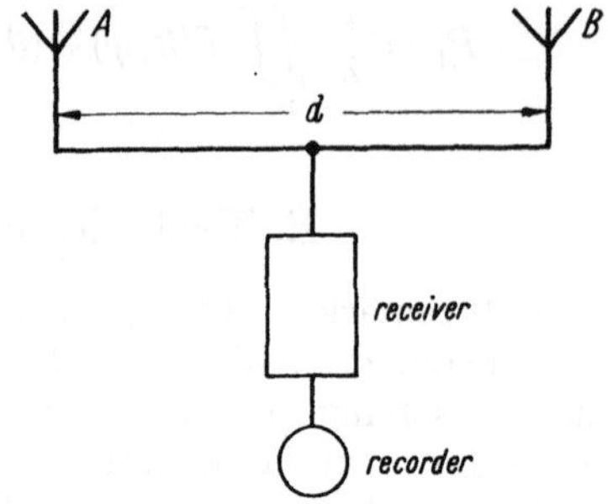

Fig. 3. A simple two-aerial interferometer.

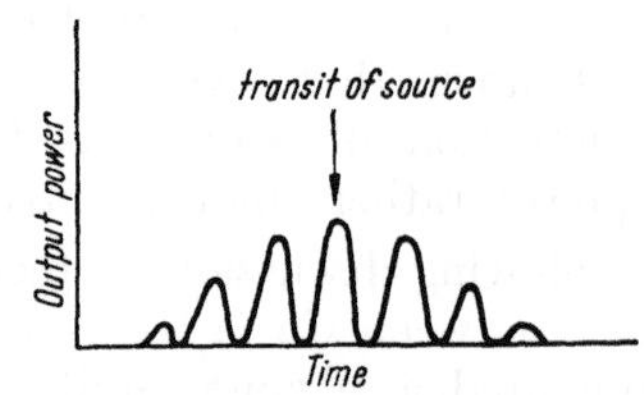

Fig. 4. The fringe pattern due to a discrete source of small angular diameter.

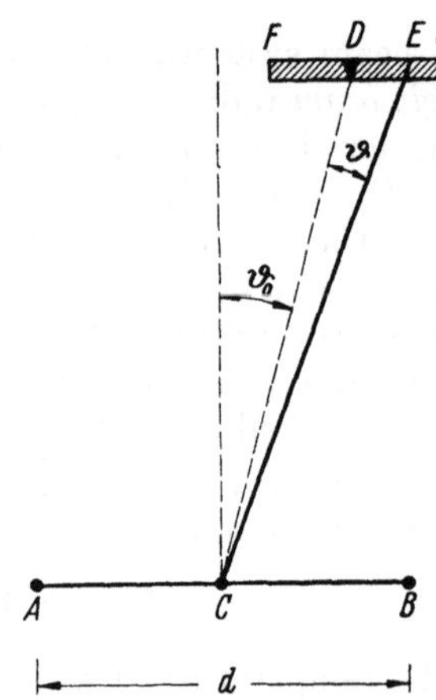

Fig. 5. The geometry of the measurement of apparent angular size.

Fig. 2. The transit of the intense source Cygnus A (19N4A) observed with a 218 ft diameter paraboloid at 158 Mc/s. The three records were taken on different nights and show different amounts of scintillation. The vertical scale is linearly proportional to power received, the horizontal scale shows time (1 division = 2 minutes).

The celestial coordinates of the sources are measured by finding the azimuth and elevation of the beam which yield the maximum deflection at a given time. The flux-density is found by calibrating the receiver from the standard source, and by calculating, or measuring, the effective area of the aerial.

The principal differences between various single beam radiometers are in the construction and design of the aerial systems. The most common type of aerial is the paraboloid which has the considerable advantage that its operating wavelength can be changed by a comparatively minor alteration; but aerials consisting of arrays of helices, horns and dipoles have also been used. One most interesting way of producing a narrow pencil-beam has been developed by MILLS[1] in Australia. He uses two long lines of dipoles in the form of a cross to produce fan-beams at right angles. The intersection between these fan-beams is a pencil-beam, and sources in this beam can be detected and displayed by recording the time average of the cross-correlation between the signals received in the two fan-beams. An aerial operating on this principle has been constructed in Sydney; it has arms 1500 feet long and, at a wavelength of $3\frac{1}{2}$ m, it has a beam width of 50 min of arc.

5. Multiple beam systems. The simplest *multiple beam* system is the two-aerial interferometer shown in Fig. 3. Two aerials A and B, separated by a baseline of length d, are joined by a cable; the centre of the cable is connected to a receiver and the output power is recorded. When a discrete source transits the plane normal to the baseline the receiver output exhibits oscillations or fringes as shown in Fig. 4. The interferometer can be regarded as having an aerial system with a polar diagram consisting of a large number of lobes, the envelope of the maxima of these lobes corresponding to the polar diagram of the component aerials. The angular width $\Delta\vartheta$ of the individual lobes, between adjacent zeros or maxima, is given by,

$$\Delta\vartheta \approx \lambda/d. \tag{5.1}$$

Thus the angular width of one lobe is half that of the beam from a single aerial of width d.

The *celestial coordinates* of a source can be measured from the fringe pattern. It is common practice to mount the aerials on an East-West baseline, and to measure Right Ascension by noting the time at which a source crosses the principal maximum of the fringe pattern which is directed along the meridian. Declination is found by a variety of methods; but the simplest is to note the period τ seconds of the fringes which are close to the maximum fringe, and deduce the declination δ from the formula,

$$\cos\delta = 13750\,\lambda/d\tau. \tag{5.2}$$

The *flux-density* of a source is measured by calibrating the receiver by means of a standard source, much in the same way as discussed above in relation to single beam technique.

The angular distribution of brightness across a source can be measured by observing the amplitude of the fringe pattern obtained with different baselines. Consider the simple case shown in Fig. 5, where the source is represented as a narrow strip FG coplanar with the baseline AB of the interferometer. The amplitude of the fringes at the receiver output K due to the source is,

$$K(d) \approx \int_{-\alpha/2}^{\alpha/2} P(\vartheta)\cdot\cos\left[\frac{\omega_0 d}{c}\sin\left(\overline{\vartheta_0+\vartheta}\right)\right]d\vartheta \tag{5.3}$$

where $P(\vartheta)\,d\vartheta$ is the power received from an element of the source of width $d\vartheta$ in direction ϑ, α is the angular extension of the source and ω_0 is the angular frequency to which the receiver is tuned. If $\alpha \ll 1$, Eq. (5.3) may be written,

$$K(d') = F_{\cos}\left(\frac{\omega_0 d'}{c}\right)\cos\left(\frac{\omega_0 d'\tan\vartheta_0}{c}\right) - F_{\sin}\left(\frac{\omega_0 d'}{c}\right)\sin\left(\frac{\omega_0 d'\tan\vartheta_0}{c}\right) \tag{5.4}$$

---

[1] B. Y. MILLS and A. G. LITTLE: Austral. J. Phys. 6, 272 (1953).

where $d' = d\cos\vartheta_0$ is the baseline projected on to a plane perpendicular to the direction of the source, and

$$\left.\begin{aligned} F_{\cos}\left(\frac{\omega_0 d'}{c}\right) &= \int\limits_{-\alpha/2}^{\alpha/2} P(\vartheta)\cos\left(\frac{\omega_0 d'\vartheta}{c}\right) d\vartheta, \\ F_{\sin}\left(\frac{\omega_0 d'}{c}\right) &= \int\limits_{-\alpha/2}^{\alpha/2} P(\vartheta)\sin\left(\frac{\omega_0 d'\vartheta}{c}\right) d\vartheta. \end{aligned}\right\} \tag{5.5}$$

Eq. (5.4) shows that the output of the receiver oscillates sinusoidally as ϑ_0 varies, and that the relative amplitude and phase of the oscillations at different baselines is given by the Fourier transform of the angular distribution of intensity across the strip. It is convenient to normalize the results obtained with any given baseline to give the *correlation coefficient* $\Gamma(d')$ where,

$$\Gamma(d') = \frac{\left[F_{\cos}^2\left(\frac{\omega_0 d'}{c}\right) + F_{\sin}^2\left(\frac{\omega_0 d'}{c}\right)\right]^{\frac{1}{2}}}{F_{\cos}(0)} \tag{5.6}$$

and $F_{\cos}(0)$ is proportional to the total power from the source and corresponds to the output observed with zero baseline.

The experimental method of measuring the angular distribution across a source consists in observing the relative amplitude, and if possible the phase, of the fringe patterns for different aerial spacings. The intensity distribution across the source, reduced to an equivalent strip, is then found by taking the Fourier transform of these readings. If only the relative amplitude is measured, and not the phase, then it is no longer possible to determine the distribution across the source uniquely and some general assumptions must be made about the shape of the source.

Many valuable modifications of the simple two-aerial interferometer shown in Fig. 3 have been introduced. For example Ryle[1] has described a method, called phase-switching, by which the gradient of the background radiation can be eliminated from the records leaving only sources of small angular diameter. Alternatively a system which eliminates the gradient of the background radiation, and also allows control of the speed at which the fringes are formed, has been described by Hanbury Brown, Palmer and Thompson[2] and is referred to as a "rotating-lobe" interferometer. The problems of determining precise coordinates have been discussed by Mills[3] and also by Smith[4], while the techniques of measuring angular diameters have been discussed by Smith[5], by Mills[6] and also by Jennison and Das Gupta[7].

6. The maximum sensitivity of a radio-telescope. The maximum useful sensitivity of a radio-telescope is frequently limited by interference of terrestrial origin; however, assuming a noisefree site and perfectly stable equipment, an inevitable limit is set by the random fluctuations in the signal itself or by thermal noise and shot noise in the receiver. Thus, if the signal itself has the characteristics of random noise, then the r.m.s. fluctuations $\Delta N_0/N_0$ in a record of the

[1] M. Ryle: Proc. Roy. Soc. Lond., Ser. A **211**, 351 (1952).
[2] R. Hanbury Brown, H. P. Palmer and A. R. Thompson: Phil. Mag. **46**, 857 (1955).
[3] B. Y. Mills: Austral. J. Sci. Res. A **5**, 456 (1952).
[4] F. G. Smith: Monthly Notices Roy. Astronom. Soc. London **112**, 497 (1952).
[5] F. G. Smith: Proc. Phys. Soc. Lond. B **65**, 971 (1952).
[6] B. Y. Mills: Austral. J. Phys. **6**, 452 (1953).
[7] R. C. Jennison and M. K. Das Gupta: Phil. Mag. **1**, 55 (1956).

total power output N_0 from the receiver have an order of magnitude given by,

$$\Delta N_0/N_0 \approx (\Delta f \tau)^{-\frac{1}{2}} \tag{6.1}$$

where τ is the post detector time constant or integration time and Δf is the pre-detector bandwidth. From Eqs. (4.1) and (6.1) it follows that the smallest change in aerial temperature ΔT_A which can be detected by simple inspection of a single trace is,

$$\Delta T_A \approx \{(F-1)\, T_0 + T_A\}\, (\Delta f \tau)^{-\frac{1}{2}}. \tag{6.2}$$

The limiting value of flux-density $S_{\min}$ which can be detected with an aerial of effective area A is therefore,

$$S_{\min} \approx \frac{2k\Delta T_A}{A} \approx \frac{2k}{A} \{(F-1)\, T_0 + T_A\}\, (\Delta f \cdot n \tau)^{-\frac{1}{2}}. \tag{6.3}$$

This limit is inversely proportional to the effective area of the aerial, and to the square root of the bandwith and time of integration. Its variation with the noise factor of the receiver is more complicated since it depends upon the relative value of the receiver noise and the signal power received by the aerial. Thus, at wavelengths greater than about 3 m, the background radiation from the sky is so great that $T_A > (F-1)\, T_0$ and the noise factor of the receiver is relatively unimportant; however, at shorter wavelengths, T_A decreases so that $T_A < (F-1)\, T_0$ and under these conditions the noise in the receiver usually sets the limit to the weakest source which can be detected.

An important consideration in the design of a receiver for a radio telescope is that it should be sufficiently stable so that its sensitivity is not limited by fluctuations of gain. For example it follows simply from the equations given above that fluctuations Δg in the gain of the receiver will not be apparent on the actual records of a source, provided that

$$\Delta g/g < (\Delta f \tau)^{-\frac{1}{2}}. \tag{6.4}$$

It is common, at metre wavelengths, to put $\Delta f \approx 1$ Mc/s and $\tau \approx 1$ sec, and it follows that the gain must be stable to at least 1 part in 1000. Stabilities of this order can be achieved by careful design using the simple arrangement in Fig. 1, but a number of other methods have been developed to overcome this difficulty. One well-known method, due to DICKE[1], involves the rapid comparison of the aerial power with the standard source. It has been elaborated very effectively by RYLE and VONBERG[2], who arranged that the power from the standard source should be held automatically equal to that from the aerial by means of a servo-loop; in this method the receiver acts as a null dectector and there are no stringent requirements for high stability in the receiver gain.

C. The radio observations.

7. The surveys. A list of the principal surveys of sources is shown in Table 1. A number of extensive surveys of the sky have been omitted from this list because they were primarily concerned with the general background radiation and give data about only a very few sources.

At the present stage of the development of radio-astronomy the published data about sources must be treated with considerable caution. There are a number of technical difficulties in making a survey and their nature depends upon the

---

[1] R. H. DICKE: Rev. Sci. Instrum. **17**, 268 (1946).

[2] M. RYLE and D. D. VONBERG: Proc. Roy. Soc. Lond., Ser. A **193**, 98 (1948).

Table 1. *The principal surveys of discrete radio sources.*

Number	Observers	Reference	Frequency Mc/sec	Declination covered	Number of sources reported	Notes on technique
1	G. J. Stanley and O. B. Slee	Austral. J. Sci. Res. A **3**, 234 (1950)	100	$+50°$ to $-50°$	22	An early survey made with an array of Yagi aerials mounted on a cliff and using sea reflections
2	M. Ryle, F. G. Smith and B. Elsmore	Monthly Notices Roy. Astronom. Soc. London **110**, 508 (1950)	81	$+7°$ to $+90°$	50	An early survey with a meridian transit interferometer using two aerials of size $20\lambda\times 1\lambda$, spaced 110λ apart
3	B. Y. Mills	Austral. J. Sci. Res. A **5**, 266 (1952)	101	$+50°$ to $-90°$	77	A meridian transit interferometer using three aerials. The maximum spacing was 270 m, and the minimum 60 m. Each aerial consisted of 24 half-wave dipoles with beamwidth 24° (E.W.) and 14° (N.S.)
4	R. Hanbury Brown and C. Hazard	Monthly Notices Roy. Astronom. Soc. London **113**, 123 (1953)	158.5	$+68°$ to $+38°$	23	A single fixed paraboloid of 218 ft aperture. The beamwidth was about 2°
5	C. A. Shain and C. S. Higgins	Austral. J. Phys. **7**, 130 (1954)	18.3	$+10°$ to $-90°$	37	A single beam from an array of thirty horizontal dipoles 0.2λ above ground, and arranged in a square. The beam width was 17°
6	J. G. Bolton, G. J. Stanley and O. B. Slee	Austral. J. Phys. **7**, 110 (1954)	100	$+50°$ to $-50°$	104	A sea-interferometer using an array of twelve Yagis. Beamwidth 12° in azimuth, 8° in vertical plane
7	J. D. Kraus, H. C. Ko and S. Matt	Astrophys. Journ. **59**, 439 (1954)	250	$+60°$ to $-40°$	207	Broadside array of 48 helical aerials. Beamwidth 1.2° in Right Ascension, 17° in declination
8	F. T. Haddock, C. H. Mayer and R. M. Sloanaker	Nature, Lond. **174**, 176 (1954)	3200	$+90°$ to $-50°$	24	50 ft aperture fully steerable paraboloid. Beamwidth $0.45°\times 0.4°$
9	J. P. Hagen, E. F. McClain and N. Hepburn	Proc. Inst. Radio Engrs. N.Y. **42**, 811 (1954)	1420	$+90°$ to $-40°$	24	50 ft aperture fully steerable paraboloid. Beamwidth 55′

10	R. X. McGee, O. B. Slee and G. J. Stanley	Austral. J. Phys. **8**, 347 (1955)	400	−17° to −49°	14	Fixed paraboloid 80 ft diameter. Beam-width 2.1°
11	J. H. Piddington and G. H. Trent	Austral. J. Phys. **9**, 74 (1956)	600	+51° to −90°	49	Paraboloid (steerable in declination only) 36 ft in diameter. Beamwidth 3.3°
12	J. R. Shakeskaft, M. Ryle, J. E. Baldwin, B. Elsmore and J. H. Thomson	Mem. Roy. Astronom. Soc. **67**, 97 (1955)	81	+83° to −38°	1936	Meridian transit interferometer with 4 aerials spaced 1900 ft apart E.W., and 168 ft N.S. Each aerial was a parabolic cylinder 320 ft × 40 ft
13	B. Y. Mills and O. B. Slee	Austral. J. Phys. **10**, 162 (1957)	86	+10° to −20° (R.A. 00^h to 08^h)	383	1500 ft Mills Cross aerial, which gives a pencil-beam of 50′ width

wavelength. For example at wavelengths of about one metre, or longer, it is likely that the limits to a survey are set, not by the difficulty of detecting sources, but by confusion between adjacent sources. The effects of such confusion, or lack of resolving power, can be serious; not only may the intensities and positions of the sources be in error, but also some of the published sources may in fact be spurious combinations of other real sources. At wavelengths less than about one metre the main difficulty is to obtain sufficient sensitivity to detect many sources, and for this reason the likelihood of confusion is less.

An intercomparison of the metre-wave surveys listed in Table 1 reveals a number of serious discrepancies between the positions and intensities of many of the sources, and it is impossible to decide what significance to attach to the observations given by any one survey without comparing them with several other surveys of the same part of the sky. In an attempt to overcome this difficulty Commission 40 of the International Astronomical Union (I.A.U.) have published "A Catalogue of reliably known Discrete Sources of Cosmic Radio Waves", see Ref. [*1*]. This short catalogue was compiled early in 1954 by a committee of radio-astronomers. It is divided into two lists. List 1, which is reproduced in Table 2, gives the data on eight sources whose existence has been definitely established, for which accurate positions have been given by a number of observers, and for which optical identifications have been secured. In some cases the identifications are supported by agreement between the radio and the optical apparent angular size. List 2 of the I.A.U. has not been reproduced here. It contains a further 30 sources for which radio positions are known only fairly accurately; these sources have been chosen because either a reasonable identification has been suggested or there is good agreement between the positions and flux densities quoted by two or more independent observers. Since the publication of the I.A.U. catalogue a great deal more observational work has been carried out and, while it seems likely that the data in Table 2 is correct, some of the data in List 2 of the I.A.U. is now known to be wrong.

A reference to the surveys listed in Table 1 shows that a variety of systems has been used by the various observers to designate the sources. In an attempt to gain a measure of uniformity the I.A.U. catalogue in Table 2 uses a system which

Table 2. *I.A.U. catalogue of discrete sources of cosmic radio waves (1954), list 1.*
(Reproduced from Astrophysical Journal, Vol. 121, 1955, by kind permission of the Editors and Dr. J. L. Pawsey.)

Constellation	I.A.U. No.	Observer's catalogue No.	Position (1950 epoch) R.A.	Position (1950 epoch) Dec.	Flux density: Frequency (Mc/sec)	Flux density: Watts m^{-2} (C/sec)$^{-1}$ $\times 10^{-26}$	Angular size: Dimensions	Angular size: Position angle	Optical identifications[1]: Object	Magnitude	Position (1950): R.A., Dec.	Angular size
Andromeda	00N4A	R 00.01 HB 2	$00^h40^m15^s\pm30^s$ $00^h42^m\ \pm6^m$	40°50′±20′ p 38° ±5° k	81.5 158	0.4 k 1.6 p			M31 NGC 224	5.0	$00^h40^m00^s$ +41°00′	
Perseus	03N4A	BSS 40 R 03.02 M 03+4 HB 6	$03^h15^m15^s\pm1\frac{1}{2}^m$ $03^h12^m\ \pm3^m$ $03^h10^m\ \pm4^m$ $03^h16^m37^s\pm4^s$	41°22′±30′ p 43°45′±30′ k 42° ±20′ b 41°25′± 6′ x	81.5 81.5 100 101 158	0.95 k 1.2 z 0.9 r 2.4 b 0.6 p			NGC 1275, colliding galaxies	13.28	$03^h16^m31^s$ 41°19′	
Taurus	05N2A	BSS 2 R 05.01 M 05+2 NRL 2	$05^h31^m29^s\pm2\frac{1}{2}^s$ $05^h31^m25^s\pm30^s$ $05^h31^m35^s\pm5^s$	22°00′± 3′ c 22°02′± 8′ a 22°04′± 5′ d	60 81.5 100 101 160 3200	18.5 a 12.5 k 18.5 a 19.0 b 18.5 a 8.0 y	$3\frac{1}{2}'\times5\frac{1}{2}'$ [2]	140° n	M1 NGC 1952 Crab Nebula	9	$05^h31^m30^s$ 21°59′.3	4′×6′ position angle 120°
Puppis	08S4A	BSS 21 M 08−4 S 08−4	$08^h20^m\ \pm4^m$	−42°30′±1° r	18.3 100 101	70 s 35 r 5.0 b	33′ >1° but <2°	E.-W.[3] b g	Galactic nebulosity		08^h20^m3 −42°48′	50′×80′ position angle 145°
Virgo	12N1A	BSS 4 R 12.01 M 12+1 S 12+1 NRL 8	$12^h28^m15\frac{1}{2}^s\pm2\frac{1}{2}^s$ $12^h28^m18^s\ \pm3^s$ $12^h28^m11^s\ \pm37^s$	12°44′± 6′ c 12°37′±10′ d 12°41′±10′ a	18.3 60 81.5 100 101 3200	110 s 25.0 a 10.5 k 12.5 a 12 b 1.4 y	$2\frac{1}{2}'\times5'$ [2]	50° n	M 87 NGC 4486	9.9	$12^h28^m18^s$ 12°40′.1	5′
Centaurus	13S4A	BSS 6 M 13−4 S 13−4 NRL 7	$13^h22^m30^s\pm4^s$ $13^h22^m24^s\pm1^m$	−42°46′± 2′ c −42°37′± 8′ a	60 100 101 3200	32.0 a 18.5 a 16.0 b 2.0 y	$3'\times6\frac{1}{2}'$ [2]	130° n	NGC 5128	6.1	$13^h22^m28^s$ −42°45′.6	20′
Cygnus	19N4A	BSS 1 R 19.01 M 19+4 HB 19 (S 20+X) NRL 3	$19^h57^m45.3^s\pm1^s$ $19^h57^m44^s\ \pm2\frac{1}{2}^s$	40°35′± 1′ d 40°35′± $1\frac{1}{2}'$ c	60 81.5 100 101 158 3200	220 a 135 k 125 a 130 b 57 p 7.0 y	2′10″×35″ 0′.9 3′.5	120°[3] e E.-W.[4] n E.-W.[5] j	Colliding gallaxies	17.9	$19^h57^m44^s.5$ 40°35′46″.3	18″×30″ position angle 150° nucleus, 3″×5″ position angle 115°
Cassiopeia	23N5A	R 23.01 HB 23 NRL 1	$23^h21^m12^s\pm1^s$ $23^h21^m36^s\pm30^s$	58°32′.1± 0′.7 d 58°35′ ±10′ p	81.5 158 3200	220 k 93 p 15.0 y	3′×4′ 5′	25°[3] e E.-W[5]. j	Galactic nebulosity		$23^h21^m11^s.4$ 58°31′52″.9	6′.3 in diameter

is based on the position of the source. For example in the I.A.U. designation 19N4A, which refers to the intense source in Cygnus, the first two digits 19 give the hours of Right Ascension ($19^h 57^m$); the letter N stands for north declination and is replaced by S for sources of southern declination; the digit 4 indicates that the declination lies between 40° and 50°; the final letter A is a serial letter.

8. Angular distribution. MILLS (survey 3, Table 1) has examined the angular distribution in galactic latitude of the 77 sources, shown in Fig. 6, which he detected with an interferometer at 101 Mc/s. He divided the survey into ten strips of equal area bounded by lines of constant galactic latitude and examined the number of sources in each strip. His results are shown in Fig. 7 where each

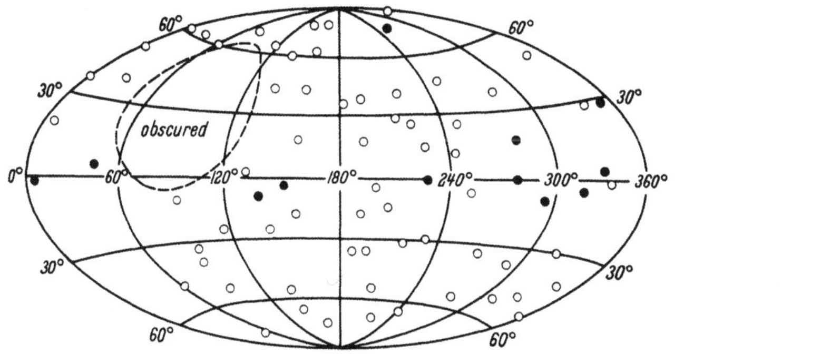

Fig. 6. The distribution in galactic coordinates of the 77 discrete sources observed by MILLS (survey 3, Table 1). Intense sources (flux-densities > 10^{-24} watt m^{-2} $(c/s)^{-1}$] are shown as black dots, the weaker sources are shown as open circles.

column represents a strip of constant galactic latitude and the individual sources are shown by symbols which denote their flux-density. The most striking feature of this diagram is the clustering of the more intense sources near the galactic

Footnotes and references of Table 2.

[1] Optical identifications are discussed by W. BAADE and R. MINKOWSKI, Astrophys. Journ. **119**, 206, 215 (1954).

[2] The with between the half-brightness isophotes along the minor and major axes.

[3] The equivalent rectangular strip of constant surface intensity.

[4] The width between half-intensity points of the measured distribution of intensity across the equivalent strip source.

[5] The equivalent circular disk of constant surface intensity (measured in the specified direction).

(a) G. J. STANLEY and O. B. SLEE: Austral. J. Sci. Res. A **3**, 234 (1950).
(b) B. Y. MILLS: Austral. J. Sci. Res. A **5**, 266 (1952).
(c) B. Y. MILLS: Austral. J. Sci. Res. A **5**, 456 (1952).
(d) F. G. SMITH: Nature, Lond. **168**, 555 (1951).
(e) R. HANBURY BROWN, R. C. JENNISON and M. K. DAS GUPTA: Nature, Lond. **170**, 1061 (1952).
(g) J. G. BOLTON, K. C. WESTFOLD, G. J. STANLEY and O. B. SLEE: Austral. J. Phys. **7**, 96 (1954).
(j) F. G. SMITH: Nature, Lond. **170**, 1065 (1952).
(k) M. RYLE, F. G. SMITH and B. ELSMORE: Monthly Notices Roy. Astronom. Soc. London **110**, 508 (1950).
(n) B. Y. MILLS: Austral. J. Phys. **6**, 452 (1953).
(p) R. HANBURY BROWN and C. HAZARD: Monthly Notices Roy. Astronom. Soc. London **113**, 123 (1953).
(r) J. G. BOLTON, G. J. STANLEY and O. B. SLEE: Austral. J. Phys. **7**, 110 (1954).
(s) C. A. SHAIN and C. S. HIGGINS: Austral. J. Phys. **7**, 130 (1954).
(x) J. E. BALDWIN and B. ELSMORE: Nature, Lond. **173**, 818 (1954).
(y) F. T. HADDOCK, C. H. MAYER and R. M. SLOANAKER: Astrophys. Journ. **119**, 456 (1954).
(z) J. E. BALDWIN and D. W. DEWHIRST: Nature, Lond. **173**, 164 (1954).

plane. Thus of 11 sources with a flux-density $> 3 \times 10^{-24}$ watt m^{-2} $(c/s)^{-1}$ 8 lie within $\pm 12°$ of the galactic equator. The relative scarcity of weak sources in the galactic plane, shown by Fig. 6 must not be regarded as significant since it may well be due to observational selection; it is common experience that it is difficult to detect weak sources in the presence of adjacent intense sources.

The concentration of the more intense sources close to the galactic plane was also found by Hanbury Brown and Hazard using a pencil-beam of 2° at 158 Mc/s (survey 4, Table 1). They surveyed an area of about 1000 square deg. within $\pm 5°$ of the galactic plane and 3000 square deg. outside these limits. They found a total of 13 sources which they classed as intense ($\geq 50 \times 10^{-26}$ watt m^{-2} $(c/s)^{-1}$); 10 of these sources lay within $\pm 5°$ of the galactic plane.

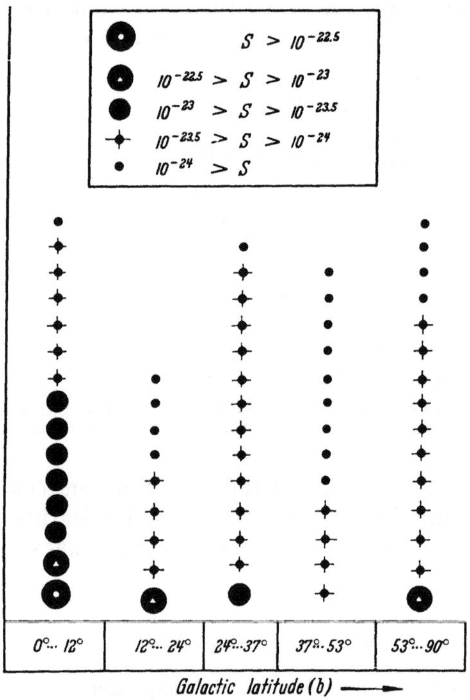

Fig. 7. The distribution in galactic latitude of the sources observed by Mills (survey 3, Table 1). To allow for the obscured region in the northern hemisphere (see Fig. 6) all sources between $l = 50°$ and $l = 130°$ have been omitted. S = flux density in watt m^{-2} $(c/s)^{-1}$.

Bolton, Stanley and Slee (survey 5, Table 1) have examined the angular distribution of 104 sources detected with an interferometer at 100 Mc/s. Their results show a marked concentration of intense sources towards the galactic plane, and also show a general increase in the number of sources near the plane irrespective of intensity. Bolton *et al.* also noted a tendency for the sources within $\pm 5°$ of the galactic plane to concentrate in galactic longitude towards the galactic centre.

More recent investigations show that the angular distribution is related to intensity and apparent angular diameter. Thus Shakeshaft *et al.* (survey 12, Table 1) have divided the 1936 sources which they observed at 81 Mc/s into two groups with angular diameters $> 20'$ and $< 20'$. Fig. 8 shows the distribution in galactic latitude of the sources with small ($< 20'$) angular diameters for galactic longitudes between $l = 90°$ and 195°. It can be seen that for sources with flux-densities $< 80 \times 10^{-26}$ watt m^{-2} $(c/s)^{-1}$ there is no evidence for a concentration in latitude. However, the results for sources of large diameter ($> 20'$) are quite different and are shown in Fig. 9. They show a pronounced concentration towards the galactic plane which is most striking for sources with an intensity $> 100 \times 10^{-26}$ watt m^{-2} $(c/s)^{-1}$.

To summarize, the present evidence shows that intense sources of large angular diameter are concentrated into the galactic plane, while sources of small angular diameter appear to be more uniformly distributed over the sky. These two types of distribution indicate that there are at least two classes of radio source present and that the members of one class, the intense sources, are in the Galaxy.

9. The number—flux-density relation. Attempts have been made to deduce the distribution in space of the sources from an analysis of the number of sources observed with different flux-densities. It is simple to show that, for a uniform spherical distribution of sources of constant space density n_0, the number of sources N_S observed with a flux-density exceeding S will be

$$N_S = \frac{4}{3}\pi n_0 \left(\frac{W_0}{S}\right)^{\frac{3}{2}} \tag{9.1}$$

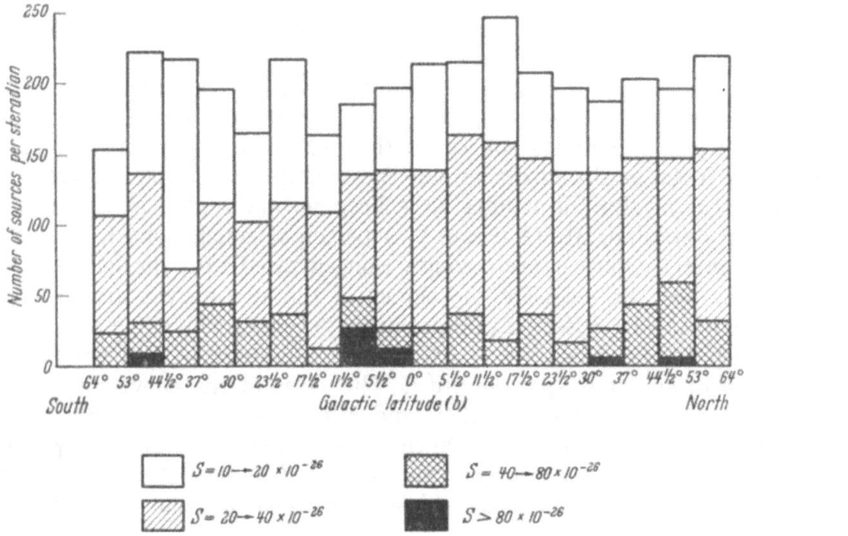

Fig. 8. The distribution in galactic latitude of the sources of small angular diameter (< 20′) observed by SHAKESHAFT *et al.* (survey 12, Table 1) between galactic longitudes $l = 90°$ to $195°$. The sources have been divided into four ranges of flux-density shown on the diagram as S watt m^{-2} (c/s)$^{-1}$.

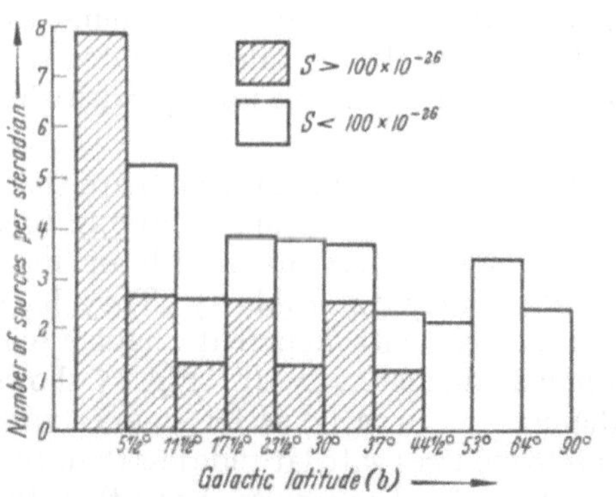

Fig. 9. The distribution in galactic latitude of the sources of large angular diameter (20′ to 180′) observed by SHAKESHAFT *et al.* (survey 12, Table 1). The ordinates represent the number of sources per steradian for two different limits of flux-density S watt m^{-2} (c/s)$^{-1}$.

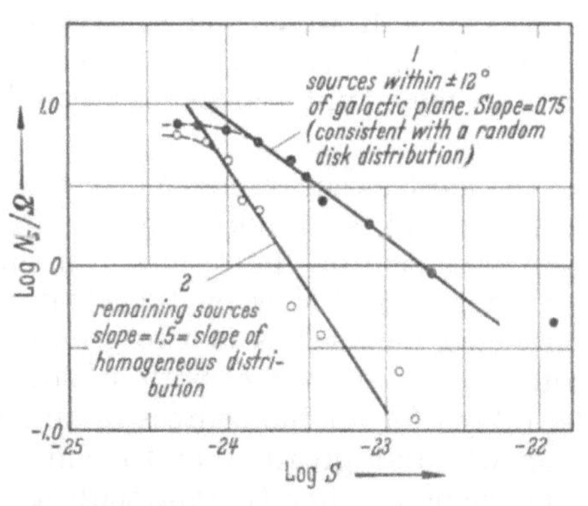

Fig. 10. The number/flux-density curve observed by MILLS for 77 sources (survey 3, Table 1). S = flux-density at 100 Mc/s in watt m^{-2} (c/s)$^{-1}$, N_s = number of sources with flux-density equal to or greater than S, N_s/Ω = number of sources observed per steradian.

where W_0 is the power per unit bandwidth per steradian radiated by each source and the effects of red-shift, curvature of space etc., are neglected. If the sources are assumed to have a dispersion in absolute magnitude such that $\log W$ is distributed as,

$$\frac{1}{\sigma\sqrt{2\pi}} \exp\{-(\log W - \log W_0)^2/2\sigma^2\} \tag{9.2}$$

then it can be shown[1] that,

$$N_S = \frac{4}{3} \pi n_0 \left(\frac{W_0}{S}\right)^{\frac{3}{2}} \exp\left(\frac{9}{8}\sigma^2\right). \tag{9.3}$$

Therefore for a simple spherical distribution of sources with constant space-density a linear relationship with a slope of -1.5 is expected between Log N_S and Log S.

Fig. 10 shows a plot of Log N_S/Ω (the number of sources per steradian) against Log S for the 77 sources observed by MILLS (survey 3, Table 1) with an interferometer at 101 Mc/s. The sources have been divided into two groups to take account of the observed concentration of intense sources into the galactic plane. One group includes sources within 12° of the galactic plane, and the other group includes all sources outside these limits. Since the two groups occupy different areas of sky the results have been normalized to give the number of sources per steradian. MILLS draws two principal conclusions from his results; firstly the intense sources are strongly concentrated into the galactic plane, as discussed in Sect. 8; secondly the slopes of the two curves are different, and indicate at least two different populations of source. The slope of the curve for sources with $b > 12°$ is about -1.5 and is consistent with a simple spherical distribution of constant space-density. The slope of the curve for sources with $b < 12°$ is significantly less (-0.75), and is consistent with a thin disc-shaped distribution in the galactic plane. On the basis of these results MILLS has proposed the division of the sources into two classes, *Class I sources* which are concentrated into the galactic plane and *Class II sources* whose angular distribution is apparently isotropic.

BOLTON, STANLEY and SLEE (survey 6, Table 1) have examined the number/flux-density relation for 104 sources observed at 100 Mc/s. For sources near the galactic plane ($b < 10°$) they found the slope of the curve to be -0.6 in reasonable agreement with MILLS; for sources at higher latitudes ($b > 10°$) they found the slope to be greater than the value of -1.5. This latter results indicates that there are more faint sources than would be expected for a simple isotropic distribution.

The most recent discussions of the number/flux-density relationship are those by RYLE and SCHEUER[2] and by MILLS and SLEE[3]. RYLE and SCHEUER base their discussion on the results of a survey of 1936 sources made with an interferometer at 81.5 Mc/s by SHAKESHAFT *et al.* (survey 12, Table 1). They have considered the sources in two groups, 1906 sources with apparent angular diameters $< 20'$, 30 sources with apparent angular diameters $> 20'$. Fig. 11 shows the results for all the sources of small diameter. Curve (2) is for sources with galactic latitude greater than 10°, curve (3) is for sources with latitude less than 10°, while the broken line (1) represents a slope of -1.5. The results shows that the slope of the curve is less for sources in the equatorial region, in qualitative agreement with the results of previous workers. The difference of slope between the two curves is not so marked as that found in previous work, however RYLE and SCHEUER point out that this may be due to the absence of sources of large angular diameter from their analysis. The slope of the curve for the non-equatorial areas ($b > 10°$) departs significantly from a slope of -1.5 and from this RYLE and SCHEUER have drawn a number of cosmological conclusions.

[1] G. WESTERHOUT and J. H. OORT: Bull. astronom. Inst. Netherl. **11**, 323 (1951).
[2] M. RYLE and P. A. G. SCHEUER: Proc. Roy. Soc. Lond., Ser. A **230**, 448 (1955).
[3] B. Y. MILLS and O. B. SLEE: Austral. J. Phys. **10**, 162 (1957).

MILLS and SLEE have published a preliminary survey of the sources in an area bounded by declinations $+10°$ to $-20°$ and right ascensions 00^h to 08^h. Fig. 12 shows the number/flux-density relation for all sources with galactic latitudes exceeding $12\frac{1}{2}°$. The black dots represent observations on a total of 311 sources found with the Mills Cross aerial and the vertical lines through each dot represent the uncertainty $\sqrt{N}$. Since the number of sources with flux-densities $> 80 \times 10^{-26}$ watt m^{-2} $(c/s)^{-1}$ in the sample area was very small, MILLS and SLEE have taken the data on 36 intense sources found in non-equa-

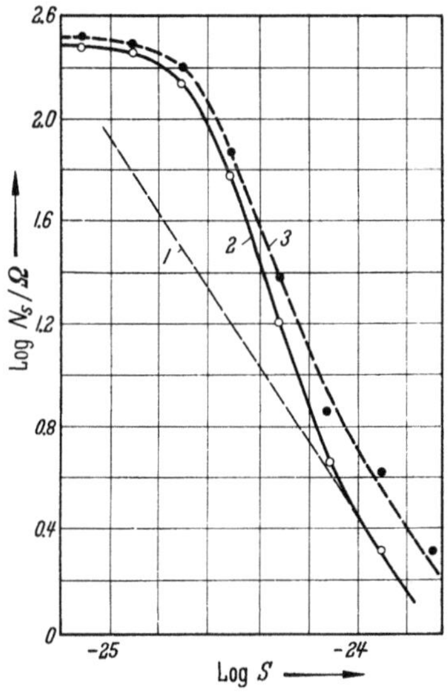

Fig. 11. The number/flux-density curve observed by SHAKESHAFT *et al.* (survey 12, Table 1) for 1906 sources of small angular diameter (< 20′) Curve (1) represents a slope of -1.5; curve (2) shows the observed results for sources with galactic latitude $> 10°$; curve (3) shows the observed results for sources with galactic latitude $< 10°$. S = flux-density at 81 Mc/s in watt m^{-2} $(c/s)^{-1}$, N_s = number of sources with flux-density equal to or greater than S, N_s/Ω = number of sources per steradian.

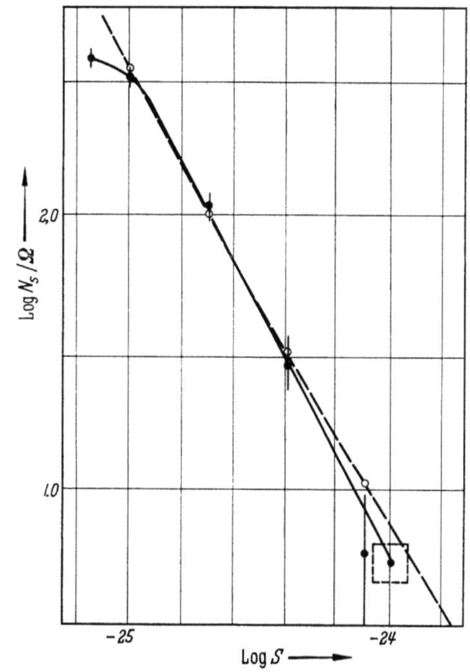

Fig. 12. The number/flux-density curve observed by MILLS and SLEE (survey 13, Table 1) for 311 sources with galactic latitudes greater than $12\frac{1}{2}°$. The rectangle encloses a point which represents 36 intense sources taken from the survey of SHAKESHAFT *et al.* The full line shows the observed results and the broken line represents these results after correction for certain instrumental errors discussed in the text. S = flux-density at 86 Mc/s in watt m^{-2} $(c/s)^{-1}$, N_s = number of sources with flux-density equal to or greater than S, N_s/Ω = number of sources per steradian.

torial regions by SHAKESHAFT *et al.* to give the point marked by a small rectangle. The full line in Fig. 12 represents the best fit to the observations. For strong sources this line is straight and has a slope of -1.8 ± 0.1 (probable error); there is a pronounced curvature for faint sources indicating an apparent reduction in the space-density of sources with flux-densities less than 10^{-25} watt m^{-2} $(c/s)^{-1}$. MILLS and SLEE state that the curvature of the line for faint sources is likely to be due to the operation of instumental effects near the limit of sensitivity. They also point out that there are two important instrumental effects which tend to increase the apparent number of sources with flux densities just above the survey limits. Thus confusion or blending effects between sources below the limit of sensitivity cause a random variation in the receiver output, and large chance excursions in this variation may be counted as sources. Secondly the rapid increase in numbers of sources with decreasing flux-density provides many more sources below the limits of visibility capable of being included due to the presence of favourable noise fluctuations, than sources above the limit which are likely to be excluded through unfavourable fluctuations. They have estimated the magnitude of these effects and the broken line in Fig. 12 shows their results after allowance has been made for the estimated instrumental effects. It is clear that these corrections have reduced the observed slope considerably; MILLS and SLEE concluded that the slope of -1.8, shown by their uncorrected observations, cannot be regarded as significant evidence that the true slope of the line differs from the theoretical value of -1.5 expected for a population of constant space-density.

A comparison of the results of MILLS and SLEE with those of RYLE and SCHEUER shows that there is significant disagreement; furthermore, as MILLS and SLEE have pointed out, the majority of the sources in the area common to both surveys do not agree in position. The principal difference between the two surveys is one of technique; thus the results analysed by RYLE and SCHEUER were obtained with a multiple-beam interferometer, while MILLS and SLEE used a Mills Cross with a pencil-beam of 50 minutes of arc. The disagreement suggests that in one or other of the surveys, or perhaps on both, the effects of confusion between the sources has been seriously underestimated. Thus, although the results of both surveys suggest that there are more faint sources than would be expected for a perfectly isotropic distribution, the shape of the number-flux-density curve cannot yet be regarded as established with sufficient certainty.

10. Apparent angular diameter. Detailed measurements of apparent angular diameter have so far been confined to a few of the principal intense sources, and there is little information on the diameters of the majority. In their survey at 81 Mc/s SHAKESHAFT *et al.* (survey 12, Table 1) found 1906 sources with angular diameters in the range 0 to 20′, and only 30 sources in the range 20 to 180′. As mentioned in Sect. 8, the intense sources of large angular diameter (20 to 180′) show a pronounced concentration into the galactic plane, while the sources of small angular diameter (0 to 20′) are distributed more isotropically. Some unpublished work of CARTER, who has surveyed about 75 sources at 101 Mc/s, indicates that the angular diameter of many of the sources is of the order of 1′.

In the absence of sufficient data to support a discussion of the properties of the sources in general, the following account will be confined to a few of the principal sources whose angular diameter is known with reasonable certainty.

Cygnus A (19N4A). The apparent angular diameter of this source was measured for the first time in 1952[1] and found to be of the order of 1′. It has been shown that, at 125 Mc/s, the apparent diameter varies from about 35″ to 2′10″ in different directions across the source, the major axis being between position angles 90 and 120°. More recently JENNISON and DAS GUPTA[2] have made observations with an interferometer in position angle 113° using aerial spacings in the range 610 to 5300 λ at 125 Mc/s. They have interpreted their results to give the distribution of brightness across the equivalent strip source shown in Fig. 13, and have concluded that it consists of two radiating centres, each being $45'' \times 35''$ and the spacing between centres being 1′25″. Their results do not, by themselves, yield a unique distribution since the technique did not take account of the relative phase of the components of the Fourier transform of the angular distribution. However subsequent measurements with another interferometer which gives both the amplitude and phase of the components has confirmed that the distribution in Fig. 13 is substantially correct.

---

[1] R. HANBURY BROWN, R. C. JENNISON, M. K. DAS GUPTA; B. Y. MILLS; F. G. SMITH: Nature, Lond. **170**, 1061 (1952).

[2] R. C. JENNISON and M. K. DAS GUPTA: Phil. Mag. **1**, 55 (1956).

The Cygnus source has been identified with two colliding galaxies and the photographic image has a diameter of a few seconds of arc. The angular size of the radio source is considerably greater and suggests that the interaction between the galaxies extends over a volume large compared with the visible nucleus. There is little positive correlation, in fact one may say there is negative correlation, between the regions of emission of light and radio waves.

Cassiopeia A (23N5A). SMITH[1] gives the diameter of this source as 5.5′. His value represents the diameter of an equivalent uniformly bright disc and is based on measurements with an east-west baseline at 215 Mc/s. His result is in reasonable agreement with the measurements at 125 Mc/s by HANBURY BROWN, JENNISON and DAS GUPTA which are shown in Fig. 14. These latter measurements indicate that the source is roughly symmetrical in outline. The source has been identified (Sect. 17) with a faint peculiar nebulosity and the diameter of the radio source is consistent with this identification.

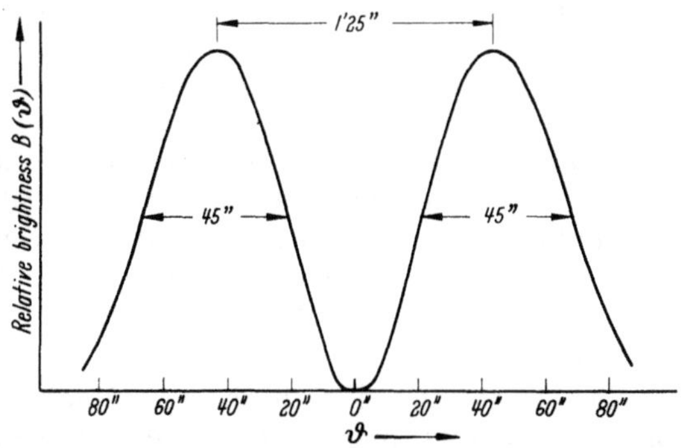

Fig. 13. The distribution of brightness across the Cygnus A source (19N4A) found by JENNISON and DAS GUPTA at 125 Mc/s. The brightness $B(\vartheta)$ in arbitrary units is plotted against ϑ the angle from the centre of the source. The distribution refers to the equivalent strip source in position angle 90°.

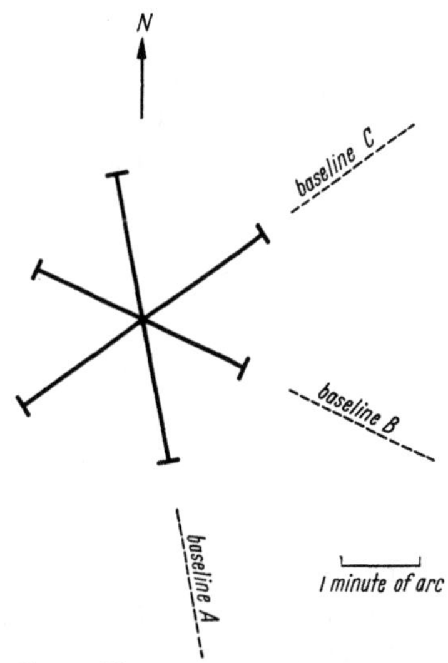

Fig. 14. The apparent angular width of the Cassiopeia A source (23N5A) found by HANBURY BROWN *et al.* at 125 Mc/s. The diagram shows the width of the equivalent strip source of uniform brightness observed in three different position angles.

Taurus (05N2A). The distribution of radio intensity across this source is of great interest since it is known to be associated with the Crab Nebula (see Sect. 17). A study of the correlation between optical and radio distributions for this object may prove to be a fruitful method of testing theories of the origin of the radiation. BALDWIN[2] has measured the distribution at 214 Mc/s using an interferometer. He concluded that the source is radially symmetrical and that the radiation extends over a disc of between 8′ and 10′ diameter.

BOISCHOT, BLUM, GINAT and LE ROUX[3] have observed an eclipse of this source by the Moon at 169 and 900 Mc/s. Their results show that at 169 Mc/s the radiation is almost entirely concentrated in the diffuse inner mass of the nebula; however, within this region there does not appear to be a detailed correlation between the brightest patches as observed optically and by radio. Their results at 900 Mc/s were not so precise; however they suggest that this frequency an appreciable fraction of the radiation arises from the region of the filaments.

[1] F. G. SMITH: Nature, Lond. **170**, 1061 (1952).
[2] J. E. BALDWIN: Observatory **74**, 120 (1954).
[3] A. BOISCHOT, E. J. BLUM, M. GINAT and E. LE ROUX: C. R. Acad. Sci., Paris **242**, 1849 (1956).

Virgo A (12N1A). MILLS[1] has studied this source at 100 Mc/s and gives its angular dimensions as an ellipse of $5\times 2\frac{1}{2}'$ with major axis in position angle 50°. BALDWIN and SMITH[2], from measurements at 81.5 Mc/s, report that this intense and comparatively small source measured by MILLS is surrounded by an extended distribution with a diameter of about 1 degree. They estimate that about 20% of the total flux from the source arises in this extended distribution. The presence of this extended distribution has not yet been confirmed and must be regarded as a tentative result, since this source lies in a region of sky where the background radiation is very irregular and difficult to interpret. This source has been identified with the elliptical galaxy N.G.C. 4486 which has the peculiar feature of a jet apparently protruding from the nucleus (see Sect. 18).

Gemini (06N2A). Auriga (04N4A). A number of sources with angular diameters of the order of 1 degree have been found close to the galactic plane. A brief summary of the data on these sources is to be found in an article by HANBURY BROWN[3], however it does not include the results of the more recent surveys by SHAKESHAFT *et al.* and by MILLS and SLEE (see Table 1). These large diameter sources appear to belong to a class of rare objects in the Galaxy and a few of them have been identified with faint nebulosities. Two examples only will be given here, further details are given in the references given above. The source in Gemini (06N2A) has been studied by BALDWIN and DEWHIRST[4] at 81 Mc/s and by RISHBETH[5] at 85 Mc/s. It has an angular diameter of about 50′ in good agreement with the peculiar nebulosity I.C. 443. with which it has been identified. The source in Auriga (04N4A) has an angular diameter of about 1.4° at 158 Mc/s[6]. This also is in reasonable agreement with a faint nebulosity in this position reported by MINKOWSKI.

Andromeda (00N4A). Source 00N4A is associated with the spiral galaxy M 31. The angular distribution of brightness across this source has been studied a number of workers[7]. BALDWIN finds that at 81.5 Mc/s, it is roughly circular in outline extending to at least 200′. The radio source is therefore considerably greater in size than the optical image of the galaxy, and BALDWIN deduces that M 31 has an extensive radio "corona" surrounding it which is responsible for about two-thirds of the total radiation.

Perseus (03N4A) has been identified with the well-known Perseus cluster of galaxies. It has been shown[8] that about 75% of the radiation from this source comes from an area of the order of 1′ and that the remainder comes from an area of the order of 2°. The small diameter component has been satisfactorily identified with the peculiar galaxy N.G.C. 1275 in the cluster: MINKOWSKI has shown that the peculiar galaxy consists of two galaxies in collision. The large diameter component is believed to be due to the aggregate radiation from the cluster.

Centaurus (13S4A) is associated with the peculiar galaxy N.G.C. 5128 which has been interpreted as two galaxies in collision. The radio source appears to

[1] B. Y. MILLS: Austral. J. Phys. **6**, 452 (1953).
[2] J. E. BALDWIN and F. G. SMITH: Observatory **76**, 141 (1956).
[3] R. HANBURY BROWN: Observatory **74**, 185 (1954).
[4] J. E. BALDWIN and D. W. DEWHIRST: Nature, Lond. **173**, 164 (1954).
[5] H. RISHBETH: Austral. J. Phys. **9**, 494 (1956).
[6] R. HANBURY BROWN, H. P. PALMER and A. R. THOMPSON: Nature, Lond. **173**, 945 (1954).
[7] R. HANBURY BROWN and C. HAZARD: Monthly Notices Roy. Astronom. Soc. London **111**, 357 (1951). — J. E. BALDWIN: Nature, Lond. **174**, 320 (1954). — J. D. KRAUS: Nature, Lond. **175**, 502 (1955).
[8] J. E. BALDWIN and B. ELSMORE: Nature, Lond. **173**, 818 (1954).

consist of a central concentration of order[1] $3' \times 6\frac{1}{2}'$, surrounded by a large diameter source[2] which at 400 Mc/s appears to extend over an area of about $5^\circ \times 3^\circ$.

11. The distance of the sources. Attempts have been made to estimate the distance of some intense sources by observing their annual parallax or proper motion. The most precise observations appear to be those of SMITH[3] who observed the differences in right ascension, over a period of about one year, between four intense sources. He concluded that there is probably no annual variation greater than 20″ in the positions of Taurus A (05N2A) and Cassiopeia A (23N5A), and 5″ in the positions of Cygnus A (19N4A) and Virgo A (12N1A). Thus the distances of the first pair must be greater than 1/20 parsec, and the latter pair must be more distant than 1/5 parsec.

A more promising technique is to observe the absorption of the radiation by interstellar neutral hydrogen. If the distribution of hydrogen in the direction of the source is known then the distance can be estimated from the hydrogen absorption lines in the spectrum of the source. This method has been exploited by WILLIAMS and DAVIES[4] to show that the distance of Cassiopeia A (23N5A) is between 2.5 and 5.5 kiloparsec, and that Cygnus A (19N4A) is at the boundaries of the Galaxy or is extra-galactic. HAGEN, LILLEY and McCLAIN[5] have used the same technique to show that the minimum distance to Cassiopeia A (23N5A) is 3 kiloparsec. Although the results of these two independent investigations are consistent, it must be noted that they do not agree with the distance of 500 parsec estimated for the Cassiopeia source by BAADE and MINKOWSKI[6] from an analysis of the motions of filaments in the nebulosity. The reasons for this discrepancy are not yet understood.

An important verification that Cygnus A (19N4A) is extra-galactic and at a very great distance has recently been provided by LILLEY and McCLAIN[7]. They observed the red-shift of the absorption line of neutral hydrogen in the source and found it to be 16700 km sec^{-1}. This value is in close agreement with the photographic red-shift reported for the colliding galaxies by BAADE and MINKOWSKI in Ref. [5].

12. The radio-frequency spectrum. The available data on the radio-frequency spectra of the sources are few and no satisfactory set of observations covering the whole radio spectrum exists. Fig. 15 shows the spectra of the four principal sources compiled by WILD in Ref. [2] from the results of different observers. For the three sources Cassiopeia A, Cygnus A and Virgo A the flux-density decreases with increasing frequency, the slope tending to diminish at the higher frequencies. The source in Taurus A, the Crab Nebula, has a much flatter spectrum. The available data suggest that the majority of the sources have spectra of the same general type as the first three given above. However sources with spectra similar to the source in Taurus have been reported and also others with intermediate types.

The best known spectrum is that of the intense source Cassiopeia A (23N5A). Table 3 shows a list of values of flux-density compiled by SEEGER; no references to the original work have been given since many of the values are as yet unpublished. There is a considerable scatter in the values shown in the table which

[1] B. Y. MILLS: Austral. J. Phys. **6**, 452 (1953).
[2] R. X. McGEE, O. B. SLEE and G. J. STANLEY: Austral. J. Phys. **8**, 347 (1955).
[3] F. G. SMITH: Nature, Lond. **168**, 962 (1951).
[4] D. R. W. WILLIAMS and R. D. DAVIES: Phil. Mag. **1**, 622 (1956).
[5] J. P. HAGEN, A. E. LILLEY and E. F. McCLAIN: Astrophys. Journ. **122**, 361 (1955).
[6] W. BAADE and R. MINKOWSKI: Reported at Jodrell Bank Symposium 1955.
[7] A. E. LILLEY and E. F. McCLAIN: Astrophys. Journ. **123**, 172 (1956).

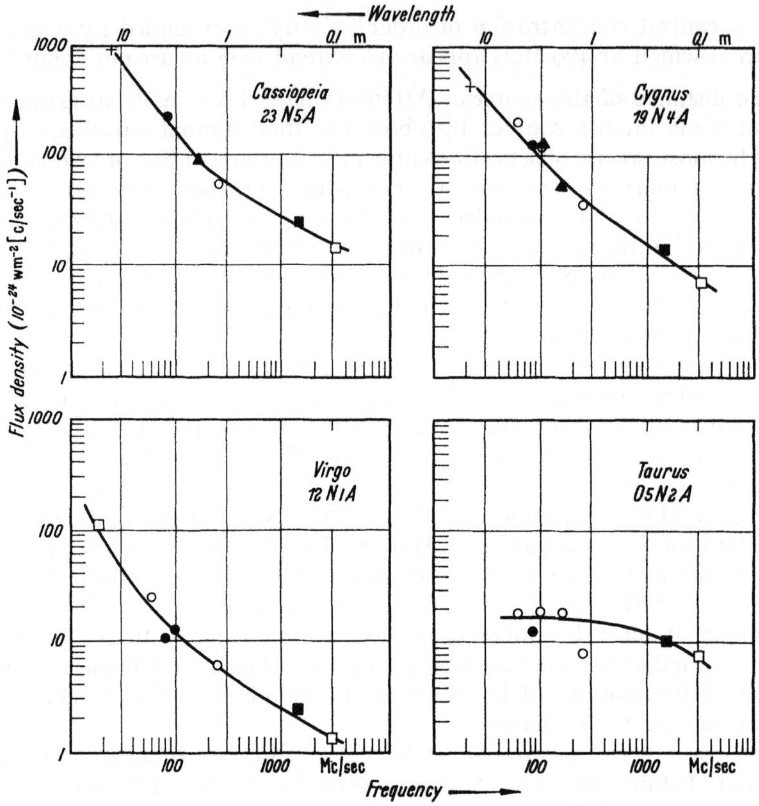

Fig. 15. Provisional radio-frequency spectra of the four principal sources as compiled by WILD in Ref. [2].

Table 3. *The flux-density versus frequency for the source Cassiopeia A (23N5A).*

Frequency Mc/sec	Flux-density watt m^{-2} $(c/sec)^{-1}$ $\times 10^{-24}$	Estimated accuracy ($\pm$%)	Observer
12.5	220	—	WELLS
15.5	240	—	WELLS
16.5	110	20	LAMDEN and LOVELL
19.0	272	20	LAMDEN and LOVELL
22.6	460	20	LAMDEN and LOVELL
22.6	940	20	HEY and HUGHES
30.0	450	20	LAMDEN and LOVELL
38.0	435	10	ADGIE
81.5	232	10	ADGIE
158	93	25	HANBURY, BROWN and HAZARD
193.5	62	25	GREBENKEMPER, MCCLAIN and HAGEN
210	121	20	ADGIE
250	57	20	KRAUS, KO, MATT
300	16	30	RAZIN and PETCHKOV
400	32	15	SEEGER
500	56	25	ADGIE
900	33—50	—	DENISSE
1420	31	20	WESTERHOUT
1420	26.6	+20 —15	HAGEN, MCCLAIN, HEPBURN
3200	12	25	HADDOCK, MAYER, SLOANAKER
9400	4.6	20	RAZIN and PETCHKOV
9500	4	25	HADDOCK and MCCULLOUGH

is apparently due to experimental error; nevertheless a mean curve drawn through these values is the most reliable spectrum which is at present available for this source.

LAMDEN and LOVELL[1] have studied the low-frequency end of the spectra of the intense sources Cassiopeia A (23N5A) and Cygnus A (19N4A). They found that, while the flux-density increased at low frequencies, it reached a maximum at about 22 Mc/s and at lower frequencies there was a rapid decrease. At frequencies less than 22 Mc/s the slope of the curve was given by $S \propto \nu^{+4.4}$ for Cassiopeia A, and $S \propto \nu^{+1.6}$ for Cygnus A. They have shown that this cut-off can be explained in terms of absorption by regions of ionized interstellar hydrogen lying between the observer and the source.

13. Polarization. There is, as yet, no evidence for the polarization of the radiation from the discrete sources and they are generally assumed to be unpolarized. RYLE and SMITH[2] have examined the intense sources Cygnus A (19N4A) and Cassiopeia A (23N5A) with an interference polarimeter at 80 Mc/s and have shown that the polarization cannot be greater than 5%. HANBURY BROWN, PALMER and THOMPSON[3] have examined the same two sources, together with the source Taurus A (05N2A), at a frequency of 158 Mc/s using a polarimeter with a bandwidth of 400 kc/s. No evidence of polarization was found and the upper limits set by this experiment are shown in Table 4.

Table 4. *Upper limits to polarisation at 158* Mc/s.

Source	Percentage of flux which is plane polarized %	Percentage of flux which is circularly polarized %
Cassiopeia A (23N5A)	<1	<4
Cygnus A (19N4A) .	<1	<4
Taurus A (05N2A) .	$<2\frac{1}{2}$	<4

WESTERHOUT[4] and also DAVIES[5] have tested the source Taurus (05N2A) for plane polarization at 1420 Mc/s. In neither of these experiments was there evidence of significant polarization. WESTERHOUT sets an upper limit of 1% to any plane polarization, while DAVIES sets an upper limit of 2.7% from his own observations and 1.5% by combining his results with those of WESTERHOUT.

It is not surprising that significant polarization has not yet been detected from the sources. If, as seems likely, a discrete source contains ionized gas in the presence of a magnetic field, then it is to be expected that any polarization in the total emission will be wiped out by rotation of the plane of polarization in the source. Thus, following OORT and WALRAVEN[6], if a polarized wave traverses a medium with electron density $N(r)$ in a field of H gauss, H making an angle ϑ to the line of sight, the plane of polarization will be turned through an angle,

$$\Omega = \frac{7490\pi}{\nu^2} \int H \cos\vartheta \, N(r) \, dr \qquad \text{radian} \tag{13.1}$$

in a distance r. Assuming, for example, that the Taurus source (the Crab Nebula) has a mass of one tenth of the Sun, concentrated in a shell 1/3 parsec thick and taking $H \cos\vartheta \approx 10^{-3}$, the rotation due to passing through the whole source is

[1] R. J. LAMDEN and A. C. B. LOVELL: Phil. Mag. **1**, 725 (1956).
[2] M. RYLE and F. G. SMITH: Nature, Lond. **162**, 462 (1948).
[3] R. HANBURY BROWN, H. P. PALMER and A. R. THOMPSON: Monthly Notices Roy. Astronom. Soc. London **115**, 487 (1955).
[4] G. WESTERHOUT: Bull. astronom. Inst. Netherl. **12**, 309 (1956).
[5] R. D. DAVIES: Jodrell Bank Annals in publication.
[6] J. H. OORT and TH. WALRAVEN: Bull. astronom. Inst. Netherl. **12**, 285 (1956).

14 radians for a 1420 Mc/s wave. Under these circumstances it seems very unlikely that any nett polarization would be observed at 1420 Mc/s or at lower frequencies.

14. Scintillation. The intensity of the radiation received from a discrete source is occasionally observed to fluctuate with periods ranging from several seconds to one or two minutes. This phenomenon is referred to as *scintillation* and an example is shown in Fig. 2. LITTLE and LOVELL and also SMITH[1] have shown that there is no detailed correlation between the scintillations observed at 81 Mc/s with receivers spaced 210 km apart. The correlation is perfect with the receivers spaced 100 m apart but decreases rapidly when the spacing is of the order of 4 km. Their experiments showed, beyond doubt, that the scintillations are not a property of the source but arise in the medium between the source and the observer.

A study of the diurnal variation of the scintillations in the Northern Hemisphere shows a strong correlation between the presence of scintillations and disturbed conditions in region F of the ionosphere[2]. In the Southern Hemisphere a correlation has been found[3] with disturbances in both region E and region F. The scintillations are believed to be caused by irregularities in the electron content of the ionosphere which are a few kilometres in size, having a length in the north-south direction which is usually about four times their width east-west[4]. They appear to move with velocities of about 100 msec^{-1}, although velocities as high as 1000 msec^{-1} have been observed. There is a close correlation between the apparent speed with which these irregularities move and disturbances in the Earth's magnetic field and auroral activity[5]. The origin of the irregularities is unknown.

A theoretical treatment of the effect of irregularities in the ionosphere has been given by LITTLE[6] and also by HEWISH[7], and a general survey of ionospheric irregularities can be found in Ref. [*3*].

15. Variable sources. An extensive search for sources of variable intensity was made by RYLE and ELSMORE[8] who examined the intensity of 100 sources nearly every day for eighteen months. They found no evidence for variability and concluded that none of the sources showed variations exceeding 10%. The more intense sources were in fact found to be constant to better than 5%.

Two variable sources have been reported in recent literature. SLEE[9] has observed that a source in Hydra (0951A) shows irregular variations in intensity at 100 Mc/s, and CARTER[10] has reported that it shows significant variations in apparent angular diameter. KRAUS, KO and STOUTENBURG[11] have reported a variable source at 242 Mc/s with coordinates R.A. $08^h\,19^m$, declination $+8°$, Epoch 1950. Neither of these sources has yet been confirmed as variable by independent observers at a different site, and the conclusion that sources of variable intensity exist must still be regarded as tentative.

---

1 C. G. LITTLE and A. C. B. LOVELL: F. G. SMITH: Nature, Lond. **165**, 422 (1950).
2 C. G. LITTLE and A. MAXWELL: Phil. Mag. **42**, 267 (1951).
3 J. G. BOLTON, O. B. SLEE and G. J. STANLEY: Austral. J. Phys. **6**, 434 (1953).
4 J. P. WILD and J. A. ROBERTS: J. Atmos. a. Terres. Phys. **8**, 55 (1956).
5 A. MAXWELL and C. G. LITTLE: J. Atmos. a. Terres. Phys. **2**, 356 (1952).
6 C. G. LITTLE: Monthly Notices Roy. Astronom. Soc. London **111**, 289 (1951).
7 A. HEWISH: Proc. Roy. Soc. Lond., Ser. A **209**, 81 (1951).
8 M. RYLE and B. ELSMORE: Nature, Lond. **168**, 555 (1951).
9 O. B. SLEE: Austral. J. Phys. **8**, 498 (1955).
10 A. W. L. CARTER: Austral. J. Phys. **8**, 564 (1955).
11 J. D. KRAUS, H. C. KO and D. V. STOUTENBURG: Nature, Lond. **176**, 304 (1955).

D. Identification.

16. General remarks. It has so far proved impossible to identify the majority of the discrete sources with known celestial objects or to associate them with any particular type of object. The present evidence shows definitely that the bulk of the sources are not associated with any well-known objects in the Galaxy such as the bright stars or nebulae, nor do they show a close correlation with bright galaxies; what little positive evidence there is suggests that they may be associated with peculiar galaxies which are relatively faint photographically.

A few discrete sources have been identified with objects observed photographically. These objects can be classified as follows:

Objects in the galaxy	Extra-galactic objects
The remnants of supernovae	Normal galaxies
Peculiar nebulosities	Peculiar galaxies
H II regions	Clusters of galaxies

A possible exception to this classification is a source in Sagittarius (17S2A) which is discussed separately under the general heading of objects in the Galaxy.

17. Objects in the Galaxy. *α) The remnants of supernovae.* Three supernovae of type I are known to have flared up in the Galaxy. In each case a radio source has been reported close to the position.

The Crab Nebula (Supernova A.D. 1054). The identification of source Taurus A (05N2A) with the Crab Nebula is well established. The position and diameter of the radio source are in reasonable agreement with the optical data; a discussion of this data is to be found in the article by BAADE and MINKOWSKI in Ref. [*4*]. It is impossible to account for the radio emission in terms of thermal radiation and its origin must be non-thermal. Recent evidence, which is discussed fully by OORT and WALRAVEN[1] shows that the light from the amorphous mass of the nebula is polarized, and it has been suggested that the light and the radio emission may have a common origin in the synchrotron radiation from electrons moving at relativistic speeds in magnetic fields (see Sect. 19).

B Cassiopeia (Tycho Brahe's nova A.D. 1572). The identification[2] of a radio source (00N6A) with the remnant of this supernova must be regarded as tentative. The proximity of the intense source in Cassiopeia (23N5A) makes it difficult to determine the position and angular diameter of the radio source reliably; furthermore no positive optical evidence for the remnant of this supernova is yet available, although BAADE and MINKOWSKI have recently reported a faint nebulosity close to the expected position.

Nova Ophiuchi (Kepler's nova A.D. 1604). A radio source in the position of this supernova has been reported by at least two independent observers[3,4] and the remnant has been observed photographically by BAADE[5].

An interesting feature of the three radio sources associated with supernovae is that the difference between their apparent radio magnitude and visual magnitude at maximum is approximately constant.

---

[1] J. H. OORT and TH. WALRAVEN: Bull. astronom. Inst. Netherl. **12**, 285 (1956).

[2] R. HANBURY BROWN and C. HAZARD: Nature, Lond. **170**, 364 (1952).

[3] J. R. SHAKESHAFT, M. RYLE, J. E. BALDWIN, B. ELSMORE and J. H. THOMSON: Mem. Roy. Astr. Soc. **67**, 106 (1955).

[4] B. Y. MILLS, A. G. LITTLE and K. V. SHERIDAN: Austral. J. Phys. **9**, 84 (1956).

[5] W. BAADE: Astrophys. Journ. **97**, 119 (1947).

A search for the radio emission from novae has been made by MILLS, LITTLE and SHERIDAN[1] who failed to detect any of them at a wavelength of 3.5 m. They concluded that the ratio of the radio emission to the light emission at maximum must be considerably less in a nova than in a supernova of type I.

β) Peculiar Nebulosities. A number of radio sources have been identified with peculiar faint nebulosities in the Galaxy. It is by no means certain that these nebulosities are all off the same type.

The Cassiopeia nebulosity. Source Cassiopeia A (23N5A), which is the most intense source in the sky at metre wavelengths, has been identified with a remarkable nebulosity. The nebulosity is faint and patches of it cover an area of about 6′ in diameter. It contains filaments of a peculiar type which exhibit a dispersion in velocities of several thousand kilometres per second. The nebulosity has been described in detail by BAADE and MINKOWSKI in Ref. [*5*]. The diameter and position of the radio source are in good agreement with the optical data and the identification is regarded as well established. The nature of this nebulosity is unknown and of considerable interest; it has been suggested that it may be the remnant of a supernova of type II.

Nebulosities in Puppis, Gemini (I.C. 443), Auriga. In each of these three cases the identification is based on reasonable agreement between the position and angular diameter of the radio source and the optical object. The nebulosities are all fairly faint photographically and show filamentary structure. Their origin is at present unknown. A discussion of the nebulosity in Puppis is given by BAADE and MINKOWSKI in Ref. [*5*]; the nebulosity in Gemini has been discussed by BALDWIN and DEWHIRST[2]; the nebulosity in Auriga has been discussed briefly by HANBURY BROWN, PALMER and THOMPSON[3].

γ) Ionised Hydrogen (H II) regions. Detectable thermal radio emission is to be expected from regions of ionized hydrogen, H II regions, surrounding intensely hot stars. Surveys of these regions have been made at 85 Mc/s by MILLS, LITTLE and SHERIDAN[4], at 3200 Mc/s by HADDOCK, MAYER and SLOANAKER[5], and a number of H II regions have also been reported as radio sources at 1420 Mc/s by HAGEN, MCCLAIN and HEPBURN[6]. Among the emission regions reported as radio sources are the well-known nebulae, Orion (N.G.C. 1976), Omega (N.G.C. 6618), Trifid (N.G.C. 6514), Lagoon (N.G.C. 6523), North American (N.G.C. 7000); altogether about fifteen hydrogen emission regions have now been measured.

The results show that the radio emission from these regions is consistent with a simple origin of the radiation in free-free transitions in an ionized gas. For example in their study of the Orion Nebula HADDOCK *et al.*[7] found the brightness temperature, as defined in Sect. 1, to be about 100° K at 3200 Mc/s. On the assumption that the electron temperature is 10^4 deg. K, the mean opacity of this nebula must be about 0.01. An opacity of this order would correspond to an average emission measure over the disc of about 2.6×10^5 (electron density$^2 \times$ thickness in parsecs), and assuming the source to be spherical, the electron density at the centre would be of the order of 400 cm^{-3}. HADDOCK *et al.* conclude that these results are in reasonable agreement with the optical data about the nebula.

---

[1] See footnote 4, p. 231.

[2] J. E. BALDWIN and D. W. DEWHIRST: Nature, Lond. **173**, 164 (1954).

[3] R. HANBURY BROWN, H. P. PALMER and A. R. THOMPSON: Nature, Lond. **173**, 945 (1954).

[4] B. Y. MILLS, A. G. LITTLE and K. V. SHERIDAN: Austral. J. Phys. **9**, 218 (1956).

[5] F. T. HADDOCK, C. H. MAYER and R. M. SLOANAKER: Nature, Lond. **174**, 176 (1954).

[6] J. P. HAGEN, E. F. MCCLAIN and N. HEPBURN: Proc. Inst. Radio Engrs. **42**, 1811 (1954).

[7] F. T. HADDOCK, C. H. MAYER and R. M. SLOANAKER: Astrophys. Journ. **119**, 456 (1954).

δ) The Source in Sagittarius (17S2A). It has been known for some years that there is a strong radio source (17S2A) in the direction of the galactic centre. This source has not yet been identified positively with any object observed photographically; it has been included here because tentative identifications have been put forward which suggest that this source may be unique in the Galaxy.

Observations of the source have been made over a wide range of frequencies, it is very prominent at high frequencies but difficult to separate from the general background radiation at low frequencies. It appears to change in angular size and shape with frequency; for example, WILLIAMS and DAVIES[1] conclude that it has angular diameter of about 1° at 1420 Mc/s, while MILLS[2] reports that at 85 Mc/s there is an extended region of several degrees, flattened in galactic latitude, with two maxima about 2° apart.

The coordinates of the source are in close agreement with the galactic centre[3] and it has been tentatively proposed[4] that the source may be identified with the nucleus of the Galaxy. On the other hand it has also been pointed out[5] that the source may be associated with a compact H II region which lies in the direction of the galactic centre at a distance of 3 kpsc from the Sun. At the present time the location of this source is controversial. Several possible models[6] have been suggested; for example the source may consist of non-thermal radiation from the galactic centre with an H II region, at 3 kpsc from the Sun, superimposed on it, or the ionized hydrogen may itself be at the galactic centre together with the non-thermal source. For a fuller discussion of these possibilities the reader must refer to the original papers.

18. Extra-galactic objects. *α) Normal galaxies.* A few radio sources have been identified with galaxies which do not appear to exhibit any peculiarity of structure and spectrum and which are believed to be normal samples of the population of galaxies. A total of about 16 *normal* galaxies have been detected up to the present; the list includes 6 *Sb*, 6 *Sc*, 1 *Sbc* and 3 of the Magellanic type. Among the well-known galaxies which have been detected are N.G.C. 224 (M 31), N.G.C. 3031 (M 81) and N.G.C. 5194-5 (M 51). No normal elliptical galaxies have yet been identified as radio sources.

Generally speaking the ratio of the light emission to the radio emission from type *Sb* galaxies, is found to be of the same order as that for our own Galaxy.

For a list of galaxies, together with a more detailed discussion, reference must be made to the companion article in the present volume by B. Y. MILLS on the "Radio-frequency radiation from external galaxies".

β) Peculiar Galaxies. A small number of radio sources have been identified with *peculiar* galaxies, and the present evidence suggests that many of the sources may be of this type.

The most spectacular case is the radio source Cygnus A (19N4A) which has been identified by BAADE and MINKOWSKI [*5*] with the collision of two spiral galaxies. It appears that, in the process of collision, the ratio of the total radio

[1] D. R. W. WILLIAMS and R. D. DAVIES: Phil. Mag. **1**, 622 (1956).

[2] B. Y. MILLS: Observatory **76**, 65 (1956).

[3] J. D. KRAUS and H. C. KO: Astrophys. Journ. **122**, 139 (1955).

[4] R. X. MCGEE and J. G. BOLTON: Nature, Lond. **173**, 985 (1954).

[5] R. D. DAVIES and D. R. W. WILLIAMS: Nature, Lond. **175**, 1079 (1955). — W. PRIESTER: Z. Astrophys. **38**, 73 (1955).

[6] R. D. DAVIES: Observatory **76**, 196 (1956). — F. G. SMITH, P. A. O'BRIEN and J. E. BALDWIN: Monthly Notices Roy. Astronom. Soc. London **116**, 282 (1956). — B. Y. MILLS: Observatory **76**, 65 (1956).

emission to the light has been increased by a factor of about 10^6 in comparison with normal galaxies. Thus, although the apparent photographic magnitude of the Cygnus source is about $+18$, it is the second most intense source in the sky at metre wavelengths. There are a few other cases in which colliding galaxies have been found to emit intense radio emission, for example N.G.C. 1275 and N.G.C. 5128.

A particularly interesting peculiar galaxy is N.G.C. 4486 in Virgo which has been identified with radio source Virgo A (12N1A). This galaxy is elliptical and appears to have a jet, predominantly blue in colour, of rapidly moving material in close proximity to the nucleus. The light from this jet is known to be polarized. So far no other similar object has been discovered in association with a radio source.

For a fuller discussion of peculiar galaxies the reader is referred to the companion article by B. Y. MILLS (p. 239 this volume).

γ) Clusters of galaxies. Most galaxies are member of clusters and it should therefore be possible to detect the aggregate radiation from a cluster. A few clusters, for example the well-known cluster in Perseus, are known to be radio sources. The reader is again referred to the companion article by B. Y. MILLS (p. 239 this volume) for further information.

19. The mechanism of radiation. The observational evidence shows that, while the radiation from a few sources, e.g. H II regions, is consistent with a thermal origin, the majority of the sources are radiating by some *non-thermal* process. Attempts to explain the non-thermal radiation have, in general, invoked either *plasma oscillations* or *synchrotron radiation.*

α) Thermal radiation. The theory of the thermal radio emission from an ionized gas has been given by a number authors[1]. For a cloud of isothermal gas with an electron temperature T_e and optical depth τ, the brightness temperature T_B (defined in Sect. 1) is given by

$$T_B = T_e[1 - e^{-\tau}]. \tag{19.1}$$

The optical depth τ at any frequency ν can be determined from an expression for the absorption coefficient $\varkappa$ which is applicable at low electron densities,

$$\left.\begin{aligned} \tau &= \int \varkappa\, ds, \\ \varkappa &= \zeta N^2 \nu^{-2} T_e^{-\frac{3}{2}} \end{aligned}\right\} \tag{19.2}$$

where ds is an element of path through the cloud, N is the electron density, and ζ is a slowly varying function of T_e given by,

$$\zeta = 9.70 \times 10^{-3} \log\left(\frac{3kT_e}{2h\nu} \times 10^{-6}\right) \quad \text{c.g.s. units.} \tag{19.3}$$

where k and h are BOLTZMANN'S and PLANCK'S constants, and it is assumed that the hydrogen is fully ionized.

If the optical depth is great ($\tau \gg 1$) then it follows from Eq. (19.1) that $T_B = T_e$, and the brightness temperature is simply equal to the electron temperature. If the optical depth is small and the cloud is uniform and of thickness t then we may write,

$$T_B = \tau T_e = \varkappa t T_e = \zeta N^2 \nu^{-2} T_e^{-\frac{1}{2}} t. \tag{19.4}$$

[1] J. H. PIDDINGTON: Monthly Notices Roy. Astronom. Soc. London **111**, 45 (1951). — Proc. Roy. Soc. Lond., Ser. A **203**, 417 (1950). — S. F. SMERD and K. C. WESTFOLD: Phil. Mag. **40**, 831 (1949).

The application of the formulae given above to the radiation from H II regions shows that the observed radiation is in reasonable agreement with that predicted from the optical data.

Attempts have been made to give a general explanation of the radiation from the Galaxy and from the sources in terms of thermal radiation; however there are at least two major difficulties which cannot be overcome. Firstly several of the discrete sources have brightness temperatures, at metre wavelengths, which exceed 10^7 deg. K. Such extreme temperatures cannot be explained, since according to Eq. (19.2) the opacity of an ionized gas varies as $T_e^{-\frac{1}{2}}$ and this leads to unreasonably high values for the mass of the sources.

A second difficulty is that the spectrum of the majority of the sources, see Sect. 12, does not correspond with that from a thermal source. Thus the flux-density S to be expected from an optically thin thermal source of angular diameter Ω steradians is, from Eqs. (1.2), (1.3) and (19.4),

$$S = \frac{2k}{c^2}\,\Omega\,\frac{\zeta N^2 t}{T_e^{-\frac{1}{2}}} \tag{19.5}$$

and is independent of the frequency. In the case of a gas which is opaque, S is given by,

$$S = \frac{2k}{c^2}\,\nu^2\,T_e\,\Omega\,. \tag{19.6}$$

For the majority of sources the flux-density decreases with increasing frequency with a spectrum of the rough form $S \propto \nu^{-1}$ which is inconsistent with a thermal origin of the radiation.

β) Plasma oscillations. Intense bursts of radio emission are occasionally observed from the Sun, for example they are often associated with solar flares. The present evidence suggests that some of these bursts may be produced by plasma oscillations possibly excited by violent disturbances in the solar corona. For a discussion the reader is referred to a review by PAWSEY and SMERD[1]. UNSÖLD[2] has suggested that the general radiation from the Galaxy may originate in great numbers of stars of low surface temperature which are characterized by powerful radio outbursts similar to those observed on the sun. This theory is not supported by the available data; since there is, as yet, no evidence for the existence of discrete sources with dimensions comparable with stars.

A general difficulty with a theory based on plasma oscillations is that it seems unlikely that in any of the extended objects, which are observed to be sources, there will be a sufficiently high density to support plasma oscillations at high radio frequencies. Thus the plasma frequency of an ionized gas is given by,

$$\nu = \frac{1}{2\pi}\sqrt{\frac{N e^2}{m\,\varepsilon_0}}\,. \tag{19.7}$$

TWISS[3] has argued that, since a density of the order of 10^{14} electrons m^{-3} is required to raise the plasma frequency to 100 Mc/s, it is unlikely that the radiation from the extended nebulosity which is associated with the Cassiopeia A source (23N5A) is due to plasma oscillations.

γ) Synchrotron radiation. The suggestion that the radio emission from the discrete sources might be due to high-energy electrons moving in magnetic

[1] J. L. PAWSEY and S. F. SMERD: Solar radio emission, p. 466, The Sun, ed G. P. KUIPER. University of Chicago 1953.

[2] A. UNSÖLD: Nature, Lond. **163**, 489 (1949) also Z. Physik **141**, 70 (1955).

[3] R. Q. TWISS: Phil. Mag. **45**, 249 (1954).

fields has been advanced by a number of authors. It has been applied to the Cassiopeia A source (23N5A) by TWISS[1], to the Crab Nebula (05N2A) by SHKLOVSKY[2] and by OORT and WALRAVEN[3], and to the peculiar galaxy N.G.C. 4486 in Virgo (12N1A) by SHKLOVSKY[4] and BURBIDGE[5].

OORT and WALRAVEN show that the total energy radiated per second by an electron of high energy is,

$$\frac{dE_{(\mathrm{Gev})}}{dt} = -A H_{\perp}^2 E_{(\mathrm{Gev})}^2 \tag{19.8}$$

where $E_{(\mathrm{Gev})}$ is the energy of the electron in units of 1 Gev $=10^9$ ev, $H_{\perp}$ gauss is the magnetic field perpendicular to the electron's path, and $A = 3.793 \times 10^{-6}$. Due to loss of energy by this process the energy of the electron will diminish to half its original value in a time,

$$T = \frac{0.00835}{H_{\perp}^2 E_a} \quad \text{years} \tag{19.9}$$

where E_a is the initial energy in units of 10^9 ev.

The radiation from the electron should be completely polarized with the electric vector parallel to the radius of curvature of the orbit; it should also be concentrated into a very sharp cone, the axis of which coincides with the velocity of the electron. For a more detailed account of this theory the reader may refer to the article by J. H. OORT on p. 100 of the present volume.

OORT and WALRAVEN have applied these formulae to calculate the magnetic field and the electron energies and densities in the Crab Nebula which would account for both the optical and the radio emission. As mentioned in Sect. 17 the light from this object is known to be polarized. They find that the magnetic field must be of the order of 10^{-3} gauss and that the median energy of the electrons must be about 2×10^{11} ev. Under these conditions the total energy stored in the electrons and the magnetic field are both of the order of 10^{49} ergs and the electrons lose half their energy by radiation in 180 years.

This theory appears to be a promising explanation of non-thermal radio emission, but the origin of the magnetic field and of the high-energy electrons both present a controversial problem. For example SHKLOVSKY has argued that the magnetic field is of interstellar origin and has been increased by turbulence in the expanding shell. He has also argued that the first electrons are accelerated by the Fermi process which depends upon successive collisions between the electrons and magnetic irregularities. OORT and WALRAVEN estimate that the total kinetic energy of expansion of the shell is only about 10^{47} ergs and do not support the view that the electrons draw their energy from that source. They conclude that the energetic electrons cannot be considered as remains of the original explosion and point to some observational evidence, moving wisps of light, which suggest that they may be continually replenished in some way by the remnant of the supernova.

At the present time no theory of the origin of the radio emission from the discrete sources can be regard as firmly established. However the discovery that the light from the Crab Nebula is polarized suggests strongly that the theory

---

[1] See footnote 3, p. 235.

[2] I. S. SHKLOVSKY: Dokl. Akad. Nauk. USSR. **90**, 983 (1953).

[3] J. H. OORT and TH. WALRAVEN: Bull. astronom. Inst. Netherl. **12**, 285 (1956).

[4] I. S. SHKLOVSKY: Astr. Zh. **32**, 215 (1955).

[5] G. R. BURBIDGE: Astrophys. Journ. **123**, 178 (1956); **124**, 416 (1956).

given above is correct in general outline, although there remain a number of difficult questions. The same mechanism has been proposed to account for the non-thermal radiation from the other discrete sources in the Galaxy, from the interstellar gas itself, and from extra-galactic nebulae.

20. Speculations on the nature and origin of the discrete sources. It is recognised that the sources must be considered as belonging to at least two classes which, following MILLS[1], we will discuss under the headings, Class I and Class II.

α) Class I sources. These sources are members of the Galaxy, they show a pronounced concentration into the galactic plane and several of them are known to have large angular diameters. Their radiation at metre wavelengths is undoubtedly non-thermal in origin. Both MILLS[1] and also HANBURY BROWN and HAZARD[2] have argued that the distance of the nearest Class I sources is of the order of 1000 parsec, and the latter authors have estimated their space-density in the neighbourhood of the Sun as about 5×10^{-8} sources parsec^{-3}. If this interpretation is correct, then the Class I sources are rare objects, like planetary nebulae, with a space density much less than that of the visible stars.

The few satisfactory identifications which have been made suggest that many of these sources are associated with optically faint extended nebulosities of a peculiar type. The nature of these objects is controversial; for example it has been proposed[3], mainly on the grounds of their enormous energy output in the radio spectrum, that they are the remnants of supernovae, possibly of type II.

As stated in Sect. 18 a number of sources, particularly at centimetre waves, are associated with H II regions and their radiation is explicable by classical thermal processes. It must be noted that these sources are not included in the category Class I as used here.

β) Class II sources. For our present purposes we shall refer to the vast majority of the sources as Class II, meaning by this distinction that they show no evidence of concentration into the galactic plane. Considered as a whole the present evidence suggests that many of these sources are extra-galactic. Thus the few which have been identified are associated with external galaxies; the apparent angular diameters of many are known to be of the order of one minute of arc, which is consistent with an origin in galaxies; their angular distribution over the sky is roughly isotropic and it is extremely difficult to make a convincing model of the sources assuming them to be in the Galaxy[4]. At one time it was argued that the majority of the sources could not be extra-galactic since they showed such poor correlation in position with the bright galaxies. This objection has now been removed by the discovery of peculiar galaxies which exhibit a vastly greater ratio of radio emission to light than normal galaxies; thus it may well be that many of the Class II sources are associated with peculiar galaxies, perhaps galaxies in collision, which are photographically faint. It is this possibility, namely that extremely distant objects may perhaps prove to be relatively easy to detect by radio, which suggests that the study of the Class II sources may lead to a significant contribution to cosmology.

In contrast to the view that the majority of the sources are extragalactic it has been argued that the Galaxy may contain a large number of ***radio stars***. This possibility cannot be dismissed, since the origin of the general background

---

[1] B. Y. MILLS: Austral. J. Sci. Res. A **5**, 266 (1952).

[2] R. HANBURY BROWN and C. HAZARD: Phil. Mag. **44**, 939 (1953).

[3] I. S. SHKLOVSKY: C. R. Acad. Sci. USSR. **94**, 417 (1954). — R. HANBURY BROWN: Observatory **74**, 185 (1954).

[4] An interesting analysis is that given by W. PRIESTER: Z. Astrophys. **34**, 295 (1954.)

radiation from the Galaxy is still open to question; nevertheless there is, as yet, no experimental evidence for the existence of a population of radio sources of stellar dimensions.

γ) The sources and the general background radiation. Isophotes of the sky show that a significant fraction, probably the major part, of the total radio emission originates in the Galaxy. It is clear that the Class I sources must contribute to this radiation, but their contribution cannot be estimated usefully since the variation of their space-density in the Galaxy in unknown. The present evidence suggests that, at high frequencies, they contribute only a few per cent of the total radiation close to the galactic plane, but it is likely that their contribution is more significant at low frequencies. The emission which is observed at high galactic latitudes cannot be explained in terms of known sources. It can therefore be stated that, at the present time, there is no evidence that the majority of the radio emission from the Galaxy arises in a population of discrete sources.

If the majority of the observed sources are extra-galactic then it is to be expected that there will be an isotropic component in the total radiation due to their integrated emission. An upper limit to this component can readily be set from an analysis of the isophotes of the sky, for example it can be shown that the brightness temperature cannot exceed about 500 deg. K at 100 Mc/s. A lower limit cannot be set reliably without a more precise knowledge of the total radiation due to the Galaxy. However it has been shown that, on the assumption that the majority of the sources are extra-galactic, the expected value of their total radiation is consistent with the rather wide limits which can be set to any such isotropic component[1].

General references.

[1] PAWSEY, J. L.: A catalogue of reliably known discrete sources of cosmic radio waves. Astrophys. Journ. **121**, 1 (1955).

[2] WILD, J. P.: Observational Radio Astronomy, Advances in Electronics and Electron Physics, Vol. 7, Editor L. Marton. New York: Academic Press Inc. 1955.

[3] Report of the Physical Society's Conference on the Physics of the Ionosphere held at the Cavendish Laboratory, Cambridge, in September 1954: The Physical Society (London) 1955.

[4] BAADE, W., and R. MINKOWSKI: On the identification of radio sources. Astrophys. Journ. **119**, 215 (1954).

[5] BAADE, W., and R. MINKOWSKI: Identification of the radio sources in Cassiopeia, Cygnus A, and Puppis A. Astrophys. Journ. **119**, 206 (1954).

[6] PAWSEY, J. L., and R. N. BRACEWELL: Radio Astronomy. Oxford: Clarendon Press 1955.

[7] COUTREZ, R.: Radioastronomie. Monographie 5: Observatoire Royal de Belgique 1956.

[8] HANBURY BROWN, R., and A. C. B. LOVELL: The Exploration of Space by Radio. London: Chapman & Hall. 1957.

[1] Examples of this type of analysis are by J. R. SHAKESHAFT: Phil. Mag. **45**, 1136 (1954) and W. PRIESTER: Z. Astrophys. **34**, 283 (1954).

Radio Frequency Radiation from External Galaxies.

By

B. Y. MILLS.

With 22 Figures.

Introduction.

The study of the radio emission of external galaxies may be said to have begun in 1946 with the observation by HEY, PARSONS and PHILLIPS[1] of fluctuations in the emission from a region in the constellation of Cygnus. They deduced that the fluctuating component "could only originate from a small number of discrete sources". This was confirmed when BOLTON and STANLEY[2] showed that the fluctuations were due to an intense emitting region less than eight minutes of arc in extent; but five years had to elapse before the extragalactic origin of this radiation was established. Meanwhile the existence of a whole class of such "discrete sources" had been discovered, and two tentatively identified with external galaxies by BOLTON, STANLEY and SLEE[3] thus suggesting that some types of galaxies might be strong radio-emitters: these tentative identifications were later confirmed and many others made. At the time of writing the number of galaxies from which radio frequency radiation has been detected is about two dozen; a large increase might be expected in the near future.

It appears that two basic classes may be recognized, "normal" and "radio" galaxies. The former radiate with the same order of intensity as, or less than, the Milky Way and the Andromeda nebula; they include the brightest and closest of the late-type galaxies. The latter have a radio emission which may be many orders of magnitude greater; they are very rare astronomical objects, but the most common form of radio source. The intense source of radiation in the constellation of Cygnus is of the latter type. This article will be principally concerned with the physically more numerous normal galaxies, although some consideration will be given in Chap. V to the radio galaxies and their possible importance in cosmological studies.

The discovery in 1951 of 21 cm line emission from interstellar atomic hydrogen opened a new field of galactic radio investigation. This line has now been observed in either emission or absorption from various external galaxies; the number of such observations is at present very limited but, because of the measurements of velocity which are possible, the results are of fundamental importance.

Unlike the H line emission, the basic mechanisms producing the continuum are not yet well understood. Two radiating processes have been identified, the free-free transitions in ionized gas clouds and the synchrotron type radiation from relativistic electrons radiating in magnetic fields; the physics of these processes are discussed in the preceding articles by HANBURY BROWN and OORT.

[1] J. S. HEY, S. J. PARSONS and J. W. PHILLIPS: Nature, Lond. **158**, 234 (1946).

[2] J. G. BOLTON and G. J. STANLEY: Nature, Lond. **161**, 312 (1948). — Austral. J. Sci. Res. A **1**, 58 (1948).

[3] J. G. BOLTON, G. J. STANLEY and O. B. SLEE: Nature, Lond. **164**, 101 (1949).

The origin of the high energy electrons has not yet been satisfactorily established, however, and, also, the possibility of other radiating mechanisms remains open.

Because of the very recent development of radio astronomy and the rapid advances that are being made at present in the study of the radio properties of galaxies, it is difficult to give an account of the subject which is not likely to require extensive modification in a few years' time, a difficulty which is accentuated by many contradictory results which have been reported. As far as possible the most reliable data are given prominence in this article; it is inevitable that the author's personal preferences will show themselves in this selection.

I. The Magellanic Clouds.

1. General remarks. The Magellanic Clouds are well suited for studying the radio properties of galaxies, since they are sufficiently large for structural details to be observed with a radio telescope having a resolution of the order of one degree. At the H line wavelength of 21 cm this is achieved with parabolic aerials of moderate size, whilst at metre wavelengths, cross-type aerials of this resolution are easily constructed. The resulting convenience of observation is somewhat offset by the comparative rarity of the galaxy type exemplified by the Clouds, that is, "late irregular", or in DE VAUCOULEURS' notation (p. 276) $SB(s)m$. Thus it may not be assumed that the results obtained are directly applicable to the more common types of spiral galaxies. Nevertheless the possibility of direct observation and comparison of the radio brightness with various stellar populations, together with the possibility of dynamical studies using the H line, are decisive factors in the importance of the Clouds. The radio observations will therefore be discussed in some detail.

Radio frequency radiation from the clouds was first observed in 1953 when observations in the H line and at 3 m wavelength were almost simultaneously successful in detecting radiation from the Large Cloud[1] [*4*]. This was the first attempt to detect the H line radiation, but several earlier unsuccessful attempts had been made at metre wavelengths, the failures resulting largely from the complexity of the galactic brightness distribution in the region and the poor resolution of the radio telescopes then available.

The H line observations have proved extremely fruitful in determining, not only the distribution and total mass of neutral hydrogen, but also the dynamics of the gas and, by inference, the total mass of each cloud and its distribution. In the matter of definite physical information, the metre wave observations have not yet proved so successful; nevertheless valuable clues as to the origin and distribution of the radio emission have been obtained, particularly when comparisons are made with the Milky Way and other galaxies.

a) The H line radiation.

2. Observational data. The H line radiation from both Magellanic Clouds was observed by KERR, HINDMAN and ROBINSON [*4*] using an 11 m parabolic aerial of beamwidth 1°.5 (incorrectly given as 1°.0 in the original paper, but corrected later). The bandwidth of the receiver, 40 kc/s between half-power points, corresponds to a Doppler velocity spread of 9 km/s. In order to remove the effects of equipment instability and background radiation, a balancing system was employed to compare the output from this narrow-band channel with the output of a 1.7 Mc/s bandwidth reference channel.

[1] B. Y. MILLS and A. G. LITTLE: Austral. J. Phys. **6**, 272 (1953).

A number of traverses of the Clouds were made with the narrow-band channel tuned to different frequencies, and from these traverses a complete set of line profiles were derived at 250 points arranged in a lattice spaced 1° in declination and 10 min in right ascension. A number of sample profiles obtained for the small cloud are shown in Fig. 1, the frequency scale having been converted to Doppler velocities.

The line profiles so obtained were described by two parameters, the "integrated brightness" and the median velocity. The "integrated brightness", B_{int}, is the total energy received from a particular direction and is obtained by integrating the line profile, thus:

$$B_{\text{int}} = \frac{2\,k}{\lambda^2} \int T_\nu \, d\nu \tag{2.1}$$

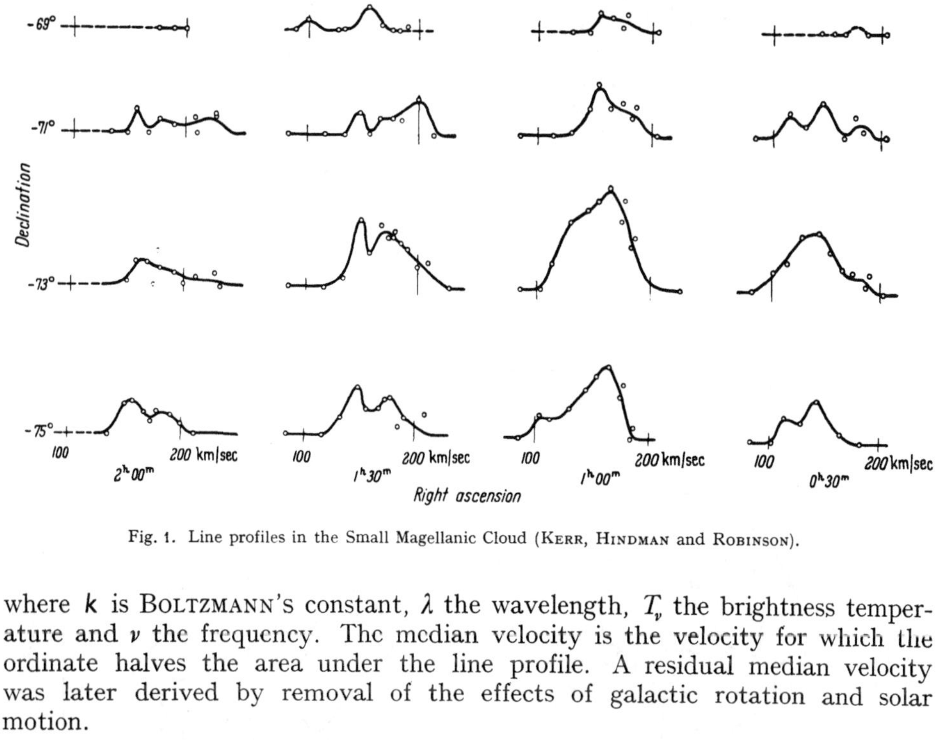

Fig. 1. Line profiles in the Small Magellanic Cloud (Kerr, Hindman and Robinson).

where k is Boltzmann's constant, λ the wavelength, T_ν the brightness temperature and ν the frequency. The median velocity is the velocity for which the ordinate halves the area under the line profile. A residual median velocity was later derived by removal of the effects of galactic rotation and solar motion.

In Figs. 2 and 3, contour plots are presented of the integrated brightness and the residual median velocities, while in Fig. 4 is shown, to the same scale, a wide angle photograph of both Clouds. The enormous extent of the distribution of neutral hydrogen, particularly in the Small Cloud, is very obvious and is one of the most surprising features of the observations. The radio and optical distributions are compared in detail in Sect. 8.

3. The total mass of neutral hydrogen. The total quantity of neutral hydrogen in each Cloud has been derived from the integrated brightness contours of Fig. 3. It is assumed that the line is unsaturated, a reasonable assumption since the maximum temperatures are only some 30° K compared with temperatures exceeding 100° K observed in the Galaxy. On this assumption we have:

$$T_\nu = T\,(1 - e^{-\tau}) \approx T\,\tau \tag{3.1}$$

where T is the excitation temperature of the line and τ the optical thickness. Now the absorption coefficient is given by Wild[1] as

$$\varkappa = 2.6 \times 10^{-15} \frac{n}{T} f(\nu) \text{ cm}^{-1} \tag{3.2}$$

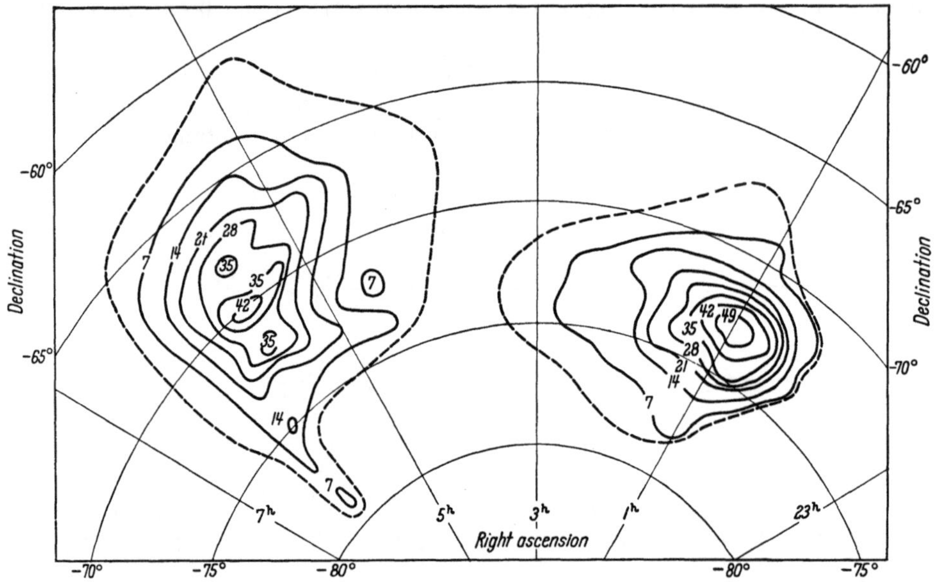

Fig. 2. The integrated hydrogen-line brightness contours of the Magellanic Clouds (Kerr, Hindman and Robinson).

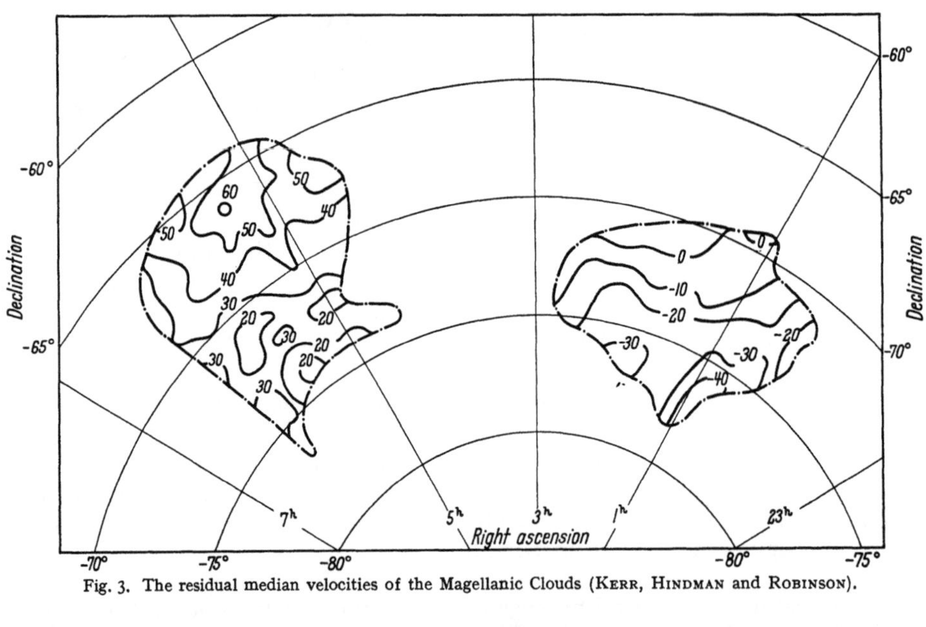

Fig. 3. The residual median velocities of the Magellanic Clouds (Kerr, Hindman and Robinson).

where n is the number of ground state atoms per cubic centimetre and $f(\nu)$ is the "line profile" function normalized so that

$$\int_0^\infty f(\nu)\, d\nu = 1 .$$

[1] J. P. Wild: Astrophys. Journ. **115**, 206 (1952).

The optical depth is therefore

$$\tau = \int \varkappa\, ds = 2.6 \times 10^{-15} \int_0^\infty \frac{n}{T} f(\nu)\, ds. \tag{3.3}$$

If it is assumed that T and $f(\nu)$ are constant along the line of sight it follows from (3.1), (3.3) and (2.1) that

$$N = 6.2 \times 10^{35} B_{\text{int}} \tag{3.4}$$

where N is the number of atoms in a projected area of 1 cm^2.

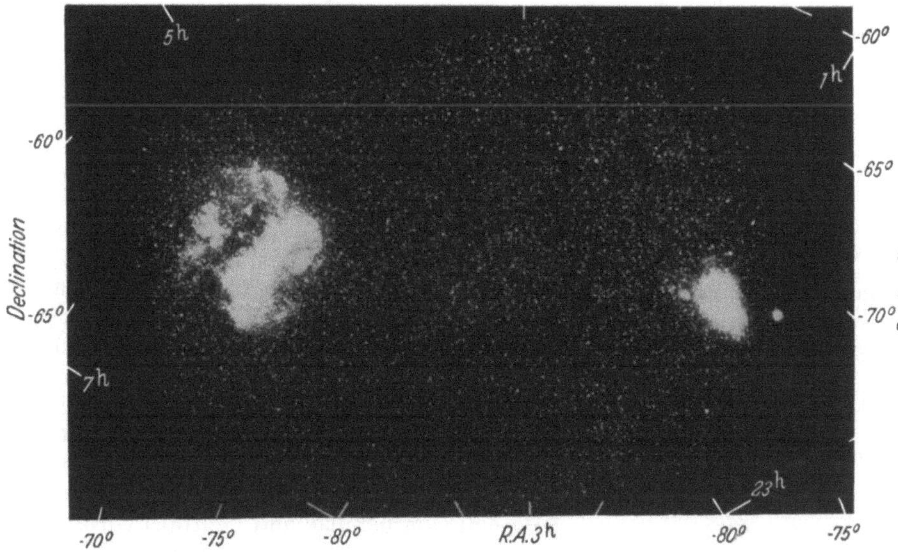

Fig. 4. A photograph of both Magellanic Clouds (DE VAUCOULEURS).

Applying this result to the Clouds and assuming a distance of 46 kpc, it is found, on integration over the contours of Fig. 3, that the total numbers of hydrogen atoms are 7×10^{65} and 5×10^{65} in the Large and Small Clouds respectively. The corresponding masses are 6×10^8 solar masses for the Large Cloud and 4×10^8 solar masses for the Small Cloud. As we shall see these represent an appreciable fraction of the total masses of the Clouds. It is difficult to estimate the accuracy of the determinations and KERR *et al.* do not attempt to do so; it would seem unlikely, however, that they could be in error by as much as a factor of two. The principal uncertainty arises in obtaining the contribution of the tenuous outer regions.

4. Motions of the Clouds and total masses. The mass of neutral hydrogen in each Cloud has been obtained directly from the integrated emission; it is possible, also, to estimate the *total* masses of the Clouds from an analysis of the internal motions. The Clouds exhibit both rotational and random motions of approximately equal energy content; a theory covering this type of motion in a gravitating system is not yet available, although the cases of pure rotation and pure random motion are well known. KERR and DE VAUCOULEURS have attempted to analyse the data by an approximate method to obtain best estimates of the masses: this is quite justified as the principal uncertainties are in the observational data rather than in the theory[1].

Considering the rotational motion first, the rotation curves shown in Figs. 5 and 6 were derived by trial and error methods in which the centres of rotation and position angles of the major axes were adjusted. It was found that, to obtain a symmetrical rotation curve the centres of rotation had to be somewhat displaced from the brightest optical regions, presumably a result of the asymmetrical Cloud structures.

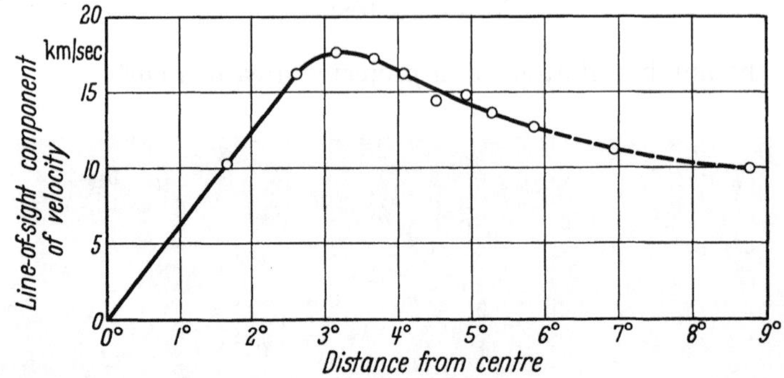

Fig. 5. Rotational curve of the Large Magellanic Cloud (KERR and DE VAUCOULEURS).

The curves are *mean* rotational curves. If curves corresponding to the peaks in the line profiles are derived, they indicate a much more rapid rotation, as also do the radial velocities of 17 emission nebulae in the Large Cloud, the only available optical velocity data. These discrepancies are interpreted by KERR and DE VAUCOULEURS as indicating differential rotation of the system, the equatorial regions, where the emission nebulae and neutral hydrogen are concentrated, rotating fastest. The large dispersion and low mean velocity of the neutral hydrogen indicates that it is much less concentrated towards the plane of the system than the ionized hydrogen: it would seem that the concentration of the latter is determined by the concentration of early type stars.

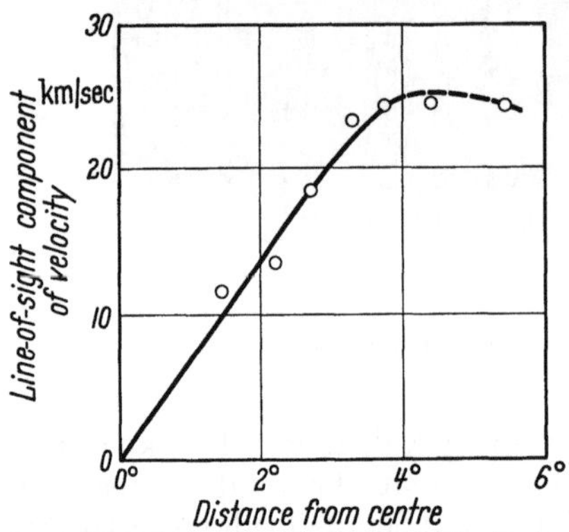

Fig. 6. Rotational curve of the Small Magellanic Cloud (KERR and DE VAUCOULEURS).

The measured velocity dispersion obtained from the line profiles was given as 25 km/sec for both Clouds, but this value is due in part to systematic mass motions and to variations of the rotational speed with depth in the Clouds. A figure of 20 km/sec was adopted as the most probable value of the random velocity dispersion alone.

Using these data KERR and DE VAUCOULEURS deduced "most probable masses" which, for the Large Cloud was 3.0×10^9 solar masses and for the Small Cloud 1.3×10^9 solar masses, with an uncertainty of about a factor of two in each case. Comparison with the results of the last section indicates that the neutral hydrogen comprises 20% of the total mass of the Large Cloud and 30% of the total mass of the Small Cloud. The uncertainties are very large.

[1] F. J. KERR and G. DE VAUCOULEURS: Austral. J. Phys. **8**, 508 (1955); **9**, 90 (1956). See also the article by G. DE VAUCOULEURS: General Physical Properties of External Galaxies, p. 311 this volume.

b) The continuum radiation.

5. Observations of the radiation. Radio frequency radiation from the clouds has now been observed at a number of wavelengths, 50 cm[1], 3 m[2], 3.5 m and 15 m. Detailed information has, however, been obtained only at 3.5 m [5]. This was a result of a preliminary survey with the 450 m cross-type radio telescope, a pencil-beam instrument of beamwidth 50′ of arc, located near Sydney, Australia. Radiation was detected from both Clouds and isophotes obtained for the whole region.

A typical recording of the large Magellanic Cloud obtained with this radio telescope is shown in Fig. 7. Radiation from the Cloud extends from about 04^h to about 06^h; the peak at about $05^h 40^m$ coincides in position with the giant emission nebula 30 Doradus, located in the Cloud. A series of such recordings was made at regular declination intervals of about 1° between declinations of

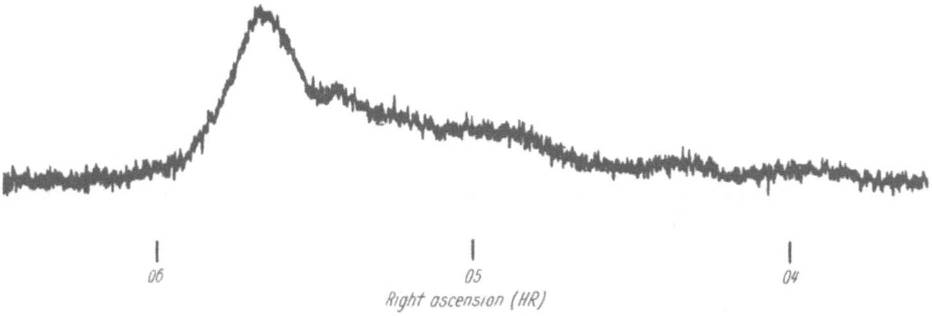

Fig. 7. A record of the Large Magellanic Cloud at a declination of −69°26′ and a wavelength of 3.5 m (MILLS).

$-62\frac{1}{2}°$ and $-75\frac{1}{2}°$, together with additional observations at intervals of about $\frac{1}{2}°$ over the central regions. Thus the central regions of the Clouds have been studied in rather more detail than the outer regions and it is possible that some fine detail has been overlooked in the latter.

The contours are shown in Fig. 8; the brightness temperatures are measured above a base temperature, T, which is about 900° K[3]. There may be some small gradients in the background radiation which would not be recorded by the method employed in reduction, but they would not be sufficiently steep to cause any serious errors. The Cloud radiation, which is a maximum at the approximate positions of $05^h 40^m$, $-69°$ for the Large Cloud and $00^h 65^m$, $-72\frac{1}{2}°$ for the Small Cloud, is quite obvious; but the full extent of the emission is difficult to estimate because of the general brightness irregularities in the area and the presence of several discrete sources.

In Fig. 9 are shown the outer boundaries derived from an examination of the original records. The dark areas include radiation connected to the main systems by strongly closed contour lines and therefore almost certainly belonging to the Clouds. The lighter areas are regions of excess radiation, often quite strong, but in which the connection with the Clouds is not so obvious. For comparison the limits of the hydrogen line emission are shown as dotted lines. It will be noted that the bright region south of $-75°$ stretching from one Cloud to the other in Fig. 8 has been omitted in Fig. 9; it appears probable that this is merely

[1] J. H. PIDDINGTON and G. H. TRENT: Austral. J. Phys. **9**, 90 (1956).
[2] U.R.S.I. Special Report No. 3; Discrete sources of extra-terrestrial radio noise. 1954.
[3] A recalibration of the equipment has resulted in increased values for all temperatures and flux densities. Here and elsewhere in the article corrected values have been used.

an irregularity in the galactic emission. On the other hand, it is considered that the possible "link" between the Clouds at a declination of $-72°$ may be real because of the evidence of the H line emission which suggests a similar "link", or at least a great extension of the Small Cloud in the direction of the Large.

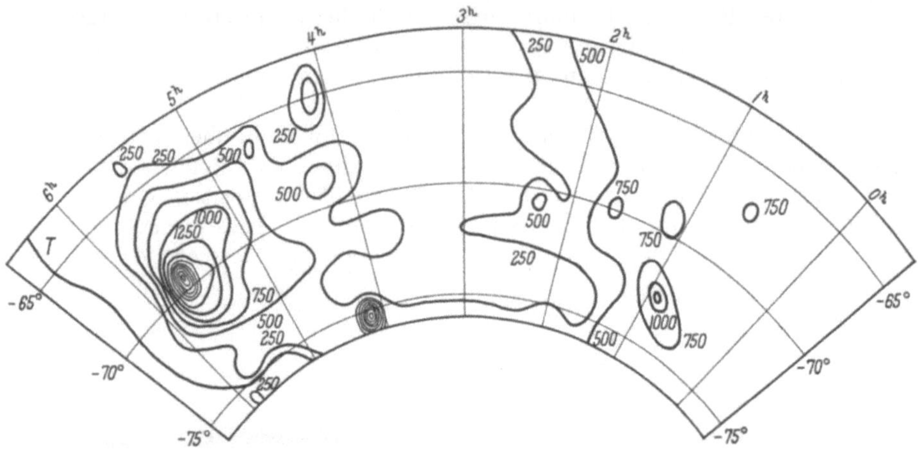

Fig. 8. Contour map of the 3.5 m isophotes in the vicinity of the Magellanic Clouds; the contour interval is 250° K (MILLS). The Large Cloud has maximum brightness at 05^h40^m, the Small Cloud 00^h54^m.

The observations at shorter wavelengths represent, at best, only estimates of total flux density and positions of the centroids of the radiation; the resolution is inadequate for the construction of contour diagrams. The 15 m observations,

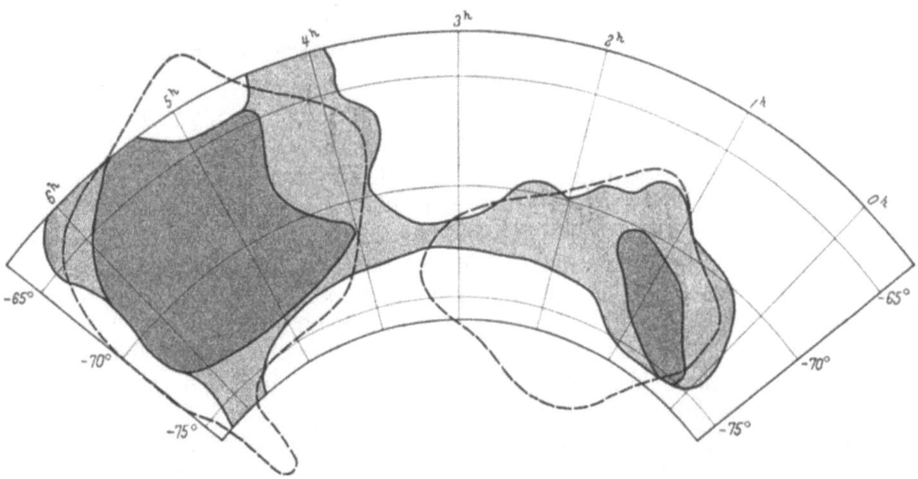

Fig. 9. The boundaries of the Clouds at 3.5 m wavelength (MILLS) compared with the neutral hydrogen boundaries (KERR, HINDMAN and ROBINSON). The dark areas represent definite regions of 3.5 m radiation and the lighter areas possible regions of the radiation. The dotted lines outline the neutral hydrogen boundaries.

on the other hand, are being carried out at the time of writing with a $1\frac{1}{2}°$ pencil beam cross-type aerial, and it is anticipated that a detailed comparison with the 3.5 m observations described above should be most informative.

6. The flux densities, radio spectra and radio magnitudes of the Clouds. The integrated flux density of the Clouds has been obtained from Fig. 7 by subtraction of a "background brightness" and direct integration from the contours according to the relation

$$S = \frac{2k}{\lambda^2} \int T \, d\Omega. \tag{6.1}$$

Difficulties arise because of the related uncertainties of the background radiation to be subtracted and the extent of the Clouds. Best estimates for the 3.5 m observations are given in Table 1, together with the two shorter wavelength observations. The estimated probable error is large for the Small Cloud because of the great uncertainty in its overall extent, resulting from the large background irregularities in its vicinity: these irregularities probably account for the high flux density of the small cloud at 3 m wavelength, as the measurement was made with a beamwidth of 8° which would have been inadequate for the separate observation of the Cloud. The 50 cm observation of the Large Cloud was made with an aerial of beamwidth 3°.3; radiation was detected only from a "point" source at the position of 30 Doradus, the more extended emission being below the level of sensitivity.

Table 1. *Flux density measurements of the Clouds at various wavelengths.*

Wavelength metres	Flux density 10^{-24} W m^{-2} (c/s)$^{-1}$	
	LMC	SMC
3.5	40 ± 4	6.0 ± 1.5
3.1	31 ± 6	9.0 ± 4.5
0.5	4.5	< 2

In each Cloud the radio emission increases between wavelengths of 0.5 m and 3.5 m. Radiation of thermal origin has either a constant spectrum, or one in which the emission decreases as the wavelength increases, depending upon the opacity of the emitting object. The principal source of emission in the Clouds at metre wavelengths therefore has a non-thermal origin, but determination of the actual spectrum law is impossible because of the large uncertainty at 3 m and the lack of sensitivity of 0.5 m. In the latter case the quoted flux density of the Large Cloud is undoubtedly too low because the weak radiation from the outer regions would not have been recorded: a large part of the remainder probably represents thermal emission from 30 Doradus and neighbouring H II regions. For comparison of the Clouds with other galaxies observed at different frequencies, the flux density at 3.5 m will be adopted and the spectra assumed to be of the form $S \propto \lambda^{0.7}$. This spectrum is consistent with the above observations and in conformity with spectra of the Milky Way and the Andromeda Nebula. Following HANBURY BROWN and HAZARD [2] it is convenient to express flux densities in a scale of "magnitudes", analogous to the magnitudes of optical astronomy: it is then possible to compare directly the relation between the radio and optical emission of different astronomical objects. Their radio magnitude scale is adopted, an arbitrary one defined by the relation

$$m_\lambda = -53.4 - 2.5 \operatorname{Log} S_\lambda. \tag{6.2}$$

Thus for the Clouds we have:

$$\text{LMC} \quad m_{3.5} = 2.6 \pm 0.1,$$

$$\text{SMC} \quad m_{3.5} = 4.7 \pm 0.3.$$

For the intercomparison of all radio observations the wavelength and flux density scale of HANBURY BROWN and HAZARD will also be adopted: in the present case this necessitates a spectral correction and a correction to equalize the flux density standards. The spectral correction is $-2.5 \operatorname{Log} (1.9/3.5)^{0.7}$, or $+0.5$ magnitude. It is difficult to obtain an accurate value for the flux density correction since there is no overlap in the coverage of the respective radio telescopes. An estimate has been made using the intense sources in Cygnus, Taurus and Virgo, the former being observed by HANBURY BROWN and HAZARD, the two latter with the Sydney cross-type radio telescope. Many measurements of the flux densities of these sources have been made at different wavelengths and it is possible to estimate their relative intensities at 3.5 and 1.9 m and thus intercompare the flux density standards: the required magnitude correction is approximately +0.3. Thus the total correction is +0.8 magnitude, with a probable error estimated to be about 0.2 magnitude. The radio magnitudes, in a form directly comparable with those of HANBURY BROWN and HAZARD, are therefore:

$$\text{LMC} \quad m_{1.9} = 3.4 \pm 0.25,$$
$$\text{SMC} \quad m_{1.9} = 5.5 \pm 0.4.$$

c) Comparisons of optical and radio data.

7. Some optical data. An enormous amount of detailed observational material on the Clouds has been collected during the last half century at the southern stations of the Harvard College Observatory; but it is only with the recent revision of the distance scale, based in part upon the Magellanic Cloud observations, that it has been possible to make real progress in understanding their basic structure and stellar content. A great deal of the recent work has been carried out at the Australian Commonwealth Observatory at Mt. Stromlo, and in Ref. [*3*] the most recent summaries of this and earlier work are given.

The Magellanic Clouds are usually classified as irregular galaxies, although a definite structure, particularly in the Large Cloud, has long been recognized. Many other galaxies in the irregular class also show the same basic structure which is characterized by a bright axial bar around which are coiled two spiral arms, one heavily populated and well defined, the other embryonic. It is suggested [*3*][1] that these galaxies represent a late-type asymmetrical barred spiral; this interpretation is given weight by the discovery of DE VAUCOULEURS that a twisted lane of absorption, which bears a remarkable similarity to the absorption features observed in many barred spirals, cuts across the axial bar of the Large Cloud[2]. These features are best observed in the Large Cloud because the tilt of the system to the line of sight is small; their presence in the Small Cloud has not been definitely established and, indeed, there is some disagreement in the matter. However, in this article, the view will be adopted that both Clouds are basically similar systems.

The total photographic magnitudes of the Clouds have been very poorly known until recently. A determined attack on the problem, made at the Mt. Stromlo Observatory by DE VAUCOULEURS and by HOGG[3,4] has led to reasonably reliable values, viz. for the Large Cloud a total photographic magnitude of 0.6 and for the Small Cloud a corresponding magnitude of 2.5.

The question of the stellar population content of the Clouds is at present a little obscure although, again, recent work has led to a much better understanding[5]. It appears that both Clouds may be regarded as predominantly Population I systems, although there is evidence that the Large Cloud population is more

[1] G. DE VAUCOULEURS: Astronom. J. **60**, 126, 219 (1955).
[2] G. DE VAUCOULEURS: Observatory **74**, 23 (1954).
[3] G. DE VAUCOULEURS: Astronom. J. **62**, 69 (1957). — Publ. Astronom. Soc. Pacific **68**, 421 (1956).
[4] A. R. HOGG: Monthly Notices Roy. Astronom. Soc. London **155**, 473 (1955).
[5] G. DE VAUCOULEURS: Irish Astronom. J. **4**, 13 (1956).

extreme (i.e. younger) than the Small Cloud. The latter view is supported by the greater quantities of ionized hydrogen and dust associated with the Large Cloud. Although globular clusters and cluster-type variables of Population II have been observed in both Clouds, the proportion of such stars is small. As an indication of the mean population of the Clouds the mass-luminosity ratio should be of significance. This has been obtained by KERR and DE VAUCOULEURS[1] from the total masses derived from the H line observations (Sect. 4) and the total magnitudes above. It is between 1 and 2 for the Large Cloud and of the order of 2 for the Small Cloud. Both Clouds therefore appear to be very late-type systems; by comparison the mass-luminosity ratio of M33, an *Sc* galaxy, has been estimated to be between 2 and 3 and that of M31, an *Sb* galaxy, to be about 12.

8. Radio-optical comparisons. The radio-emitting efficiency of a galaxy is defined most simply by the difference of radio and optical magnitudes. For the Clouds we have:

$$\text{LMC} \quad m_{1.9} - m_p = 2.8,$$

$$\text{SMC} \quad m_{1.9} - m_p = 3.0.$$

The values are rather uncertain, particularly for the Small Cloud, but they are consistent with the assumption that both Clouds have a similar efficiency of radio emission. Comparisons with other galaxies are made in Sect. 10.

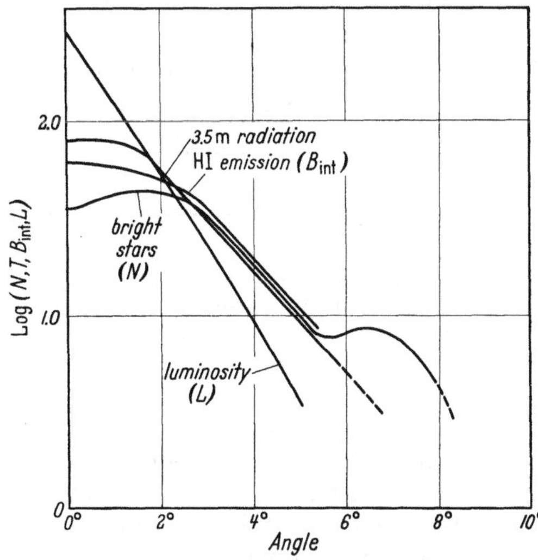

Fig. 10. A comparison of the average radial distributions of light, 3.5 m radiation, 21 cm hydrogen line radiation and bright stars in the Large Magellanic Cloud.

The radio resolution is sufficient to compare in detail the distribution of various elements of the Large Cloud [*5*]. The mean radial distribution of the light emission, the number density of bright stars ($m < 14$), the neutral hydrogen and the 3.5 m radio emission, as measured from the optical centre, are compared in Fig. 10 as logarithmic functions with arbitrary zeros. It is clear that the distributions of neutral hydrogen, 3.5 m radio emission and bright stars are all quite similar; in particular the logarithmic decrements between radii of 2° and 6° are identical. On the other hand the light distribution is much more sharply concentrated towards the centre and between 2° and 6° the logarithmic decrement is steeper.

The detailed distributions of these elements have been intercompared by DE VAUCOULEURS[2]: his results are summarized in Table 2. He divides the Cloud into various characteristic regions to each of which is assigned a qualitative estimate of the concentration of each element.

Inspection of the table reveals that the closest correspondence is between the neutral hydrogen and the 3.5 m emission, with the bright stars also showing some correspondence with the latter. The correspondence between light and 21 cm emission is also good, but between light and both the 3.5 m emission and bright stars is very poor. One might almost say that the neutral hydrogen occupies an

[1] F. J. KERR and G. DE VAUCOULEURS: Austral. J. Phys. **9**, 90 (1956).
[2] G. DE VAUCOULEURS: Ref. [7], p. 244.

intermediate position with light at one extreme and bright stars at the other, the 3.5 m emission fitting somewhere between the neutral hydrogen and bright stars.

Similar detailed comparisons cannot be made convincingly with the Small Cloud because of the greater tilt of the main system and the great amount of distortion exemplified by the prominence. In addition, the 3.5 m emission is much weaker and its separation from the background irregularities more difficult. The "prominence" is particularly interesting. It appears most pronounced in neutral hydrogen: on the northern side of the hydrogen prominence there is a possible prominence of the 3.5 m emission and on the southern a concentration of bright stars forming the visible prominence discovered by Shapley. The concentration of "light" in this region is very small.

Table 2. *The concentration of various elements in the Large Cloud (after de Vaucouleurs).*

Scale: 4 very strong concentration; 3 strong concentration; 2 some concentration; 1 present but no concentration; 0 very weak or absent.

Region in *LMC*	Light	Bright stars	HI	3.5 m emission
Axial bar	4	0	3	1
Region around 30 Doradus	3	4	3	4
Bright inner spiral arms	3	4	3	2
Dark inter-arm spaces	1	0	3	2
Faint outer arms	3	1	1	0

Summing up these comparisons it appears that, while none of the populations investigated show a one-to-one correspondence, and therefore do not have a direct physical relationship, the 3.5 m emission has a distribution more akin to the interstellar gas and the bright stars than the visible light. It thus appears to be related to Population I.

II. Neighbouring bright galaxies.

9. The Andromeda nebula. Practically all the neighbouring bright galaxies with magnitudes brighter than 9, and some fainter ones, have now been investigated for the presence of continuous radio emission[1]. In the majority of cases the only information is an indication of the presence or absence of emission and, if present, a rough estimate of its intensity; but for two of the brightest, M 31 (the Andromeda nebula) and NGC 5128, there have been extensive observations of brightness distributions and spectra. The latter galaxy is, however, quite abnormal, both in photographic appearance and in the amount of radio emission: it is discussed later in Chap. V under "Radio Galaxies" p. 267).

The Andromeda nebula was the first of the "normal" galaxies to be detected and also the first discrete radio source for which radio brightness distributions were obtained. This double observation, a landmark in the science of radio astronomy, was made in 1950 by Hanbury Brown and Hazard with a 67 m diameter parabolic radio telescope[2]. They found that the radiation from the

[1] Very recently observations have also been reported of H-line radiation from the bright galaxies M 31, M 33, M 51 and M 81. H.C. van de Hulst and E. Raimond: Afd. Naturrk. **65**, 157 (1956). — N.H. Dieter: Publ. Astronom. Soc. Pacific **69**, 356 (1957). — D.S. Heeschen: Astrophys. Journ. **126** (1957) (in press).

[2] R. Hanbury Brown and C. Hazard: Monthly Notices Roy. Astronom. Soc. London **111**, 357 (1951).

nebula, although weak, was quite easily detectable[1] and extended over an angle greater than their beamwidth of 2°.

In order to obtain the distribution across the nebula, traverses were made at seven separate declinations separated by about $\frac{1}{2}$°. It was found that the galaxy is situated in a region where the galactic gradient is rather high so that it was difficult to obtain an accurate brightness distribution. However, by making a correction for the gradient in declination, they were able to construct the brightness contours shown in Fig. 11: the dotted ellipse represents approximately the optical dimensions of the main body of the nebula. It would seem from the diagram that the radio emission is distributed quite differently from the optical, although the obvious distortion due to galactic irregularities reduces the significance of this difference. HANBURY BROWN and HAZARD concluded that the

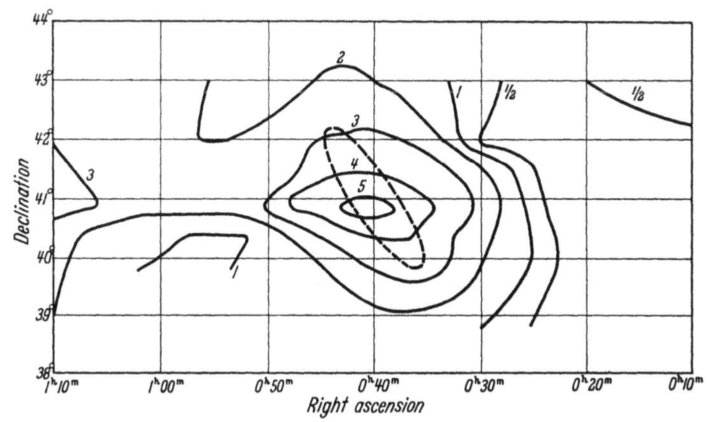

Fig. 11. A contour map of the 1.9 m isophotes in the vicinity of the Andromeda Nebula (HANBURY BROWN and HAZARD). The dotted ellipse represents approximately the optical dimensions of the main body of the nebula.

angular dimensions of the nebula appeared roughly equal in right ascension and declination but, owing to the uncertainties in the observations, were unwilling to infer that the radio and optical distributions differed. However, in 1952 SHKLOVSKII, as the result of an analysis of Milky Way surveys, suggested that the Galaxy was surrounded by a vast quasi-spherical "corona" of radio emission and pointed to the observations of M 31 as additional evidence for the presence of such a corona in the radio emission of a galaxy[2]. In view of this correspondence it seems safe to conclude that the large East-West extension of the nebula is real.

Two other radio observations of the Andromeda Nebula have subsequently been made with somewhat lower resolution. Both support the above results in suggesting that the radio emission is distributed much more widely than the optical, but galactic irregularities are likely to affect both more seriously for quantitative measures. BALDWIN used the Cambridge interferometer and measured some Fourier components of the radio brightness distribution along various axes: he deduced an East-West brightness distribution and compared it with distributions of light and mass, concluding that the radio emission was more widely distributed than either and approximately spherical in shape[3].

[1] It is possible that radiation from the Andromeda Nebula had been detected earlier in a survey of radio sources although the position of the source in question was about 3° different from that of the Nebula. M. RYLE, F. G. SMITH and B. ELSMORE: Monthly Notices Roy. Astronom. Soc. London **110**, 508 (1950).

[2] I. S. SHKLOVSKII: Astr. J. Moscow **29**, 418 (1952).

[3] J. E. BALDWIN: Nature, Lond. **174**, 320 (1954).

KRAUS used a fan beam (1°.2×8°) and also concluded that the radio emission was more widely distributed than the optical, but did not attempt to derive the axial ratio[1].

All observers estimated the flux density of M 31; their results are given in Table 3. Within the expected errors of about 20% these observations are consistent and may be fitted roughly to the spectral law assumed for the Magellanic Clouds, i.e. $S \propto \lambda^{0.7}$.

The flux density of HANBURY BROWN and HAZARD at 1.9 m will be adopted, which leads to a radio magnitude, $m_{1.9}$, of 6.0.

Table 3. *Radio observations of the Andromeda Nebula.*

Observers	Wavelength m	Flux density 10^{-24} W m^{-2} (c/s)$^{-1}$
BALDWIN	3.7	2.0
HANBURY BROWN and HAZARD	1.9	1.6
KRAUS	1.2	1.0

10. Other bright galaxies. A further 21 bright galaxies have been examined for radio emission and, of these, 13 have been detected. The observations have been made in England by HANBURY BROWN and HAZARD using the 67 m fixed parabolic reflector which has a limited coverage between declinations of +38° and +68° [3] and in Australia by MILLS using the 450 m cross-type radio telescope with a coverage between about +10° and −80° [5].

There is a region between +10° and +39° not covered by either instrument, and here unfortunately lies the fourth brightest galaxy, M 33. A Cambridge survey of radio sources includes this region but no source is listed at the position of the galaxy[2]. The omission is not significant, however, as the detection of a source of the expected intensity (~10^{-24} Wm^{-2} (c/s)$^{-1}$) and angular size (~1°) would not necessarily be expected with the Cambridge instrument which has a much reduced sensitivity for sources of large angular size.

It is found that all 16 of the bright galaxies of a late type (*Sb* and later) which have been investigated are detectable radio emitters. On the other hand none of the 8 early-type galaxies (up to *Sab*) have been detected. It is necessary to differentiate here between these observations, in which the galaxies have been chosen for separate study in the basis of their optical brightness, and surveys of radio sources in which lists of sources are prepared and afterwards investigated for coincidences with galaxies. In the latter case there will be a selection of the galaxies of abnormally great radio emission, which should not be used for establishing the mean emission of the class of galaxies to which they belong. The majority of such "radio galaxies" manifest optical abnormalities which support their exclusion from the class of "normal" galaxies, in others the abnormality is not so obvious, e.g. NGC 1316, and their exclusion is largely a matter of common sense; finally there may sometimes be a real doubt as to whether the radio emission of a galaxy detected in a source survey is normal or not but here, for consistency, such cases will not be considered.

All the observations of bright galaxies, including the Clouds and M 31 are summarized in Table 4. Those of particular interest noted in the table are descri-

[1] J. D. KRAUS: Nature, Lond. **175**, 502 (1955).

[2] J. R. SHAKESHAFT, M. RYLE, J. E. BALDWIN, B. ELSMORE and J. H. THOMSON: Mem. Roy. Astronom. Soc. **67**, 97 (1955).

bed in detail below. The galaxies are arranged in order of increasing radio magnitude, and the flux densities quoted refer to the wavelength at which the observation was made, 3.5 m for the southern galaxies and 1.9 m for the northern galaxies; the radio magnitude is given at 1.9 m, a spectral and calibration adjustment being applied for the longer wavelength observation as in Sect. 6. Photographic magnitudes are the *total* magnitudes taken from compilations of DE VAUCOULEURS[1].

Table 4. *A summary of observations of bright galaxies.*
The standard classification is given in column (a), that of DE VAUCOULEURS (p. 276) in column (b).

Galaxy	Classification a	Classification b	Flux density W m^{-2} (c/s)$^{-1}$ $\times 10^{-26}$	$m_{1.9}$	m_p	$m_{1.9} - m_p$	Remarks
LMC	*I*	*SB(s) m*	4000	3.4	0.6	2.8	
SMC	*I*	*SB(s)* m_p	600:	5.5:	2.5	3.0:	
M 31	*Sb*	*SA(s) b*	160[2]	6.0	4.0	2.0	
NGC 4945	*SBc*	*SB(s) c*	30	8.7	7.8:	0.9:	see below
NGC 1068 (M 77)	Sb_p	*(R) SA (rs) b*	30	8.7	9.6	−0.9	see below
NGC 5236	*SBc*	*SAB(s) c*	28	8.8	7.5:	1.3:	
NGC 253	*Sc*	*SB(s) c*	23	9.0	7.5	1.5	sample record, Fig. 12b
NGC 3031	*Sb*	*SA(s) ab*	12[2]	8.9	7.75	1.1	
NGC 891	*Sb*	*SA(s)* b_{sp}	10:[2]	9.0:	10.7	−1.7:	see below
NGC 5194/5 (M 51)	*Sc*	*SA(s) bc*	5.5[2]	9.7	8.5	1.2	detected by averaging records
NGC 4258	*Sb*	*SAB(s) b*	5.0[2]	9.8	9.1	0.7	detected by averaging records
NGC 300	*Scd*	*SA(s) cd*	10:	9.9:	8.5:	1.4:	see below
NGC 6744	*SBbc*	*SAB(r) bc*	7.5	10.2	9.1	1.1	
NGC 2841	*Sb*	*SA(r?) b*	[2]	10.4:	10.2	0.2:	see below
NGC 55	*I*	*SB(s)* m_{sp}	4.0	10.9	7.8	3.1	see below
NGC 5457 (M 101)	*Sc*	*SAB(rs) d*	< 10[2]	> 9.1	8.1	> 1.0	see below
NGC 205	*E 5*	E^+Sp	[2]		8.9	> 0.1:	not detected, see below
NGC 221	*E 2*	*dE*[2]			9.1	> 0.6	not detected, see below
For	*dE*	*dE*	< 9.0	> 10.0	7.5:	> 2.5:	not detected
IC 5267	*Sa*	*SA(s) a*	< 4.0	> 10.9	10.8	> 0.1	not detected, see below
Scul	*dE*	*dE*	< 3.8	> 11.1	7.1:	> 4.0:	not detected
NGC 1291	*SBO*	*(R) SB(s)*O^+	< 3.4	> 11.2	9.4	> 1.8	not detected
NGC 4594	*Sab*	*SA(s) ab*	< 2.8	> 11.3	8.9	> 2.4	not detected
NGC 3115	*E 7*	E^+7	< 2.5	> 11.4	10.15	> 1.2	not detected

NGC 4945. This is a late-type galaxy seen edge on. Because of its low galactic latitude (+12°) absorption effects are uncertain and possibly large.

NGC 1068. The photographic magnitude of this galaxy is known accurately, but there is some uncertainty in the radio magnitude because of a possible contribution from the neighbouring galaxy NGC 1055 which is not separately resolved by the radio telescope. Assuming the radio emission is divided between the two galaxies in the ratio of their optical brightness, the radio magnitude of NGC 1068

[1] G. DE VAUCOULEURS: A revision of the Harvard Survey of Bright Galaxies. Australian National University Mimeograph, 1952/53. — Astronom. J. **61**, 430 (1956). — Ann. Obs. Houga II, No. 1 (1957).

[2] Measurements made at 1.9 m wavelength: unlabelled flux densities were measured at 3.5 m wavelength.

would be raised from 8.6 to 8.9. The efficiency of radio emission, defined by $m_{1.9}-m_p$, would still be rather high compared with the other galaxies observed. This is not altogether unexpected as it is a galaxy of rather unusual type displaying strong and broad emission lines in the nucleus[1]: there is clearly some doubt as to whether it should be classified as "normal".

NGC 891. The radio observation is not completely reliable because of the proximity of a strong source (source No. 5 in the list of Hanbury Brown and Hazard[2]) and because the galaxy is located on the edge of a cluster which itself may be a radio source. It appears that the radio emission is surprisingly large. However the galaxy is seen edge-on with a pronounced absorbing band crossing the nucleus and self absorption may reduce the light by as much as one magnitude. The final value of $m_{1.9}-m_p$ cannot be accorded very much weight.

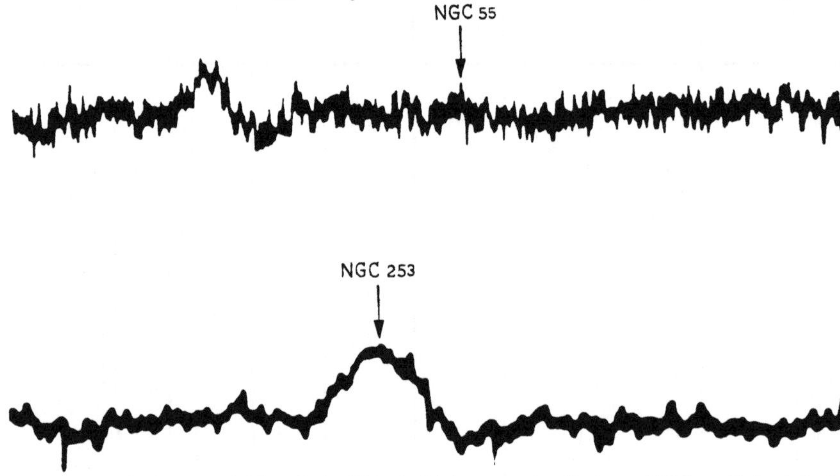

Fig. 12. Records obtained on some bright galaxies (Mills). (a) A very weak radio source, NGC 55. (b) A relatively strong radio source, NGC 253.

NGC 300. This galaxy is in a region of many weak radio sources and it is therefore difficult to estimate the radio emission or even to be sure of the identification. The radio source appears superimposed on a weaker source of large angular size and its apparent position is about 10′ east of the galaxy, but the displacement could well be due to insufficient resolution. The photographic magnitude is also uncertain as the galaxy has a very low surface brightness and the published measures differ widely.

NGC 2841. There appears to be no detailed description of the radio measurements of this galaxy, the only reference being a statement of its radio magnitude[3]. Since it is a very weak source on the Manchester scale, it was presumably detected by averaging of records. Resolution effects are probably becoming very important at this level with the 67 m parabola.

NGC 55. This galaxy was near the limit of detection with the Sydney instrument and several observations were required to be sure of its existence: a sample record is shown in Fig. 12a. There is some difficulty in the classification of the galaxy, as it is seen nearly edge-on, but the photographic evidence for a Magellanic type appears quite strong and this classification has therefore been adopted.

[1] C. K. Seyfert: Astrophys. Journ. **97**, 28 (1943).

[2] R. Hanbury Brown and C. Hazard: Monthly Notices Roy. Astronom. Soc. London **113**, 123 (1953).

[3] U.R.S.I. Special Report No. 3; Discrete sources of extraterrestrial radio noise. 1954.

NGC 5457. HANBURY BROWN and HAZARD remark that a radio source was observed near the position of this galaxy but it was so close to a more powerful source (IAU 14N5A) that measurement of the flux density was impossible and only an upper limit could be given.

NGC 205, 221. No indication of radio emission has been found from these two companions of the Andromeda nebula. Actual limits of flux density have not been published, only values of $m_p - m_{1.9}$[1]. The quoted values of $m_{1.9} - m_p$ were obtained by correcting to the present photographic magnitude scale.

IC 5267. This was earlier listed [5] as an uncertain but possible identification. However it was later found that the radio source was about $\frac{1}{2}°$ distant from the galaxy and therefore probably unassociated. The observation of such a faint galaxy was attempted in the first place because of a similarity in appearance to NGC 5128 and NGC 1316 (both "radio galaxies") which had been noted by EVANS[2].

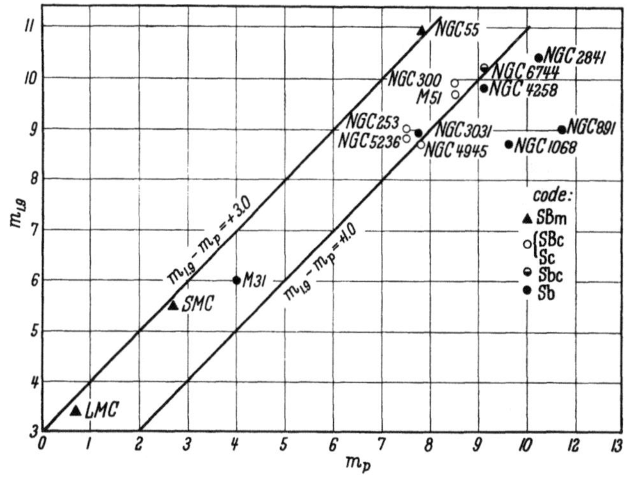

Fig. 13. A comparison of the radio and optical emission for all the normal galaxies detected. Lines defined by $m_{1.9} - m_p = 3$ and $m_{1.9} - m_p = 1$ are shown, representing approximately the detectable categories.

III. The radio emission of normal galaxies.

11. The dependence of emission on galaxy type. There are, at present, insufficient observations to determine accurately the radio emission of different types of normal galaxies. However, it does appear that they may be divided into three main categories; (i) The magellanic types, of relatively low emission, which have values of $m_{1.9} - m_p$ around 3; (ii) the *Sb* and *Sc* galaxies, including barred spirals which have a mean $m_{1.9} - m_p$ of 0.7 ± 0.2 if all such galaxies are included, and 1.1 ± 0.1 if one excludes the galaxies NGC 1068 and NGC 891 on the grounds of optical abnormality and uncertain identification respectively; (iii) the early type galaxies which are undetectable. Previously a distinction had been made between *Sb* and *Sc* galaxies since there appeared to be a small difference between their radiating efficiencies [5]: with the correction of the earlier radio magnitude scale for the southern galaxies this difference is no longer significant. In Fig. 13 the radio magnitudes are plotted against the photographic for all the galaxies detected; lines defined by $m_{1.9} - m_p = 3$ and $m_{1.9} - m_p = 1$ are shown, representing approximately categories (i) and (ii).

[1] See footnote 2, p. 245.
[2] D. S. EVANS: Monthly Notices Roy. Astronom. Soc. London **109**, 94 (1949).

The failure to detect radiation from *any* early-type galaxy is very significant although some of the individual observations have little weight themselves, because of insufficient sensitivity. When coupled with the failure to detect radiation from *any* globular cluster (two crucial observations are 47 Tucanae, $m_{1.9}-m_p>7.5$, and NGC 362, $m_{1.9}-m_p>4.2$ [5], we may conclude that objects of Population II probably do not contribute to the cosmic radio emission by any measurable extent.

To summarize, it is found that, among those investigated, only galaxies with a substantial proportion of Population I are detectable radio emitters. Also, it was concluded in Sect. 8 that the distribution of radio brightness in the Large Magellanic Cloud displayed characteristics of this population. The galaxy NGC 4594, an undetectable *Sab*, has some Population I, as dust is conspicuous and emissions regions present, but the Population II in the huge nucleus contributes practically all the light: it would seem to be a marginal case where greater sensitivity is required. The evidence appears to point to the necessary presence of Population I in a galaxy for radio emission to occur in normal amounts. However, there is no direct correlation, for the magellanic types which have a very much higher proportion of Population I than the *Sb* galaxies are substantially weaker radio emitters and, in the Large Cloud, the distribution of radio brightness cannot be directly correlated with the visible components of the population.

Before discussing these problems in detail, let us first review briefly some of the characteristics of radio emission from the Milky Way.

12. Radio emission from the Milky Way. As in optical astronomy, an understanding of the radio properties and structure of external galaxies requires a study of our own galaxy, the Milky Way. The process works both ways because the Milky Way, although it is unique in the amount of detail which can be observed, is too close to obtain easily an adequate overall picture. The radio Milky Way is discussed at length in OORT's article: here, discussion is confined to those features which are particularly relevant to the observations of external galaxies.

The early surveys of galactic radio emission resulted in much confusion, due to the lack of appreciation of the effects of finite resolution in a radio telescope; these effects were particularly marked at the longer wavelengths. Now, however, with the advent of high resolution pencil-beam aerials at all wavelengths the picture is much clearer. It is apparent that there are two *principal* subsystems of radio emission, one flattened and confined closely to the plane of the Galaxy, the other forming a very extensive quasi-spherical "corona". These components were first recognized by SHKLOVSKII in 1952[1], but their presence was not conclusively demonstrated until several years later: the exact nature and structure of the subsystems, however, has not yet been elucidated. The possibility of further subdivision of the coronal component, mentioned by OORT (p. 100), is not relevant to the present discussion.

The two subsystems are demonstrated in Fig. 14, which shows some galactic sections at longitudes near the galactic centre. These are provisional sections obtained with the cross-type radio telescope at Sydney[2]: this instrument has a pencil beam response of 50′ of arc between half-power points so that the narrow belt of radiation close to the galactic plane is completely resolved. Minor irregularities in the coronal distribution have been smoothed out since at this stage it is uncertain whether they may not be the result of small calibration differences between observations at different declinations. The asymmetry in the coronal

[1] I. S. SHKLOVSKII: Astr. J. Moscow **29**, 418 (1952).
[2] B.Y. MILLS, E.R. HILL and O.B. SLEE: Observatory **78**, 116 (1958).

component is a feature of all surveys. Because the distribution of radiation along the galactic plane is very irregular, reliable estimates of the width of the belt cannot be obtained by examination of a few such galactic sections. However, a detailed map of an area near the galactic centre has been prepared[1], and this suggests a width between half brightness contours of about three or four degrees, corresponding to a distance between half density points of about 500 pc near the galactic nucleus. It is difficult to obtain the thickness of the wider distribution because of the unknown value of the integrated extragalactic radiation which presumably forms a uniform background. It has been estimated that the width between half-brightness contours is of the order of 60° or 70° corresponding to a thickness near the galactic centre of about 10 kpc; the emission undoubtedly extends much further in an attenuated form [5].

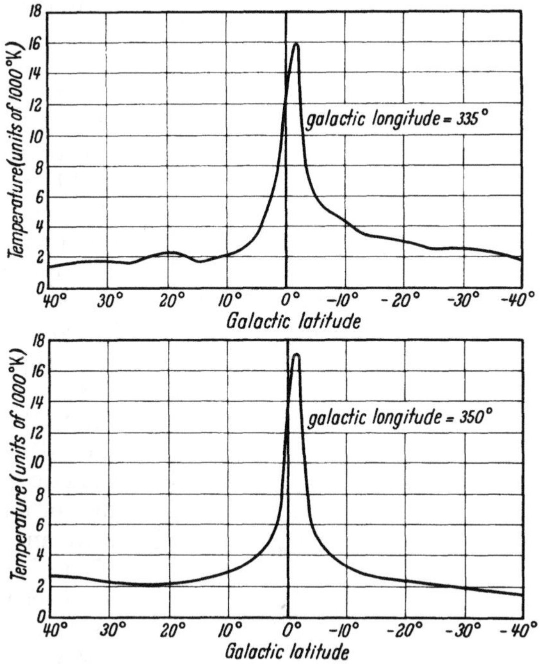

Fig. 14. Some provisional sections near the galactic centre at 3.5 m wavelength obtained with the Sydney cross-type radio telescope.

There is general agreement that the spectrum of this wide distribution, or corona, is non-thermal, the index defining the brightness spectrum, $B \propto \lambda^n$, lying between 0.5 and 1.0 at wavelengths of a few metres. At longer wavelengths, however, it appears that the rate of increase slows down and perhaps eventually reverses. In Fig. 15 are shown some observations by ELLIS of the brightness near the galactic centre and the south galactic pole[2]: they were made in Tasmania where the critical frequencies of the ionosphere are very low, and reliable observations can sometimes be made at frequencies as low as one megacycle. The brightness is constant between wavelengths of 30 and 60 m.

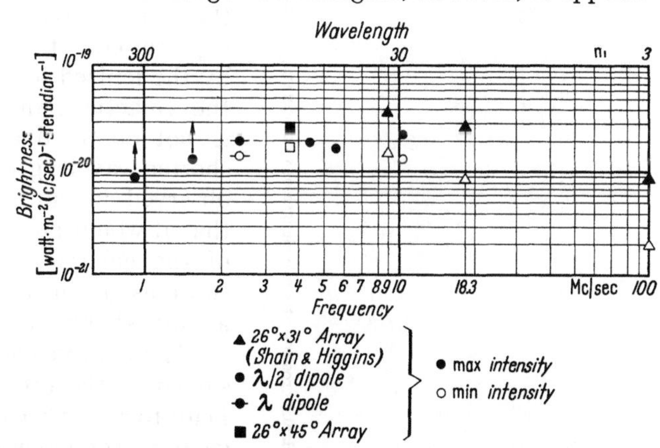

Fig. 15. The spectra of the sky brightness near the galactic centre and the south galactic pole (ELLIS). Maxima correspond to observations near the centre, minima to observations near the pole.

The situation regarding the spectrum of the "disk" subsystem is not so clear because individual measurements depend critically on the resolution of the radio

[1] B. Y. MILLS: Observatory 76, 65 (1956).
[2] G. R. ELLIS: J. Geophys. Res. 62, 229 (1957).

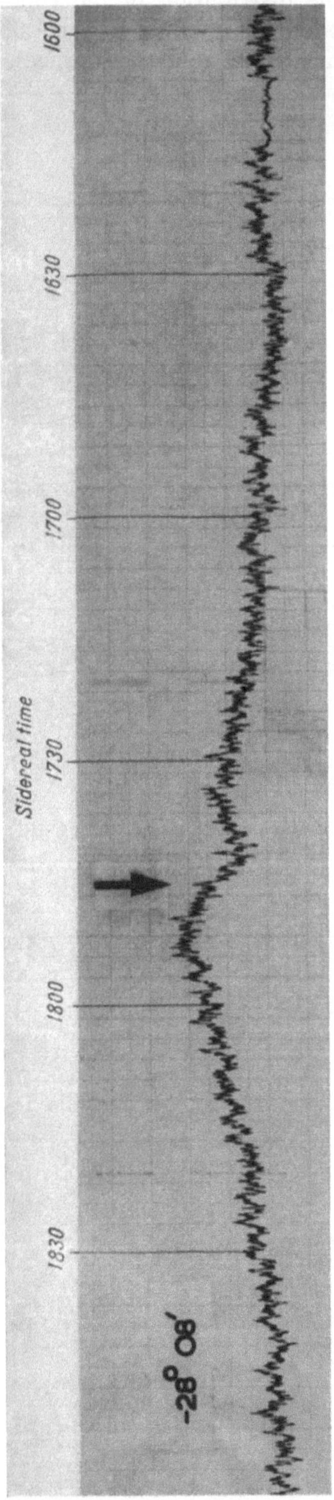

Fig. 16. A record of a galactic crossing at a wavelength of 15 m (SHAIN). Heavy absorption due to ionized hydrogen may be seen at a right ascension of approximately $17^h 40^m$. The arrow indicates a galactic latitude of $-1°.3$.

telescope. It was originally suggested by SHKLOVSKII[1] that the emission in this belt arises in thermal emission of the ionized hydrogen distributed throughout the Galaxy: this would have a spectrum which is either constant or a decreasing function of wavelength. However, results obtained with the Sydney cross-type radio telescope suggested that the emission at 3.5 m wavelength is too high for this to be true [*5*]: also the emission appears to be much more widely distributed than the ionized hydrogen. This question now appears to have been resolved in favour of a non-thermal origin by some measurements of SHAIN using a cross-type radio telescope with a beamwidth of 1°.4 at 15 m wavelength[2]. At this wavelength the general background radiation from the galactic corona plus the integrated extragalactic emission has a brightness temperature of more than 10^5 °K: ionized hydrogen with an electron temperature of the order of 10^4 °K is therefore observed everywhere in absorption. A typical record of a galactic crossing obtained by SHAIN is shown in Fig. 16; it appears that a narrow absorption feature is superposed on a wider region of *excess* emission. The latter corresponds in width at the base to the narrow peaks of the disk distribution of Fig. 13 which must therefore be non-thermal in origin. The former is presumably the result of absorption caused by the ionized hydrogen clouds very close to the galactic plane; in general this feature is not observed on the 3.5 m sections. At shorter wavelengths the hydrogen should be observed in emission, but SHAIN concludes that it would not be a significant component of the galactic emission at wavelengths as short as 0.5 m, although it may become so at still shorter wavelengths.

The relative contributions of the two subsystem to the total galactic emission are difficult to assess because of the unknown extragalactic component, but there is no doubt that, because of the great difference in spatial extent, the coronal contribution far exceeds that of the disk. Using "best estimates" for the parameters of the two distributions the emission may be calculated

[1] See footnote 1, p. 256.
[2] C. A. SHAIN: Austral. J. Phys. **10**, 195 (1957).

roughly. It is easily shown that the observed flux density of a uniformly emitting transparent spheroid at distance D is given by

$$S = \frac{\pi}{3} \cdot \frac{kT}{\lambda^2} \left(\frac{d}{D}\right)^2 \alpha \tag{12.1}$$

where λ is the wavelength, k is BOLTZMANN's constant, α is the axial ratio, d is the thickness of the spheroid measured along the axis of rotation, and T the brightness temperature observed through an equatorial diameter.

Consider the Milky Way at the distance of the Andromeda Nebula, i.e. 500 kpc, and assume that the distribution of volume emissivity is uniform to the half density regions and zero beyond. Best estimates for the disk parameters at a wavelength of 3.5 m are, $T=15000°$ K, $\alpha=15$, $d=500$ pc, yielding a flux density of 2.7×10^{-25} W m^{-2} (c/s)$^{-1}$. For the corona the corresponding "best estimates" are $T=4000°$ K, $\alpha=1.5$, $d=10^4$ pc, yielding a flux density of 2.8×10^{-24} W m^{-2} (c/s)$^{-1}$. Thus the ratio of the two components is of the order of 10, and the total flux density about 3×10^{-24} W m^{-2} (c/s)$^{-1}$, comparable with that of the Andromeda Nebula itself.

13. Interpretation of the normal galaxy observations. In view of the paucity of the data, any interpretation of the foregoing observations must be rather speculative, but it is natural to assume that the two subsystems of radio emission present in the Milky Way find their counterparts, perhaps in different proportion, in other galaxies. Neither of these subsystems appears to be *directly* related to any optical component of a galaxy. The corona, although superficially resembling the corona of globular clusters of Population II, displays much less central concentration; moreover, early-type galaxies rich in Population II are not detectable as radio emitters. The disk resembles somewhat a Population I distribution but it appears rather too thick, the thickness being about 500 pc compared with about 100 pc for the O and B stars and 250 pc for the low-velocity neutral hydrogen. It is known, however, that high-velocity hydrogen exists in a much wider distribution.

In the Large Magellanic Cloud the distribution of the 3.5 m radio emission has similarities to that of the neutral hydrogen and bright stars, suggesting a general relationship to the Population I component of the galaxy. Since the total emission is less than that of Sb and Sc galaxies by about 2 magnitudes, i.e. by about the ratio of the emission of the coronal and disk subsystems in the Milky Way, it seems possible that the main emission is of the disk type, the corona being weak or absent; this picture is consistent with a relationship between the disk and Population I components. In the Small Cloud the distributions are obviously very distorted by the large prominence, but the 3.5 m emission boundaries agree closely with the neutral hydrogen boundaries over wide areas and nowhere do they significantly exceed them. The total emission is also much less than for Sb and Sc galaxies and, again, a reasonable interpretation seems to be that the coronal component is weak or absent.

The contours of M31 obtained by HANBURY BROWN and HAZARD show no evidence of a flattened subsystem so that it appears the spherical subsystem must predominate, as in the Milky Way: the similarity in the galaxies is further enhanced by the near equality of their total emissions demonstrated in the last section. From the similar values of $m_{1.9}-m_p$ for all the Sb and Sc galaxies it would appear that their main source of emission is a similar type of corona. In early-type galaxies the presence of the flattened subsystem, with its Population I characteristics, would not be expected, but the lack of observable emission in *any* such galaxy investigated suggests that the coronal emission is also weak or absent. It is clear that the total radiation of a galaxy at the longer wavelengths cannot be ascribed to a subclass of objects belonging to Populations I or II.

Following his suggestion of the existence of a radio corona, SHKLOVSKII, in 1953, advanced the theory that the emission originated in a vast quasi-spherical distribution of tenuous matter and magnetic fields in which high energy electrons ($\sim 10^9$ eV) were radiating by the synchrotron mechanism[1]. This theory received substantial support when a similar suggestion of SHKLOVSKII[2], concerning the origin of both the optical and radio emission of the Crab nebula, was confirmed by DOMBROVSKII[3] and by OORT and WALRAVEN[4]. As the result of a tentative analysis of physical conditions in the corona based on equilibrium concepts, MILLS suggested in 1955 a possible maximum in the radio emission at a wavelength of 60 m [5]: this is just the range in which ELLIS has now found the sky brightness ceases in increase and perhaps begins to decrease (see Fig. 15). Several other possible models have been advanced, however, and at present no firm conclusion can be drawn concerning the physical conditions, or the origin of the high energy electrons[5]. Further discussion is given in the articles by OORT and HANBURY BROWN.

The origin of the flattened subsystem is even more obscure, although it appears that it cannot originate in the thermal emission of ionized hydrogen. The basic process is also, very probably, that of synchrotron type emission of energetic electrons, but it is not known if the band of radiation is caused by the integrated emission of many individual radio sources or whether it arises in a diffuse, more or less homogeneous medium. It can be shown that the integrated emission of type I supernovae (Population II objects) is inadequate to account for the observed total emission, but there seems a distinct possibility that the much more numerous type II supernovae, perhaps exemplified by the intense source in Cassiopeia, IAU 23 N 5 A, are the basic sources[6]. These objects, being of Population I, presumably occur frequently in the midst of interstellar gas concentrations where the conditions would appear promising for strong radio emission. They are probably more concentrated towards the galactic plane than the observed radio emission and perhaps correspond to the band of Class I radio sources[7] discussed in the article by HANBURY BROWN: one might then interpret the wider band of emission as the result of the dispersion through the galactic disk of energetic electrons produced by the outbursts. The problem may be finally solved by the analysis of data obtained from the many high resolution radio telescopes now in operation.

IV. Radio emission from clusters of galaxies.

14. Observations of localized clusters. As spiral galaxies are known to be radio emitters, it might be expected that dense clusters of galaxies, which usually contain an appreciable proportion of spirals, should be detectable radio sources at distances far exceeding the limit of detectability of their individual members. HANBURY BROWN and HAZARD were prompted by their observations of individual galaxies to look for such radiation from some dense localized clusters. Two

[1] I. S. SHKLOVSKII: Astr. J. Moscow **30**, 15 (1953).
[2] I. S. SHKLOVSKII: Dokl. Akad. Nauk USSR. **90**, 983 (1953).
[3] V. A. DOMBROVSKII: Dokl. Akad. Nauk USSR. **94**, 1021 (1954).
[4] J. H. OORT and TH. WALRAVEN: Bull. astronom. Inst. Netherl. **12**, 285 (1956).
[5] G. R. BURBIDGE: Astrophys. J. **123**, 178 (1956). — L. SPITZER: Astrophys. J. **124**, 20 (1956).
[6] B. Y. MILLS, A. G. LITTLE and K. V. SHERIDAN: Austral. J. Phys. **9**, 84 (1956).
[7] B. Y. MILLS: Austral. J. Sci. Res. A **5**, 266 (1952).

were detected, the Perseus and Ursa Major II clusters[1] [2] a third source close to the cluster centred on NGC 911, which was originally identified with the cluster, was later found to be more appropriately identified with the galaxy NGC 891; it is discussed in Sect. 10.

The two clusters detected both radiate substantially more than expected on the assumption that all the cluster members have a radio emission similar to the normal spirals, that is $m_{1.9}-m_p=+1.0$. The value of $m_{1.9}-m_p$ for the Perseus cluster is -1.4, and for the Ursa Major II cluster, -1.9[2]. The discordance is further increased when it is remembered that the emission of the early-type galaxies, which usually predominate in a dense cluster, is much less or absent.

After the definite identification of a number of intense sources with galaxies radiating many orders of magnitude more than the "normal" galaxies, it appeared that this excess cluster emission could well arise in one or more abnormal members. It was pointed out by BAADE and MINKOWSKII that in the Perseus cluster the brightest member, NGC 1275, had long been recognized as highly abnormal, suspected of representing a direct collision between galaxies: they suggested that the major emission of the cluster was arising in this one galaxy [1]. This suggestion was confirmed when BALDWIN and DEWHIRST were able to measure the position of the radio source more accurately and to demonstrate that it coincided closely with NGC 1275, and further, that the radio source had a small angular size compared with the cluster as a whole[3]. Making the assumption that the source corresponding to the galaxy NGC 1275 had a negligible angular size, they were able to determine the relative contributions from this one galaxy and the remainder of the cluster. They found that the cluster accounted for about $\frac{1}{5}$ of the total radiation, an upper limit in view of the assumption made. The value of $m_{1.9}-m_p$ for the cluster is therefore $+0.3$ or higher.

To supplement these limited northern observations, a survey of the most likely southern clusters has been made with the 450 m Sydney cross-type radio telescope[4]. Five were examined and radio sources found close to four, but in only two cases does it seem possible that the integrated radiation from the cluster is being detected. The observations are discussed below.

Cluster in Cetus. The position of the centre of this cluster is given by SHAPLEY[5] as $01^h06^m.2$, $-15°54'$ (1950). A radio source of flux density 5.1×10^{-25} $\mathrm{W\,m^{-2}\,(c/s)^{-1}}$ was found at a position of $01^h05^m.0$, $-16°19'$ (1950): the source was unresolved indicating an angular size less than about $\frac{1}{3}°$ compared with a size of about 1° for the cluster. Since the probable errors in the radio position are only a few minutes of arc and the discordance nearly half a degree, it is clear that the source cannot be identified with the integrated emission of the cluster. There are no obvious cluster members within the limits of error of the radio position so that an association between the cluster and the source does not appear very likely.

Cluster in Dorado. This cluster, approximately a degree in diameter[6], has its centre at $04^h28^m.6$ $-53°45'$ (1950), whilst a radio source, of flux density 5.8×10^{-25} $\mathrm{W\,m^{-2}\,(c/s)^{-1}}$ is located at $04^h27^m.8\pm0^m.2$, $-53°59'\pm3'$ (1950) and is of small angular size. Thus an association with the integrated emission of the

---

[1] R. HANBURY BROWN and C. HAZARD: Monthly Notices Roy. Astronom. Soc. London **113**, 123 (1953).

[2] The photographic magnitudes used by HANBURY BROWN and HAZARD have been corrected by -0.7 magnitude for consistency with the different standards used elsewhere in this article.

[3] J. E. BALDWIN and D. W. DEWHIRST: Nature, Lond. **173**, 164 (1954).

[4] B. Y. MILLS: Unpublished data.

[5] H. SHAPLEY: Proc. Nat. Acad. Sci. U.S.A. **26**, 41 (1940).

[6] H. SHAPLEY: Proc. Nat. Acad. Sci. USA. **21**, 587 (1935).

cluster again appears unlikely. However, the galaxy IC 2082, a cluster member which SHAPLEY has suggested may represent an example of colliding galaxies[1], is practically coincident with the radio position. An identification between this galaxy and the radio source therefore appears to be a distinct possibility.

Cluster in Grus. This cluster[2] at a position of $22^h24^m.5$, $-49°03'$ (1950) has no associated radio source within the sensitivity limits of existing equipment. The integrated-photographic magnitude is estimated from SHAPLEY'S data as 11.2 and the upper limit of the radio emission corresponds to 11.3; thus the mean $m_{1.9}-m_p$ is greater than 0.1. All that can be said is that there is no substantial excess of radio emission over that expected from the "normal" galaxies of the cluster.

Cluster I in Reticulum. A radio source of flux density 1.9×10^{-25} W m^{-2} (c/s)$^{-1}$ and angular size of the order of $\frac{1}{2}°$ is found at a position of $03^h44^m.0$, $-53°23'$ (1950). This agrees with the cluster position, which is given only to the nearest degree[3], and the angular size suggests that an identification may be made. Because of the uncertain optical position, however, the identification is also uncertain. The integrated photographic magnitude of the cluster[4] is estimated at 10.0 and the radio magnitude of the source is 9.2; thus the mean value of $m_{1.9}-m_p$ is -0.8, a radio emission apparently too high to be produced by the normal galaxies of the cluster.

Cluster II in Reticulum. The optical position of this cluster is also given to the nearest degree[3]. Within this uncertainty there is a radio source at position $03^h27^m.5$, $-53°30'$ (1950) of flux density 8×10^{-25} W m^{-2} (c/s)$^{-1}$: because of the low intensity an estimate of angular size was not possible. The integrated photographic magnitudes is estimated at 11.2 and the radio magnitude is 10.0; thus the mean value of $m_{1.9}-m_p$, -1.2, again suggests an abnormally high radio emission if the identification is correct.

Table 5. *A summary of the observations of clusters of galaxies.*

Cluster	Apparent $m_{1.9}-m_p$	Remarks
Perseus	-1.4	After correcting for radiation from NGC 1275 the cluster has $m_{1.9}-m_p \gtrsim +0.3$
Ursa Major II	-1.9	
Cetus	—	Source observed but probably not associated with the cluster
Dorade	—	Source observed but probably associated with IC 2082
Grus	$>+0.1$	Cluster not observed
Reticulum I	-0.8	Angular size of source $\approx\frac{1}{2}°$
Reticulum II	-1.2	

The important results of the cluster observations are summarized in Table 5; it is apparent that dense clusters tend to exhibit a much more powerful radio emission than would be expected from their component galaxies. In some cases it seems that a radio galaxy in the cluster is responsible, in others no definite interpretation is possible. The faintness of the radio emission, and the small angular size of the clusters themselves, limits the information which may be derived from present equipment. However, there remains one agglomeration of galaxies to which neither of these difficulties obtain, that is, the "Local Supergalaxy".

[1] H. SHAPLEY: Private communication.
[2] See footnote 5, p. 261.
[3] See footnote 6, p. 261.
[4] A correction of -0.7 magnitude has been applied to SHAPLEY'S photographic magnitudes for consistency with the other data used in this article.

15. Radio emission from the "Local Supergalaxy". For more than 30 years a distinct belt of bright galaxies has been recognized in the north galactic hemisphere, stretching around a great circle nearly at right angles to the plane of the Milky Way. It has been suggested that this agglomeration of galaxies may represent an enormous organization of individual galaxies and whole clusters forming a galaxy of the second order, or "supergalaxy". Recently DE VAUCOULEURS has investigated this system and suggests that the concentration extends also to the south galactic hemisphere in much attenuated form; he concludes that the Milky Way galaxy is situated in an outlying region[1]. While this concept of a supergalaxy has not met with general acceptance because of the possibility that the concentration is the result merely of a chance alignment of a few clusters, there is no doubt concerning the reality of the concentration itself.

In 1953 two observations were made of radio emission apparently emanating from this belt of galaxies. One, with the 67 m, Manchester parabola[2], was limited to the declination range $+70°$ to $+40°$; the other, with KRAUS's fan-beam antenna[3] ($1.°2 \times 17°$) extended the range of declinations to about $-30°$. An almost continuous ridge of emission, of width about 10°, was found over the whole range. In Fig. 17 are shown the comparisons made by KRAUS between the belt of emission he observed and the distribution of bright galaxies; the similarity is obvious. Examination of his actual records, however, shows that the separation of the

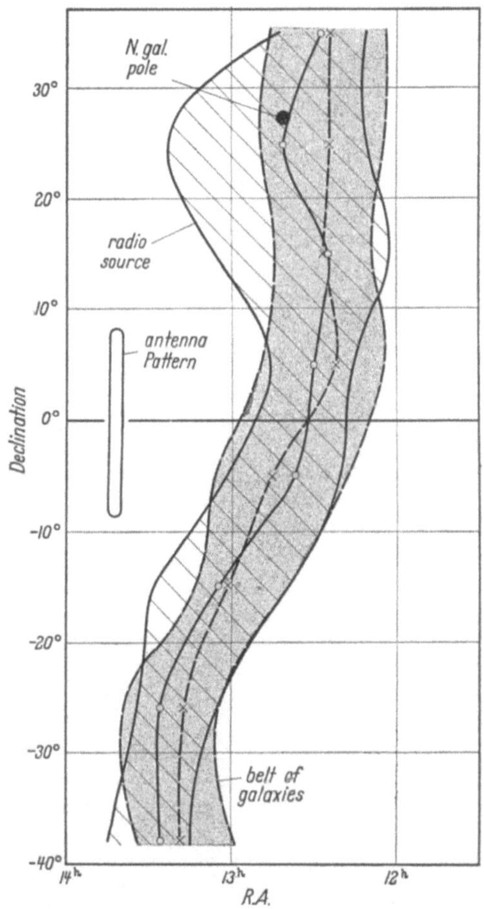

Fig. 17. Comparisons between the belt of radio emission and the belt of bright galaxies in the supergalaxy (KRAUS).

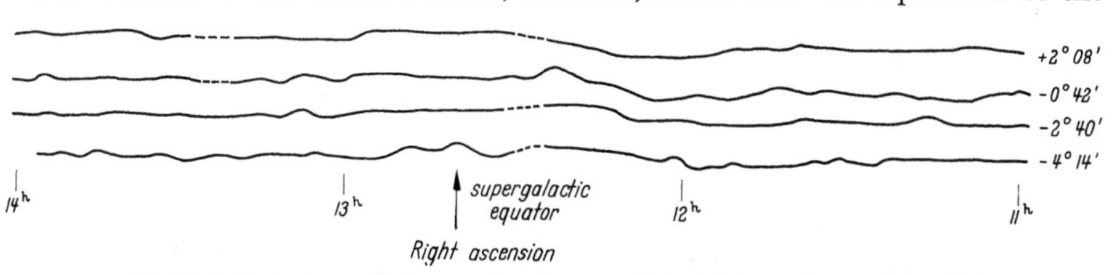

Fig. 18. Some sample records of supergalaxy traverses obtained at 3.5 m with a 50′ beamwidth.

[1] G. DE VAUCOULEURS: Astronom. J. **58**, 30 (1953). — Vistas in Astronomy, Vol. II, Sect. 15. London and New York: Pergamon Press 1955.

[2] R. HANBURY BROWN and C. HAZARD: Nature, Lond. **172**, 997 (1953).

[3] J. D. KRAUS: Astronom. J. **59**, 113 (1954).

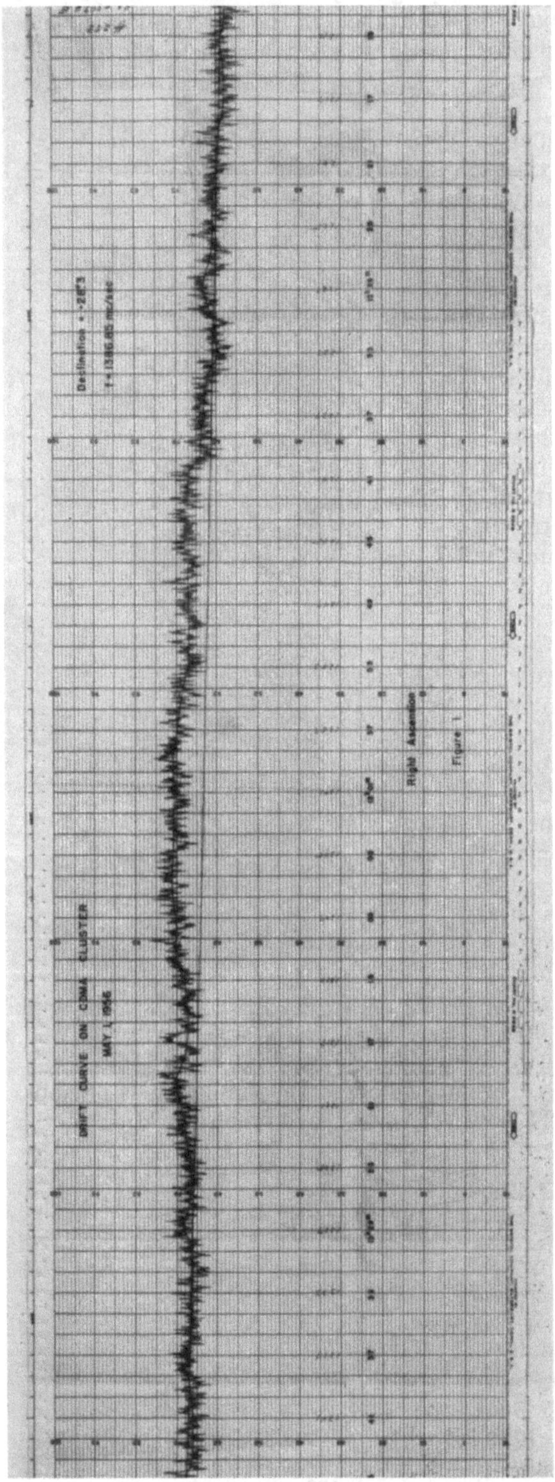

Fig. 19. A record obtained of the hydrogen line emission from the Coma cluster (HEESCHEN).

supergalaxy radiation from the general irregularities in the background is not always easy with an aerial of only moderate resolution; a detailed analysis is therefore being made of the high resolution Sydney survey at 3.5 m in an attempt to obtain more reliable data. Some sample traverses of the supergalaxy region obtained in this survey are shown in Fig. 18 [1].

It has been concluded by all who have discussed the question that the integrated radio emission of the normal galaxies in the supergalaxy falls far short of the observed brightness ($\sim 10^{-25}$ W m^{-2} (c/s)$^{-1}$ per square degree at 1.2 m wavelength). In fact the radiation from normal galaxies alone would probably be undetectable with present-day equipment. The cause of the excess emission must lie either in the existence of a large number of "radio galaxies" (a large number is required to account for the relative smoothness of the isophotes) or in emission arising in the intergalactic space. There are several difficulties associated with the former interpretation, principally that on present evidence radio galaxies do not appear to exist in sufficient numbers to produce

---

[1] *Note added in proof.* These observations have trown doubt on the existence of the southerly part of the supergalactic radiation observed by KRAUS: it appears, in particular, that the maxima near the supergalactic equator in Fig. 18 are associated with a galactic feature. — E. R. HILL: Austral. J. Phys. **11**, 580 (1958).

the observed radiation over the whole emitting region. The second alternative has been discussed at length by SHKLOVSKII[1] who suggests the synchrotron type radiation from relativistic electrons as a mechanism of emission. This requires the existence of very weak intergalactic fields of the order of 10^{-7} gauss, in conjunction with electrons of 10^9 eV having a space density of about 10^{-13} cm^{-3}. If correct, this model suggests strongly the existence of the supergalaxy as a physical entity. It seems possible that a similar mechanism could be operative in some of the clusters of Sect. 14 which display an excess of radio emission.

16. Hydrogen line emission from clusters of galaxies. Just as clusters of galaxies at great distances might be expected to have detectable radio emission in the continuum, so also it might be expected that the hydrogen line emission from dense clusters should be detectable. This has been verified for the Coma cluster by HEESCHEN using the 8 m Harvard radio telescope[2]: one of his original records showing a drift curve through the cluster is reproduced in Fig. 19. He made 80 such drift curves at different frequencies and was able to construct curves representing the distribution in right ascension of the integrated brightness, and the frequency line profile. The line profile indicates a mean red-shift of $+7000$ km per sec which, in view of the large uncertainties, is in good agreement with the optical figure of $+6680$ km/sec. The total mass of neutral hydrogen is estimated as 8×10^{13} solar masses compared with SCHWARZSCHILD's value of 3×10^{14} for the total mass of the cluster[3]: the proportion of hydrogen is therefore about 25%.

Although one cannot draw general conclusions from the observation of a single cluster, it would appear that this will prove a powerful method of studying their structure and dynamics when larger radio telescopes come into general use[4].

V. Radio galaxies.

17. Identifications of radio galaxies. In the I.A.U. catalogue of reliably known radio sources [*6*] there are six sources with galactic latitude greater than $12\frac{1}{2}°$ (Class II sources) and with flux densities at 100 Mc/s greater than 2×10^{-24} W m^{-2} (c/s)$^{-1}$. All but one of these have now been identified, at various levels of confidence, with galaxies which are emitting at radio wavelengths many times the power of the bright galaxies discussed earlier. Such galaxies have been called "radio galaxies" and the majority display some easily recognized optical peculiarity which may be related to their radio emission. The observations are sumarized in Table 6 and the individual identifications discussed below. Two other identifications have been added to the table, IAU 19N4A, the strongest extragalactic source, which escapes classification as a Class II source because of its low galactic latitude and IAU 03 N4A which, although of low flux density, has been reliably identified as a radio galaxy. There are many more *possible* identifications with weaker sources which do not have the same degree of reliability as those listed; because some are undoubtedly wrong they are not included. The seven examples listed in the table comprise the principal evidence for the existence of this class of object.

19N4A (Cygnus A). This was the first discrete source discovered and is the strongest radio galaxy: it is the second strongest source in the sky. Nevertheless the final identification of the source was a long and arduous task which

[1] I. S. SHKLOVSKII: Astr. J. Moscow **31**, 533 (1954).

[2] D. S. HEESCHEN: Astrophys. J. **124**, 660 (1956).

[3] M. SCHWARZSCHILD: Astronom. J. **59**, 273 (1954).

[4] The observation of H-line emission from two more clusters, those in Hercules and Corona Borealis, has recently been reported. D. S. HEESCHEN: Publ. Astronom. Soc. Pacific **69**, 350 (1957).

Table 6. *Observations of Radio Galaxies.*

For uniformity with Table 4, the flux densities are those measured with the Sydney cross-type radio telescope at 3.5 m wavelength or the Manchester 67 m parabola at 1.9 m; wavelength.

Radio source (IAU designation)	Flux density at 3.5 m (10^{-24} W m^{-2} (c/s)$^{-1}$)	$m_{1.9}$	Associated Galaxy			$m_{1.9}-m_p$
			NGC	m_p	Reference	
19N4A	57[1]	2.1	anon	15.1	[*1*]	−13.0
13S4A	88	2.6	5128	6.5:	2	− 3.9:
12N1A	24.3	3.9	4486	9.95	2	− 6.0
03S3A	9.5	4.9	1316	9.5	2	− 4.6
16N0A	8.9	5.0	anon	18:	4	−13:
09S1A	6.7	5.3	anon	16:	3	−11:
03N4A	0.65[1]	7.0	1275	12.0:	2	− 5.0:

finally required photographs and spectra obtained with the 200-inch Hale telescope to demonstrate the extreme abnormality of the galaxy. The identification was first suspected in 1949, but a direct photograph with the 100-inch Mt. Wilson telescope revealed no abnormalities in the galaxy[5] and the radio position was not considered sufficiently accurate to justify the use of the larger telescope. It was not until a long series of observations by SMITH[6] led to a much more accurate position that a photograph was made by BAADE with the Hale telescope; this showed immediately that the galaxy had a double nucleus, suggesting a collision was in progress between two separate galaxies. This view was confirmed with a spectrum obtained by MINKOWSKI which revealed a startling and unique collection of strong and fairly broad emission lines, identified with the lines produced by the interstellar gas in a highly excited state, such indeed as could be produced by the suggested collision. The lines observed included those of [NeV], [NeIII], [OIII], [OII], [OI], [NII] and Hα: the width of the lines corresponds to a velocity dispersion of 400 km/sec. The red-shift can be measured very accurately and is equal to 16830 km/sec, corresponding to a distance of 9×10^7 pc in the new scale. This work of BAADE and MINKOWSKI is described in a classic paper in Ref. [*1*].

The exact nature of the mechanism producing the radio emission in the Cygnus source has not yet been established. It must be very powerful, however, because the total radio power is of the same order as the total power emitted in the optical range by the stars in the galaxies and the total power emitted by the interstellar gas in the visible emission lines. It is not unreasonable to suppose that synchrotron type emission is the basic mechanism, for the turbulent ionized gas masses would undoubtedly produce strong magnetic fields, while at the same time acting as powerful accelerators of electrons by a Fermi type process. Other puzzling features, however, are the great extent over which the radio emission occurs, $\frac{1}{2}' \times 2'$, which is several times the visible size of the galaxies and, more especially, an apparent paucity of emission in the central regions where the collision appears most active[7]. The angular size measurements are discussed more fully by HANBURY BROWN in Sect. 10 of his article (p. 224). Other observations of the source are listed in Ref. [*6*].

[1] Measurement made at 1.9 *m*.
[2] See footnote 1, p. 253.
[3] R. L. MINKOWSKI: Unpublished data.
[4] R. L. MINKOWSKI: Ref. [7], p. 108.
[5] B. Y. MILLS and A. B. THOMAS: Austral. J. Sci. Res. A **4**, 158 (1951).
[6] F. G. SMITH: Nature, Lond. **168**, 962 (1951).
[7] R. C. JENNISON and M. K. DAS GUPTA: Phil. Mag. **1**, 65 (1956).

13S4A (Centaurus A). The identification of this radio source with the galaxy NGC 5128 was first suggested by BOLTON, STANLEY and SLEE in 1949[1]; the galaxy is very bright and has an obvious anomalous feature consisting of a dark bar cutting across an apparently regular *Eo* galaxy. Later measurements confirmed the identification and showed that the radiation is emitted with two markedly different distributions, one concentrated in a very small region approximately coincident in position and shape with the dark bar of the galaxy and the other forming a very extensive corona[2,3]. The flux density quoted in Table 6

Fig. 20. A photograph of the radio galaxy NGC 5128. (Mount Wilson and Palomar Observatories.)

refers to the integrated emission of both components; the former contributes about one third of the total at wavelengths of a few metres.

A photograph of the galaxy is shown in Fig. 20, and in Fig. 21 are shown the radio contours obtained with the 450 m Sydney cross-type radio telescope at 3.5 m: the resolution of the instrument is inadequate to resolve the central bright source which has a size estimated to be $3' \times 6\frac{1}{2}'$ [3].

On long exposure plates the optical limits of the nebula may be followed much further than in Fig. 20. DE VAUCOULEURS has found that, in its outer regions, the nebula is elliptical in position angle 30° with an axial ratio of about 1.5; he has traced the emission to a distance of about 0°.5 along an East-West axis[4]. The outer parts of the radio emission, however, are in position angle 10°

[1] J. G. BOLTON, G. J. STANLEY and O. B. SLEE: Nature, Lond. **164**, 101 (1949).
[2] B. Y. MILLS: Austral. J. Sci. Res. A **5**, 456 (1952).
[3] B. Y. MILLS: Austral. J. Phys. **6**, 452 (1953).
[4] G. DE VAUCOULEURS and K. V. SHERIDAN: Ref. [7], p. 169.

and have an axial ratio of about 4: the maximum diameter along a major axis is about 8°. Correspondence between the visual and radio emission is therefore not at all close.

The galaxy does not fall into any easily defined category: only one other is known which resembles it, NGC 1947, but this is not detectable as a radio source[1,2]. It has been suggested by BAADE and MINKOWSKI [*1*] that NGC 5128 represents a collision between a late-type spiral, seen edge-on and responsible for the dark obscuring band, and an *S*0, principally responsible for the light from the nebula. Faint emission lines suggesting an excitation of the interstellar gas lend some support to this suggestion.

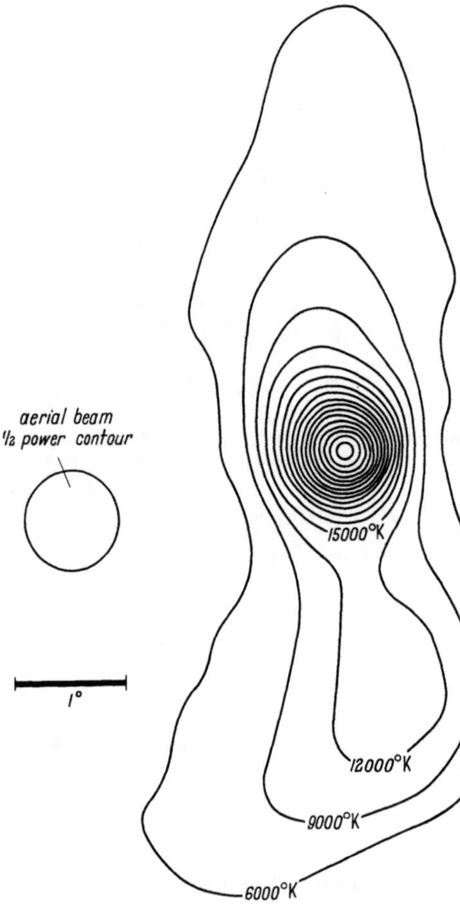

Fig. 21. Contours of 3.5 m radiation from the radio source Centaurus A (Sheridan), identified with NGC 5128; the beam-width is 50′ of arc, the wavelength 3.5 m.

12N1A (Virgo A). The identification with the galaxy NGC 4486 was also suggested by BOLTON, STANLEY and SLEE[3]. Like NGC 5128, the galaxy has a very anomalous optical feature, in this case a very blue "jet" or series of bright condensations extending from the nucleus of the galaxy. The radio emission is not directly connected with this feature, however, for the length of the jet is but 20″ arc in position angle 290°, whereas it has been found that the radio emission arises in a vastly greater volume. Interferometer measurements, when interpreted in terms of a symmetrical elliptical model, give an object of size $5' \times 2\frac{1}{2}'$ in position angle 50°[4].

It was suggested by SHKLOVSKII[5] that the optical emission from this "jet" and the radio emission both result from the synchrotron mechanism applied to high energy electrons and magnetic fields in the galaxy, that is, an explanation very similar to the one he successfully proposed for the optical and radio emission of the Crab nebula discussed in the article by HANBURY BROWN. This explanation has received substantial confirmation by the observation by BAADE[6] of strong polarization, of the order of 30% in the optical emission from the "jet". The cause of this abnormality is, however, completely unknown.

---

1 G. DE VAUCOULEURS: Observatory **73**, 252 (1953).
2 B. Y. MILLS: Observatory **74**, 248 (1954).
3 See footnote 1, p. 267.
4 B. Y. MILLS: Austral. J. Phys. **6**, 452 (1953).
5 I. S. SHKLOVSKII: Astr. J. Moscow **32**, 209 (1955).
6 W. BAADE: Astrophys. J. **123**, 550 (1956).

03S3A (Fornax A). The identification with NGC 1316 has had rather a chequered history. Photographs of the nebula with telescopes of poor resolution had suggested a distinct similarity to the radio galaxy NGC 5128, consequently there was a temptation to identify the radio source with the former galaxy, which was fairly close to the radio positions then available. However the agreement in position was not good and, when better photographs of the nebula showed clearly that the similarity in appearance was purely fortuitous and that the nebula, an *S*0, displayed no real abnormalities, the fate of the identification seemed assured. Surprisingly, though, observations with the Sydney cross-type radio telescope showed a large source with complex brightness distribution whose centroid is practically coincident with the centre of the galaxy[1]. The angular size, of the order of 1°, is about three times the maximum extent of the nebula, in conformity with other identified sources, and there would therefore seem to be little doubt the identification. By analogy with NGC 4486 it seems possible that emission is caused by the synchrotron process, but that electron energies and/or magnetic fields do not reach sufficiently high values for optical emission.

16N0A (Hercules A). Recent accurate radio positions of this source have led to a probable identification, again using the Hale telescope[2]. The two most accurate radio positions (1950 coordinates) and their associated probable errors are $16^h 48^m\ 41^s \pm 2^s$, $+5°\ 06' \pm 5'$[3] and $16^h 48^m\ 46^s \pm 3^s$, $+5°\ 05' \pm 2'$[4]: the position of the galaxy is $16^h\,48^m\,49^s$, $+5°\ 01'.8$ in reasonable agreement. The galaxy displays a double nucleus and rather strong emission lines of [O III] and [O II] indicating that an active collision may be in progress. The angular size of the galaxy is about $\frac{1}{2}'$ and that of the radio source about $2\frac{1}{2}'$[5]. The red-shift is 25050 km/sec corresponding to a distance of 1.4×10^8 pc in the new scale. All told, the identification appears very probable.

09S1A (Hydra A). This source may also probably be identified with a faint double galaxy[6]. The most accurate source position is $09^h 16^m 43^s \pm 3^s$, $-11°\ 52'.5 \pm 2'$ (1950)[4] compared with the galaxy position of $09^h 16^m 42^s$, $-11°\ 53'$ (1950). The angular size of the radio source is $1\frac{1}{2}'$ compared with about $\frac{1}{2}'$ for the galaxy[7].

Despite the very good agreement in position, the identification is not as reliable as the others for two reasons. Firstly, the spectrum of the galaxy is not obviously peculiar; the λ 3727 line of [O II] is detectable but not unusually strong, and so there is no real evidence from the optical data for an active collision. Secondly, it is reported by SLEE[8] that the apparent intensity of the source varies substantially in a time less than one day: such a variation is inconsistent with radiation from a whole galaxy. However, the apparent angular size shows variations also, uncorrelated with those of the intensity and it therefore appears likely that both variations are locally imposed[7]. Neither of these objections appear to be of great consequence and in view of the close correspondence between position and angular size the identification is considered probable.

---

[1] B. Y. MILLS: Observatory **74**, 248 (1954).
[2] R. L. MINKOWSKI: Ref. [7], p. 108.
[3] J. R. SHAKESHAFT, M. RYLE, J. E. BALDWIN, B. ELSMORE and J. H. THOMSON: Mem. Roy. Astronom. Soc. **67**, 97 (1955).
[4] B. Y. MILLS: Unpublished data.
[5] A. W. L. CARTER: Unpublished data.
[6] R. L. MINKOWSKI: Unpublished data.
[7] A. W. L. CARTER: Austral. J. Phys. **8**, 564 (1955).
[8] O. B. SLEE: Austral. J. Phys. **8**, 498 (1955).

03N4A (Perseus A). The identification with the galaxy NGC 1275 in the Perseus cluster is discussed in Sect. 14. Recent work by MINKOWSKI has fully confirmed that two galaxies are colliding[1]. The galaxies involved are a spiral of early type, very luminous with $M_p \approx -19$, and a strongly distorted late type spiral. Because of the brightness of the galaxies this object appears the most suitable for studying the process of a collision between galaxies.

MINKOWSKI was able to construct a detailed picture of the collision process by examination of the emission line spectra at different places in the system. On the northern side two sets of lines are observed with velocities of $+5200$ km/sec and $+8200$ km/sec; in the nucleus and to the south one set only is observed, very wide and asymmetrical, with a velocity approximately equal to $+5000$ km per sec. The interpretation given is that the galaxies are inclined to each other and the collision started in the south and proceeded to the north as they interpenetrated each other. The two velocities to the north correspond to the velocities of the individual galaxies and the single wide lines to the south to the resultant mixture of turbulent gas; the mean velocity is close to that of the early type galaxy because of the much greater mass of gas associated with it.

Because of the weakness of the radio source, it has not yet been possible to study the distribution of radio emission in detail. However, the angular size has been given as 2.2′[2], compared with a size of the order of $\frac{1}{2}'$ for the visible parts of the galaxies.

18. Some characteristics of radio galaxies. Of the seven radio galaxies listed, four or perhaps five appear to represent a collision or interaction between two very close galaxies: in the others the abnormality causing the radio emission is present in a single galaxy. The question arises whether there is any optical peculiarity which enables a galaxy, or pair of galaxies, to be recognised as an abnormal radio emitter.

An investigation of four known pairs of interacting bright galaxies with the Sydney cross-type radio telescope showed that none has a markedly abnormal emission although one pair, NGC 4038/39, emits about 2 magnitudes more than an average spiral galaxy. The observations are summarized in Table 7.

Table 7. *Observations of colliding or interacting galaxies.*

Galaxy NGC	Flux density at 3.5 m (10^{-26} W m^{-2} (c/s)$^{-1}$)	$m_{1.9}$	m_p	$m_{1.9}-m_p$
4038/39	10	9.9	10.6	-0.7
3256	<3	>11.2	10.6	$>+0.6$
1487	<5	>10.7	11.9	>-1.2
520	<9	>10.0	11.6	>-1.9

It appears that interacting and colliding galaxies are not necessarily radio sources. The degree of interaction may be the key, as none of the above galaxies represents a particularly violent collision; on the other hand only two of the colliding radio galaxies (the Cygnus galaxy and NGC 1275) have spectra suggesting a great abnormality in the excitation of the interstellar gas. It is therefore not yet possible to come to any conclusions regarding the necessary condition for radio emission to occur in a collision of galaxies.

In the case of single galaxies the situation is even more obscure. In M87 with its bright "jet" of, presumably, high energy electrons, an optical abnormality is obvious; however, if the synchrotron mechanism is involved and if magnetic fields or electron energies were smaller in the region of the "jet", the radio emission would be unaffected but there would be no recognizable optical pe-

[1] R. L. MINKOWSKI: Ref. [7], p. 111.
[2] A. W. L. CARTER: Unpublished data.

culiarity. No physical reason has been discovered yet for the presumed existence of relativistic electrons throughout the galaxy. Thus the existence of radio galaxies without any abnormality detectable by optical means appears probable; NGC 1316 would seem to be of this type.

In every case the distribution of radio emission appears to be several times wider than that of the stars, with the possible exception of NGC 4486, but even here the existence of an extensive corona has been suggested[1]. This wide distribution appears also in the normal galaxies, although probably to a smaller extent, and is possibly a fundamental property: it is quite consistent with the explanation in terms of synchroton emission from high energy electrons. The spatial extent of the emission is of interest: it ranges from 10^5 pc for the Hercules source to 10^4 pc for the Virgo source. The range of 10:1 is remarkably small and may be smaller still if the existence of a corona around the Virgo source is confirmed.

The observations in Table 6 suggest that it might be possible to group the galaxies into two classes, one radiating with $m_{1.9}-m_p$ of the order -5 and the other with a value around -13. The numbers are insufficient to establish such a conclusion, however, and it seems more probable that the radio emission of a galaxy may be anywhere in, and probable outside, the observed range from $m_{1.9}-m_p$ equal to $+4$ to $-13\frac{1}{2}$. The absolute emission is greatest for the Cygnus colliding galaxies with an output of approximately 2×10^{28} watt $(c/s)^{-1}$ at 3.5 m. For the Andromeda nebula the output is about 10^{22} watt $(c/s)^{-1}$, or more than a million times less, while for the undetected Sculptor dwarf galaxy the emission is less than 5×10^{18} watt $(c/s)^{-1}$ at the same wavelength, a ratio of more than 10^9 between the strongest and weakest galaxies.

The radio spectra of the galaxies in Table 6 are poorly determined, although it has been found that, in all cases, the emission increases with increasing wavelength up to a wavelength of at least 30 m. In only two cases, IAU19N4A (Cygnus A) and IAU12N1A (Virgo A), have sufficient independent observations been made to construct the spectrum with any accuracy; the other sources are either too weak for easy observation or have too large an angular size for reliable flux density measurements. The spectra of the two former sources were measured first by STANLEY and SLEE[2] who concluded that, at metre wavelenghts, the flux density of each was approximately proportional to the wavelength. This law still appears to be a reasonable approximation to their spectra even at centimetre wavelengths[3], although some recent measurements have suggested that the variation of flux density with wavelength may be somewhat less steep[4].

Two additional spectral features have been noted, a decrease in the flux densities of Cygnus A and Centaurus A at wavelengths greater than about[5] 15 m and a possible flattening of the spectra of all radio sources in the range of wavelengths between about[6] 20 cm and 1 m. The decrease in flux density at long wavelengths appears to be well established although it is at present uncertain whether it is a result of absorption by ionized hydrogen, in the radio galaxies themselves or in the Milky Way, or whether it represents a real feature of the emission spectrum. The suggested flattening of spectra in the decimetre range

---

[1] J. E. BALDWIN and F. G. SMITH: Observatory **76**, 141 (1956).
[2] G. J. STANLEY and O. B. SLEE: Austral. J. Sci. Res. A **3**, 234 (1950).
[3] J. P. HAGEN, E. F. MCCLAIN and N. HEPBURN: Astronom. J. **59**, 323 (1954).
[4] R. ADGIE and F. G. SMITH: Observatory **76**, 181 (1956).
[5] R. J. LAMDEN and A. C. B. LOVELL: Phil. Mag. **1**, 725 (1956). — H. W. WELLS: J. Geophys. Res. **61**, 535 (1956). — R. G. ELLIS: J. Geophys. Res. **62**, 229 (1957).
[6] N. G. ROMAN and F. T. HADDOCK: Astrophys. J. **124**, 35 (1956).

is rather suspect as this is just the range where there are marked changes in the techniques of measurement, and since it apparently applies to *all* radio sources investigated. However, if it is established that such a spectral feature exists and is related to the emission process in a radio source, there is no doubt of its importance; it should prove a clear indicator of the red-shift of very distant radio galaxies.

19. Cosmological aspects. The identification of the strongest Class II sources with external galaxies has pointed the way to a new approach to the problem of cosmology; some of the implications were discussed first in 1952 when the possi-

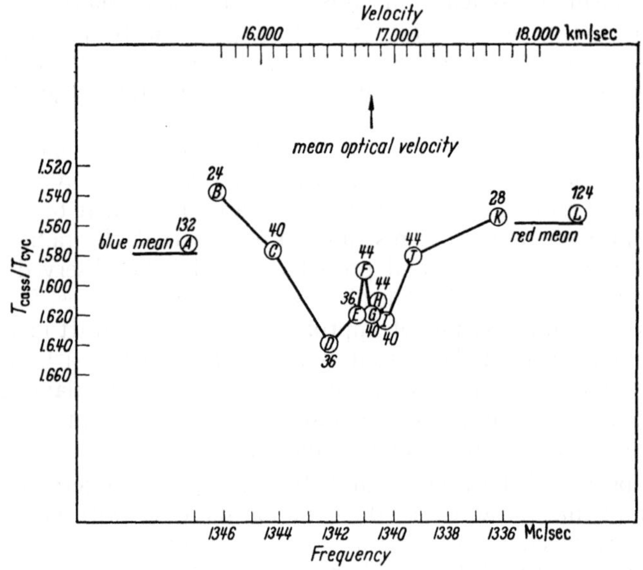

Fig. 22. Absorption of radiation from the Cygnus colliding galaxies by associated neutral hydrogen (LILLEY and MCCLAIN). The red-shift is identical with that deduced from the optical emission lines.

bility of an extragalactic origin of the Class II sources was advanced[1]. The importance of the radio measurements is indicated by the simple consideration that present equipment would be able to detect radiation from galaxies of the Cygnus type at distances corresponding to recession velocities of about $0.8\,c$ where cosmological effects must be very pronounced.

A classic experiment was performed at the American Naval Research Laboratory in 1955 when LILLEY and MCCLAIN observed the absorption of continuum radiation by neutral hydrogen associated with the Cygnus colliding galaxies[2]. The absorption curve they obtained is reproduced in Fig. 22. The hydrogen produced maximum absorption at the wavelength expected if the red-shift was a true Doppler effect, i.e. $\Delta\lambda/\lambda$ was identical for optical and radio emission. This showed firstly, that the same cosmological equations can be used in analysis of radio and optical data and secondly, that the observed red-shift is almost certainly a Doppler shift due to a real recession. If other radio galaxies have a similar envelope of neutral hydrogen, a distance scale is available without the necessity of making identifications with visible galaxies. Such measurements are likely to be extremely difficult however, for, in the case of the Cygnus source, the absorption amounts to only about 5%; if other sources are similar, very

[1] B. Y. MILLS: Austral. J. Sci. Res., Ser. A **5**, 266 (1952).
[2] A. E. LILLEY and E. F. MCLAIN: Astrophys. J. **123**, 172 (1956).

high signal-to-noise ratios are required and any given radio telescope will be able to measure the red shift of only a small proportion of the sources it can detect.

One important aspect of radio cosmology is the possibility of measuring the total apparent emission of all the very distant galaxies. The radio brightness in the direction of the galactic poles includes both galactic and extra-galactic contributions and, while it is easy to set upper limits to the latter by equating the former to zero (giving $T < 500°$ K at 3 m wavelength), a measurement of the actual contribution is difficult. It may be inferred from the distribution of galactic emission, but such an inference is necessarily uncertain and usually very dependent on the galactic model assumed[1]. At long wavelengths a direct method of measuring the galactic contribution seems possible, using very high resolution aerials to observe extremely distant H II regions which are opaque and do not transmit the radiation beyond them. The nebula 30 Doradus in the Large Magellanic Cloud should be well clear of the galactic corona and a suitable object for such a measurement.

The simplest application of radio astronomy to cosmology is the counting of radio sources down to various limiting magnitudes, a method which was tried and virtually abandoned in optical astronomy. In a uniform static Euclidean universe and with a radio telescope of infinite resolution and sensitivity the number of sources, N, with flux densities greater than S, is given by the relation

$$\operatorname{Log} N = C - 1.5 \operatorname{Log} S. \tag{19.1}$$

This relation is independent of any dispersion in emission or clustering of the radio sources and requires only that N should be sufficiently large for chance effects to be negligible. Other model universes give different and more complicated relationships between N and S, so that by a simple counting process it seems possible that the form of the actual universe could be derived. Unfortunately, however, these other relationships depend upon dispersion and evolutionary effects in the radio galaxies themselves, so that this method of analysis is unlikely to yield definitive results without extensive supplementary data, except probably for a distinction between steady-state and evolutionary universes. Even here, however, many uncertainties arise because the form of the observed relation between N and S may also be profoundly affected by departures of the instrument from the ideal assumed. Departures in the form of finite resolution and random noise both tend to increase the factor 1.5 in Eq. (19.1)[2]. Two recent surveys in which large numbers of radio sources were counted gave very different values for this slope factor[2,3], the survey of higher resolution giving the lower value (1.8 compared with 3.0). It was concluded by MILLS and SLEE that neither of these results gave a clear indication of any significant departure from uniformity in the distribution of radio sources. Details of the surveys are given in HANBURY BROWN'S article.

At levels where the resolution is inadequate for reliable source counting it is possible to analyse the statistics of the output fluctuations of a radio telescope These statistics will be different for different model universes, which may therefore be tested against the observations. It is not possible, however, to derive useful information about the form of the universe directly from measurements of this type because such factors as clustering and the angular sizes of the radio galaxies may affect the statistics profoundly and cannot be separated from the

[1] J. E. BALDWIN: Monthly Notices Roy. Astronom. Soc. London **115**, 691 (1955).
[2] B. Y. MILLS and O. B. SLEE: Austral. J. Phys. **10**, 162 (1957).
[3] M. RYLE and P. A. G. SCHEUER: Proc. Roy. Soc. Lond., Ser. A **230**, 4, 8 (1955).

more basic cosmological effects without supplementary information. In the two recent surveys mentioned above opposite conclusions were drawn from an analysis of background fluctuations.

Since the simple counting of sources and measurement of background fluctuations are not adequate for a complete attack on the cosmological problem it is necessary to consider other possibilities. Two have been mentioned, red-shift measurements using the hydrogen line, and a measurement of the total extragalactic emission. The former, however, although very direct and powerful in its application, is probably limited to relatively few galaxies, not very distant cosmologically speaking, even with the largest radio telescopes planned at present; the latter has clearly a very limited usefulness in deciding between competing cosmologies, although very useful for fixing *one* parameter in any assumed model. Angular size measurements may be the key to the problem, for we have seen that, unlike the total emission, the spatial extent of identified radio galaxies shows comparatively little variability, and therefore the angular size is a reasonably accurate measure of the "distance". Such measurements, although not easy, may be carried out almost at the sensitivity limit of a radio telescope; the methods are discussed in the article by HANBURY BROWN.

General references.

[1] BAADE, W., and R. MINKOWSKI: Identification of the Radio Sources in Cassiopeia, Cygnus A and Puppis A. Astrophys. Journ. **119**, 206 (1954). — On the Identification of Radio Sources. Astrophys. Journ. **119**, 215 (1954).

[2] BROWN, R. HANBURY and C. HAZARD: Extragalactic Radiofrequency Radiation. Phil. Mag. **43**, 137 (1952).

[3] BUSCOMBE, W., S. C. B. GASCOIGNE and G. DE VAUCOULEURS: Problems of the Magellanic Clouds. Austral. J. Sci., Suppl. **17**, No. 3 (1954).

[4] KERR, F. J., J. F. HINDMAN and B. J. ROBINSON: Observations of the 21 cm Line from the Magellanic Clouds. Austral. J. Phys. **7**, 297 (1954).

[5] MILLS, B. Y.: The Observation and Interpretation of Radio Emission from some Bright Galaxies. Austral. J. Phys. **8**, 368 (1955).

[6] PAWSEY, J. L.: A Catalogue of Reliably Known Discrete Sources of Cosmic Radio Waves. Astrophys. J. **121**, 1 (1955).

[7] VAN DE HULST, H. C.: Radio Astronomy. Symposium No. IV of the Int. Astron. Union. Cambridge: Cambridge University Press 1957.

Classification and Morphology of External Galaxies.

By

G. DE VAUCOULEURS.

With 7 Figures and 11 Plates.

Introduction.

The earliest systems of classification of "nebulae" by W. HERSCHEL and others have been described by H. D. CURTIS in his article of the "Handbuch der Astrophysik" [*A*][1]. The *spiral* form of some nebulae was first detected visually by Lord ROSSE and his assistants between 1845 and 1850[2] and the appellation *elliptical* first used in a purely descriptive way was applied to non-spiral tub presumably external stellar systems by S. ALEXANDER in 1852[3].

The great abundance of the spiral type was brought to light by the photographic surveys of the end of the XIX-th century. The spiral structure of the Andromeda nebula was detected photographically in 1888 by I. ROBERTS. These surveys also revealed the great abundance of smaller objects of the smooth, structureless elliptical type and the very elongated objects described as "spindles". In 1918 CURTIS [*9*] isolated a special type of spiral, the so-called "Φ" type, characterized by a diametral bar across the nucleus and a general ring-like structure; he also identified the "spindle" nebulae with a longitudinal dark lane as edgewise spirals with peripheral dark matter [*10*].

An early alphabetical classification of apparent shapes, introduced by M. WOLF[4], has been extensively used at Heidelberg by REINMUTH [*62*] and at Lund by LUNDMARK [*44*], HOLMBERG, REIZ, DANVER and others[5]. Its relation to the standard classification has been given by SHAPLEY and Miss AMES [*68*]. For faint nebulae which show little or no structure, a descriptive classification, based on "concentration" and "elongation", introduced by SHAPLEY[6] in 1929, has been in use at Harvard for some years; it bears little relation to the standard classification.

Rather similar systems of morphological classification were introduced, about 30 years ago, by HUBBLE [*25*] and by LUNDMARK [*43*] with three main types, viz. elliptical, spiral (normal and barred) and irregular. This scheme, extensively used and later further developed by HUBBLE, has been accepted as standard up to the present time.

---

[1] The main references are listed at the end of the chapter "General physical properties of external galaxies", p. 311.

[2] Earl of ROSSE: Phil. Trans. Roy. Soc. Lond. **1850**; **1861**.

[3] S. ALEXANDER: Astronom. J. **2**, 95.

[4] M. WOLF: Veröff. Königstuhl-Heidelberg **3**, Nr. 5 (Plate V).

[5] See Ann. Lund Obs. **6** (1937); **9** (1941); **10** (1942).

[6] H. SHAPLEY: Harvard Bull. No. 849.

I. Classification.

1. Standard classification. This is the Mt. Wilson classification as used by HUBBLE between 1925 and 1935; it has been so often described [*A*, *B*, *D*] that only a short summary will suffice here. It is illustrated by HUBBLE's well known "tuning-fork" diagram (Fig. 1).

a) Description of types.

α) Elliptical nebulae (*E*), range from circular or globular objects, such as NGC 3379, to elongated, lenticular objects, such as NGC 3115. As a rule they show no structural details, besides a small, bright and strongly condensed nucleus around which the textureless nebulosity decreases smoothly outwards in all directions to an indefinite edge where it fades into the general luminosity of the night sky.

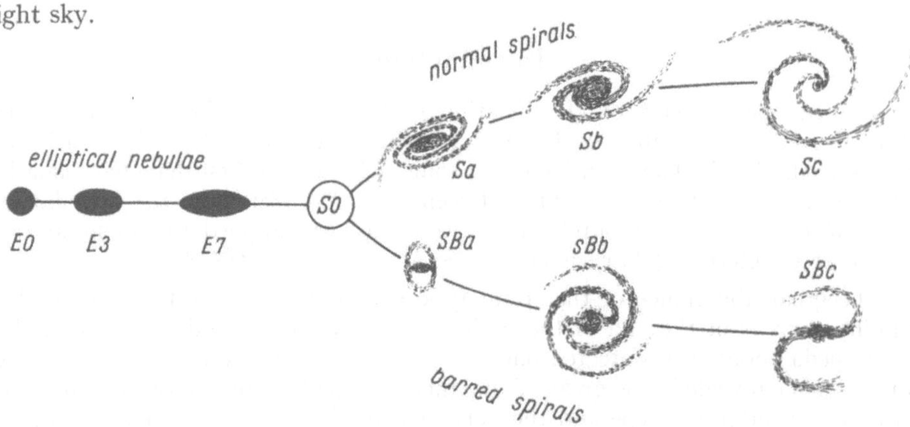

Fig. 1. Standard classification: HUBBLE's tuning-fork diagram (1925).

Sub-types are defined by the index $n = 10\,(1 - b/a)$, if a, b are the apparent major and minor axes measured on photographs. The most strongly elongated objects, type *E* 7, such as NGC 3115, depart notably from a geometrical "elliptical" shape, being pointed near the ends of the major axis.

β) Normal spirals (*S*) show the characteristic spiral arms when seen pole-on and, as a rule, a "spindle" shape with heavy absorption lanes of dark matter when seen edge-on. In the normal spirals the arms emerge tangentially from a bright central nucleus at opposite points on its indefinite edge and vanish after about one complete turn of the best fitting logarithmic spiral [*12*]; (cf. Sect. 6). In the more regular or classical examples only two main arms, very nearly symmetrical with respect to the nucleus, are present. In most cases, however, additional or secondary arms may exist and the spiral pattern is often far from regular.

Sub-types, noted *a*, *b*, *c*, are defined by the relative importance of the nucleus (decreasing from *a* to *c*) and the degree of unwinding and resolution of the arms (increasing from *a* to *c*). According to HUBBLE [*25*]: "the arms appear to build up at the expense of the nuclear regions and unwind as they grow; in the end the arms are wide open and the nuclei inconspicuous. Early in the series the arms begin to break up into condensations, the resolution commencing in the outer regions and working inwards until in the final stages it reaches the nucleus itself" ([*5*], p. 326). The resolution referred to is into blue supergiants and emission objects characteristic of a Type I population. The gradual decrease of the axial ratio nucleus/spiral arms is best seen in edgewise systems (see Plate VIII), while face-on systems show more clearly the increasing resolution and irregularity from "early" types (*Sa*) to "late" types (*Sc*) (Plate V)[1].

Intermediate types: HUBBLE also introduced the notion of "lateral" extension or width of the classification sequence in its intermediate section, giving M 81 with "large nuclear region and thin, rather open arms" and M 94 "having smaller nuclear region with closely coiled arms" as extreme cases. This distinction has been little used in practice.

Of more importance was the recognition of objects intermediate between normal and barred spirals, such as M 83 and M 61 which have been classified alternatively as *Sc* or *SBc*. Their intermediate characteristics first noticed by HUBBLE ([*B*], p. *46*) and LUNDMARK [*43*], [*44*] have been discussed by LINDBLAD and LANGEBARTEL [*41*][2]. They more or less fill the gap between the two branches of the tuning-fork diagram.

γ) Barred spirals (*SB*), include the "pin-wheel" or "*Φ*-type" first described by CURTIS [*9*]. In it a very bright central nucleus is crossed diametrically by a bar at the extremities of which spiral arms start at right angles (in "late" sub-types) or tangentially from the rim of a continuous ring of which the bar is a diameter (in "early" sub-types). Additional or secondary arms may exist, but as a rule the symmetry of the pattern is more regular than in normal spirals (see Plate VI).

Sub-types, noted *a, b, c,* are defined as for the normal spirals by the relative size of the nucleus and the degree of resolution and opening of the spiral structure. In HUBBLE's original system the ring, closed in the *SBa* and *SBb* sub-types, opens at *SBc*, producing the aspect sometimes described as "*S*-shaped" spirals. *SBa* objects observed under various angles give rise to singular "Saturn-like" shapes (Plate X).

δ) Irregulars (*I*), were described originally by HUBBLE [*25*] as a class of objects "lacking both dominating nuclei and rotational symmetry" and of which "the Magellanic Clouds are the most conspicuous examples" ([*25*], p. 328). These wered termed more specifically "Magellanic nebulae" by LUNDMARK [*43*], [*44*]. However, the class was broadened by HUBBLE to include peculiar or chaotic objects which "do not find a place in the sequence of classification" since "the remaining irregulars might be arbitrarily placed in the regular sequence as highly peculiar objects, rather than in a separate class ... Others, such as M 82, are merely nondescript" ([*B*], p. 47).

In fact the symbol *I* has often been used as almost equivalent to the subscript *p* for "peculiar"; such an extension of the notation is both confusing and unwarranted. As a result the relation of "irregulars" to other sections of the classification sequence was not clear, some of them being clearly related to late-type spirals and others to early-type spirals

b) Frequency of types.

From 600 bright galaxies in the Lick and Mt. Wilson plate collections HUBBLE [*25*] found the following apparent relative frequencies:

Type	*E*	*Sa, SBa*	*Sb, SBb*	*Sc, SBc*	*I*
Frequency	17%	19%	25%	36%	2.5%

[1] In accordance with establish custom, the words "early" and "late" are used in connection with the position in the spiral sequence but have no temporal connotation.

[2] See also in [*38*] a discussion by LINDBLAD of the barred spiral characteristics in the Andromeda nebula.

Hence the spirals constituted about 80% of the sample, with $(S+SB)/E=4$ or 5 and, further, a ratio $S/SB=2$ or 3.

However, SHAPLEY and AMES[1] found that in the Coma-Virgo region for $m<12$ spirals comprise only 46% of the population and for $m<14$, $S=48\%$, $E=47\%$, $I=5\%$.

These early statistics, although not necessarily inconsistent, were affected by selection effects arising from clustering and different interpretations of some nebular types not included in the standard classification. Frequencies based on revised types are given in Sect. 3c.

2. Revisions and additions. About 1935 Hubble undertook a systematic morphological study of the approximately 1000 brighter galaxies listed in the Shapley-Ames Catalogue, north of −30° declination, with a view to refining his original classification scheme. This work begun with the 100-inch reflector and later continued with the 200-inch reflector was practically completed at the time of HUBBLE's death in 1953; his notes collected and organized by SANDAGE will be published in the near future with a collection of about 175 photographs illustrating the morphological features characterizing the various galaxy types. Through the courtesy of Dr. A. R. SANDAGE the main revisions introduced by HUBBLE to his system can be briefly described here.

a) The most important addition was the introduction of the SO and SBO types regarded as transition stages between the ellipticals and spirals at the branching off point of the tuning-fork.

SO objects have the smooth appearance of ellipticals, but a luminosity distribution more like that of spirals, although no spiral arms are visible. They are characterized by a sharp, bright nucleus in the centre of a more or less uniform disc or "lens" having a rather sharp outer rim and surrounded by a faint, diffuse "envelope" with indefinite boundaries; the diameter of the lens in which dark crescents or rings of obscuring matter are often observed is usually about one-third of that of the envelope (see Plate III).

There is apparently a continuous transition from "late" ellipticals, such as NGC 4270, 4958, etc. and "early" SO's such as NGC 3630, 4564, etc., then through various stages of the SO sequence to the "earliest" regular spirals.

HUBBLE distinguished two groups of SO objects:

$SO(1)$: smooth lens and envelope; early examples are NGC 1201, 1332, and late examples NGC 3065, 4684.

$SO(2)$: some structure in the envelope in the form of a dark zone and ring; examples are NGC 4459 and 4111, the latter seen edgewise.

In edgewise objects the presence of rings manifests itself by the appearance of "ansae" simulating Saturn's ring, such as in NGC 4215, 7332, etc. (Plate IX). The transition stage SO/a between SO and Sa shows incipient spiral structure in the envelope.

SBO objects are characterized by a bar through the central lens, sometimes broad and hazy, sometimes narrow and sharp; the envelope may form faint outer rings, sometimes conspicuous, sometimes vague or imperceptible (Plate IV).

HUBBLE distinguished three groups of SBO objects:

$SBO(1)$: a bright lens, with broad, hazy bar and no ring, surrounded by a larger, fainter envelope, e.g. as in NGC 3384, 4262, etc., some have circular envelopes, e.g. NGC 4203.

$SBO(2)$: a broad, weak bar across a primary ring, with faint outer secondary rings, e.g. NGC 2859.

[1] H. SHAPLEY and A. AMES: Harvard Bull. **1926**, No. 838, 3; **1930**, No. 876, 39.

SB0(3): well developed bar and ring pattern, with the bar stronger than the ring, e.g. NGC 4643, 5101.

b) Many of the objects originally classified as *SBa* were then reclassified as *SB0* and the *SBa* sub-division had to be redefined as follows:

SBa: smooth bar and lens, with poorly developed, closely coiled arms in envelope and either massive and structureless or filamentary and partially resolved.

HUBBLE further distinguished two or three groups of barred spirals, one in which the arms extend from the rim of a ring crossed by a bar, e.g. NGC 2217, 5566, 5701 (*SBa*), NGC 4999, 5950 (*SBb*), and one in which the arms start at the ends of the bar without ring, e.g. NGC 2798, 4290, 7743 (*SBa*), NGC 1300, 5430, 6951 (*SBb*). In another group still the ring is formed of closely coiled filamentary arms, e.g. NGC 3185, 4037, 4385, 4389. This distinction remains well marked at the *SBb* stage, showing well developed, partially resolved arms going through more than one revolution, but there is no spiral structure in the lens; primary rings consist of spiral arms, secondary rings are seldom found. It was not followed up into the *SBc* sub-division, characterized by wide open, well resolved spiral arms with absorption lanes and spiral structure in the lens, but no ring pattern.

SANDAGE has also recognized two types of normal spirals, one in which the arms start at the rim of a ring structure, the other in which they start from a central nucleus. Some objects of the ringed type had previously been described by SHAPLEY and PARASKEVOPOULOS[1] as "plate spirals" and an investigation of ring structures was made by RANDERS [*57*].

There are further differences in the *multiplicity* of the spiral pattern, some objects showing only two main regular arms, others having a great many tightly coiled whorls. REYNOLDS had already pointed out[2] that some spirals have "massive" arms, e.g. M 31, M 33, while others have "filamentary" arms, e.g. M 81, M 101. The importance of this distinction was acknowledged by HUBBLE [*B*] but could not be incorporated in the classification.

c) An extension of the normal spiral sequence beyond the stage *Sc* was proposed by SHAPLEY[1] who used the notation *Sd* for objects such as NGC 7793 (Plate V, 20), showing a very small, bright nucleus and many knotty irregular spiral arms. This notation could also, and perhaps more appropriately, be applied to highly disorganized and complex spirals of low surface brightness, such as NGC 4395—4401 (see Plate VII).

A parallel extension of the barred spiral sequence beyond the stage *SBc* was introduced by de VAUCOULEURS[3] through the recognition of spiral structure in the Magellanic Clouds and objects of similar type, such as NGC 1313 (Plate VI, 32), 4027, 4618, etc., which may be noted as *SBd* or *SBm*.

Still "later" stages extending the spiral sequence into the irregular types may be represented by objects such as IC 2574 or NGC 2366, IC 4662, etc. These objects are characterized by low surface brightness, high degree of resolution and sometimes outstanding emission nebulosities similar to 30 Doradus in the Large Cloud. They always show an abundance of blue supergiants and strong emission lines in their spectrum.

---

[1] H. SHAPLEY and J. S. PARASKEVOPOULOS: Proc. Nat. Acad. Sci. U.S.A. **26**, 31—36 (1940) = Harvard Rep. 184.

[2] J.H. REYNOLDS: Observatory **50**, 185—189 (1927); with comments by HUBBLE, Observatory **50**, 276—281 and by REYNOLDS, Observatory **50**, 308.

[3] G. DE VAUCOULEURS: Observatory **74**, 23—31 (1954). — Astronom. J. **60**, 126—140, 219—230 (1955).

An important characteristic of the group of irregulars related to the magellanic type $I(m)$ is their small diameter and low luminosity which marks them as dwarf galaxies. Typical objects of this group are NGC 6822, IC 1613, the Sextans system, the Wolf-Lundmark-Melotte nebula, etc.[1] (Plate VII).

In the poorest and smallest of them emission objects may be few or absent, but when their distance is small enough they are always well resolved into blue supergiants and giants.

d) The existence of dwarf ellipticals (*dE*) of very low surface brightness was first brought to notice in 1938 through the discovery of the Fornax and Sculptor systems by SHAPLEY[2]. These are close enough to be resolved on blue plates at $m \approx 18$; except for their very low density they seem to share all other population characteristics of normal ellipticals. Because of their very low surface brightness such systems are difficult to detect and might be much more abundant in space than their belated discovery suggests; nevertheless a special search for them on Harvard plates failed to disclose other examples. A few more "Sculptor type" systems have, however, been found since 1949 with the 48-inch Palomar Schmidt[3].

Dwarf galaxies of a possibly related type have been investigated in 1952 by REAVES [*58*] on plates of the Virgo cluster taken by C. D. SHANE with the Lick 20-inch astrograph. These objects of which the brightest example in the Virgo cluster is IC 3475 have the same smooth, circular or little elongated appearance as the dwarf ellipticals and likewise show little central condensation. HUBBLE and REAVES suggested that they may be related to a group of dwarf spirals of low surface brightness exemplified by NGC 3299 because their colour was at first thought to be blueish; however, recent observations indicate them to be reddish and this strengthens the similarity with the Sculptor-type systems.

e) After all such additional types or variants have been weaned out, there remains a hard core of "irregular" or "peculiar" objects which do not seem to fit in any of the recognized types. One group of such objects consists of strongly interacting or colliding systems such as NGC 1275, 4038—39, 5128, etc.[4] which can often be identified by their distorted structure, abnormal spectrum and strong radio-emission; these systems are discussed in the article by B. Y. MILLS. Double or multiple systems showing moderate interaction in the form of connecting links, distorted spiral arms or irregular filamentary extensions are discussed by F. ZWICKY. Only isolated galaxies or weakly interacting systems are considered in the present article.

Even among them, however, there remains a small number of objects not clearly assignable to any of the previous types; one possible group, exemplified by NGC 3034, NGC 3077, etc. is characterized by an early-type spectrum (A, F) contrasting with a reddish colour ($C \approx +0.8$), irregular absorption patches and filaments and a smooth, unresolved nebulous structure indicating an absence of blue supergiants and of discrete emission nebulosities (see Plate XI). Another possible group, exemplified by NGC 5253[5] is characterized by fairly strong

[1] See E. HUBBLE: Astrophys. Journ. **62**, 409—433 (1925) = M.W.C. 304. — W. BAADE: Astronom. Nachr. **234**, 407 (1929). — F. ZWICKY: Phys. Rev. **58**, 478 (1940). — K. LUNDMARK: V.J.S. **68**, 382 (1933) = Lund. Medd. (I), Nr. 135.

[2] H. SHAPLEY: Bull. Harvard Coll. Obs. **1938**, No. 908.

[3] R. G. HARRINGTON and A. G. WILSON: Proc. Astr. Soc. Pacific **62**, 118—120 (1950). — A. G. WILSON: Proc. Astr. Soc. Pacific **67**, 27—29 (1955).

[4] See W. BAADE and R. MINKOWSKI: Astrophys. Journ. **119**, 215—231 (1954). — F. ZWICKY: Ergebn. exakt. Naturw. **29**, 344—385 (1956).

[5] D. S. EVANS: Observatory **72**, 164—166 (1952).

emission lines and a complex structure unlike that of the magellanic irregulars or late-type spirals. Still other puzzling objects such as MAYALL's nebula[1], also show strong emission lines but have a fairly smooth structure[2].

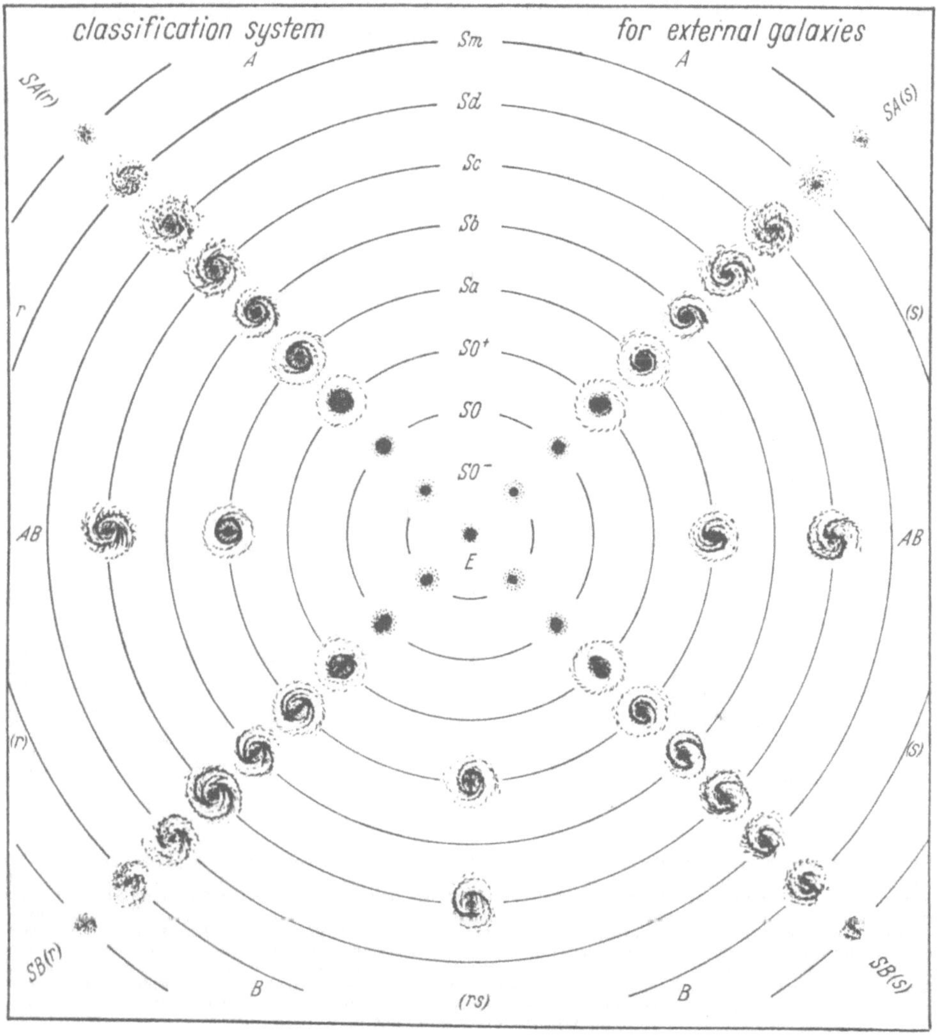

Fig. 2. Revised classification: a plane projection of the classification volume. Compare with Fig. 3. The ordinary spirals *SA* are in the upper half of the figure, the barred spirals *SB* in the lower half. The ring types (*r*) are to the left, the spiral types (*s*) to the right. Ellipticals and lenticulars are near the center, magellanic irregulars near the rim. The main stages of the classification sequence from *E* to *Im* through *SO*⁻, *SO*, *SO*⁺, *Sa*, *Sb*, *Sc*, *Sd*, *Sm* are illustrated, approximately on the same scale, along each of the four main morphological series *SA* (*r*), *SA*(*s*), *SB*(*s*), *SB*(*r*). A few mixed or "intermediate" types *SAB* and *S*(*rs*) are shown along the horizontal and vertical diameters respectively. Structures predominantly of Type I population are dashed, of Type II dotted. This classification scheme, as used in the Mount Stromlo survey of southern galaxies, is superseded by the slightly revised and improved system illustrated in Fig. 3 and Plates I to XI.

3. Revised classification. In the course of a survey of bright southern galaxies with the 30-inch Reynolds reflector at Mount Stromlo [*80*] a classification and

[1] R. T. SMITH: Proc. Astr. Soc. Pacific **53**, 187 (1941); also Astrophys. Journ. **119**, 225 (Fig. 14) (1954).

[2] Some early-type spirals are known to show abnormally wide emission lines in the spectra of their nuclei, although the origin of such high random velocities (≈ 5000 km/sec) is not known, this need not be considered as an indication of a distinct type, but merely of a peculiarity.

notation system has been developed to include in a consistent scheme all or most of the recent revisions and additions to the standard classification. The classification follows as closely as possible the revised Mt. Wilson-Palomar scheme, but includes the additional types and sub-types suggested by the Harvard, Lick and Mt. Stromlo work; the notation is based on the original Mt. Wilson system, but has been modified and supplemented as required by the intro-

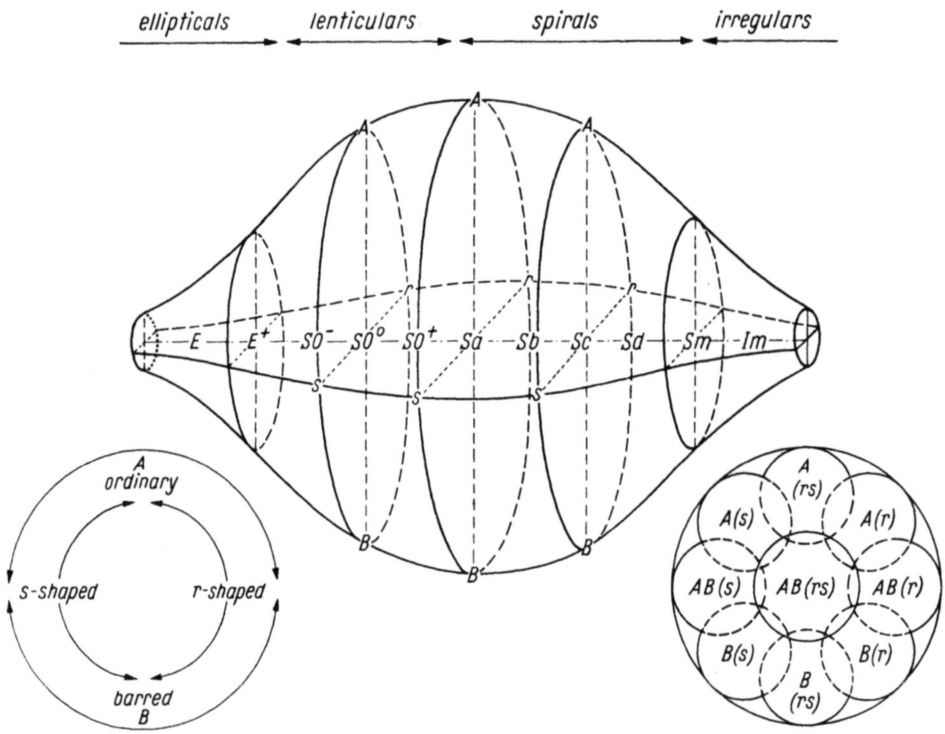

Fig. 3. A 3-dimensional representation of the revised classification scheme and notation system. From left to right are the four main *classes*: ellipticals E, lenticulars SO, spirals S irregulars I. Above are the ordinary *families* SA, below the barred families SB; on the near side are the S-shaped *varieties* $S(s)$, on the far side the ringed varieties $S(r)$. The shape of the volume indicates that the separation between the various sequences $SA(s)$, $SA(r)$, $SB(r)$, $SB(s)$ is greatest at the transition stage SO/a between lenticulars and spirals and vanishes at E and Im. A central cross-section of the classification volume illustrates the relative location of the main types and the notation system. There is a continuous transition of mixed types between the main families and varieties across the classification volume and between stages along each sequence; each point in the classification volume represents potentially a possible combination of morphological characteristics. For classification purposes this infinite continuum of types is represented by a finite number of discrete "cells." Compare with Plates I and II.

duction of new types and sub-types and so as to permit a finer description of structural features. Classification and notation are illustrated in Fig. 2 which may be regarded as an extension of Hubble's tuning-fork diagram.

Actually Fig. 2 may be considered as a plane projection of a three dimensional representation, the idea of which was first conceived during conversations with Dr. Sandage in August 1955. It is schematically illustrated in Fig. 3 which may be consulted in conjunction with Fig. 2 to follow the detailed description of the adopted classification and notation (compare also with Plates I and II). The three-dimensional classification appears necessary to represent objects of mixed characteristics in correct relation to the main sequences as well as the progressive divergence of the various sequences from E to SO/a through the SO stages and their ulterior convergence from SO/a to I through the S stages.

a) *Isolated galaxies.*

The present classification scheme and notation system rest on the following principles:

Four *classes* are retained: ellipticals *E*, lenticulars *SO*, spirals *S*, irregulars *I*, which coincide with the main divisions introduced by HUBBLE.

The two *families* of lenticulars and spirals, originally denoted "normal" *S* and "barred" *SB* by HUBBLE, are now designated "ordinary" *SA* and "barred" *SB*, so as to permit the use of the compound symbol *SAB* for "intermediate" objects of mixed characteristics. The symbol *S* alone is used when a spiral object cannot be more accurately classified as either *SA* or *SB* because of poor resolution, unfavorable tilt, etc. The "ordinary" spirals are not more "normal" than those of the "barred" family which are at least as common (cf. Sect. 3c).

In the new class of lenticulars the two families are denoted *SAO* and *SBO*, depending on the absence or presence of a bar structure across the central lens; "intermediate" objects with a very weak bar are noted *SABO*. The symbol *SO*, used for *SAO* in HUBBLE's notation, is now used for a lenticular object which cannot be more precisely classified as either *SAO* or *SBO*; this is often the case for edgewise objects[1].

Two main *varieties* are recognized in each of the lenticular and spiral families, the "annular" or "ringed" type, denoted (*r*), and the "spiral" or "*S*-shaped" type, denoted (*s*). Intermediate types are noted (*rs*). In the "ringed" variety the structure includes circular (sometimes elliptical) arcs or rings (*SO*) or consists of spiral arms or branches emerging tangentially from an inner circular ring (*S*). In the "spiral" variety two main arms start at right angles from a globular or little elongated nucleus (*SA*) or from an axial bar (*SB*).

The distinction between the two families *A* and *B* and between the two varieties (*r*) and (*s*) is most clearly marked at the transition stage *SO/a* between the *SO* and *S* classes. It vanishes at the transition stage between *E* and *SO* on the one hand, and at the transition stage between *S* and *I* on the other (cf. Fig. 3).

Four *sub-divisions* or *stages* are distinguished along each of the four spiral sequences *SA*(*r*), *SA*(*s*), *SB*(*r*), *SB*(*s*), viz. "early", "intermediate" and "late" denoted *a*, *b*, *c* as in the standard classification, with the addition of a "very late" stage, denoted *d*. Intermediate stages are noted *Sab*, *Sbc*, *Scd*. The transition stage towards the magellanic irregulars (whether barred or not) is noted *Sm*, e.g. the Large Magellanic Cloud is *SB*(*s*)*m*.

Along each of the non-spiral sequences the signs + and — are used to denote "early" and "late" subdivisions; thus E^+ denotes a "late" *E*, the first stage of the transition towards the *SO* class[2]. In both the *SAO* and *SBO* sub-classes three stages, noted SO^-, SO°, SO^+ are thus distinguished; the transition stage between *SO* and *Sa*, noted *SO/a* by HUBBLE, may also be noted Sa^-. Notations such as Sa^+, Sb^-, etc. may be used occasionally in the spiral sequences, but the distinction is so slight between, say, Sa^+ and Sb^-, that for statistical purposes it is convenient to group them together as *Sab*, etc. Experience shows that this makes the transition subdivisions, *Sab*, *Sbc*, etc. as wide as the main sub-divisions, *Sa*, *Sb*, etc.[3].

---

[1] The symbol (*S*) may be used when even the distinction between *S* and *SO* is no longer possible, as in the case of very small, poorly resolved objects.

[2] An example is NGC 3115 which, although selected as the type object for *E*7 in HUBBLE's original scheme, shows an absorption marking and a beginning of nuclear differentiation on 200 inch plates.

[3] No E^-, nor I^- or I^+ are at present clearly recognized, but such notations may later prove useful for a still finer classification based on accurate surface photometry.

The classification of "irregulars" is still somewhat in doubt; objects obviously related to the magellanic type, but which do not show clearly the characteristic spiral structure, are noted $I(m)$; those with an elongated core and asymmetical branches are probably later stages of the SB sequences, e.g. NGC 4449, 6822, etc., while those more nearly symmetrical and without bar-like core are probably later stages of the SA sequences, e.g. IC 1613, IC 2574, etc. (see Plate VII) the distinction vanishes in the ultimate, chaotic dwarf stages of low surface brightness, e.g. the Wolf-Lundmark nebula or the Sextans system[1].

A faint, *outer* ring-like structure is often observed in lenticulars and early-type spirals which is not clearly related to a specific type, i.e. with some variants it appears about equally in all four sequences near the transition stage SO/a. This particularity which seems therefore more characteristic of a certain stage of evolution than of any definite line of evolution, is denoted by an (R) *preceding* the symbol of the class. Examples of $(R)\,SA$ are NGC 1068 (M 77), NGC 4736 (M 94), NGC 7217, etc. of $(R)\,SB$: NGC 1291, 1326, 2859, etc. Some objects have both an outer (R) structure and an inner (r) pattern; a good example is NGC 6753, noted $(R)\,SA(r)\,ab$ (Plate V, 22).

In the spiral sequences supplementary data of some interest are the *multiplicity* of the spiral pattern and the *character*, "massive" or "filamentary" of the spiral arms. These are included whenever possible as subscripts respectively after the family and variety symbols. Thus M 33 is noted $SA(s)_2\,c_m$ and described as "an ordinary, late-type, S-shaped spiral with two main massive arms". When one or more additional and weaker arms are present this is noted $2+1$ (e.g. M 99), $2+2$ (e.g. M 51), etc. An asymmetrical, regular spiral with one arm stronger than the other as is often the case among late-type barred spirals (e.g. NGC 7479) is noted $1+1$ (not 2). Branching arms not clearly assignable to a definite multiplicity are noted 1^+, 2^+, etc. as the case may be; this is a frequent occurrence among latetype spirals. A complicated spiral pattern of multiplicity higher than 4, as often observed in late-type spirals, is simply noted n (e.g. NGC 2903, 7793); this, however, occurs also among early-type spirals of the ringed variety (e.g. NGC 4736, 7217).

Edgewise systems which can usually be classified only as $E7$, SO, $S(a, b, c, d)$ or I are noted (sp) for "spindle". For ellipticals the traditional notation $E0$, $E1, \ldots E7$ may be preserved, although it is not homogeneous with the rest of the classification scheme and notation system used for the other classes; it is really almost superfluous when measured dimensions and flattening are given.

An advantage of the present notation system is that it is possible to retain significant information on objects poorly resolved by dropping one or more symbols which from right to left refer to stages, varieties, families and classes of increasing generality. Thus S may be either SA or SB, SO either SAO or SBO, SA is either $SA(r)$ or $SA(s), \ldots,$ Sa is either SAa or $SBa, \ldots,$ etc. Hence the nature and number of symbols used can be selected according to the amount of detail visible in any given object with any particular telescope. Conversely the degree of completeness in the symbolism gives an indication of the relative degree of resolution available; this, of course, depends on both the size of the telescope and the size of the nebula.

[1] Objects of low surface brightness (when it is not obviously due to local obscuration, e.g. as for IC 10, IC 342 etc.) are almost invariably dwarf systems; this particularity may be noted by a (d) *preceding* the symbol of the class which is usually either E or I. The existence of dwarf lenticulars and spirals remains to be demonstrated. Criteria are still lacking for the assignment of objects to the giant and supergiant groups on purely morphological grounds.

b) Pairs and multiplets.

These are discussed in detail by ZWICKY. It is nevertheless of some importance for the present discussion to determine the separation between neighboring objects within which their mutual interaction begins to cause significant structural distortion and thus sets a limit to the validity of the classification system established in the first instance for isolated galaxies. A rough classification of pairs is as follows:

$P(a)$: pairs of widely separated objects showing no clear sign of interaction (occasional optical pairs are included in this group);

$P(b)$: close pairs showing clear signs of interaction, such as tidal distortion of their outlying parts;

$P(c)$: colliding pairs where the "main bodies" are in actual contact and show extensive disruption of their internal structure.

Many parameters are obviously involved, such as separation, diameters, masses, velocities, etc. in the interaction of two galaxies. Only the first two are more less directly accessible in most cases and more precisely the ratio x' of the projected separations $s'_{1,2}$ between the nuclei G_1, G_2 of a pair to the sum of their apparent photographic radii r_1, r_2. The true separation s is not known in individual cases, but for a random distribution of orientations of $\overline{G_1 G_2}$ with respect to the line of sight the mean value of s'/s is $\pi/4 \approx 0.79$. For 40 pairs south of $-35°$ observed with the Reynolds reflector on Mt. Stromlo[1] the following mean values were obtained:

$$P(a) \quad \bar{x}' = 3.65 \quad (\text{range}: 1.6 - 9:) \quad n = 30,$$

$$P(b) \quad \bar{x}' = 1.30 \quad (\text{range}: 0.6 - 1.6) \quad n = 8,$$

$$P(c) \quad \bar{x}' = 0.3: \quad (\text{range}: 0.2: - 0.4:) \quad n = 2.$$

Allowing for scale factors the average projected and actual separations are:

$$P(a) \quad \bar{s}' = 18 \text{ kpc} \quad (\text{range 8 to 45 kpc}) \quad \bar{s} = 23 \text{ kpc},$$

$$P(b) \quad \bar{s}' = 6.5 \text{ kpc} \quad (\text{range 3 to 8 kpc}) \quad \bar{s} = 8 \text{ kpc}$$

on a distance scale in which HUBBLE's expansion parameter is 180 km/sec per Megaparsec. The mean critical separation ($x' \approx 1.6$, $x \approx 2.0$) within which tidal distortion becomes notable is then of the order of 10 kpc. Among the bright galaxies south of $-35°$ only 10 out of more than 400, or less than 2.5%, are within this limit. Hence, the revised classification should be applicable to at least 97 or 98% of the external galaxies, of which about 95% can be included in the normal scheme of Fig. 2 and the remaining 2 or 3% must still be regarded as special or "peculiar" cases reserved for further analysis.

c) Frequency of revised types.

A classification of over 200 southern galaxies in the revised system [*80*] yields the following apparent relative frequencies for a sample essentially selected according to apparent magnitude and surface brightness (all objects listed in the Shapley-Ames Catalogue south of $-35°$).

---

[1] The photographic diameters used in the Mt. Stromlo survey [*80*] are corrected for tilt effects, i.e. reduced to "face-on", and define a system of dimensions intermediate between those of REINMUTH [*62*] and of SHAPLEY [*66*].

The regular spirals SA, SB make up just over fifty percent of the sample, ellipticals and lenticulars close to a quarter each. The discrepancies between earlier statistics (cf. Sect. 1b) may therefore have hinged mainly on whether

Table 1. *Frequency of revised types.*

Class/Family	*E*	*SO*	*SA*	*SB*	*I*	*Pec*
Frequency (%) . .	23.4	21.0	24.4	26.3	3.4	1.5

the lenticulars were counted as ellipticals or spirals the ratio S/E changing correspondingly from 1/1 to 3/1. Note also that the ordinary spirals appear no longer more abundant than the barred spirals despite the inclusion of intermediate types (SAB) with the SA group in the table.

Table 2. *Frequency of sub-divisions.*

Type	*E*	*E/SO*	*SO*	*SO/a*	*Sa*	*Sab*	*Sb*	*Sbc*	*Sc*	*Scd*	*Sd*	*Sm*
Frequency	22.0	9.5	10.5	9.0	4.5	6.5	7.5	7.5	10.5	8.5	2.0	2.0

The detailed frequency in each of the sub-divisions along the regular classification sequence is given in Table 2, in which $E/SO = E^+$, SO^- and $SO/a = SO^+$, Sa^-.

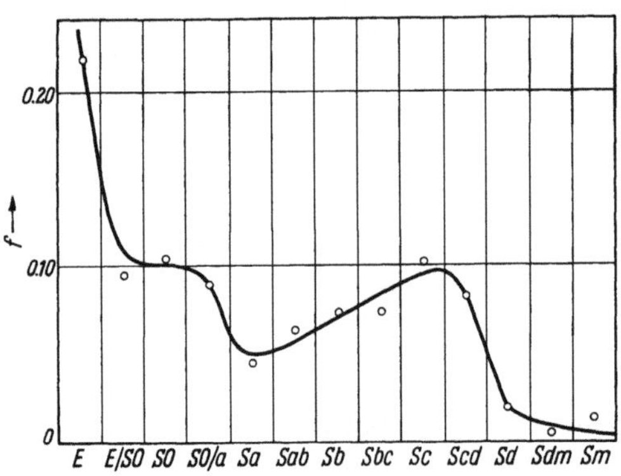

Fig. 4. Frequency distribution of revised types in the Mount Stromlo survey.

A plot of the frequencies (Fig. 4) gives a fairly smooth curve indicating maxima at Sc and E separated by a minimum at Sa. The low apparent frequency of the Sd, Sm stages is certainly due to a large extent to the low absolute luminosity and surface brightness of galaxies at these stages, but it is not yet possible to correct the apparent frequencies for such selection effects. On the other hand the high frequency of ellipticals cannot be due in general to an abnormally high absolute luminosity, so that it probably reflects the actual great abundance of this class[1].

[1] It should be emphasized that these provisional data based on a rather small sample of 200 objects are still subject to further revision; more definite and detailed results will eventually be forthcoming through the current reclassification of 1250 bright galaxies in the Shapley-Ames Catalogue.

II. Morphology.

a) Qualitative morphology.

4. Description of typical examples. The following eleven plates illustrate the revised classification system and will serve to support the description of the main morphological characteristics used as classification criteria. Since the revised three-dimensional classification recognizes at least 16 stages along the sequence from *E* to *Im* and 9 or more pure or mixed types at each stage across it (at least in the spiral sequences), a fully illustrated description, allowing also for tilt and orientation effects, would require several hundred examples. However, the main criteria used to define the position of any normal galaxy along and across the classification volume can be illustrated with only a fraction of the total number of possible cases.

Plates I and II describe the main criteria used to distinguish between families and varieties in a central cross-section near the stage *Sb*.

Plates III and IV illustrate the ellipticals and principal stages of the lenticular sequences.

Plates V and VI illustrate the four main stages along each of the four principal sequences of ordinary and barred spirals.

Plate VII shows the (magellanic) irregular later stages of the spiral sequences.

Plates VIII and IX illustrate the appearance of edgewise system of the lenticular, spiral and irregular classes.

Plate X illustrates the effects of tilt and orientation on the appearance of *SB*(*s*)*a* or *SB*(*r*)*a* spirals.

Plate XI shows some peculiar lenticulars and (non-magellanic) irregulars of the M 82 type.

In order to minimize as far as possible the complications arising from differences in scale and resolution, several sources of illustrations were used as follows:

1. Mt. Palomar 200-inch reflector (P 200″) and 48-inch telescope (P 48″).
2. Mt. Wilson 100-inch and 60-inch reflectors (W 100″; W 60″)[1].
3. Mt. Stromlo 74-inch and 30-inch reflectors (S 74″; S 30″).
4. Isaac Roberts 20-inch reflector (IR 20″)[2].

For technical reasons all photographs are reproduced here as inverted negative prints, i.e. are mirror images of the objects as they appear on the sky, and the orientation with respect to the celestial coordinates is arbitrary.

The classification criteria are described in relation to the three main parts of a galaxy defined by HUBBLE:

a) the *nucleus*, i.e. the very small, very bright central condensation often sharply defined as the centre of symmetry of the structure; it is round in *SA*, elliptical in *SB*.

b) the *lens*, smooth, bright and sharply defined in the lenticulars and early spirals, being crossed by the bar in *SB* and brighter at the edge in *S*(*r*).

c) the *envelope*, generally faint and smooth with indefinite outer boundary in the lenticulars, brighter and occupied by the spiral arms in spirals. The outer (*R*) structure appears often in its outer parts near the stage *SO*/*a*.

The distinction between these three regions is most definite near the middle of the classification sequence (*SO*/*a*) and vanishes at *E* and *Im*.

[1] The Mt. Wilson and Palomar photographs are reproductions from the "Hubble Memorial Volume" generously communicated in advance of publication by Dr. A. R. SANDAGE.

[2] The Isaac Roberts photographs are from direct prints of the original plates, now stored at the Paris observatory and kindly made available some years ago by Prof. P. COUDERC.

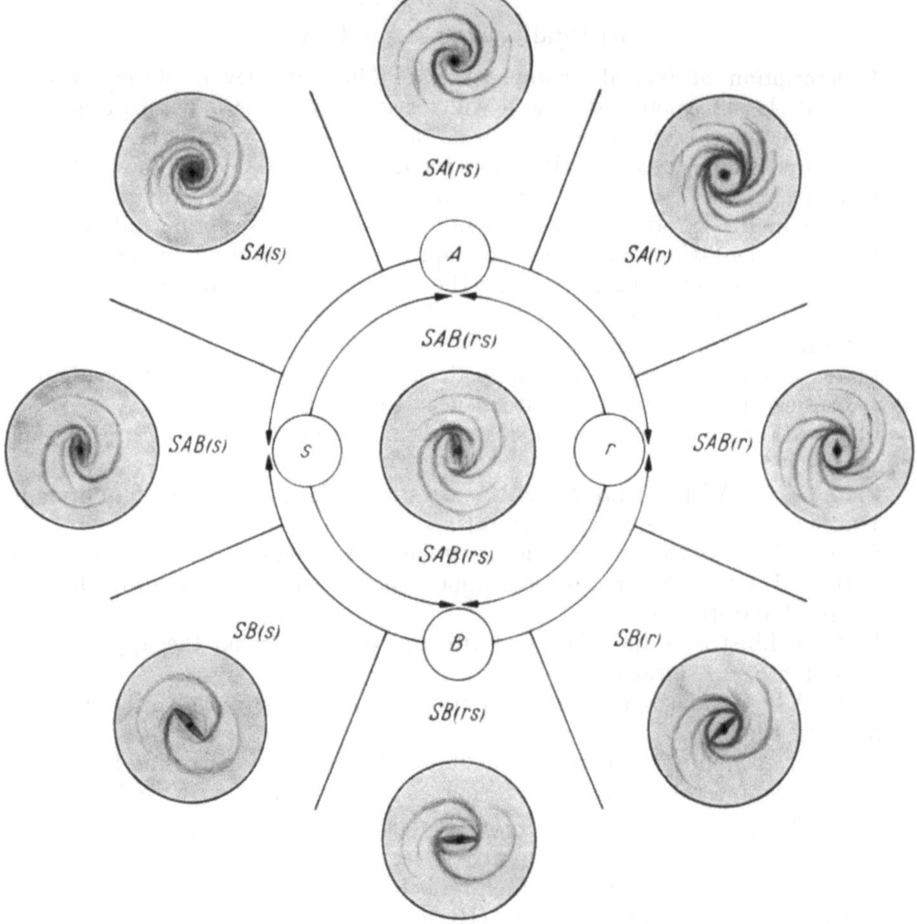

Plates I and II. Cross-section of classification volume near stage Sb.

The spirals of the ordinary family *SA* are above, of the barred family *SB* below; the ringed variety *S* (*r*) is to the right, the *S*-shaped variety *S* (*s*) to the left. The mixed types *SAB* (*rs*) are in the centre. There is a continuous transition between each type and only the main forms are illustrated; actual objects can occupy any intermediate position in the plane (volume).

The main classification criteria are as follows:

SA (*r*): has a small, sharp, very bright and round nucleus isolated in the centre of a circular ring at the edge of which emerge many tightly wound filamentary spiral arms or arcs; the nucleus and ring often merge and disappear in the over-exposed image of the central bulge (as illustrated), but the high multiplicity of the spiral pattern is usually sufficient for identification; a weak outer ring (*R*) made up of many closely coiled spiral arcs is often present in early stages of this sequence. Examples are NGC 488 (illustrated from P 200″), 7217, 6753 (Plate V, 22); other examples: NGC 4736 (M 94) (Plate V, 21), *Sab* and 5055 (M 63), *Sbc*.

SA (*rs*): has a fairly small, fairly sharp, bright and round nucleus in the centre of a diffuse lens or bulge out of which two main arms and two or more additional, weaker arms emerge tangentially; the two main arms simulate an incomplete ring around the lens; weak outer arms simulate an (*R*) structure in the early stages. Examples are NGC 1068 (M 77), 3147, 4237, possibly NGC 3521, 4800, 7079, 7590. NGC 5194 (M 51) (illustrated from IR 20″) is not a very good example as it is more nearly *Sbc* and intermediate between *SA* (*rs*) and *SA* (*s*).

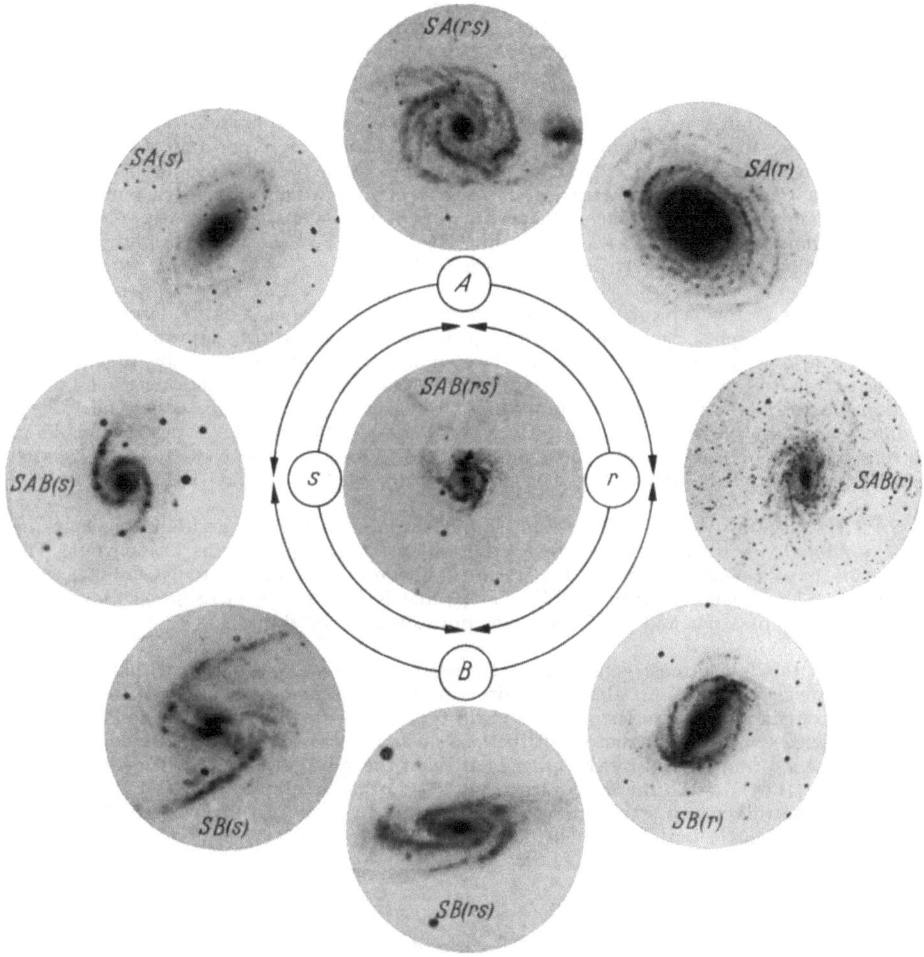

SA(s): This is the typical, regular logarithmic spiral; it has a fairly large, diffuse, round nucleus extending smoothly into the circular or little elongated lens which shows some spiral pattern of dark matter; two main spiral arms with occasional branching or weaker secondary arms start at the rim of the lens. Examples are NGC 3031 (M 81), *Sab* (illustrated from IR 20″), 4569, 7205 (Plate V, 18), 7331. One diameter of the lens is occasionally slightly brighter and longer and the outer arms may have a tendency to "return" to the lens, indicating a transition toward the *SAB(s)* type; examples are NGC 224 (M 31), *Sb* and 4321 (M 100), *Sbc*.

SAB(s): Has a fairly small, bright, elongated nucleus crossed by a weak, twisted dark lane in the centre of a weak and broad bar marking the major axis of a faint, diffuse and elongated lens with much spiral pattern of dark matter having a tendency to run parallel to the bar. Two main arms emerge near the extremities of the bar along a smooth, curved path turning sharply just outside the lens; their faint extremities tend to return to the lens along an almost circular outer loop, simulating an (*R*) structure in the early stages. Examples are NGC 1566 (illustrated from S 30″), NGC 5236 (M 83), *Sbc*, 4579, *Sab* and 7392.

SB(s): This is the typical "*S*-shaped" barred spiral; it has a small, very bright and elongated nucleus distorted by a strong, twisted dark lane as it crosses over from one side to the other of the strong, narrow bar marking the major axis of a much elongated lens. Two strong, main arms start sharply at right angles to the bar at both ends and return faintly to it after completing a turn of a quasi-circular loop. Examples are NGC 1097,

1300, 1365 (illustrated from S 30″), 5383. Later stages in this sequence are markedly asymmetrical as NGC 7479, *Sc* and 7741, *Scd* (Plate VI, 31). Some, like NGC 1300, have a third, fainter arm forming a half-ellipse close to the lens and indicating a transition toward the *SB*(*rs*) type.

SB(*rs*): has a fairly small, very bright and elongated nucleus crossed by a twisted dark lane, as in *SB*(*s*), in a very strong, very narrow bar marking the major axis of an elongated lens whose rim is brighter, especially near the extremities of the bar; this produces a characteristic "dash-dot in brackets" pattern, thus (— o —), from which emerge two main spiral arms with faint additional branches near the rim of the lens. Examples are NGC 4548, 4593, 7124 (illustrated from S 74″). Some, like NGC 4349, *Sab*, having a brighter lens and more continuous rim mark the transition toward *SB*(*r*).

SB(*r*): This is the typical "*Φ*-shaped" barred spiral. It has a fairly large, elongated nucleus with weak spiral dark lane in a strong, narrow bar along the major axis of an elliptical ring marking the edge of the lens. Two main arms start tangentially from the ring near the extremities of the bar (i.e. at right angles to the bar); one or two fainter arms branch out from breaks in the ring near its minor axis. In early stages the main arms tend to form an outer, circular (*R*) structure, as in NGC 1433 (Plate VI, 25); in later types the breaks in the ring tend to produce slightly hexagonal shapes. Examples are NGC 1433 (illustrated from S 74″), *Sa* or *Sab*, 3185, 3351, *Sab* (Plate VI, 26), 2523, *Sbc* (Plate VI, 27).

SAB(*r*): has a fairly small, little elongated nucleus in a fairly broad and faint bar marking the major axis of a little elongated ring from which several spiral arms branch out. The main arms have a slight tendency to "return" inwards. Examples are NGC 1832, 7531, *Sab*, 6744, *Sbc* (illustrated from S 30″). Some, like NGC 6902, 6935, *Sa*, 6937, have only very faint traces of a bar and mark the transition towards *SA*(*r*).

SAB(*rs*): this is the most general mixed type involving all possible transitions between the main typical patterns. An almost infinite variety is possible here, but for classification purposes the main characteristics of this hybrid type are: a small bright nucleus in a broad, diffuse bar with some spiral structure in the lens. The bar crosses a nearly circular or often hexagonal pseudo-ring formed by the inner sections of the spiral arms. A very good example is NGC 4303 (M 61), *Sbc* (illustrated from IR 20″); other examples are NGC 3145, 6814. At a later stage this mixed structure is well illustrated by the central regions of NGC 5457 (M 101) and NGC 6946, both *Scd*.

Plate III. Ellipticals and ordinary lenticulars.

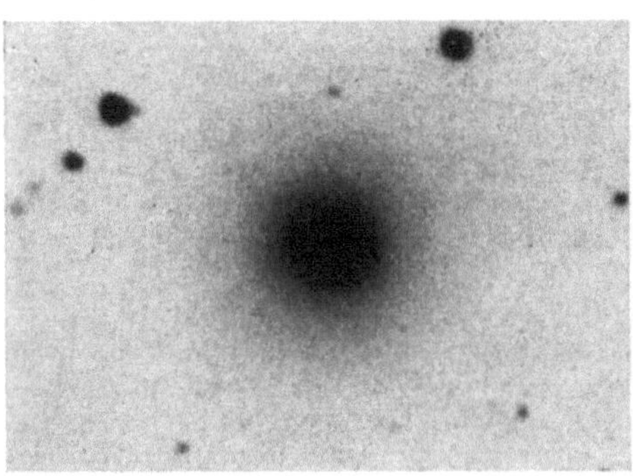

1. NGC 7144 (S 74″): *E0*. The appearance of a distinct nucleus is due to a photographic effect; there is actually a smooth, continuous decrease of luminosity from the centre outwards.

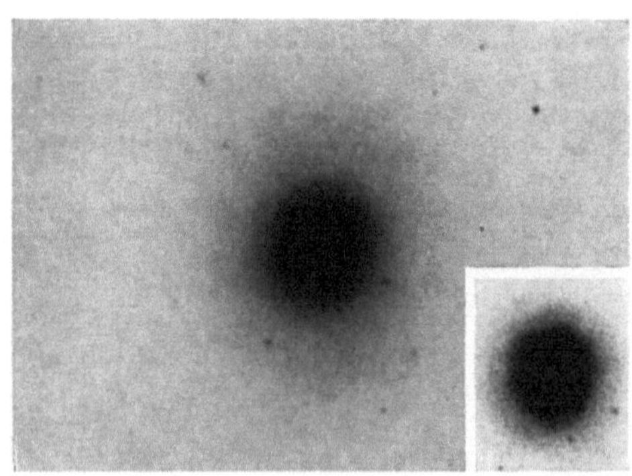

2. NGC 5273 (P 200″): *SA*(*s*)O^-. Traces of spiral structure in the smooth lens and envelope surrounding the large nucleus; little or no dark matter. With lower resolution lens and envelope would simulate a late elliptical; compare with 1 and 5.

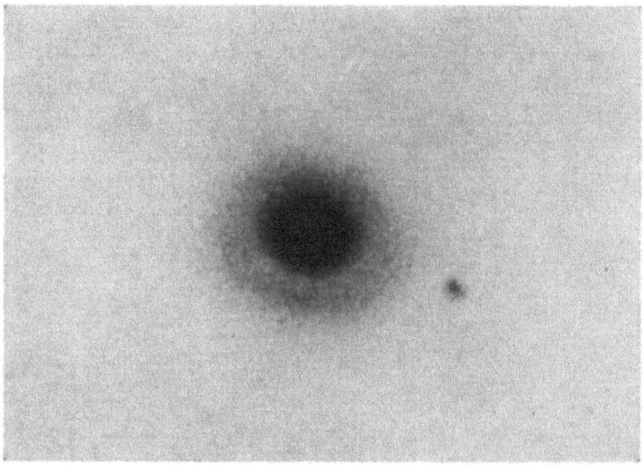

3. NGC 4459 (W 100″): $SA(s)O^{\circ}$. A weak ring or whorl of dark matter appears in the lens close to the large nucleus. The envelope does not show up on this view taken with a short exposure. Compare with 4 and 7.

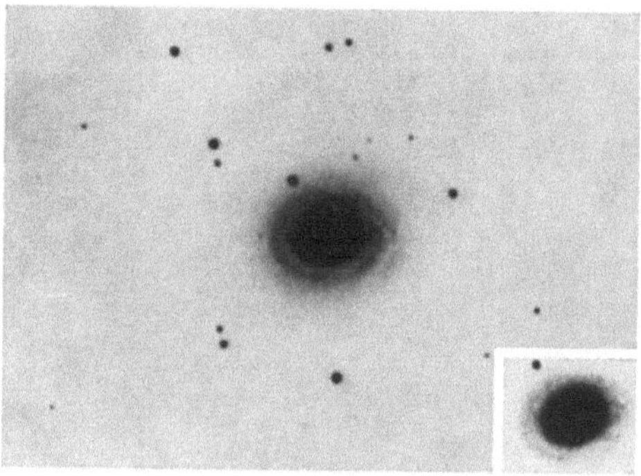

4. NGC 2855 (P 200″): $SA(s)O^{+}$. The ring of dark matter outlines a weak smooth spiral structure around the large and bright nucleus. Compare with 3 and 8.

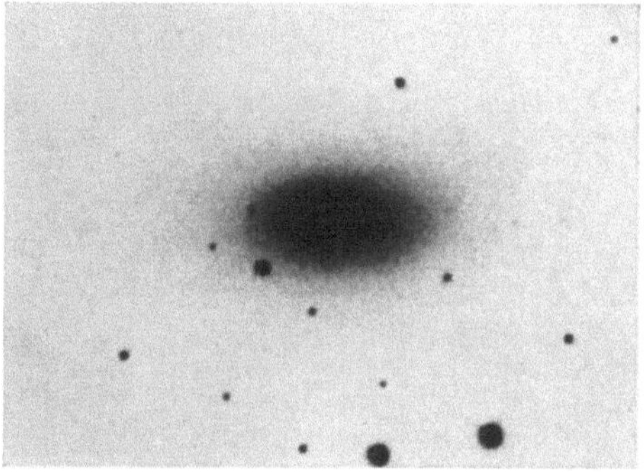

5. NGC 7457 (W 100″): $E^{+}5$. The merest traces of differentiation between nucleus and lens appear at the first transition stage between *E* and *SO*. Compare with 1, 2 and 6.

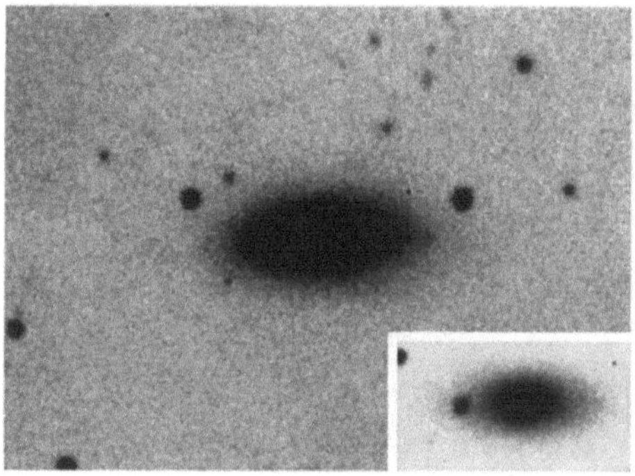

6. NGC 7166 (S 74″): $SA(r)O^{-}$. The lens and envelope still simulate an elliptical, but a small nucleus appears in the lens. Note size of nucleus; compare with 2 and 5.

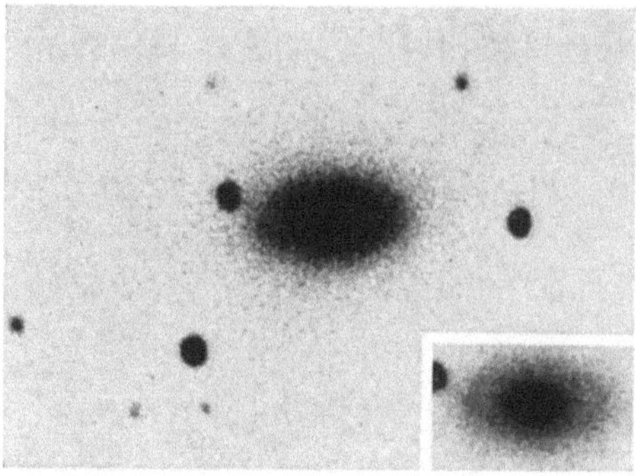

7. NGC 1553 (S 30″): $SA(r)O^{\circ}$. A weak trace of ring appears at the edge of the lens. Note small, distinct nucleus in lens and faint, outer envelope. Compare with 3, 6 and 8.

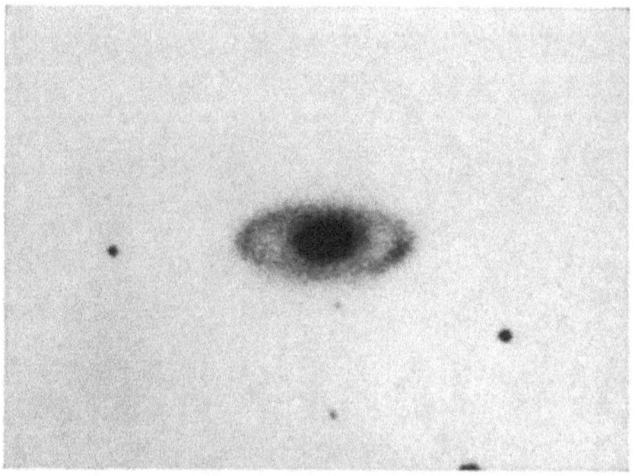

8. NGC 7702 (S 74″): $SA(r)O^{+}$. A definite ring, well separated from the small nucleus appears at the edge of the lens. The very faint envelope does not show up on this image of short exposure. Note asymmetry of nucleus and ring near minor axis indicating the presence of dark matter in the equatorial plane. Compare with 4 and 7.

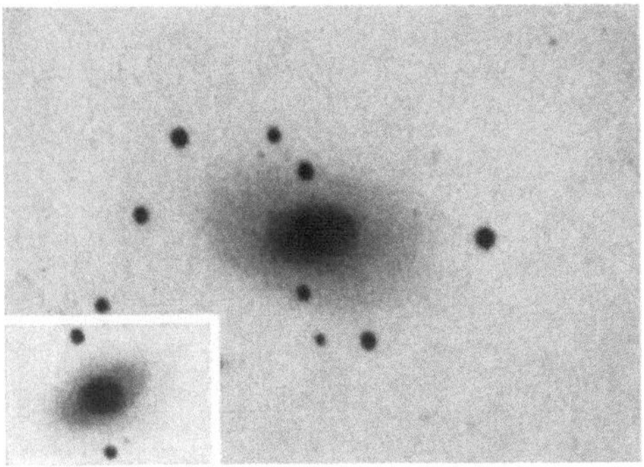

9. NGC 7079 (S 74″): *SB(s)O°*. Small bright nucleus in short and bright bar surrounded by very faint traces of spiral arcs in lens; extensive envelope. Compare with 12 and 13.

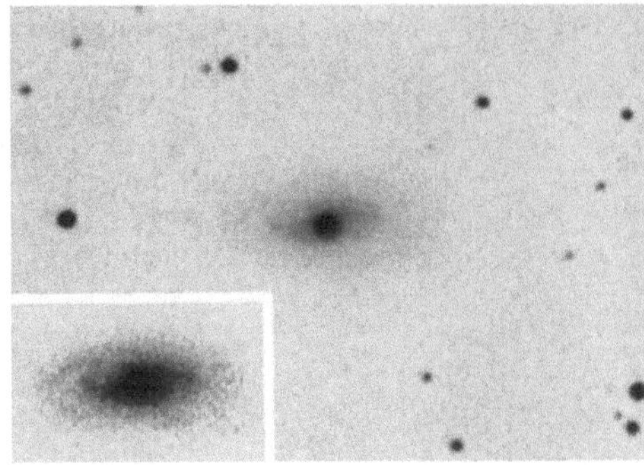

10. NGC 1617 (S 30″): *SB(s)O⁺*. Very small nucleus in short and fainter bar curving into spiral arms or arcs in lens and envelope. The spiral pattern is blurred by insufficient resolution here. Compare with 9 and 14.

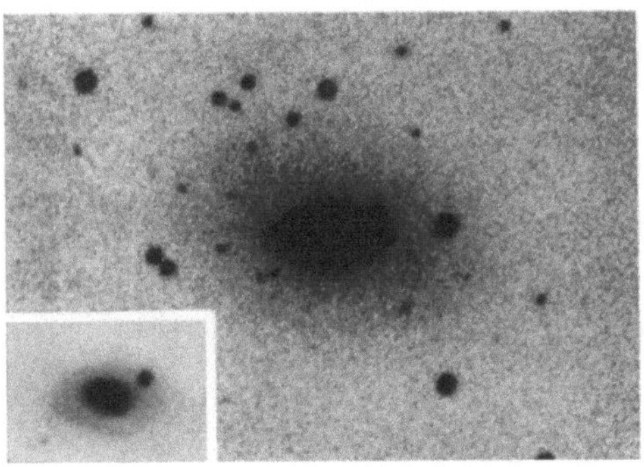

11. NGC 6873 (S 74″): *SAB(rs)O°*. Fairly small nucleus in elongated lens with dark ring or crescent, surrounded by extensive, bright envelope. Compare with 3, 7 and 9.

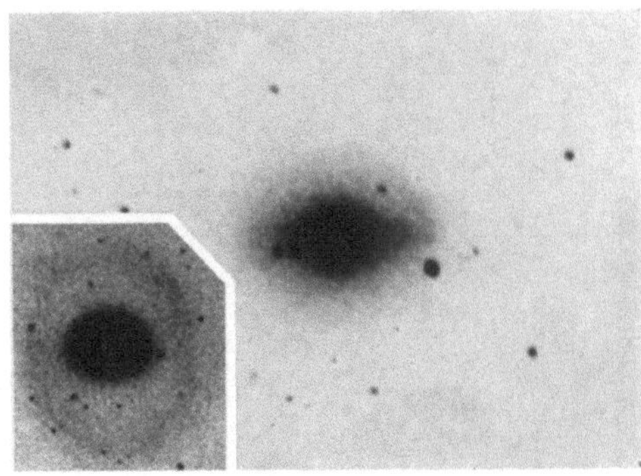

12. NGC 1291 (S 30″): *(R) SB(s)O⁺*. Fairly large nucleus on weak bar marking major axis of little elongated bright lens. Faint outer ring structure from which emerge two very weak spiral arms or arcs. No ring structure in lens. Compare with 9, 10 and 16.

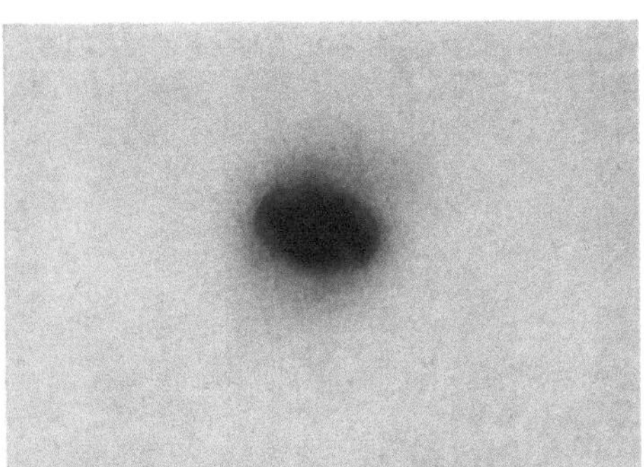

13. NGC 4262 (P 200″): *SB(r)O°*. Small nucleus in short, stubby bar (stronger near extremities) in bright diffuse lens. The nucleus appears large on this image taken in yellow light. Compare with 9, 15 and 16.

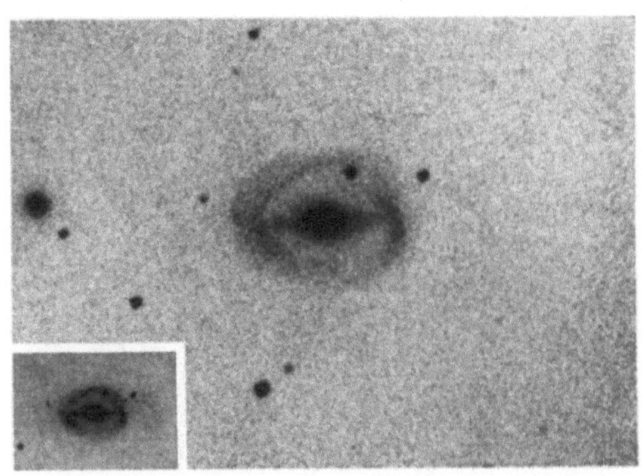

14. NGC 1512 (S 30″): *SB(r)O⁺*. Small, elongated nucleus on narrow, bright bar marking the major axis of elliptical ring; the ring is stronger near the extremities of the major axis. Traces of spiral arms or arcs in weak envelope outside ring. Compare with 8 and 25; NGC 1510 (*EO*) is at left near edge of field.

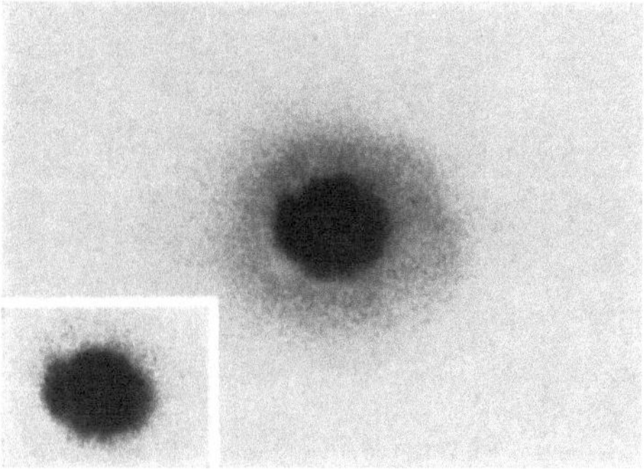

15. NGC 3032 (P 200″): *SAB(r) 0°*. Fairly large, elongated nucleus on short, bright stubby bar, surrounded by dark ring or arcs in bright lens. The envelope does not show up on this print. Compare with 3, 7 and 13.

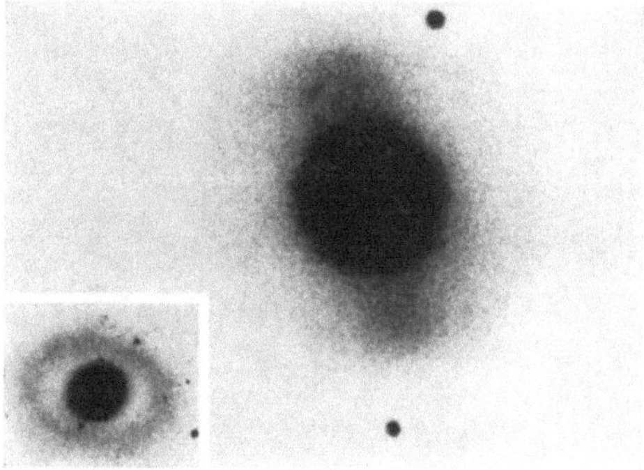

16. NGC 2859 (P 200″; inset P 48″): *(R) SB(r) 0°*. Fairly large nucleus on weak bar marking major axis of little elongated lens, slightly stronger near rim where bar appears brighter. Faint outer ring structure from which emerge two very weak spiral arms or arcs. No spiral structure in lens. Compare with 12, 14 and 15.

Plate V. The two main sequences of ordinary spirals.

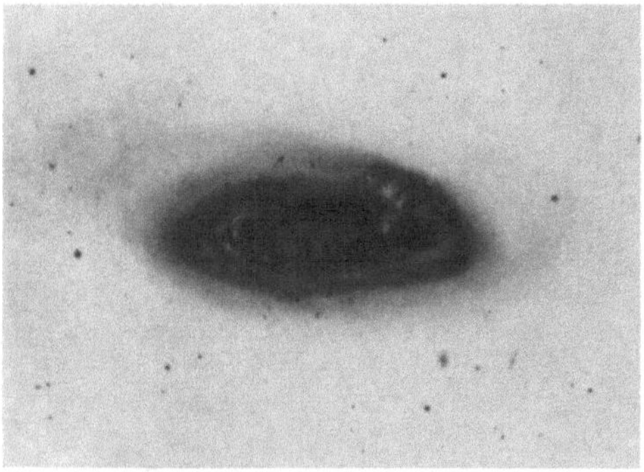

17. NGC 4569 = M 90 (P 200″): *SA(s) ab*. A large central bulge from which emerge two smooth and massive regular arms which are closely wound and separated by dark lanes. Faint outer spiral extensions showing tendency to return towards central lens. Compare with 4, 21 and 39.

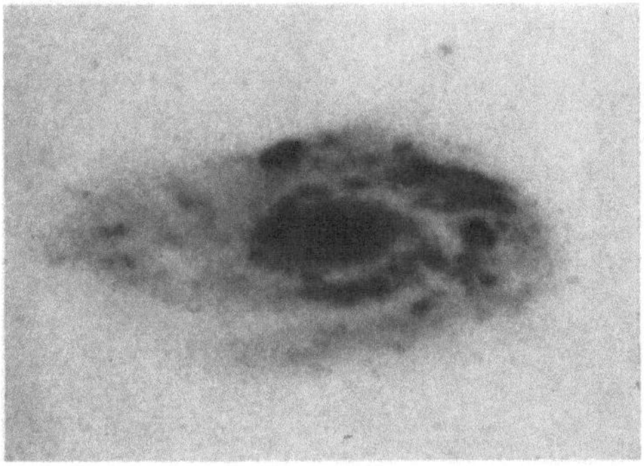

18. NGC 7205 (S 74″): *SA(s) bc*. Smaller central bulge with inner nucleus from which emerge two main massive and knotty regular arms; these are less closely wound and show some branching. Compare with 17, 18 and 40 or 41.

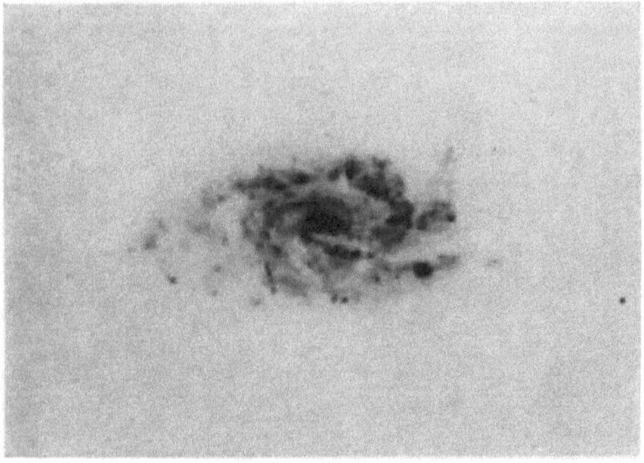

19. NGC 1084 (W 100″): *SA (s) c*. A small central nucleus from which start two main arms with much irregular structure, bright knots and emission objects. Only the brighter, inner regions are shown in this short-exposure image. Compare with 18, 20, 23 and 42.

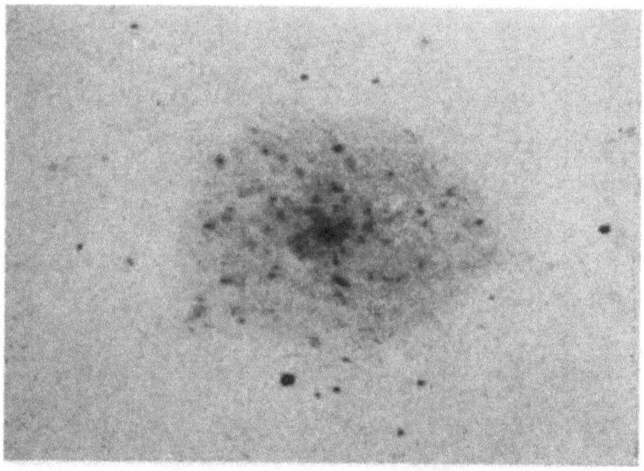

20. NGC 7793 (S 30″): *SA(s) d*. A very small central nucleus, resembling a globular cluster, is surrounded by several irregular and broken spiral arms with very many knots. Note that the arms still start from the nucleus. The mean surface brightness decreases steadily from *S a* to *S d*. Compare with 19, 24 and 33.

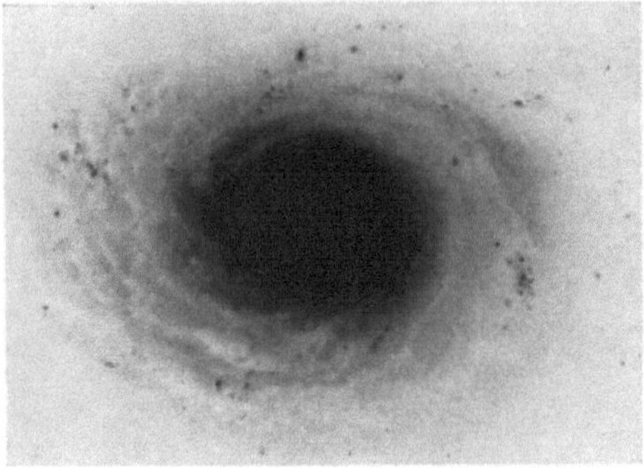

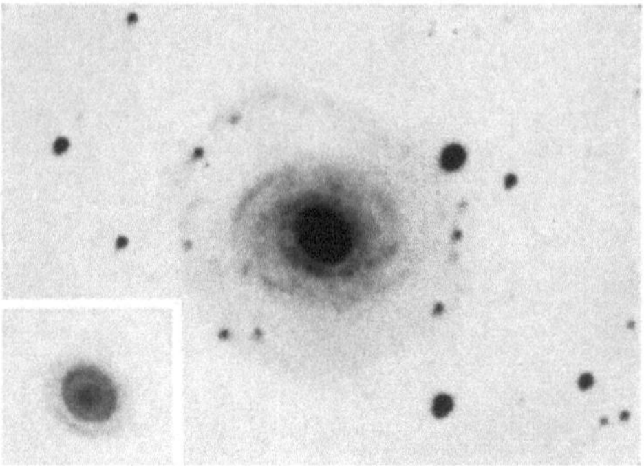

21. NGC 4736 = M 94 (P 200″): (*R*) *SA* (*r*) *ab*. Only the complex spiral pattern of high multiplicity in the lens appears in this reproduction. The small nucleus and inner ring (*r*) surrounding it are lost in the overexposed central bulge; the outer, smooth ring structure (*R*) extending beyond the limits of the figure is not shown either. Compare with 8, 17 and 22.

22. NGC 6753 (S 74″): (*R*) *SA* (*r*) *b*. The inset shows the nucleus and bright inner ring (*r*) from which emerge the complex spiral pattern in the lens, only partly resolved here; the outer ring (*R*), which begins to break up into spiral arcs with bright knots is faintly connected to the inner spiral pattern in two diametrically opposite points. Compare with 8, 21 and 23 (allowing for differences in scale and resolution).

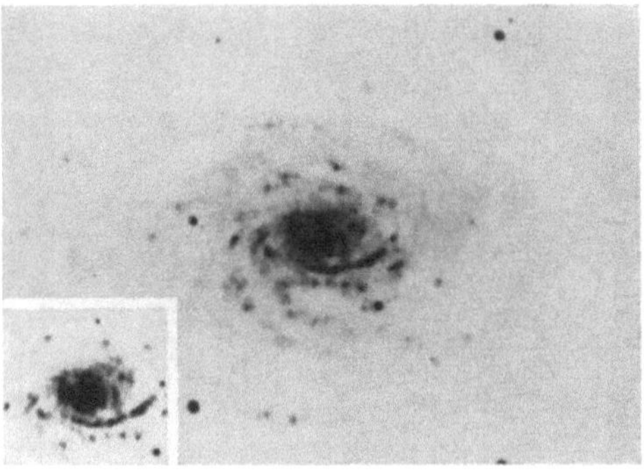

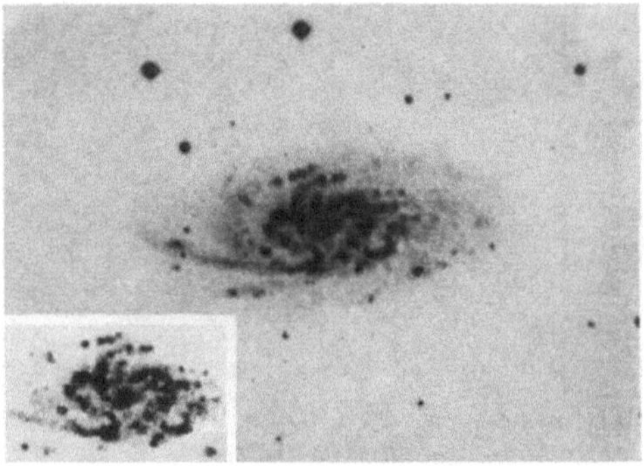

23. NGC 5962 (W 100″): *SA* (*r*) *c*. The inset shows the nucleus and inner ring which begins to break up at this stage and from which extend the many knotty and filamentary arms; only weak traces remain of the outer ring structure. Compare with 19, 22 and 24.

24. NGC 6643 (W 60″): *SA* (*r*) *d*. The inset shows the small nucleus separated from broken remnants of the inner ring extending into the complex, irregular spiral pattern. Compare with 20, 23 and 33.

Plate VI. The two main sequences of barred spirals.

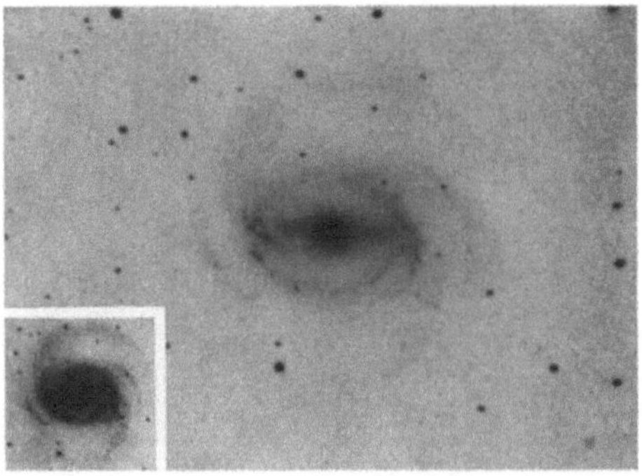

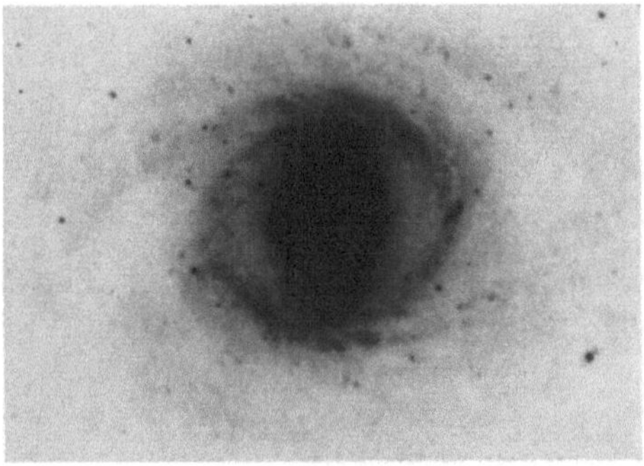

25. NGC 1433 (S 74″): (*R′*) *SB* (*r*) *a*. A small, elongated nucleus whose major axis is tilted at 45° to a long, narrow bar along which run filaments of dark matter. The bar marks the major axis of an elliptical ring from which emerge two faint main arms near the extremities of the bar and two weaker branches forming an incomplete outer ring (inset). Compare with 14, 26 and 29.

26. NGC 3351 = M 95 (P 200″): (*R′*) *SB* (*r*) *ab*. The nucleus and bar are enhanced on this photograph taken in yellow light; the elliptical ring appears circular or slightly hexagonal by projection foreshortening; note the two main arms emerging from ring near the extremities of the bar and weaker branches. Compare with 25, 27 and 30.

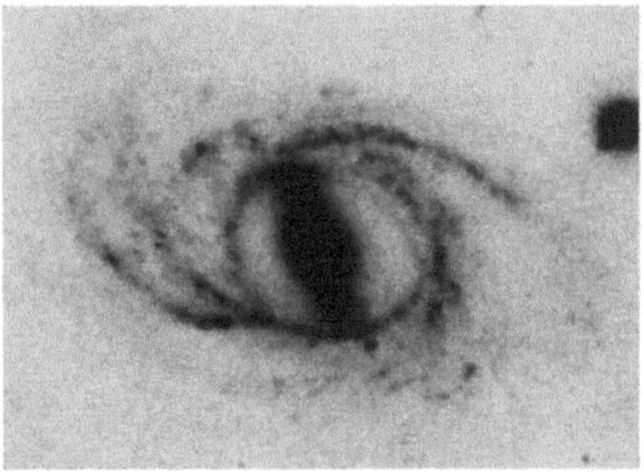

27. NGC 2523 (P 200″): *SB(r)bc*. The nucleus and bar are still strong, but the ring is narrower and knotty; two main spiral arms, narrow and knotty (one with branching) emerge from the ring near the extremities of the bar; additional, weaker arms start near minor axis of elliptical ring (here appearing circular in the projection). Note slight asymmetry. Compare with 26, 28 and 31.

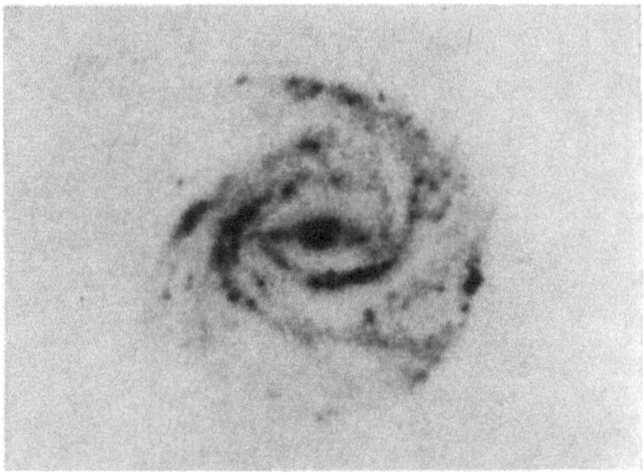

28. NGC 3367 (P 200″): *SB(r)cd*. The nucleus and bar are smaller and fainter; the elliptical ring is weaker and begins to break up into knots; the arms are branching and filamentary; note strong asymmetry. Compare with 24, 27 and 31.

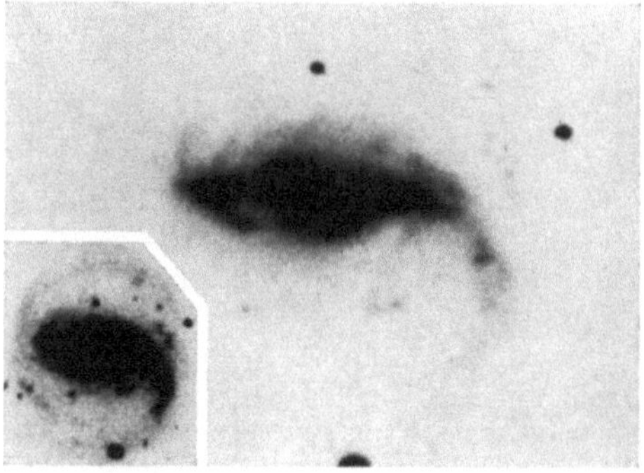

29. NGC 7552 (S 74″): *(R′) SB(s)a*. A small, elliptical nucleus in a bar-like, elongated lens with much internal structure marked by dark lanes; two main spiral arms (one stronger)emerge from the extremities of the bar and return each to the opposite extremity after a halfturn, thus simulating an outer ring structure. Compare with 10, 23, 30 and Plate X.

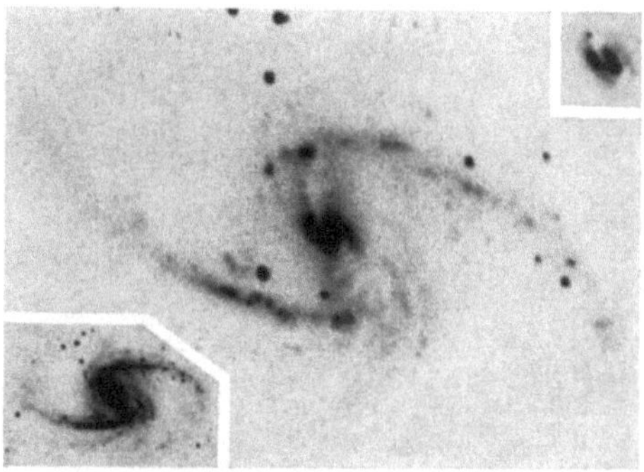

30. NGC 1365 (S 30″): *(R′) SB(s)b*. The upper right inset shows the complex structure of the nucleus due to the "crossing-over" of the main dark lane running along the bar which is seen here foreshortened in the projection; the two bright and knotty main spiral arms emerge at right angles to the bar and return to it after a half-turn (lower inset), thus simulating an outer ring. Compare with 27, 29, 31 and Plate X.

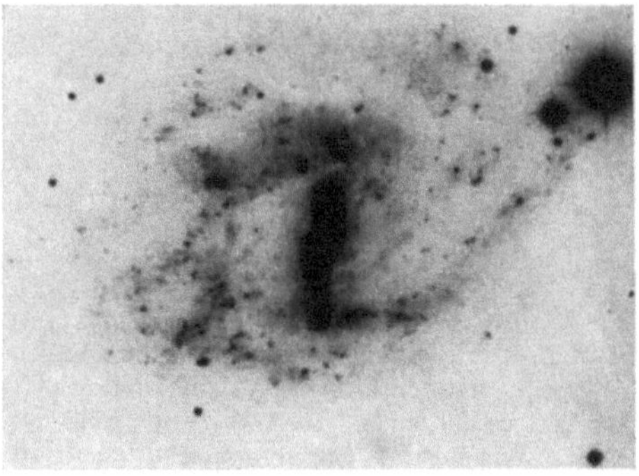

31. NGC 7741 (P 200″): *SB(s)cd*. The nucleus becomes inconspicuous in the bar (foreshortened in the projection); the two main, heavy arms show much branching and clumpiness; one is much stronger than the other, producing a characteristic asymmetry. Compare with 27, 30 and 32.

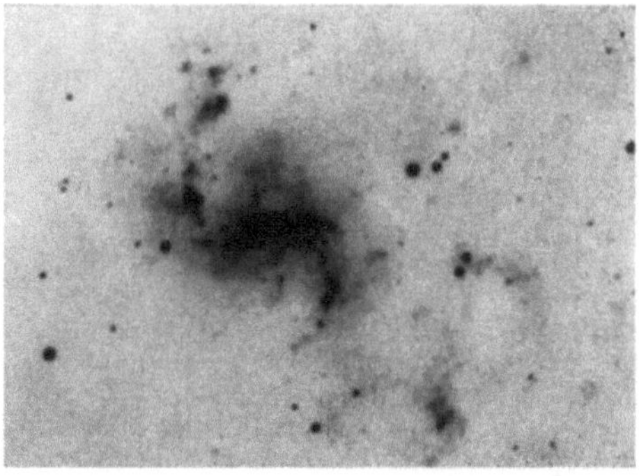

32. NGC 1313 (S 74″): *SB(s)d*. Only a very small nucleus remains in the center of the bar from which emerge two short arms broken up into clumpy sections. The rest of the spiral pattern is weak, very irregular and strongly asymmetrical; note also asymmetrical halfring on one side of bar. This is the transition stage towards the magellanic irregulars. Compare with 31, 36 and 37.

Plate VII. The irregular later stages of the spiral sequences.

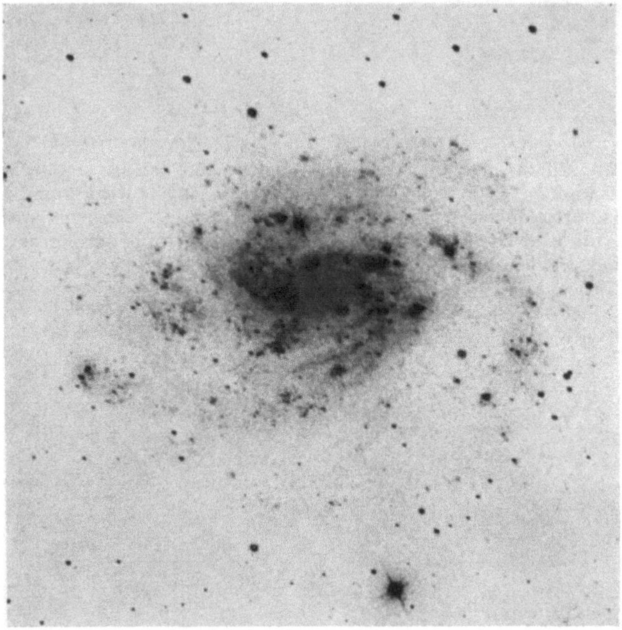

33. NGC 300 (S 30″): *SA (s) d.* A very small, round nucleus, similar to a globular cluster, lies in the middle of the central region from which emerge several, highly resolved and branching arms of low surface brightness. Compare with 20 and 34 NGC 300 is very similar to M 33 but of slightly later type.

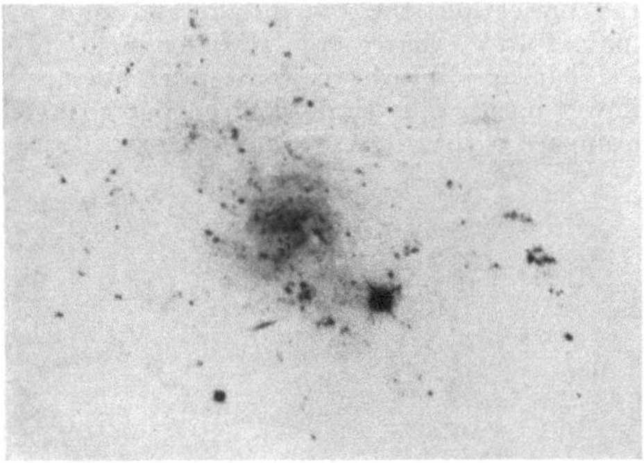

34. NGC 45 (P 200″): *SA (s) dm.* Only the central regions of this object of very low surface brightness are visible on this print; the nucleus is very small and faint in a central, amorphous mass from which emerge very faint, irregular and branching arms. Compare with 33 and 35.

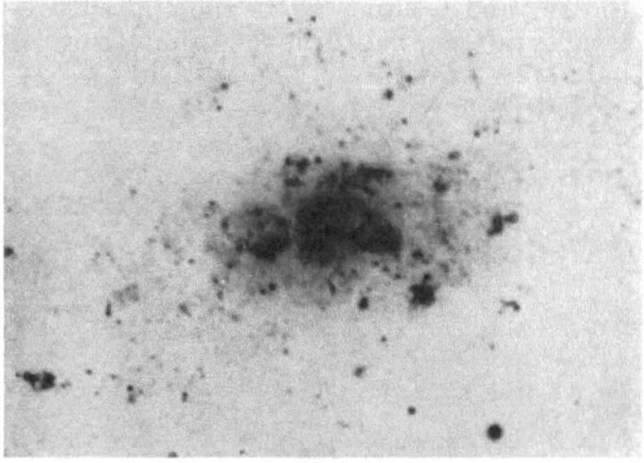

35. NGC 5204 (P 200″): *SA (s) m.* Only a weak spiral pattern remains in the highly resolved, irregular, but symmetrical distribution of supergiant stars and emission knots surrounding the central condensation. Compare with 34 and 37.

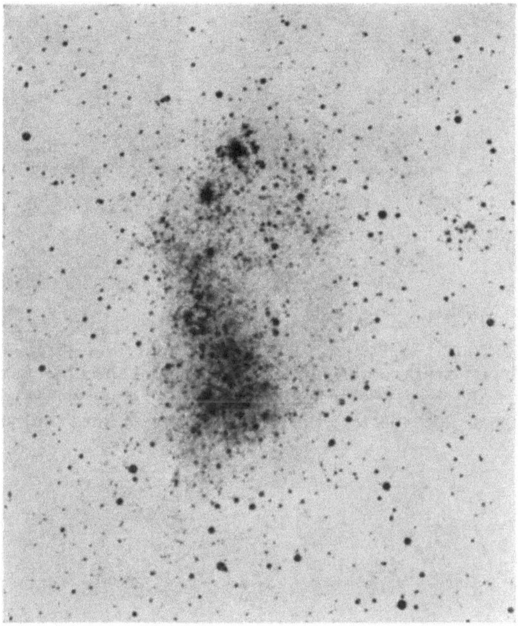

36. Small Magellanic Cloud (AE 2″): $SB(s)m_p$. The spiral pattern is reduced to the axial bar from which emerge one bright, knotty and asymmetrical arm (above, right)—including many supergiants—and one faint, smooth arm (lower left, invisible on print) including no bright supergiant. Traces of embryonic arms appear at the other extremities of the bar. The peculiarity is the asymmetrical distorsion towards the Large Cloud (to right). Compare with 32 and 37.

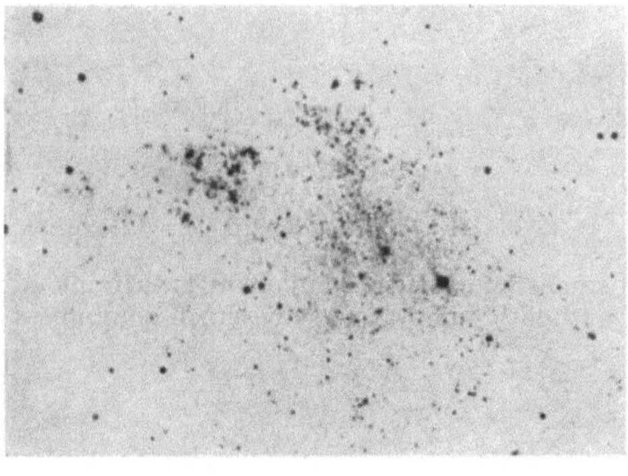

37. IC 1613 (P 48″): *IBm*. The residual structure includes the well resolved axial bar and asymmetrical bright patch rich in supergiants and emission objects. Note characteristic asymmetry. Compare with 36 and 38.

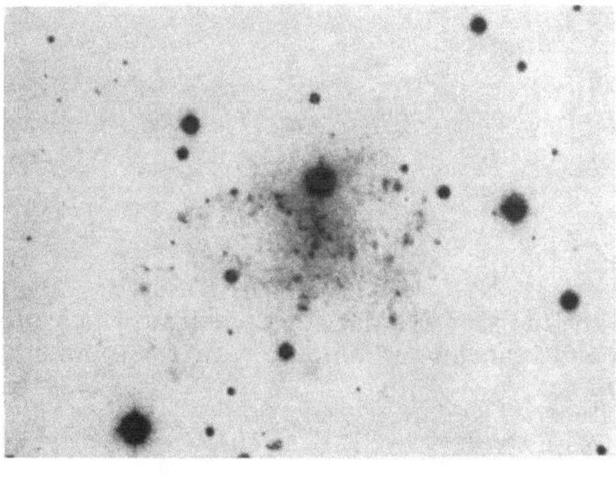

38. Dwarf I in Andromeda (2^h31, $+40°19'$) (P 200″): $dIB(s)m$. The characteristic structure of the late-type $SB(s)$ spirals is still recognisable, including axial bar (foreshortened in the projection), with embryonic spiral arms emerging at its extremities; one of the arms can be faintly traced to the left and lower part of the field. Compare with 36 and 37.

Plate VIII. Edgewise systems: Ordinary spirals.

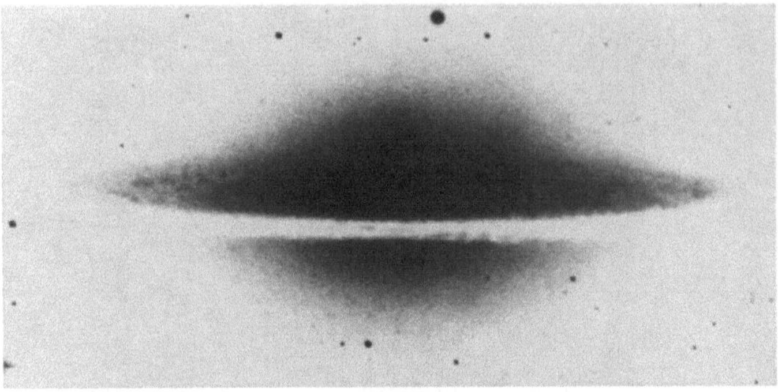

39. NGC 4594 (P 200″): *SA*(*s*:)*a*. The very bright and large spheroidal component, made up of high-velocity stars, emits most of the light of the object; the thin and weak flat component includes only very few bright Population I supergiants; it is surrounded by a heavy peripheral ring of dark matter; the spiral pattern is faint and smooth. Compare with 17 and 21.

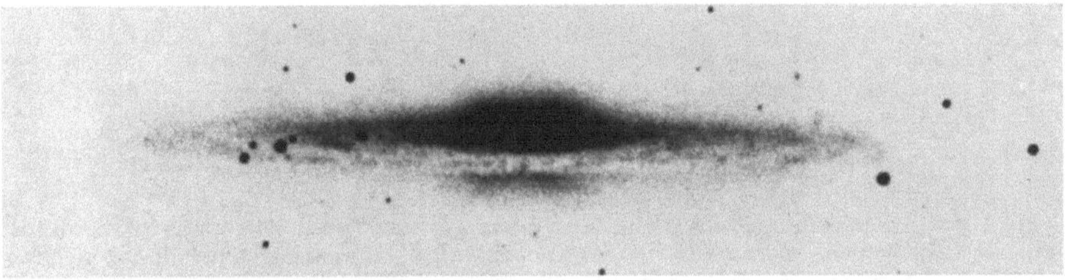

40. NGC 5746 (W 60″): *SA*(*s*:)*b*. The spheroidal component is very much reduced, but still contributes a large fraction of the light; the spiral pattern of bright and dark arms is well defined (save for the projection foreshortening), regular and still fairly smooth, although it begins to show some clumpiness. Compare with 18 and 22.

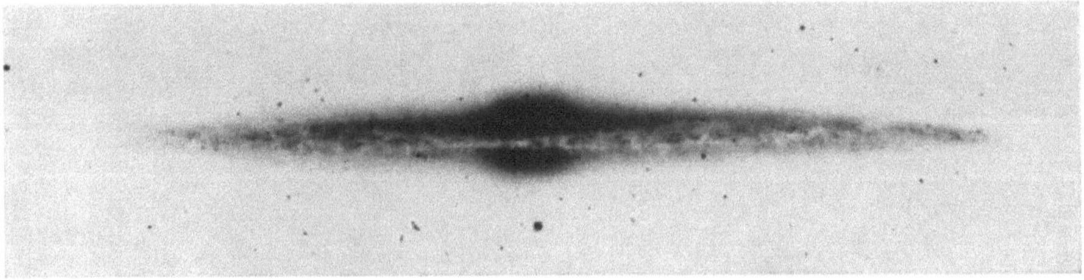

41. NGC 4565 (P 200″): *SA*(*s*:)*bc*. The still smaller spheroidal component is surrounded by an extensive flat component in which bright and dark spiral arms show much clumpiness and branching. Compare with 18, 19 and 23.

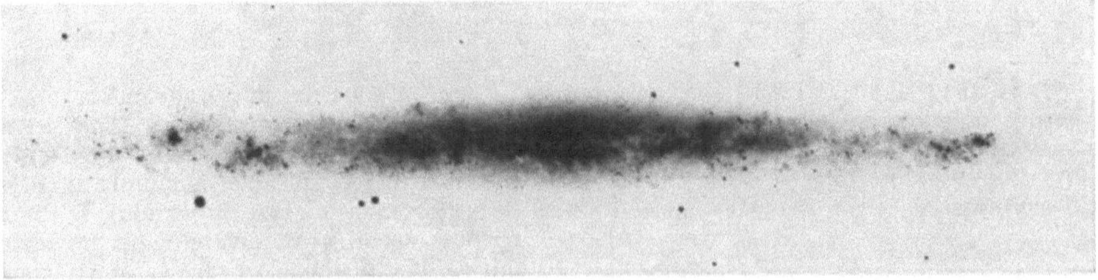

42. NGC 4244 (P 200″): *SA*(*s*:)*cd*. The spheroidal component is now inconspinuous in the center of a very extensive flat component in which the spiral pattern of bright and dark matter shows much irregularity and very many well resolved knots. Compare with 20 and 24.

Plate IX. Edgewise systems: Lenticulars and magellanic irregulars.

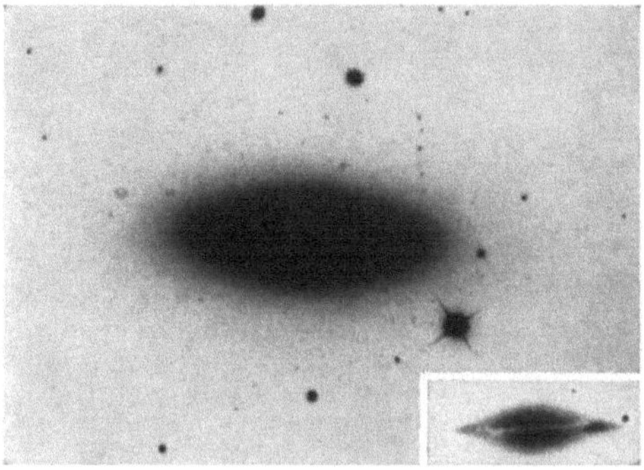

43. NGC 5866 (P 200″): *SA*(*s*)*O*°:(*sp*). The outer envelope simulates an elliptical, but the lens (inset) shows a narrow ring of dark matter surrounding the large nucleus. Compare with 3, 44 and 45.

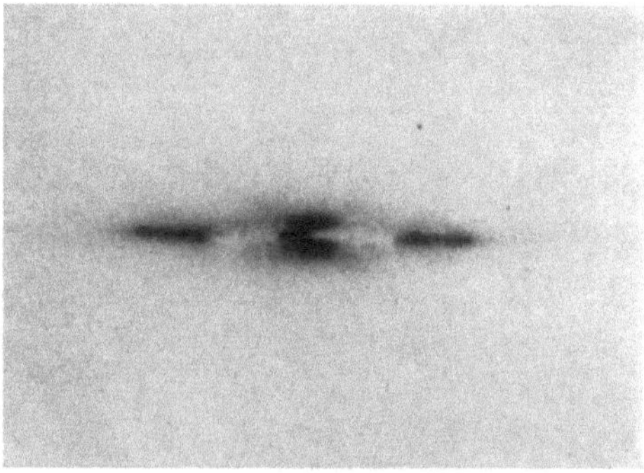

44. NGC 4710 (W 100″): *SA*(*rs* ?)O^+(*sp*). This short exposure shows only the lens and fairly small nucleus surrounded by a heavy ring of dark matter and bright ansae, being the edge-on view of an incomplete ring or incipient spiral arms.

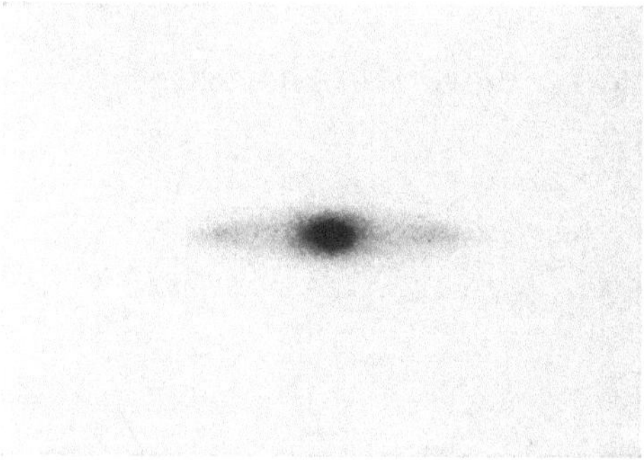

45. NGC 4215 (W 100″): *SA*(*r*)*O*°: (*sp*). This short exposure shows only the small nucleus and the weak ansae, being the edge-on view of the incipient ring structure in the lens. Compare with 7, 8 and 44.

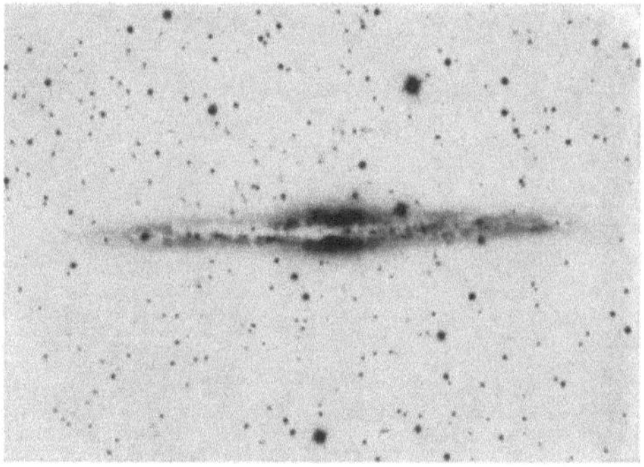

46. NGC 891 (P 200″): *SA*(*s*:)*b*(*sp*). This edge-on view of a typical ordinary spiral is for comparison with 43 and 44. Note how the dark matter, limited to a ring close to the nucleus in the lenticulars, spreads to the whole of the intermediate and outer parts of the spiral pattern. Compare also with 39 and 42.

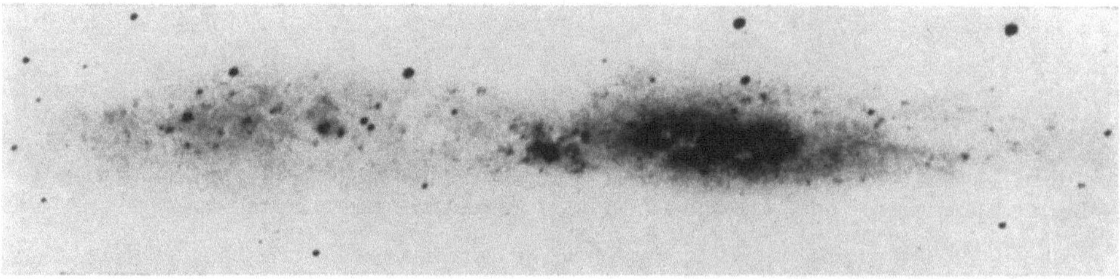

47. NGC 55 (S 30″): *SB*(*s*)*m*(*sp*). The brighter part which is not resolved, except for a few bright emission knots and complex dark patches, represents a "rear" end-on view of the axial bar; the fainter part (left) which is partly resolved represents the end-on view of the main asymmetrical spiral arm. Between them dark clouds obscure the distant parts of the arm; weak embryonic arms can be identified to the right; compare with 36, 37, 38 and 48.

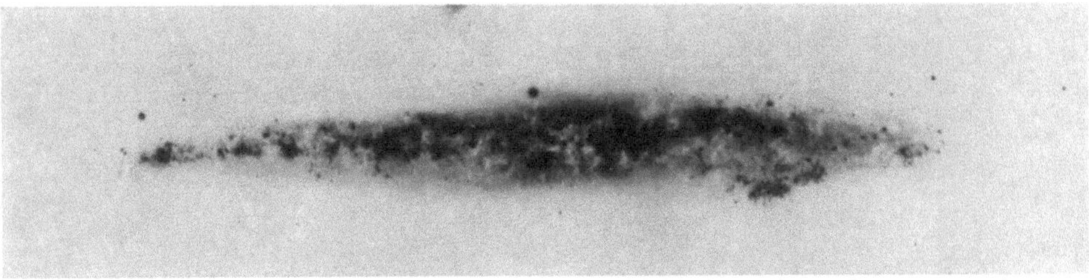

48. NGC 4631 (P 200″): *SB(s)d* or *m(sp)*. The nearly central brighter part which is not resolved and against which complex dark clouds are projected represents a "front" end-on view of the axial bar; from it emerge two well-resolved, irregular spiral arms spreading out in opposite directions. Note characteristic asymmetry. Compare with 32, 38, 42 and 47.

Plate X. Orientation effects on appearance of SB(s)a or SB(r)a spirals.

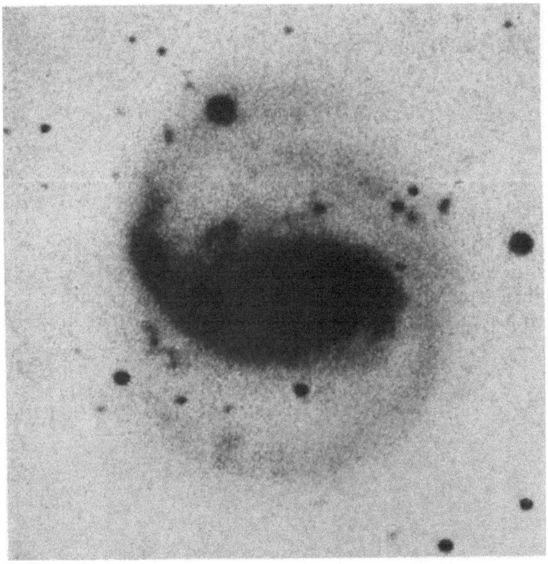

49. NGC 7552 (S 74″): *(R′) SB(s)a*. Face-on view; same as 29; note outer loops. Details of bar and lens, seen from "above" are in 52; note pattern of dark matter.

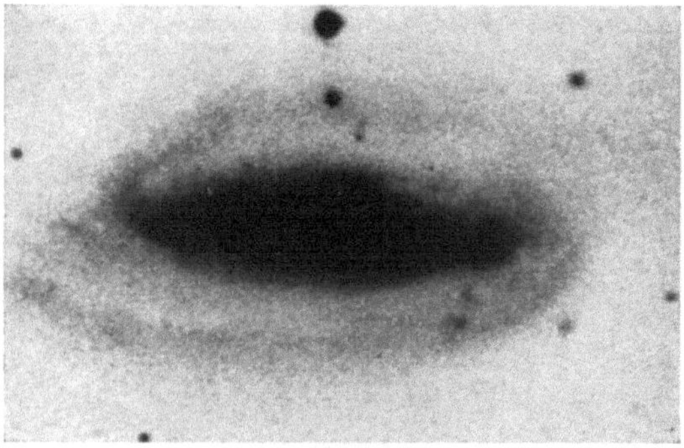

50. NGC 7582 (S 74″): (*R′*) *SB* (*s*) *a*. Broadside view; details of bar and lens, seen from "across", are in 53; note pattern of dark matter.

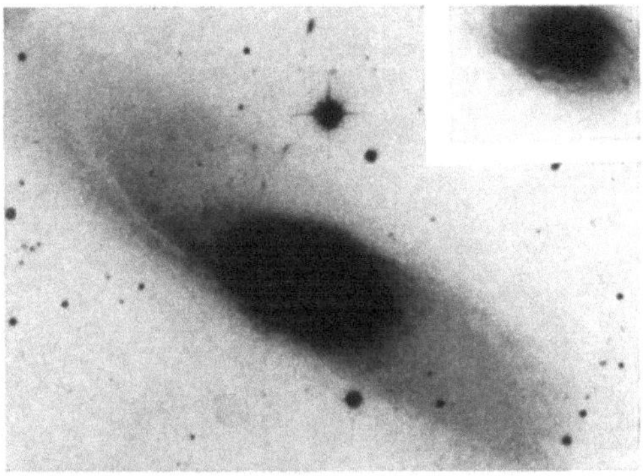

51. NGC 5566 (P 200″): *SB* (*rs*) *a*. End-on view; details of bar and lens, seen from "along", are in inset. There is an (*r*) structure in the lens which does not show up on this print. Note lane or layer of dark matter splitting outer arms.

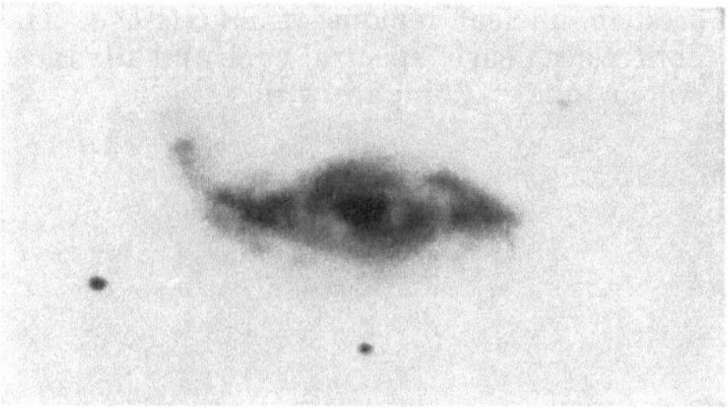

52. NGC 7552. Detail of bar and lens.

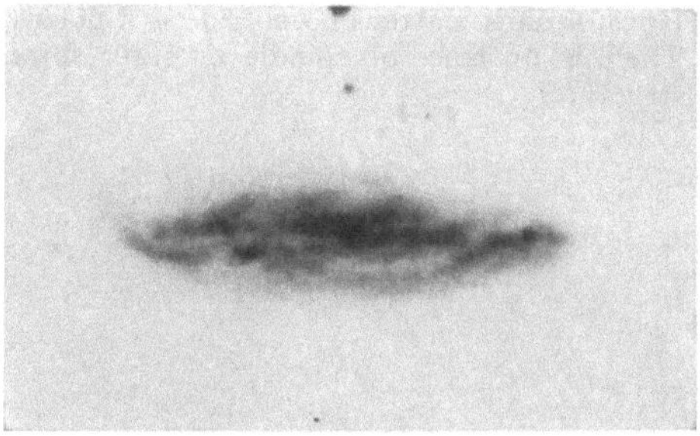

53. NGC 7582. Detail of bar and lens.

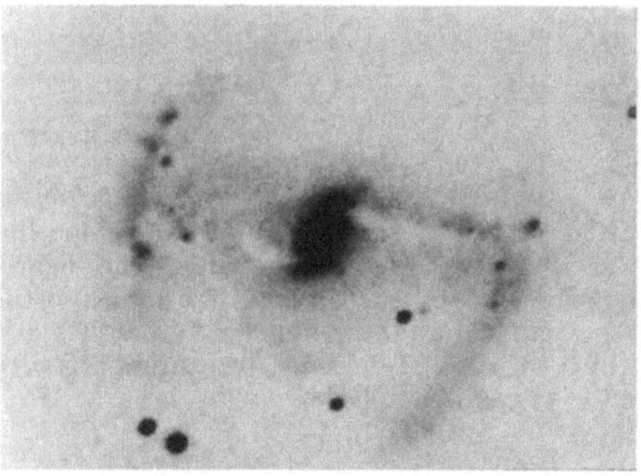

54. Lens of NGC 5383 (W 60″): *SB* (*s*) *b*. Seen nearly "face-on" as in 49; note structure of nucleus and pattern of dark matter; compare with 30.

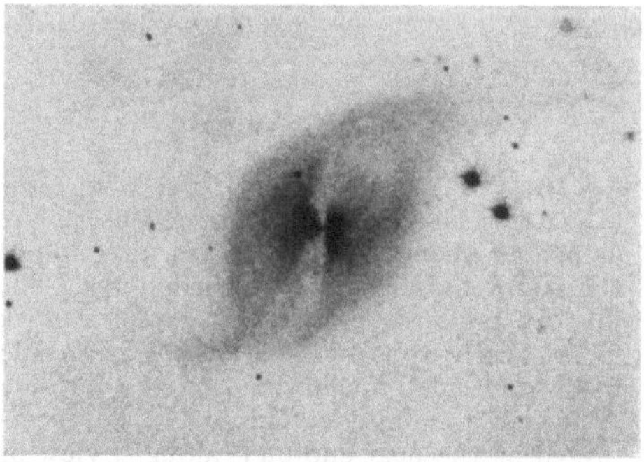

55. Lens of NGC 3718 (W 60″): *SAB* (*s*) *a*. The bar is seen nearly "end-on" as in 51. Compare also with 30 and 54.

Very peculiar-looking structures can be produced in the lens and nucleus of early *SB* (*s*) galaxies depending on tilt and orientation of the bar with respect to the line of sight.

Plate XI. Peculiar lenticulars and non-magellanic irregulars.

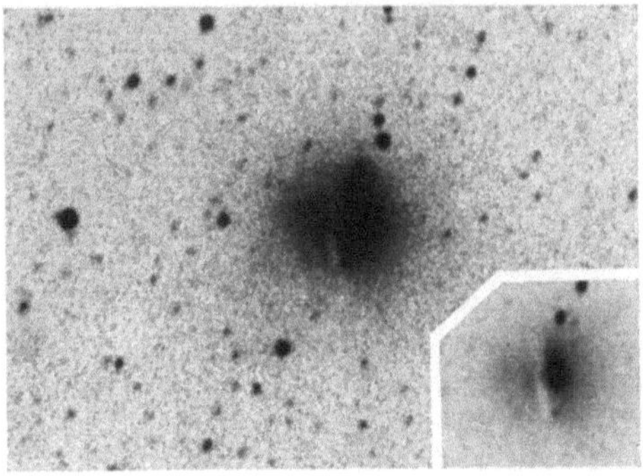

56. NGC 1947 (S 74″): *SA0p*. Little elongated, smooth nebulosity, rather similar to elliptical galaxy or to spheroidal bulge of early spiral, cut by two slightly curved lanes of dark matter on one side of nucleus. There is no trace of spindle or outer spiral structure.

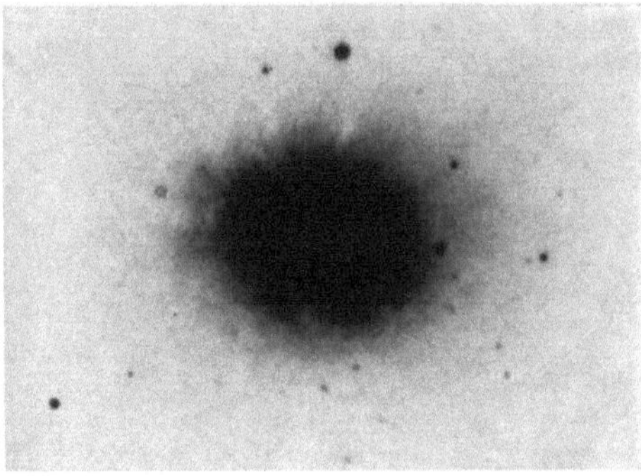

57. NGC 3077 (P 200″): *I*(*0*) or *SAB0p*. Little elongated, smooth nebulosity marked by irregular patches of dark matter; two of them emerging on both sides near minor axis simulate embryonic spiral pattern as seen in nuclear regions of *SAB*(*s*)*0/a*. Has contrasting early spectral type and advanced colour index. Compare with 15.

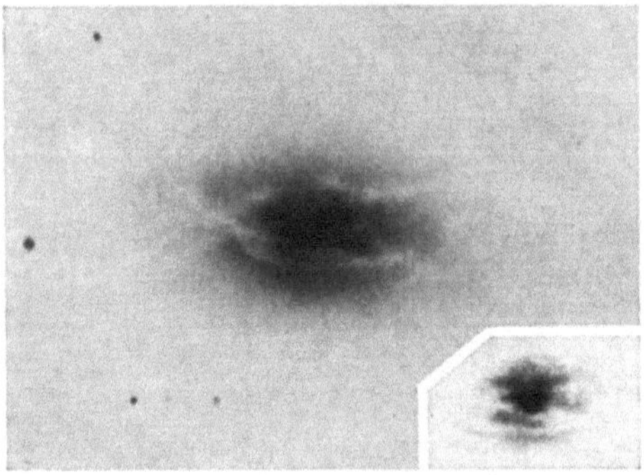

58. NGC 4753 (W 100″): *I*(*0*) or *SB*? (*s*:) *0⁺p*. The inner isophotes (inset) are elongated at right angles to the direction of the major axis of the outer isophotes. The nuclear region is crossed by complex dark lanes which appear to be on the near side below the small nucleus and on the far side above it (see inset). This could be interpreted as the near end-on ("along") view of the bar and lens of an armless *SB*(*s*)*0/a* object. Compare with 55 and 56.

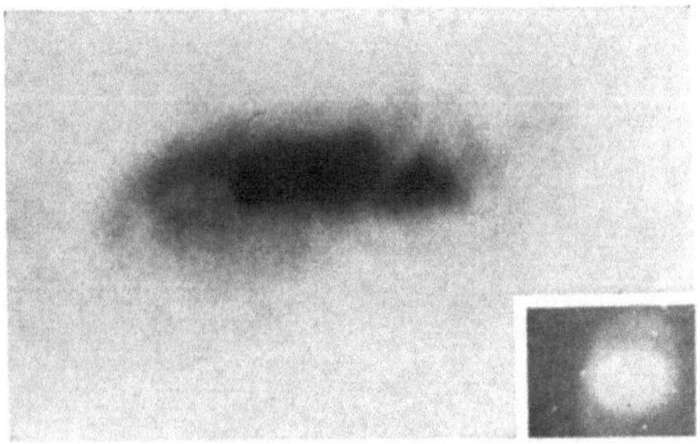

59. NGC 4691 (P 200″): (*R*) *SB*(*s*:)*0/ap*. The weak outer whorls in the envelope (inset) show that this system is seen nearly face-on; note the complex structure of the bar crossed and surrounded by patches and curved lanes of dark matter. Compare with 52, 57, 58 and 60/61.

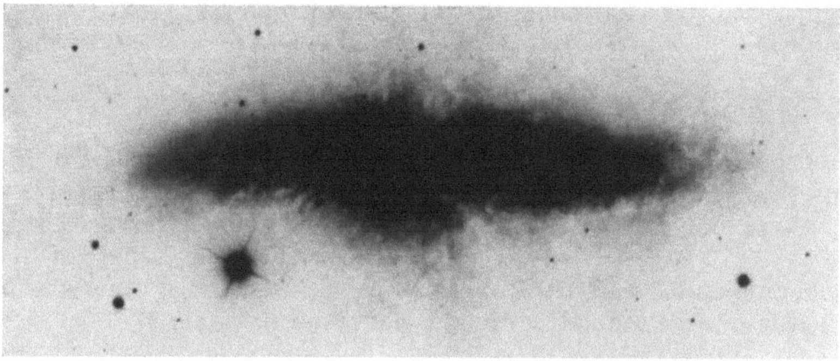

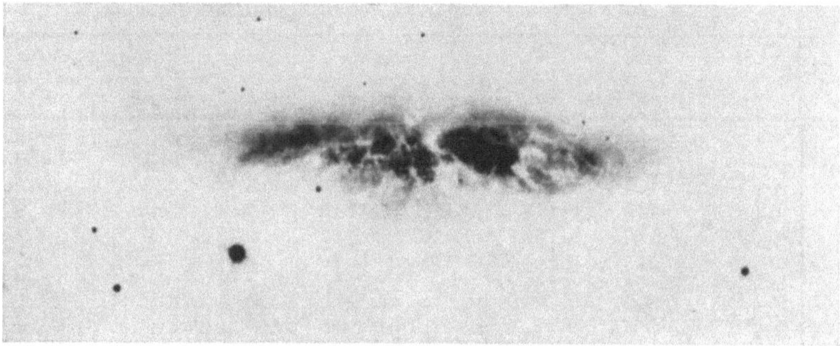

60/61. NGC 3034 = M 82 (P 200″): *I(O)* or *SB(s:)Op*. This is interpreted tentatively as the "broad-side", nearly edge-on view of the bar and lens of an armless *SB(s)O/a* object. Has contrasting early-type spectrum and advanced colour index. Compare with 53, 57 and 59. Note especially the great similarity between 59 and 61 and the characteristic "crossing-over" of the main dark lane in the nucleus[1].

b) Quantitative morphology.

5. Apparent and true flattening. α) *Ellipticals.* The distribution of apparent and true flattening among elliptical galaxies has been investigated by HUBBLE [*25*], MACHIELS [*45*] to [*47*], TEN BRUGGENCATE [*8*], DE VAUCOULEURS [*80*] and others.

The elementary theory is simple, but its application to the physical situation is complicated by selection and other systematic effects and no more elaborate theory has yet been developed.

Consider an ellipsoid of revolution of equatorial diameter a_0 and polar axis b_0 whose equatorial plane is tilted at an (acute) angle i on the line of sight and which appears projected on the tangent plane to the celestial sphere as an ellipse of apparent axes $a = a_0$ and $b \leq b_0$, the true flattening is $e = 1 - (b_0/a_0) = 1 - q_0$, the apparent flattening is $\varepsilon = 1 - b/a = 1 - q$, and

$$\sin i = \sqrt{(q^2 - q_0^2)/(1 - q_0^2)} \tag{5.1}$$

For a random space orientation of the minor axes of the ellipsoids the probability that the inclination on the line of sight of the equatorial plane of any system lies between i and $i + di$, is $\cos i$. $di = d(\sin i)$, so that the probability

[1] This interpretation of the non-megallanic irregulars is consistent with their luminosities, colours and dimensions (cf. p. 315, Fig. 1 and 2). It also makes it possible to understand the rapid, solid-body rotation observed along the major axis of M 82 (N. U. MAYALL, in "Problems of Cosmical Aarodynamics", U.S. Central Air Documents Office, p. 181, 1951).

for an ellipsoid of true flattening e_m to appear of type En, i.e. with an apparent flattening ε_n, is

$$\Delta_{n-\frac{1}{2}}^{n+\frac{1}{2}} (\sin i) = \sqrt{\frac{q_{n+\frac{1}{2}}^2 - q_{n-\frac{1}{2}}^2}{1 - q_0^2}} . \tag{5.2}$$

HUBBLE [25] has computed for the eight groups of ellipticals $E0$, $E1$, ... $E7$ the theoretical frequencies of apparent flattening for each of the corresponding groups of true flattening. The limits of the flattening intervals chosen were 0.00—0.05—0.15—0.25— ... —0.65—0.75; note that the range of flattening for the $E0$ groups (0.00—0.05) is only half as wide as for the other groups. This subdivision leads to the relative frequencies listed in Table 3.

Table 3. *Frequencies of apparent flattening among ellipsoids.*

Actual	Apparent								Total
	E0	E1	E2	E3	E4	E5	E6	E7	
E7 . . .	0.055	0.111	0.114	0.116	0.121	0.132	0.164	0.187	1.000
E6 . . .	0.059	0.123	0.126	0.133	0.148	0.187	0.224		1.000
E5 . . .	0.067	0.140	0.148	0.166	0.216	0.263			1.000
E4 . . .	0.079	0.169	0.190	0.250	0.312				1.000
E3 . . .	0.100	0.225	0.299	0.376					1.000
E2 . . .	0.145	0.378	0.477						1.000
E1 . . .	0.300	0.700							1.000
E0 . . .	1.000								1.000
Total . .	1.805	1.846	1.354	1.041	0.797	0.582	0.388	0.187	8.000
Frequency	0.226	0.231	0.169	0.130	0.100	0.073	0.049	0.023	1.000

The totals at the bottom of the table give the theoretical distribution of apparent flattening in a population of ellipsoids randomly oriented and equally distributed over the whole interval of e (0.00—0.75). HUBBLE used this table to analyse the observed distribution of 87 bright ellipticals among the eight groups $E0$ to $E7$; excluding 2 doubtful cases the relative frequencies for 85 objects were as follows:

E0	E1	E2	E3	E4	E5	E6	E7
20%	15%	17%	12%	15%	7%	8%	6%

By comparison with the last line of Table 3 there is an excess of strongly flattened shapes $E6$ to $E7$ and a deficiency of circular forms $E0$ to $E1$, but Hubble noted that much of the divergence could be due either to fluctuations in the small sample or to departures of the most flattened ellipticals from true ellipsoids and also to the inclusion in the $E7$ group of some unrecognized early-type spirals (actually lenticulars) seen edgewise.

The same data were again discussed by TEN BRUGGENCATE [8] and by MACHIELS [45] without significant modification of HUBBLE'S conclusions. MACHIELS [47] and later WYATT[1] also analyzed the larger sample of rather more than 200 objects classified as E in the Shapley-Ames Catalogue and for which ε can be computed from the published dimensions; MACHIELS concluded that there is no statistically significant excess or deficiency of any group compared with a uniform distribution of true flattening[2], while WYATT is of the opinion that

[1] S. P. WYATT: Astronom. J. **55**, 187 (1950) (Abstract).

[2] See also discussion by MACHIELS [46] of the objects classified as "elliptical" in the Harvard Catalogue of the Coma-Virgo cluster (Harvard Ann. **88**, No. 1 (1930)] in which he emphasizes the errors introduced by the definition and unequal widths of the flattening intervals.

there is a significant excess of highly oblate systems over nearly spherical ones and further that there is a definite deficiency of systems of intermediate (true) flattening with a minimum near $e \approx 0.3$.

Because of the mixture of types in earlier statistics none of these conclusions can be accepted without further discussion and HUBBLE's reasonable warning of possible systematic errors appears fully justified. In particular, it is now clear that the shapes of the isophotes of the most flattened "ellipticals" depart notably from similar geometrical ellipsoids (cf. p. 323, Fig. 7) and that many objects of the other classes, mainly lenticulars, have been erroneously included in the early statistics of ellipticals. As noted in Sect. 3 this is a most serious cause of confusion in the old data.

More significant results can probably be obtained from a discussion of 48 objects classified as E or E^+ in the Mt. Stromlo survey [*80*]. The observed relative frequencies in flattening intervals limited at 0.00—0.07—0.17— ... —0.67—0.77 are given in Table 4 for both the "estimated" types $E\,0$ (including $E\,0$ to 1), $E\,1$ (including $E\,1$ to 2), etc. and for the "measured" types derived from actual measurements of the axes of the photographic images[1].

Table 4. *Observed relative frequencies of estimated and measured flattening of ellipticals.*

Flattening ε	0.00—0.07	0.07—0.17	0.17—0.27	0.27—0.37	0.37—0.47	0.47—0.57	0.57—0.67	0.67—0.77	Total
Type	$E\,0$	$E\,1$	$E\,2$	$E\,3$	$E\,4$	$E\,5$	$E\,6$	$E\,7$	
Frequency of ε_{est}	23	17	17	13	16	8	4	2	100
Frequency of ε_{meas}	25	19	15	19	6	10	4	2	100
Mean	24	18	16	16	11	9	4	2	100

The mean frequencies agree closely with the theoretical values for a uniform distribution of true flattening and random orientation (Table 3). This is shown by the following summary of the relative frequencies observed and computed

Types	$E\,0$—$E\,1$	$E\,2$—$E\,3$	$E\,4$—$E\,5$	$E\,6$—$E\,7$
Observed (%)	42	32	20	6
Computed (%)	42	33	19	6

In view of the smallness of the sample on which this result rests, the conclusion must, however, remain provisional and subject to verification. A final discussion should be possible after the completion of the current revision of the Shapley-Ames galaxies for the whole sky.

β) Lenticular galaxies. The distribution of apparent and true flattening among lenticular galaxies as a separate class, distinct from the ellipticals, has not yet been studied in any detail. The distribution of apparent flattening ε_m in a sample of 33 lenticulars included in the Mt. Stromlo survey [*80*] is given in Table 5.

Table 5. *Frequency of apparent flattening among lenticulars.*

ε_m	0.0—0.2	0.2—0.4	0.4—0.6	0.6—0.8
Observed (%) .	31	27	12	30
Random (%) . .	40	20	20	10

[1] The mean systematic difference for the 48 objects is $\varepsilon_m - \varepsilon_e = +0.016$ and the average deviation (excluding nine $E\,0$ objects) is ± 0.034. The internal probable error of the measured flattening is about ± 0.025.

For comparison the last line gives the theoretical frequencies for a random distribution under the same assumptions as in Table 3. The deficiency in the interval 0.4 to 0.6 is probably accidental, but the deficiency of nearly circular objects ($\varepsilon < 0.2$) is probably real and the excess of elongated objects ($\varepsilon > 0.6$) is certainly significant. The observed distribution merely confirms that lenticulars are strongly flattened systems and behave in this respect more like spirals than ellipticals; a larger sample would be required for the derivation of the frequency distribution of true oblateness.

γ) Spirals. The problem of the true space orientation of the equatorial planes of spirals as derived from the distribution of apparent flattening has been the subject of many investigations with conflicting results. Some early studies had suggested that there is an excess of elongated shapes compared with expectations for a random (uniform) distribution of rotation axes[1]; this has not been confirmed by other, more recent investigations[2], but the discussion is probably not yet closed[3].

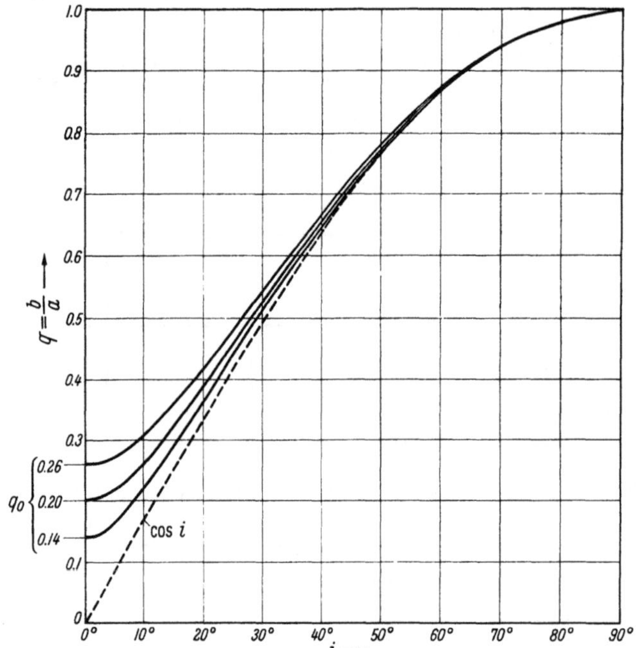

Fig. 5. Apparent flattening $q = b/a$ as a function of true flattening q_0 and tilt angle i for ellipsoids.

Consider a large number of flat circular discs whose axes are randomly oriented and which are observed from a distant point or one such disc observed from a large number of points uniformly distributed over the sphere; the probability for a given observer to be located between latitudes i and $i + di$ is proportional to the area of the zone defined by these angles, i.e. $\cos i.\ di = d(\sin i)$. Hence a statistically uniform distribution of $\sin i$ characterizes a random orientation of the planes. For geometrical ellipsoids of apparent axial ratio $q = b/a$ and true axial ratio $q_0 = (b_0/a_0)$, $\sin i$ is given by (5.1). The relation between i and q is shown in Fig. 5 for a number of values of q_0 between 0.14 and 0.26. According to a discussion by HOLMBERG [*21*] the mean axial ratio of spirals seen exactly edge-on is $q_0 = 0.20$.

For the physical systems the situation is complicated by systematic errors and selection effects which have vitiated most—perhaps all—discussions. In

[1] e.g. J. H. REYNOLDS: Monthly Notices Roy. Astronom. Soc. London **81**, 129 (1920); **82**, 510 (1922). — F. G. BROWN: Monthly Notices Roy. Astronom. Soc. London **98**, 218 (1938); **99**, 14 (1938); see also [*46*].

[2] e.g. H. KNOX-SHAW: Monthly Notices Roy. Astronom. Soc. London **69**, 72 (1908); **98**, 587 (1938). — C. C. L. GREGORY: Monthly Notices Roy. Astronom. Soc. London **84**, 456 (1924); see also [*12*], [*47*].

[3] e.g. E. ÖPIK: Observatory **46**, 51, 65 (1923). — R. O. REDMAN: Observatory **61**, 216 (1938). — F. G. BROWN: Observatory **61**, 250 (1938). — A. G. WALKER: Monthly Notices Roy. Astronom. Soc. London **100**, 623 (1940). — S. P. WYATT and F. G. BROWN: Astronom. J. **60**, 420 (1955).

the first place spiral galaxies depart strongly from ideal ellipsoids and the departure depends on the exact type, late-type spirals are more flattened than early-type objects; for instance the flat component including the spiral arms has a mean axial ratio of about 12:1 in $Sb - Sc$ edgewise spirals[1] while the "spherical" component in the central bulge has a mean axial ratio of about 2 to 1 in $Sa - Sb$ edgewise spirals ([*12*], p. 82 to 86); then the ratio between the spherical component and the flat component varies greatly along the spiral sequence. Another difficulty is the presence in the measured axial ratios of systematic errors which have been thoroughly investigated experimentally by HOLMBERG [*21*]; in general the flattening derived from micrometric measurements of the plates is exaggerated as compared with photometric determinations of the isophotes; such measuring errors must be corrected before any conclusion can be derived concerning the distribution of true and apparent flattening. A third effect is the projection or tilt effect: a given galaxy will be assigned a larger diameter seen edgewise than when seen face-on; for a transparent oblate spheroidal system of true axial ratio q_0 and projected ratio q, in which the radial intensity distribution in the face-on view is $I_1(x, 0)$, it can be shown[2] that the indensity distribution along the major axis of the projected image is given by

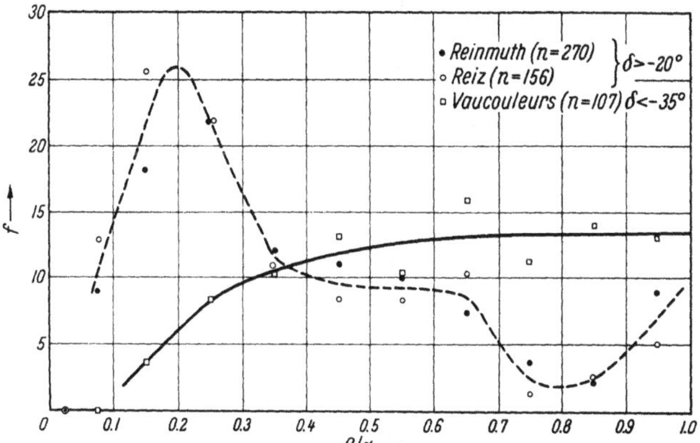

Fig. 6. Frequency distribution of apparent axial ratio among spirals in the Heidelberg and Mount Stromlo surveys.

$$I_i(x, 0)/I_1(x, 0) = 1/q = a/b \gtrsim 1 \tag{5.3}$$

in particular, for the edge-on view ($i = 0°$)

$$I_0/I_1 = 1/q_0 > 1 .$$

Hence, in a sample selected according to apparent diameter, edgewise systems will be included that would not have been included had they been oriented face-on; this introduces a spurious excess of elongated shapes compared with a sample from a given volume of space. The exact amount of this selection effect cannot be computed theoretically and it is very difficult to determine empirically.

From a discussion of 270 spirals north of $\delta = -20°$ listed in REINMUTH's catalogue and whose major diameters are not less than 3′0, after correction for measuring errors and projection effects, HOLMBERG [*21*] concluded that the distribution of $\sin i$ is quasi uniform and therefore that the axes are randomly oriented. A similar study of 156 spirals larger than 3′0 listed in REIZ' catalogue led to the same conclusion. However, the basic material (the Heidelberg Bruce collection) is common to both catalogues and the correction for projection effect was more or less arbitrarily selected so as to produce the desired result. Furthermore, the basic statistics of the frequency distribution of apparent axial ratio among spirals are in serious disagreement as shown in Table 6 and Fig. 6 where

[1] A. B. WYSE and N. U. MAYALL: Astrophys. Journ. **95**, 39 (1942).
[2] S. P. WYATT and F. G. BROWN: Astronom. J. **60**, 420 (1955).

the two sets of Heidelberg data (uncorrected and corrected for measuring errors only)[1] are listed.

Table 6. *Frequency of axial ratio and tilt among spirals.*
Absolute frequency of apparent axial ratio (uncorrected).

β/α	1.0	0.9	0.8	0.7	0.6	0.5	0.4	0.3	0.2	0.1	0.05	Sum
REINMUTH	24	6	10	20	27	30	33	59	49	12		270
REIZ	8	4	2	16	13	13	16	30	40	10		156
VAUCOULEURS	14	15	12	17	11	14	11	9	4	0		107

Relative frequency of tilt angle (corrected for measuring errors).

$\sin i$	1.0–0.8	0.8–0.6	0.6–0.4	0.4–0.2	0.2–0.0	Remarks	
REINMUTH	0.16_5	0.15	0.21_5	0.23_5	0.24_5	$\bar{q}=0.20$	$a \gtrsim 3'0$
REIZ	0.13	0.20	0.18_5	0.20_5	0.28		
VAUCOULEURS	0.27	0.26	0.21_5	0.18	0.07_5	$q=0.14$	$m_H \lesssim 13.0$

The Heidelberg data indicate a conspicuous deficiency of nearly round shapes ($0.7<\beta/\alpha<0.9$) and a large excess of elongated shapes ($0.2<\beta/\alpha<0.4$), while the Mt. Stromlo data indicate a nearly constant distribution of flattening for $b/a>0.4$ and a gradual deficiency of elongated shapes for $b/a<0.4$ (comparison with other data indicates that measuring errors are negligible in the Mt. Stromlo data). The selection rule for the Heidelberg data is $a \gtrsim 3'0$, while it is $m_{\text{Harvard}} \lesssim 13.0 \pm$ for the Mt. Stromlo data; this, however, does not account for the discrepancy since a segregation of the Mt. Stromlo data according to corrected diameter (reduced to face-on) indicates a still greater excess of round shapes for the larger objects[2].

A possible reason for the excess of elongated shapes in the Heidelberg data is the mixture of types and in particular the inclusion of edgewise lenticulars with spirals; as noted by HOLMBERG [*21*] the deficiency of nearly round shapes is probably partly subjective (e.g. decimal error). The statistical treatment may account for the deficiency of elongated shapes in the Mt. Stromlo data. A final discussion will be possible when the revised types and dimensions of the brighter spirals over the whole sky are published.

6. Shape of spiral arms. The shape of spiral arms has been studied by VON DER PAHLEN[3], GROOT[4], REYNOLDS[5] and in greater detail by DANVER [*12*] who has determined the parameters of the best fitting logarithmic spiral for 98 systems.

The shape of the arms in the equatorial plane of the spiral can be found empirically by projecting a photograph on an inclined plane whose tilt and orientation are adjusted by trial and error until the projected outline appears roughly circular. It can be also determined analytically.

[1] For the observers who measured the Heidelberg Bruce Collection the mean relation between measured (β/α) and corrected (b/a) axial ratio is as follows:

b/a	1.00	0.8	0.6	0.4	0.2
β/α	1.00	0.76	0.53	0.31	0.125
$\beta/\alpha : b/a$	1.00	0.95	0.88	0.77	0.63

[2] Median axial ratio: 0.72 for $D>4'0$; 0.63 for $3'0<D\lesssim 4'0$; 0.47 for $2'0<D\lesssim 3'0$; for a random distribution Fig. 5 shows that the median ratio is in the range 0.52 to 0.55.

[3] H. VON DER PAHLEN: Astronom. Nachr. **188**, 249 (1911).

[4] H. GROOT: Monthly Notices Roy. Astronom. Soc. London **85**, 535—541 (1925); **86**, 146 (1926).

[5] J. H. REYNOLDS: Monthly Notices Roy. Astronom. Soc. London **85**, 1014—1020 (1925).

Let r, φ be the polar coordinates of a point P of an arm in the plane of the spiral and ϱ, ϑ its coordinates in the tangent plane to the celestial sphere, if $\omega = \frac{\pi}{2} - i$ is the angle between the plane of the spiral and the tangent plane, then

$$r = \varrho \sqrt{1 + \tan^2\omega \cdot \sin^2\vartheta} = \varrho \cdot c\,, \tag{6.1}$$

$$\tan\varphi = \tan\vartheta / \cos\omega\,. \tag{6.2}$$

$c = 1$ for $\vartheta = 0°$ and $180°$, $c = \sqrt{1 + \tan^2\omega}$ for $\vartheta = 90°$ and $270°$. If the spiral is logarithmic, as suggested by the empirical results,

$$\operatorname{Log} r = A - B\varphi \tag{6.3}$$

the measured values of $r = \varrho$ for $\vartheta = 0°$ and $180°$ give A and B, then r is computed and ϱ measured for $\vartheta = 90°$ and $270°$, whence $\cos\omega = \varrho/r$. If circular symmetry is assumed in the plane of the spiral for the outer outline of the system another determination is $\cos\omega = b/a$, neglecting the small thickness of the flat component.

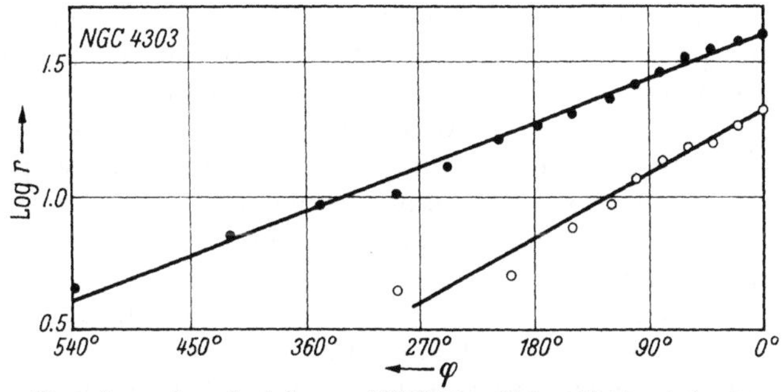

Fig. 7. Comparison of spiral arms of NGC 4303 with logarithmic spirals, after C. G. DANVER.

In practice the empirical method is quicker and just as accurate; for moderately tilted spirals ω can be determined to within 2° or 3° according to DANVER [12]. Usually the best fit is given by a logarithmic spiral, if

$$r = r_0 \cdot \exp(\varkappa\,\alpha) \quad \text{with} \quad \alpha = \pi\,\varphi/180°\,, \tag{6.4}$$

$$\operatorname{Log} r = \operatorname{Log} r_0 + \varkappa\, c\, \varphi \quad \text{where} \quad c = \frac{\pi}{180} \operatorname{Log} e = 0.00758 \tag{6.5}$$

and

$$\varkappa = \cot\mu\,, \tag{6.6}$$

μ being the characteristic angle between the tangent and the radius vector. The best fitting spiral is obtained through a least-square solution in $x = \varphi$ and $y = \log r$ coordinates which gives $\log r_0$ and $\varkappa c$. The origin $\varphi = 0$ may be taken at the visible extremity of an arm. A plot of $\operatorname{Log} r$ vs. φ for the two main arms of NGC 4303, type $SAB(rs)bc$, measured by DANVER on a Lick photograph is shown in Fig. 7. The mean characteristic angle for 190 arms in 98 objects is $\bar{\mu} = 73.°4 \pm 0.°4$ (p.e.) with a dispersion $\sigma = 6.°1 \pm 0.°3$ (p.e.); the range of individual values is from $\mu = 54°$ to $\mu = 86°$. Within the same object μ is not the same for different arms and the corresponding dispersion is $\sigma = 4.°9 \pm 0.°3$. Along a given spiral arm there is a slight tendency for μ to increase outwards, at least in the outer parts. The mean values of μ decrease along the morphological sequence from Sa to Sc, but the change is only slight between Sb and Sc (Table 7).

Table 7. *Mean geometrical parameters of spiral arms.*

Type	SAb	SAc	SBb	SBc	All
μ	75°2 ± 0°9	72°5 ± 0°6	75°2 ± 1°0	73°0 ± 2°3	73°4 ± 0°4
σ_μ	5°9 ± 0°7	6°4 ± 0°4	5°0 ± 0°7	8°3 ± 1°6	6°1 ± 0°3
n	41	112	24	13	190
$\bar{L}$	1.80 ± 0.09	1.74 ± 0.05	1.62 ± 0.08	1.61 ± 0.09	1.76 ± 0.04
σ_L	0.58 ± 0.06	0.51 ± 0.03	0.38 ± 0.05	0.34 ± 0.06	0.54 ± 0.03
W	328° ± 20°	317° ± 10°	277° ± 17°	211° ± 24°	310° ± 11°
σ_W	125° ± 14°	105° ± 7°	82° ± 12°	88° ± 17°	115° ± 6°

The angular winding W of the arms may be measured by the value of φ for which they apparently reach the nucleus; their linear development L by the ratio of their total length to the apparent major axis a. The mean angular winding for 190 arms is $W = 310° \pm 11°$ (p.e.), i.e. the arms can be followed over slightly less than one complete turn, with a large dispersion $\sigma_W = 115° \pm 6°$ (p.e.); their average length is $L = 1.76 \pm 0.04$, with $\sigma_L = 0.54 \pm 0.03$; one arm is usually longer than the other, on the average $\Delta W = 58° \pm 5°$, $\Delta L = 0.45 \pm 0.05$. The length of the arms, measured in both W and L decreases along the sequence, late-type spirals having shorter arms (Table 7). There is a slight correlation between apparent diameter a and W, $r_{a,W} = +0.32 \pm 0.07$, probably indicating the presence of some systematic errors in the measurement of W as the arms are more easily traced in the larger objects ($W \approx 250°$ for $a < 3'$, $W \approx 420°$ for $a > 21'$); there is, however, no perceptible correlation between a and L.

Note that in any study of the geometrical shapes of spiral arms there is a strong selection of the more regular and simple systems; in many objects whose spiral structure has a multiplicity greater than 3 or 4 it is practically impossible to follow individual whorls for more than a small fraction of a turn. Further, in many systems the spiral arms are made up of a large number of shorter arcs more strongly inclined on the radius vector (i.e. having smaller μ), than the overall pattern. Such structure is also observed in some ringed types for which part of the structure $\mu \approx 90°$. Finally in many barred spirals the faint extensions of the arms seem to return toward the bar and follow a general path more like a circle or an ellipse than a logarithmic spiral (cf. Sect. 4).

Index of illustrated galaxies.

NGC	Tpye	Plate and Figure	NGC	Type	Plate and Figure
45	*SA(s) dm*	VII, 34	4594	*SA(s:) a*	VIII, 39
55	*SB(s) m*	IX, 47	4631	*SB(s) dm*	IX, 48
300	*SA(s) d*	VII, 33	4691	*(R)SB(s:) O/ap*	XI, 59
488	*SA(r) b*	II	4710	*SA(rs?) O+*	IX, 44
891	*SA(s:) b*	IX, 46	4736	*(R)SA(r) ab*	V, 21
1084	*SA(s) c*	V, 19	4753	*I(O)*	XI, 58
1291	*(R)SB(s) O+*	IV, 12	5194	*SA(rs) bc*	II
1313	*SB(s) d*	VI, 32	5204	*SA(s) m*	VII, 35
1365	*(R')SB(s) b*	II; VI, 30	5273	*SA(s) O−*	III, 2
1433	*(R')SB(r) a*	II; VI, 25	5383	*SB(s) b*	X, 54
1512	*SB(r) O+*	IV, 14	5566	*SB (rs) a*	X, 51
1553	*SA(r) O°*	III, 7	5746	*SA(s:) b*	VIII, 40
1566	*SAB(s) b*	II	5866	*SA(s) O°*	IX, 43
1617	*SB (s) O+*	IV, 10	5962	*SA(r) c*	V, 23
1947	*SAOp*	XI, 56	6643	*SA(e) d*	V, 24
2523	*SB(r) bc*	VI, 27	6744	*SAB(r) bc*	II
2855	*SA(s) O+*	III, 4	6753	*(R)SA(r) b*	V, 22
2859	*(R)SB(r) O°*	IV 16	6873	*SAB (rs) O°*	IV, 11
3031	*SA(s) ab*	II	7079	*SB(s) O°*	IV, 9
3032	*SAB(r) O°*	IV, 15	7124	*SB(rs) b*	II
3034	*I(O)*	XI, 60/61	7144	*EO*	III, 1
3077	*I(O)*	XI, 57	7166	*SA(r) O−*	III, 6
3351	*(R')SB(r) ab*	VI, 26	7205	*SA(s) bc*	V, 18
3367	*SB(r) cd*	VI, 28	7457	*E+ 5*	III, 5
3718	*SAB(s) a*	X, 55	7552	*(R')SB(s) a*	VI, 29; X, 49/52
4215	*SA(r) O°:*	IX, 45	7582	*(R')SB(s) a*	X, 50/53
4244	*SA(s) cd*	VIII, 42	7702	*SA(r) O+*	III, 8
4262	*SB(r) O°*	IV, 13	7741	*SB(s) cd*	VI, 31
4303	*SAB(rs) bc*	II	7793	*SA(s) d*	V, 20
4459	*SA(s) O°*	III, 3	SMC	*SB(s) mp*	VII, 36
4565	*SA(s:) bc*	VIII, 41	I 1613	*IBm*	VII, 37
4569	*SA(s) ab*	V, 17	And I	*dIB(s) m*	VII, 38

General Physical Properties of External Galaxies.

By

G. DE VAUCOULEURS.

With 36 Figures.

Introduction.

The early investigations into the structure and general physical properties of external galaxies up to 1935 have been reviewed by CURTIS [*A*] and by HUBBLE [*B*]. The order of magnitude of their dimensions, luminosities and masses was established over thirty years ago by HUBBLE [*25*] and by LUNDMARK [*44*]. In the following decade more detailed studies of their colours, spectra and dynamics were undertaken principally at the Harvard, Lick, Mt. Wilson, Lund and Stockholm observatories. The main results available at the end of the war have been summarized by SHAPLEY [*C*] and by VOGT [*D*].

During the last decade, however, a considerable body of new information based on more accurate photographic and photoelectric work has accumulated which supersedes much of the pre-war data. The revision of the extragalactic distance scale[1] has also contributed to an improved assessment of the relative and absolute properties of external galaxies in relation to the Galaxy [*E*].

In the present article only the general physical properties of individual galaxies are considered at the exclusion of both their detailed internal structure or composition and their statistical or collective properties; neither the stellar luminosity function, stellar population characteristics, etc., nor the general distribution functions (mass, luminosity, etc.) of the galaxy population are discussed[2]. For similar reasons only the mass determinations based on rotational methods applied to individual galaxies of known morphological type will be considered at the exclusion of the order-of-magnitude estimates based on statistical methods applied to pairs, groups or clusters of galaxies of various types.

I. Optical properties.

a) Photographic dimensions.

The relative dimensions of galaxies of different types have been discussed by HUBBLE [*25*], SHAPLEY [*66*], DE VAUCOULEURS [*80*] and others using apparent photographic diameters measured directly by visual inspection of the plates, and by SHAPLEY [*67*] and HOLMBERG [*23*], using the maximum diameters detected on microphotometer tracings of the plates.

1. Micrometric diameters. The initial discussion by HUBBLE in 1926 [*25*] of apparent diameters measured on early reflector plates at Lick and Mt. Wilson had suggested a steady and large variation along the classification sequence,

[1] See K. LUNDMARK: Medd. Lunds Obs. (I) **1946**, Nr. 163; **1954**, Nr. 187. — W. BAADE: Trans. Internat. Astronom. Union **8**, 397 (1952).

[2] These are in fact still so poorly known that any attempt to summarize the conflicting views of recent years would retain hardly more than historical interest in a few years hence.

the mean apparent diameter at $m=10.0$ (HOLETSCHECK, visual) increasing as follows:

$E0$	$E4$	$E6$	$E7$	Sa, SBa	Sb, SBb	Sc, SBc
2′	3′	4′	5′	6′	7′	9′

According to these data the average ratio of diameters is $S/E=2.4$, but the range of variation from $E0$ to late S approaches 5 to 1. A ratio of 10 to 1 was even suggested by LUNDMARK[1] from a discussion of galaxies associated in 60 pairs or groups measured mainly on Franklin-Adams plates.

However, from a discussion of over 400 bright galaxies [*66*] measured on long-exposure photographs taken with the Bruce 24-inch refractor of the Harvard southern station, SHAPLEY obtained later[2] a mean ratio $S/E=1.7$. The ratio actually increases from about 1.5 at $m=10$ to 11 to about 2.5: at $m=13$ (Harvard, photographic), an indication of selection effects and/or mixture of types depending on luminosity and/or apparent diameter. Further, a special discussion of 85 objects in the Virgo cluster—and therefore all practically at the same distance—indicated a still smaller range of variation as follows:

$E0-3$	$E4-5$	$E6-7$	S	Sa	Sb	Sc	SB	SBa	SBb	SBc
4′6 (17)	3′6 (7)	2′4 (6)	4′2 (19)	4′4 (12)	6′2 (5)	6′0 (8)	2′5: (1)	3′9 (4)	4′3 (3)	7′3: (2)

The increase from early to late spirals found by HUBBLE was still present, but the trend from $E0$ to $E7$ was reversed and the mean diameters for the two main classes were $S=4'8$ $(n=55)$, $E=3'9$ $(n=30)$, a ratio $S/E=1.23$. If the distance of the Virgo cluster is $\Delta=6.5$ Mpc (assuming $H=180$ km/sec per Mpc) the corresponding diameters are $\bar{S}\approx 9$ kpc and $\bar{E}\approx 7.5$ kpc. The disagreement with the Mt. Wilson results called for further comparisons based if possible on more abundant and more homogeneous data and allowing for the new types isolated during the last two decades.

The results of such comparisons, based on diameters of over 200 southern galaxies in the Mt. Stromlo survey [*80*], corrected for tilt effects and reduced to unit distance (about 10 Mpc in the revised scale), are collected in Table 1 and

Table 1. *Mean corrected diameters reduced to unit distance.*

Type	E	$S0$	$S0/a$	Sa	Sab	Sb	Sbc	Sc	Scd	Sdm
$\overline{D(0)}$	1′17	1′22	(2′4:)	1′38	1′71	1′98	2′25	2′17	2′25	1′87
n	(42)	(27)	(15)	(8)	(13)	(14)	(13)	(19)	(10)	(6)

shown in Fig. 1. Ordinary and barred spirals SA, SB have been grouped together since the mean ratio in the sample is close to unity ($\overline{SA}/\overline{SB}=0.96$); no distinction is made here between the (r) and (s) varieties although for barred spirals an uncertain difference of the order of 20% appears indicated: $\overline{SB(s)}/\overline{SB(r)}=1.2$.

The substantial equality of dimensions of ordinary and barred spirals confirms the earlier results of HUBBLE [*25*] and SHAPLEY [*66*]. The range of variation

[1] K. LUNDMARK: Medd. Astr. Obs. Uppsala **1926**, No. 9.
[2] H. SHAPLEY: Harvard Obs. Bull. **1934**, No. 895, 24.

from E to late S is less than 2 to 1 in this sample selected according to apparent magnitude and surface brightness ($m_{pg} \leqq 13 \pm$, Harvard); at the chosen unit distance $\Delta = 10$ Mpc, $1' = 2.9$ kpc and the range of mean diameters is from about 3.5 kpc at $E/S0$ to 6.5 kpc at Sc; magellanic spirals Sm average 5.5 kpc and the few magellanic irregulars Im are still smaller.

The range of diameters within each class is at least 5 to 1 for spirals, 4 to 1 for lenticulars and 3 to 1 for ellipticals, but this latter range is certainly reduced by the magnitude/surface brightness restriction in the Shapley-Ames catalogue from which the sample is drawn. The mean ratio $S/E = 1.7$ agrees well with its value derived from similar Harvard data (H.A. **88**, 4), but does not support the approaching equality suggested by the Harvard work on the Virgo cluster (H.B. 895) quoted above.

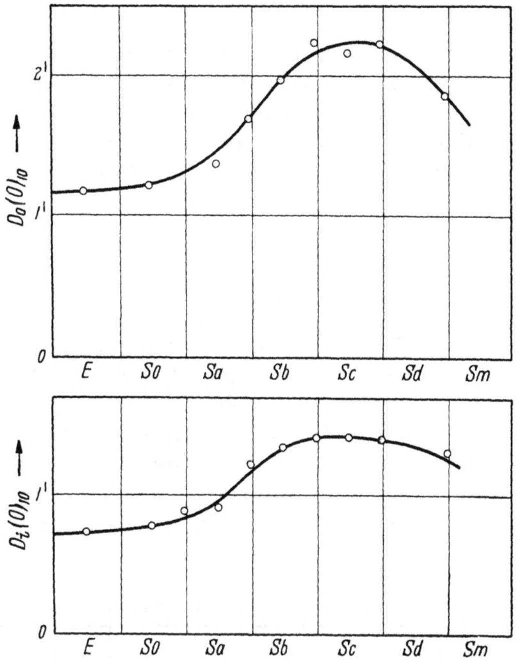

Fig. 1. Mean apparent diameters vs. galaxy type. The mean apparent photographic diameters of the Shapley-Ames galaxies south of $-35°$ (Mount Stromlo survey) corrected to "face on" and reduced to unit distance (10 Mpc) are plotted as a function of galaxy type along the classification sequence. Dimensions are given for the bright, inner regions (D_i) and for the faint, outer regions (D_0).

Part of the discrepancies may be attributed to the different observing and measuring conditions, since the scales of dimensions of different surveys are not only different but depend also on galaxy type[1]; the absence of correction for tilt effects and mixture of types are other contributing factors.

The change of scale is even greater between dimensions measured by mere visual inspection of photographs or on microphotometer tracings of the plates. A comparison of diameters measured at Mt. Wilson and Harvard by both methods for a number of galaxies of different types was made by SHAPLEY[2]; it indicated a very large increase in apparent dimensions through the use of the microphotometer, amounting to 2.3 times for spirals and 5.5 times for ellipticals as compared with the early Mt. Wilson data and still to 1.4 and 1.7 times respectively compared with the Harvard data.

Such factors applied to the Harvard data on the Virgo cluster lead to a vanishing difference between ellipticals and spirals, but their application to the more general Harvard and Mt. Stromlo statistics still leaves a margin of the order of one-third in favour of the spirals.

2. Microphotometric diameters. There are only few published determinations of dimensions from microphotometer tracings and they are variously affected by selection effects. An extensive series of photometric measurements of 112 bright

---

[1] For instance the relations between the "standard" dimensions in the Mount Stromlo survey and the dimensions in the catalogues of SHAPLEY [*66*] and of REINMUTH [*62*] are

Harvard/Stromlo $= 1.25$ (E) and 1.18 (S)

Heidelberg/Stromlo $= 0.72$ (E, $S0$) and 0.84 (SA, SB).

[2] H. SHAPLEY: Monthly Notices Roy. Astronom. Soc. London **94**, 806 (1934) = Harvard Reprint 105.

galaxies by Miss PATTERSON [53a] at Harvard was never published in full; a summary of the maximum dimensions detected on the tracings, published in 1942 by SHAPLEY [67], gives only the absolute diameters in kiloparsecs for assumed distances which are not specifically stated. The mean diameters (based on the 1936 distance scale in which $H = 550$ km/sec per Mpc) are as follows:

$$E: \quad 4.88 \pm 0.44 \text{ (m.e.) kpc} \quad (n = 13),$$
$$S0, Sa, Sb: \quad 4.35 \pm 0.23 \text{ (m.e.) kpc} \quad (n = 30),$$
$$Sc: \quad 4.67 \pm 0.25 \text{ (m.e.) kpc} \quad (n = 69).$$

The selection effect was estimated through a comparison of the mean apparent magnitudes of objects in the Virgo cluster (H.A. **88**, 2 magnitudes), showing that on the average the brighter galaxies ($\Delta m = 0.9$ for E, 0.5 for $S0$, S) were selected for diameter measurements; the differential effect in favour of the ellipticals amounts to $\Delta m = 0.4$ mag. or $\Delta \operatorname{Log} D = -0.2\,\Delta m = -0.08$, a factor 0.83 in the diameters; the mean diameters corrected for selection are then

$$E: \ 4.05 \text{ kpc}, \quad S: \ 4.57 \text{ kpc}, \quad S/E = 1.13.$$

The trend toward equality of spiral and elliptical diameters as defined by progressively fainter brightness levels is confirmed. However, this equality refers to samples selected according to apparent magnitude, i.e. refers substantially to giant systems only, and while dwarf ellipticals are definitely known, there is no clear case of dwarf spirals of comparable dimensions.

In a photometric study of nearby galaxies HOLMBERG [23] gives the maximum recorded dimensions of 27 members of the Local group and of the M 81 and M 101 groups; the mean values based on revised distance moduli (+1.6 mag. in the Local Group; +3.2 mag. in the other groups) and galaxy types are collected in Table 2 and shown in Fig. 2.

Table 2. *Mean absolute magnitudes and dimensions of nearby galaxies.*

Type	Objects	n	Magnitudes (M_{pg}) mean	Magnitudes (M_{pg}) range	Dimensions (kpc) mean	Dimensions (kpc) range
dE	NGC 147, 185, 205, 221	4	−14.3	−13.5/−15.1	2.1	1.7/3.5
I(0)	NGC 3034, 3077, 5195	3	−17.1	−16.6/−18.0	7.1	6.2/9.5
gSb	NGC 224, 3031	2	−19.5		25.0	
Sc	NGC 598, 2403, 2976, 5194, 5204, 5457, 5474, 5585	8	−17.2	−15.5/−19.0	9.3	4.8/19.5
Sm	LMC, SMC, NGC 2366, IC 2574, Ho IV	5	−16.2	−14.3/−18.1	7.4	4.8/10.5
dIm	NGC 6822, IC 1613, W.L.M. Ho I, Ho II	5	−14.1	−12.8/−16.1	2.3	1.7/3.1

The mean value for the ellipticals, based only on the four dwarf companions of the Andromeda nebula, is not representative of the E class as a whole. The only two Sb galaxies (M 31, M 81) are both supergiant systems. The mean values for Sc, Sm, dIm are probably more representative and may constitute a fair sample; the decrease in dimensions from Sc to dIm agrees fairly well with the data in Table 1 (note that the effect of tilt is not corrected in HOLMBERG'S data). The three non-magellanic irregulars are all giant systems of the peculiar M 82 type[1]. The data are plotted in Fig. 2 which, in conjunction with Fig. 1, gives the

[1] HOLMBERG designates them as "Type II" irregulars because of their red colour; however the spectra indicate early types and morphological characteristics suggest instead a relation to lenticulars; see Plate XI in the preceding article.

best available evidence on the relative and absolute dimensions of galaxies of various types.

The overall range of diameters among galaxies appears to be at least 15 to 1, but the range does not seem to exceed 5 to 1 among late-type spirals. *Dwarf* ellipticals and irregulars are 4 to 5 times smaller than the average spiral; since the ratio for all ellipticals is only 1.5 or less, the existence of giant ellipticals approaching in size the largest spirals is indirectly confirmed, but such systems may be restricted to the great clusters[1].

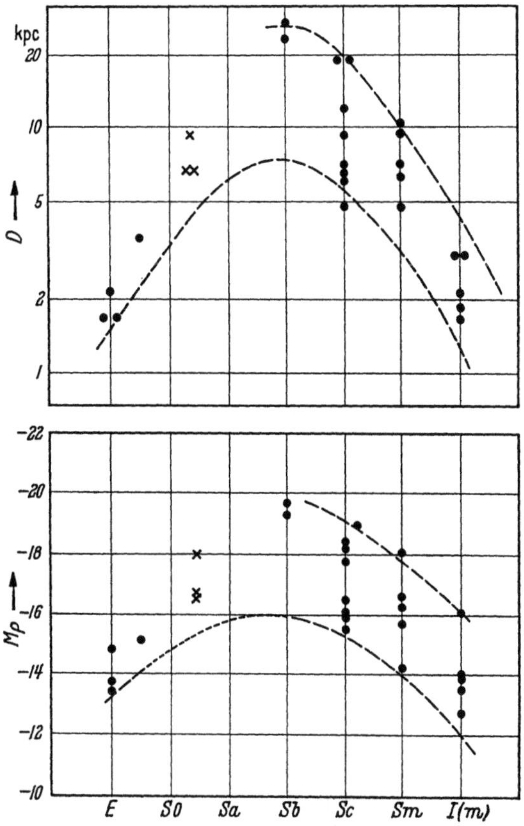

Fig. 2. Absolute dimensions and magnitudes of 27 bright galaxies. The microphotometric diameters and photographic magnitudes are plotted as a function of galaxy type along the classification sequence. Note the position of the non-magellanic irregulars $I(0)$ marked by crosses.

b) Integrated luminosities and colours.

3. Luminosities. Measurements of the integrated luminosities of external galaxies have been often attempted in the past, but it is only in recent years that reliable results have been achieved by photographic and photoelectric methods. The main photographic contributions are due to BIGAY [*4*], [*5*] using the Fabry method and to HOLMBERG [*20*], [*23*] using the slow, but accurate method of integration of isophotes or luminosity profiles. The main photoelectric contributions are due to WHITFORD [*85*], STEBBINS and WHITFORD [*70*], [*72*] and PETTIT [*56*] using the Fabry method. Lists of integrated P magnitudes of bright galaxies based on reductions of these data to a standard system have been compiled by SANDAGE [*28*] and DE VAUCOULEURS [*82*], [*83*].

The two main results of interest here refer to the mean surface brightness and to the mean absolute magnitude as a function of morphological type.

α) The mean surface brightness has no absolute meaning as it depends on the particular system of dimensions used in the comparison; for the dimensions given in the Shapley-Ames catalogue and corrected total magnitudes [*81*], [*82*], the following mean values may be quoted:

	Mean ($m_{pg}/\square'$)	Dispersion
E, *S0*	11.75	0.90
SA, *SB*	13.15	0.75

[1] There is no such object in the immediate neighbourhood of the galaxy; examples in the Virgo cluster, at a distance $\Delta = 6.5$ Mpc, are NGC 4486 (M 87) and NGC 4472; the Harvard microphotometric diameters are respectively $11'5 = 22$ kpc and $16'3 = 31$ kpc which is an extreme case.

The conventional surface brightness is here defined by the expression $m_1 = m_T + 2.5 \text{ Log } D d$; if the galaxies are treated as ellipses of axes D, d the numerical value of m_1 should be decreased (the surface brightness increased) by $2.5 \text{ Log} \pi/4 = -0.26$ mag.; if instead of the dimensions listed in HA **88**, 2 those of any other catalogue are used the values of m_1 should be corrected by $5 \text{ Log } K$, where K is the scale factor of the other catalogue compared with HA **88**, 2.

In the absence of a satisfactory system of dimension data the present results must be regarded as giving only a preliminary indication and they are also affected by selection effects in an unknown manner[1].

β) The mean absolute magnitudes can be derived, in principle, from the constants in the magnitude/red-shift relations

$$P_c = 5 \text{ Log } c z + B \tag{3.1}$$

where P_c is the photographic apparent magnitude corrected for latitude and selective effects (negligible for the brighter galaxies) and $z = \Delta\lambda/\lambda$ is the red-shift. The most complete and recent data, based on 474 galaxies in the Lick-Mt. Wilson-Palomar list of red-shifts discussed by SANDAGE [*28*], are collected in Table 3.

Table 3. *Differential mean absolute photographic magnitudes $\Delta M = M - \overline{M}$ of galaxies observed for red-shift.*

Type	*E*	*SA* 0	*SB* 0, *SB a*	*SA a*	*SB b*	*SA b*	*SA c, SB c*	All
ΔM	−0.14	+0.17	+0.29	−0.12	−0.33	−0.16	+0.21	0.00
n	117	67	36	54	27	76	90	474

The integrated magnitudes used in the table are in a system which comes very close to the total or asymptotic magnitudes extrapolated to infinity. According to this table the *SB b* galaxies are statistically slightly brighter than average while the *SB* 0 and *SB a* galaxies appear slightly fainter. However, the large scatter in the magnitude/red-shift relation results in probable errors of the order of ± 0.3 mag. for the mean values in Table 3; hence the differences between different types are probably mainly illusory. Within the observational errors the mean absolute magnitudes of galaxies of all classes *in this sample* appear therefore practically equal. The mean value of B for all classes is -4.235 ± 0.128 (p.e.); if the expansion coefficient is $H = 180$ km/sec per Mpc [*28*] the corresponding absolute magnitude corrected for 0.25 mag. absorption in the galaxy[2] is $\overline{M}(m) = -18.2$, while the brightest galaxies reach up to $M \approx -20$.

It must be emphasized, that the sample of galaxies observed for radial velocity is strongly affected by observational selection and the mean absolute magnitude per unit volume of space is much fainter than indicated by the figures above. For a Gaussian luminosity function of dispersion σ the relation between the mean absolute magnitude $\overline{M}(m)$ in a sample selected according to apparent magnitude and the mean per unit volume $\overline{M}_0$ is[3]

$$\overline{M}(m) = \overline{M}_0 - 1.382\sigma^2. \tag{3.2}$$

[1] For instance low luminosity dwarfs (I and E) are grossly under-represented in all existing catalogues so that the mean surface brightness of, say, the ellipticals is certainly not homogeneous and probably too high compared with the spirals.

[2] E. HUBBLE: Astrophys. Journ. **79**, 8–76 (1934) = M.W.C. 485

[3] K. G. MALMQUIST: Medd. Lunds Astr. Obs., **2**, No. 22 (1920).

The value of σ is not well known but appears to be between 1 and 2 magnitudes [*23*] so that the corrected mean magnitude $\overline{M}_0$ may be of the order of −15 or −16.

From a photometric study of 28 galaxies in the Local Group and the M 81 and M 101 groups, during which care was taken to include all known dwarf members, HOLMBERG derived in 1950 [*23*] rather different values which may refer more directly to a given volume of space but are not as representative of the total population in view of the smallness of the sample. The results, revised and corrected[1] are collected in Table 2 and Fig. 2.

The general mean, about −16 or −17 depending on weights, is in fair agreement with the red-shift data, but large systematic differences are indicated between dE, S and $dI\,(m)$ galaxies. The sample is too small and certainly not representative for ellipticals and early type spirals; for late-type spirals a mean absolute magnitude per unit volume of $-16\frac{1}{2}$ or −17 indicated by both the redshift data and the nearby galaxies may be more representative.

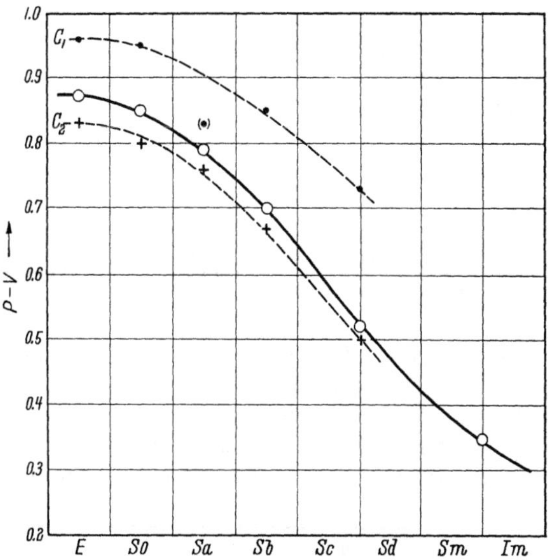

Fig. 3. Mean integrated colour indices of bright galaxies. The mean colours of the nuclear regions (dots, C_1) and outer regions (crosses, C_2) are also indicated.

The overall range of absolute magnitudes among galaxies of all types is at least 7 magnitudes from the faintest dE, dI, to the brightest gSb systems. Among spiral galaxies of a given type the range is probably less, perhaps not more than 4 magnitudes, but here again the small sample and selection effects may vitiate the statistics. From a comparison of Tables 2 and 3 the existence of giant ellipticals is again indicated; the brightest cluster galaxies reach up to about $M_{pg} \approx -20$ according to the red-shift data [*28*].

4. Colours. The integrated colours of galaxies have been measured by HOLMBERG [*20*], [*23*] using accurate photographic methods[2] and by STEBBINS and WHITFORD [*70*], [*72*], BIGAY *et al.* [*6*], [*7*] and PETTIT [*56*] using photoelectric techniques.

α) *The mean colour indices* for the various morphological types are collected in Table 4; the considerable scatter among individual values within each types is shown in Fig. 3 by the mean and range in PETTIT's data (reduced to $P-V$) for galaxies brighter than the 13th magnitude.

---

[1] Cf. Sect. 2; revised photographic magnitudes used for some objects are LMC: 0.7, SMC: 2.2, NGC 2366: 11.5, NGC 5204: 11.7. The system Ho IV ($m = 12.9$) whose velocity [*28*] indicates membership in the M 101 group has been added. The Sculptor and Fornax elliptical dwarfs whose magnitudes are very uncertain and which would not normally be observed in radial velocity surveys because of their low surface brightness were rejected.

[2] Early photometric measurements by WHIPPLE (Harvard Circ. **1935**, 404), SEYFERT [Harvard Ann. **105**, 209 (1937)], Miss SCHATTSCHNEIDER [Astronom. Nachr. **264**, 165 (1937)], VASHAKIDSE [Bull. Abastumani **5** (1940); **6** (1942); **8** (1945); **13** (1953)] and others do not compare in accuracy with the more recent results.

The colour systems and regions measured differ slightly from one series to another, but the adopted mean values should be very close to the $P-V$ system[1]. It is clear that on the average the integrated colour grows redder along the classification sequence from the magellanic irregulars to the ellipticals. The small group of irregulars having red colours includes peculiar and misclassified objects as well as non-magellanic irregulars of the M 82 type. The large range of colours at each stage (Fig. 3) indicates, however, that integrated colour is at best only a crude classification index and is affected by many complicating factors.

Table 4. *Mean colour indices of normal galaxies.*

Type	E	$S0$	Sa	Sb	Sc	$I(m)$	(I)	Remarks
S and W (1937)	+0.90 (30)	—	+0.86 (13)	+0.84 (12)	+0.59 (36)	—	—	C_p
S and W (1952)	+0.86 (32)	+0.83 (4)	+0.83 (4)	+0.82 (11)	+0.47 (11)	+0.27 (3)	+0.74 (2)	C_p
HOLMBERG	+0.77 (8)	—	+0.81 (1)	+0.84 (3)	+0.48 (8)	+0.33 (4)	+0.82 (3)	$P_g - P_v$
PETTIT	+0.94 (64)	+0.92 (50)	+0.85 (49)	+0.76 (79)	+0.57 (85)	+0.44 (6)	+0.80 (6)	CI, $m_{pg} < 13$
Adopted	+0.87	+0.85	+0.80	+0.75	+0.55	+0.35	+0.8:	$P-V$

β) The effect of emission lines is indicated by the following comparison, according to PETTIT's data [*56*], of the mean colour indices of galaxies listed by HARO [*19*] as having strong emission lines in their spectra (cf. Sect. 11) with the mean normal colours of galaxies of the same morphological types in whose spectra emission lines are weak or absent. The effect is apparently most pronounced at the stage Sa

Type	E, $S0$	Sa	Sb	Sc
Normal	+0.93	+0.85	+0.76	+0.57
Emission	+0.89 (3)	+0.64 (4)	+0.66 (13)	+0.58 (8)
Em.-Norm.	−0.04	−0.21	−0.10	+0.01

where galaxies with unusually strong emission lines are known to occur; it is small for lenticulars and ellipticals whose emission lines are usually much weaker and also for late-type spirals (and magellanic irregulars) whose mean colour differs probably but little from the integrated colour of their common emission lines.

γ) The effect of the red-shift may be statistically represented, according to PETTIT [*56*], by the relations

$$C = +0.84 + 0.04\,(m_{pg} - 9.0) \qquad (E)\,, \tag{4.1}$$

$$C = +0.50 + 0.08\,(m_{pg} - 9.0) \qquad (SA) \tag{4.2}$$

based on several hundred objects in the magnitude range $m = 9$ to $m = 17$. Since the red-shift is also very nearly a linear function of apparent magnitude in this range [*28*], these relations express the effect of the redward displacement of the

[1] The relation between PETTIT's colours and $P-V$ as represented by STEBBINS and WHITFORD [*72*] is CI = $(0.018 \pm 0.027) + (1.056 \pm 0.033)\ C_p$, according to SANDAGE [*28*].

spectral energy curves; it is not clear why this effect should be so much larger for spirals (if anything the opposite might have been expected), but, here again, observational selection and other systematic effects must be involved which prevent a straightforward interpretation of the results.

The colour-velocity relation for elliptical galaxies, according to STEBBINS and WHITFORD[1], is

$$C = +0.84 + 0.0133 \cdot 10^3 V \qquad (V \text{ in km/sec}). \tag{4.3}$$

For a discussion of colour effects, real or apparent, related to the red-shift, see the chapters on Cosmology.

δ) *The effect of tilt* of the equatorial planes on the line of sight is nil or negligible for ellipticals and early lenticulars which are substantially transparent. For spirals and irregulars a small effect may be expected, but the empirical evidence is not clear, the change from the face-on to the edge-on position is probably less than ± 0.1 mag.[2].

c) Luminosity and colour distribution.

5. Ellipticals. The distribution of luminosity (surface brightness) in elliptical galaxies has been investigated by HUBBLE [*26*], REDMAN [*59*], REDMAN and SHIRLEY [*61*], OORT [*52*], Miss PATTERSON [*53a*], DE VAUCOULEURS [*76*], EVANS [*17*], DENNISON [*13*], Miss HAZEN [*88*] and others (see references to Table 5) using photographic methods and by VAN HOUTEN, OORT and HILTNER [*29*] and BAUM[3] using photoelectric methods. The technical problems of the photographic method have been discussed by REDMAN [*59*], and by DE VAUCOULEURS [*75*]; the theoretical interpretation of the results has been discussed by HUBBLE [*26*], TEN BRUGGENCATE [*8*], OORT [*52*], BELZER, GAMOW and KELLER [*3*], DE VAUCOULEURS [*78*] and others[4].

A finding list of ellipticals measured photometrically is given in Table 5.

Table 5. *Ellipticals analysed for luminosity distribution.*

NGC	Type	References	NGC	Type	References
205	$E^+ 5p$	[*59*], [*61*], [*73*], [*69*][5]	4278	$E1$	[*26*]
221	$E2$	[*26*], [*59*], [*61*], [*29*], [*73*]	4283	$E0$	[*26*]
410	$E2$	[*26*]	4406	$E3$	[*26*]
584	$E3$	[*26*]	4472	$E1$	[*26*]
1344	$E4$	[*17*]	4486	$E0p$	[*26*]
1399	$E0$	[*17*]	4494	$E1$	[*52*]
1404	$E1$	[*17*]	4552	$E0$	[*26*]
1549	$E0-1$	[*17*]	4621	$E5$	[*26*]
2672	$E1$	[*59*], [*61*]	4699	$E2$	[*26*], [*59*], [*61*], [*76*]
2693	$E2$	[*59*], [*61*]	4976	$E4$	[*17*]
3115	$E^+ 7$	[*26*], [*52*], [*76*], [*29*][6]	5557	$E1$	[*59*], [*61*]
3377	$E6$	[*59*], [*61*]	7454		[*59*], [*61*]
3379	$E1$	[*26*], [*59*], [*61*], [*76*], [*13*]	**I** 1459	$E3$	[*17*]
3605	$E4$	[*59*], [*61*]	**I** 4296	$E0$	[*17*]
3608	$E1$	[*59*], [*61*]			

[1] J. STEBBINS and A. E. WHITFORD: Astrophys. Journ. **108**, 413 (1948) = M.W.C. 753.

[2] See E. HOLMBERG: Medd. Lund (II) **1947**, No. 120.

[3] W. A. BAUM: Publ. Astronom. Soc. Pacific **67**, 328–330 (1956).

[4] See e.g. the discussion of globular clusters by R. v. D. R. WOOLLEY, Monthly Notices Roy. Astronom. Soc. London **114**, 191–209 (1954).

[5] J. H. REYNOLDS: Monthly Notices Roy. Astronom. Soc. London **94**, 519 (1934).

[6] See also S. P. WYATT: Astronom. J. **58**, 50 (1953) (Abstr.). — R. C. WILLIAMS and W. A. HILTNER: Astrophys. Journ. **98**, 47 (1943) (uncalibrated isophotes).

α) Luminosity law. The most significant result of these investigations is the discovery of a common luminosity law to which all typical ellipticals seem to obey; this suggests that all elliptical galaxies are essentially built on the same basic model, individual objects differing only in scale, density and flattening. There is, however, no general agreement as yet on the precise formulation of the luminosity law.

HUBBLE [*26*] found that the photometric profiles measured along the minor and major axes could be represented by a formula of the type

$$\left.\begin{aligned} B &= \frac{B_0}{(r+a)^2} \\ \text{or}\quad \log\frac{B}{B_0} &= -2\log\left(\frac{r}{a}+1\right) \end{aligned}\right\} \tag{5.1}$$

where B_0 is the central brightness and a a scale parameter; two independent parameters are therefore available to fit the observed profiles[1]. In a sample of 15 ellipticals of total magnitudes 8 to 13 and types $E0$ to $E7$ the range of values of a was from 2″ to 10″. The mean luminosity distribution for the 15 objects was very accurately represented by this formula down to $r/a \approx 15$ and fairly accurately to $r/a \approx 30$; the formula breaks down for $r/a < 0.3$ and does not apply at $r=0$; hence the range of distances to which it applies accurately is about 50 to 1 and approximately 100 to 1; the corresponding luminosity ratios are about 10^2 and 10^3.

The validity of the Reynolds-Hubble formula in the inner and intermediate regions of other ellipticals has been verified by all subsequent studies, but its applicability to the outer regions of low luminosity has been questioned by REDMAN [*59*] and by DE VAUCOULEURS [*76*], [*78*]; in fact the results of DE VAUCOULEURS [*78*], DENNISON [*13*] and HAZEN [*88*] indicate that at large distances from the centre the observed intensities fall short of those predicted by formula (5.1) when the parameters are adjusted to fit the brighter inner regions (Fig. 4). This is a reasonable result since the asymptotic slope of relation (5.1) is -2 and the integral diverges for $r \to \infty$ while in a physical system it appears more plausible that most of the light (and mass) is concentrated in the inner and intermediate regions rather than diluted in the outermost fringes.

An alternative formula whose domain of validity appears wider has been proposed by DE VAUCOULEURS [*76*], [*78*]; it may be written

$$\operatorname{Log}\mathcal{B} = -3.33\,(\alpha^{\frac{1}{4}} - 1) \tag{5.2}$$

where $\alpha = r/r_e$ if r_e is the *effective* radius (or semi-major axis in a flattened object), i.e. by definition the radius within which is emitted half the total luminosity[2], and $\mathcal{B} = B/B_e$ the surface brightness along this particular isophote; note that there is no other arbitrary parameter than the scale factor r_e since B_e is uniquely defined by r_e. The applicability or otherwise of this formula to any object can be tested apart from the interpretations of r_e and B_e by plotting the *observed*

---

[1] This formula was originally introduced in the form $L = C/(x+1)^2$ by REYNOLDS [Monthly Notices Roy. Astronom. Soc. London **74**, 132 (1913)] to represent the luminosity distribution along the major axis in the nuclear regions of the spiral nebula M 31 (cf. Sect. 7).

[2] An important application of the formula is the computation of the integrated luminosity within any given radius and, in particular, the total or asymptotic magnitude [*82*], [*83*] since the integral of (5.2) converges for $r \to \infty$. Another application is the definition of objective and physically significant dimensions of galaxies, such as the effective diameter $2r_e$ [C. R. Acad. Sci. Paris **226**, 1692 (1948); **227**, 548 (1948)]. This is not possible with a luminosity law whose integral diverges for $r \to \infty$.

Log B as a function of $r^{\frac{1}{4}}$; if the plot is a straight line the law applies and the slope of the line is $-3.33/r_e^{\frac{1}{4}}$, then Log B_e is read off at $r=r_e$ [1].

Although this formula was originally based on measurements of only 3 ellipticals of types $E\,0$ to $E\,7$ measured by the author [*76*] and later of 5 objects accurately measured by several other observers [*78*] (see Fig. 5), it has since been applied and found to represent satisfactorily all ellipticals for which measurements of sufficient accuracy have been published. In the range covered by HUBBLE's measurements, in particular, it is indistinguishable observationally

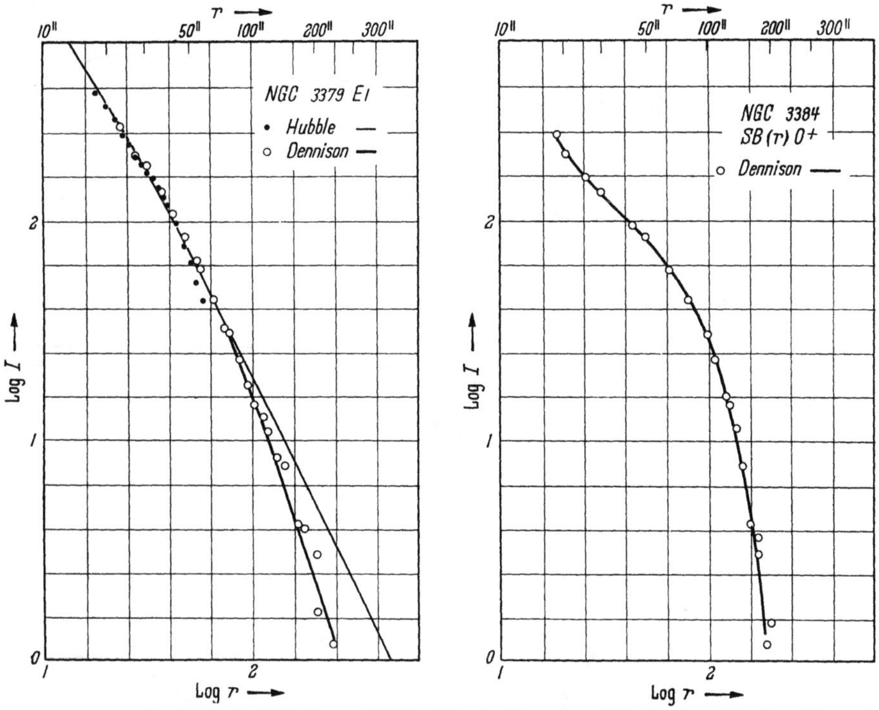

Fig. 4. Luminosity distribution in NGC 3379 and NGC 3384, after DENNISON. Note the departure from HUBBLE's rule in the outer parts. Compare with Fig. 8.

from formula (5.1), but it gives a much better representation of the observations at greater distances (Figs. 4 and 5), at least up to $r/r_e \approx 5$ and perhaps as far as $r/r_e \approx 15$. Formula (5.2) breaks down for $r/r_e < 0.03$ and does not apply at $r=0$ (HUBBLE's formula is better in this range); hence the range of distances to which the formula applies is at least 150 to 1 and possibly 500 to 1. Over this range the surface brightness varies by a factor of at least 10^4 and possibly in excess of 10^5 [*78*] [2].

[1] The coefficient in formula (5.2) first obtained in a purely empirical manner was 3.25 [*76*], but it is easy to show that if the formula is exactly obeyed from $r=0$ to infinity its form requires that the coefficient be 3.33.

[2] Formula (5.2) has been found to apply also to the distribution of galaxies in spherical clusters of galaxies of the Coma type [DE VAUCOULEURS, C. R. Acad. Sci. Paris **227**, 586 (1948); C. D. SHANE and C. A. WIRTANEN, Astronom. J. **59**, 285—304 (1954)] and to the distribution of stars in the outer regions of large globular clusters [GASCOIGNE and BURR, Monthly Notices Roy. Astronom. Soc. London **116** 570—582 (1956)]. The theoretical interpretation in terms of a truncated isothermal (MAXWELL) distribution has been outlined by the author [C. R. Acad. Sci. Paris **227**, 548 (1948); Monthly Notices Roy. Astronom. Soc. London **113**, 134—161 (1953)] and developed more rigorously by WOOLLEY and collaborators [Monthly Notices Roy. Astronom. Soc. London **114**, 191—209 (1954)].

Still another empirical formula has been advanced by BAUM[1] which may be written

$$B = L_0/2r\,(r + r_0) \tag{5.3}$$

where L_0 and r_0 are a "characteristic" luminosity and a "characteristic" radius. It is said to represent the luminosity distribution out to very great distances from the nucleus in a dozen galaxies (in particular M 87) measured by photon counting techniques, but details of the measurements and tests of the goodness of fit have not been published yet. Note that again two arbitrary parameters are involved and that the integral diverges at infinity; furthermore the formula breaks down in the centre since $B \to \infty$ when $r \to 0$.

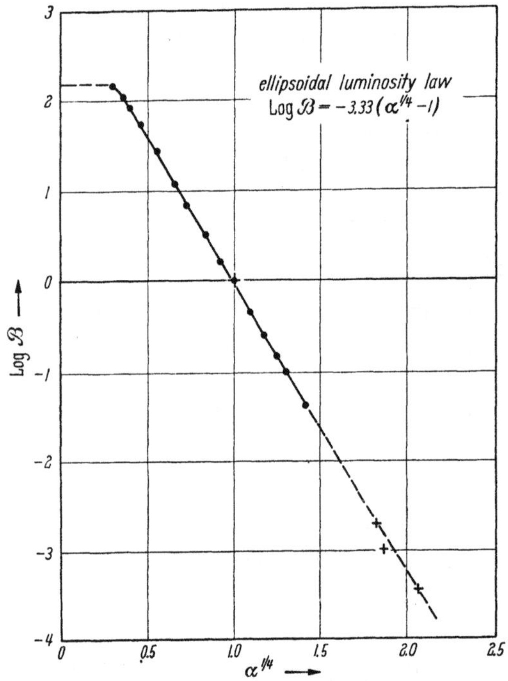

Fig. 5. Ellipsoidal luminosity law. The mean reduced brightness $\mathcal{B} = B/B_e$ for five well observed ellipticals is plotted against the reduced radius (or semi-major axis) $\alpha = a/a_e$. Except in the centre Log $\mathcal{B}$ is a linear function of $\alpha^{\frac{1}{4}}$.

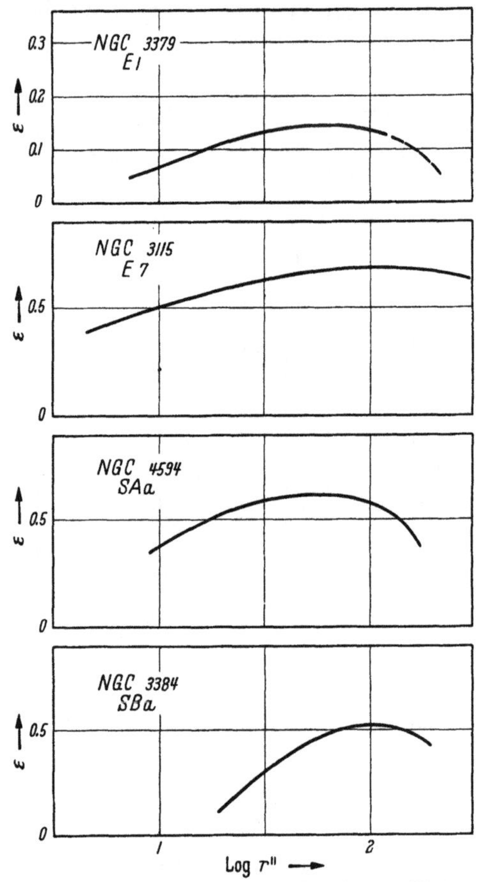

Fig. 6. Flattening curves of ellipticals and spirals. The flattening $\varepsilon = 1 - b/a$ of the isophotes is plotted against the radius or semi-major axis (in seconds of arc).

β) The apparent ellipticity of the isophotes can be derived from accurate determinations of the luminosity profiles along the major and minor axes [*52*], [*76*] or better from complete tracings of the luminosity contours with an isophotometer [*13*]. The results are conveniently represented as a plot of the measured flattening $\varepsilon = 1 - b$ as a function of Log r, if r is the radius (semi-major axis) of any particular isophote. Examples are shown in Fig. 6. There is general agreement, at least among the more recent and accurate determinations, to show that the flattening increases from the nucleus outwards up to a maximum corresponding very nearly with the apparent flattening estimated by direct inspec-

[1] W. A. BAUM: Publ. Astronom. Soc. Pacific **67**, 328 (1956).

tion of the photographs and beyond which a slow decrease takes place in the faint outer regions, generally beyond the limits of visual detection of the nebular image[1].

Further, there is a conspicuous departure of the shape of the isophotes from geometrical ellipses, at least in the most elongated objects (Fig. 7). It follows that the isophotal surfaces in the elliptical galaxies are not similar, concentric ellipsoids and the derivation of space light densities can only be achieved through approximate numerical procedures. The analytical treatment [*8*] constitutes, however, a useful guide and may serve as an introduction to theoretical models.

γ) Colour distribution. Apart from the integrated colour indices of elliptical galaxies discussed in Sect. 4, very few direct observations are available on the colour distribution as a function of distance to the centre; some information may be obtained, however, by comparing the integrated colours of a given object measured through field diaphragms of various apertures.

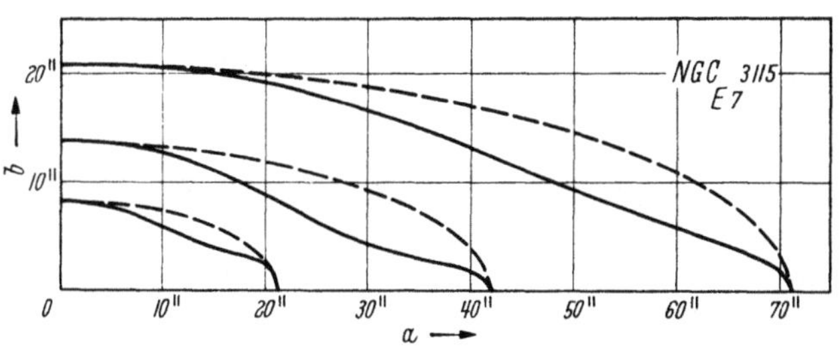

Fig. 7. Some isophotes of NGC 3115, after Oort. Note the departures from ellipses.

It is usually stated that there is little or no change in colour from the nucleus outwards; for instance, according to Pettit [*56*], out of 58 ellipticals measured through two or more diaphragms, 52% were found redder toward the centre, 32% appeared bluer and 16% nearly neutral. However, rough, overall statistics may be misleading; if one restricts the discussion to the largest and brightest systems ($m \leq 13.0$) and compare only the colours measured through the smallest and largest diaphragms, all objects appear redder in the nucleus; the mean colour difference (nucleus-integrated) $= +0.09 \pm 0.02$ (p.e.) (range: $+0.02$ to $+0.22$) for the mean diaphragm apertures 0′45 (range: 0′3 to 0′7) and 2′8 (range: 1′3 to 4′1) corresponding to a magnitude difference (nucleus-integrated) $= 1.2$ mag. (range: 0.66 to 2.14 mag.). By reference to Table 4 this indicates for the nuclear region, contributing about $\frac{1}{3}$ of the integrated light, a mean colour index $(P-V)_1 = +0.96$ and for the surrounding annular zone contributing the remaining $\frac{2}{3}$, a mean colour index $(P-V)_2 = +0.83$.

6. Lenticulars. A number of galaxies now recognised as lenticulars have been occasionally measured along with bona fide ellipticals in some of the photometric studies on luminosity distribution. A finding list of these objects is given in Table 6.

[1] In the nuclear regions the observed flattening is falsified by the instrumental smoothing which (for perfect guiding) tends to give circular images whatever the true shape of the isophotes; this effect is, however, practically negligible at distances greater than a few seconds of arc from the centre depending on telescope size, seeing conditions, etc. The two dimensional problem of the correction of the observed luminosity distribution for instrumental smoothing has been solved by Burr [Austral. J. Phys. **8**, 30–53 (1955)], Bracewell [Austral. J. Phys. **8**, 54–60 (1955)] and others. A detailed application to luminosity profiles of elliptical galaxies has been made by Burr and de Vaucouleurs (unpublished).

Table 6. *Lenticulars analysed for luminosity distribution.*

NGC	Type	References	NGC	Type	References
1023	$SB0$	[29]	3610	$SB0$	[59], [61]
1291	$(R)SB(s)0^+$	[17], [79]	4111	$S0(sp)$	[59], [61]
2768	$SA0$	[59], [61], [29]	4374	$S0$	[26]
3384	$SB(r)0^+$	[13]	4382	$S0$	[26]
3607	$SA0$	[59], [61]	5102	$SA0$	[17]

α) Luminosity distribution. In several cases only the bright nucleus was measured and mistaken for an elliptical to which HUBBLE's law was found to apply within the small range of intensities measured. In the earliest lenticular stage $E/S0$ or $S0^-$ it is possible that this or one of the other laws which govern the luminosity distribution in typical ellipticals may still apply to a first approximation. But it certainly does not apply to intermediate and late lenticulars, especially those of the barred variety. As an example the direct isophotes of NGC 3379, type $E1$, and of NGC 3384, type $SB(r)0/a$, after DENNISON [13] are shown in Fig. 8 and the luminosity profiles in Fig. 4 where it can be compared with HUBBLE's law.

Two other examples, one of the intermediate $SA0$ galaxy NGC 1553, after EVANS [17], the other of the late (transition) type $(R)SB(s)0^+$ system NGC 1291, after DE VAUCOULEURS [79] are given in Fig. 9. Obviously there is no general law applicable to all lenticulars and there is as yet not enough observational data for the derivation of special laws applicable to the various sub-types. Nevertheless, the indications are that in the nuclear regions the luminosity distribution approximates that in the genuine ellipticals, while in the lens and in the envelope it tends to follow the quasi exponential decrease characteristic of spirals (cf. Sect. 7). This is consistent with the accepted position of the lenticulars in the classification sequence. There is a great need for more detailed studies of the luminosity distribution in lenticular galaxies.

β) Colour distribution. Direct data on the colour distribution are very few and perhaps not too reliable, but again some indication of the colour change between the inner and outer regions may be obtained from a comparison of integrated colour indices measured by PETTIT [56] through diaphragms of various sizes. For $SA0$ and $SB0$ objects brighter than $m=13.0$, the following mean colour differences (nucleus-integrated) are obtained:

$$SA0\colon\ +0.19\pm0.03\ \text{(p.e.)}\ n=8,\ AD=\pm0.13\ \text{(range: }-0.03\text{ to }+0.54),$$

$$\text{or}\qquad +0.16\pm0.03\ \text{(p.e.)}\ n=6,\ AD=\pm0.06\ \text{(range: }+0.06\text{ to }+0.30),$$

$$SB0\colon\ +0.10\pm0.04\ \text{(p.e.)}\ n=5,\ AD=\pm0.09\ \text{(range: }+0.01\text{ to }+0.32).$$

The nuclei of the lenticulars are therefore decidedly redder than the whole systems and by about the same amount as in ellipticals. Allowing for the mean apertures ($SA0$: 0'38/3'1, $\Delta m=1.55$; $SB0$: 0'52/2'7, $\Delta m=0.76$) as in Sect. 5, the following corrected colour indices of the inner and intermediate regions are obtained:

$$\begin{aligned}
SA0\colon&\ \text{nuclear region,}\quad \tfrac{1}{4}\ \text{of integrated light,}\ (P-V)_1=+1.0\colon\\
&\ \text{lens and envelope,}\ \tfrac{3}{4}\ \text{of integrated light,}\ (P-V)_2=+0.8\colon\\
SB0\colon&\ \text{nuclear region,}\quad \tfrac{1}{2}\ \text{of integrated light,}\ (P-V)_1=+0.9\colon\\
&\ \text{lens and envelope,}\ \tfrac{1}{2}\ \text{of integrated light,}\ (P-V)_2=+0.8\colon
\end{aligned}$$

in close agreement with the corresponding values for ellipticals.

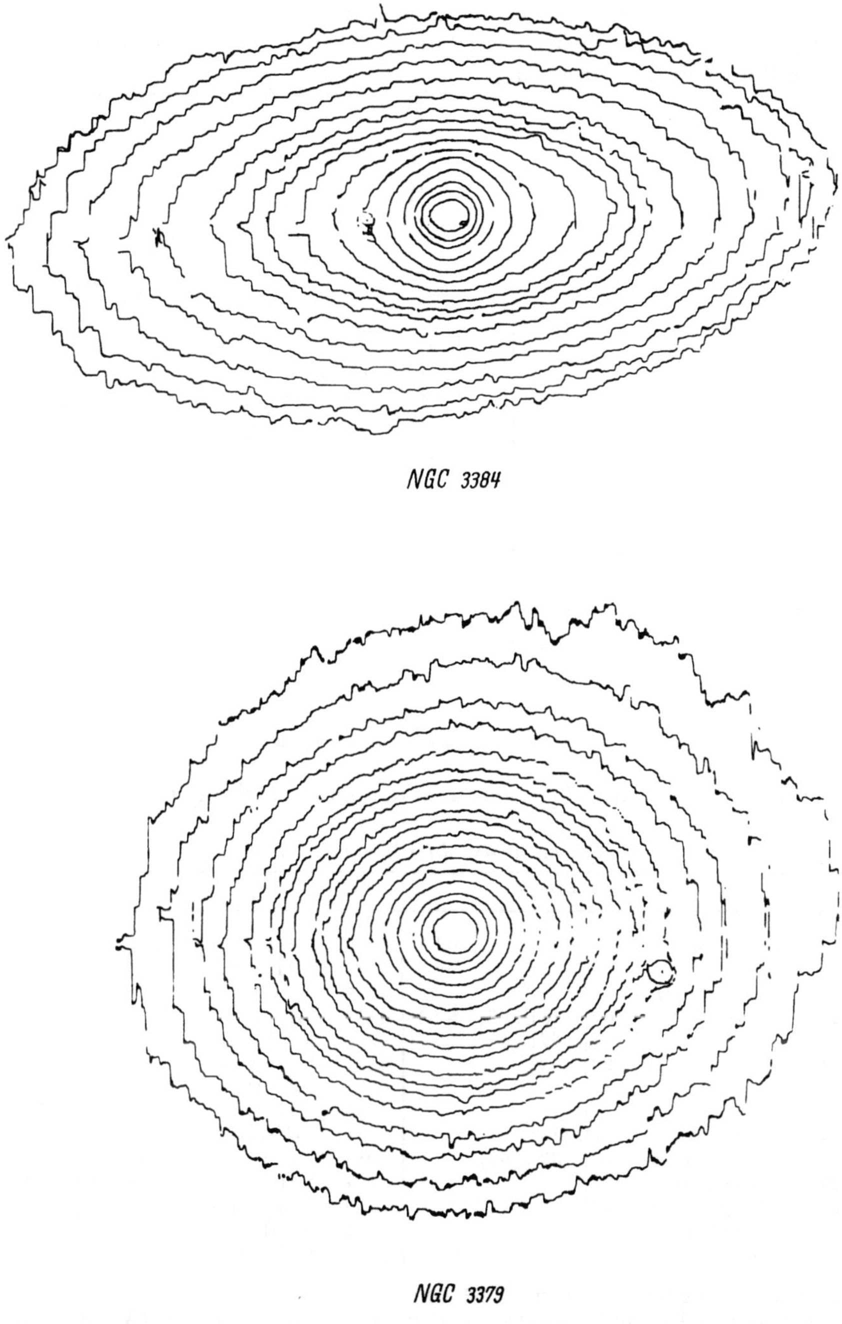

Fig. 8. Direct isophotes of NGC 3379 and NGC 3384, after DENNISON. From tracings with the automatic isophotometer of the University of Michigan Observatory.

The colour increase towards the nucleus in the E and $S0$ classes may be related to a segregation of stellar masses as if, for instance, stars of the red giant branches of the H—R diagram which have above average mass were more concentrated than the faint dwarf contributing the bulk of the mass, but a small fraction of the integrated luminosity.

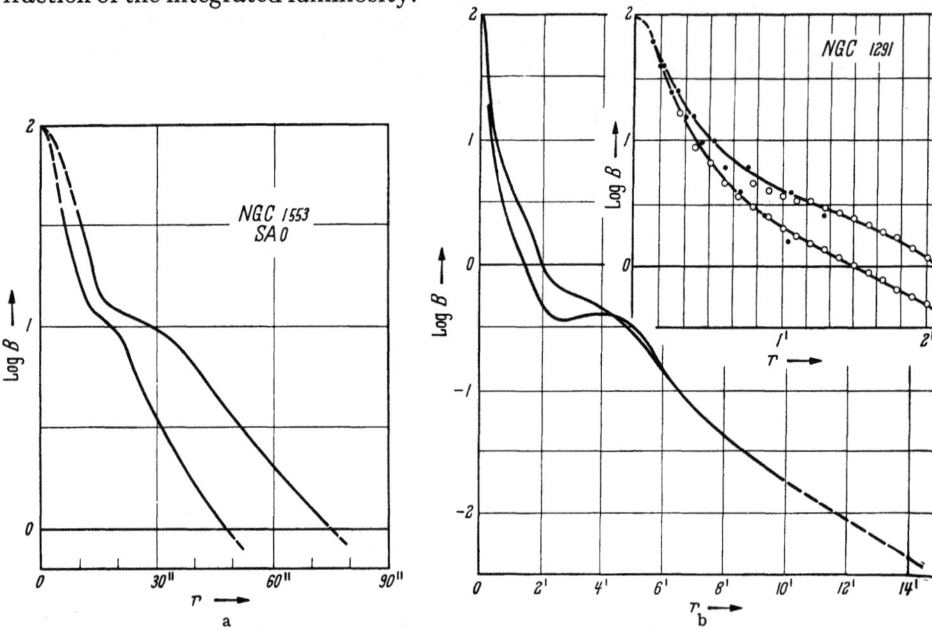

Fig. 9a and b. Luminosity profiles of lenticulars: NGC 1553, after Evans, and NGC 1291 after de Vaucouleurs. Compare with direct photographs shown on p. 291/292.

7. Spirals. A finding list of spirals measured photometrically is given in Table 7.

Table 7. *Spirals analysed for luminosity distribution.*

NGC	Type	References	Notes	NGC	Type	References	Notes
LMC	*SB(s)m*		1, 2, 3	4192	*SA:bc*	[69]	
SMC	*SB(s)mp*		1, 3	4216	*SAB(r)bc*	[16]	
55	*SB(s)m(sp)*	[79]		4217	*Sb (sp)*	[69]	
224	*SA(s)b*	[60], [18], [69]	4, 5, 6	4254	*SA(s)bc*	[69]	
598	*SA(s)c*	[53], [64], [69]		4258	*SAB(s)ab*	[69]	
891	*SA(s)b:(sp)*	[20], [64]		4321	*SAB(s)bc*	[69]	
2841	*SA(r:)$_n$ b*	[29]		4565	*SA(s:)bc(sp)*	[69]	
2976	*SA$_n$d*	[13]		4594	*SA(s:)a*	[29], [76]	8, 5, 10, 11
3031	*SA(s)b*	[64], [13]		4631	*SB(s)?d*	[69]	
3623	*SAB(rs)ab*	[20], [29], [69]		5055	*SA(r)c*	[40]	
3627	*SAB(s)bc*	[69]		5194	*SA(s)c*	[64], [69]	8, 9
3628	*Sb* pec (*sp*)	[69]		5457	*SAB(rs)cd*	[64], [69]	
4027	*SB(s)m*		7	7331	*SA(s)bc*	[64]	
4038—4039	*S* pec.		7	7814	*SA(s:)a(sp)*	[29]	

[1] G. de Vaucouleurs: Astronom. J. **62**, 69 (1957).
[2] O. J. Eggen and G. de Vaucouleurs: Publ. Astronom. Soc. Pacific **68**, 421 (1956).
[3] A. R. Hogg: Monthly Notices Roy. Astronom. Soc. London **115**, 473 (1955).
[4] R. C. Williams and W. A. Hiltner: Publ. Michigan **8**, No. 7 (1941).
[5] G. Thiessen: Mém. Soc. Roy. Sci. Liège (4) **15**, 411 (1955).
[6] N. N. Mikhelson: Pulkovo Bull. **19**, No. 151, 93 (1953).
[7] F. S. Patterson: Harvard Bull. 913, 13 (1940).
[8] A. Velghe: Bull. Astronom. Belg. **3**, 326 (1945).
[9] F. I. Lukazkaya: Russ. Astronom. J. **20**, No. 3, 1 (1943).
[10] J. H. Oort: Monthly Notices Roy. Astronom. Soc. London **106**, 159 (1946).
[11] R. C. Williams and W. A. Hiltner: Astrophys. Journ. **98**, 47 (1943).

α) Luminosity distribution. In view of the great variety of structures among spirals there is no simple general law governing the luminosity distribution. There is good evidence, however, that at least in ordinary spirals the smoothed radial luminosity distribution is approximately exponential in the outer parts. The mean luminosity distribution in M 33, type *SA* (*s*) *c*, measured in blue light by Miss PATTERSON [*53*] from six central sections is shown in Fig. 10; the abscissae are for the major axis; the mean axial ratio is $b/a = 0.87$.

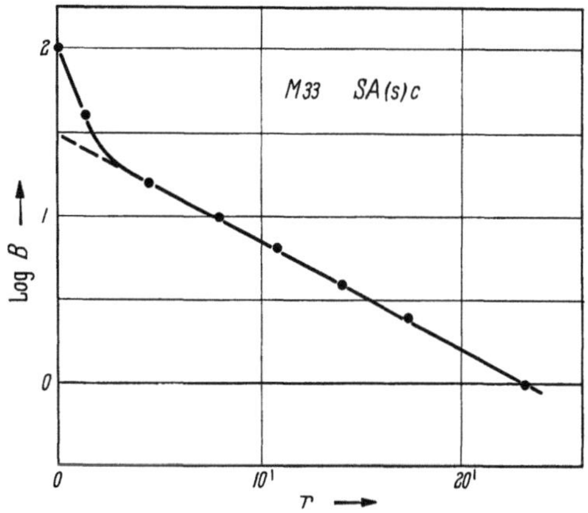

Fig. 10. Mean luminosity profile of Messier 33, after Miss PATTERSON. Note exponential decrease outside central region.

The relative importance of the nuclear bulge (spherical component) and spiral structure (flat component) and its variation along the spiral sequences is readily indicated by the departure from the exponential law in the central regions; a comparison of the luminosity profiles along the major axes of NGC 4594, type *SAa*, and NGC 2841, type *SAb*, after VAN HOUTEN *et al.* [*29*], and of NGC 5643, type *SAc*, after the writer (unpublished) is shown in Fig. 11. The progressive reduction of the spherical component from *Sa* to *Sc* is in evidence and again the exponential luminosity law in the flat component. It is conceivable that a precise classification along the spiral sequence from *S*0/*a* to *Sm* and perhaps *Im* could be made to depend on the ratio between the integrated luminosities of the spherical and flat components; i.e. on the fraction of the total luminosity contributed by the central bulge producing the excess of light above the exponential component.

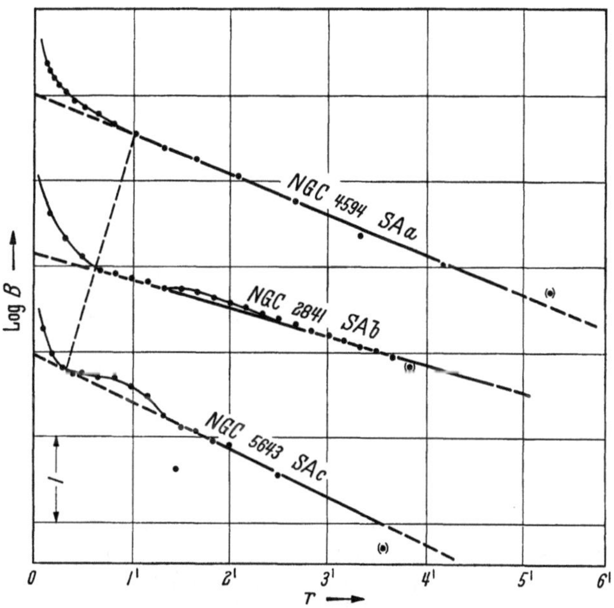

Fig. 11. Typical luminosity profiles of ordinary spirals. Note progressive reduction of spheroidal component from *Sa* to *Sc*; compare with direct photographs shown on p. 298.

The luminosity distribution in the spherical component appears to approximate that in elliptical galaxies. REYNOLDS in his early study[1] of the nuclear region of the Andromeda nebula, type *SA* (*s*) *b*, found that within 7′ from the centre the luminosity profile along the major axis could be represented by $L = C/(x+1)^2$, a formula subsequently applied by HUBBLE to ellipticals (cf.

[1] R. H. REYNOLDS: Monthly Notices Roy. Astronom. Soc. London **74**, 132 (1913).

Sect. 5). EVANS in his studies of southern "elliptical" nebulae [17] found that the luminosity distribution in the nuclei of NGC 1291, type *SB*(s)0/*a*, and of NGC 6744, type *SB*(*r*) *bc*, also follow rather well the Reynolds-Hubble rule. Obviously there is a transition zone where neither the ellipsoidal nor the exponential law apply. This suggests that a law of the form

$$B = A\,[f(r) + 1]\,\exp(-k\,r)\,, \tag{7.1}$$

where $f(r)$ is any one of the ellipsoidal laws discussed in Sect. 5, may represent the general features of the luminosity distribution in spirals. As an example Fig. 12 shows the decomposition of the mean luminosity profile along the major

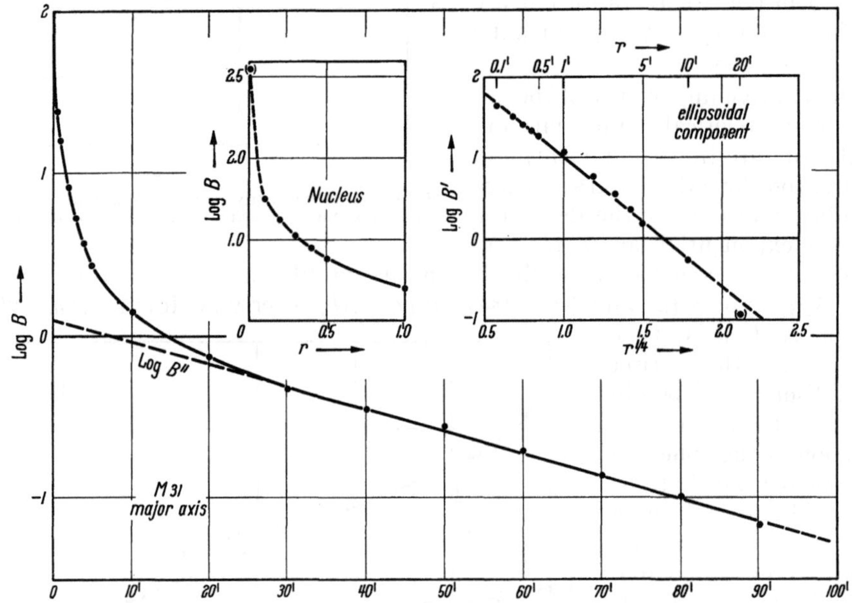

Fig. 12. Decomposition of luminosity profile of Messier 31. Note applicability of ellipsoidal luminosity law to spheroidal component in central regions.

axis of M 31, according to the data of REDMAN and SHIRLEY [60]; the outer parts for $r > 30'$ are well represented by Log $B'' = +0.10 - 0.014\,r$ (r in minutes of arc), whence Log $[B'(r) + 1] = $ Log $B -$ Log B''; the inset gives a plot of Log B' as a function of $r^{\frac{1}{4}}$; the ellipsoidal law (5.2) is verified within the experimental errors for $r < 20'$.

The *spiral arms* are more or less prominent on the unsmoothed luminosity profiles, their importance increases from *Sa* to *Sc*-*Sd* and is greater in blue or ultra-violet light than in yellow or red (cf. this section, below). Their contribution to the integrated light can be estimated by subtracting from the measured brightness maxima a smooth continuous background interpolated between the minima; according to HOLMBERG [23] the arms contribute about 20% (0.2 mag.) to the integrated luminosity of the *Sb* spirals M 31, M 81, in blue light. On isophotic maps the arms are generally indicated only by minor ripples of the contours and tend to disappear (cf. similar effect for the coronal streamers); as an example the direct isophotes of M 81, type *SA*(s) *b*, in blue light are shown in Fig. 13 after DENNISON [13].

The *apparent flattening* derived from the luminosity profiles or better from direct isophotes has been studied for only a small number of edgewise spirals.

A plot of the measured flattening $\varepsilon = 1 - (b/a)$ as a function of Log r (r: semi-major axis of isophote) is given in Fig. 6 for NGC 4594, type $SA(s)a$, after

a

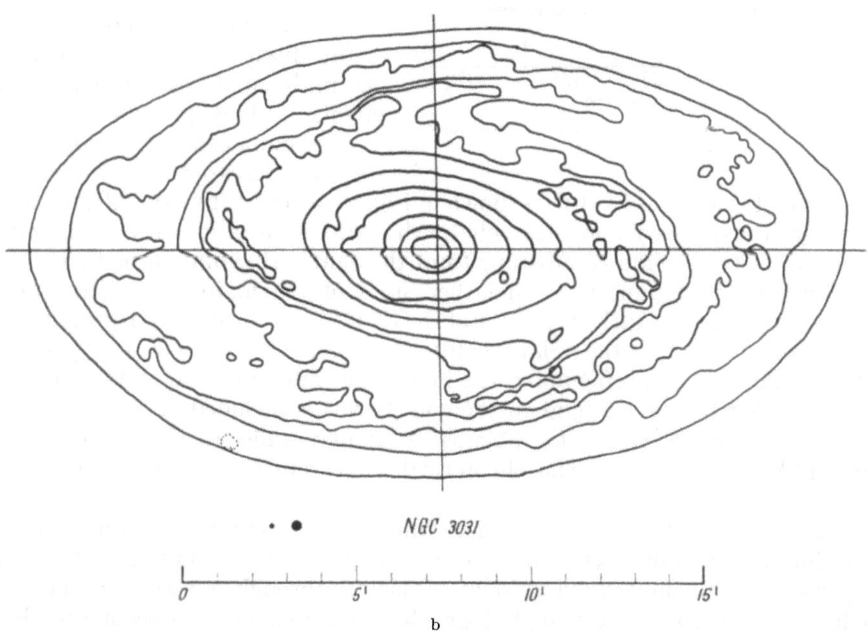

b

Fig. 13 a and b. Direct isophotes of Messier 81, after DENNISON. From tracings with the automatic isophotometer of the University of Michigan Observatory.

DE VAUCOULEURS [76]. As in ellipticals the flattening increases at first from the nucleus outwards to a flat maximum, then decreases in the faint outer parts.

The apparent flattening depends on the wave-length; in the short wave-lengths the flat component, including the spiral structure, is more prominent and the isophotes are more elongated and depart strongly from ellipses; in the long wave-lengths the spherical component, including the nuclear bulge and the outer "corona", is enhanced and the isophotes are less flattened and approximate more closely to ellipses. The mean or maximum flattening ε depends on the sub-type or stage along the spiral sequence, ranging from about 0.6 at *Sa*, to 0.7 at *Sb* and 0.8 at *Sc*; this is simply another expression of the progressive change in the relative importance of the spherical component.

When the system is not seen exactly edgewise or when there is much internal structure the flattening profile has no immediate interpretation in terms of true flattening and tilt angle.

The *asymmetry* of the luminosity profile along the minor axis is an important feature of systems whose planes are tilted at small angles to the line of sight ($i < 40°$). For instance in 200 spirals whose tilt angle and apparent dimensions were measured by DANVER [*12*], the ratio between the lengths b_1, b_2 of the two "halves" of the minor axis on either side of the nucleus varies as a function of i as indicated in Fig. 14; for comparison the ratios a_1, a_2 for the major axis are also given. While a_1/a_2 is nearly constant[1], b_1/b_2 decreases sharply for systems seen nearly edge-on[2].

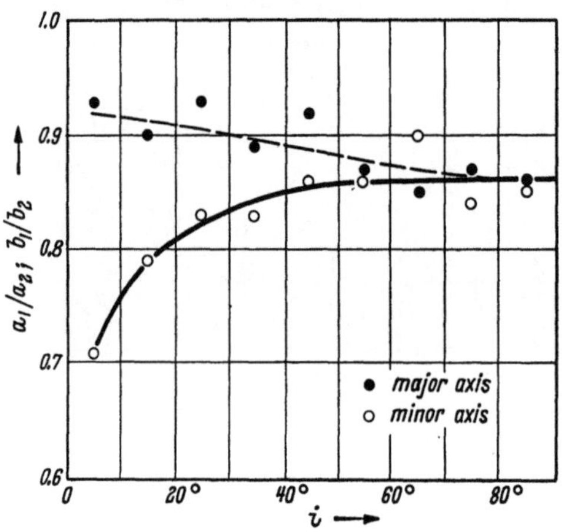

Fig. 14. Mean apparent asymmetry of spirals as a function of tilt angle, after DANVER.

This phenomenon has been extensively discussed in connection with the problem of absorption and other effects in tilted spirals (cf. Sect. 8, 9).

The luminosity distribution along the minor axes of the *SAb* spiral NGC 7331, whose tilt angle is $i = 30°$, is shown in Fig. 15, after LINDBLAD [*34*]; the asymmetry can be expressed quantitatively by the ratio of the intensities (or magnitude difference) in two points equidistant from the centre (LINDBLAD) or by the ratio of the integrated luminosities on either sides of the major axis (HOLMBERG). For instance in the early-type spiral NGC 3623, studied by HOLMBERG [*20*] and for which $i = 15°$, the relative difference $(A_1 - A_2)/(A_1 + A_2)$ of the areas under the luminosity profile along the minor axis amounts to 6 or 11%, depending on whether the excess light in the spiral arms is included or not in the integration; the average values for 12 cross sections parallel to the minor axis are 7 and 14%, respectively. The theoretical interpretation of the asymmetry is discussed in Sect. 8.

β) The colour distribution can be determined either directly through a comparison of luminosity profiles or contours in several colours, or indirectly through a comparison of the integrated colours measured through a series of apertures. The first method applies best to photographic measurements, the second to photoelectric determinations.

[1] The ratio is slightly less than 1 because of the unavoidable accidental errors of measurement and of some real structural asymmetry.

[2] The ratio may be expected to increase again almost to unity for systems seen exactly edge on.

From 69 spirals brighter than the 13th photographic magnitude, measured by PETTIT [56] through small and large diaphragms, the following values are obtained for the colour difference ΔC (nucleus-integrated).

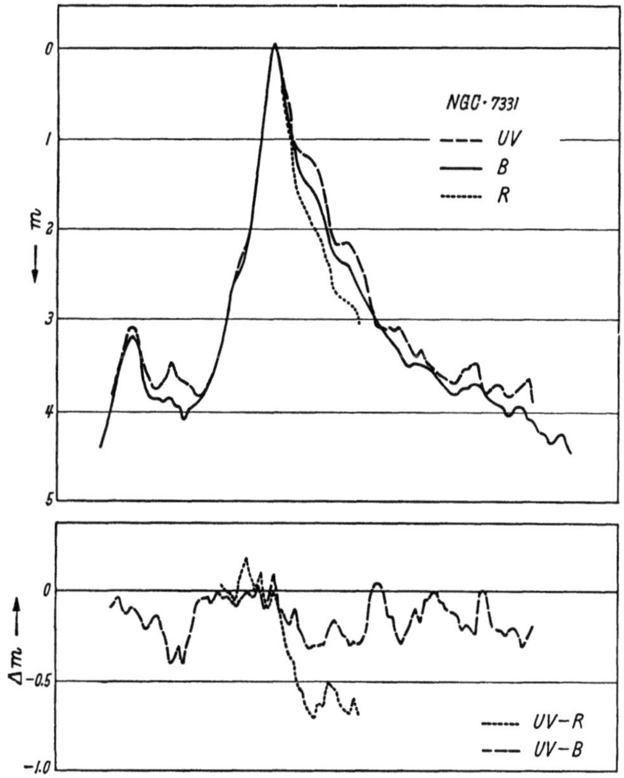

Fig. 15. Luminosity and colour profiles along minor axis of NGC 7331, after LINDBLAD. Note characteristic asymmetry of tilted spiral discussed in text.

The colours of the nuclear regions of spirals come close to the mean integrated colour of ellipticals and lenticulars $P-V=+0.85$, but are perceptibly bluer than the corresponding regions of these systems for which $(P-V)_1=+0.95$

Table 8. *Colour difference (nucleus-integrated) in spirals.*

Type	*SAa, SBa*	*SAb, SBb*	*SAc, SBc*
ΔC	$+0.05\pm0.03$	$+0.16\pm0.02$	$+0.22\pm0.03$
n	(10)	(37)	(22)
A.D.	±0.10	±0.12	±0.15
Range	$(-0.18,\ +0.32)$	$(-0.13,\ +0.60)$	$(-0.01,\ +0.50)$
Apertures . . .	0′36/2′85	0′38/3′85	0′46/5′32
Δm	1.46	1.90	2.72
C_1, C_2 (PETTIT).	$+0.90/+0.83$	$+0.92/+0.73$	$+0.79/+0.55$
$(P-V)_1$, $(P-V)_2$	$+0.83/+0.76$	$+0.85/+0.67$	$+0.73/+0.50$

(Sects. 5, 6). The mean colour outside the nuclear regions of spirals is notably bluer than in the corresponding regions of ellipticals and lenticulars for which $(P-V)_2=+0.8$; actually the colour index $(P-V)_2$ decreases along the sequence from *Sa* to *Sc* in keeping with the corresponding changes in population characteristics. Note that the change in $(P-V)_2$ is much larger than in $(P-V)_1$; even in late-type spirals the nuclei seem to have the colour characteristics of Population II or "old" Population I. The overall variation of the mean colours $(P-V)_1$ and $(P-V)_2$ as a function of morphological type is shown in Fig. 3.

The detailed colour distribution along the principal diameters of a number of spirals has been determined by SEYFERT [*64*], LINDBLAD and collaborators [*34*], [*35*], [*40*], FRICKE [*18*], SHCHEGOLEV [*69*] and others; it confirms the redder colour of the nucleus and indicates that the bluest regions are in the spiral arms[1]. As examples the luminosity distribution in blue and red light and colour differences along two rectangular diameters of M 33 and M 101, after SEYFERT [*64*],

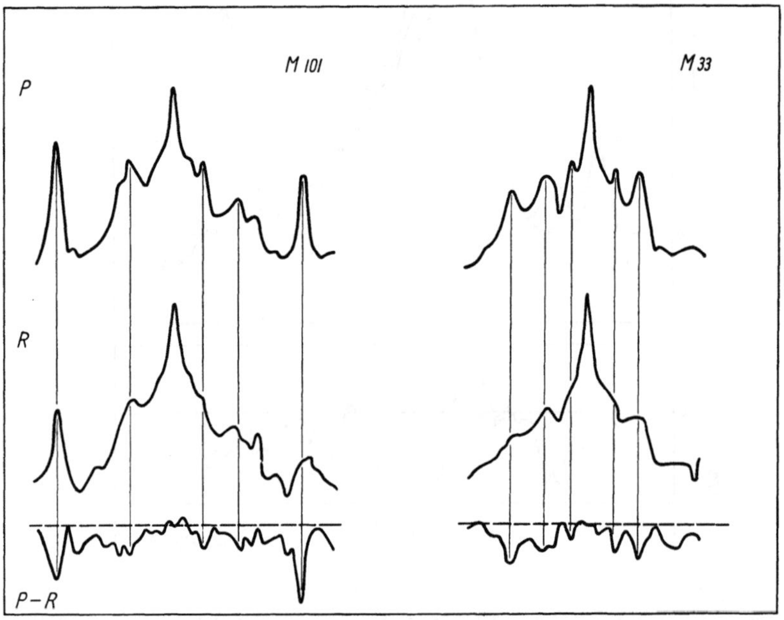

Fig. 16. Luminosity and colour profiles of Messier 101 and Messier 33, after SEYFERT. Note minima of colour index in spiral arms and nearly constant colour of background.

are shown in Fig. 16. The bright "knots" including supergiant *OB* stars and associated emission nebulosities are the bluest regions in the arms. Table 9 gives, according to SEYFERT's data, the mean "blue-red" colour difference $\Delta C_k =$ (nucleus-knots) and the corresponding mean colours of the knots in this system (λ 4250, λ 6300 ±)[2].

Table 9. *Colour of bright knots in spirals.*

NGC	Type	C_k	C_k	Range	n
598	*SA c*	0.66	+0.24	−0.24/+0.60	9
2403	*SA cd*	0.68	+0.12	−0.31/+0.56	10
4631	*SB d*: (*sp*)	0.64[3]	+0.16	−0.03/+0.35	8
5194	*SA c*	0.74	+0.26	+0.14/+0.48	8
5457	*SAB cd*	0.76	+0.03	−0.20/+0.43	10
Mean		0.70	+0.16		

[1] This has been known qualitatively for many years, see e.g. F. H. SEARES: Proc. Nat. Acad. Sci. U.S.A. **2**, 553 (1916). — Proc. Astronom. Soc. Pacific **28**, 191 (1916). — Popular Astr. **25**, 34 (1917). — E. F. CARPENTER: Publ. Astronom. Soc. Pacific **43**, 294 (1931), and is shown in striking manner by composite blue-red photographs, see F. ZWICKY, Publ. Astronom. Soc. Pacific **67**, 232 (1955).

[2] Because of the variable effect of emission lines there is no simple relation to the $(P-V)$ system.

[3] Refered to NGC 4627.

The integrated light of the arms themselves outside conspicuous knots is only slightly bluer than the nuclear regions, by about 0.2 mag. at *Sb-Sc* (M 81, M 33) and 0.3 mag. at *Sc-Sd* (M 51, M 101) on the red index scale; if, however, the smooth background light is subtracted, the corrected colours of the arms average 0.6 or 0.7 mag. bluer than the nuclei at *Scd*, indicating a mean (red) index $\approx +0.2$, very nearly equal to the mean colour of the knots.

The colour distribution in the central bulge and the nucleus proper has seldom been determined in detail with high resolution; according to some preliminary photoelectric measurements of M 31 by THIESSEN[1], the colour index increases right up to the centre where it reaches $(P-V)_0 = +1.02$ as measured through an aperture of radius 3″9. The mean colour index 1′0 from the centre is $(P-V)_1 \approx +0.93$. However, according to STEBBINS and WHITFORD [72] the mean colour indices $C_p \approx (P-V)$ measured through diaphragms of $r = 0'35$, 1′0, 2′05 are +0.98, +1.01, +1.00 respectively.

d) Absorption, diffraction and polarisation.

8. Pure absorption. The theoretical asymmetry of the colour and luminosity profiles in spirals has been computed by HOLMBERG [22] for a number of models of galaxies treated as oblate spheroids in which the light intensity I of luminous matter follows a tri-dimensional Gaussian distribution of axial ratio f and the optical density D of the absorbing material a similar distribution of axial ratio pf; thus

$$I(x, y, z) = \frac{1}{f\sqrt{(2\pi)^3}} \exp\left(-\frac{x^2}{2} - \frac{y^2}{2} - \frac{z^2}{2f^2}\right), \tag{8.1}$$

$$D(x, y, z) = \frac{u}{pf\sqrt{(2\pi)^3}} \exp\left(-\frac{x^2}{2} - \frac{y^2}{2} - \frac{z^2}{2f^2}\right). \tag{8.2}$$

By numerical integration the luminosity and colour distributions along the minor axis were computed as a function of the tilt angle i in the case of $f = 0.2$ and for different values of the optical thickness a and physical thickness p of the absorbing layer; the relation between tilt, absorption and asymmetry was obtained as shown in Fig. 17 for $p = 0.4$ and in Table 10 for $i = 15°$. In the case of NGC 3623 for which $i = 15°$ and $p = 0.4$, the observed asymmetry could be accounted for if $a = 0.5$ to 1.0 mag., a reasonable value. The distribution of colour excess is also shown in Fig. 17. According to this model which allows only for

Table 10. *Asymmetry of tilted spirals for $i = 15°$.*

a (mag)	p = 0.0	p = 0.4	p = 1.0
0.5	17.8%	18.6%	0.0%
1.0	23.9%	28.3%	0.0%
2.0	27.9%	37.9%	0.0%

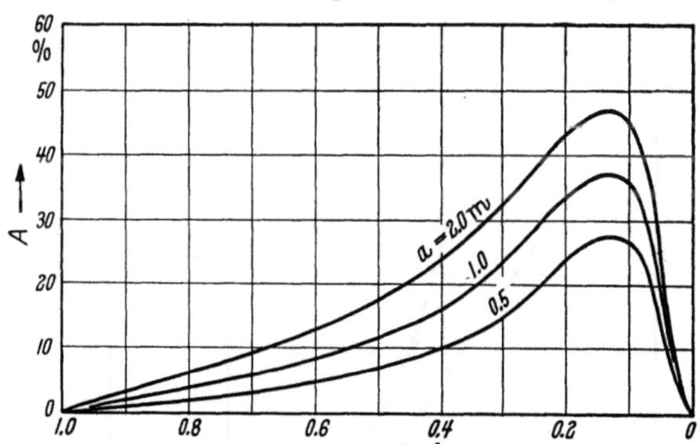

Fig. 17. Relation between asymmetry, absorption and tilt in oblate spheroids for pure absorption, after HOLMBERG. The relative physical thickness of the absorbing layer is $p = 0.4$ and its optical thickness is given the values $a = 0.5$, 1.0 or 2.0 magnitudes.

[1] G. THIESSEN: Mém. Soc. Roy. Sci. Liège (4) **15**, 411–413 (1955).

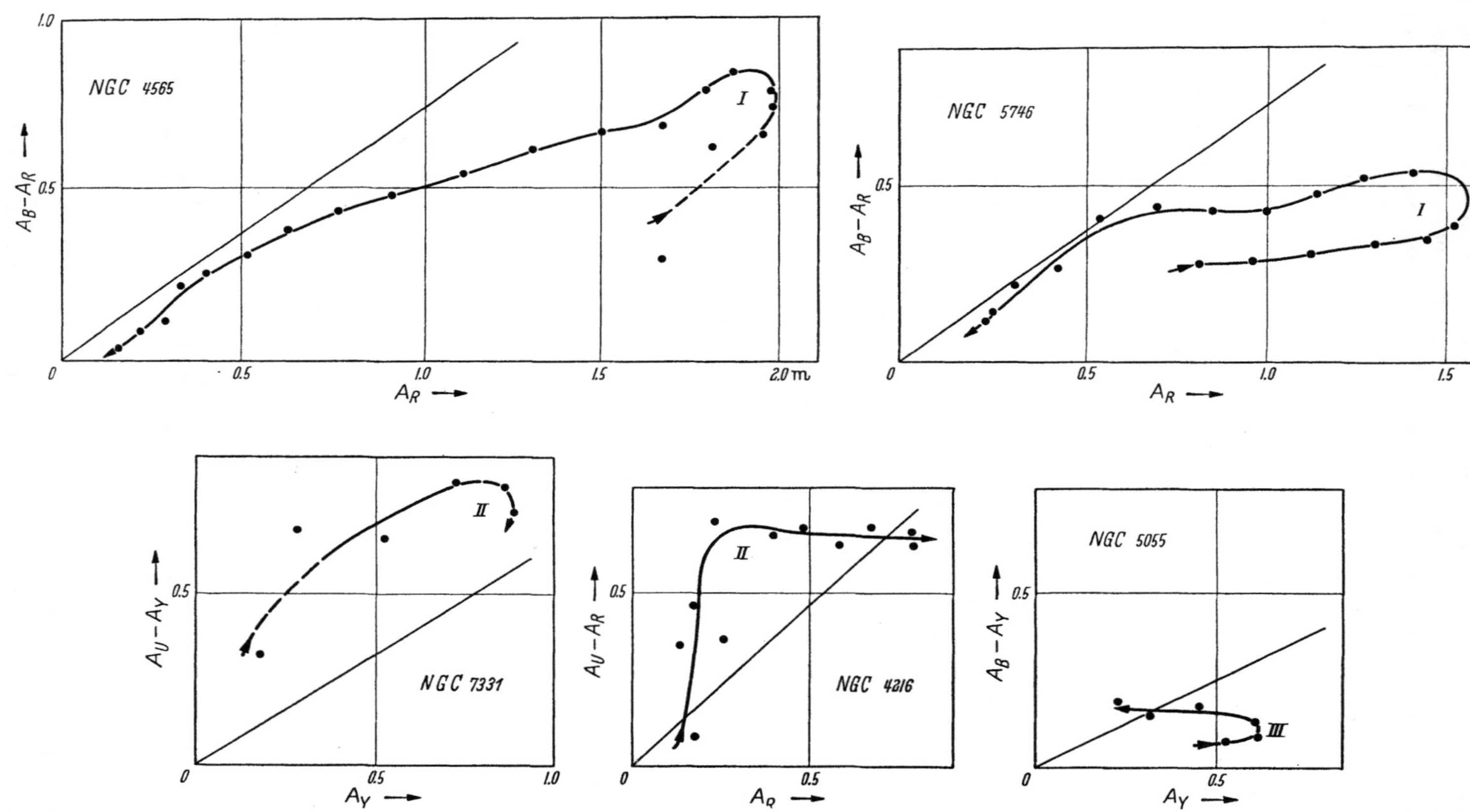

Fig. 18. Colour-luminosity asymmetry curves of tilted spirals, after LINDBLAD and ELVIUS. The differential asymmetry between short and long wavelengths is plotted against the asymmetry of the luminosity profile at the longer wavelength. The asymmetry is expressed by the magnitude difference between the brighter and fainter half of the minor axis at equal distances from the nucleus. The directions of increasing distances are indicated by the arrows. The straight line shows the theoretical asymmetry relation for pure absorption.

pure absorption the fainter and redder side of the minor axis is the near side; the maximum colour excess, about +0.2 mag., varies but little when the (photographic) optical thickness of the absorbing layer increases from 0.5 to 2.0 mag. This is very nearly equal to the maximum reddening on the faint side of M 31 as measured by STEBBINS[1].

9. Absorption and diffraction. A different interpretation has been put forward by LINDBLAD [*35*] and developed by Mrs. A. ELVIUS [*16*] which relies on detailed studies of the relation between the asymmetries in the colour and luminosity profiles along the minor axis for spirals tilted at various angles to the line of sight. The asymmetry is here measured by the magnitude difference A between points on the faint and bright sides at equal distances from the nucleus; the difference in asymmetry between two colours, say $A_B - A_R$, is plotted against the asymmetry in one colour, say A_R (Fig. 18). Different types of asymmetry curves are obtained for spirals of similar types (*Sb*, *Sc*) but variously inclined on the line of sight; the asymmetry curve—a straight line—expected for a simple "absorption" model, such as HOLMBERG'S (with $p=0$) is shown for comparison.

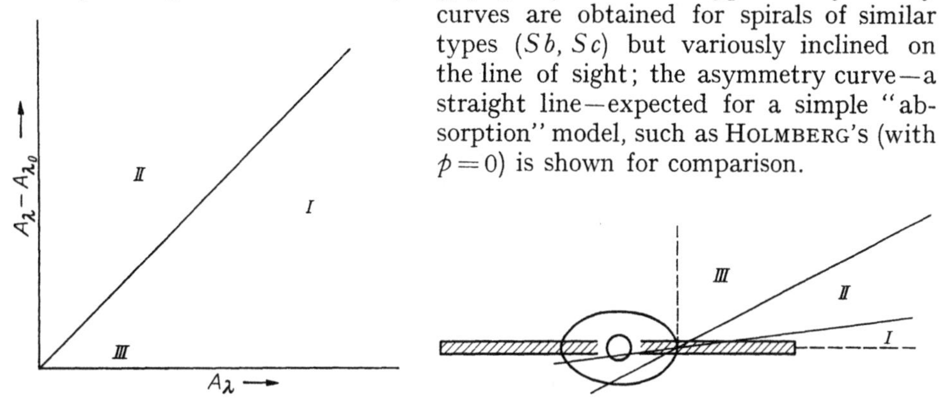

a b

Fig. 19 a and b. Relation between maximum asymmetry and tilt, after Mrs. ELVIUS. The general location of the maximum of the asymmetry curves for three intervals of tilt angle is indicated. The "pure absorption" line is shown. Compare with Fig. 18.

(I) For spirals such as NGC 4565, NGC 5746, type *SA*(*s*)*bc*, seen very nearly edge-on ($i < 10°$) the asymmetry curve is below the absorption line and the maximum asymmetry reaches high values (1.5 to 2 mag. in yellow or red light).

(II) For spirals such as NGC 224, NGC 7331, type *SA*(*s*)*b*, or NGC 4216, type *SAB*(*r*)*c*, whose tilt angle is somewhat larger ($10° < i < 30°$), the asymmetry curves are generally above the absorption line and the maximum asymmetry is moderate (about 1.0 mag. in yellow and red).

(III) Finally, for spirals such as NGC 5055, type *SA*(*r*)*c*, whose tilt angle is fairly large ($i > 30°$), the asymmetry curve lies generally below the absorption curves again, but the maximum asymmetry is small (about 0.5 mag. in yellow or red). The general relation between tilt angle and maximum asymmetry are shown in Fig. 19.

The interpretation involves a detailed assessment of the relative importance of absorption and diffraction in the scattering of the light from the bright nucleus by interstellar grains located in the region of the spiral arms. Since diffraction by interstellar particles is selective, the diffracted light being bluer than the source, and because it is strongly concentrated in the direction of propagation (forward scattering) it appears possible that under suitable conditions of tilt, nuclear brightness, etc. the near side of the minor axis may be both brighter and bluer than the far side.

[1] J. STEBBINS: Monthly Notices Roy. Astronom. Soc. London **110**, 422 (1950).

Consider two points P_1, P_2 in the plane of a galaxy and equidistant from the nucleus on the projected minor axis (Fig. 20); let I_B be the light intensity of the central bulge (per unit area) projected in front of P_2 (or beyond P_1) and I_D the intensity of light from the nucleus diffracted in the forward direction by the interstellar layer near P_1, suppose further that the intensity diffracted backwards in P_2 is negligible, then if $(1-k)$ is the fraction of I_B transmitted through the interstellar layer in the direction of observation

$$I_1 - I_2 = I_D - (1-k)\, I_B \tag{9.1}$$

the near side is the brighter if $I_D > (1-k)\, I_B$.

Numerical computations of expected diffraction effects have been made by Mrs. ELVIUS [*16*] for a simple galaxy model and interstellar particles of plausible properties. It appears that asymmetry curves of type I are generally ascribable to the absorption effects predominating at small tilt angles; curves of type II

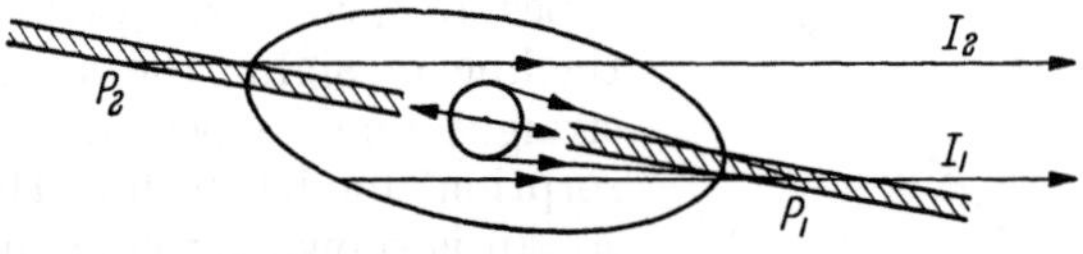

Fig. 20. Geometry of diffraction problem in tilted spirals. At point such as P_1 located on the near side of the minor axis the observed luminosity is a complicated function of the luminosity of the spheroidal component reduced by absorption in the flat dust layer and of the intensity of light from the central nucleus diffracted near P_1 in the dust layer.

can be only explained by the predominant diffraction effects at moderate angles; curves of type III represent residual absorption effects when diffraction becomes negligible at large angles[1].

10. Polarisation. The first, inconclusive attempts to observe spirals in polarised light were due to REYNOLDS[2] and MEYER[3]. Measurable polarisation was first detected by ÖHMAN [*51*] in some dark clouds of M 31.

Detailed studies of the polarisation effects in external galaxies are still limited to a very small number of spirals (NGC 4565, 5055, 7331) observed by Mrs. ELVIUS [*14*], [*15*] using ÖHMAN's birefringent polarigraph.

The distribution of polarisation in NGC 7331, type *SA(s)b*, is shown in Fig. 21, on an isophotic map of the system; the orientation of the segments gives the direction of the electric vector, their length the proportion of polarised light. The degree and direction of polarisation is not the same on both sides of the major axis of the asymmetrical image; on the bright side the mean polarisation is $p = -2.0\%$ and the direction of the strongest vibration is approximately parallel to the minor axis; on the faint side it is $p = +4.5\%$ and the direction of the strongest vibration is approximately parallel to the direction of the isophotes or of a spiral arm.

The general interpretation proposed by Mrs. ELVIUS involves considerations on the absorption and diffraction effects of interstellar grains; the "positive" polarisation observed on the fainter (and redder) side of the spiral is regarded as an absorption effect of the same nature as the Hiltner-Hall effect in the galaxy and in agreement with the theory of DAVIS and GREENSTEIN; the electric vectors are parallel to the dust lanes in the spiral and mark the direction of the magnetic field; the "negative" polarisation observed on the brighter (and bluer) side of the spiral is then attributed to the diffracted light of the nucleus in regions

[1] See details and results of an analysis of NGC 4216 for diffraction effects in [*16*].
[2] J. H. REYNOLDS: Monthly Notices Roy. Astronom. Soc. London **72**, 553 (1952).
[3] W. F. MEYER: Lick Obs. Bull. **10**, 68 (1926).

where the tilt angle and the phase function are favourable to forward scattering (see Sect. 9, above), here again the interpretation requires that the brighter half of the minor axis be the nearer.

For spirals such as NGC 5055 tilted at greater angles where diffraction effects are no longer important, only the "positive" polarisation is observed. More observations of this kind and in spirals of various types and orientations are needed.

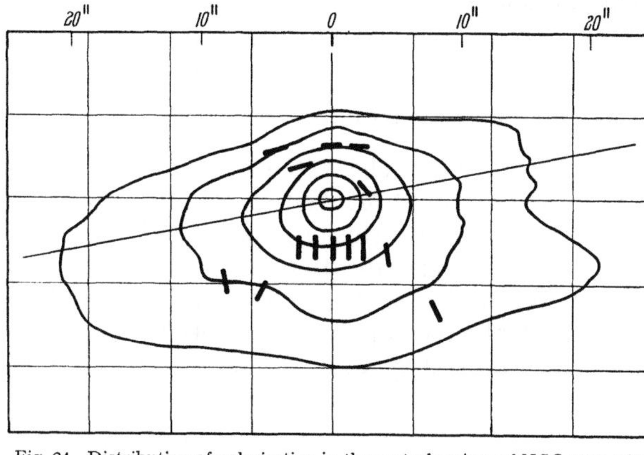

Fig. 21. Distribution of polarisation in the central regions of NGC 7331, after Mrs. A. Elvius. The direction of the electric vector is shown by the thick segments on a rough isophotic map of the inner regions. The major axis of the outer spiral structure is marked. Note that the electric vector lies parallel to the isophotes on the obscured side and perpendicular to the isophotes on the bright side.

Very few data are available as yet on the mean polarisation of the light of galaxies of various types. According to some recent photographic observations of 12 bright galaxies by Vashakidze [74] a measurable proportion of polarised light can be detected, above and in excess of the observational errors, at least for the late-type spirals and magellanic irregulars[1]. The results are collected in Table 11 and shown in Fig. 22 where the zero line has been placed at $P_0 = 4\%$, the estimated amplitude of the experimental errors.

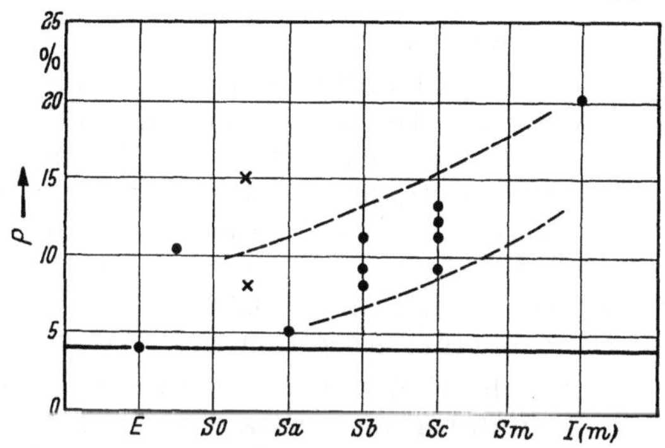

Fig. 22. Mean polarisation of 13 bright galaxies, after Vashakidze. The non-magellanic irregulars $I(0)$ are marked by crosses. The level of experimental errors is about 4%.

There is a fairly definite indication that the proportion of polarized light increases steadily from *E*, *S*0, *Sa* where it is small or negligible, to *Sc*, *I*(*m*) where it reaches and possibly exceeds 10%. This seems to agree well with the parallel trend in colour, spectral type, emission lines, etc. noted earlier and the interpretation is generally the same. Note the marked polarisation of the non-magellanic irregulars

Table 11. *Mean polarisation of galaxies.*

Type	Objects	n	P	$P-P_0$
dE	NGC 205, 221	(2)	7%	—
I(0)	NGC 3034, 3077	(2)	11%	7%
SBa	NGC 4203	(1)	5%	—
Sb	NGC 224, 3031, 4558	(3)	9%	5%
Sc	NGC 598, 2976, 5195, 5457	(4)	11%	7%
I(*m*)	NGC 4449	(1)	20%	16%

[1] This is not the polarisation of the integrated light but the mean proportion of polarised light measured in numerous places along one or several diameters, and whatever the orientation of the plane of polarisation.

$I(0)$ of the M 82 type. Polarisation measurements may eventually yield a useful index of the abundance of dust and/or of the strength of magnetic fields in galaxies.

e) Spectra and energy distribution.

11. Spectral types[1]. The early spectral observations have been described by CURTIS [*A*] and by HUBBLE [*B*].

α) Mean types. The extensive series of nebular spectra accumulated since 1936 by HUMASON at Mt. Wilson-Palomar and MAYALL at Lick [*28*] have hardly changed the main conclusion, viz. that the mean spectral type changes but little throughout the classification sequence. The results of HUMASON, based on 546 Mt. Wilson-Palomar spectra are collected in Table 12; note that as a rule these spectral types apply to the bright nuclear regions, little is known of the integrated spectral types of the faint outer regions.

Table 12. *Mean spectral types of galaxies.*

Type	*E*	*S* 0	*S a*	*S b*	*S c*	All
Spectra	*G* 3.7	*G* 2.2	*G* 1.4	*F* 9.6	*F* 6.1	*G* 1.4
n	(178)	(117)	(84)	(102)	(65)	(546)
$(P-V)_0$	0.55	0.52	0.50	0.45	0.37	0.50
C E	+0.32	+0.33	+0.30	+0.30	+0.18	

The trend from "early" (*F*) type to "late" (*G*) type in the spectra agrees qualitatively with the trend noted in the colours along the sequence $I \to S \to S0 \to E$. However, the scatter of individual spectral types at each stage is large and parallels the scatter of the colour indices. The mean colour indices $(P-V)_0$ corresponding, after EGGEN[2], to the observed spectral types for dwarf stars of luminosity class V in the Yerkes system have been added for comparison with the observed colours listed in Table 4. There is a considerable colour excess, about +0.3 mag. for *E*, *S*0, *S a*, *S b* and +0.2 mag. for *S c*, which may be due to the combined effects of internal absorption and mixture of spectral types[3].

Although nebular spectra are generally of solar type with dwarf characteristics, high dispersion spectra of the nuclei of the brighter galaxies indicate that the lines are wider and shallower than in the corresponding stellar spectra. This must result in part from the effect of mixture of stellar spectra of different types, as was first shown by WHIPPLE in 1935 [*84*], but there is good evidence that at least in some galaxies, this is also caused by a large dispersion of stellar velocities. For instance, although the central regions of M 32 (*dE* 2) and M 31 (*gS b*) have very similar spectral types, respectively *G* 3 and *G* 5 according to HUMASON [*28*], the absorption lines in the spectrum of M 31 are ill defined and approximately twice as wide as in the spectrum of M 32. However, it is not yet possible to disentangle precisely the effects of mixture of types and of velocity dispersion in any given galaxy; this would be important for the testing of theoretical models.

β) Early types. Some galaxies exhibit abnormally early spectral types, sometimes in sharp disagreement with the type inferred from colour. The best known

---

[1] Important new investigations ([*87*] to [*91*]) have been published after this article was written. A summary has been added in proof, p. 363.

[2] O. J. EGGEN: Astrophys. Journ. **113**, 663–671 (1951).

[3] This ecxess has now been removed by the more accurate spectral classifications of MORGAN and MAYALL [*90*]; see p. 363.

example is NGC 3034 (M 82), the brightest non-magellanic irregular in the M 81 group [23] which shows strong emission lines corresponding to spectral type *A* 5 according to HUMASON [28] or *A* 6 to *A* 7 according to MAYALL[1], while the colour index is about $P-V=+0.8$ [23], [56], [70] as is usually observed in *E*, *S* 0 or *Sa* galaxies of spectral types *G* 0 to *G* 5. Another example, also in the M 81 group [23], is NGC 3077 which shows "an early type spectrum with N_1, N_2, Hβ, Hγ and 3727 in emission" according to MAYALL [28] and has a colour index $P\text{-}V=+0.7$ [23], [56]; a third possible case is NGC 5195, the non-magellanic irregular companion to NGC 5194 (M 51) whose spectral type is *F* 5 according to HUMASON [28] and has a colour index $P-V=+0.7$ [23], [56]. This seems to indicate that contrasting spectral type (*A* 5 to *F* 5) and colour (0.7 to 0.8) are related characteristics of this small, but definite group of non-magellanic irregulars whose main morphological characters are an absence of supergiant stars and a smooth texture diversified only by strong, irregular absorption markings. It is doubtful that the abnormal colour-spectrum relation can be entirely explained by absorption effects, nor, when present, by the emission lines.

γ) *Emission lines.* An important characteristic of many nebular spectra is the presence of emission lines superimposed on the generally predominant continuum. This has been known for a long time with respect to late spirals and magellanic irregulars where the strongest emission lines of galactic nebulosities including Hα, Hβ, ..., N_1, N_2 [O III], λ 3869 [Ne III], λ 3727 [O II] are almost always observed in spectra including sections of the spirals arms; this, of course, is a natural consequence of the presence of emission nebulosities associated with the supergiant *OB* stars located in the arms.

Much less expected was the discovery by MAYALL with the Crossley UV spectrograph at Lick [48] of the frequent occurence of λ 3727—either alone or in association with other emission lines—in the nuclear regions of many galaxies of all types; successive reports by MAYALL and HUMASON[2,3] have indicated the generality of this phenomenon. The most recent and complete information based on 278 Mt. Wilson spectra by HUMASON [28] is collected in Table 13; earlier data based on 222 Lick spectra by MAYALL[3] which include magellanic irregulars are also given for comparison.

Table 13. *Frequency of occurence of λ 3727 emission.*

Type	*E*	*S* 0	*Sa*	*Sb*	*Sc*	*Sm, Im*	All
Mt. Wilson	0.18	0.48	0.62	0.80	0.85	—	0.54
n	(82)	(52)	(37)	(66)	(41)		(278)
Lick	0.19	0.27	0.25	0.54	0.78	0.93	0.58
n	(31)	(11)	(20)	(52)	(94)	(14)	(222)

The lower frequencies in the Lick data are explained by the smaller dispersion which makes it more difficult to detect weak emission lines. Both series agree to indicate a steady and large variation in the frequency of occurence of the line along the classification sequence from *E* to *Im* (Fig. 23) and again in agreement with the various trends suggested by morphological and photometric data. In broad terms emission of λ 3727 is correlated with the abundance of type I population (gas, dust, supergiants); it is generally present and strong where type I

[1] N. U. MAYALL: Sky and Telescope **8**, 4—5 (1948).
[2] M. L. HUMASON: Publ. Astronom. Soc. Pacific **59**, 180 (1947).
[3] W. BAADE and N. U. MAYALL: Problems of cosmical Aerodynamics, p. 165—184. 1951.

is abundant (I, late S) and usually weak or absent where type I is poorly represented or missing (E, $S0$); in this latter case when emission is present it is found as a rule only in the nuclear regions.

Generally speaking λ 3727 emission is accompanied by other emission lines, especially Hα; there are, however, differences in intensities and presence or absence of correlation between 3727 and Hα which are not yet understood. For instance Page[1] found emission lines in 17 out of 24 components of double galaxies, but Hα was observed without λ 3727 in 4 cases and λ 3727 without Hα in 2 cases; further, λ 3727 appeared to extend to greater distances from the nuclei than Hα.

In some rare objects studied by Seyfert [*65*] the emission lines localised in the nuclear regions are exceedingly strong and broad (Fig. 33); these objects are listed below together with their revised morphological types and integrated colour indices after Pettit [*56*].

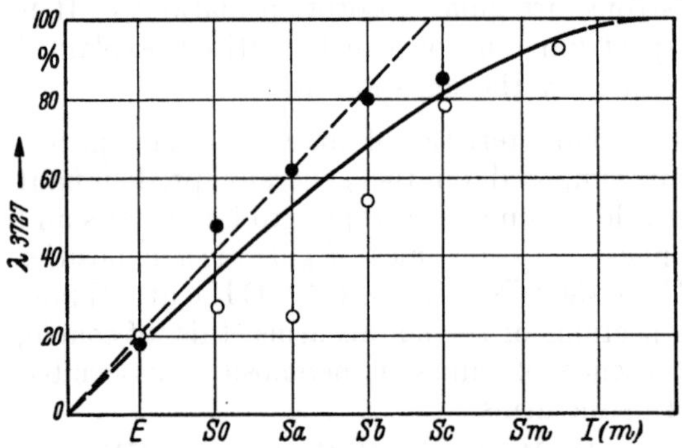

Fig. 23. Frequency of occurrence of λ 3727 emission, after Humason and Mayall. Dots: Mt. Wilson-Palomar data; circles: Lick data.

NGC 1068,	(R) $SAab$,	+0.65
NGC 1275,	Pec,	+0.78
NGC 2782,	Sa,	+0.65
NGC 3077,	I(0),	+0.78
NGC 3227,	Sb,	+0.79
NGC 3516,	$SB0^+$,	+0.83
NGC 4051,	Sb,	—
NGC 4151,	SAa,	—
NGC 4258,	$SABb$,	+0.75
NGC 5548,	Sa,	—
NGC 6814,	$SABb$,	+0.89
NGC 7469,	(R) $SBab$,	+0.61.

The line profiles indicate turbulent velocities of several thousands ot kilometers per second and ranging up to 8500 km/sec for the total width of the hydrogen lines in NGC 3516 and 7469. The emission lines include all the stronger lines observed in planetary nebulae, viz. Hα to Hζ, [O I] λ 6300—6364, [O II] λ 3726 to 3730, [O III] λ 4363, 4959, 5007, [Ne II] λ 5755, 6548, 6584, [N I] λ 5199, [S II] λ 4068, 4076, 6717, 6731, [S III] λ 6310, [Ne III] λ 3869, 3968, [A IV] λ 4711, 4740, [Fe VII] λ 5158, 5276, 5720, 6086. The nuclei of these objects are exceedingly bright, contributing from 6% (NGC 1068) to 48% (NGC 4151) of the total (photographic) luminosity of the galaxies and of these fractions 10 to 20% are in the emission lines. There seems to be a positive correlation between the width of the H lines on the one hand and the absolute magnitudes of the nuclei or the ratio nucleus/total luminosity. Most of these objects are early-type spirals (Sa, Sb), but NGC 1275 is now known to be a colliding pair[2] and NGC 3077 is one of the non-magellanic irregulars with absorption markings (see preceding article, Plate XI).

The great width of the lines shows that the emission cannot be attributed to an abnormal concentration of individual H II regions such as commonly found in late-type spirals and magellanic irregulars in which the lines are always narrow. No satisfactory explanation of this nuclear emission has been proposed.

A method for the rapid detection of galaxies with strong emission lines, especially λ 3727 and λ 3869, by means of three-colour photography with a Schmidt camera has been described by Haro [*19*]. These galaxies are characterised by

[1] T. Page: Astronom. J. **54**, 47 (1948); **55**, 179 (1950).

[2] W. Baade and R. Minkowski: Astrophys. Journ. **119**, 215 (1954).

abnormally low colour indices which may not be wholly attributable to the effect of the emission lines transmitted by the filters. According to preliminary estimates by HARO, their colour indices are probably less than +0.2 and possibly about 0.0 on the average. Hence these objects must be regarded as belonging to a different category from the spirals with broad emission lines which have an average colour index of about +0.8 (see above). Examples of these "blue" galaxies noted by HARO are NGC 2415, 3353, 3991a, b, 4234, 4670, 5253; other possible examples are NGC 244, 263, 1345, 3510, 4218, IC 3600.

It is obvious that much further work will be required to clarify the spectral classification of galaxies.

12. Energy distribution. *α) Ellipticals.* Up to the present the energy distribution based on multicolour photometry has been published for only one object, M 32, whose central regions were measured by STEBBINS and WHITFORD [*71*] through six wide-band colour filters; their results are given in Table 14; the second line (Δm) gives the relative magnitudes with an arbitrary zero point compared to the mean of six *dG*6 stars[1].

Table 14. *Six-colour photometry of bright galaxies.*

Object	Type	*Sp*	λ / $1/\lambda$	3530 2.83	4220 2.37	4880 2.05	5700 1.75	7190 1.39	10300 0.97	Remarks
M 32	*E*2	*G*3	Δm	+0.54	+0.45	+0.27	+0.03	−0.29	−0.91	
M 31	*SAb*	*G*5	Δm	+0.84	+0.52	+0.27	+0.03	−0.28	−0.83	Nucleus
M 77	(*R*) *SA*(*rs*) *ab*	*G*3	Δm	+0.21	+0.26	+0.18	+0.02	−0.19	−0.67	Nucleus
M 94	(*R*) *SA*(*r*) *ab*	*F*8	—	0.00	+0.13	+0.14	+0.06	−0.19	−0.58	
M 64	*SA*(*s*) *ab*	*G*3	—	+0.17	+0.21	+0.17	+0.03	−0.20	−0.60	
Mean	*Sab*	*G*1	Δm	+0.13	+0.20	+0.16	+0.04	−0.19	−0.62	
M 33	*SA*(*s*) *c*	*A*7−*G*0	Δm	−0.36	0.00	−0.01	+0.04	−0.02	−0.20	Nucleus
M 51	*SA*(*s*) *c*	*F*8	—	−0.30	−0.05	+0.06	+0.06	−0.11	−0.58	
M 101	*SAB*(*rs*) *d*	*G*0	—	−0.18	−0.06	+0.04	+0.10	−0.15	−0.21	
Mean	*Scd*	*F*5	Δm	−0.28	−0.04	+0.03	+0.07	−0.09	−0.33	

This energy curve affords a useful, though not very sensitive test of models of the stellar population (H—R diagram, luminosity function) in elliptical galaxies. Such comparisons have been made by DE VAUCOULEURS [*78*] and by ROBERTS[2] for some plausible population models with results in satisfactory general agreement with the observed energy curve[3].

Incidental to a study of Hα emission in galaxies (cf. Sect. 11) the colour temperature of the continuum between λ 3727 and λ 6730 was measured by PAGE[4] for NGC 4486, type *E*0*p*, $T_0 = 3800°$ K and NGC 3077, type *I*(0), $T_0 = 5500°$ K.

[1] Some recent and as yet unpublished observations of WHITFORD throw some doubt on the reliability of these data and this unique example may not be typical of elliptical galaxies in general.

[2] M. S. ROBERTS: Astronom. J. **61**, 195 (1956).

[3] The model computed by the author and based on the scanty information available in 1951—1952 on "Type II" systems is superseded by the recent model of ROBERTS (l. c.) based on more complete information now available. However, the earlier model still gives a fair approximation of a plausible system it it is regarded as describing an "old" Population I. It is well known that the integrated colours of Population II and "old" population I systems are practically identical [cf. C. V. REDDISH, Observatory **76**, 68, (1956)] and this confirms the relative insensitivity of the colour criterion for distinguishing between alternative models ofs tellar population.

[4] T. PAGE: Astronom. J. **55**, 179 (1950).

β) Spirals. The energy distribution in the continuum has been measured by SEYFERT [65] for six emission spirals (cf. Sect. 11); the energy curves determined by comparison with standard stars on the Greenwich system were found to be representable by a single value of the gradient from λ 4000 to 6700 Å and the corresponding black-body colour temperatures, computed with $\Phi_0 = 1.10$, range from 4750° K (NGC 4151) to 8000° K (NGC 7469) with the others (NGC 1275, 1068, 3516, 4051) at 5250° K.

As a by-product of a spectrophotometric study of the night sky H. W. BABCOCK and J. J. JOHNSON[1] determined the spectral energy curve of the nuclear regions of M 31 (uncorrected for the much fainter sky background). In the visible range (λλ 4100 to 6000 Å) the colour temperature (corrected to no atmosphere) was found to be only 3650° K, an unusually low value. For comparison similar determinations by GREENSTEIN[2] and by V. A. DOMBROVSKY[3] gave 4200 and 4650° respectively.

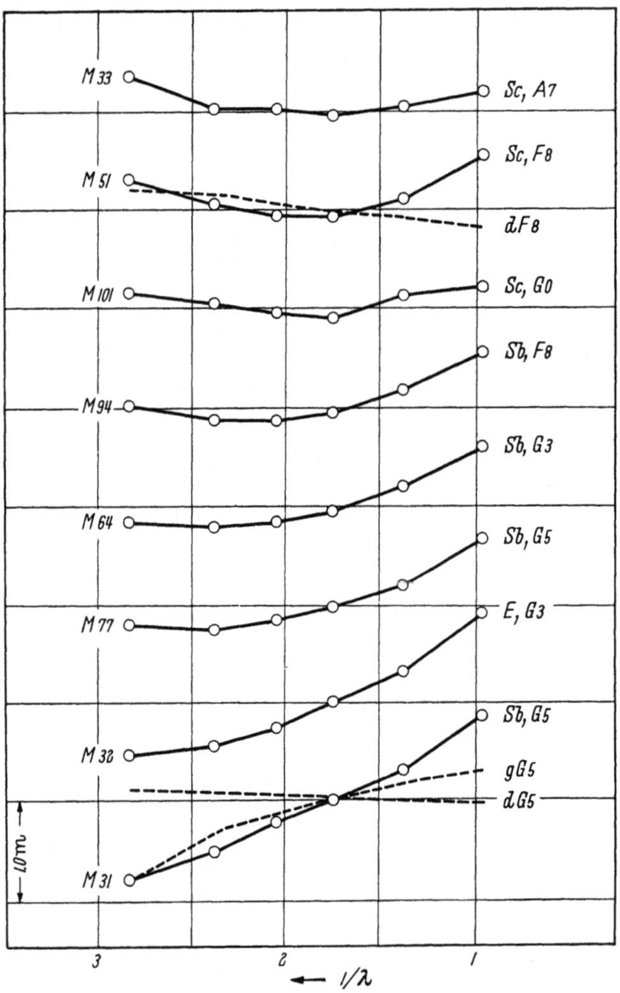

Fig. 24. Spectral energy curves of bright galaxies referred to the average of *dG*6 stars, after J. STEBBINS and WHITFORD.

Except for objects with strong emission lines the energy distribution is probably better determined by photometry through narrow-band colour filters such as the six-colour photometry of STEBBINS and WHITFORD [71]. The results are collected in Table 14 and Fig. 24; the band-width transmitted by the filters is too large to show such details as the intensity drop near 3900 Å and the curvature of the gradient curves reduces the significance of colour temperature estimates, unless restricted to a small range. However, between λλ 4200 and 7200 Å, the mean colour temperature of M 31 (central region) is $T_0 = 4250$°K, in substantial agreement with the mean of the spectrophotometric determinations. It will be recalled that the mean colour index of this region is $P - V = +1.0$ (Sect. 7) in fair agreement, the spectral

[1] H. W. BABCOCK and J. J. JOHNSON: Astrophys. Journ. **92**, 271 (1940).
[2] J. L. GREENSTEIN: Astrophys. Journ. **88**, 505 (1938).
[3] V. A. DOMBROVSKI: Publ. Leningrad Astr. Obs. **15**, 166—203 (1950).

type $G5$ [28] however, is much earlier than either gradient or colour suggest, an effect qualitatively attributable to the effect of mixture of stellar types, but of which there is as yet no quantitative theoretical estimate.

II. Mechanical properties.

a) Rotation.

13. Direction of rotation. The direction of rotation of spirals, especially of the spiral arms, is of importance for checking theoretical models. In principle a spiral galaxy can rotate either in the direction of "winding" of the arms (convex side preceding, arms "trailing") or of "unwinding" (concave side preceding, arms "leading"). In a spiral tilted on the line of sight spectrographic observations indicate which end of the major axis of the projected image is approaching ($v - v_s < 0$) and which is receding with respect to the centre of the spiral of systemic velocity v_s. Direct photographs usually indicate the direction of the spiral pattern: "right-handed" if a point moving inwards along a spiral arm rotates clockwise, "left-handed" if counterclockwise. In order to determine the direction of rotation of the spiral pattern it is also necessary to know which part of the minor axis of the image is on the near side (see Fig. 25); i.e. the true space orientation of the spiral.

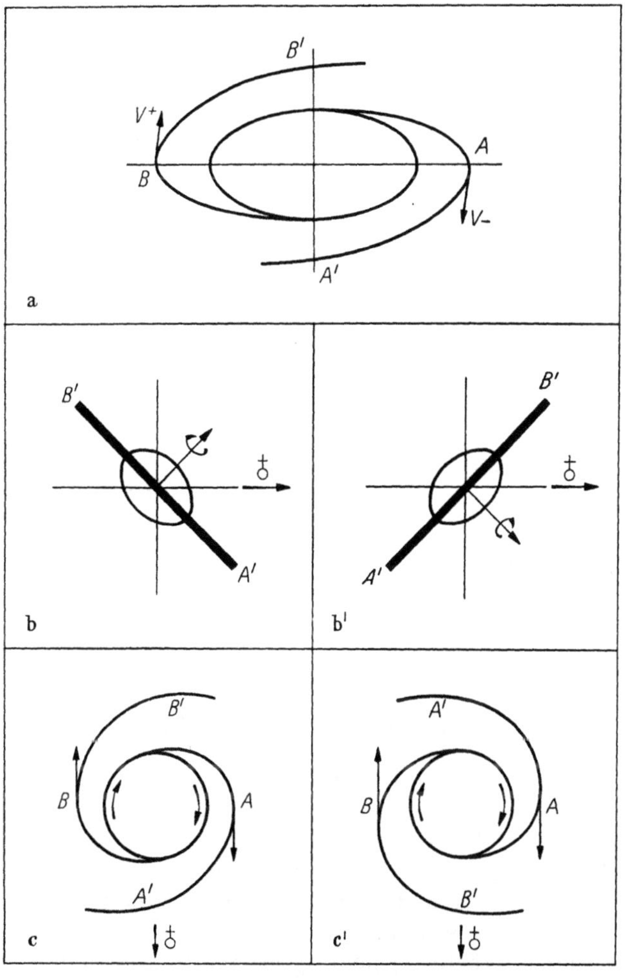

Fig. 25 a–c. Relation between sense of tilt and direction of rotation of spirals. For a given sense of spectroscopic rotation in a tilted spiral (a), the arms may be either "leading" (c) or "trailing" (c'), depending on the sign of the tilt angle on the line of sight (b, b').

After the initial studies of V.M. SLIPHER[1] and H.D. CURTIS[2], this problem has been discussed by HUBBLE [27], LINDBLAD and his collaborators [32], [33], [36], [37], [39], HOLMBERG [20], [22], DANVER [11], [12], DE VAUCOULEURS [77],

[1] V.M. SLIPHER: Proc. Amer. Phil. Soc. **56**, 403 (1917). — Publ. Amer. Astronom. Soc. **4**, 232 (1920). — Popular Astr. **29**, 272 (1921).

[2] Publ. Lick Obs. **13**, Part. 2, 45–54 (1918).

IRWIN, FRICKE and others[1]. There is as yet no general agreement on the interpretation of the various criteria. Table 15 gives a finding list of spirals studied for rotation and tilt.

Table 15. *Spirals studied for rotation and tilt.*

NGC	Type	Spiral pattern	Major axis	Side obscured	End approach.	i	References
224	*SA(s)b*	*L*	NE–SW	NW	SW	15°	[27], [39], [36] [2]
598	*SA(s)c*	*R*	NE–SW	NW(?)	NE	33°	[27], [39], [36]
1068	*(R) SA(rs)b*	*L*	NE–SW	?	NE	70° ±	[27] [2]
2146	$SA(s{:})b_p$	*L*	NW–SE	SW	SE	30° ±	[77]
2613	$SA(s)_n bc$	*R*	NW–SE	SW	NW		[27]
2683	*SA(s):c*	*R*	NE–SW	NW	NE	8°	[27] [2]
2841	*SA(?)b*	*L*	NW–SE	NE	NW	26°	[27]
3031	*SA(s)b*	*L*	NW–SE	SW	SE	31°	[27]
3169	$SA(?)ab_p$	*L*	NE–SW	SE	NE		[27] [2]
3190	$SA(s{:})a_p$	*L(?)*	NW–SE	SW	SE		[27], [35] [2]
3432	*SB(s)cd*	*R(?)*	NE–SW	NW	NE		[27]
3556	*SAB:(s)cd*	*R*	E–W	N	E	7°	[2]
3623	*SAB(rs)ab*	*L*	N–S	E	N	15°	[27] [2]
3627	*SAB(s)bc*	*L*	N–S	E(?)	N(?)	33°	[27] [2]
4088	$SAB(s)cd_p$	*L*	NE–SW	SE	NE	21°	[27], [39]
4216	*SAB(r)b*	*L*	NE–SW	SE	NE	9°	[27], [39] [2]
4244	*SA(s:)cd*	*R(?)*	NE–SW	NW	NE	4°	[3]
4258	*SAB(s)b*	*L*	NW–SE	SW	SE	35°	[27], [39] [2]
4527	*SA(?)bc*	*L*	NE–SW	NW	SW	20°	[2]
4565	*SA(s:)bc*	*R(?)*	SE–NW	NE	SE	5°	[39], [36] [2]
4594	*SA(s)a*	?	E–W	S	W	(11° ?)	[27] [2]
4736	*(R) SA(r)ab*	*R*	NW–SE	?	SE	55°	[27]
4826	*SA(rs)a*	*R*	NW–SE	NE	SE	30°	[27] [2]
5005	*SA(s)ab*	*L*	NE–SW	NW	SW(?)	23°	[27] [2]
5033	*SA(s)c*	*L*	NE–SW			32°	[2]
5055	$SA(r)_n bc$	*R*	SE–NW	SW	NW	32°	[40] [2]
5746	*SA(s)b*	*R*	N–S	E	S	6°	[27] [2,3]
5907	*SA(s)c*	*L*	NW–SE	SW	SE	3°	[3]
6503	*SAB?(s)cd*	*L*	NW–SE	SW	SE		[27]
7331	*SA(s)bc*	*R*	N–S	W	N	31°	[27], [35], [39] [2,3]

The alternative interpretations of the main criterion, viz. the asymmetrical luminosity and colour distributions near the minor axis, are given in Sects. 8 and 9. According to most investigators, following SLIPHER and CURTIS, the obscured and redder side is always the near side and with this interpretation all spirals investigated have "trailing" arms. According to LINDBLAD'S theory of spiral galaxies[4] the arms must "lead" in the rotation and the distribution of dark matter is interpreted so as to agree with this prediction which requires that, in general, the obscured and redder side is the far side.

Actually there is general agreement that for spirals seen almost exactly edge-on, say for $i < 10°$, the dark lane marks the near side ("primary" absorption in LINDBLAD'S terminology), but then, in most cases, the spiral pattern cannot be traced unambiguously; for spirals tilted at a somewhat greater angle, say $i \approx 10$ to 30°, in which the sense of the spiral pattern can be followed clearly, the "primary" dark lane is no longer projected against the central bulge and the "secondary" absorption lanes are differently interpreted either as still marking the near

[1] See references to Table 15.
[2] J. B. IRWIN: Astronom. J. **57**, 15–16 (1952), and unpublished manuscript.
[3] G. DE VAUCOULEURS: Astrophys. Journ. **127**, 487 (1958).
[4] See e.g. Stockholm Obs. Ann. **13**, No. 10 (1941); also [33], [36], [41].

side (SLIPHER, CURTIS, HUBBLE) or as indicating the far side (LINDBLAD). The general requirements of the two interpretations concerning the distribution of dark matter in relation to the bright spiral arms is shown in Fig. 26; in the "trailing" case the dark matter is regarded as concentrated in a more or less continuous, thin layer permeating the equatorial plane of the system in the region of the spiral arms; in the "leading" case the dark matter is supposed to be narrowly limited to the (inner) concave edge of the bright arms and with a very small vertical spread so as to be hidden from view on the near side. Colour and

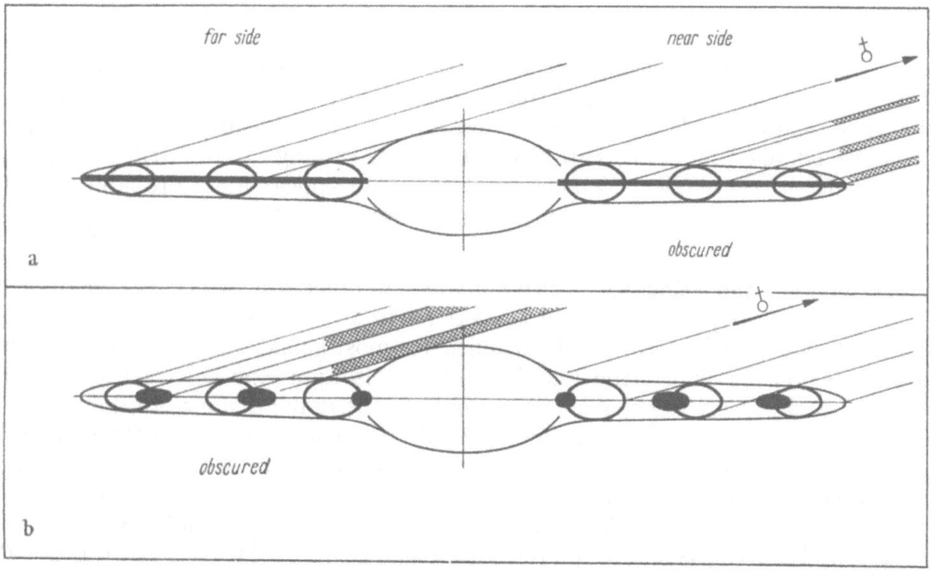

Fig. 26 a and b. Alternative hypotheses on the distribution of dark matter in spirals. (a) Usual interpretation: if the dark matter is more or less uniformly spread in the equatorial plane of the system the obscured side is the near side. (b) LINDBLAD's interpretation: if the dark matter is localized in the concavity of the bright arms the obscured side is the far side.

polarisation effects have been variously interpreted by LINDBLAD and his colleagues in support of this point of view (cf. Sect. 9).

IRWIN[1] has studied a criterion related to the asymmetry in the luminosity profiles or isophotes of the nuclear regions, viz. the location of the innermost "nucleus" or peak intensity in the central bulge; in four nearly edgewise systems with primary absorption lanes the "nucleus" is closer to the edge of the absorption lane, i.e. displaced in projection towards the obscured (near) side which has the steeper luminosity gradient (cf. Fig. 15). This same relation holds for 12 other systems of greater tilt and in which the spiral arms can be traced; this leads to the conclusion that the arms are trailing in all systems, but it may depend again on the interpretation of the secondary absorption lanes in the 12 systems.

Attempts have been made by HUBBLE [*27*] and DE VAUCOULEURS [*77*] to find unequivocal cases of primary absorption where the sense of the spiral pattern can still be traced; in two such objects (NGC 3190, NGC 2146) the arms were found to be trailing, but because in both cases one arm is tilted on the principal plane of the system by interaction with another nearby galaxy the interpretation may be and has been questioned. LINDBLAD has even suggsted [*36*], [*39*] that in some early type systems, such as NGC 2681, the faint outer spiral pattern may be of opposite sense to the stronger inner pattern. Other attempts have

[1] J. B. IRWIN: Astronom. J. **57**, 15–16 (1952). — Sky and Telescope **11**, 116 (1952) and private communication.

been made by LINDBLAD to "open" the spiral structure of NGC 4565 [*36*] and especially NGC 4594 [*37*] seen almost exactly edgewise to show that in these systems the arms are leading; the tilt angle is so small, however, that the results are not convincing and the data can be re-interpreted differently[1]. The spiral structure of several spirals of types $SA(s)b$ or c (NGC 4244, 4565, 5746, 5907) which, according to LINDBLAD and Mrs. ELVIUS [*16*], [*35*], [*39*] show unquestionable "primary" dark lanes ($i<6°$) has been traced by the writer [*91*], in all cases the observed velocities were found to agree with the sign predicted for trailing arms[2].

In view of these conflicting interpretations of all general criteria and individual cases it is not yet possible to come to a final conclusion with complete assurance; it is, however, generally conceded that in most, if not all, *ordinary* spirals studied, the same relation holds between direction of rotation and sense of spiral pattern, i.e. all well defined spiral arms are either leading or trailing. In the writer's opinion the balance of the evidence makes the latter case appear more probable and it is also in agreement with the recent radio results on rotation and spiral structure in our own galaxy.

14. Periods and velocities. Except for a few objects in which the velocities of emission-line objects located in distant parts of the system can be observed the rotational velocities can be determined for most galaxies only in their bright central regions in which the angular velocity is approximately constant (see below), i.e. the linear velocity is proportional to the radial distance to the nucleus and it follows that the spectral lines (usually absorption lines) are straight and tilted with respect to the comparison lines in curved-slit spectra when the slit is oriented along the major axis of the system. For instance PEASE[3] found that in the central, regions of M 31, $v=-316-0.48x$ (x in seconds of arc, v in km/sec) and in NGC 4594, $v=+1180-2.78x$ ($x<150''$). At $x=2'$ the rotational velocities reach 58 km/sec in M 31, 330 km/sec in NGC 4594. The early results on rotational periods and velocities have been reviewed by CURTIS [*A*], HUBBLE [*B*] and VOGT [*D*].

Great progress has been achieved in this field since 1935, through the use of emission lines, especially λ 3727 (cf. Sect. 11), which often extend to much greater distances from the nucleus than the absorption lines; the accuracy of the measurements is also improved by the better definition of the lines. The linear relationship between velocity and distance to the nucleus in the central parts of the galaxies—which therefore rotate apparently like solid bodies—has been verified for a rather large sample including objects of all classes from E to I. It has been checked also that measurements on individual condensations showing λ 3727 emission lead to rotational velocities in good agreement with those derived from the average inclination of absorption lines[4].

---

[1] Cf. G. DE VAUCOULEURS: La Nature, Paris **1952**, No. 3210, 303—308 and [*91*].

[2] Other criteria have been discussed with inconclusive results. See C. G. DANVER, Medd. Lund (I) **1940**, No. 157; Ann. Obs. Lund. **1942**, No. 10; B. LINDBLAD and R. BRAHDE, Astrophys. Journ. **104**, 211 (1946); A. ELVIUS, Stockholm Obs. Ann. **18**, No. 9 (1956); G. DE VAUCOULEURS, Observatory **69**, 150 (1949); **74**, 23 (1954); Astronom. J. **60**, 126 (1955); Mrs. V. MCKIBBEN-NAIL and H. SHAPLEY, Proc. Nat. Acad. Sci., Washington **41**, 685 (1955) = Harvard Reprint 426; H. C. ARP, Astronom. J. **61**, 31 (1956).

[3] F. G. PEASE: Mt. Wilson Comm. **1916**, No. 32; **1918**, No. 51.

[4] The constancy of the angular velocity in the inner parts of galaxies has been considered sometimes as a surprising and unexpected property calling for special explanations; see e.g. F. ZWICKY [Astrophys. Journ. **86**, 217 (1937)] or E. HOLMBERG [Monthly Notices Roy. Astronom. Soc. London **99**, 650 (1939)]. This, however, is not necessary as it can be readily explained by various plausible models of density distribution (cf. Sect. 15); it is also well known that in an homogeneous ellipsoid rotational velocity is proportional to radial distance.

If the distances of the systems are known the rotation periods can be derived: Let i be the tilt angle between the equatorial plane and the line of sight, and v'_r the apparent velocity at the angular distance r from the centre along the apparent major axis of the nebula, the rotational velocity in the equatorial plane of the galaxy is

$$v_r = v'_r \sec i = k\, r \tag{14.1}$$

Since

$$v = c\,\frac{\Delta\lambda}{\lambda} \quad \text{and} \quad r = \frac{R}{d},$$

if R is the linear radius and d the distance of the galaxy,

$$k = c\,\frac{\Delta\lambda}{\lambda}\,\frac{d}{R}. \tag{14.2}$$

The rotation period is

$$P = 2\pi R/v_r = 2\pi\, d/k. \tag{14.3}$$

If v_r is expressed in km/sec, r in seconds of arc, and d in parsecs, the rotation period in years is given by

$$P = 29.7\, d \cdot r/v_r = 29.7\, d/k \tag{14.4}$$

or

$$P = 6.16 \cdot 10^9/\omega$$

if the angular velocity ω is expressed in km $\cdot$ sec^{-1} $\cdot$ kpc^{-1}.

According to some provisional data for about 30 objects discussed by MAYALL in 1948[1], the rotation periods (corrected for the revision of the distance scale) in the regions where the angular velocity is constant are as follows:

$E7, S0$	5 to 10×10^6 years,
Sa	10 to 20×10^6 years,
Sb	10 to 40×10^6 years,
Sc	20 to 80×10^6 years.

There is, however, considerable scatter, part of which arises from the uncertainties on individual distances; however, some late Sc (both ordinary and barred) indicate periods well in excess of 100 million years.

More reliable results are available for only a few individual bright nearby galaxies which have been studied in greater detail, viz. the Galaxy, the Magellanic Clouds, M 31, M 33 and M 81 (cf. Sect. 15) to which may be added NGC 4594 and NGC 3115 although the distances of these latter are rather uncertain. The results are collected in Table 16; as far as such a small sample may be regarded as representative, the indication is in agreement with the statistical data to suggest a steady increase in the rotation periods along the classification sequence, from a few million years among ellipticals, to about one hundred million years near the middle of the spiral stage, and up to several hundred million years among late-type irregular spirals of the magellanic type. Note, however, that all objects in Table 16 are giant or supergiant systems and the sample is too small to indicate the probable correlations between period of rotation, dimensions and mass or density.

[1] N. U. MAYALL: Sky and Telescope **8**, 3–5 (1948).

Table 16. *Galaxies analyzed for rotation and mass.*

Object	Type	Data	k	P_0	P_m	R_m	D	References
				(in 10^6 years)		(kpc)	(Mpc)	
L.M.C.	$SB(s)m$	H I, H II	—	—	200:	3.0	0.05	[*30*], [*31*][1,2,3]
S.M.C.	$SB(s)mp$	H I	—	—	350	3.0:	0.05	[*30*], [*31*][1]
M 33	$SA(s)c$	H II	—	—	100	2.5	0.5	[*31*], [*42*], [*49*], [*63*], [*86*]
M 31	$SA(s)b$	HII, Abs.	0.48	25	200}	9	0.5	[*1*], [*2*], [*42*], [*50*], [*63*], [*86*][4]
M 81	$SA(s)b^-$	H II	—	—	100:	5?	2.5	[5]
NGC 4594	$SA(s)a^+$	Abs.	2.78	40:	—	?	4:	[*49*][7]
NGC 4111	S0	Abs.	35:	(3—4)	—	?	4:	[6]
NGC 3115	E^+7	Abs.	9.8	10:	—	?	4:	[*52*], [*63*][6]

Data: Abs.: velocities from inclined absorption lines of unresolved central part. H I: velocities from 21 cm line of neutral hydrogen substratum. H II: velocities from emission lines of discrete gaseous nebulosities.

$k = v_r/r$: angular velocity constant near centre ($R \to 0$) in km/sec per second of arc; for M 31: $r < 150''$; NGC 4594: $r < 150''$; NGC 3115: $r < 45''$; NGC 4111: $r < 6''$, keplerian branch observed out to $50''$.

P_0: period of rotation of inner region ($R \to 0$).

P_m: period of rotation at $R = R_m$.

b) Masses of individual galaxies.

15. Rotational methods. The rotational analysis of large galaxies gives the most direct approach and probably the most reliable data for mass determinations. The theoretical problem has been discussed by BABCOCK [*2*], OORT [*52*], WYSE and MAYALL [*86*], PEREK [*54*], [*55*] and others. Observational data have been supplied and/or analyzed by BABCOCK [*2*], OORT [*52*], MAYALL and ALLER [*49*], SCHWARZSCHILD [*63*], LOHMANN [*42*], KERR and DE VAUCOULEURS [*30*], [*31*], HUMASON, G. MÜNCH and others[8].

A finding list of galaxies analyzed for rotation and mass is given in Table 16.

The most extensive rotation curves have been secured for the Magellanic Clouds by means of optical and radio observations of H I and H II emission lines, and for M 31, M 33 and M 81 by optical observations of the brightest H II regions located in the outer spiral arms. In all cases the rotation curve consists of a straight inner part in the region of constant angular velocity (cf. Sect. 14) up to a maximum at $R = R_m$ beyond which the rotational velocity decreases with increasing distance to the centre and tends asymptotically towards KEPLER's third law (Fig. 27)[9].

α) Point mass approximation: Keplerian branch. The simplest way to estimate the total mass is from a consideration of the Keplerian branch of the velocity

[1] F. J. KERR, J. V. HINDMAN and B. J. ROBINSON: Austral. J. Phys. **7**, 297—314 (1954)

[2] R. E. WILSON: Publ. Lick Obs. **13**, 187—190 (1918).

[3] M. W. FEAST, A. D. THACKERAY and A. J. WESSELINK: Observatory **75**, 216—221 (1955).

[4] N. U. MAYALL and O. J. EGGEN: Publ. Astronom. Soc. Pacific **65**, 24—29 (1953).

[5] G. MUNCH: Amer. Astr. Soc. Meeting, Aug. 1956 and private communication.

[6] M. L. HUMASON: Ann. Rep. Mt. Wilson Obs. **1936/37**, 31.

[7] F. G. PEASE: Proc. Nat. Acad. Sci. U.S.A. **2**, 520 (1916).

[8] See references to Table 16.

[9] In a few other objects, mainly "spindles" of either E, $S0$ or Sa types, such as NGC 3115, NGC 4111 and NGC 4594 and in the central bulge of Sb, Sc spirals, the solid-body rotation of the inner regions only can be observed as described in Sect. 14. Other tilted system showing inclined lines have been briefly reported on by MAYALL [Sky and Telescope **8**, 3—5 (1948) and "Problems of Cosmical Aerodynamics" 180—183 (1951)]. A larger list and a very complete discussion of such systems will be published shortly by N. U. MAYALL and P. O. LINDBLAD (private communication).

curve, by approximating the motion of an outlying object to that of a test particle moving in a circular orbit in the field of a central mass of negligible dimensions. The mass is then given immediately by

$$\mathcal{M} = \frac{R \times v_R^2}{G} \tag{15.1}$$

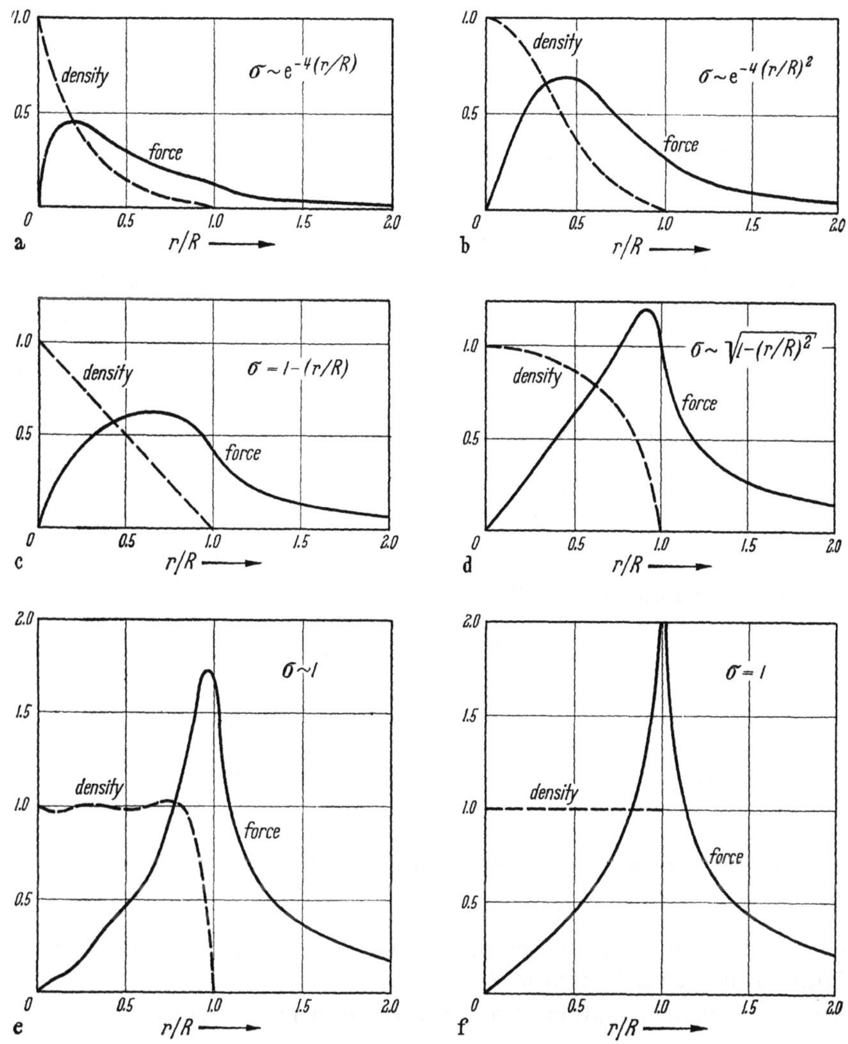

Fig. 27 a—f. Typical density and force curves in the thin disc approximation, after WYSE and MAYALL. Abscissae: $\alpha = r/R$ (R = radius of disc). Ordinates: arbitrary units, mutually consistent for density and focre.

where G is the gravitational constant, v_R the circular velocity, R the distance to the centre, or in solar units

$$\mathcal{M}_\odot = 1.13 \times 10^{-3} \cdot d \cdot r \cdot v_r^2 = 3.81 \times 10^{-5} \cdot P \cdot v_r^3 \tag{15.2}$$

if the distance d of the system is expressed in parsecs, the radius r of the orbit in seconds of arc, the period of rotation P in years and v_r in km/sec.

If $R \gg R_m$, this may be a fairly close approximation, especially if it can be checked that the computed velocity curve agrees closely with the observations over a sufficient range of distances.

β) Method of Wyse and Mayall: thin disc approximation. In a strongly flattened system such as a spiral galaxy outside the central bulge the random velocities are much smaller than the rotational velocity so that a simple two-dimensional model with circular orbits gives probably a fair approximation. This case has been discussed in detail by WYSE and MAYALL [*86*].

Let v_r be the rotational velocity at distance r from the centre, the central force is given by

$$F = \frac{v_r^2}{r} \tag{15.3}$$

and it is also, by definition, the derivative of the potential V, i.e.

$$F = \frac{\partial V}{\partial r} = \frac{\partial}{\partial r} \int \frac{dm}{\Delta} \tag{15.4}$$

where dm is a mass element at distance Δ from the point considered.

For a *homogeneous* circular disc of radius R and density σ, the potential at any point inside the disc is

$$V = 4R\,\mathsf{E}\,\sigma \qquad (r < R) \tag{15.5}$$

and at a point outside the disc and in its plane it is

$$V = 4R\left[\frac{r}{R}(\mathsf{E} - \mathsf{K}) + \frac{R}{r}\mathsf{K}\right] \qquad (r > R) \tag{15.6}$$

where K, E are LEGENDRE's complete elliptic integrals of the first and second kinds with the modulus r/R or R/r. Hence

$$F = -4\frac{R}{r}(\mathsf{K} - \mathsf{E})\,\sigma \qquad \text{for } r < R, \tag{15.7}$$

$$F = -4(\mathsf{K} - \mathsf{E})\,\sigma \qquad \text{for } r > R. \tag{15.8}$$

For an *inhomogeneous* disc of density σ_0 at $r = 0$, σ_r at $r = r$, σ_R at $r = R$

$$F = +4\frac{R}{r}(\mathsf{K} - \mathsf{E})\,\sigma_R + 4\int_{\sigma_R}^{\sigma_r} \frac{R}{r}(\mathsf{K} - \mathsf{E})\,d\sigma + 4\int_{\sigma_r}^{\sigma_0} (\mathsf{K} - \mathsf{E})\,d\sigma \quad \text{for } r < R. \tag{15.9}$$

In the thin disc model $\sigma_R = 0$ and σ_r is taken as the projected mass of matter in a column of unit cross-section at $r = r$.

If the density distribution is represented by a power series

$$\sigma = A + B\left(\frac{r}{R}\right) + C\left(\frac{r}{R}\right)^2 + D\left(\frac{r}{R}\right)^3 + E\left(\frac{r}{R}\right)^4 + F\left(\frac{r}{R}\right)^5 \tag{15.10}$$

with $\sigma_R = A + B + C + D + E = 0$, Eq. (15.9) can be written

$$\left.\begin{aligned} -\frac{F}{4} &= B \cdot M(\alpha) + C \cdot N(\alpha) + D \cdot O(\alpha) + E \cdot P(\alpha) + F \cdot Q(\alpha) \text{ if } \alpha = \frac{r}{R} < 1, \\ -\frac{F}{4} &= B \cdot m(\beta) + C \cdot n(\beta) + D \cdot o(\beta) + E \cdot p(\beta) + F \cdot q(\beta) \text{ if } \beta = \frac{R}{r} < 1 \end{aligned}\right\} \tag{15.11}$$

where the coefficients, $M, N, \ldots, m, n, \ldots$, computed by WYSE and MAYALL are given in Table 17.

Table. 17. *Coefficients of series expansion of force curve, Eq.* (15.11).

α	$M(\alpha)$	$N(\alpha)$	$O(\alpha)$	$P(\alpha)$	$Q(\alpha)$
0.00	0.000000	0.000000	0.000000	0.000000	0.000000
0.05	0.152244	0.078467	0.059270	0.052508	0.049181
0.10	0.250306	0.156490	0.120110	0.105891	0.098935
0.15	0.327413	0.233622	0.183255	0.161000	0.149893
0.20	0.390836	0.309407	0.249048	0.218626	0.202764
0.30	0.488414	0.455024	0.388685	0.344138	0.317438
0.40	0.556362	0.589266	0.537178	0.486420	0.449627
0.50	0.600194	0.707442	0.689528	0.646259	0.605285
0.60	0.622188	0.803799	0.837191	0.819431	0.786914
0.70	0.622469	0.870658	0.966765	0.993740	0.988674
0.80	0.598718	0.896406	1.056628	1.143057	1.186751
0.90	0.543543	0.859079	1.066319	1.210724	1.314476
0.95	0.497264	0.800152	1.012205	1.171750	1.297028
1.00	0.415966	0.666667	0.842961	0.977778	1.086418

β	$m(\beta)$	$n(\beta)$	$o(\beta)$	$p(\beta)$	$q(\beta)$
1.0	0.41597	0.66667	0.84296	0.97778	1.08642
0.9	0.27962	0.43440	0.53523	0.60718	0.66162
0.8	0.20301	0.31170	0.38055	0.42853	0.46409
0.7	0.14680	0.22373	0.27164	0.30454	0.32864
0.6	0.10344	0.15684	0.18970	0.21205	0.22830
0.5	0.06961	0.10516	0.12684	0.14150	0.15212
0.4	0.04351	0.06556	0.07892	0.08793	0.09445
0.3	0.02406	0.03618	0.04351	0.04846	0.05208
0.2	0.01057	0.01587	0.01911	0.02136	0.02311
0.1	0.00262	0.00394	0.00492	0.00591	0.00698
0.0	0.00000	0.00000	0.00000	0.00000	0.00000

The observed rotational velocity curve is first transformed through (15.3) into a force curve from which the coefficients A, B, C, D, E, F are computed through (15.11) by means of the tabulated factors so as to fit exactly five suitably spaced points of the empirical force curve. This gives a density distribution curve (15.10) in arbitrary units; if the rotational velocities are expressed in km/sec and the radial distances in *minutes* of arc, the density σ is obtained in solar masses per square minute of arc by multiplying by the factor $0.0678d$, and the density in solar masses per square parsec by multiplying by the factor $8.01 \times 10^5/d$, when the distance d of the galaxy is expressed in parsecs.

It is then easy to obtain the total mass of the system in solar units

$$\mathcal{M}_\odot = 0.0678d \cdot 2\pi \int_0^R r\,\sigma\,dr \tag{15.12}$$

or, substituting the power series approximation (15.10)

$$\mathcal{M}_\odot = 0.426 R^2 \cdot d\left(\frac{A}{2} + \frac{B}{3} + \frac{C}{4} + \frac{D}{5} + \frac{E}{6} + \frac{F}{7}\right) \tag{15.13}$$

(R: minutes of arc, d: parsecs).

In practice, however, the quantitative derivation of the density law (15.10) from a given velocity curve is complicated not only by the unavoidable accidental errors in the velocity measurements and the resulting uncertainty in the interpolated, smooth velocity curve, but also to a considerable extent by the alternating character of the series involved so that the quantities of physical significance are obtained as differences between much larger coefficients of opposite signs and the results are largely indeterminate.

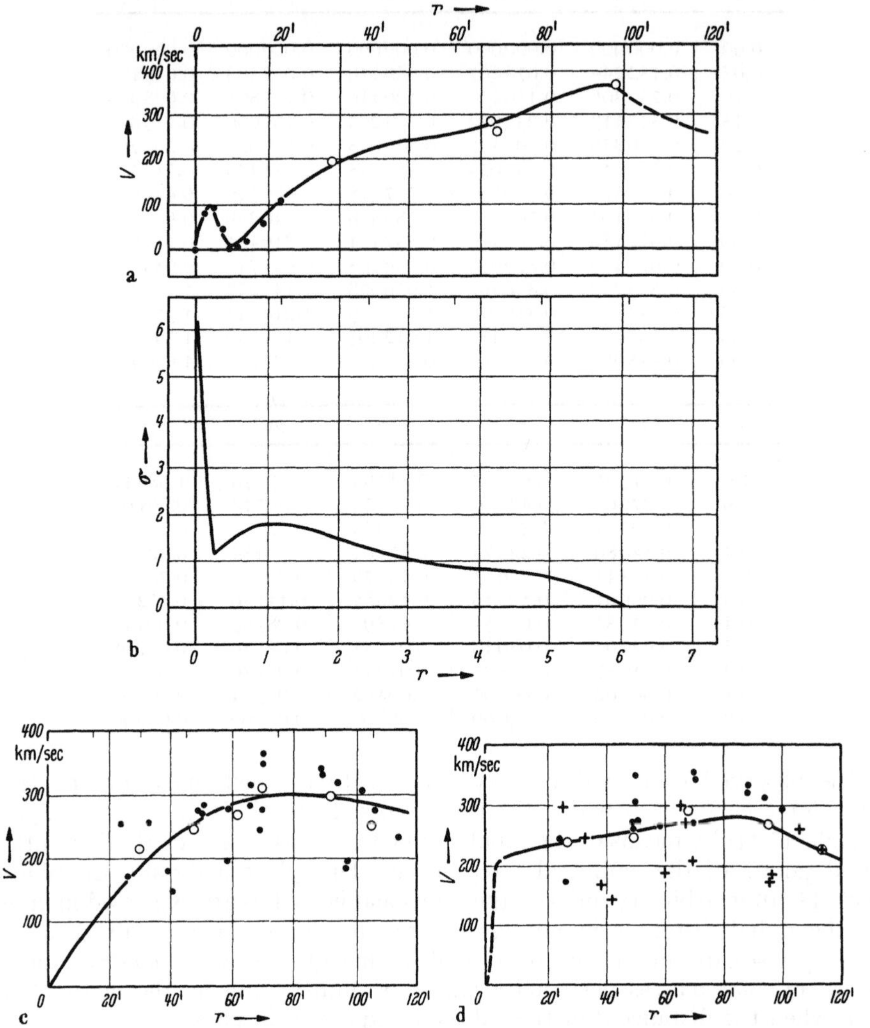

Fig. 28 a—d. Observed and computed velocity curves of Messier 31. (a) Velocity curve after BABCOCK (1939); dots: from absorption spectrum; circles: from emission regions. (b) Projected density distribution after WYSE and MAYALL (1942); ordinates: thousands of solar masses per square parsec, abscissae: kiloparsecs, for a distance $D = 0.21$ Mpc (old scale). (c) Velocity curve after LOHMANN (1954); dots: velocities of emission regions observed by MAYALL and HUMASON (1951), circles: mean points. (d) Velocity curve after SCHWARZSCHILD (1954); dots: velocities of emission regions on north-following side, crosses: velocities of emission regions on south-preceding side, circles: mean points.

A semi-empirical approach is therefore to be preferred in most cases. WYSE and MAYALL have computed the coefficients $A, \ldots, F$ in the power series approximations to six simple models of density distribution, as follows:

$$\text{I. } \sigma \sim \exp(-4\alpha); \quad \sigma = 1.0 - 3.97500\alpha + 7.66042\alpha^2 - 8.89636\alpha^3 + 6.02084\,\alpha^4 - 1.80990\alpha^5,$$

$$\text{II. } \sigma \sim \exp(-4\alpha^2); \quad \sigma = 1.0 + 0.17105\alpha - 5.87265\alpha^2 + 6.93605\alpha^3 - 1.46637\alpha^4 - 0.76818\alpha^5,$$

$$\text{III. } \sigma = 1 - \alpha,$$

IV. $\sigma \sim 1 - \alpha^2$; $\quad \sigma = 1.0 - 0.2530\alpha + 2.1460\alpha^2 - 9.3284\alpha^3 + 13.3833\alpha^4 - 6.9479\alpha^5$,

V. $\sigma \sim 1$; $\quad \sigma = 1.0 - 1.0\alpha + 10.41667\alpha^2 - 36.45833\alpha^3 + 52.08333\alpha^4 - 26.04167\alpha^5$,

VI. $\sigma = 1$.

These expressions apply only to $0 \leq r/R \leq 1$; for $r/R > 1$, $\sigma = 0$ by definition. The first two density laws are approximately exponential and approximately Gaussian; the third is exactly linear; the fourth approximates the (projected) density distri-

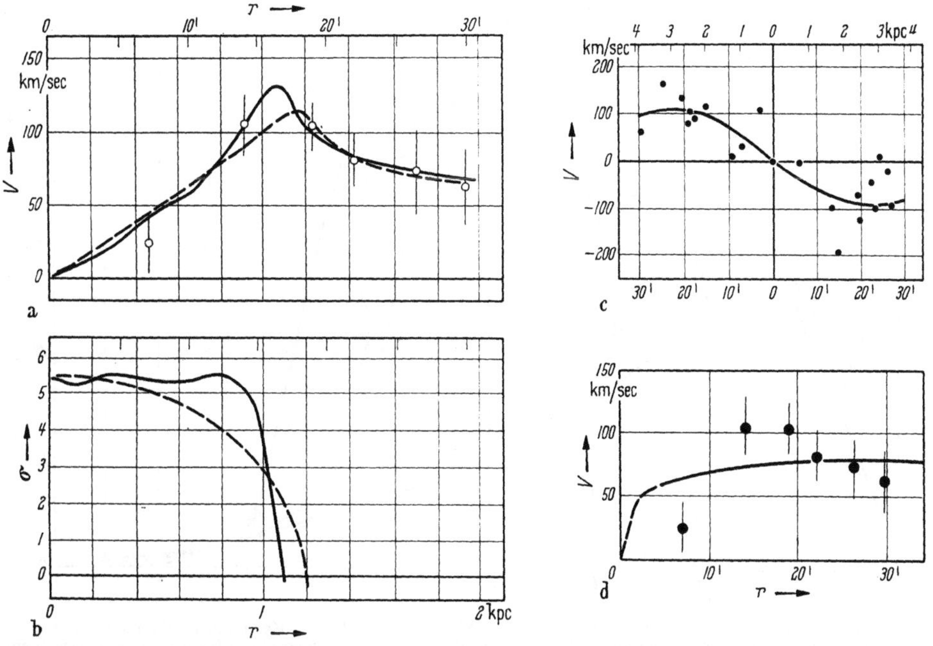

Fig. 29 a—d. Observed and computed velocity curves of Messier 33. (a) Velocity curve after MAYALL and ALLER (1942); circles are mean velocities from emission regions with probable error shown by segments. (b) Projected density distribution after WYSE and MAYALL (1942); continuous curve: quasi-homogeneous disc, dashed curve: projected oblate spheroid. The corresponding theoretical velocity curves are shown in (a). Ordinates: hundreds of solar masses per square parsec, abscissae: kiloparsecs for a distance $D = 0.22$ Mpc (old scale). (c) Velocity curve after LOHMANN (1954); dots: velocities of individual emission regions observed by MAYALL and ALLER. (d) Velocity curve after SCHWARZSCHILD (1954); dots: mean points of MAYALL and ALLER.

bution in an oblate spheroid; the fifth is an approximation of the constant density of an homogeneous disc exactly represented by the sixth. The density distributions and corresponding force curves are illustrated in Fig. 27.

It is then a fairly simple matter to determine by trial and error which model or combination of models gives a satisfactory fit, within the observational errors, to any observed velocity curve. Note that it is preferable to test the goodness of fit on the observed velocities rather than on the derived force curve. This method has been applied to M 31 and M 33 by WYSE and MAYALL [*86*] (Fig. 28 and 29) and to the Large Cloud by KERR and DE VAUCOULEURS [*31*] (Fig. 30). The results, corrected as required for changes in the distance scale, are collected in Table 18.

A serious disadvantage of this method is that it neglects the thickness of the system and thus prevents the derivation of space densities; another weakness is that the velocity curve is very in sensitive to the mass distribution in the outer parts of low space density but of large volume which contribute an appreciable fraction of the total mass. A closer approximation may be obtained by the following methods.

γ) Perek's method: oblate spheroid approximation. The forces curves of nonhomogeneous spheroids of axial ratio c/a have been discussed by PEREK [*54*] when the density law is of the form

$$\varrho = \varrho_c (1 - m^2)^n \tag{15.14}$$

where

$$m^2 = \frac{x^2 + y^2}{a^2} + \frac{z^2}{c^2} \tag{15.15}$$

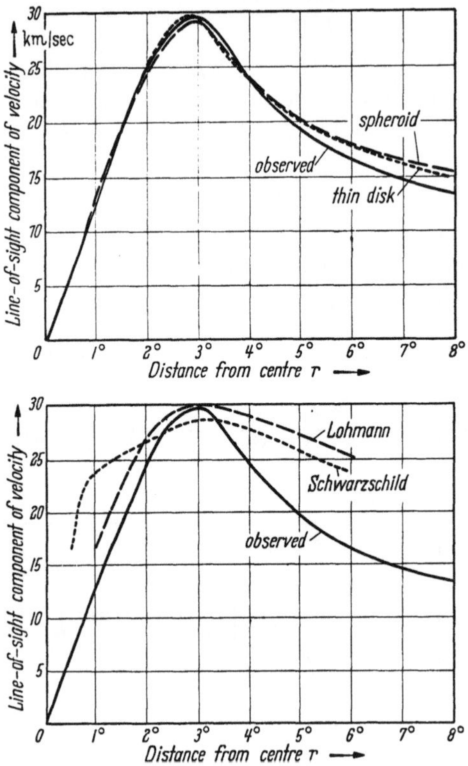

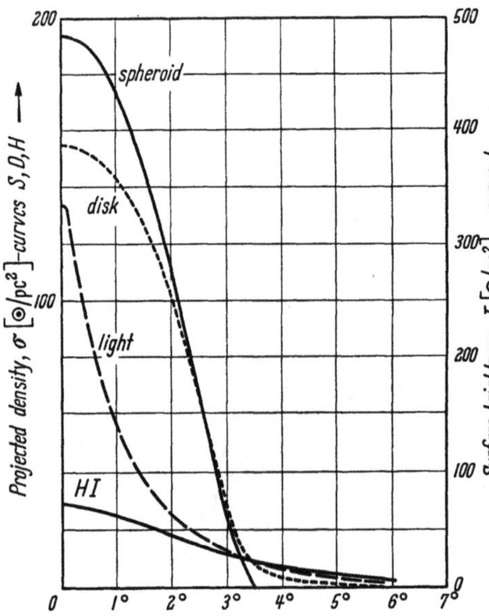

Fig. 30. Observed and computed velocity curves of Large Magellanic Cloud, after KERR and DE VAUCOULEURS (1956). The observed velocity curve, corrected for resolving power, of the equatorial component of the neutral hydrogen 21 cm emission line (continuous curve) is compared with the theoretical velocity curves for the best fitting thin disc and spheroid models (dots and dashes) to which correspond the projected density distributions shown to the right. The velocity curves derived according to LOHMANN's and SCHWARZSCHILD's formulae are shown for comparison. Compare with Fig. 29.

Table 18. *Masses of galaxies computed by different methods* (unit = 10^9 ⊙).

Method	Object Data Dist. Tilt	L.M.C. H I 0.046 (Mpc) 65°	M 31 H II 0.42 (Mpc) 15°	M 33 H II 0.44 (Mpc) 33°
	a) *Rotational mass*			
1. KEPLER		1.5	170	3.5
2. WYSE-MAYALL		1.7	190	3.5
3. PEREK		1.9	—	4
4. LOHMANN		(3—4)	270[1]	10
5. SCHWARZSCHILD		(2—3)	140[2]	5[3]
	b) *Extrapolated mass*			
1. Rotation		2.5	—	5.5
2. Rotation + velocity dispersion		3.0	—	—

[1] Corrected figure from a least-square solution (private communication) superseding published value ($330 \cdot 10^9$).

[2] For $D = 0.46$ Mpc.

[3] For $D = 0.48$ Mpc.

ϱ_c being the central space density. The potential V at the point (x, y, z) is

$$V = \pi \varrho_c G \frac{a^2 c}{n+1} \int \left(1 - \frac{r^2}{a^2+u} - \frac{z^2}{c^2+u}\right)^{n+1} \frac{du}{(a^2+u)\sqrt{c^2+u}} \tag{15.16}$$

where G is the gravitational constant and $r^2 = x^2 + y^2$. The limits of the integral are $0, \infty$ for internal points and λ, ∞ for external points. λ is the greatest positive root of the equation

$$\frac{r^2}{a^2+\lambda} + \frac{z^2}{c^2+\lambda} = 1\,. \tag{15.17}$$

Putting

$$\left.\begin{aligned} \cos\psi_0 &= \frac{c}{a}\,, & r_0 &= \frac{r}{a}\sin\psi_0 \\ \cos\psi &= \sqrt{\frac{c^2+\lambda}{a^2+\lambda}}\,, & z_0 &= \frac{z}{a}\sin\psi \end{aligned}\right\} \tag{15.18}$$

and

($\psi = \psi_0$ for internal points), Eq. (15.16) and (15.17) may be written

$$\left.\begin{aligned} V = \pi G \varrho_c a^2 \cot\psi_0 \Big[&\frac{1}{n+1} w_{10} - (w_{20} r_0^2 + w_{11} z_0^2) + \\ &+ \frac{1}{2} n (w_{30} r_0^4 + 2 w_{21} r_0^2 z_0^2 + w_{12} z_0^4) + \cdots \Big] \end{aligned}\right\} \tag{15.19}$$

and

$$r_0^2 \sin^2\psi + z_0^2 \tan^2\psi = 1 \tag{15.20}$$

where the numerical coefficients w have been tabulated by PEREK in the range $0 \leq \psi \leq 1.50$ of $\psi = \frac{1}{2} w_{10}$. The component of the attractive force parallel to the equatorial plane[1] is then

$$\left.\begin{aligned} F = -2\pi G \varrho_c a \frac{\cos\psi_0}{\sin^2\psi_0} r_0 \Big[&w_{20} - n (w_{30} r_0^2 + w_{21} z_0^2) + \\ &+ \frac{n(n-1)}{2} (w_{40} r_0^4 + 2 w_{31} r_0^2 z_0^2 + w_{12} z_0^4) + \cdots \Big] \end{aligned}\right\} \tag{15.21}$$

which for $n = 0$ (homogeneous spheroid) reduces to

$$F_r = -2\pi G \varrho \frac{\cos\psi_0}{\sin^2\psi_0} w_{20} r_0 \tag{15.22}$$

and for $n = 1$ and points in the equatorial plane of the system reduces to

$$F_r = -2\pi G \varrho_c a \frac{\cos\psi_0}{\sin^2\psi_0} r_0 \left[w_{20} - w_{30} r_0^2 + w_{40} r_0^4 + \cdots \right] \tag{15.23}$$

PEREK [55] has constructed a convenient nomogram, reproduced in Fig. 31, for the rapid resolution of (15.20); Table 19 gives only the values of w_{20}, w_{30}, w_{40} needed to compute the force curve in the cases $n = 0$ and $n = 1$ for $z = 0$ according to (15.22) and (15.23).

Once a suitable value of c/a has been chosen, say $\frac{1}{5}$ to $\frac{1}{10}$ (cf. Sect. 7), $\cos\psi_0$ and r_0 are obtained for a given r through (15.18), then (15.20) is solved for ψ by means of the nomogram and the coefficients w found in the table for the argument ψ give the force curve through (15.22) or (15.23).

[1] For further details concerning the force perpendicular to the equatorial plane, the velocity of escape, etc. and application to the Galaxy, see PEREK [54] and Contr. Ast. Inst. Masaryk Univ. **1**, No. 8 (1951), and also M. SCHMIDT, Bull. Astr. Inst. Netherl. **12**, No. 468, 15–41 (1956).

Table 19. *Coefficients of attractive force in Eq.* (15.23).

ψ	w_{20}	w_{30}	w_{40}	ψ	w_{20}	w_{30}	w_{40}
0.00	0.000	0.000	0.000	1.00	0.545	0.248	0.131
0.50	0.079	0.011	0.002	1.10	0.696	0.361	0.216
0.60	0.134	0.026	0.006	1.20	0.862	0.500	0.332
0.70	0.207	0.053	0.016	1.30	1.04	0.662	0.478
0.80	0.300	0.097	0.036	1.40	1.23	0.843	0.650
0.90	0.413	0.160	0.073	1.50	1.43	1.04	0.841

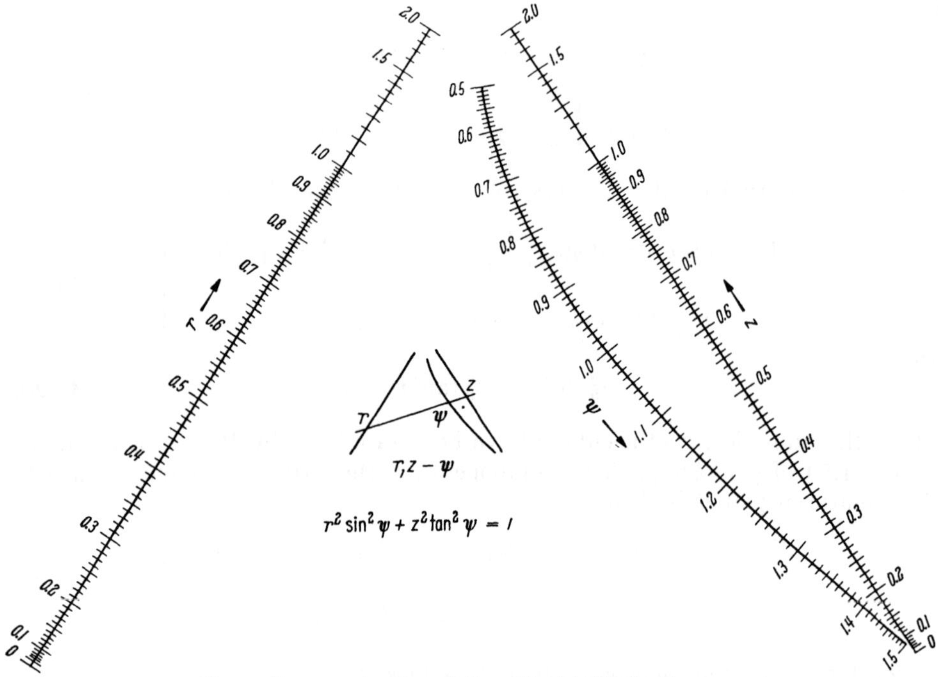

Fig. 31. Nomogram for the resolution of Eq. (15.20), after PEREK.

The *total mass* of the spheroid is

$$\mathcal{M} = 4\pi a^3 \sqrt{1-e^2} \int_0^1 m^2 \varrho\, d m = \varkappa \pi a^3 \sqrt{1-e^2} \cdot \varrho_c \tag{15.24}$$

where $\sqrt{1-e^2} = c/a$ and

$$\varkappa = 2^{2n+3} \frac{\Gamma(n+1)\,\Gamma(n+2)}{\Gamma(2n+4)}. \tag{15.25}$$

For $n=0$, $\varkappa=\frac{4}{3}$, for $n=1$, $\varkappa=\frac{8}{15}$. Further, $\varkappa=\frac{1}{3}$ if $\varrho=\varrho_c(1-m)$ and $\varkappa=\frac{2}{15}$ if $\varrho=\varrho_c(1-m)^2$.

The *projected density* on the equatorial plane is

$$\sigma = \varrho_c\, a \sqrt{1-e^2}\,(1-m_0^2)^{n+\frac{1}{2}} \cdot 2^{2n+1} \frac{[\Gamma(n+1)]^2}{\Gamma(n+2)} \tag{15.26}$$

if $m_0 = r/a$; for $n=1$ and $r=0$, this reduces to

$$\sigma_c = \tfrac{4}{3} \varrho_c\, a \sqrt{1-e^2} = \tfrac{4}{3} \varrho_c \times c. \tag{15.27}$$

This method has been applied to the Large Cloud by KERR and DE VAUCOULEURS [31] (Fig. 30). The results are in good general agreement with those from method (β), but, as could be expected for a model allowing for the spread of matter outside the equatorial plane, the total mass and central projected density are somewhat higher. For the Large Cloud, with $n=1$ and $c/a=\frac{1}{5}$, $\mathcal{M}$ is about 10% and σ_c about 25% greater than in the thin disc approximation. A comparison of the observed and computed velocity curves and projected density distributions according to the two models is given in Fig. 30.

The oblate spheroid approximation, although it gives a satisfactory fit to the observed velocity curve, still takes little or no account of the outer regions of low density where both optical and radio observations indicate the presence of an appreciable density of matter, stellar and interstellar. Several empirical methods may be used to remedy this deficiency.

δ) Method of Lohmann-Bottlinger. BOTTLINGER has proposed several empirical formulae to represent the gravitational force in the galaxy. LOHMANN [42] has assumed that one of these gives an exact representation of the force curve in the equatorial plane of a spiral galaxy, it is

$$F = a\,r/(1 + b\,r^3) \tag{15.28}$$

so that

$$v^2 = a\,r^2/(1 + b\,r^3) \tag{15.29}$$

from which the total mass is obtained analytically

$$\mathcal{M} = a/b\,G \quad (G\text{: gravitational constant}). \tag{15.30}$$

In terms of the coordinates r_m, v_m of the maximum of the velocity curve $a=3\,(v_m/r_m)^2$, $b=2/r_m^3$ so that

$$\mathcal{M} = \frac{3}{2}\,\frac{r_m v_m}{G} \tag{15.31}$$

This method has been applied by LOHMANN [42] to M 31 and M 33, and by KERR and DE VAUCOULEURS [31] to the Large Cloud with the results shown in Table 18. The theoretical velocity curve (15.29) fits the observed curves for M 31 and M 33 fairly well, but the fit is much poorer for the more extensive Keplerian branch of the rotation curve of the Large Cloud (Fig. 30). Note that while the total mass thus obtained allows automatically for the matter distributed in the outer regions, in so far as (15.28) fits the observed force curve, the method as such does not imply any definite law for the density distribution.

ε) Schwarzschild's method. In this method, on the contrary, the basic hypothesis is a specific assumption concerning the projected density distribution $\sigma(r)$ which is taken as strictly proportional everywhere to the optical surface brightness $I(r)$, i.e. it is assumed that the mass/luminosity ratio f is constant throughout the system. For computation purposes it is further assumed that the luminosity distribution can be approximated by a series of straight segments, so that

$$I(r) = \sum_n a_n (1 - r/R_n) \tag{15.32}$$

and

$$\sigma(r) = \sum_n A_n (1 - r/R_n) \tag{15.33}$$

where $A_n = f \times a_n$, with $f=$ const.

The thin disc approximation, following the method of WYSE and MAYALL (see above) for a linear distribution (III), then leads to a simple solution of the velocity curve and to a total mass

$$\mathcal{M} = \tfrac{1}{3}\pi \sum_n A_n R_n^2 \tag{15.34}$$

which is expressed in solar units if distances are in parsecs and a_n, A_n in solar units per square parsec. The circular velocity is given by

$$v_c^2(r) = 59.0 \frac{f}{D} r \sum_n a_n g(r/R_n) \tag{15.35}$$

where $g(r/R_n)$ is the function $M(\alpha)$, $m(\beta)$ of WYSE and MAYALL (Table 17) and D the distance of the galaxy (in parsecs if r is expressed in the same unit).

The method was applied by SCHWARZSCHILD [63] to M 31 and M 33 and by KERR and DE VAUCOULEURS [31] to the Large Cloud with the results given in Table 18. The theoretical velocity curve gives an acceptable fit of the observations in M 31 (Fig. 28), but the agreement is poor for M 33 and for the Large Cloud (Figs. 29 and 30). It seems, therefore, probable that the assumption of a constant mass/luminosity ratio throughout a galaxy is not justified, at least for late-type systems in the range *Sc* to *Sm*.

ζ) Extrapolation to infinity. An attempt has been made by KERR and DE VAUCOULEURS [31] to allow for the mass of stellar and interstellar matter spread in the outer regions beyond the limiting radii R of the models in the thin disc and oblate spheroid approximations by an extrapolation method similar to that used in the determination of the total (integrated) luminosities of galaxies (cf. Sect. 3).

The fraction of the total luminosity (or H I mass) within any given radius is computed by direct integration of the isophotes; if it is assumed that the mass given by the thin disc or oblate spheroid approximations refers to the limiting radii of the models only and that the combined mass (gas + stars) is distributed as the mean of the hydrogen and light, it is possible to estimate, at least approximately, the factor by which the mass given by the models ought to be multiplied to obtain the total mass of the system.

This method has been applied to the Large Cloud, the Small Cloud and M 33 with the results shown in Table 18. Unless the correction is very small the computed velocity curve departs from the observed curve in the same direction as in SCHWARZSCHILD'S method and probably for the same reason. In all the previous treatments the neglect of random motions is, however, another significant defect which may be responsible for at least some of the discrepancies.

16. Velocity dispersion. The rotational methods presuppose that the observed velocities give a close approximation to the pure circular velocities that would prevail in a system where random motions are negligible. This may be a reasonable approximation in late-type spirals of extreme flattening. The opposite extreme is the case of a spherical system in which random motions (or radial oscillations) are predominant and general rotation negligible. This is probably a good approximation in the case of globular clusters and spheroidal galaxies; since, however, velocities of individual stars cannot be observed in galaxies and because the overall velocity dispersion is difficult if not impossible to determine[1], the method of estimating total masses from random motions developed for globular

[1] The velocity dispersion which, in principle, could be derived from the width of the spectral lines is, in practice, confused by the broadening due to the mixture of stellar types in the composite spectrum of a galaxy (cf. Sect. 11).

clusters of stars or galaxies have not been applied yet to individual galaxies. In most cases actual galaxies are in an intermediate situation where the velocity dispersion although generally smaller than the rotational velocity (except near the centre) is not entirely negligible (see additions, p. 366).

For a system in a dynamically steady state, the following general relation between the circular velocity in the absence of random motions v_c, and the mean rotational velocity observed when random motions are present v_0, has been given by OORT [52], after JEANS[1]

$$v_c^2 - v_0^2 = \overline{u^2}\left[-\frac{\partial \log \varrho}{\partial \log r} - \frac{\partial \log \overline{u^2}}{\partial \log r} - \left(1 - \frac{\overline{v^2}}{\overline{u^2}}\right)\right] \qquad (16.1)$$

where u, v, w, are the x, y, z components of the random velocity and ϱ the space density[2]. The third term of the bracket is probably small and may be neglected, but there is at present no way of estimating the density gradient $\partial \log \varrho/\partial \log r$, nor the mean square velocity and still less its gradient with sufficient accuracy for practical application to any external galaxy.

The difficulty of estimating the velocity dispersion has been noted already; the density gradient could be derived in an edgewise elliptical system ($E\,7$) from the luminosity distribution only if the assumption $f=$ const. is made, for which there is no clear support in the observations. Nevertheless the method has been tentatively applied by OORT [52] and by SCHWARZSCHILD [63] to HUMASON'S rotation data on NGC 3115 by making plausible assumptions on the factors involved. The results of SCHWARZSCHILD are collected below from which follows

Distance from centre	Circular velocity	Rotational velocity	Velocity dispersion
8″1	510 km/sec	79 km/sec	320 km/sec
16 1	580 km/sec	158 km/sec	340 km/sec
36 2	530 km/sec	316 km/sec	240 km/sec
42 5	480 km/sec	447 km/sec	100 km/sec

$f=100$ and $\mathcal{M}=9\times10^{10}\,\odot$, if the distance of NGC 3115 is $D=2.1$ Mpc a value which is subject to a large relative uncertainty[3]. In this case $c/a\approx\frac{1}{3}$ and the root mean square velocity is of the same order or even larger than the rotational velocity in the inner regions; the computed circular velocity is accordingly very uncertain.

When the random motions are not so large and/or if the system is not in a steady state a simplified treatment may give a sufficient approximation, at least for the derivation of the total mass. Such a treatment was applied by KERR and DE VAUCOULEURS [31] to the Large Cloud for which $c/a\approx\frac{1}{5}$ and the total mass is thereby increased by about 20% only. For typical spirals of greater flattening the correction is probably negligible.

A comparison of mass estimates of the Large Cloud, M 31 and M 33 by different methods is given in Table 18.

17. Perturbations. SCHWARZSCHILD [63] has attempted to derive the mass of M 32 from an asymmetry in the shape and velocity distribution of the spiral

[1] J. JEANS: Monthly Notices Roy. Astronom. Soc. London **82**, 122 (1922).

[2] Note that for strict dynamical equilibrium $\overline{u^2}=\overline{v^2}=\overline{w^2}$ [JEANS, Monthly Notices Roy. Astronom. Soc. London **76**, 81 (1915)] which is not the case in a flattened system where in general $\overline{w^2}<\overline{u^2}\approx\overline{v^2}$.

[3] The same data, neglecting random motions, give through (7.12) $\mathcal{M}\gtrsim 2.2\times10^{10}\,\odot$.

pattern of the Andromeda nebula in its south-preceding half compared with the north-following half. Between 40′ and 90′ from the centre of M 31 the rotational velocities in the *sp* and *nf* sections differ by about $\Delta v = 80$ km/sec, this and a corresponding structural asymmetry may be caused by the attraction of M 32 if the elliptical companion is close to the equatorial plane of the spiral. Assuming that the interaction has been effective during only a quarter of the period of rotation P of M 31 (if it had been much longer a quasi permanent, symmetrical figure of equilibrium would have been set up with respect to the radius vector M 31 to M 32), the simple relations

$$\Delta v = g\,t, \quad g = \frac{G\mathcal{M}}{d^2}$$

may be written, where $\mathcal{M}$ is the mass of M 32 and d its distance to the perturbed region of M 31; with $t = P/4 = 5 \times 10^7$ years, $d = 8000$ pc, $\Delta v = 80$ km/sec: $\mathcal{M} \approx 2.5 \times 10^{10}\,\odot$. This figure, although very uncertain, appears to be of the right order of magnitude.

c) Mass luminosity ratio.

18. Summary of results. Table 20 gives a summary of the most probable estimates of masses and integrated photographic luminosities of galaxies of different types in the current distance scale ($H = 180$ km/sec/Mpc). In view of

Table 20. *Masses and luminosities of bright galaxies.*

Objects	Type	m_P	$m-M$	D (Mpc)	Log $\mathcal{L}$ ($\odot=1$)	Log $\mathcal{M}$ ($\odot=1$)	Log f ($\odot=1$)
L.M.C.	*SB(s)m*	0.7	19	0.05	9.4	9.6:	0.2:
S.M.C.	*SB(s)mp*	2.6:	19	0.05	8.7:	9.1:	0.4:
M 33	*SA(s)c*	6.2	24	0.5	9.2	9.8	0.6
M 31	*SA(s)b*	4.2	24	0.5	10.2	11.3	1.1
M 81	$SA(s)b^-$	7.8	27	2.5	9.9	11.0	1.1
NGC 4594	$SA(s)a^+$	9.0	28:	4:	9.8:	11.2:	1.4:
NGC 3115	gE^+7	10.1	28:	4:	9.3:	11.0:	1.7:
M 32	*dE2*	9.1	24	0.5	8.2	10.4:	2.2:

the inherent uncertainties only round figures are given for the distances and distance moduli; the estimated probable errors of $m-M$ are ± 0.4 mag. for NGC 3115 and NGC 4594 and ± 0.2 mag. for all others. Note that if $m-M$ is increased by 0.2 mag., D increases by 10%, $\mathcal{L}$ by 20%, $\mathcal{M}$ by 10% and $f = \mathcal{M}/\mathcal{L}$ decreases by 10%. The apparent magnitudes m are on the same P system as in Table 2; asymptotic magnitudes (extrapolated to infinity) are 0.1 to 0.2 mag. brighter; the probable errors range from ± 0.05 to ± 0.1 mag. The adopted absolute magnitude of the Sun is $P_\odot = +5.37$[1]. The mass estimates have been adjusted for changes in the distance scale as required and corrected where necessary to allow for velocity dispersion and outlying regions (NGC 3115, 4594); the mass of the Large Cloud was revised to give more weight to optical determinations and computed for $i = 70°$ (instead of 65°). The probable errors of Log $\mathcal{M}$ range from about ± 0.1 (M 31, M 33, M 81) to about ± 0.3 (M 32, NGC 3115, NGC 4594); the probable errors of Log f from ± 0.1 to ± 0.2.

19. Mass luminosity ratio. A plot of the data (Fig. 32) shows a close correlation between mass/luminosity ratio and morphological type, the mean f ratio

[1] J. STEBBINS and G. E. KRON: Sky and Telescope **16**, 56 (1956). — Astronom. J. **61**, 327 (1956).

ranging from $f \approx 100$ at E to $f \approx 1$ at Im. The values of f at either end of the classification sequence are still very uncertain, however, and only the average logarithmic value $f \approx 10$ near Sb appears fairly well established.

The steady increase of f along the sequence $I \rightarrow S \rightarrow E$ is consistent with similar trends in colour, spectrum, etc. (cf. Sects. 5 to 12). Any such observable characteristic which varies monotonely along the classification sequence will therefore be correlated with the mass/luminosity ratio. In recent years particular significance has been attached to the correlation involving the colour index C [24]; assuming that definite values may be attached to hypothetical "pure" Type I and "pure" type II populations, the correlation was regarded as a manifestation of the changing ratio of Type I to Type II along the classification sequence; however, if the five-fold increase of f in the range Sa to E indicated by Fig. 32 is real, the negligible variation of C in the same range (cf. Fig. 3) detracts from the significance of the correlation. Then it is now realised that supposedly "pure" Type II systems, such as globular clusters, exhibit a range of colour indices[1] and have much lower mass-luminosity ratios than elliptical galaxies [63], while "old" Type I systems have very nearly the same average colour as Type II systems[2]. Finally, because of absorption effects, emission lines, etc. (cf. Sects. 4 and 11) the colour index is a poor indicator of population type.

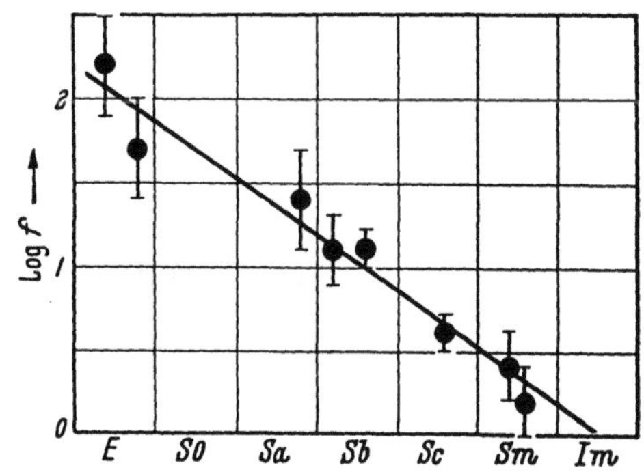

Fig. 32. Logarithmic mass to luminosity ratio as a function of galaxy type.

Alternatively one might assume that it is possible to assign definite values of f to "pure" Type I and Type II systems and derive the composition of a mixed system from the observed intermediate value of f [63]. Thus, Fig. 32 suggests $f_1 \approx 1$, $f_2 \approx 100$, so that if one writes

$$\mathcal{M} = \mathcal{M}_1 + \mathcal{M}_2, \tag{19.1}$$

$$\mathcal{L} = \mathcal{L}_1 + \mathcal{L}_2 \tag{19.2}$$

the observed value $f \approx 10$ near Sb leads to

$$\mathcal{M}_1/\mathcal{M} = 0.1, \quad \mathcal{M}_2/\mathcal{M} = 0.9,$$

$$\mathcal{L}_1/\mathcal{L} = 0.9, \quad \mathcal{L}_2/\mathcal{L} = 0.1.$$

Hence one might conclude that while nine tenths of the mass of an Sb spiral resides in its Type II component, nine tenths of its light comes from its Type I component.

However, this is probably an oversimplified picture and the results are of doubtful significance; it ignores the contribution of interstellar matter to the total mass, the probable change of f-ratio of a Type I population from "young" to "old" and, again, the great difference between the f-ratios of elliptical galaxies on the one hand and of globular clusters on the other, both supposedly of "pure" type II according to BAADE. It seems obvious that great variations exist in the luminosity functions and H–R diagrams of stellar systems and present

[1] G. E. KRON and N. U. MAYALL: Astronom. J. **61**, 327 (1956).

[2] C. V. REDDISH: Observatory **76**, 68 (1956).

ideas will certainly require considerable modifications and refinements before a satisfactory interpretation of the observed correlation between mass-luminosity ratio and morphological type can be evolved. There is, above all, a great need for more observational and theoretical work on the dynamics of galaxies.

Fig. 33. Typical spectra of galaxies (N. U. Mayall, Lick Observatory).

Note concerning Sect. 11, added to proofs.

(December 1957)

A new era in the analysis of spectra of galaxies has been opened by a recent investigation of low and medium-dispersion spectra of bright galaxies by W. W. MORGAN and N. U. MAYALL [91] which indicates both a larger spread of spectral types than previous studies suggested and a greater complexity in the composite spectra than was heretofore suspected. The main results are illustrated in Figs. 33 and 34 showing some typical spectra and the correlation between revised spectral types and revised morphological types.

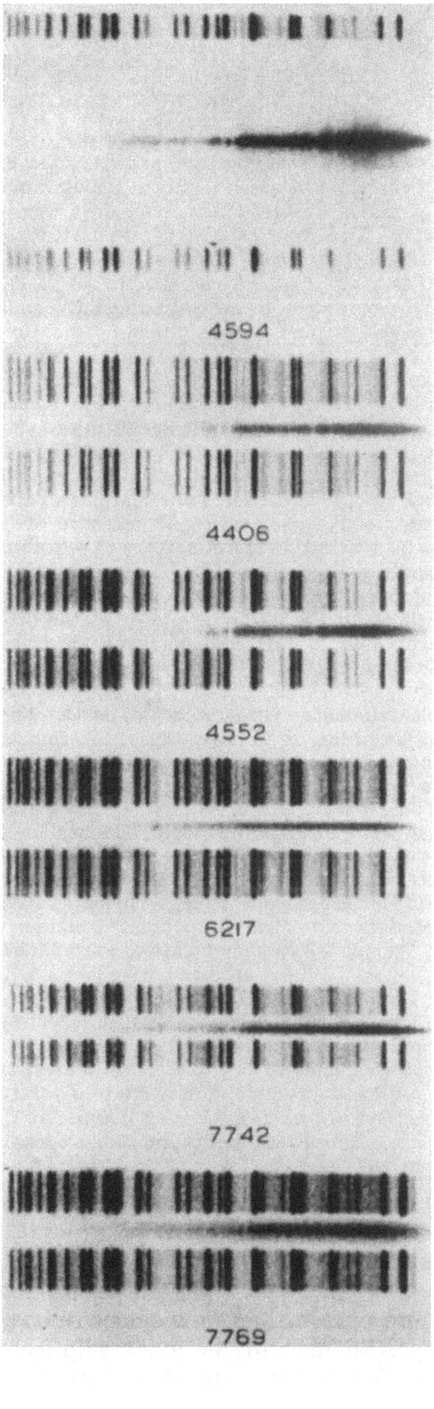

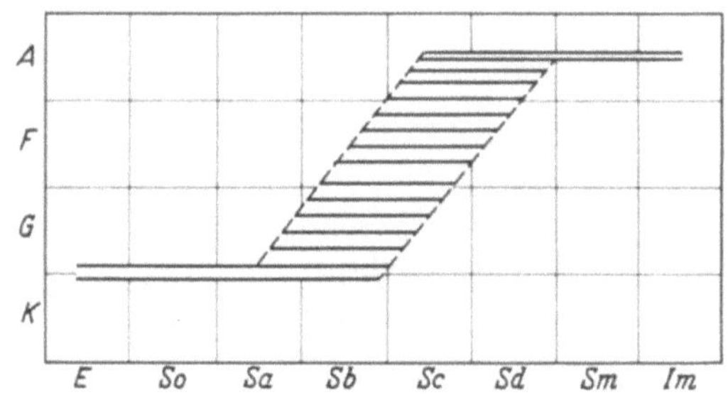

Fig. 34. Correlation between spectral type and morphological type

Because of the composite character of the spectra and their dependence on spectral range (see below) the spectral type is estimated in the narrow interval λ 3850 to 4100 Å which includes the *H* and *K* lines of Ca II, the Balmer lines of hydrogen beyond Hγ and the cyanogen bands allowing to distinguish between dwarf and giant characteristics in the later types. In order to avoid giving an illusory accuracy to estimates of spectral type only *A*, *AF*, *F*, *FG*, etc. groups are indicated.

A-type spectra are observed among magellanic irregulars (Fig. 36) and late-type spirals *Sd*, *Sm*; the major contributors to light in the short wavelength range are *A*-type stars. However, in the blue, around λ 4340 Å the spectral type derived from the strength of the *G*-band and Hγ, is about *F* 8, indicating the presence of *F* and *G* type stars which probably belong to the main sequence of Population I.

AF-spectra are generally observed in spirals of types *Sc*, *Sd* having little nuclear concentration of light, i.e. a small spherical component; their mean type is *F*0 to *F*2 near λ 4000 Å and *F*8 near λ 4340 Å. The stellar population is similar to that of the more advanced types, but the proportion of *A*-type stars must be less and the main contribution to blue-violet light comes from *F*-type stars. *Sc* spirals with a stronger nuclear concentration of light give *F*-type spectra and intermediate *Sb* spirals give *FG*-type spectra. This, of course, indicates an increasing proportion of *F*- and *G*-type stars. The early spirals of types *Sa* and *Sb* with a prominent spherical component and most lenticulars and ellipticals give *G*- and *K*-type spectra.

A rather unexpected result is that the *K*-type systems present spectral characteristics indicating that in light of short wavelengths the main contributors are *K* giants of Population I and not main sequence dwarfs or Population II giants as was generally assumed in recent years. This was clearly demonstrated by MORGAN through a study of the cyanogen bands near λ 4200 Å in spectra of M 31; furthermore, the presence of the TiO bands near λ 5900 and 6200 Å discloses the presence of a substantial proportion of *M*-Type giants. The conclusion is that most of the light of the nuclear regions of M 31 comes from stars of types *gG* 8 to *gK* 3 in the violet, from stars of types *gK* 4 or *gK* 5 in the blue and from *M*-type giants in the red and infra-red; nevertheless, some features in the ultra-violet suggest also the presence of a sizable population of ordinary main sequence dwarfs of types *G*0 to *G*5. The inferred H-R diagram of the central regions of M 31 is outlined in Fig. 35; it is similar to the diagram of an "old" Population I, such as that of "evolved" open clusters, and is distinctly different from that of a typical Population II, such as on globular clusters.

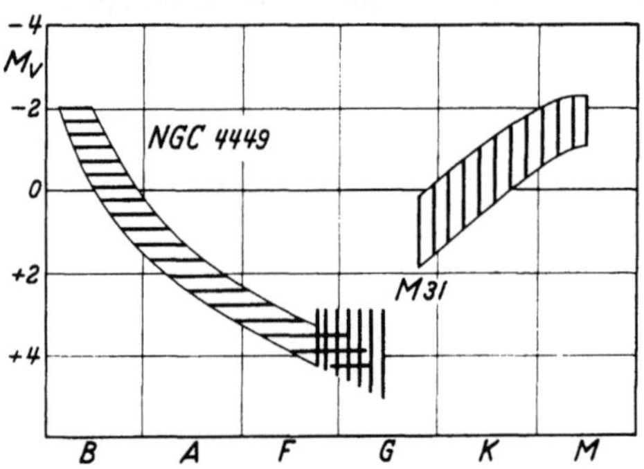

Fig. 35. Schematic H-R diagrams of M 31 and NGC 4449, after MORGAN and MAYALL.

The central regions of M 33 show considerable differences with respect to M 31. The Balmer lines of hydrogen are very prominent and, according to MAYALL, the small central nucleus gives an *A*-type spectrum, while the surrounding regions give a *F*-type spectrum near λ 4000 Å and a *G* 0 spectrum near λ 4340 Å; this indicates a large proportion of main sequence *A* and *F* stars that do not appear in M 31.

The magellanic irregular NGC 4449 shows also very strong ultra-violet Balmer lines corresponding to type *A* 7, while near λ 4340 Å the type is about *F* 5; to this absorption spectrum are superimposed the strong emission lines characteristic of H II regions, including λ 4959 and 5007 Å or [O III] and 3727—29 of [O II]. This leads to the schematic H-R diagram shown in Fig. 35 which includes many *B* stars and a heavily populated main sequence in the *A* and *F* types, similar to that of a "young" Type I population, exemplified by the expanding associations in the Galaxy.

An unpublished study of the spectrum of the brightest region of the bar of the Large Magellanic Cloud by the writer at Mt. Stromlo (1956) leads to similar conclusions: the spectral type indicated by the ultraviolet Balmer lines and the *K* line of Ca II is *B* 5 to *A* 5, while the strength of H γ and that of the *G*-band corresponds to *F* 5 to *F* 8; of interest is the strength of the 3820 Å line of He I indicating a substantial contribution of *B* 0 to *B* 5 stars in the ultra-violet; however, the only emission observed in a section clear of definite H II regions is λ 3728 Å of [O II]; because of the low surface brightness of the system superimposed emission bands of the airglow are faintly visible (Fig. 36).

An effect of considerable interest noted by MORGAN in spectra of medium dispersion (150 Å/mm) is the probable existence of a correlation between the widths of the spectral lines and the absolute magnitudes of galaxies giving a *K*-type spectrum. It has been known for a long time that the absorption lines in the spectrum of M 31 are broader and more diffuse than in the spectrum of M 32; MORGAN noticed, further, that the spectral lines of the giant ellipticals NGC 4486 (M 87) and NGC 4649 (M 60) in the Virgo cluster are still wider than

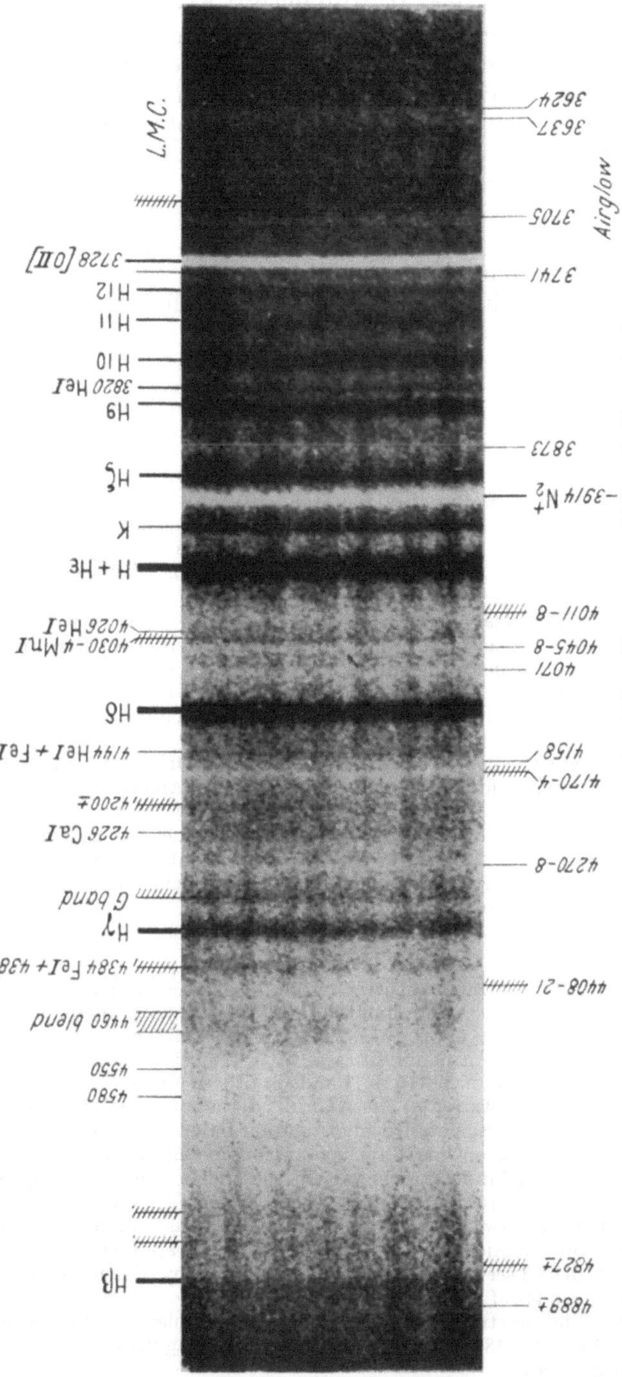

Fig. 36. Spectrum of Large Magellanic Cloud (G. DE VAUCOULEURS, Mt Stromlo Observatory).

the corresponding lines in the spectra of M 31 and M 81 and that the width of the lines decreases as one considers elliptical galaxies of decreasing luminosity. Again, spectra of NGC 4594 and M 31 show clearly the greater width of the lines in the former.

In an original investigation, R. MINKOWSKI[1] has estimated the velocity dispersion in the central regions of M 31 and M 32 through a comparison of their spectra with stellar spectra secured through "diffuse" slits of Gaussian profiles corresponding to various values of the velocity dispersion. The spectrum of M 31 in the range λ 3900 to 4400 Å corresponds to that of a star of spectral type $G8$ affected by a velocity dispersion of the order of 225 km/sec; the spectrum of M 32 to $K0$ and 100 km/sec.

This effect which, through the virial theorem, is clearly associated with the different masses of the systems (cf. Sect. 16), opens up a promising avenue for the spectroscopic determination of either the masses themselves or of elements likely to be correlated with them, such as the absolute magnitudes.

Bibliography.

General references.

[*A*] CURTIS, H. D.: The Nebulae. In Handbuch der Astrophysik, Vol. V/2, p. 833—936; Vol. VII, p. 550—563. Berlin: Springer 1933, 1936. — A General survey of the early work in the field with extensive bibliography up to 1935; includes finding lists of MESSIER, W. HERSCHEL and J. HERSCHEL numbers, tabular descriptions of various classification systems and a list of published reproductions of drawings and photographs of nebulae.

[*B*] HUBBLE, E. P.: The Realm of the Nebulae. Oxford: University Press 1936. — The standard semi-popular account of the pre-war Mt. Wilson work with fairly complete references to modern American sources up to 1935.

[*C*] SHAPLEY, H.: Galaxies. Philadelphia: Blakiston Co. 1943. — A popular account with emphasis on the Harvard work, especially on the Magellanic Clouds and nearer galaxies. No references to original papers.

[*D*] VOGT, H.: Die Spiralnebel. Heidelberg: C. Winter 1946. — A thorough semi-popular summary with very complete references, mainly to American and German sources, up to 1944.

[*E*] DE VAUCOULEURS, G.: L'exploration des galaxies voisines par les méthodes optiques et radio-électriques. Paris: Masson 1958. — A semi-popular account of optical and radio studies of nearby galaxies with emphasis on southern hemisphere studies at Sydney and Mount Stromlo. No references to original papers.

Special references.

[*1*] BAADE, W., and N. U. MAYALL: Distribution and motions of gaseous masses in spirals, in Problems of Cosmical Aerodynamics (Paris Symposium 1949), Central Air Documents Office, Dayton, 1951, pp. 165—184. — A preliminary report on the Mt. Wilson and Lick work on emission nebulosities in M 31 and other large spirals, occurrence of λ 3727 emission, rotation of M 31, M 33 and six other spirals.

[*2*] BABCOCK, H. W.: The rotation of the Andromeda nebula. Lick Obs. Bull. **19**, No. 498, 41—51 (1939). — The first detailed investigation based on both absorption lines in the central region and emission lines in a few gaseous nebulosities.

[*3*] BELZER, J., G. GAMOW and G. KELLER: On the stellar dynamics of spherical galaxies. Astrophys. Journ. **113**, 166—180 (1951). — Model of a spherical ($E0$) galaxy based on HUBBLE's luminosity law and an assumed constant mass/luminosity ratio.

[*4*] BIGAY, J. H.: Introduction à la photométrie photographique des nébuleuses extragalactiques. J. des Observ. **34**, 89—104 (1951). — Survey of methods used in measuring integrated magnitudes of galaxies.

[*5*] BIGAY, J. H.: Photométrie photographique des nébuleuses extragalactiques. Ann. d'Astrophys. **14**, 319—382 (1951). — Photographic magnitudes by Fabry photometry of 175 bright galaxies.

[*6*] BIGAY, J. H.et al.: Photométrie photoélectrique des nébuleuses extragalactiques. Ann. d'Astrophys. **16**, 133—138 (1953).

[*7*] BIGAY, J. H., et R. DUMONT: Déterminations photoélectriques de magnitudes globales de nébuleuses extragalactiques (2ème série). Ann. d'Astrophys. **17**, 78—84 (1954).

[1] See Annual Report of the Director of the Mt. Wilson and Palomar Observatories for 1953/54, p. 26.

[*8*] Bruggencate, P. ten: Die Helligkeitsverteilung im Innern elliptischer Nebel. Z. Astrophys. **1**, 275—299 (1930). — Re-discussion of Hubble's data on frequency of flattening and luminosity distribution in ellipticals. See corr. in Z. Astrophys. **2**, 83 (1930).

[*9*] Curtis, H. D.: Descriptions of 762 Nebulae and Clusters photographed with the Crossley reflector. Publ. Lick Obs. **13**, Part I (1918). — Includes a description of the "Φ-type" (p. 12 and Plate I).

[*10*] Curtis, H. D.: A Study of Occulting Matter in the Spiral Nebulae. Publ. Lick Obs. **13**, Part II (1918). — Includes many photographs of tilted spirals and "spindles" showing dark markings; mainly of historical interest as the first systematic study of absorption effects in galaxies.

[*11*] Danver, C. G.: The inclination of the galaxies towards the tangential plane. Medd. Lund Obs. (I) **1940**, No. 157. — A brief discussion of tilt criteria.

[*12*] Danver, C. G.: A Morphological Investigation of some Near Galaxies. Ann. Lund. Obs. **1942**, No. 10. — A very complete discussion of the orientations and shapes of 202 bright galaxies with a catalogue of descriptive parameters for 98 spirals.

[*13*] Dennison, E. W.: Isophotometry of extragalactic nebulae. Astronom. J. **59**, 320, 348 (1954) (Abstract) and Univ. Michigan Thesis (private communication).

[*14*] Elvius, Mrs. A.: A polarigraphic study of the spiral nebula NGC 5055 (M 63). Stockholm Obs. Ann. **17**, No. 4 (1951). — The first detailed investigation of polarisation effects in a spiral; it is restricted to the innermost, bright, ring-like structure.

[*15*] Elvius, Mrs. A.: Polarization of light in the spiral galaxy NGC 7331 and the interpretation of interstellar polarization. Stockholm Obs. Ann. **19**, No. 1 (1956). — A more detailed discussion of polarisation in the central regions of an ordinary spiral.

[*16*] Elvius, Mrs. A.: Diffraction of light by interstellar particles in spiral galaxies. Stockholm Obs. Ann. **18**, No. 9 (1956). — A detailed theoretical discussion of absorption and diffraction effects in spirals with a photometric analysis of NGC 4216.

[*17*] Evans, D. S.: Surface photometry of southern elliptical nebulae. Monthly Notices Roy. Astronom. Soc. London **111**, 526—536 (1951); **112**, 606—613 (1952). — Rough photographic isophotes of central regions of objects south of declination $-30°$ that are brighter than $m_{pg} = 12.0$ and are classified as ellipticals in the Shapley-Ames Catalogue [*68*].

[*18*] Fricke, W.: Die Intensitäts- und Farbverteilung im Andromeda-Nebel. Z. Astrophys. **34**, 137—167 (1954). — Luminosity profiles (U, B, V) along a number of cross sections in M 31; interpretation in terms of pure absorption effects.

[*19*] Haro, G.: Preliminary note on blue galaxies with emission lines. Bol. Obs. Tonantzintla y Tacubaya **1956**, No. 14, 8—18. — Method of detection for galaxies with strong UV emission by means of three filters.

[*20*] Holmberg, E.: A photographic photometry of the spiral nebulae NGC 891 and NGC 3623. Medd. Lund Obs. (II) **1945**, No. 114. — Approximate luminosity profiles (B, V) of a number of cross sections parallel to minor axis; definition of asymmetry; integrated magnitudes.

[*21*] Holmberg, E.: On the apparent diameters and the orientation in space of extragalactic nebulae. Medd. Lund Obs. (II) **1946**, No. 117. — Experimental study of measuring errors in micrometer photographic dimensions and statistical discussion of several catalogues (Harvard, Heidelberg).

[*22*] Holmberg, E.: On the absorption in the spiral nebulae. Medd. Lund Obs. (II) **1947**, No. 120. — A very thorough theoretical and empirical discussion of (pure) absorption effects for various models of galaxies with or without nuclear condensation.

[*23*] Holmberg, E.: A photometric study of nearby galaxies. Medd. Lund Obs. (II) **1950**, No. 128. — Dimensions and integrated magnitudes (P, V) of 40 bright galaxies in the Local Group and in the M 51 and M 81 groups; an attempt to determine the dimensions and luminosity functions for a representative sample.

[*24*] Holmberg, E.: On the Masses and Luminosities of Extragalactic Nebulae. Medd. Lunds Obs. (I) **1952**, No. 180. — An attempt to correlate mass/luminosity ratio with colour index and to derive the mass function of the nearby galaxies discussed in [*23*].

[*25*] Hubble, E.: Extragalactic nebulae. Astrophys. Journ. **64**, 321—369 (1926). — A classical paper including the original description of the standard classification system and a preliminary analysis of the statistical properties of the "extragalactic nebulae" as external galaxies. It is now mainly of historical interest.

[*26*] Hubble, E.: Distribution of luminosity in elliptical nebulae. Astrophys. Journ. **71**, 231—276 (1930). — A classical paper giving the photographic luminosity profiles of 15 ellipticals (some now known as lenticulars) and the formulation of Hubble's standard luminosity law. Includes a useful table of the truncated isothermal distribution.

[27] HUBBLE, E.: The direction of rotation in spiral nebulae. Astrophys. Journ. **97**, 112—118 (1943). — Includes a survey of tilt criteria based on absorption effects and a discussion of a number of critical cases indicating "trailing" arms; has a systematic tabulation of all tilt and rotation data up to 1943.

[28] HUMASON, M. L., N. U. MAYALL and A. R. SANDAGE: Redshifts and magnitudes of extragalactic nebulae. Astronom. J. **61**, 97—162 (1956) = Mt. Wilson and Palomar Obs. Rep. No. 181 = Lick Obs. Bull. No. 542. — A fundamental paper including all red-shift data to date for over 800 galaxies observed at Mt. Wilson-Palomar (620 objects) and Lick (300 objects) between 1935 and 1955, or 63% of the Shapley-Ames objects brighter than $m_H = 13.0$, north of declination $-30°$. The Mt. Wilson list includes spectral types, the Lick list has notes on spectral and other peculiarities. Includes a list of integrated P magnitudes for 576 bright galaxies derived by SANDAGE from PETTIT's and STEBBINS and WHITFORD's photoelectric observations [56], [70], [72]. Provisional derivation of the Hubble redshift parameter $H = 180$ km/sec per Mpc. which defines the distance scale in current use beyond the Local Group and nearby groups.

[29] HOUTEN, C. J. VAN, J. H. OORT and W. A. HILTNER: Photoelectric measurements of extragalactic nebulae. Astrophys. Journ. **120**, 439—453 (1954). — Luminosity profiles (U, B, V) along main axes and additional cross sections of 9 galaxies (2 ellipticals, 3 lenticulars, 4 early spirals) from direct recordings; of interest as the first successful attempt to by-pass photographic techniques.

[30] KERR, F. J., and G. DE VAUCOULEURS: Rotation and other Motions of the Magellanic Clouds from Radio Observations. Austral. J. Phys. **8**, 508—522 (1955). — Discussion of radial velocity observations secured in the first survey of 21 cm line emission in external galaxies; determination of rotation curves in both Clouds; comparison with optical velocities.

[31] KERR, F. J., and G. DE VAUCOULEURS: The Masses of the Magellanic Clouds frome Radio Observations. Austral. J. Phys. **9**, 90—111 (1956). — Derivation of masses from rotation curves obtained in [30] with a discussion of various methods and comparison of their results. Includes a survey of mass determinations for other galaxies and a brief discussion of possible evolutionary trends along the classification sequence.

[32] LINDBLAD, B.: The orientation of the planes of spiral nebulae inferred from the dark lanes of occulting matter. Ark. Mat., Astronom. Fys., Ser. A **24**, No. 21 (1934) = Stockholm Obs. Medd. No. 14. — Proposes an interpretation of the distribution of dark matter such that the obscured side of a tilted spiral is the far side, leading to a direction of rotation opposite to that advocated by SLIPHER and CURTIS. The first of a long series of papers in support of a theoretical prediction that spiral arms are "leading" in the rotation.

[33] LINDBLAD, B.: On the interpretation of spiral structure in the nebulae. Astrophys. Journ. **92**, 1—26 (1940) = Stockholm Obs. Medd. No. 56. — Summary of the theory of spiral structure developed by the author since 1926 and discussion of tilt criteria favoring the predicted direction of rotation.

[34] LINDBLAD, B.: On the distribution of light-intensity and colour in the spiral nebula NGC 7331. Stockholm Obs. Ann. **13**, No. 8 (1941). — Luminosity and colour (U, B, V) profiles along the main axes and interpretation of the asymmetry to show that the obscured side is the near side.

[35] LINDBLAD, B.: On the absorption of light in the central regions of the spiral nebula NGC 7331 and related subjects. Stockholm Obs. Ann. **14**, No. 3 (1942). — A more elaborate discussion of luminosity and colour asymmetry in spirals; introduces the idea of diffraction besides absorption effects by interstellar particles.

[36] LINDBLAD, B.: On the dynamical theory of spiral structure. Stockholm Obs. Ann. **15** No. 4 (1948). — An elaboration of the earlier theoretical work of the author; includes a discussion of inner and outer spiral structure in NGC 2681 and an attempt to "open" the spiral pattern of NGC 4565, both in support of the view that the arms are "leading".

[37] LINDBLAD, B.: On the run of the spiral arms and the direction of rotation in NGC 4594. Publ. Astronom. Soc. Pacific **63**, 133—136 (1951) = Stockholm Obs. Medd. No. 69. — An attempt to "open" the spiral structure through a restitution of the distribution in the plane of the spiral of bright knots measured on a 200-inch UV photograph. The plot is interpreted so as to give "leading" arms, but the opposite interpretation is just as plausible (cf. [91]).

[38] LINDBLAD, B.: On a barred spiral structure in the Andromeda nebula. Stockholm Obs. Ann. **19**, No. 2 (1956). — Restitution of the structure of M 31 in its own plane and comparison with the typical *SBb* spiral NGC 7723. An excellent discussion of what appears to be a transition structure between the typical *SA(s)b* and *SAB(s)b* types of the revised classification scheme. The interpretation of the pattern in support of the "leading" arms is, however, less convincing.

[*39*] LINDBLAD, B., and R. BRAHDE: On the direction of rotation in spiral nebulae. Astrophys. Journ. **104**, 211—225 (1946) = Stockholm Obs. Medd. No. 58. — A critical discussion of the objects selected by HUBBLE in [*27*] as indicating unequivocally that spiral arms are "trailing" in the rotation of all spirals, with a review of tilt criteria involving the distribution of novae, globular clusters, etc, in M 31.

[*40*] LINDBLAD, B., and J. DELHAYE: On the distribution of light-intensity and colour in the spiral nebula Messier 63. Stockholm Obs. Ann. **15**, No. 9 (1949). — Luminosity and colour profiles (*B*, *V*, *R*) along several cross sections and interpretation of the asymmetry in terms of a distribution of dark matter such that the far side is fainter and redder, hence that the spiral arms are "leading" in the rotation. This discussion applies only to the innermost, bright region of the spiral.

[*41*] LINDBLAD, B., and R. G. LANGEBARTEL: On the dynamics of stellar systems. Stockholm Obs. Ann. **17**, No. 6 (1953). — A theoretical discussion of barred spirals and transition types; see especially pp. 58—59 and Plates I, II illustrating *SAB*(*s*) and *SAB*(*rs*) types of the revised classification scheme.

[*42*] LOHMANN, W.: Die Masse des Andromeda- und Dreieck-Nebels. Z. Astrophys. **35**, 159—164 (1954). — Mass estimates based on an assumed force curve after BOTTLINGER. The published value for M 31, viz. $3.3 \cdot 10^{11}$ suns, is superseded by the result of a least-square solution of the same data, viz. $(2.7 \pm 0.9) \cdot 10^{11}$ suns (private communication).

[*43*] LUNDMARK, K.: A preliminary classification of nebulae. Ark. Math., Astronom. Fys., Ser. B **19**, No. 8 (1926) = Medd. Obs. Upsala No. 7. — Classification by shape and concentration; introduces the Magellanic Cloud type.

[*44*] LUNDMARK, K.: Studies of anagalactic nebulae. Nova Acta Reg. Soc. Sci. Upsala, vol. extraord. **1927** = Medd. Obs. Upsala, No. 30. — A comprehensive survey of the early studies of non-galactic nebulae and a summary of the author's work, mainly statistical, at Heidelberg, Greenwich, Mt. Hamilton and Wt. Wilson. This important volume (124 pp. and 10 plates) is now valuable mainly for its historical interest.

[*45*] MACHIELS, A.: Sur la répartition des aplatissements des nébuleuses elliptiques. Bull. Astr. Paris (2) **6**, 317—324 (1930). — A rediscussion of HUBBLE's data given in [*25*].

[*46*] MACHIELS, A.: Classification des nébuleuses extragalactiques par leurs formes. Bull. Astr. Paris (2) **6**, 405—416 (1930). — A discussion of the Harvard data on the Virgo cluster (H.A., 88, Nol).

[*47*] MACHIELS, A.: Classification, formes et orientations des nébuleuses extragalactiques Bull. Astr. Paris (2) **9**, 471—478 (1933). — A discussion of the Shapley-Ames Catalogue [*68*] and other Harvard lists.

[*48*] MAYALL, N. U.: The occurrence of λ 3727 [OII] in the spectra of extragalactic nebulae. Lick Obs. Bull. **19**, No. 497, 33—39 (1939). — The first systematic discussion of the frequency of occurrence of line emission in galaxies; see also Publ. Astronom. Soc. Pacific **51**, 282—286 (1939).

[*49*] MAYALL, N. U., and L. H. ALLER: The rotation of the spiral nebula Messier 33. Astrophys. Journ. **95**, 5—23 (1942) = Contrib. Lick Obs. (II) No. 1. — A classical paper reporting on rotational velocities of 20 emission line objects with a derivation of a mean rotation curve and a brief comparison with rotation data in other galaxies. See discussion of the data by WYSE and MAYALL in [*86*].

[*50*] MAYALL, N. U.: Comparison of rotational motions observed in the spirals M 31 and M 33 and in the Galaxy, in: The Structure of the Galaxy. Publ. Obs. Univ. Michigan **10**, 19—24 (1951). — A survey of the Lick work on rotational motions in spirals, with a progress report on current work on M 31, supplementing an earlier report by BAADE and MAYALL [*1*]. Includes reproductions of spectra of NGC 5005 and 5055 showing inclined lines; see also MAYALL, Sky and Telescope **8**, 3—5, 17 (1948).

[*51*] OHMAN, Y.: A polarigraphic study of obscuring clouds in the great Andromeda nebula M 31. Stockholm Obs. Ann. **14**, No. 4 (1942). — The first successful attempt to detect polarisation in external galaxies; the observations refer to small isolated clouds in the central bulge of N 31 and are interpreted to support the conclusion that the obscured side is the far side of the nebula and that the arms are "leading" in the rotation. The interpretation has been changed since 1948 without changing the conclusion, however.

[*52*] OORT, J. H.: Some Problems concerning the structure and dynamics of the galactic system and the elliptical nebulae NGC 3115 and 4494. Astrophys. Journ. **91**, 273—306 (1940). — An important paper including data on luminosity distribution in two ellipticals (*E*0, *E*7) and a preliminary discussion of the dynamics of NGC 3115; see the rediscussion by SCHWARZSCHILD in [*63*].

[*53*] PATTERSON, F. S.: The luminosity gradient of M 33. Harvard Bull. **914**, 9—10 (1940). — Mean luminosity profile (*Pg*) indicating exponential decrease outside the nuclear region.

[*53a*] PATTERSON, F. S.: Surface photometry of external galaxies. Harvard University Thesis (unpublished), May 1941. — This fundamental and unfortunately unpublished

paper was received in private communication from its author (now Mrs. JONES) after the present chapter had been completed. It gives a very thorough study of the luminosity distribution in 123 galaxies, mainly in the Virgo cluster, from plates taken with the 12-inch Metcalf refractor of the Oak Ridge Station.

[54] PEREK, L.: Distribution of Mass in the Galactic system. Contrib. Astr. Inst. Masaryk Univ., Brno **1**, No. 6 (1948). — An important paper introducing non-homogeneous spheroids in the interpretation of rotational velocity curves of galaxies; general theory, numerical data for special density laws; application to the Galaxy. See table and nomogram in [55].

[55] PEREK, L.: Table for Computing the Potential and Attractive Force of Spheroids. Bull. Astr. Inst. Czechosl. **2**, No. 5, 75—79 (1950). — Summary of formulae developed in [54]; table and nomogram for non-homogeneous spheroids of mass density law $\varrho = \varrho_c (1 - m^2)^n$.

[56] PETTIT, E.: Magnitudes and color indices of extragalactic nebulae determined photoelectrically. Astrophys. Journ. **120**, 413—438 (1954). — An important list of magnitudes (P, V) and colours for 558 galaxies observed for redshift at Lick and Mt. Wilson-Palomar. A list of integrated magnitudes of bright galaxies partly based on PETTIT's data is given by SANDAGE in [28].

[57] RANDERS, G.: A note on the evolution of extragalactic nebulae. Astrophys. Journ **92**, 235—246 (1940) = Contrib. Mt. Wilson Obs. No. 634. — A preliminary discussion of ring structures in spirals.

[58] REAVES, G.: Dwarf galaxies in the Virgo cluster. Astronom. J. **61**, 69—76 (1956) = Lick Obs. Bull. No. 540 and Univ. of California Thesis, 1953 (private communication). — Discussion of dwarf galaxies of low surface brightness; description of the IC 3475 type.

[59] REDMAN, R. O.: Photographic photometry of elliptical nebulae. Monthly Notices Roy. Astronom. Soc. London **96**, 588—604 (1936). — Luminosity profiles of 6 objects; derivation of integrated magnitudes; discussion of technical problems involved.

[60] REDMAN, R. O., and E. G. SHIRLEY: Photometry of the Andromeda nebula M 31. Monthly Notices Roy. Astronom. Soc. London **97**, 416—423 (1937). — Luminosity profiles and integrated magnitude.

[61] REDMAN, R. O., and E. G. SHIRLEY: Photographic photometry of the elliptical nebulae (Second Paper). Monthly Notices Roy. Astronom. Soc. London **98**, 613—623 (1938). — Luminosity profiles of 9 objects; comparison of integrated magnitudes with HUBBLE's results and the Harvard (Shapley-Ames) data.

[62] REINMUTH, K.: Die Herschel-Nebel. Veröff. Sternw. Heidelberg **1926**, No. 9. — A major photographic survey of the General Catalogue nebulae north of declination $-20°$; includes data on dimensions, type in WOLF's system and condensed description of 4200 objects.

[63] SCHWARZSCHILD, M.: Mass distribution and mass-luminosity ratio in galaxies. Astronom J. **59**, 273—284 (1954). — A rediscussion of observational data on the rotation of M 31, M 33, NGC 3115, M 32 to test the assumption of identical space distribution of mass and luminosity. Includes additional evidence from velocity dispersion in clusters on the average mass of ellipticals and spirals.

[64] SEYFERT, C. K.: The distribution of color in spirals. Astrophys. Journ. **91**, 528—545 (1940). — Luminosity and color (P, V) profiles of 5 large spirals indicating a background of uniform colour between the bluer arms.

[65] SEYFERT, C. K.: Nuclear emission in spiral nebulae. Astrophys. Journ. **97**, 28—40 (1943) = Mt. Wilson Contr. 671. — Discussion of broad emission lines in the nuclei of some early-type spirals.

[66] SHAPLEY, H.: The angular diameters of bright galaxies. Harvard Ann. **88**, No. 4 (1934). Micrometric diameters of 447 bright galaxies measured on long-exposure photographs taken with the 24-inch Bruce refractor.

[67] SHAPLEY, H.: Note on the comparative diameters of spheroidal and spiral galaxies. Proc. Nat. Acad. Sci. U.S.A. **28**, 186—191 (1942) = Harvard Rep. 238. — Diameters of 112 bright galaxies from microphotometer tracings of long-exposure photographs; details of this work have not been published yet; see ref. [53a].

[68] SHAPLEY, H., and A. AMES: A survey of the external galaxies brighter than the 13-th magnitude. Harvard Ann. **88**, No. 2 (1932). — An important survey and reference catalogue of 1249 bright galaxies including dimensions and types in HUBBLE's standard system from a variety of sources and photographic magnitudes from homogeneous small-scale, in focus plates. For a provisional revision of types see [81] and for a discussion of the magnitude system see [82], [83].

[69] SHCHEGOLEV, D. E.: Photometritcheskoe Issledovanie 15 spiralnikh galaktik. Izvestia Glav. Astr. Obs. Pulkovo **20**, No. 156, 87—109 (1956). — An important investigation

of photometric profiles and colours (*P*, *V*) of 15 bright spirals on an absolute magnitude scale; see abstract in Astronom. J. USSR. **32**, 16 (1955).

[*70*] Stebbins, J., and A. E. Whitford: Photoelectric magnitudes and colors of extragalactic nebulae. Astrophys. Journ. **86**, 247—273 (1937) = Contrib. Mt. Wilson Obs. 577. — An important list of magnitudes and colours of 165 bright galaxies measured through a series of diaphragms of small aperture; the measured magnitudes are far short of the integrated magnitude but provide the basic material for the standardization of other series; see discussion in [*82*], [*83*].

[*71*] Stebbins, J., and A. E. Whitford: The colors of the extragalactic nebulae. Astrophys. Journ. **108**, 413—428 (1948) = Contrib. Mt. Wilson Obs. 753. — Six-colour photometry of 8 bright galaxies providing data on the spectral energy curve from λ 3650 to λ 10000 Å. Includes a discussion of the colour-excess of distant galaxies.

[*72*] Stebbins, J., and A. E. Whitford: Magnitudes and colors of 176 extragalactic nebulae. Astrophys. Journ. **115**, 284—291 (1952) = Mt. Wilson-Palomar Rep. No. 64. — A very important series of photometric data for the standardization of other series; see discussion in [*82*], [*83*].

[*73*] Sytinskaja, N. N.: Absolute photometry of extragalactic nebulae NGC 205 and 221. Astronom. J. USSR. **20**, 54—57 (1943). — Luminosity profiles (*P*) on an absolute magnitude scale.

[*74*] Vashakidze, M. A.: Determination of the degree and plane of polarisation of the light of extragalactic nebulae. Bull. Abastumani Astrophys. Obs. **1955**, No. 18, 15—28. — Polarisation measurements in many points along a number of cross sections in 13 bright galaxies; a measurable degree of intrinsic polarisation is detected in spirals and magellanic irregulars.

[*75*] Vaucouleurs, G. de: Introduction à l'analyse microphotométrique des nébuleuses extragalactiques. J. des Observ. **31**, 113—128 (1948) = Contrib. Inst. Astrophys. Paris No. B-23. — Survey of earlier work and discussion of the technical problems of the detailed photographic photometry of galaxies.

[*76*] Vaucouleurs, G. de: Recherches sur les nébuleuses extragalactiques. I. Sur la technique de l'analyse microphotométrique des nébuleuses brillantes. Ann. d'Astrophys. **11**, 247—287 (1948) = Contrib. Inst. Astrophys. Paris No. B-26. — Includes a detailed discussion of sources of error and methods of correction in the surface photometry of bright galaxies; photometric profiles of 3 ellipticals and one spiral; comparison with the results of Hubble, Oort and Redman. Definition of the "effective" dimensions; standard law of luminosity distribution for ellipticals. See also abstracts in C. R. Acad. Sci. Paris **226**, 1692—1694; **227**, 548—550 (1948).

[*77*] Vaucouleurs, G. de: Orientation spatiale et sens de rotation de la nébuleuse spirale NGC 2146. Ann. d'Astrophys. **13**, 362—366 (1950) = Contrib. Inst. Astrophys. Paris No. B-57. — Discussion of tilt and direction of rotation of a three-armed spiral suggesting that the arms are "trailing". See abstract in C. R. Acad. Sci. Paris **231**, 32—33 (1950).

[*78*] Vaucouleurs, G. de: On the distribution of mass and luminosity in elliptical galaxies. Monthly Notices Roy. Astronom. Soc. London **113**, 134—161 (1953). — Summary of best available data on luminosity distribution law in ellipticals; description of semi-empirical model of globular galaxy, based on the truncated isothermal sphere, accounting for luminosity distribution and other observational data. This model, based on early data on Type II populations, is superseded by more recent works.

[*79*] Vaucouleurs, G. de: Extragalactic studies in the southern hemisphere. Occasional Notes Roy. Astronom. Soc., London **3**, No. 18, 118—142 (1956). — Mainly historical, but includes otherwise unpublished luminosity profiles of NGC 1291 and isophotes of NGC 55; also tabular data on brightest southern galaxies and 10 photographs of 9 objects. Useful as an introduction to [*80*].

[*80*] Vaucouleurs, G. de: A survey of bright galaxies south of $-35°$ declination with the 30-inch Reynolds reflector (1952—1955). Mem. Commonwealth Obs. **1956**, No. 13. — An extensive photographic survey of bright southern galaxies (210 Shapley-Ames objects, 120 NGC and IC objects, 130 uncatalogued objects). Includes a preliminary description of the revised classification scheme and notation system, statistical discussions of type and dimension data, axial ratios, ring-structures as geometric distance indicators, etc. Detailed tables of revised types and dimensions; with 8 plates giving photographs of 44 galaxies.

[*81*] Vaucouleurs, G. de: A provisional revision of the Harvard Survey of bright galaxies (Shapley-Ames Catalogue). Austral. Nat. Univ. Mimeogram **1953**. — Includes revised data on types in Hubble's system, dimensions and corrected magnitudes for 1250 bright galaxies from published sources available in 1952 with individual references. The corrected (total) magnitudes are derived from the original Harvard magnitudes according to the principles discussed in [*82*], [*83*]. A final revision is in preparation

including the material in [80] and other Mt. Stromlo data for the southern zone together with more recent Lick and Mt. Wilson-Palomar data for the northern zone.

[82] VAUCOULEURS, G. DE: Photographic magnitudes of the brighter external galaxies. Astronom. J. **61**, 430—437 (1956). — Discussion of 10 basic photographic and photoelectric series of nebular magnitudes; derivation of standard total P magnitudes of 100 bright galaxies; discussion of the Harvard magnitudes in the Shapley-Ames catalogue [68]; derivation of correction formulae used in [81]; statistical discussion of completeness of Shapley-Ames Catalogue.

[83] VAUCOULEURS, G. DE: Etudes sur les galaxies brillantes, I: Magnitudes photographiques. Ann. Obs. du Houga **2**, Part I (1957). — This is a more complete discussion of the questions summarized in [82] with full details, tables, etc. Comparison with the integrated magnitudes of SANDAGE [28] indicates excellent agreement.

[84] WHIPPLE, F.: The color and spectra of external galaxies. Harvard Obs. Circ. **1935**, No. 404. — The first systematic discussion of the effect of mixture of stellar types in the integrated colours and spectra of galaxies; interpretation of colour excess; computed typical line profiles.

[85] WHITFORD, A. E.: Photoelectric magnitudes of the brightest extragalactic nebulae. Astrophys. Journ. **83**, 424—432 (1936) = Contrib. Mt. Wilson Obs. No. 543. — The first photoelectric measurements of quasi-total magnitudes of 11 bright galaxies.

[86] WYSE, A. B., and N. U. MAYALL: Distribution of mass in the spiral nebulae Messier 31 and Messier 33. Astrophys. Journ. **95**, 24—47 (1942) = Contrib. Lick Obs. **2**, No. 2. — A fundamental paper including the theory and detailed tables for the derivation of mass distribution from rotational velocity curves in the thin disc approximation; with an application to the observations of BABCOCK on M 31 [2] and of MAYALL and ALLER on M 33 [49].

[87] DIETER, N. H.: Observations of neutral hydrogen in M 33. Publ. Astronom. Soc. Pacific **69**, 356 (1957). — Preliminary note on the detection of the 21 cm line in M 33; rotational velocity; mass estimate (10^{10} suns).

[88] HAZEN, M. L.: The distribution of intensity in elliptical galaxies of the Virgo cluster. Thesis, University of Michigan 1957. Includes important new data on the luminosity distribution in 24 galaxies derived from 36-inch and 48-inch Schmidt plates analyzed with the Williams and Hiltner isophotometer.

[89] HULST, H. C. VAN DE: Rotatie en waterstofdichtheid van de Andromeda Nevel afgeleid uit waarnemingen van de 21 cm lijn. Kon. Nederl. Akad. Wetensch., Amsterdam **65**, 157—160 (1956). — Preliminary communication of observations of the 21 cm line emission in M 31; intensity and velocity distribution; mass estimates: HI $= 1.9 \times 10^9$ suns, total mass $= 1.5 \times 10^{11}$ suns. Comparison with Galaxy.

[90] MORGAN, W. W., and N. U. MAYALL: A Spectral Classification of Galaxies. Publ. Astronom. Soc. Pacific **69**, 291—303 (1957). — A fundamental paper on the spectral classification of galaxies in the MK system; discussion of luminosity criteria in composite spectra; tentative H-R diagrams. See also the discussion of spectra of globular star clusters by MORGAN, in Publ. Astronom. Soc. Pacific **68**, 509—516 (1956).

[91] VAUCOULEURS, G. DE: Tilt criteria and direction of rotation of spiral galaxies. Astrophys. Journ. **127**, 487—503 (1958). — A review of the observational evidence on the directionof rotation of ordinary and barred spirals; discussion of tilt criteria, both classical and new. List of critical objects indicating that in all cases the arms are trailing in the rotation.

Multiple Galaxies.

By

F. Zwicky

With 11 Figures.

I. Historical.

1. Investigators of extragalactic objects have occupied themselves successively with the study of individual galaxies, multiple galaxies, clusters of galaxies and, most recently, with intergalactic matter. While all of these studies, in the past, have gone their individual ways, it is now not only appropriate but necessary to consider the interrelations and the transitions between the various forms of the large scale aggregations of matter. These aggregations are in fact not at all as distinct units as are for instance the stars and atoms. The decision of whether or not a specific individual star, dust cloud or gas cloud belongs definitely to any given galaxy, group of galaxies or to an intergalactic formation is indeed at the present state of our knowledge an almost impossible one to make. In order to shed some light on this problem, we therefore propose in this article to outline directives for the future investigation of multiple galaxies. For this purpose we shall sketch some possible morphological and structural aspects of characteristic double and multiple galaxies. Referring to the history of the subject at hand, it should be mentioned that good photographs of double galaxies were probably first obtained by F. G. Pease with the 60-inch reflector of the Mount Wilson Observatory. Pease published his results in two classic papers [*1*] in 1917 and 1920.

Knut Lundmark [*2*], during his early pioneering work on extragalactic nebulae was the first to point out the importance of the study of double galaxies for the evaluation of the absolute masses and of the luminosity function of stellar systems. Following him, E. Holmberg [*3*] continued and extended Lundmark's work in several important papers. Generally speaking, however, the study of multiple galaxies has been sadly neglected for several decades and it is only recently being reactivated through the work of F. Zwicky [*4*], [*5*] on double and multiple galaxies which are interconnected by faint luminous intergalactic formations. The main reason for the neglect of double galaxies as an object of study is to be found in the fact that all efforts of the large reflectors had to be concentrated on the observation of individual galaxies as well as of large clusters of galaxies. The latter, as is well known, served as stepping stones for the construction of the redshift-distance relation to remote regions of cosmic space.

A review of a large part of the knowledge available on multiple galaxies has recently been given by Zwicky [*5*]. In the present short account we shall only present those aspects of multiple galaxies which have not been treated in the Ergebnisse, to which the readers are herewith referred for additional information. Also, it should be mentioned that clusters of galaxies are treated separately in the following article in this volume.

II. Morphology of multiple galaxies.

2. A preliminary review of the structural and kinematic aspects of double galaxies leads to the recognition of the following types of significant features. We find in the first place that there exist double galaxies of considerable permanence whose components cannot escape from one another. On the other hand, two galaxies may temporarily appear to form a double system, although their relative velocities are high enough to lead to an ultimate and progressively increasing separation. Permanent multiple galaxies are *dynamic units* analogous to physical double or multiple stars.

The *components of permanent multiple galaxies* may be *clearly separated*, or they may be *coexisting*. Revealing information on two or more types of galaxies coexisting within the same volume of space has recently been obtained through the application of the *method of analytical composite photography* [*6*], [*7*]. The structural features revealed by this method have in fact led to the recognition not only of the universality of the phenomenon of coexistence of many different structural types of galaxies, but of the coexistence of different dynamic formations within gaseous nebulae of the type of the Crab Nebula [*8*] as well.

In many cases the components of permanent groups of multiple galaxies are clearly *interconnected* by luminous and probably also by dark intergalactic formations. Such connections were of course recognized very early and are clearly in evidence in many of the systems photographed originally by PEASE [*1*]. That there are many spectacular luminous *intergalactic formations which link very widely separated galaxies* and groups of galaxies was in recent years shown by ZWICKY's systematic discoveries and investigations [*4*], [*5*].

The multiple galaxies which are of a temporary nature and which cannot be considered as stationary dynamic units may be of the following characteristic types. In the first place the galaxies may pass around each other, the encounter starting from a state of complete separation and finishing again in a state of complete separation. There is a change of velocity and direction of motion of each of the galaxies with respect to a universal inertial system of reference [*9*], but only insignificant internal rearrangement takes place within the two galaxies. Secondly the encounter may take place at very close range with induced tidal effects, internal rearrangement of the structure of the galaxies. This case involves an appreciable loss of the total translational energy of the galaxies involved, with transfer of the lost energy to internal energy of these galaxies. Finally, the encounter may be so close or even head-on that either considerable disruption of both systems results, or perhaps even total mutual capture.

The general systematics of the morphological features of double and multiple galaxies will of course need many refinements through the consideration of the physical character of the individual galaxies which constitute the groups in question. Obviously, the results of close encounters will be quite different, depending on whether the galaxies involved contain much dust and free interstellar gas or if they are composed essentially only of stars. In the first case the disruption not only will be of a quite different overall geometrical character than in the second, but it may also lead to the generation of radio waves, cosmic rays and other radiations which will not occur during the collision of two purely stellar systems which do not contain any finite dispersed matter. Unfortunately, our observational data, at the present time, are entirely insufficient to allow many detailed statements about the physical conditions within interacting groups of galaxies. Most of the results which we present here are therefore merely indicative. They are not only incomplete, but in many cases quite uncertain and should be considered merely as starting points for a future thoroughly comprehensive program of investigation.

III. Permanent multiple galaxies.

3. Coexisting galaxies. Several individual galaxies obviously may form a permanent multiple dynamic system, if their relative velocities are too small to result in an ultimate separation. Actually the members of a permanent group can coexist, occupying the same volume of space, although this remarkable fact has not in the past been duly emphasized. In the case of our own galaxy, for instance, it has long been recognized that it is made up of sub-systems which, dynamically and historically seem to have gone through independent evolutions. Three of these subsystems are (1) the Milky Way system proper, which is thought to be a spiral galaxy, (2) the large globular system in which the spiral system is imbedded and whose outskirts reach far outside of the flattened spiral, and (3) the globular clusters, most of which are generally being associated with the globular subsystem of our galaxy. From their appearance and behaviour, we may, however, consider the globular star clusters as independent stellar systems, especially if statistical mechanical problems of the type relating to the theory of the luminosity function of galaxies are discussed [*3*], [*9*]. Globular clusters of stars are also known to be associated with many other galaxies, such as the *Sb* spiral Messier 31 (the Andromeda nebula), the more irregular Magellanic clouds, the globular galaxy NGC 4486 and dwarf galaxies such as the Fornax system [*10*].

Fig. 1. Coexistence of a blue normal and a yellow-green barred spiral galaxy within the brighter member (NGC 5194) of the Whirlpool nebula, Messier 51. The barred spiral and the "elliptical" companion NGC 5195 are indicated by solid line contours within which the dotted yellow-green population is distributed with remarkable uniformity. The blue normal spiral is sketched by means of the black shaggy spots.

Most remarkable among the coexisting galaxies is a discovery made with the help of the so-called analytical composite photography, recently developed by F. Zwicky [*6*], [*7*] for the study of the differently coloured luminous populations of stellar systems. An analysis of the Whirlpool nebula (Messier 51) by means of this type of photography revealed that, while the blue stars form a disrupted and very poorly organized spiral structure, the yellow-green stars have arranged themselves in a highly streamlined spiral with smoothly distributed surface luminosity [*9*]. Furthermore, this latter spiral is of the barred type, while the blue population structure is a normal spiral. In Fig. 1 we give a schematic representation of these startling facts.

Astronomers, until recently, following Hubble's classification [*11*], have assumed universally that normal spirals (*S*) and barred spirals (*SB*) represent two entirely different evolutionary sequences, and numerous theoretical investigations have attempted to explain these sequences. With the discovery that normal and barred spiral structures can coexist within the same galaxy, the whole problem of the structure and of the evolution of stellar systems must be viewed from an entirely new perspective.

As a consequence of the recognition of the possible coexistence of widely different structural types of galaxies one naturally suspects that such coexistence might be a rather general phenomenon. Indeed, a preliminary search indicates, that probably all of Hubble's original types [*11*] can coexist in any combination one may wish to choose. While this poses entirely new and unsuspected problems for the theoretical astronomers, it shows before all, that Hubble's classification of stellar systems must now be considered as rather obsolete in the sense that it will have to be replaced by a far more refined morphology of galaxies, if such a morphology is to serve as a sound observational basis for future theories. One of the first fascinating problems of such theories will be to analyse just how long two stellar systems of different structures can coexist without markedly influencing one another or without entirely loosing their individuality. One of the simplest problems in this category is that related to the perturbations which the various material components and the gravitational and electromagnetic fields of our own galaxy exert on the globular and on the open star clusters which are located within its confines. It will be worthwhile to search for observable effects of this type on the structure of the star clusters in the Milky Way system.

4. Clearly separated systems. Multiple systems which consist of clearly separated individual galaxies may be of two types. The structures of the individual component galaxies may more or less clearly reveal disturbances such as tidal actions caused by the proximity of the other galaxies in the system, or, no effects of this kind may be readily in evidence. In this latter case the proximity in space of two galaxies can be proved only through rather elaborate investigations, resulting in accurate distance determinations of the individual components of the suspected double galaxy. The present writer is not convinced that sufficient proof has been furnished in any particular case that any two optically close and structurally undisturbed galaxies are really close neighbors in space and that they actually form a dynamically stable unit.

Among the members of the Local Group of galaxies, for instance, Messier 31 (the great nebula in Andromeda), Messier 32 and NGC 205, 147 and 185 are supposed to constitute a dynamic unit. While some tidal distortions possibly exist in NGC 205 no effects of this kind are certainly discernible in the other four galaxies. The view that we here deal with a dynamic unit of five stellar systems therefore derives its main support from the fact that the distances of the individuals, as so far determined [*12*], seem to be roughly the same. The small dispersion in radial velocities indicates that the systems cannot escape from one another, although this conclusion might of course be vitiated by large tangential components of the velocities. The radial velocities for four of the galaxies, as given by Humason [*13*] are $V_s = c \cdot \Delta\lambda/\lambda = -266$, -214, -239 and -266 km/sec for Messier 31, Messier 32, NGC 205 and NGC 185 respectively.

(c = velocity of light, $\Delta\lambda$ = displacement of a spectral line whose unshifted wavelength is λ).

A similar problem exists with respect to some of the small stellar systems on the far outskirts of the Milky Way system. Again, the globular cluster NGC 2409 and the dwarf Capricorn stellar system, of the type of a medium open cluster, may be independent dwarf galaxies, or their velocities may be so small that together with our own galaxy they constitute a dynamically stationary multiple galaxy.

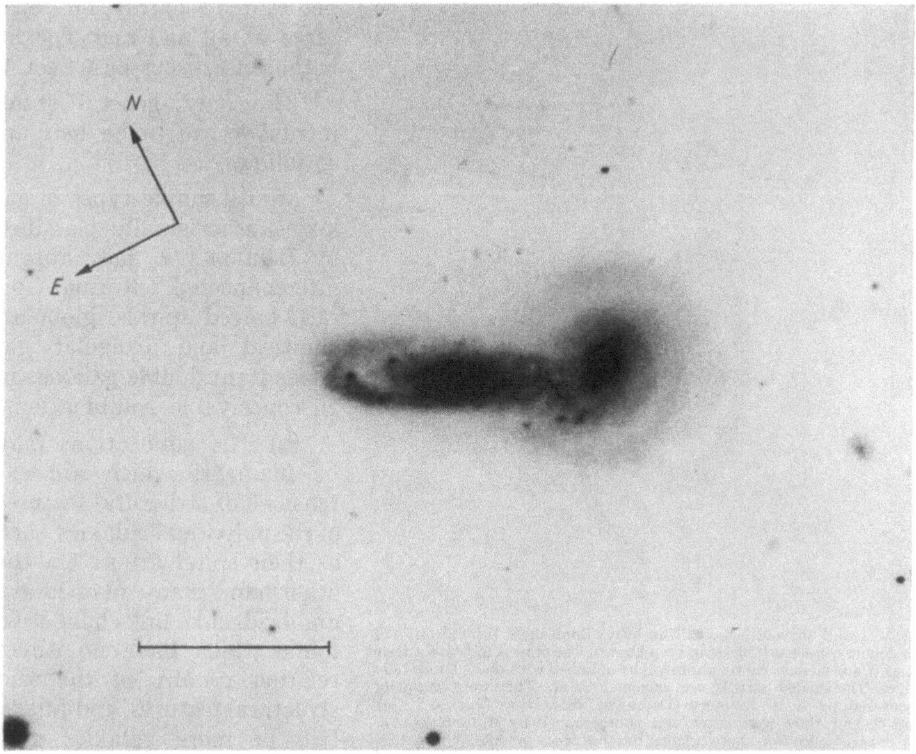

Fig. 2. Optical Double Nebulae NGC 5544 to 5545 at R. A. = $14^h 14^m 53^s$, Decl. + 36°41′20″ (1950). Somewhat out of context it may be added that the field back of the double galaxy contains an extremely distant and richly populated medium compact cluster of galaxies, which is one of the few discovered with the 200-inch telescope and not readily visible on plates taken with the 48-inch Schmidt telescope. Scale indicates one minute of arc.

The optical double galaxy is another type of system the interpretation of whose real nature is problematic. One of these cases is shown in Fig. 2.

According to TH. PAGE [*14*] the first member, NGC 5544, of the double system in Fig. 2 has a redshift corresponding to a symbolic velocity of recession equal to $V_s = (3090 \pm 70)$ km/sec, while its companion NGC 5545 has $V_s' = V_s + \Delta V_s$, where $\Delta V_s = (10 \pm 100)$ km/sec. PAGE erroneously lists both galaxies as of the elliptical type, whereas they are smoothly wound spirals. The practical equality of the values of V_s for the two galaxies suggests that they are close neighbors in space, although there is no other check available so far for this conclusion, since no tidal action between the two systems can be seen. The author has, however, in similar cases succeeded in photographing very faint tidal extensions, and he has also attempted to obtain such evidence for NGC 5544 to 5545. Unfortunately, the night sky has recently been so bright, because of solar activity, that the search for extremely faint tidal extensions must be postponed until a later date.

5. Interconnected galaxies. α) *Close pairs or groups.* Here a perfectly bewildering multitude of morphological types exist which it is quite impossible to classify at the present time. The sky survey with the 48-inch Schmidt telescope has brought to light tens of thousands of interconnected galaxies with separations up to say three diameters of the main bodies of the individual galaxies involved. In the limit, these bodies may hardly be separated at all and may coexist in the sense discussed in Sect. 3.

Fig. 3. The Whirlpool Nebula. The heavy lines show the structure of the double nebula as it has long been known. The crosses indicate a flare (I) as it appears clearly on photographs obtained with the Schmidt telescopes. The shaded parts II are extremely faint. They were originally discovered by G. W. Ritchey [Decennial Publ. Univ. Chicago 8, 397 (1904)], and they were later also photographed by M. de Kerolyr. For reasons unknown, these original findings were practically forgotten by most of the experts in the field until Zwicky [7], [15] rediscovered the formation II and showed by his method of composite photography that it is presumably this very broad and faint spiral arm which connects NGC 5194 and NGC 5195, rather than the very much more pronounced narrow and luminous spiral arm *S*. The reality of the faint dotted cloud (III) is still in doubt. The seven large dots represent some of the foreground stars.

The only general statements we can make here are as follows.

(a) All simple types of galaxies, as originally classified by Hubble [11] may appear interconnected. Normal spirals, barred spirals, globular, elliptical and irregular galaxies form double galaxies in all conceivable combinations.

(b) The connections may be filaments which are extensions of structural features of the individual galaxies, such as their spiral arms. On the other hand many filamentary and bridgelike links have been found which have no direct relation to any of the said structural features, and finally two or more galaxies may appear imbedded in a more or less diffuse luminous cloud.

(c) The luminous connections between galaxies seem to consist most frequently of stars. In a few special cases the existence of gases has been traced from emission lines. From certain striking features such as annular absorption rings it is practically certain that some of the bridges between neighboring galaxies also contain dust, a result which is naturally to be expected if the bridges are formed as a result of close encounters of stellar systems.

(d) The discovery, discussed in Sect. 3, that different structural types of galaxies can coexist within the same volume of space, makes us suspect that the morphological possibilities of the interconnections between two galaxies are far more complex than was stated above in paragraph (a). Indeed, we must now inquire which particular structure or which structures within one system of coexisting galaxies is interconnected with which structure in the companion galaxy. In this sense it would for instance be of the greatest interest to unravel the detailed morphological features of the Whirlpool Nebula Messier 51, a sketch of which is shown in Fig. 3 [reprinted from Physics Today 6, No. 4, 7—11 (1953)].

The Whirlpool Nebula is being photographed by the author in five colour ranges, infrared, red, yellow green, blue and ultraviolet with the intention of using all combinations between these ranges for the construction of composite photographs [*9*], [*15*]. These composites are expected to reveal much concerning the individual structures of the two components of the Whirlpool Nebula and how they are interconnected. So far, only the combination of blue and of yellow-green is available. From this combination it was deduced that NGC 5194 is a poorly organized normal spiral in the blue population of stars while it appears as an exceedingly smooth open barred spiral structure in the yellow-green population. On the other hand, NGC 5195 seems to be irregular in the blue and perhaps an elliptical or Theta barred spiral in the yellow-green. As stated before, the brighter spiral arm clearly seems to miss the companion NGC 5195, at least as far as the blue population is concerned. From this we conclude that the two spiral arms of NGC 5194 *do not lie in one plane*. We may expect the structural features of the whole system to become still more complicated if it should be found that the individual spiral arms of the different colour populations do not really coincide.

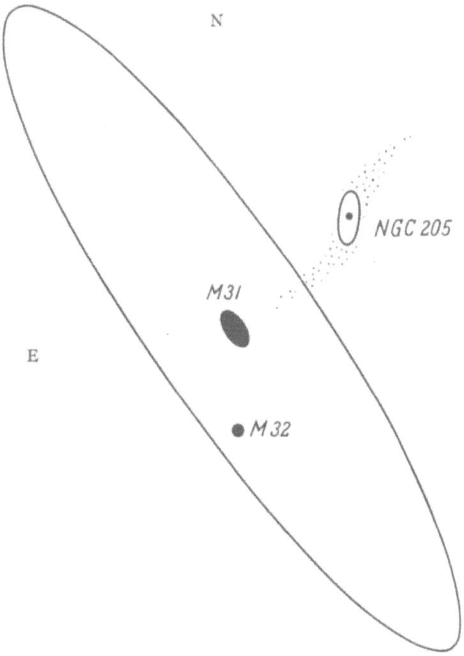

Fig. 4. Schematic drawing of the structural relation between the Great Nebula in Andromeda and its fainter companion NGC 205.

Starting from our own immediate neighborhood, the problem of whether or not the Milky Way system and the two Magellanic clouds are all interconnected and in what manner, presents one of the most fascinating and difficult problems. As Professor KNUT LUNDMARK first pointed out to the author in 1952, Sir J. F.W. HERSCHEL had already watched in South Africa, in 1838, for a bridge between the Milky Way and the Large Magellanic Cloud and had written as follows [*16*]:

"Entirely without telescopic aid, when seated at a table in the open air, in the absence of the moon I scanned the southern skies from a South African Observatory at the Cape of Good Hope and found that no branch of the Milky Way whatsoever forms any certain and conspicuous junction with the Nubecula Major; though on very clear nights I have sometimes fancied a feeble extension of the nearer portion of the Milky Way in Argo (where it is not above 15° or 20° distinct) in the direction of the Nubecula".

The present author subsequently contacted a number of Australian astronomers and radio astronomers about the matter. While Dr. G. DE VAUCOULEURS attempted by various optical methods to explore interconnections between the three neighboring systems and also searched for a countertide on the opposite side of the Milky Way, the radio astronomers derived intensity contour lines which give some support for the assumption that extensions of the three systems towards one another exist. Because of the intervening effects of complicated and rich more nearby clouds of stars, gases and dust, much more and painstaking work will be needed to clear up the question of any interconnecting formations between the Milky Way and the Magellanic Clouds.

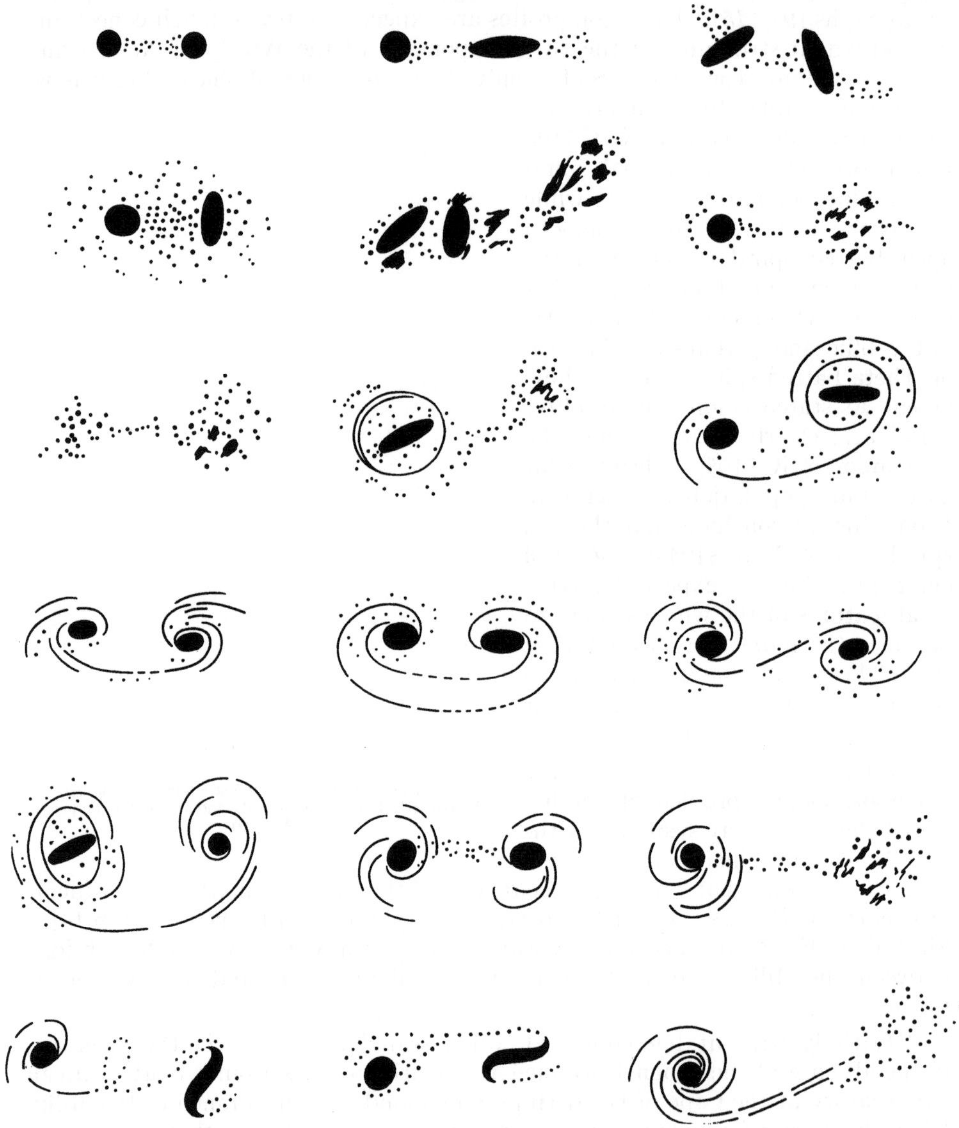

Fig. 5. Some of the many types of bridge and filament connections observed between neighboring galaxies.

As far as the next nearest optically visible group of galaxies is concerned, we already mentioned that among the group of five stellar systems around the great nebula in Andromeda only NGC 205 shows any structural features which might be interpreted as being due to the gravitational field of Messier 31 itself. On plates obtained with the 48-inch Schmidt telescope, NGC 205 clearly appears as an open barred spiral, one of whose extensions is pointing towards the center

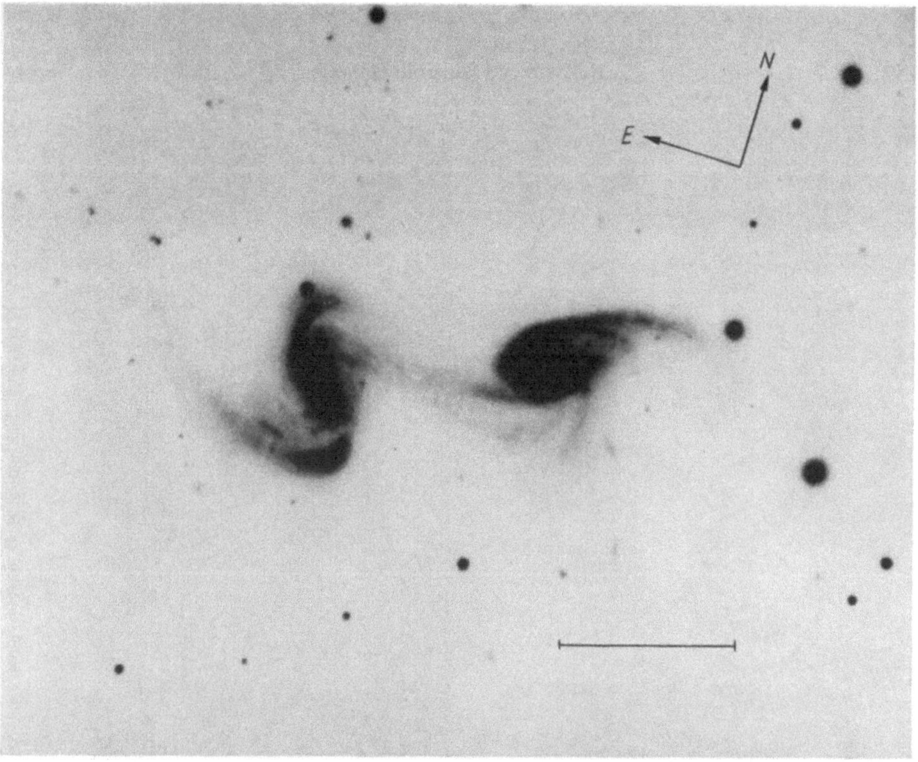

Fig. 6. NGC 5257 to 5258 at R.A. $13^h37^m12^s$, Decl. $+1°0'13''$ (1950.0), photographed on Eastman 103 *a*-0 emulsion, exposure 40 min, with the 200-inch telescope. According to TH. PAGE [*14*] the symbolic velocities of recession for the two galaxies are respectively $V_s = 6600$ and $V_s' = 6485$ km/sec. This is one of the systems originally observed by F. G. PEASE [*1*]. Scale indicates one minute of arc.

N

E

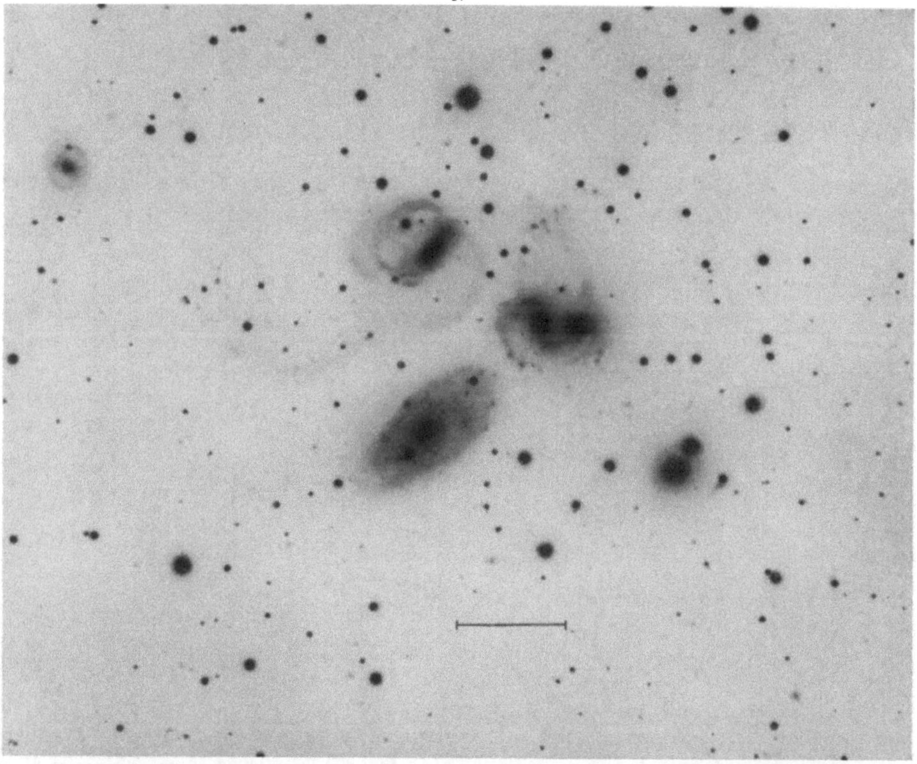

Fig. 7. Photograph with the 200-inch telescope on Eastman 103 *a*-*D* plate behind *GG* 11 Schott glass filter of STEPHAN'S Quintet, NGC 7317, 7318a, 7318b, 7319, 7320 at R.A. $22^h33^m43^s$ and Decl. $+33°42'$ (1950.0). The symbolic velocities of recession of the four first mentioned galaxies, which are respectively of the types *E* 4, *E* 2, *SB b* and *SB b* are $V_s = 6736$ 6638, 5638, and 6657 km/sec. Notice that there is a difference of 1100 km/sec between the first and the third members of the group. If the group is a permanent dynamic unit, its total mass must be large enough to retain NGC 7318b. If we adopt HUBBLE'S old distance scale, corresponding to a value of $V_s = 550$ km/sec per million parsecs the distance of the group would be about twelve million parsecs, its diameter of the order of 16000 parsecs and its mass, if the group is permanent, is greater than 10^{12} times the mass of the Sun. Adopting any of the more recent expanded distance scales, the mass of the system would have to be greater still, if it is not eventually to loose any of its members. Scale = 1′ arc.

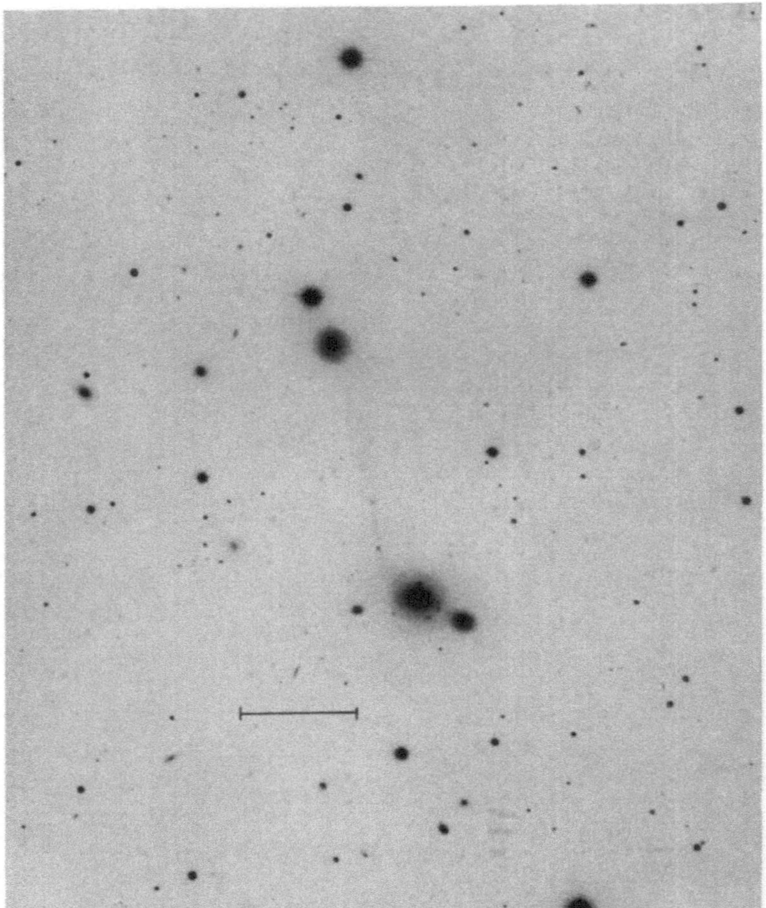

Fig. 8. Interconnected triple galaxy at R.A. $16^{h}48^{m}.0$ and Decl. $+45°35'$ (1950). Going from South to North the members are an *SO*, an *Sa* and an *SO* or *SBO* galaxy, respectively. Photograph on an Eastman 103*a*-O emulsion with the 200-inch elescope. Scale indicates one minute of arc.

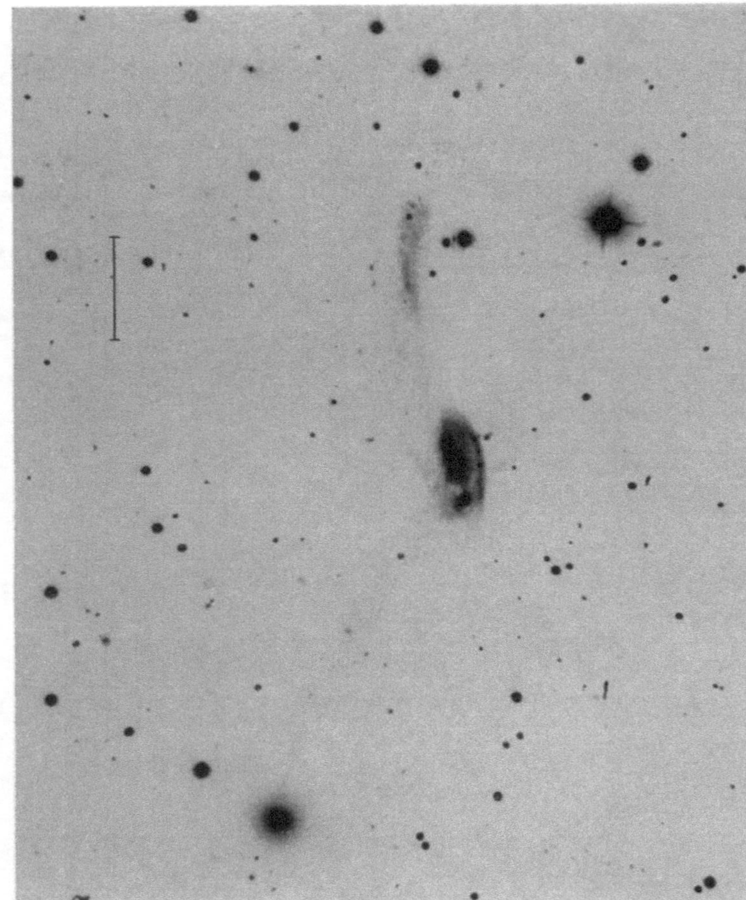

Fig. 9. Interconnected double galaxy at R.A. $17^{h}18^{m}6^{s}$ and Decl. $+49°7'36''$ (1950). The Southern member is globular, the Northern is a spiral with a long ejected spiral arm. 200-inch telescope photograph on Eastman 103*a*-O emulsion. Scale indicates one minute of arc.

of Messier 31 and which might be interpreted as a direct tide, while the symmetrical countertide points in the opposite direction, as shown in Fig. 4.

In Fig. 5 we give sketches of a few of the very many types of connections between neighboring galaxies. It is seen for instance that for normal and barred

N

E

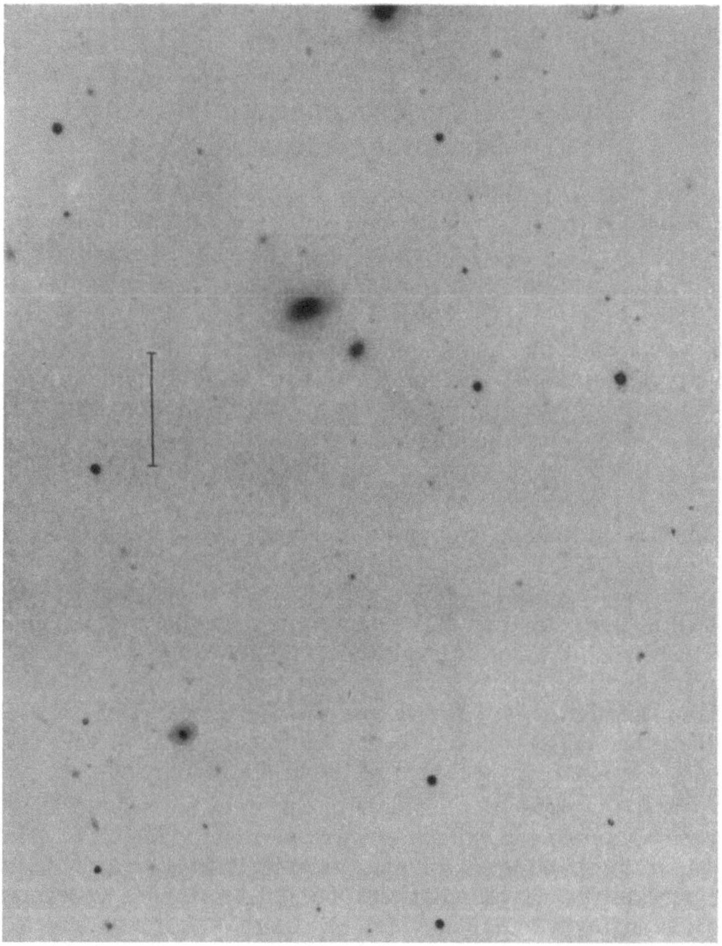

Fig. 10. Interconnected double galaxy at R.A. $9^h 55^m 49^s$ and Decl. $+29°6'.0$ (1950). The northern galaxy is an *SO* type, the southern is elliptical. The very long luminous ejected cloud toward the south is exceedingly faint and has been slightly touched up to make it show in the reproduction. 200-inch telescope photograph on Eastman 103*a-D* emulsion behind yellow *GG* 11 Schott glass filter. Scale indicates one minute of arc.

spirals all conceivable relative configurations occur, a fact which indicates strongly that these configurations are the result of double or multiple encounters.

Finally, in conclusion of this section, two photographs obtained with the 200-inch telescope are reproduced in Figs. 6 and 7, showing one interconnected pair and the famous Stephan Quintet.

β) Groups of widely separated interconnected galaxies. With wide separation we mean those groups in which the individual galaxies are either separated by more than three diameters or which show filaments, luminous clouds or tidal spurs which are many times as extended as the main bodies of the galaxies proper. The discovery and the subsequent exploration of these interesting structures is mainly due to the present author [*4*], and the results, as available at the present time, have been discussed by him in an extended paper [*5*]. In order to avoid unnecessary duplication, we therefore refer the interested reader to this article as well as to a recent review of the subject by Professor OTTO STRUVE in "Sky and Telescope" Vol. XVI, 162—166 (1957). Since, however, good photographs are not easily obtained of extended luminous intergalactic formations, we reproduce in the Figs. 8, 9 and 10 a few systems not included in the original article in the Ergebnisse Vol. 29 [*5*].

IV. The kinematics and dynamics of multiple galaxies. Gravitational lenses.

6. All of the discussions which follow are based on HUBBLE'S old distance scale, corresponding to a symbolic velocity of recession (universal redshift) equal to $V_s = 550$ km/sec per million parsecs. The investigations by TH. PAGE and E. HOLMBERG, mentioned in the following, use $V_s = 526$ km/sec per million parsecs.

Various values for the masses of individual galaxies have been obtained from the study of their absolute luminosities, from internal motions and rotations, from the velocity dispersion in clusters of galaxies and from the differences in radial velocities of the components of double galaxies. Radio astronomy and the possible future discovery of gravitational lens effects promise to offer further approaches to the masses of galaxies.

The masses of individual galaxies were originally derived from either absolute luminosities [*11*], assuming that these masses are, in solar units, equal to a few times the luminosities of these galaxies, also measured in solar units. This approach is thus based on the assumption that the mass-luminosity ratio for a whole galaxy is, in order of magnitude, the same as the corresponding ratio averaged over all of the known stars in the solar neighborhood. This of course disregards the possibility of large masses being concentrated in dark bodies, gases and dust. The masses of even the most luminous galaxies, on these assumptions, are only a few billion times that of the Sun ($10^9 \cdot M_\odot$).

ZWICKY, in 1933, made the first determination of masses of galaxies on the basis of the virial theorem and the velocity dispersion in clusters of galaxies [*17*]. For instance, for the six hundred brightest galaxies in the Coma cluster he derived an average value for the individual masses of about $2 \times 10^{11} M_\odot$, or several hundred times the value estimated from luminosities. Although it was later found that clusters of galaxies contain very many dwarf galaxies and probably much dark dispersed intergalactic matter, such as gases and dust [*9*], the whole discrepancy probably cannot be entirely described to the existence of these components. This conclusion is strengthened by the results of PAGE and HOLMBERG on the masses of the components of double galaxies. PAGE [*14*], from the investigation of 35 double systems, obtained a mean mass of $8 \times 10^{10} M_\odot$ for the component galaxies and a mass luminosity ratio of 348 in the above mentioned solar units. E. HOLMBERG, in his most recent studies [*18*], derived for the components of 26 double galaxies a mean mass of $6.5 \times 10^{10} M_\odot$ and a mass luminosity ratio equal to 127. Although these investigations are subject to rather large errors of both observational and theoretical origin, they nevertheless indicate, in confirmation of the results derived from the velocity dispersions in clusters, that the masses of galaxies are very much larger than could be inferred from their luminosities. The methods used by PAGE and HOLMBERG can be strengthened, if more accurate differences in radial velocities of multiple galaxies are obtained, and if not only average differences are used but the whole fields of the radial velocities over the full extent of the galaxies involved are studied. Furthermore, one must make sure that one does not deal with optical doubles or with galaxies which are clearly escaping from one another. Also, it will be necessary to obtain accurate distances of the double galaxies studied.

The requirements mentioned for trustworthy determinations of the masses of the components of double galaxies are obviously severe and not easily satisfied. Independent approaches are therefore most desirable. Unfortunately, the determination of the mass of a galaxy from its rotation and its internal motions is beset by equally severe difficulties [*9*]. On the other hand, the discovery of gravitational lens effects among galaxies would do away with most of the uncertainties [*17*], [*9*]. For details on the theory of gravitational lenses the reader is referred to "Morphological Astronomy", by F. Zwicky, Springer Verlag 1957.

What features of multiple galaxies reflect phenomena which took place during the creation of stars and galaxies and what other features are due to evolution or to the effects of rather recent encounters is a most fascinating problem. Far too little, however, is known at the present time about these subjects to permit us to make any statements here, lest we wish to run the danger of indulging in idle speculations. We only recommend that, an attack on multiple galaxies with all of the means at our disposal should certainly prove most fruitful to unravel the mysteries surrounding the origin and the evolution of the large scale aggregates of matter in the universe.

V. Colliding galaxies as radio sources.

7. During the past decade the methods of radio astronomy have opened up new vistas on the field of multiple galaxies, and a few most significant results have been achieved already. Data have been given in the literature [*19*], [*20*] on the following extraordinary galaxies: NGC 1275, a double galaxy in Cygnus [Cygnus A radio source], NGC 2623, NGC 4486, NGC 4038 to 4039 and NGC 5128. From direct photographs, as well as from spectral analyses in the photographic range, the first two mentioned systems seem to be definitely galaxies in collision. NGC 2623 and NGC 4038 to 4039 also have the appearance of colliding galaxies, while the nature of NGC 4486 and NGC 5128 is more uncertain.

Table 1. *Apparent photographic magnitudes m_p and radio magnitudes m_r of a few strong extragalactic radio sources. By definition* [*20*] *it is* $m_r = -53.4 - 2.5 \log S$, *where S is the flux density in watts per square meter and per cycle per second at a wavelength of two meters. The constant* (-53.4) *was adjusted so as to make* $m_r = m_p$ *for average S_b or S_c galaxies, as they are listed in the Shapley-Ames catalogue.*

Object	m_p	m_r	$m_p - m_r$	Ratio r_0
NGC 5128	6.1	3.7[1]	2.4	~ 10
NGC 4486	9.9	4.7	5.2	~ 100
NGC 1275	13.3	7.2	6.1	~ 300
Cygnus A	17.9	2.2	15.8	$\sim 2 \times 10^6$

Among the few abnormally powerful extragalactic radio sources there seems to exist a bewildering variety of physical conditions, so that no systematic classification is as yet possible. For instance, the ratio of the luminosities of these objects in the radio regions and the optical regions covers a most astounding range, as is seen from the entries in Table 1, which were kindly supplied me by my colleague at the California Institute of Technology, Professor J. G. Bolton.

[1] Includes an allowance for extended envelope. r_0 is defined by the relation $m_p - m_r = 2.5 \operatorname{Log}_{10} r_0$ and therefore represents the ratio of the radio intensity of the abnormal galaxies to that of normal S_b or S_c galaxies, where the intensities are measured in terms of unit radiations in the photographic range.

Dr. BOLTON also submits that the sources Hydra A and Hercules A have values m_r of about +6.0 and photographic magnitudes of about +17. These are extragalactic objects with double nuclei and, if the identifications can be accepted, are objects intermediate between Cygnus A and NGC 1275.

NGC 1275. This double galaxy, which is shown in Fig. 11, is in the center of the large Perseus Cluster and is located at R.A. $3^h\,16^m\,30^s$ and Decl. $+41°\,19'.6$ (1950).

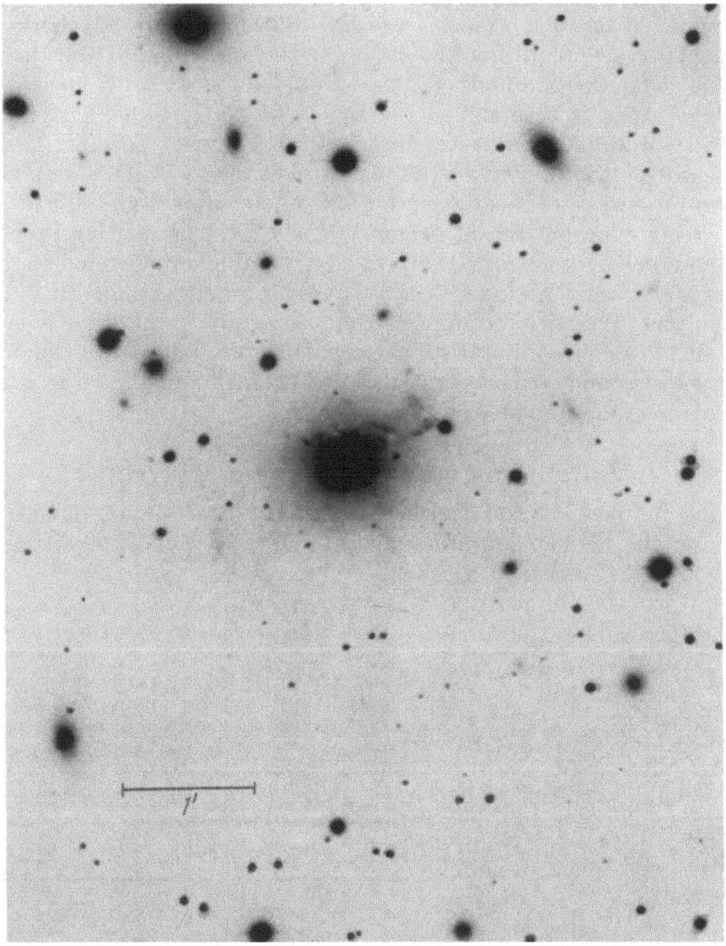

Fig. 11. NGC 1275. Photograph taken with the 200-inch telescope, emulsion Eastman 103a-0, exposure time 10 min. Scale indicates one minute of arc.

For a description of some of the structural and spectroscopic characteristics of NGC 1275 we quote from the director's report for the Mount Wilson and Palomar Mountain Observatories in the Year Book No. 54 (1954—1955) of the Carnegie Institution of Washington, where we read on p. 25:

"Although colliding galaxies are safely established as radio sources, we have at present little information about the details of such collisions. The strongest radio source of this kind, Cygnus A, is not very suitable for detailed investigation because of its great distance. A much better subject is the colliding pair NGC 1275 in the Perseus cluster of galaxies. Here an early-type spiral is in collision with a late-type spiral and the pair is close enough to show much structural detail on plates taken at the 200-inch telescope. An examination of the existing plates convinced Baade that it should be possible to localize the regions of the colliding gases on exposures with suitable plate-filter combinations if in such 'hot spots' the total light emitted by the colliding gases were of the same order as the continuum of the unresolved stars. Attempts thus to localize the hot spots by their H_α, O III, and other emissions were successful, and a detailed picture of the emission regions was obtained.

The spectroscopic investigation of these spots by MINKOWSKI led to the following results. In the northern part of the object the emission lines are double, showing a velocity of + 5200 km/sec for the early-type spiral and of + 8200 km/sec for the late-type spiral. The emission spectra of both nebulae consist of strong O II lines and weak lines of hydrogen. In the southern part an entirely different type of spectrum appears, showing one set of asymmetrical lines of considerable width, indicating a velocity of + 5200 km/sec. H lines and forbidden lines of O I, O II, O III, Ne III, and S II appear. This spectrum appears superposed on the nucleus of the early-type spiral where it was first observed by HUMASON many years ago. But it is not restricted to the nuclear region, as happens in certain spirals with bright semistellar nuclei, and it appears with somewhat less intensity farther to the south.

The appearance of the object and the spectroscopic observations suggest the interpretation that the late-type spiral is inclined against the early-type spiral in such a way that in the north the late-type spiral is in front. Moving toward the early-type spiral, it has penetrated the other system in the center and south of it, where now highly excited gas with large internal motions shows the aftermath of the collision. The velocity of the mixed gas is close to the velocity of the early-type system; this requires that the early-type system be the more massive one, a conclusion which is supported by the general appearance of the two galaxies. The actual collision is in progress to the north, where the gas masses of the two spirals can still be seen separately. A more detailed discussion of the collision suggests a duration of the order of a million years; more than half this time has passed, and the effects of the collision should now be past their maximum."

One of the most important results of this investigation is [*5*] that the two spirals involved in the collision have a relative radial velocity of 3000 km/sec. This is the highest value so far known, except for the difference of 7000 km/sec found by ZWICKY for the apparently interconnected galaxies IC 3481 and IC 3483. This large velocity difference led many astronomers to doubt that the IC 3481 and IC 3483 could really be members of a triple system of galaxies in the process of a mutual encounter [*4*]. The large velocity difference observed in NGC 1275, however, make this assumption appear less improbable than it originally was.

Cygnus A. This radio source seems to be related to two late type galaxies located at R.A. $19^h\,57^m\,45^s$ and Decl. $+40°\,35'\,46''$ whose redshift corresponds to a symbolic velocity of recession of approximately $V_s = 16830$ km/sec. The distance of the system, on HUBBLE's old distance scale is 3.3×10^7 parsecs. For a description of the photographic appearance of the double galaxy and some of its spectroscopic properties we refer to the article by W. BAADE and R. MINKOWSKI [*19*]. We only mention here that some high excitation lines of Ne III, Ne V and O III are observed in high intensity, indicating the violence and the effectiveness of the collision of the gas clouds involved, although their radial velocity differences are only of the order of 500 km/sec, much smaller than in NGC 1275. The Cygnus A double galaxy seems to be a giant system with a total optical emission estimated at 5.6×10^{42} ergs/sec $\approx 2\times10^9$ times the luminosity of the Sun and an even greater total emission in the radio region of 8×10^{42} ergs/sec. BAADE and MINKOWSKI remark that "the source of energy for the radio emission may be the relative kinetic energy of the colliding galaxies, which is of the order of 10^{59} ergs for a relative velocity of 500 km/sec". At the present rate of radio emission the total energy radiated in the radio region in a period of one million years would be of the order of 2×10^{56} ergs. The collision may last a few million years. We again mention that all estimates made refer to HUBBLE's old distance scale, assuming an average value $V_s \approx 550$ km/sec per million parsecs.

NGC 4486 (Messier 87). This globular galaxy, which is a member of the Virgo cluster and is located at R.A. $12^h 28^m 18^s$ and Decl. $+12°\,40.1'$, shows a jet like extension originating in its center. According to spectra obtained by HUMASON, the jet is rather blue in colour but has a continuous spectrum. Recently some of the prominent condensations in the jet were found by Baade to be partially polarized. The globular galaxy itself has a normal spectrum of type *G*, but superposed on the nucleus appears a strong emission line of O II at λ 3727 which is shifted relative to the nuclear *G*-type spectrum by -295 ± 100 km/sec. Contrary to the cases of Cygnus A and NGC 1275, where the radio emission may be ascribed to the gas clouds within two colliding galaxies impinging on one another, the origin of the radio emission in NGC 4486 is not cleared up. This origin becomes still more puzzling if the observations, recently announced [*21*] by Brown and Smith, can be confirmed that a sizeable part of the radio emission comes from a disk 50′ of arc in diameter, as contrasted with the visual disk of NGC 4486 which is only a few minutes of arc across.

NGC 5128. The origin of the radio emission from NGC 5128, located at R.A. $13^h 22^m 28^s$ and Decl. $-42° 45\overset{'}{.}6$ (1950), is as little understood as that from NGC 4486. Baade and Minkowski [*19*] have suggested that NGC 5128 is a double galaxy consisting of an unresolved E_0 galaxy and a second stellar system of much later type in front of it or interpenetrating it, thus accounting for the unusually strong and wide absorption lane. The apparent radial velocity of the system is 450 km/sec, with superposed velocity differences which are not very conspicuous. Because of this fact, the interaction of the two assumed galaxies would be of an entirely different type from that in the NGC 1275 and Cygnus A radio sources. Zwicky, with the 48-inch Schmidt telescope has recently found some evidence of polarized light being emitted from certain condensations of NGC 5128. These observations, which must be checked with the 200-inch telescope, are not necessarily related to the radio emission, which seems to come from two regions, namely, a central ellipsoidal part a few minutes of arc in extent and a very much larger area about two degrees in diameter [*19*].

8. Conclusions. From the very scanty data available so far, it is clear that most interesting results may be expected from the study of multiple galaxies. For the present we cannot be sure, however, about very many of the structural and physical features of these systems. For instance, for the groups shown in the Figs. 8, 9 and 10, no spectroscopic data of any kind are available on the main bodies constituting these groups, not to speak about the spectra of the connecting links. The fact that from photographic observations in the visual range the radio emissive properties can in no way be predicted, suggests that it would be of the greatest importance, as in many other fields of astronomy, to extend, if possible, the spectral explorations into the ranges from one micron to one centimeter wavelength, as well as to the far ultraviolet. Efforts which are now being made in these directions will have to make use both of entirely novel recording instruments and of high flying rockets to get away from the absorption in the atmosphere.

As to the dynamic interpretation of the structural features of multiple galaxies, one aspect clearly begins to stand out which indicates that some of the views by present day astronomers about stellar systems may be in need of a radical revision. These views are for instance maintained by Baade and Minkowski in their discussion [*19*] of the radio source Cygnus A, where they say:

"This suggests that we are dealing with the exceedingly rare case of two galaxies which are in actual collision. The main features of such a collision have been discussed by Spitzer and Baade [*22*]. On the cosmical time scale, collisions of galaxies are a rather frequent phenomenon in the rich clusters of galaxies. As far as the stars of the colliding systems are concerned, such a collision is an absolutely harmless affair. The average distance between two stars is so large that the two galaxies penetrate each other without any stellar collision."

The study of the features of multiple galaxies shows immediately that there must be something radically wrong in the above conclusions by Spitzer and

BAADE. Indeed, it is at once obvious from the observations that it is clouds, filaments and jets of *stars* which are ejected massively from galaxies in collision. As far as the very extended bridges and filaments between largely separated galaxies are concerned, we have not, in fact, succeeded so far to prove that they contain gases and dust, although this is undoubtedly the case. The conclusions by SPITZER and BAADE, as well as by others, concerning the consequences of the encounters of galaxies, are erroneous because of their neglect to evaluate the effects of large scale tidal effects from system to system, as well as the interactions between gas and dust clouds with individual stars and large groups of stars. Furthermore, the electromagnetic actions within the dispersed clouds may materially contribute to the internal viscosity of stellar systems which, by the testimony of the structural features of the connecting links of galaxies, must be very much greater than is usually admitted.

Unfortunately, because of the present relatively high intensity of the night sky, caused by prolonged solar activity, it will be some time before we can hope to explore efficiently the very faint extensions in multiple galaxies As soon as feasible, it is proposed, however, to pursue such studies with the utmost vigor. This is important not only because of the information to be gained on the structure and evolution of the galaxies themselves. The analysis of the extended projections into intergalactic space will ultimately give us new clues on the population of intergalactic space and on the average density in the universe, which, from other lines of reasoning recently given, seems to be so great as to rule out any of the cosmological theories so far proposed which are based on the general theory of relativity and which assume that the universal redshift observed in the spectra of distant galaxies corresponds to an actual expansion of the universe [*9*], [*23*].

Bibliography.

[*1*] PEASE, F. G.: Astrophys. Journ. **46**, 24—55 (1917); **51**, 276—308 (1920).
[*2*] LUNDMARK, K.: Kgl. Sv. Vetensk. Handl. **60**, No. 8 (1920). — Upsala Medd. **1926**, No. 8; No. 16 = VJS **61**, 254 (1926); **1927**, No. 30; No. 41 = VJS **63**, 350 (1928).
[*3*] HOLMBERG, E.: Ann. Obs. Lund **1937**, No. 6. — Medd. Lunds Astr. Observ., Ser. I **1954** No. 186.
[*4*] ZWICKY, F.: Experientia (Basel) **6**, 441—445 (1950). — Publ. Astronom. Soc. Pacific **64**, 242—246 (1952). — Physics Today **6**, No. 4, 7—11 (1953). — Phys. Blätter **9**, 406 to 415 (1953).
[*5*] ZWICKY, F.: Ergebn. exakt. Naturw. **29**, 344—385 (1955).
[*6*] ZWICKY, F.: Griffith Observer **17**, No. 12 (1953).
[*7*] COUDERC, P.: L'Astronomie **68**, 406—415 (1954).
[*8*] ZWICKY, F.: Publ. Astronom. Soc. Pacific **68**, 121 (1956).
[*9*] ZWICKY, F.: Morphological Astronomy. Heidelberg: Springer 1957.
[*10*] BAADE, W., and E. P. HUBBLE: Publ. Astronom. Soc. Pacific **51**, 40—44 (1939).
[*11*] HUBBLE, E. P.: The Realm of the Nebulae. New Haven, Conn.: Yale University Press 1936.
[*12*] BAADE, W.: Astrophys. Journ. **100**, 147 (1944).
[*13*] HUMASON, M. L., N. U. MAYALL and A. R. SANDAGE: Astronom. J. **61**, 97—162 (1956).
[*14*] PAGE, TH.: Astrophys. Journ. **116**, 63 (1952).
[*15*] ZWICKY, F.: Publ. Astronom. Soc. Pacific **67**, 232 (1955).
[*16*] HERSCHEL, Sir J. F. W.: Results of astronomical observations made during the years 1834 to 1838 at the Cape of Good Hope. Cornhill: Smith, Elder & Co. 1847.
[*17*] ZWICKY, F.: Helv. phys. Acta **6**, 110—127 (1933). — Astrophys. Journ. **86**, 217 (1937).
[*18*] HOLMBERG, E.: Medd. Lunds Astr. Observ., Ser. I **1954**, No. 186.
[*19*] BAADE, W., and R. MINKOWSKI: Astrophys. Journ. **119**, 206, 215 (1954).
[*20*] BOLTON, J. G.: Publ. Astronom. Soc. Pacific **68**, 477 (1956).
[*21*] Radio halo around galaxy Messier 87. Sky and Telescope **16**, 116 (1957).
[*22*] SPITZER, L., and W. BAADE: Astrophys. Journ. **113**, 413 (1951).
[*23*] ZWICKY, F.: Proceedings of the Third Berkeley Symposium on Mathematical Statistics and Probability, Vol. III, p. 113—143. Berkeley, Cal.: University of California Press 1956.

Clusters of Galaxies.

By

F. Zwicky.

With 5 Figures.

I. Introduction.

1. Historical. The term *nebulae* means on the one hand clouds of gas and dust which are distributed among the stars. Within the Milky Way system, there are perhaps a few thousand of these *galactic* nebulae, their distribution exhibiting a decided preference for low galactic latitudes. The remaining objects, numbering hundreds of millions within the reach of the 200-inch telescope, are stellar systems distributed through extragalactic space. These systems, which, because of the absorption of light through the dust near the galactic plane, are mostly found in higher galactic latitudes, have been variously called *island universes* (Herschel), *extragalactic nebulae* or *external galaxies*. We shall refer to them simply as *galaxies*.

The *original catalogues* included both types of nebulae, as well as star clusters located in our own galaxy. It is remarkable that, if we map those objects from these catalogues which we now know to be extragalactic nebulae, we find many of them bunched together, indicating thus the presence of several of the well known *clusters of galaxies*. For instance, of the 107 bright objects included in the famous list [*1*] by Messier (1730 to 1817) there are 35 galaxies. About 16 among these are probably members of *the Virgo Cluster of Galaxies* (see Fig. 1, where the distribution over the sky of the 35 galaxies of Messier's list is plotted).

Later on Sir William Herschel, from 1786 until 1802, published several catalogues in which he listed the positions and the structural features of several thousand nebulae. From plots of these nebulae the locations, and outlines of about a dozen of the nearest clusters and clouds of galaxies readily emerge. It is still of interest to read the discussion of the results of the two Herschel's by Alexander von Humboldt in his "Cosmos" (New York 1866, Harper Brothers, Publishers, Vol. IV, p. 25 and following). He writes for instance "This zone which has been termed the nebulous region of Virgo contains one third of all the nebulous bodies in a space embracing the eighth part of the surface of the celestial hemisphere."

Sir John Herschel's general catalogue (GC) was published in 1864. It represents the first systematic survey of the entire sky to a fairly uniform limit of apparent brightness and contains about 4630 nebulae and clusters of stars observed by his father and himself, together with about 450 discovered by others. This catalogue was replaced by Dreyer's New General Catalogue in 1890 (referred to as NGC). Two supplements, the Index Catalogues (IC) bring the list up to 1907. The great majority of the 7840 NGC objects and the 5386 IC objects, are extragalactic nebulae. Professor KNUT LUNDMARK constructed charts over the whole sky plotting all of these nebulae. These charts, which contain much information concerning the nearby clusters of galaxies have unfortunately never been published.

In more recent times, many lists have been made for special purposes, but only one that covers the entire sky in a homogeneous manner. This latter is

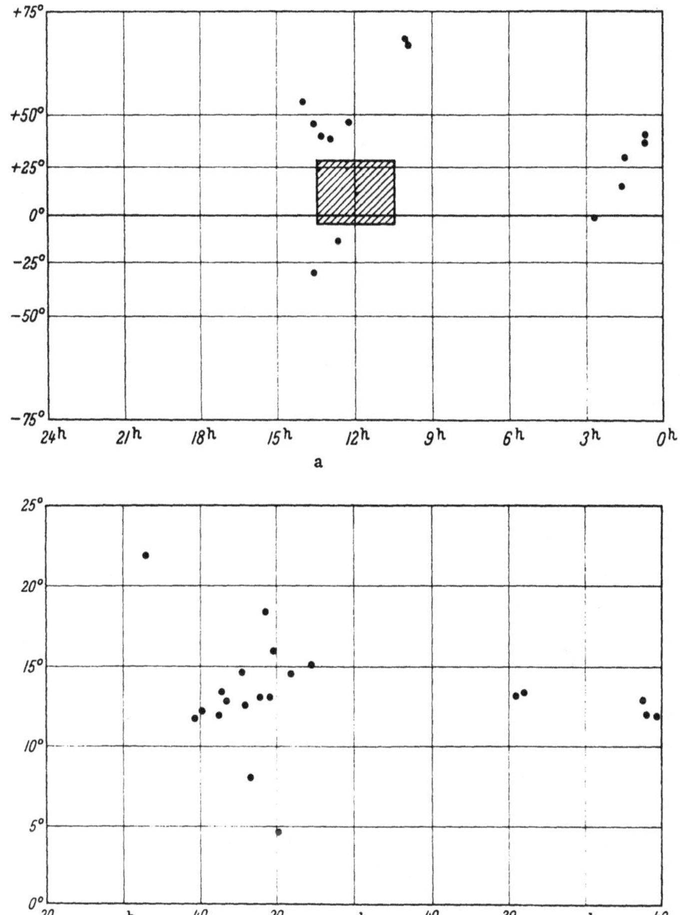

Fig. 1 a and b. Distribution of the galaxies which are included in MESSIER's list of 107 nebulous objects and star clusters.

the Harvard Survey [2] of galaxies brighter than thirteenth magnitude, which contains 1249 objects (1188 NGC nebulae, 48 IC nebulae and 13 others). The apparent photographic magnitudes of the galaxies in this catalogue will need some slight revisions, both because of newer more accurate determinations and because of the fact, that the faint outskirts of many galaxies contribute more to the total luminosity than was originally realized.

A chart of equal population contour lines or isopleths, constructed on the basis of the data in the Shapley-Ames catalogue is to be found in the book "Morphological Astronomy" by F. ZWICKY [3]. In the light of the 1250 brightest galaxies in the sky, only a few of the nearest clusters are clearly discernible.

These include in particular the Virgo cluster, with its southern extension, the still nearer but more open Ursa Major Cloud and the two clusters in Fornax.

2. Surveys of galaxies presently being in preparation. For a short description of the two most recent and most extensive surveys of galaxies and of clusters of galaxies I refer to an article by J. NEYMAN, E.L. SCOTT and C.D. SHANE [*4*] on the "Statistics of Images of Galaxies with Particular Reference to Clustering". These authors write:

"C. D. SHANE and C. A. WIRTANEN [*5*] are currently conducting a survey to the eighteenth magnitude based on astrometric plates taken with the Carnegie 20-inch astrograph of the Lick Observatory. The survey covers the sky from the north celestial pole to 23° south declination. The plates are systematically spaced and overlap at least one degree on each side. Galaxies are counted in areas 10′ square to the plate limit of about magnitude 18.4."

"FRITZ ZWICKY has under way a project to determine from images obtained by schraffier methods with the 18-inch Mount Palomar Schmidt telescope the magnitudes of all galaxies to the 15-th. When completed, the resulting some 15000 objects should provide an exceptionally complete catalogue of individual galaxies."

The plates for the survey by SHANE and WIRTANEN have all been taken and some of the counts have already been published [*5*].

The data for about 5000 galaxies of ZWICKY's survey are ready at the present time, and some of these have already been made available [*6*] through the Office of Naval Research (U.S. Navy), which is currently partly supporting this program.

ZWICKY's catalogue (with P. WILD and E. HERZOG collaborating) will consist of the following tables and charts:

1. Tables giving the positions and apparent photographic magnitudes of all galaxies brighter than the 15th and of many galaxies somewhat fainter than the 15th. The magnitudes are determined by the Schraffier method with the 18-inch Schmidt telescope and using squared images $1' \times 1'$ in size.

2. Tables listing the positions, diameters, populations and estimated distances of all clusters of galaxies clearly discernible on panchromatic plates taken with the 48-inch Schmidt telescope.

3. Charts, originally drawn on the scale of the plates of the 48-inch Schmidt ($1\ \text{mm} = 67''.2$) show the positions of all galaxies measured, using various symbols for the different ranges in photographic magnitudes. The contours of all of the clusters of galaxies investigated are also drawn.

As an example the chart and the tables for one special field of 36 square degrees area are shown in the following (Table 1, 2 and Fig. 2).

In Fig. 2 the outlines of the clusters drawn are approximate equal population contours or isopleths along which the number of galaxies per square degree is roughly double the average number of field galaxies in the surrounding field.

Once the whole sky has been explored after the manner shown in the Tables 1 and 2, as well as in the chart of Fig. 2, we shall be able to derive much statistical information on the general distribution of galaxies and clusters of galaxies, the partial segregation of galaxies of different brightness, the structure and population of clusters, the relative numbers of cluster galaxies and field galaxies, the structural features of clusters of galaxies, intergalactic obscuration and many other aspects of the large scale distribution of matter in the universe.

Table 1.

Positions (epoch 1950.0) and apparent photographic magnitudes m_p of the 95 brightest galaxies in the square field of 36 square degrees area which is centered at R.A. = $10^h 4^m$ and Decl. = $-0°30'$ (Epoch 1950) respectively.

n = number of independent determinations of right ascension, declination and m_p.

The fifth column gives the NGC or IC numbers of the respective galaxies.

The parenthesis ( ) means that the magnitude in question is somewhat less certain than those given for other galaxies.

The bracket ⟨ ⟩ means that the respective magnitude was estimated from the direct image on a plate (Eastman 103 $a-0$ emulsion) photographed with the 48-inch Schmidt telescope.

R = image of the galaxy was close to the border of the field.

D = magnitude was estimated from a direct image photographed with the 18-inch Schmidt telescope, which lies too close to the image of another galaxy or star to be measured by the Schraffier method.

F = galaxy is too faint for safe measurement from the Schraffier image.

L = galaxy with diameter greater than one minute of arc, and its image is therefore larger than the Schraffier square used.

N = magnitude is that of the nucleus of the galaxy only.

(D) = direct image was estimated in addition to the Schraffier image because the latter was of poor quality.

R.A.	Decl.	n	NGC/IC	m_p	n	Remarks
$9^h 52^m 17^s$	$-1°43'.5$	1 1		⟨15.5⟩		
9 53 08	−1 35.2	8 8		(15.4)	8	*F*, double
9 53 16	−1 38.0	8 8		15.4	13	(*D*)
9 53 33	−1 35.5	1 1		⟨15.5⟩		
9 53 58	+2 05.2	4 4		(15.7)	4	*F*
9 54 01	+1 40.0	6 6	NGC 3062	15.0	8	
9 54 47	−2 10.7	2 2		(15.5)	6	(*D*), *B*
9 55 59	−2 35.0	2 2		15.3	4	
9 56 06	−0 41.8	1 1		⟨15.5⟩		
9 56 09	−2 50.0	2 2		(15.7)	4	*F*
9 56 53	−0 00.9	8 7		15.0	8	
9 57 10	−1 54.6	4 4		15.5	4	
9 57 18	−2 38.3	2 2	NGC 3083	14.5	4	
9 57 35	+2 23.8	4 4		14.7	5	*D*
9 57 39	−2 44.2	2 2	NGC 3086	14.9	4	
9 57 48	+0 44.6	1 1		⟨15.5⟩		
9 57 50	+0 56.7	4 4		15.1	4	
9 57 58	−2 43.8	2 2	NGC 3090	14.2	6	(*D*)
9 57 58	−1 55.4	4 4		14.9	4	
9 58 00	+0 47.3	1 1		⟨15.5⟩		
9 58 13	+1 28.6	4 4		15.4	7	(*D*), *L*
9 58 15	−2 46.4	2 2	NGC 3092 ?	15.1	4	
9 58 21	−2 43.9	2 2	NGC 3093	15.1	4	
9 58 36	+0 09.6	4 4		(15.6)	4	*F*
9 58 38	+0 34.2	4 3		15.4	4	
9 58 38	+0 27.9	4 4		15.2	4	
9 59 03	−2 45.3	2 2	NGC 3101	(15.6)	4	*F*
9 59 57	−2 47.8	2 2		(15.6)	8	(*D*), *F*
9 59 57	+0 32.9	6 6		15.1	5	*D*
10 00 32	−2 09.6	2 2		14.9	4	
10 00 54	−1 18.3	2 2		15.6	4	double
10 01 22	+1 45.3	5 5		15.2	4	
10 01 52	−2 11.1	2 2		14.9	4	
10 01 58	+2 01.8	4 4		(15.6)	4	*F*
10 02 03	−1 30.1	2 2		15.5	4	

Table 1. (Continued.)

R. A.	Decl.	*n*	NGC/IC	m_p	*n*	Remarks
$10^h\,02^m04^s$	$+2°07'.9$	4 4		(15.6)	4	*F*
10 02 50	+0 47.2	4 4		15.1	8	
10 03 12	+2 29.3	2 2		15.1	4	
10 03 16	+0 52.6	4 4	IC 590	14.4	8	double
10 03 39	−1 18.5	3 3		15.3	9	
10 03 40	−0 57.8	2 2		15.4	4	
10 04 22	−0 59.3	4 4		15.2	8	
10 05 00	+0 31.3	2 2		(15.4)	5	*F*, (*D*)
10 05 18	+0 32.7	4 4		15.1	9	
10 05 19	−3 14.7	6 6		15.2	4	
10 05 22	+0 31.0	2 2		15.2	4	*D*
10 05 26	−2 15.2	4 4	IC 592	14.8	8	
10 05 46	−2 16.9	4 4	IC 593	14.8	12	(*D*)
10 05 46	−3 31.4	4 4		14.9	4	
10 05 55	+0 36.8	4 4		14.8	8	
10 05 59	−0 25.3	6 6		14.8	8	
10 06 14	−0 01.4	6 6		15.3	8	
10 07 00	−1 56.2	2 2		(15.5)	4	
10 07 17	+1 28.2	4 4		15.4	8	
10 07 26	−1 10.2	2 2		(15.4)	5	*F*, (*D*)
10 07 26	−2 35.3	1 1		⟨15.5⟩		
10 07 28	−2 13.0	4 4		15.1	8	
10 07 53	+2 28.5	6 6		15.0	8	
10 07 58	−1 33.3	2 2		15.2	4	
10 08 06	−1 47.9	4 4		15.5	11	*F*, (*D*)
10 08 08	−2 11.2	4 4		15.2	13	(*D*)
10 08 24	−0 15.4	6 6		15.4	8	
10 08 31	+1 28.2	2 2		(15.7)	4	*F*
10 08 44	+0 41.4	6 6		14.3	12	*L*, (*D*)
10 08 46	+0 13.4	4 4		(15.6)	7	*F*, (*D*)
10 09 12	−2 29.1	2 2		15.1	4	
10 09 20	+0 11.2	6 6		15.4	8	
10 09 27	−2 53.8	2 2		15.4	4	
10 09 30	−1 54.0	4 4		15.1	8	
10 09 55	−1 34.2	2 2		(15.6)	4	*F*
10 10 07	−3 30.5	2 2		(15.8)	4	*F*
10 10 07	−2 35.2	2 2		15.3	4	
10 10 11	−3 15.1	4 4		15.2	8	*F*
10 10 14	−2 23.4	4 4		15.3	8	
10 10 16	−2 26.4	4 4		15.0	8	
10 10 28	−2 24.8	4 4		15.3	8	
10 10 44	−2 17.0	4 4		15.2	8	*D*
10 10 57	−1 38.2	4 4		(15.5)	4	*F*
10 11 02	−2 54.7	1 1		⟨15.5⟩		
10 11 05	−0 40.6	6 6		14.6	8	
10 11 09	−3 07.6	2 2		(15.4)	5	*F*, (*D*)
10 11 13	−2 33.8	2 2		15.3	4	*F*, (*D*)
10 11 14	−0 39.9	2 2		15.3	4	
10 11 18	+0 48.0	8 8		14.9	11	*L*
10 11 37	−2 48.2	2 2		(15.4)	5	*F*, (*D*)
10 11 43	−2 33.6	1 1		⟨15.5⟩		
10 11 52	−0 37.4	6 6		(15.5)	9	*F*
10 12 06	−0 34.9	6 6		15.2	13	double, (*D*)
10 12 12	−0 05.5	8 8		15.3	3	*F*
10 12 17	−1 08.2	6 6		15.0	8	
10 13 19	−0 09.2	2 2		15.3	4	
10 14 24	−1 08.1	1 1		⟨15.5⟩		
10 14 39	−3 14.8	4 4	IC 600	13.9	10	*F*, (*D*)
10 15 13	−2 07.3	4 4		15.4	8	*L*, double
10 15 52	−3 21.2	1 1		⟨15.5⟩		

Table 2.

Square field of 36 square degrees area centered at R.A. 10^h4^m and Decl. $-0°30'$ (1950).

Positions, populations, diameters and estimated distances of clusters of galaxies, as identifiable on limiting panchromatic plates obtained with the 48-inch Schmidt telescope. The diameters are averages derived from the equal population contours which are bounding the clusters in Fig. 2. The distances are estimated on HUBBLE's old distance scale (corresponding to an apparent velocity of recession of distant galaxies equal to about 550km/sec per million parsecs). The distance ranges are designated as follows:

Near = smaller than 70 million light years (for instance the clusters Virgo, Cancer, Hydra I, Perseus, Coma).

Medium Distant = MD for the approximate range from 70 to 150 million light years (for instance the Corona Borealis cluster).

Distant = D, Very Distant = VD and Extremely Distant = ED describe successively the ranges from 150 to 250, 250 to 350 and greater than 350 million light years, respectively.

The clusters have been divided into three classes, namely:

a) *Compact clusters* with one single major central concentration of galaxies containing a dozen or more galaxies "contacting" each other. Many of the compact clusters such as those in Perseus, Coma and Corona Borealis show remarkable spherical symmetry.

b) *Medium compact clusters* have either two or more major concentrations of galaxies, or, if there is only one dominant major concentration, the galaxies in it are not visibly contacting each other, but are separated by distances two or three times those which are taken as the "classical" diameters of the galaxies.

c) *Open clusters* with the appearance of dispersed clouds do not show any prominent concentrations of the member galaxies. The "nearby" and very extended Ursa Major-cloud is an example of this class.

x and y are arbitrary coordinates increasing toward East and North and measured in cm. On the plate used the Boss Catalogue Star No. 13916 (R.A. $10^h5^m23^s$, Decl. $-0°7'.6$, epoch 1950) is located at $x = 18.7$ cm and $y = 69.5$ cm.

No.	Population	Character	Diameter cm	Distance	E ← x_{cm}	y_{cm} ↑N
1	107	Very open	2.2	MD	9.7	84.4
2	40	Medium compact	0.8	VD	10.3	82.5
3	66	Compact	0.5	ED	20.7	82.3
4	44	Medium compact	0.8	D	31.9	81.7
5	117	Open	1.3	VD	14.8	81.6
6	80	Open	1.3	VD	17.5	80.9
7	64	Medium compact	0.7	VD	10.8	80.0
8	68	Compact	0.6	VD	9.2	80.3
9	80	Medium compact	1.2	VD	26.9	78.6
10	43	Compact	0.5	ED	25.8	77.8
11	150	Open	2.2	D	7.1	78.1
12	111	Medium compact	1.2	VD	7.3	76.9
13	523	Open	4.6	D	19.2	73.3
14	93	Open	1.7	VD	16.3	74.2
15	81	Very compact	0.6	ED	15.5	71.9
16	310	Open	3.0	VD	8.2	73.1
17	98	Open	1.4	VD	22.4	73.1
18	82	Open	1.2	D	27.3	72.9
19	91	Open	1.3	VD	29.4	71.3
20	68	Open	1.0	VD	8.8	69.3
21	125	Open	1.4	VD	18.5	68.3
22	185	Medium compact	1.8	VD	24.1	69.1
23	152	Medium compact	1.8	D	14.7	67.7
24	487	Medium compact	5.0	Near	27.0	66.7
25	64	Medium compact	0.9	VD	28.9	66.7
26	68	Compact	1.0	VD	14.4	63.7
27	188	Compact	1.5	D	23.1	60.4
28	94	Open	1.5	D	13.6	60.1
29	154	Open	4.0	Near	8.7	55.5
30	74	Compact	1.0	VD	14.8	57.3
31	128	Open	2.7	Near	25.3	56.8
32	82	Compact	0.9	VD	21.4	52.0

Total of 4117 galaxies.
Average population per cluster 129 galaxies.
Medium distant and nearby galaxies are scattered more or less uniformly over the whole field.

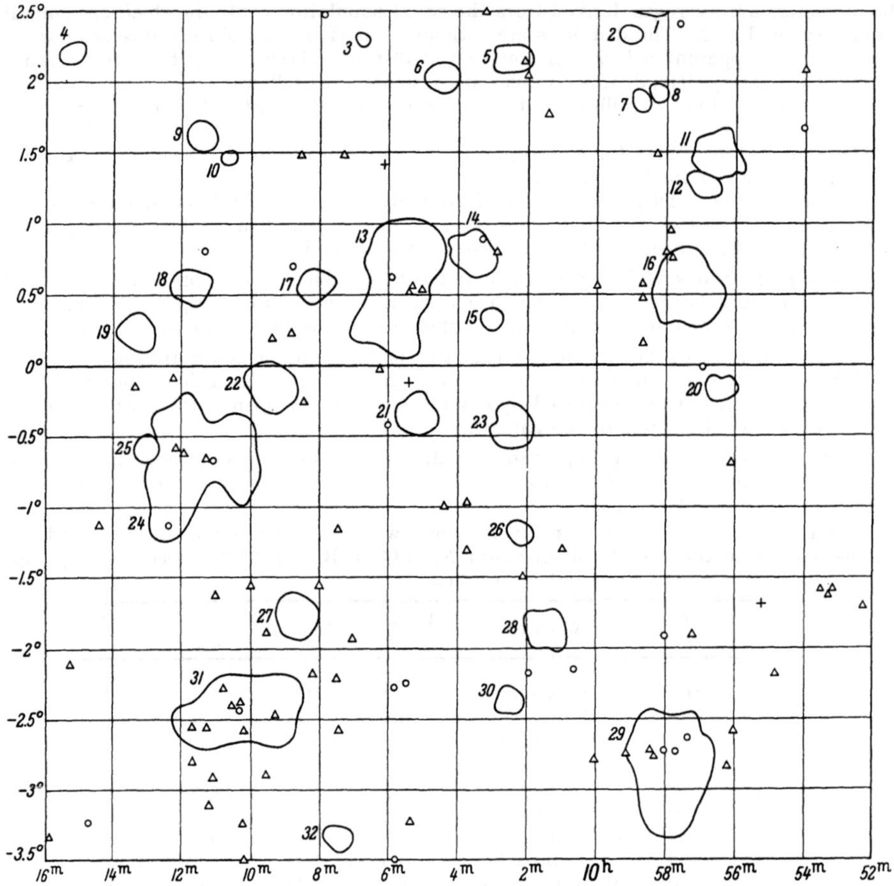

Fig. 2. Plot of all galaxies listed in Table 1 and of all clusters of galaxies shown in Table 2. The following symbols for apparent photographic magnitudes have been used. ■ $m_p \leq +12.0$. ⊡ $+12.0 < m_p \leq +13.0$. ● $+13.0 < m_p \leq +14.0$. ○ $+14.0 < m_p \leq +15.0$. △ $+15.0 < m_p \leq +16.0$.

+ GC stars Nos.	R.A.	Decl. 1950.0
13712	$9^h 55^m 1^s$	$-1°42'11''$
13916	10 4 23	−0 7 35
13932	10 6 7	+1 24 21

II. Well known clusters of galaxies.

3. As mentioned before, the existence of some of the nearest clusters was known to the HERSCHEL's over a century ago. These clusters were naturally investigated first when the large modern refractors and reflectors became available. Other clusters, much more distant, have more recently assumed a prominent place in the astronomical literature because they served HUBBLE and HUMASON as stepping stones for their exploration of the universal redshift to distances where the shifts $\Delta\lambda$ have values which make the symbolic apparent velocity of recession $V_s = c \cdot \Delta\lambda/\lambda_0$ as great as 65000 km/sec.

For a more detailed discussion of some of these clusters we refer to the literature [3], [7], [8]. We wish to call special attention here only to the possibility, suggested originally by ZWICKY [9], that the Milky Way system and the whole so-called Local Group of Galaxies [7], [8] form actually a part of the Virgo cluster, being located in its very outskirts. G. DE VAUCOULEURS [10] has investigated this suggestion more closely and has concluded that the Local Group of Galaxies is to be considered as a minor irregularity or secondary agglomeration near the border of the system of galaxies which is ellipsoidally distributed around the center of the Virgo cluster. Following a nomenclature introduced by H. SHAPLEY [11] this system has often been called the "Local Supergalaxy". According to DE VAUCOULEURS [10], radio observations made at the Ohio State University and at the University of Manchester, gave equal radiation intensity contour lines which closely follow the isopleths (equal numbers of galaxies contour lines) in the Local Supergalaxy. DE VAUCOULEURS has also started an investigation of what he calls the "Southern Supergalaxy", a supergalactic system, seen almost edge-on, and extending from Cetus, through Fornax, Eridanus and Horologium, to Dorado. This supergalaxy is probably the closest and most sharply defined of the large systems of galaxies beyond the Local Supergalaxy. Its apparent nucleus is marked by the dense group of more than a dozen bright galaxies including NGC 1365, 1374, 1379, 1380, 1381, 1387, 1389, 1399, 1404, 1427, 1437 and many fainter objects. With a distance modulus $m - M = 27.0$ (m = apparent magnitude, M = absolute magnitude of a star or galaxy), as estimated from redshifts and magnitude distribution, it is only slightly farther away than the Virgo cluster. The linear dimensions of both the Local and the Southern Supergalaxies on HUBBLE's old distance scale[1], are of the order of six to seven megaparsecs. This comes very close to ZWICKY's original estimate [9] that an average "cluster cell" has a volume of the order of 400 Mpc^3. It also strengthens his contention that these cluster cells, each containing one cluster, are space fillers.

In Table 3 we list some of the data on the best known large nearest clusters of galaxies, whose symbolic velocities of recession V_s are smaller than 10000 km/sec.

The locations and the mean symbolic velocities of recession V_s of the Virgo Cluster and of the Ursa Major Cloud are such that they perhaps both are part of the Local Supergalaxy mentioned previously [10].

III. Structure of individual clusters.

As suggested in the caption for Table 2, it is convenient to use a preliminary classification of clusters into three classes, namely compact, medium compact and open clusters. From a large number of investigations, the following general features of clusters of galaxies begin to delineate themselves. (For details consult in particular Refs. [3] and [21].)

4. Population of clusters. The total population of the richest clusters amounts to about ten thousand member galaxies in the brightness range $m_{\max}$ to $m_{\max} + 6$, where $m_{\max}$ is the photographic magnitude of the brightest member galaxy. There are of course many much smaller groups, but it is not certain at the present, to what extent they can be considered as independent dynamic units which are capable of traversing freely the humps of gravitational potential between the large clusters. Likewise it is not clear at the present time, just how many field galaxies actually exist. HUBBLE [7] originally estimated that field galaxies are

[1] Since no new distance scale for clusters has as yet been definitely established, we shall use throughout this article HUBBLE's original scale corresponding to a reshift equivalent to an apparent symbolic velocity of recession $V_s = c \cdot \Delta\lambda/\lambda = 550$ km/sec per megaparsec. c = velocity of light, $\Delta\lambda$ = shift of wavelength toward the red, λ = original wave length of the shifted line.

Table 3. *Some characteristics of the*

Cluster	1950 R.A.	1950 Decl.	$\mathcal{N}$ (Messier)	$\mathcal{N}$ Shapley-Ames	$\mathcal{N}$ 18-inch Schmidt	$\mathcal{N}$ 48-inch Schmidt	Brightest galaxy
Pisces	$1^h\ 0^m$	$+30°00'$	—	—	120	—	NGC 315
Subgroups: C_1	$1^h\ 21^m$	$+33°00'$	—	—	—	—	NGC 499
C_2	$1^h\ 6^m$	$+32°00'$	—	—	—	—	NGC 410
C_3	$0^h\ 39^m$	$+29°00'$	—	—	—	—	—
C_4	$0^h\ 14^m$	$+30°00'$	—	—	—	—	—
Perseus	$3^h\ 16^m$	$+41°19'$	—	1	360	—	NGC 1270
Fornax I	$3^h\ 30^m$	$-36°00'$	—	24	100	—	NGC 1316
Fornax II	$3^h\ 30^m$	$-21°00'$	—	30	—	—	NGC 1309
Cancer	$8^h\ 17^m.6$	$+21°06'$	—	—	84	300	NGC 2563
Hydra I	$10^h\ 34^m$	$-27°16'$	—	3	220	—	NGC 3309
Ursa Major Cloud	11^h to 13^h	$+20°$ to $+60°$	5	211	—	—	NGC 4826 (?)
Virgo: main body	12^h to 13^h	$0°$ to $+20°$	16	ca. 150	—	—	NGC 4486
Southern extension	$12^h.5$ to $13^h.5$	$0°$ to $-20°$	2				NGC 4594
Coma	$12^h\ 56^m.5$	$+28°13'.5$	—	—	654	10724	NGC 4489
Pegasus I	$23^h\ 18^m$	$+7°56'.5$	—	—	62	369	NGC 7619

$\mathcal{N}$ = number of cluster member galaxies included respectively in the Messier and Shapley-Ames catalogues or identified with the 18-inch and 48-inch Schmidt telescopes.

n = number of cluster member galaxies for which the values of V_s have been measured. $\overline{V}_s$ and $(\overline{\Delta V_s^2})^{\frac{1}{2}}$ were determined as averages from these n measures. Thus $\overline{V}_s = \sum V_s/n$ and $\overline{\Delta V_s^2} = \frac{1}{n}\sum (V_s - \overline{V}_s)^2$. All values of V_s are taken from the paper by Humason, Mayall and Sandage in the Astr. J., **61**, 97 (1956).

in the majority and that cluster galaxies do not account for more than a few percent of the total number of extragalactic stellar systems. From the extensive investigations with the Schmidt telescopes on Palomar Montain, Zwicky [*9*] concluded that cluster galaxies are the rule, rather than the exception and that field galaxies account at most for a few percent of the total. A field galaxy is defined as a system of high enough kinetic energy to escape the gravitational field of the large clusters of galaxies.

5. Types of galaxies in the clusters. The open and medium compact clusters, such as the Virgo cluster and the Ursa Major Cloud contain all types of galaxies, that is, globular, elliptical and irregular systems, as well as normal and barred spirals. The very rich and compact clusters seem to be made up mostly of elliptical and S_0 systems, with very few spirals in appearance. The explanation of this fact is to be sought for in the much more frequent close encounters of galaxies in compact clusters, as a consequence of which the loose parts of the individual galaxies are severed from them and are dispersed as intergalactic matter. Zwicky [*18*] on this basis, predicted the existence of observable amounts of both dark and luminous (stars) ***intergalactic matter.***

6. Types of intergalactic matter. Stimulated by the mentioned theoretical considerations, Zwicky undertook a systematic search for intergalactic matter with the Schmidt telescopes and the large reflectors of the Mount Wilson and Palomar Observatories. Interesting luminous formations were discovered which connect many widely separated galaxies [*3*], [*19*]. Likewise, enormous clouds of luminous intergalactic matter were found in the central regions of the Coma cluster the A-cluster and other clusters. From the fact that so far none of these luminous clouds has been observed to emit any emission lines or to be polarized, it is concluded that they consist of stars. It is remarkable, that most of the *best known nearby clusters of galaxies.*

m_p of brightest galaxy	$\bar{V}_s = c \cdot \overline{\Delta\lambda}/\lambda$ in km/sec		$(\overline{\Delta V_s^2})^{\frac{1}{2}}$	Extension	Distance (old scale) in Mpc	Character	Discoverer	Literature
+13.2	n	—	—	18°×5°	7	Medium compact	HUBBLE	[12], [13]
+13.5	—	4373	354	2°			HUMASON	
+13.3	—	5076	503	1°			ZWICKY	
—	—	—	—	1°				
—	—	6560	410	1°				
+12.7	5	5302	437	8°	11	Compact spherical	M. WOLF	[14], [15]
+10.1	5	1860	152	6°	3.3	Medium compact	SHAPLEY	[11]
+10.1	17	1692	482	16°×12°	—	Open	—	—
+13.8	4*	4854	427	7°	9		CARPENTER	[16]
+12.7	—	—	—	10°	(7.3)	Medium compact	ZWICKY	[17]
+8.0	115	1040	471	30°×40°	2	Open	—	—
+10.7	73	1222	643	15°×40°	2.3	Medium compact	—	—
+9.2								
+13.2	23	6645	1037	12°	—	Compact spherical	M. WOLF	[18]
+12.54	5	3606	273	2°	7.25	Medium compact	—	—

* In addition to NGC 2562 and 2563, the outlying galaxies NGC 2535 and 2623 were also included as possible members of the Cancer cluster.

luminous intergalactic formations are rather blue, corresponding to a colour index roughly equal to that of the outer parts of the spiral galaxies.

The existence of dark intergalactic matter within the large clusters of galaxies could be proved by ZWICKY [20], [31], [3] through counts of very distant galaxies and distant clusters of galaxies in the rear of nearby large clusters of galaxies. Thus, near the north galactic pole, the average numbers of galaxies per square degree and of clusters of galaxies, as counted with the 48-inch Schmidt telescope, are respectively only about one third and one sixth of the corresponding numbers of galaxies and of clusters of galaxies in certain regions as far as 30° from the galactic pole. This deficiency in the number of faint galaxies and of clusters of galaxies can be accounted for if we assume a local absorption of light of the order of six tenth of a magnitude, because of its passage through intergalactic dust selectively accumulated within the Coma cluster. The same effect holds true, although to a lesser extent, for all regions covered by other nearby large clusters, such as the Virgo cluster.

7. Segregation of galaxies of different brightness and of different structural types within the large clusters. It is generally found that the bright galaxies show a greater tendency toward clustering than the faint ones. While the former are relatively more concentrated toward the center, the latter are comparatively more numerous toward the outskirts of a cluster. This fact was overlooked by the original investigators of clusters, who, from the statistics of the central regions of clusters, derived a completely erroneous luminosity function [22] for the member galaxies, which, instead of increasing monotonely with decreasing brightness of the galaxies, showed a sharp maximum at the absolute photographic magnitude $M_p = -14.2$ and a dispersion of only 0.85 magnitudes. With the subsequent discovery of many dwarf stellar systems, the maximum in the luminosity function was pushed more and more toward smaller luminosities. At the same time it became apparent, that the relative number of irregular galaxies, originally estimated by HUBBLE [7], [8] at only 2.5%, increased enormously.

The degree of segregation is illustrated by the following data obtained with the 18-inch and the 48-inch Schmidt for the Coma Cluster. The total numbers of cluster galaxies counted with these two telescopes are respectively n^t (18-inch) $= n^t_{18} = 654$ and n^t (48-inch) $= n^t_{48} = 10724$, corresponding to the ranges in ap.parent photographic magnitude $+13.2 < m_p \leq +16.5$ and $+13.2 < m_p \leq +19.0$. It therefore is $\bar{R} = n^t_{18}/(n^t_{48} - n^t_{18}) = 0.065$. On the other hand, if we consider only the center of the cluster, we count respectively per square degree with the two Schmidts within 5′ of arc from the center $\bar{n}_{18} = 1431$ and $\bar{n}_{48} = 2578$. Therefore R(center) $= 1.25$. As we go to clusters poorer in population the values of R(center) *decrease*, and the values of $\bar{R}$ *increase*.

The segregation of types has so far only been investigated for the brighter galaxies of the Virgo cluster, as they are listed in the Shapley-Ames catalogue [*2*]. The spirals among these galaxies show a lesser tendency toward clustering than the ellipticals, as was shown by ZWICKY [*23*] through a novel statistical method of *regularization* of the cluster.

Table 4. *Number of galaxies per square degree at various distances from the center of the Coma I cluster, as observed with the 18-inch Schmidt telescope.*

Approximate range of apparent photographic magnitudes: $+13.2 \leq m_p < +16.6$.

Distance from center in minutes of arc	Numbers per square degree observe	Calculated
2.5	2777	2692
5	1431	1763
10	643	795
20	285	281
30	193	158
40	120	103
50	95	74
60	66	55
80	15	33
100	23	19.5
120	12	10.5
140	1	4.5
160	2	1.0

8. The radial distribution of galaxies in the rich compact clusters. Among the rich compact clusters, remarkably many show almost perfect spherical symmetry. Among these are, for instance, the Hydra I, the Coma I and the Perseus cluster. Furthermore, the radial distribution curve of the number of the brighter galaxies per square degree, as a function of the distance from the center can be very closely represented by the density distribution of a projected gravitational isothermal gas sphere, composed of atoms or molecules of one mass only. Since R. EMDEN [*24*] originally calculated the properties of these spheres, we shall call them Emden isothermal spheres. For instance, in the Coma I cluster, from the center $r = 0$ to a distance $r = 160$ minutes of arc, the number of galaxies per square degree falls from about 3000 to 1. The agreement between the observed numbers of galaxies and those calculated from the Emden isothermal sphere is shown in Table 4.

From the agreement of the observed with the calculated numbers we can draw two very important conclusions, namely, (1) the Coma I cluster is stationary in the sense of statistical mechanics, and (2) NEWTON's law of gravitation must be valid over distances of several millions of light years to bring about the mentioned agreement [*25*]. These conclusions are strengthened by the fact of the partial segregation of bright and faint (massive and less massive) galaxies. A quantitative check of the observed degree of segregation with the theory is, however, not possible at the present time, for the simple reason that no one has as yet developed any such theory. In the absence of definite and satisfactory ideas about the statistical mechanics of gravitating systems (difficulty of divergent integrals and non-availability of any thermodynamic equations of state and functions of state) it might be suggested that we obtain some ideas about clustering through the use of modern computing machines, following step by step the changes in configuration of an n-body system of masses which react according to NEWTON's law of gravitation ($n \gg 1$).

One further uncomfortable conclusion, which can be drawn from the stationary character of the large spherical clusters of galaxies, concerns the time necessary for the formation of these clusters. This time would seem to be very much longer than the 10^{10} years admissible in an expanding universe [26].

In this connection the very important question arises if the radial distribution of galaxies in rich clusters is not only functionally the same in all clusters but also in absolute dimensions. A study of this problem can conveniently be made through the determination of the so-called *structural index* or the structural length α of the equivalent Emden isothermal sphere. If this sphere consists of particles of one mass only, α is theoretically given by

$$\alpha = (\overline{v^2}/12\pi \Gamma \bar{\varrho}_0)^{\frac{1}{2}}, \tag{8.1}$$

where $\overline{v^2}$ is the square of the velocity dispersion in the cluster, Γ is the universal gravitational constant and $\bar{\varrho}_0$ is the average central density of the cluster. The "surface density" $q = q_0$ in the center of the Emden sphere as projected on a plane is given by

$$q_0 = n_{r=0} \times \overline{\mathcal{M}} = 6.056\,\alpha\bar{\varrho}_0 = 6.056\,[\bar{\varrho}_0\,\overline{v^2}/12\pi\Gamma]^{\frac{1}{2}}, \tag{8.2}$$

where $\overline{\mathcal{M}}$ = average mass of the $n_{r=0}$ galaxies counted per unit area (cylinder) in the center of the cluster. The structural length α of a cluster can be determined through counts of its member galaxies. If the number counted per square degree, for instance, and as a function of r, ie $\bar{n}(r)$, then the function $\bar{n}(r)$ has its inflexion point at $r_i = 1.75\alpha'$, where α' is the structural index in angular measure. The absolute structural length is consequently obtained as $\alpha = D\alpha'$. Provided, therefore, that the distance D is known, and $\overline{v^2}$ has been determined through a study of the radial velocities of a sufficient number of member galaxies of the cluster, the counts of the brighter members allow us to determine the central density $\bar{\varrho}_0$ and, for that matter, the density $\varrho(r)$ at any distance from the center of the cluster.

As an illustration we give some data on the Coma cluster which were derived from the analysis of the distribution of the 654 member galaxies counted with the 18-inch Schmidt. The structural index derived from this distribution is $\alpha' = 2$ minutes of arc. Assuming a value for the distance $D = 13.8$ million parsecs (HUBBLE's old distance scale), we obtain $\alpha = 8000$ parsecs $= 2.48 \times 10^{22}$ cm and $\bar{z}(r=0) = 1.06 \times 10^6$ bright galaxies per cubic megaparsec in the center of the cluster. The spatial velocity dispersion, as determined from radial velocities of 21 galaxies is $(\overline{v^2})^{\frac{1}{2}} = 1800$ km/sec. It therefore follows from Eq. (8.1) that $\bar{\varrho}_0 = 2.1 \times 10^{-23}$ gram/cm³, a remarkably high central density. Furthermore, the average mass of the 654 brightest galaxies is $\overline{\mathcal{M}} = 5.98 \times 10^{44}$ gram $= 3 \times 10^{11}\,\mathcal{M}_\odot$ and the total mass of the cluster is $\mathcal{M}_{\mathrm{Cl}} > 2 \times 10^{14}\,\mathcal{M}_\odot$.

The same result was deduced by ZWICKY [27] as long ago as 1932 through an application of the Virial theorem to the Coma Cluster. The Coma Cluster, as well as all other clusters for which velocity dispersions are known, appears to be therefore several hundreds of times more massive than would be judged on the basis of estimates of the masses of the brighter component galaxies as derived from their luminosities. Whether these galaxies are much more massive than the conventional methods of mass determination indicate or whether dwarf galaxies and intergalactic matter account for the greatest part of the total mass of clusters remains to be seen. Recent observations of the intensity of the 21 cm radio wave from neutral hydrogen in the Coma Cluster have furnished a first most important clue. They indicate indeed [28] that the total mass of the neutral atomic hydrogen alone in the Coma Cluster is of the order of $10^{14}\,\mathcal{M}_\odot$, a result which, if correct, closely checks ZWICKY's original deductions and predictions [27]. It was also announced recently that the 21 cm wave is shifted by an amount corresponding to $\bar{V}_s \sim 7000$ km/sec which is in close agreement with the optically observed value of $\bar{V}_s$ (see Table 3). From the width of the 21 cm wave a velocity dispersion $(\overline{\Delta V_s^2})^{\frac{1}{2}} \sim 500$ km/sec is deduced, which is about half of the corresponding value obtained from the optically observed spectra of 23 member galaxies. If correct, these radio telescopic observations greatly strengthen ZWICKY's contention [27], [4], [3] that the average density of matter throughout the universe is far greater than was previously thought possible by most astronomers [7].

Structural indices have so far been well determined for the globular clusters in Pegasus I, Cancer, Fornax, Hydra I, Perseus, Coma, Corona Borealis and Hydra II. Assuming HUBBLE's old distance scale, the absolute values of these indices lie all in the remarkably narrow range from 1.42×10^{22} cm to 3.30×10^{22} cm. If spherical clusters of roughly equal populations in the first three magnitude ranges from $m_{\max}$ to $m_{\max}+3$ are chosen, the structural lengths are almost identical ($m_{\max}$ = apparent magnitude of the brightest cluster galaxy). Thus the structural lengths of the clusters in Coma and Corona Borealis are closely the same [3], [4], [29]. It will be of the greatest interest to pursue this subject further and to determine structural indices for a few dozen clusters in order to establish in this way an independent absolute yardstick for the measurement of very large distances. Such a yardstick will be particularly reliable if it can be shown by independent arguments that the richest spherical clusters at all distances within the reach of the 200 inch telescope are of the same structure and the same population.

9. Clusters near and far. If there exists an expansion or an evolution of the universe, or both, one should expect clusters of galaxies at great distances to be different from clusters nearby. For instance, differences in structure, in population, in the frequency of supernovae, in degree of segregation of faint and of bright member galaxies, in the colour range and in the characteristic spectra of the member galaxies should be expected. Feasible tests for such differences were formulated by ZWICKY [30], and data for several of these tests are now being gathered with the Schmidt telescopes and the 200-inch reflector on Palomar Mountain. One of the simplest of all tests is based on the counts of the member galaxies of a cluster in different approximate magnitude ranges [3], [21], [30], and of the number of member galaxies which are in "contact" with one another. For instance, in the richest spherical clusters, about a dozen of the most luminous galaxies (range $m_{\max}$ to $m_{\max}+3$, where $m_{\max}$ is the apparent magnitude of the brightest galaxy) are in virtual contact, and this number is independent of the distance, as far as the data indicate which were obtained with the large telescopes on Palomar Mountain.

Instead of studying the total numbers and the concentration of the cluster member galaxies in a definite brightness range, for instance from $m_{\max}$ to $m_{\max}+3$, we may also attempt to derive criteria for the distances of clusters through an analysis of all the member galaxies $\mathcal{N}_{\mathrm{Cl}}$ countable with a given telescope and of the total number $\bar{n}_0$ which represents the concentration of galaxies per square degree extrapolated to the center of the cluster. ZWICKY [27], [21], [3] found from extensive counts with the 48-inch Schmidt that

$$\mathcal{N}_{\mathrm{Cl}} = A(\gamma - \gamma_0) + \Delta P(\gamma) \tag{9.1}$$

where γ is the diameter of the cluster, defined as the average diameter of the population contour line or isopleth along which the number of galaxies per square degree is twice that of the average number in the surrounding field. A is a constant. The identifiability of a cluster as such vanishes on the average when its diameter becomes smaller than $\gamma = \gamma_0$. This means that at this point so many member galaxies are beyond the reach of the telescope used that the cluster cannot be recognized as such, although some of the brighter member galaxies may still be recorded. If all clusters are included in the analysis, the data show a considerable scatter [3] around the relation (9.1). If only the richest spherical clusters are used, the relation (9.1) should represent the observations very closely. The function $\Delta P(\gamma)$, which contains higher order terms in $(\gamma - \gamma_0)$, becomes appreciable only for nearby clusters, for which γ is of the order of one degree. For the central density $\bar{n}_0$ per square degree of various clusters as photographed consistently with the *same* telescope we therefore have the following relation:

$$\bar{n}_0 = B(\gamma - \gamma_0)/\gamma^2, \tag{9.2}$$

where B is a constant, or, as long as $\gamma \gg \gamma_0$,

$$\bar{n}_0 = B/\gamma. \tag{9.3}$$

For photographs taken with the 48-inch Schmidt and using Eastman 103a-E emulsions behind red Plexiglass filters it is: $\gamma_0 = 1'.79$.

Example: For the Coma cluster (Table 4) and the Corona Borealis Cluster [21], [3], as counted with the 18-inch Schmidt we have respectively $\bar{n}_0(\text{Coma}) = 2777$ and $\bar{n}_0(\text{Cor. Bor}) = 10887$ galaxies per square degree within circles of $2'.5$ and $1'$ radii. Therefore $\bar{n}_0(\text{Cor.Bor})/\bar{n}_0(\text{Coma}) = 3.92$, while the ratio of the symbolic velocities of recession (in km/sec) is $V_s(\text{Cor.Bor})/V_s(\text{Coma}) = 22000/6600 = 3.33$. This ratio probably also represents the relative distances of the two clusters.

On the other hand, from counts with the 48-inch Schmidt it was found [21], [25], [27] that within a circle of $5'$ radius for the Coma Cluster $\bar{n}_0(\text{Coma}) = 2748$ and for a circle of $2'$ radius of the Corona Borealis Cluster $\bar{n}_0 = 12865$. Thus $\bar{n}_0(\text{Cor.Bor})/\bar{n}_0(\text{Coma}) = 4.68$.

The greatest concentration of galaxies so far counted by Zwicky (unpublished result) is in the center of an extremely distant cluster discovered by M. L. Humason at R.A. $0^h 24^m$ and Decl. $+16°54'$. Within a circle of 0.5 minutes of arc radius of the center of the cluster the density of occupation amounts to $\bar{n}_0 = 220032$ galaxies per square degree as counted on plates obtained with the 200-inch telescope. The corresponding density for the Hydra II cluster, whose symbolic velocity of recession $V_s = 61000$ km/sec is the largest so far published by Humason, is $\bar{n}_0(\text{Hydra II}) = 91700$ galaxies per square degree. The distance of the Cluster 0024+1654 may thus be estimated to be 2.4 times the distance of the Hydra II cluster.

Both the study of the *structural indices* α' (in angular measure) and of the *total populations* of clusters of galaxies, as seen with a given telescope, lead thus to two entirely novel methods for the determination of the relative distances of rich spherical clusters. Both methods are based on the fundamental principles of *dimensionless morphology*, since they make only use of the operations of identification of objects and of counting them. The distance obtained from α' is thus independent of any possible interference of intergalactic obscuration or of any changes of the apparent magnitudes and colours of the cluster member galaxies Unfortunately the determination of the apparent structural index may become difficult when exceedingly distant clusters are involved in which fewer than one hundred members can be counted with the 200-inch telescope In this case the study of the so-called *distribution index* [29] may be profitably substituted for the analysis of the structural index.

Suppose that we intercompare the distribution of galaxies in very rich spherical clusters for each of which three exposures with the 200-inch or the 48-inch Schmidt are available. For the first exposure of any cluster we adjust the characteristic exposure time t_i so that only the brightest two or three member galaxies are recorded. The two other exposure times may be chosen as $3t_i$ and $9t_i$, or, if the sky glow interferes in a shorter time than $9t_i$, we may chose $2t_i$ and $4t_i$. We now find on the second and third exposures of all clusters those circles of radii ξ for which the distribution index DI

$$\mathrm{DI} = \mathcal{N}(\xi)/[100\,\mathcal{N}(\xi/10) - \mathcal{N}(\xi)] \tag{9.4}$$

is equal for all clusters. Here $\mathcal{N}(\xi)$ and $\mathcal{N}(\xi/10)$ are the total number of cluster galaxies within circles of the radii ξ and $\xi/10$ respectively, after $\bar{n}_\xi$ galaxies for each square degree have been subtracted as the maximum possible number of background galaxies within the circle of radius ξ. $\bar{n}_\xi$ is the average number of galaxies per square degree along the circle of radius ξ. $\mathcal{N}(\xi)$ is therefore the minimum possible population of the clusters. In Table 5 we list some data [29] for those five most distant clusters for which HUMASON has determined the symbolic velocities of recession V_s. The values of ξ are in minutes of arc and have been chosen inversely proportional to the values of the corresponding V_s. If the symbolic velocities of recession V_s are actually proportional to the distances D of the clusters, then ξ is proportional to $1/D$.

Table 5. *Intercomparison of the populations in the first three brightest magnitudes and of the distribution indices (DI) of five distant clusters of galaxies and of the nearer clusters in Coma and in Corona Borealis.*

Cluster	V_s in km/sec	$(\overline{\Delta V_s^2})^{\frac{1}{2}}$	ξ	$\mathcal{N}_{\min}$	DI
Hydra II	61000	—	8′.6	451	0.110
1253 + 4422	59300	—	8.8	394	0.115
0925 + 2044	57600	—	9.1	391	0.129
1304 + 3110	54900	—	9.6	332	0.107
0138 + 1840	51700	—	10.0	461	0.115
Corona Borealis	22000	1208	24.0	551	0.103
Coma	6645	1037	78.3	467	0.084

Clusters are being designated by two numbers referring to right ascension and declination. Cl 1253 + 4422 for instance is located at R.A. 12^h53^m and Decl. 44°22′ (epoch of 1950). The coordinates of the Hydra II clusters are R. A. $8^h55^m38^s$, Decl. + 3°21′6″: Corona Borealis cluster, R. A. 15^h20^m3, Decl. + 27°53′12″, Coma cluster R.A. 12^h56^m5, Decl. + 28°13′.5.

It is seen that the average value of the distribution index DI of five distant clusters is $\overline{\mathrm{DI}} = 0.115$ and that the largest deviation from the mean is $\Delta(\mathrm{DI}) = 0.014$. The values of DI for the Corona Borealis cluster and the Coma cluster, as obtained from counts with the 48-inch and the 18-inch Schmidt telescopes respectively, have been added for comparison, although it is difficult to adjust these counts exactly to the counts obtained with the 200-inch telescope for the five more distant clusters. In any event, it is seen that for rich spherically symmetrical clusters the inverse of the angular radius ξ of the circles corresponding to equal distribution indices DI promises to give fairly reliable values of the relative distances of these clusters. Notice also that all of the clusters in Table 5 have a minimum total population $\mathcal{N}(\xi) = \mathcal{N}_{\min}$ which is independent of the distance. The total population, to the distances reached, does not therefore show any effects of evolution.

A study of the distribution index of the previously mentioned exceedingly distant cluster 0024 + 1654 leads to a value of ξ which is about half that of the

Hydra II cluster. Cluster 0024+1654 may therefore be expected to lie at about twice the distance of the Hydra II cluster. This conclusion which checks that previously obtained from a study with one and the same telescope of the number of galaxies per square degree in the centers of clusters as a function of the distance of the clusters. A photograph of the central regions of Cluster 0024+1654 is reproduced in the Fig. 3.

In Fig. 4 we give a plot of the distribution of galaxies in and around Cl 0024+1654, as it was derived from the original plate of Fig. 3.

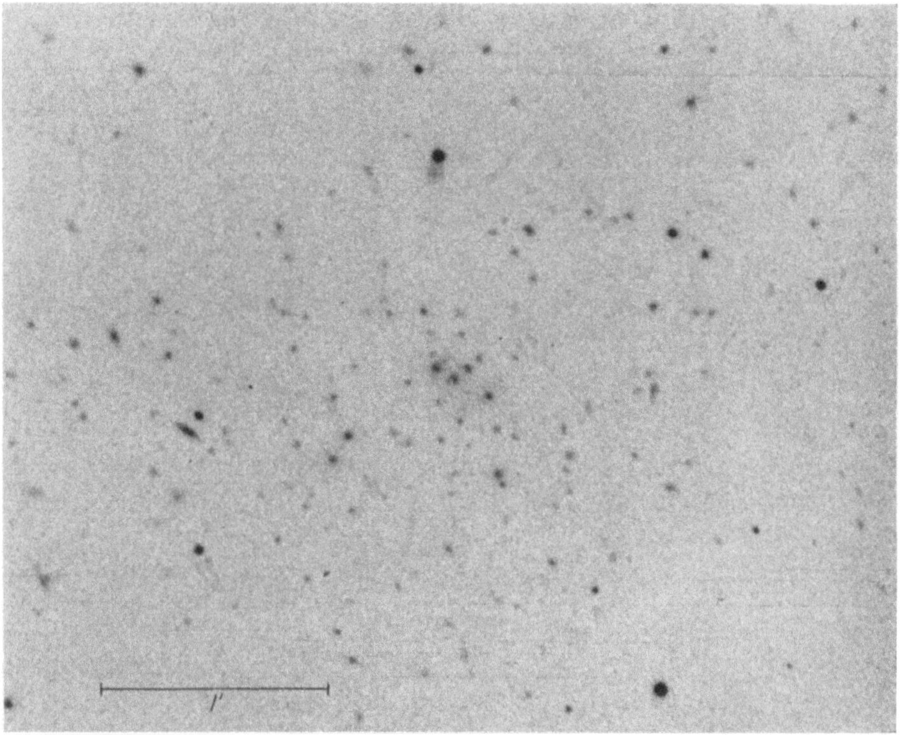

Fig. 3. 200-inch telescope photograph of the central parts of the cluster of galaxies Cl 0024+1654, obtained on Eastman 103*a-D* emulsion behind *GG* 11 Schott filter: exposure time 45 min, seeing 2. The scale indicates one minute of arc.

One of the most important conclusions of these investigations is that rich spherical clusters of galaxies, at all distances which can be reached with the 200-inch telescope, seem to contain approximately the same total populations as far as galaxies in the first three brightest magnitude classes are concerned. Also, their structures, radial distribution curves, relative segregation of galaxies of different brightness seem to be essentially the same at all distances.

10. The luminosity function of cluster galaxies. Disregarding the non-linear term in Eq. (9.1), and assuming, in close agreement with the data available for the more distant clusters, that their total populations $\mathcal{N}_{\mathrm{Cl}}$, as seen with a given telescope, are $\mathcal{N}_{\mathrm{Cl}} = A(\gamma - \gamma_0)$ we may derive a luminosity function for the brighter cluster member galaxies. It follows [*20*], [*3*] that

$$\nu(M)\,dM = B \times 10^{(M - M_{\mathrm{max}})/5}\,dM \tag{10.1}$$

where $\nu(M)\,dM$ is the number of galaxies in a given volume of space whose absolute photographic magnitudes lie in the range M to $M+dM$, and $M_{\max}$ is the magnitude of the brightest cluster galaxy.

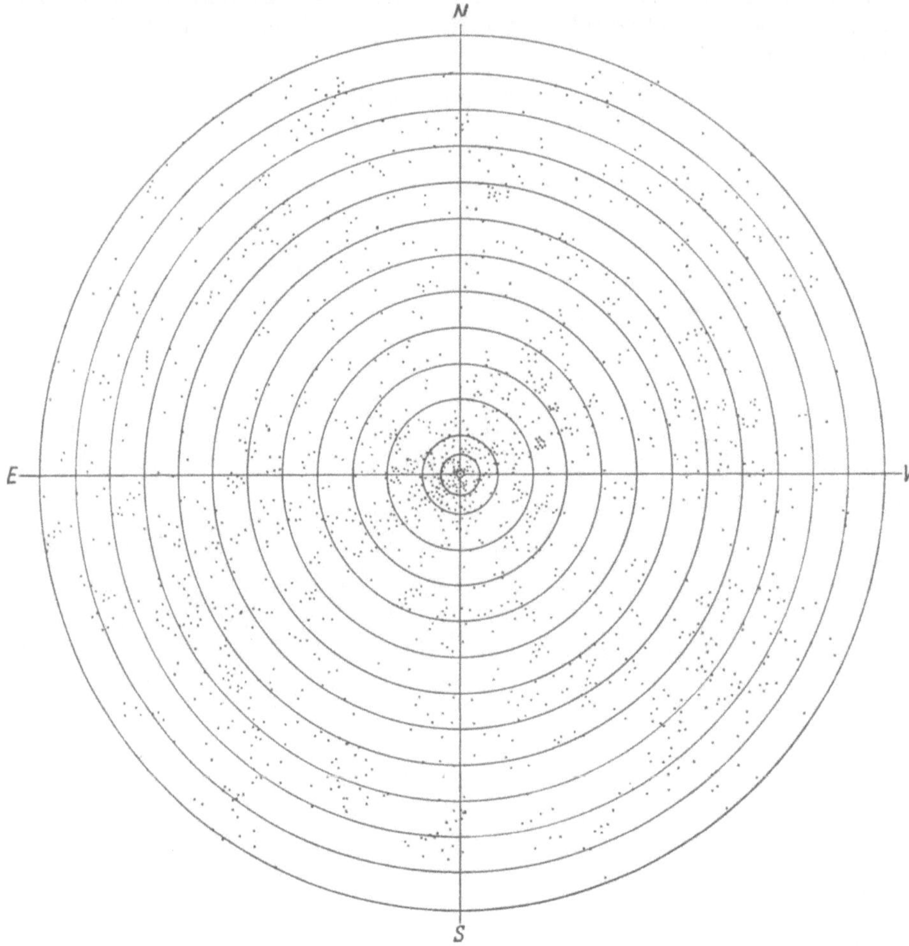

Fig. 4. Distribution of galaxies in and around Cl 0024 + 1654. The number of galaxies in the very center tends to a value 250000 per square degree, the highest as yet counted in any part of the sky. The circles shown have respectively the radii 0.5, 1, 2, 3, 4, ..., 11, 12 min of arc.

IV. Kinematics and dynamics of clusters of galaxies.

11. A very extensive report [*31*] of many years of work on the redshifts of galaxies at the Mount Wilson, Palomar and Lick Observatories has only recently been published. This report contains the redshifts of 620 galaxies observed at Mount Wilson and Palomar. Included in these data are the redshifts for 26 clusters of galaxies. In addition, the redshifts of 300 galaxies are given as observed at the Lick Observatory. Among all of these galaxies there are 114 which were observed both at Lick and at Mount Wilson-Palomar. The enormous wealth of data contained in the mentioned report will no doubt lead to many theoretical analyses, but the time has been too short for any of them to appear in the literature. We therefore indicate here only a few of the preliminary conclusions which these redshift data allow us to draw about the physical conditions which prevail within and between the clusters of galaxies.

α) The velocity dispersion within clusters of galaxies. It may be stated as a preliminary conclusion that the velocity dispersion increases with the size and the population of a cluster. For field galaxies and members of small groups the dispersion in radial velocities $(\overline{\Delta V_s^2})^{\frac{1}{2}}$ is of the order of 250 to 500 km/sec. For the largest clusters like those in Coma and Corona Borealis it may be as high as 1200 km/sec, corresponding to a dispersion of the actual peculiar space velocities of the order of 2000 km/sec.

β) Translation, rotation or oscillation of a cluster of galaxies as a whole. There is no indication that either translation or rotation of one cluster of galaxies relative to the surrounding neighboring clusters is of a magnitude which is easily observable. Except for the systematic apparent relative velocities which must be ascribed to the universal redshift, there are no detectable random motions of clusters of galaxies relative to one another. This is a most surprising and important fact, indicating either a radical change in the law of gravitation over great distances or the apparent elimination of gravitation because of the finiteness of the speed of propagation of gravitational forces [*21*].

γ) The masses of galaxies and of clusters of galaxies. Using the virial theorem [*27*], [*3*], the masses of clusters of galaxies may be derived from the observed velocity dispersions. An analysis of the Coma Cluster by ZWICKY [*27*], [*3*] thus produced the result that the average mass of one of the 670 brightest member galaxies of this cluster should be $\mathcal{M}_{\mathrm{Cl}} = 10^{11} \mathcal{M}_{\odot}$ to $2 \times 10^{11} \mathcal{M}_{\odot}$. This is clearly an average mass for galaxies which is much greater than the values obtained by any other method. Actually, if the masses of compact globular galaxies should be as great as $10^{11} \mathcal{M}_{\odot}$, such galaxies acting as gravitational lenses should have been discovered long since [*32*], [*3*]. We must therefore conclude that only a small fraction $a \mathcal{M}_{\mathrm{Cl}}$ of the mass of the Coma cluster derived from the virial theorem is actually accounted for by the 670 brightest galaxies and that the remainder $(1-a) \mathcal{M}_{\mathrm{Cl}}$ resides in dwarf galaxies and in intergalactic matter concentrated within the confines of the cluster. This conclusion actually prompted the search for dwarf galaxies and for intergalactic matter, both of which have been subsequently found [*27*], [*3*], [*28*]. If, as may be reasonably assumed, a good fraction of the mass of the cluster resides not only in dispersed intergalactic matter but also in dwarf galaxies, it can be shown, that the luminosity function of the cluster member galaxies must increase with algebraically increasing absolute magnitude M roughly as constant $\times 10^{\beta M}$ where β lies somewhere in the range from 0.2 to 0.4. This means that the number of galaxies per unit volume of space is proportional to $\mathcal{M}_g^{-2.5\beta}$ where $\mathcal{M}_g$ designates the mass of a galaxy. This is in agreement with the conclusions about the luminosity function which we have discussed in section (*Cg*).

δ) Intercomparison of the kinematics of galaxies and of clusters of galaxies. Neighboring galaxies, relative to one another, may have velocities of 100 to 5000 km/sec, or even more [*3*], [*19*], [*21*] while stars within galaxies have velocities usually inferior to 100 km/sec. The energy per gram residing in the translational motion of galaxies is therefore very much greater than the energy per gram residing in the translational motion of stars within galaxies. On the other hand the translational velocities of neighboring clusters of galaxies relative to one another are *small* compared with the relative motions of the individual galaxies. These are most important facts, the significance of which is not yet clear, as we have already mentioned in the preceding.

V. Counts of clusters of galaxies in depth; numbers as a function of angular size.

12. One of the most powerful methods which has recently been developed for the analysis of observations on the large scale distribution of matter in the universe is that of *dimensionless morphology* [*20*], [*21*], [*3*]. This method makes use only ot the operations of identification of various objects, of counting them and of measuring angles as seen from the site of the observer. Most gratifying and puzzling results have been obtained in particular from the application of this method to the study of the statistical distribution of clusters of galaxies in dependence on angular size. It can be shown, that if globular clusters of galaxies of the same character were distributed uniformly and randomely throughout a non-expanding Euclidean universe, the number of such objects, within a given solid angle of the sky which subtend linear angular diameters lying in the interval γ to $\gamma + \Delta\gamma$ would be equal to

$$N_\gamma \cdot \Delta\gamma = \text{const} \cdot \Delta\gamma/\gamma^4 \qquad (12.1)$$

provided that no optical obscuring effects or fading because of distance effects remove any of the objects involved from visibility. For an expanding universe of the type described by EINSTEIN and DE SITTER [*33*] we should expect

$$N'_\gamma \cdot \Delta\gamma = \text{const} \cdot \Delta\gamma/\gamma^4 \left[1 - \overline{V}_s(\gamma)/c^3\right], \qquad (12.2)$$

where $\overline{V}_s(\gamma)$ is the average apparent symbolic velocity of recession for globular clusters of galaxies (or any other objects under investigation) whose average angular diameters are equal to γ. The details of how this test for the expanding universe was applied to the data on 706 clusters of galaxies of the type listed in Table 2 is to be found in the literature [*21*]. We here give only the final results as represented by the data in Table 6 and in Fig. 5.

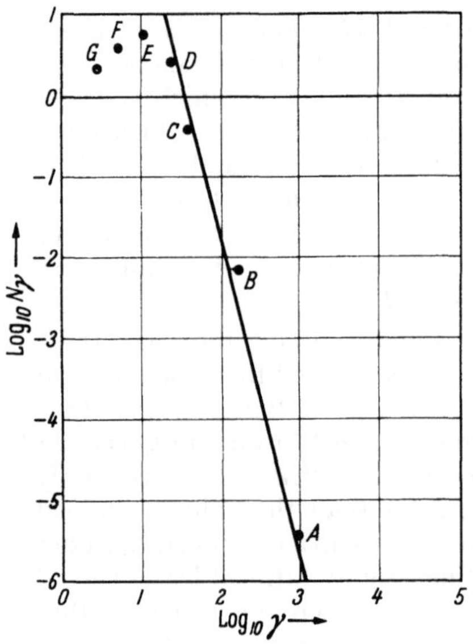

Fig. 5. Distribution of the 706 clusters of galaxies of Table 6 in dependence on their angular diameters. $N_\gamma \cdot \Delta\gamma$ is the number of clusters per one hundred square degrees whose angular diameters lie in the range γ to $\gamma + \Delta\gamma$. One minute of arc is used as the unit for measuring the angles.

Table 6. *Distribution of the Angular Diameters γ of 706 Clusters of Galaxies located in areas totalling 600 square degrees in regions around the North Galactic Pole.*

Among many good plates available from the 48-inch Schmidt telescope those were chosen which contain the largest numbers of distant faint clusters, thus indicating regions of relative minima of intergalactic obscuration. The first column gives the ranges $\Delta\gamma$ of the diameters in minutes of arc. In the second column are the numbers $\mathfrak{N}$ of clusters of the whole area of 600 square degrees which lie in the various ranges $\Delta\gamma$, as identified from 48-inch Schmidt plates. The final column lists the function $N_\gamma = \mathfrak{N}/6 \cdot \Delta\gamma$ which gives the average number of clusters per one hundred square degrees in a range of angular diameters equal to one minute of arc around the midpoint of the range.

$\Delta\gamma$	$\mathfrak{N}$	N_γ	$\Delta\gamma$	$\mathfrak{N}$	N_γ
1800′—300′	(see text)	(3.3×10^{-6})	15—7.5	273	6.07
300 — 60	10	6.96×10^{-3}	7.5—4.0	80	3.81
60 — 30	67	0.372	4.0—2.0	27	2.25
30 — 15	250	2.78			

The point A of Fig. 5 is naturally quite uncertain, since there are only a few clusters of galaxies of large apparent diameters. The possible relative fluctuations to be expected for such small numbers are of course considerable. In agreement with the operational definition of clusters of galaxies and of their diameters (see text following Fig. 2) we have admitted for our purposes within the unobscured 20000 square degrees around the two galactic poles only the Virgo cluster into the range γ from 1800 to 300 minutes of arc. The value of N_γ chosen for the point A is consequently equal to 3.33×10^{-6}.

In analysing Fig. 5 we notice in the first place that the points A, B, C and D lie as closely on the theoretical straight line for a non-expanding Euclidean universe [Eq. (12.1)] as may be reasonably expected. It is also noteworthy that the analysis which was carried out some time ago [*20*] on the basis of only 156 clusters checks very closely with the fuller analysis of the 706 clusters presented here. In both cases the rise of N'_γ with decreasing values of γ, above the straight line shown, which we should expect from the theory of the expanding universe, does not materialize. This is in agreement with the fact that, except for the spectral redshift itself, none of the effects predicted for the case of an expanding universe has as yet been found to exist. The newest data [*31*] on the deviations of $V_s = C \cdot \Delta\lambda/\lambda$ from the linearity with distance make it in fact also doubtful if these deviations, if real, can be explained on any of the presently current models of an expanding universe. The investigations on the number of clusters of galaxies as a function of angular size are being extended to include several thousand clusters, and a definite decision between the concepts of expansion and non-expansion of the universe should then become possible. The functions N_γ and N'_γ in the Eqs. (12.1) and (12.2) show differences at great distances which are greater than any which have so far been pointed out for conceivable tests deviced to decide between expansion and non-expansion of the universe. For an apparent velocity of recession of the order $V_s = c/6$, where c is the velocity of light, the ratio N'_γ/N_γ is equal to about 2. In relation to the data shown in Fig. 5 this means that to the left of the point B the points C, D and E should rise above the straight line shown, and somewhere between C and D the observational point should lie by an amount Log 2 above this line.

The location of the points E, F and G is of course due to the failing power of the telescope and may also be caused partly by the obscuration through intergalactic dust which seems to be thinly and somewhat unevenly spread throughout the whole visible universe [*20*].

VI. Distribution of clusters of galaxies in breadth.

13. Both the distribution of galaxies and of clusters of galaxies in some ten thousand square degrees around the north galactic pole, as counted on panchromatic plates obtained with the 48-inch Schmidt telescope is surprisingly *non-uniform*. The unevenness of the distribution of galaxies can only partly be explained by clustering. The striking fact that at the galactic pole itself far fewer galaxies are counted than in some regions thirty or forty degrees away from it was interpreted by Zwicky [*20*], [*3*] as indicating a local relative accumulation or intergalactic dust within the Coma cluster, as well as other nearby clusters. The data on counts of very distant clusters of galaxies lead to the same conclusion. In regions covered by nearby extended clusters of galaxies few very distant clusters are found. Thus, on a 48-inch Schmidt plate covering forty square degrees centered on the Coma cluster only eight other clusters are found, while in other fields as many as eighty distant clusters can be identified [*21*]. The dust in the Virgo cluster seems to be less dense than that in the Coma cluster, there being about 15 distant clusters visible on 48-inch Schmidt plates covering 40 square degrees each in various regions of the sky occupied by the extended Virgo cluster. The obscuration through the central parts of the Coma cluster may be estimated to dim the light passing through it by about 0.6 photographic magnitudes [*21*], [*3*].

Because of the work done with the large Schmidt telescopes our views on numbers of clusters of galaxies have radically changed during the past thirty years. HUBBLE [*7*] originally estimated that at the limit of the 100-inch reflector, that is, including galaxies of the apparent magnitude as faint as $m_p = 20.5$, one might expect about one cluster of galaxies per fifty square degrees. According to our present data there are, however, already as many as one cluster per hundred square degrees and one cluster per *one* square degree identifiable with the 18-inch and 48-inch Schmidt telescopes respectively. To the limit of $m_p = 19.5$ for the photographic magnitude of the faintest galaxies distinguishable from stars on 48-inch Schmidt plates there are thus roughly one hundred times more clusters than HUBBLE had estimated. Going to plates taken with the 200-inch telescope the great increase in the number of clusters of galaxies which should be expected has not materialized. Surprisingly enough there have been only a very few clusters found which are not identifiable on 48-inch Schmidt plates [*34*]. This means, in agreement with the results from counts of galaxies and of clusters of galaxies, that there is enough intergalactic dust, probably both within the large clusters of galaxies and in smaller concentration between them to interfere with the apparent structure and therefore with the identifiability of very distant clusters of galaxies, although individual galaxies within these clusters may still be visible.

VII. Superclustering non-existent.

14. Because of the gravitational interactions between particles of matter, we have in the universe agglomerations of gases in gas clouds and in stars, agglomerations of dust particles in dust clouds, agglomerations of stars in clusters of stars and in galaxies and finally, the galaxies bunch together in clusters of galaxies. But here the systematic agglomeration of various bodies of matter seems to stop. There are no well delineated and organized clusters of distinct clusters of galaxies. To be sure, clouds of galaxies can be found which contain various local concentrations in the numbers of galaxies, resembling a congregation of a small number of ill defined washed-out clusters of galaxies. But there is *no such thing as a globular supercluster* of well defined individual clusters of galaxies or any system remotely resembling such a supercluster. This important fact has become evident from ZWICKY's recent survey [*3*] of the distribution of the centers of clusters of galaxies, which in unobscured regions is of the character of a uniform and random distribution of non-interacting objects. The finding that there are no superclusters may be closely related to the result discussed in Sect. 11β that there are no observable random motions of any cluster of galaxies relative to its spatial neighbors. As suggested before, these two facts indicate that either the character of the gravitational forces between two bodies materially decreases when these bodies are separated by distances greater than about twenty million light years (on the old distance scale), or the absence of superclusters of clusters of galaxies is due to the fact that such superclusters cannot get organized because of the finiteness of the speed of transmission of gravitational interactions.

VIII. The universal redshift, extragalactic distances and the methodology of the study of clusters of galaxies.

In conclusion a short review may be given of the methods which have been used by astronomers for the study of the distribution and the internal structure of clusters of galaxies.

1. Photographic studies of the apparent photographic and of the photovisual magnitudes of the brightest galaxies in clusters of galaxies (Sect. 15).
2. Photoelectric determinations of magnitudes and colours of the brightest cluster member galaxies in six colour ranges [*35*] (Sect. 16).
3. Study of the spectra of some of the brightest member galaxies of distant clusters [*31*] (Sect. 17).
4. The application of the methods of *dimensionless morphology*, introduced by ZWICKY [*20*], [*21*], [*3*] to the internal structure of clusters of galaxies (Sect. 18).

15. Studies of the apparent magnitudes. The study of the apparent magnitudes of the brightest galaxies in clusters was introduced originally by HUBBLE [*7*] for the determination of the absolute distances of these clusters. In the course of time it has become apparent that this method is subject to many objections. Since it does not use any absolute length comparison standard, relative to which the distances of the clusters might be measured, many assumptions must be introduced to take the place of such a standard. HUBBLE and some of his successors [*31*] assumed that the twenty brightest member galaxies in rich clusters have on the average the same absolute magnitudes in all clusters. This of course disregards all differences between individual clusters as well as possible effects of evolution. Nevertheless it must be said that from all indications so far available the brightest galaxies in rich clusters actually seem to be closely of the same absolute magnitude, although great caution is indicated if this preliminary result is to be used as a basis for far reaching conclusions.

Furthermore, the study of the apparent magnitudes of the brightest cluster member galaxies can only lead to a determination of the distances of clusters if there is no intergalactic obscuration and, most important of all, if the corrections are accurately known which must be introduced because of the effects of the universal redshift itself on the apparent magnitudes of the galaxies involved. These corrections can only be calculated if the redshifts for the galaxies in question have been measured and if their intrinsic spectral energy distribution curve is known. How difficult these problems are has become glaringly evident from the history of the Colour Excess Effect in distant galaxies, the existence of which was announced by STEBBINS and WHITFORD [*35*] in 1948. After much work and discussion this effect appears to have now been proved non-existent by WHITFORD [*36*] and CODE. In addition, many purely observational difficulties arise, such as the uncertainty of which galaxies are actually cluster members, the indefiniteness of the boundaries of galaxies and the effects of background and night sky radiation on the measurements of magnitudes.

We therefore conclude that, although the observation and the analysis of the apparent magnitudes of the brightest galaxies in clusters can in principle lead to the determination of the relative distances of these clusters, we are in practice still far from having achieved this goal. After the relative distances have been established the problem of course still remains of getting absolute standards of length in order to establish an absolute extragalactic distance scale.

16. Multicolor photometry of distant cluster member galaxies. This method which was introduced by STEBBINS and WHITFORD, in principle could give the relative distances *and the redshifts* or symbolic velocities of recession $V_s = c \cdot \Delta\lambda/\lambda$ of distant clusters of galaxies. In its practical application it is subject, however, to all of the difficulties mentioned in the preceding Sect. 15. In the determination of V_s an additional difficulty arises if intergalactic obscuration should be found to be strong enough not only to cause absorption of light, but if this absorption is selective, causing relative changes of color.

If the difficulties mentioned can be overcome, the multicolor photometry of Stebbins and Whitford may ultimately lead to the determination of redshifts for clusters too distant to be investigated spectroscopically with the 200 inch telescope.

17. The spectra of distant cluster member galaxies. This of course is the direct method to determine the redshifts of distant clusters. Only $V_s = c \cdot \Delta\lambda/\lambda$ is obtained, where λ is the wave length of the unshifted line and $\Delta\lambda$ is the shift in wave length. The distance D cannot of course be derived from the study of the spectra alone.

In addition to giving the values of V_s, the study of the spectra may become of importance to detect possible changes due to evolution. It will also be of importance to analyse the width of the spectral lines as a function of the distance, there being some indications that a progressive widening with the distance may exist, and that this widening is not the same for absorption and for emission lines.

Unfortunately, for the present, the march into greater depths of space by this method seems to be stopped. The solar activity for a few years to come may indeed be expected to be such as to produce a brightness of the sky glow too great to make spectral investigations of clusters of galaxies with values of V_s greater than 100000 km/sec profitable.

18. The methods of dimensionless morphology. These methods were introduced by Zwicky [*20*], [*21*], [*3*] in order to avoid the difficulties which plague the approaches to observational cosmology discussed in the preceding Sects. 15 to 17. Only the operations of identification of galaxies (distinguishing them from stars) and counting them in areas of given size are used. In Sects. 9 and 12 of the present review we have discussed, how the methods of dimensionless morphology lead to a relative distance scale for clusters of galaxies which is not affected by any of the uncertainties inherent in the other methods. We call attention again to the fact, that the application of the methods of dimensionless morphology has so far failed to uncover any differences in the structure and population of the most distant clusters as compared with the analogous properties of nearby clusters One further investigation along these lines, which is now in progress, endeavors to establish the frequency of supernovae in clusters. If there is any systematic evolution in the universe, whether this universe be expanding or not, the average frequency of supernovae per individual galaxy in distant clusters may be expected to be different from the corresponding frequency in nearby clusters.

The methods of dimensionless morphology can be slightly generalized in order to derive not only relative distances of clusters of galaxies but their redshifts as well. It is only necessary to count the populations of the clusters in various colour ranges $C_1, C_2, \ldots, C_k$. One may for instance obtain photographs of a cluster in the color range C_i using a number of exposure times t_k. The ratios of the numbers of galaxies $n(C_i, t_k)$ are then studied for different clusters. After certain calibrations, these ratios can give accurate information on the redshifts of extremely distant clusters whose characteristics cannot be studied by any of the methods proposed so far, such as for instance the methods discussed in the Sects. 15 to 17. Naturally, as all other indirect determinations of the redshift, the results of the method just described can also appear falsified if reddening of galaxies due to intergalactic dust should interfere.

If a reliable redshift-distance relation $\Delta\lambda/\lambda =$ function of D is to be established, the methods of dimensionless morphology must be used to determine D, and $\Delta\lambda$ must be measured by *direct* recording and analysis of the spectra of distant galaxies. All other methods known today are subject to the as yet unresolved difficulties discussed in the preceding.

The Cluster 1448+2617. In order to push his researches to clusters more remote than those listed in Table 5, HUMASON during the past few years has scanned 48-inch Schmidt survey plates. Clusters whose symbolic velocities of recession are likely to be greater than $V_s = 100000$ km/sec appear as faint granulated smudges on limiting panchromatic (Eastman 103 *a-E* emulsions behind red filters) plates, while on blue sensitive plates they may be hardly visible at all. HUMASON has selected a few of these clusters as the most promising candidates and has already spent considerable time investigating them with the powerful nebular spectrograph of the 200-inch reflector. Particular attention has so far been paid the Cluster 0024+1654, shown in Figs. 3 and 4, as well as the Cluster 1448+2617. While HUMASON, with long exposures (ca. 50 hours), attempted to obtain spectra of member galaxies of these clusters, WHITFORD applied his six colour photometry to determine the distance and indirectly the redshift, while ZWICKY, from his methods based on dimensionless morphology derived the distance of Cl 1448+2617. The results of these investigations, which were informally coordinated by Dr. HUMASON, at the present are as follows.

a) HUMASON made two attempts at photographing the spectra of member galaxies of Cl 1448+2617 which unfortunately did not show any pronounced emission lines. The identification of absorption lines (H and K) is most uncertain in this case because of the spectrum being badly blurred by the light of the sky glow. HUMASON feels that there are some features which *might possibly be the H and K absorption lines* and that they show a redshift $\Delta\lambda$ corresponding to a symbolic velocity of recession $V_s = C\cdot\Delta\lambda/\lambda = 118000$ km/sec, but that this value cannot be accepted as a fact until more data are obtained and preferably, if member galaxies showing emission lines can be observed.

b) A. R. SANDAGE has photographed the cluster both directly and with the schraffier method. His plates are quite suitable for the determination of the apparent magnitudes of the brightest member galaxies. The material available has, however, so far not been analyzed.

c) ZWICKY determined the distance of Cl 1448+2617 both from the analysis of the distribution index (Substitute for the structural index) and the central density of the member galaxies per square degree. This analysis gives for the distance a value D[Cl 1448+2617] $= 1.5\,D$[Hydra II Cl] of V_s[Cl 1448+2617] $=$ 91500 km/sec, if the velocity-distance relation were linear[1].

d) A. E. WHITFORD[2] also worked on Cl 1448+2617 with his general method of six color photometry and at the request of Dr. HUMASON estimated

[1] On the simple theory of a non-expanding universe and a gravitational drag [37] of light as the cause for the universal redshift, one should expect as a first approximation $V_s/c = \varkappa D + \varkappa^2 D^2/2$. The quadratic deviation $(\Delta\lambda/\lambda)^2/2$ from linearity would in this case rise up the V_s for the Cluster 1448+2617 to a value $V_s = 91500 + 14000 = 105500$ km/sec.

[2] Dr. WHITFORD discussed the problems involved and his preliminary data at a meeting of the Midwestern Astronomers at the Yerkes Observatory in the spring of 1956.

the redshift of this cluster to correspond to $V_s \approx 110000$ km/sec. Recently W. A Baum [38] repeated the work of Whitford and published his opinion that the redshift of Cl 1448+2617 corresponds to a symbolic velocity of recession $V_s \approx 120000$ km/sec. Baum also states that up to this limit the velocity distance relation is strictly linear. Cosmologists should be cautioned, however, that any such statement far exceeds what can be justified on the basis of observational data presently available. In the first place Humason's final spectral data will have to be awaited until one can have any ultimate confidence in the values of V_s of these very distant clusters. As far as their relative distances are concerned, some exceedingly knotty problems remain to be solved. The solution of these problems will probably only be achieved through the application of such analyses as the Stebbins-Whitford six colour photometry and the methods of dimensionless morphology [3], [20], [21]. In addition to the difficulties mentioned before, six colour photometry must contend with the problem of placing the multicolour band passes on an absolute system of intensities. With only a handful of observers having access to the large telescopes and working on the problems related to the universal redshift one must be doubly careful in drawing any too precipitous conclusions. How even good observers can go wrong in these matters is amply exemplified through the emergence of the extra colour excess [35], [36] of distant galaxies (Stebbins-Whitford effect) and its disappearance as a result of the most recent investigations. For the time being, those of us who are engaged in the exploration, by optical means, of most distant clestial objects are also temporarily greatly hindered in their activities by the recent intensification of the sky glow due to the increased solar activity, which may be expected to last for several years.

Bibliography.

[1] See Duncan, John C.: Messier's Nebulae and Star Clusters. Leaflet No. 240 of the Astronomical Society of the Pacific, March 1949.

[2] Shapley, H., and A. Ames: A Survey of the External Galaxies Brighter than the Thirteenth Magnitude. Harvard Observatory Annals, **88**, No. 2 (1932).

[3] Zwicky, F.: Morphological Astronomy. Heidelberg: Springer 1957.

[4] Proceedings of the Third Berkeley Symposium on Mathematical Statistics and Probability, Vol. III, p. 75—111. Univ. California Press 1956.

[5] Shane, C. D., and C. A. Wirtanen: Astronom. J. **59**, 285—304 (1954).

[6] Zwicky, F.: Photometry of Galaxies Brither than the Fifteenth Photographic Magnitude with the 18-inch Schmidt Telescope on Palomar Mountain. Final Report to the Office of Naval Research, Feb. 5, 1955. Contract Nonr-449(00).

[7] Hubble, E. P.: The Realm of the Nebulae. Yale University Press 1936.

[8] Vogt, H.: Die Spiralnebel. Heidelberg: Carl Winter 1946.

[9] Zwicky, F.: Publ. Astronom. Soc. Pacific **50**, 218 (1938).

[10] Vaucouleurs, G. de: Sci. Amer. **191**, No. 1, 30—35 (1954).

[11] Shapley, H.: Galaxies. Philadelphia: P. Blackiston Son & Co. 1943.

[12] Hubble, E. P., and M. L. Humason: Astrophys. Journ. **74**, 43 (1931).

[13] Zwicky, F.: Proc. Nat. Acad. Sci. U.S.A. **23**, 251 (1937).

[14] Wolf, M.: Astronom. Nachr. **170**, 211 (1905).

[15] Zwicky, F.: Proc. Nat. Acad. Sci. U.S.A. **28**, 317 (1942); **28**, 355 (1942).

[16] Carpenter, E. F.: Report to the AAAS Meeting in Pasadena, June 1931. — Zwicky, F.: Publ. Astronom. Soc. Pacific **62**, 196, 256 (1950).

[17] Zwicky, F.: Proc. Nat. Acad. Sc. U.S.A. **27**, 264 (1941).

[18] Wolf, M.: Astronom. Nachr. **155**, 127 (1901). — Zwicky, F.: Astrophys. Journ. **86**, 217 (1937). — Publ. Astronom. Soc. Pacific **63**, 61 (1951).

[19] Zwicky, F.: Experientia, Basel **6**, 441 (1950). — Physics Today **6**, No. 4, 7—11 (1953). — Ergebn. exakt. Naturw. **29**, 344—385 (1955).

[20] Zwicky, F.: Helv. phys. Acta **26**, 241—254 (1953).

[21] Zwicky, F.: Proceedings of the Third Berkeley Symposium on Mathematical statistics and Probability, Vol. III, p. 113—144. Univ. California Press 1956.

[22] HUBBLE, E. P.: Astrophys. Journ. **84**, 158 (1936).
[23] ZWICKY, F.: Astrophys. Journ. **95**, 555 (1942).
[24] EMDEN, R.: Gaskugeln. Leipzig: B. G. Teubner 1907.
[25] ZWICKY, F.: Leaflet No. 163 of the Astronomical Society of the Pacific, Sept. 1942.
[26] ZWICKY, F.: Proc. Nat. Acad. Sci. U.S.A. **25**, 604 (1939); **26**, 332 (1940).
[27] ZWICKY, F.: Helv. phys. Acta **6**, 110 (1933). — Astrophys. Journ. **86**, 217—246 (1937).
[28] HEESCHEN, D. S.: Meeting of the Amer. Astr. Soc. and the Astr. Soc. of the Pacific, August 24—28, 1956 in Berkeley, California. See also the Scientific American, October 1956, p. 66. Also Astrophys. Journ. **124**, 660 (1956).
[29] ZWICKY, F.: Publ. Astronom. Soc. Pacific **68**, 331—338 (1956).
[30] ZWICKY, F.: Publ. Astronom. Soc. Pacific **63**, 17—25 (1951).
[31] HUMASON, M. L., N. U. MAYALL and A. R. SANDAGE: Astronom. J. **61**, 97—162 (1956).
[32] ZWICKY, F.: Phys. Rev. **51**, 290, 679 (1937).
[33] EINSTEIN, A., and W. DE SITTER: Proc. Nat. Acad. Sci. U.S.A. **18**, 213 (1932).
[34] ZWICKY, F.: Publ. Astronom. Soc. Pacific **65**, 215 (1953).
[35] STEBBINS, J. L., and A. E. WHITFORD: Astrophys. Journ. **108**, 413 (1948). — WHITFORD, A. E.: Astronom. J. **58**, 49 (1953). — Astrophys. Journ. **120**, 599 (1954).
[36] WHITFORD, A. E.: Astronom. J. **61**, 352 (1956).
[37] ZWICKY, F.: Proc. Nat. Acad. Sci. U.S.A. **15**, 773 (1929).
[38] See "Sky and Telescope", Vol. XVI, No. 2, p. 60, December 1956.

Large Scale Organization of the Distribution of Galaxies[1].

By

JERZY NEYMAN and ELIZABETH L. SCOTT.

With 7 Figures.

I. Introduction.

1. Historical. In reading the astronomical literature concerned with the distribution of galaxies in space, one encounters a combination of two basic ideas. One of these ideas is that the galaxies are clustered, and the other is that the large-scale organization of the distribution of galaxies is statistically uniform. Occasionally, the possibility of higher order clustering of galaxies is also mentioned.

As early as 1811, HERSCHEL[2] recognized that a uniform distribution cannot account for the many double and triple galaxies he catalogued. While he did not realize the tremendous distances of these objects, his conclusion is supported by contemporary astronomers.

A clear idea of the possibility of a hierarchy of clustering was expressed in 1908 by CHARLIER[3], "Suppose N_1 stars together form a galaxy G_1, N_2 galaxies form together a galaxy of second order G_2, N_3 galaxies of the second order form a galaxy of the third order $G_3, \ldots$". However, no mathematical theory was developed from these general ideas, nor was there then available sufficient observational material to support them.

Beginning in the 1930's, extensive surveys of the position, magnitude and other characteristics of the brighter galaxies were published. These were followed by counts of galaxies in regularly placed squares on photographic plates reaching as far out into space as the telescopes of the Harvard Observatory, of the Mt. Wilson Observatory and, now, of the Lick Observatory would permit. A listing of the existing surveys with a suggested program for the future is given by SHANE in [*8*].

Using a portion of these data, BOK[4] showed that the over-all distribution of galaxies is not uniform and SHAPLEY[5] indicated that the irregular absorbing material within our own Galaxy does not account for the irregularities in the distribution of galaxies. By this time the relative importance of the ideas of large scale uniformity and of large and small scale clustering in the distribution of galaxies was often discussed. The combination of these ideas is seen clearly in the following passage published in 1936 by HUBBLE [*3*], to whom extensive surveys and studies of galaxies are due.

"While the large-scale distribution (of galaxies) appears to be essentially uniform, the small-scale distribution is very appreciably influenced by the well-known tendency towards clustering. The phenomena might be roughly represented

---

[1] This paper was prepared with the partial support of the National Science Foundation, U.S.A.

[2] Sir WILLIAM HERSCHEL: Phil. Trans. Roy. Soc. Lond. **101**, 269–336 (1811).

[3] C. V. L. CHARLIER: Ark. Mat., Astronomi Fys. **4**, No. **24**, 1–15 (1908).

[4] B. J. BOK: Harvard Obs. Bull. **1934**, No. 895, 1–8.

[5] H. SHAPLEY: Harvard Obs. Bull. **1932**, No. 890, 1–3.

by an originally uniform distribution from which nebulae have tended to gather about various points until now they are found in all stages from random scattering, through groups of various sizes, up to occasional great clusters ... ."

The present article presents an account of the efforts made since 1950 to embody the combination of ideas described above into an exact probabilistic-statistical theory capable of quantitative confrontation with observational data. It would be natural to base such a theory on some dynamical considerations. Unfortunately, problems of this kind are extremely difficult and their investigation is still in its infancy. Thus, while the subject matter of the work reported is indeed indeterministic in character, it is descriptive consisting of the construction of a hypothetical chance mechanism which, if allowed to function in space and time, could produce, with due allowance for random variation, the various statistical characteristics of the observational data that are already available or might be expected in the near future. The original chance mechanism conceived, termed the theory of simple clustering of galaxies, refers to the general ideas summarized in the quotation from HUBBLE. The more recent work, on clustering of higher order, refers to the general ideas of CHARLIER, published in 1908.

II. Dynamical problem of infinite mass in infinite space.

2. ULAM's problem of motion of infinitely many randomly distributed particles subjected to Newtonian gravitation. In 1954 ULAM[1] described his work on the following problem. For each point on the real line with integer coordinate $k=0$, ± 1, $\pm 2, \ldots$, a random experiment E_k is performed which has probability of success equal to one-half. All the experiments are completely independent. If the experiment E_k yields success, then a particle of unit mass is placed at the point k. At time $t=0$ all the experiments are completed, and the particles begin to be subjected to Newtonian gravitation. Their motions are studied and the problem considered is the distribution of particles at any time $t>0$.

ULAM's treatment of this problem consisted in replacing the infinite real line by a long segment, in replacing the differential equations of motion of the particles by second order difference equations, and in using a high speed computer to evaluate the particles' displacements. The work was repeated a number of times, each time starting with an independent initial distribution of particles. A similar study was done in two dimensions, with the interior of a large circle replacing the segment of the real line. In both cases, as the time t was increased, the particles tended to form clusters.

The material offered in the following pages is devoid of dynamical considerations. The developments presented are predominantly statistical in nature, supplemented with some kinematical considerations; with reference to the results described immediately above, they are based on the presumption that, in general, galaxies are clustered in space.

III. Theory of simple clustering of galaxies.

3. Quasi-uniform distribution of particles. In this section we introduce the notion of a quasi-uniform distribution of particles. The space considered will always be the three-dimensional Euclidean space E and the term "region" will be used to denote a Borel subset of E. We shall say that a (random or stochastic or statistical) distribution of particles in the space E is quasi-uniform [5] if it satisfies the following conditions.

[1] S. ULAM: Infinite models in physics (mimeographed). New Mexico: Los Alamos Scientific Laboratory 1954.

(i) To every region R in E there corresponds a non-negative integer-valued random variable $\gamma(R)$ representing the number of particles located in R.

(ii) The distribution of $\gamma(R)$ depends only on the volume $V(R)$ (on the measure) of R.

(iii) If $\{R_i\}$ represents a countable sequence of disjoint regions, then the corresponding random variables $\gamma(R_i)$ are completely independent.

It will be noticed that these three conditions refer to the *mechanism* creating the distribution of particles rather than on the distribution itself. This way of speaking is a convenient abbreviation that will be used constantly, and it is hoped that it will not cause any misunderstanding.

In order that a distribution of particles be quasi-uniform it is necessary and sufficient that the probability generating function $G_{\gamma(R)}(t)$ of $\gamma(R)$, defined for $|t| \leq 1$ as

$$G_{\gamma(R)}(t) = E(t^{\gamma(R)}), \tag{3.1}$$

be of the form

$$G_{\gamma(R)}(t) = e^{V(R)\,h(t)}, \tag{3.2}$$

with

$$h(t) = h_0 - \sum_{k=1}^{\infty} h_k t^k, \quad \sum_{k=1}^{\infty} h_k = h_0 \tag{3.3}$$

where $h_0 > 0$ and $h_k \geq 0$ for $k = 1, 2, \ldots$.

The particular case of a quasi-uniform distribution corresponding to $h_1 = h_0$ and, consequently, $h_k = 0$ for $k > 1$ is called the statistically uniform distribution of particles. In this case $\gamma(R)$ is a Poisson variable with expectation $V(R)\,h_0$. In the general case, formulae (3.2) and (3.3) indicate that the number of particles $\gamma(R)$ in R can be presented in the form

$$\gamma(R) = \sum_{k=1}^{\infty} k\,\xi_k, \tag{3.4}$$

where $\xi_1, \xi_2, \ldots, \xi_k, \ldots$ are all completely independent Poisson variables and the expectation of ξ_k is $h_k V(R)$. Further analysis indicates that ξ_k stands for the number of k-tuples of particles whose positions coincide.

4. Basic assumptions of the theory of simple clustering of galaxies. In building the theory of simple clustering of galaxies the authors started with an extrapolation of the Copernican idea known as the *cosmological principle* [*1*]. The ordinary, broad formulation of this principle is that, except for local irregularities, the general characteristics of the universe are the same, independent of the point from which the universe is observed. In the following this principle is understood to mean that the observable universe is a single realization of a stochastic process which is stationary with respect to displacements in space but, possibly, nonstationary in time. In other words, for any fixed time T it is postulated that the distribution of mass in space is a result of a fixed chance mechanism the functioning of which within a region Λ of a fixed shape and size is independent of where this region is located in space, close to the observer or far away.

The second point of departure is the general idea that galaxies are clustered, as indicated in Sect. 1.

The first attempt to put into formulae a combination of the cosmological principle, on the one hand, and of the presumption that the galaxies are clustered, on the other, resulted in the theory of simple clustering. Here, the chance mechanism supposed to have produced the actual distribution of galaxies in space is composed of three separate mechanisms, as follows.

First, imaginary points C, called *cluster centers*, are distributed in space quasi-uniformly (see Sect. 3). Second, to every cluster center a random number ν of galaxies is attached. The random variables ν are positive, integer-valued, completely independent among themselves and independent of all other variables of the system. The distribution of ν is characterized by a probability generating function $G_\nu(t)$. Finally, when the position of any particular cluster center C' and the number ν' of galaxies attached to it are determined, a third chance mechanism is envisaged, determining the positions of the ν' galaxies. Let $u = (u_1, u_2, u_3)$ be the triplet of coordinates of the cluster center C' and let $X = (X_1, X_2, X_3)$ be a triplet of random variables interpreted as coordinates of a galaxy belonging to the cluster centered at u. It is postulated that this triplet X has a conditional (given u) probability density function depending upon the distance

$$\eta = \sqrt{\sum_{i=1}^{3} (x_i - u_i)^2} \tag{4.1}$$

between any point $x = (x_1, x_2, x_3)$ and the cluster center C' located at u. This probability density will be denoted by $f^*(\eta)$.

The above assumptions regarding the position of any particular galaxy are supplemented by further assumptions that, given the positions of all the cluster centers, the triplets of coordinates of particular galaxies are completely independent. Also, it is assumed that each triplet of coordinates of a galaxy is independent of all the other random variables of the system.

The above assumptions determine the properties of the spatially stationary stochastic process representing the distribution of galaxies in space. In particular, given a system of disjoint regions $R_1, R_2, \ldots, R_s$, and denoting by N_i the number of galaxies included in R_i, for $i = 1, 2, \ldots, s$, a formula can be written, in terms of the unspecified functions $h(t)$, G_ν and f^*, representing the joint probability generating function of the s random variables $N_1, N_2, \ldots, N_s$. This formula is given in Sect. 9 as a specialization of a more general result.

An interesting property of the formula in question should be noted. This is that, whatever the quasi-uniform distribution of cluster centers, characterized by the function (3.3), whatever the distribution of the number of galaxies per cluster characterized by $G_\nu(t)$, and whatever the probability density $f^*(\eta)$, there exist a uniform distribution of cluster centers, characterized by the constant $h_0 = h_0^*$, and a corresponding distribution of ν, characterized by the probability generating function $G_\nu^*(t)$, such that the resulting distribution of variables $N_1, N_2, \ldots, N_s$ will be the same in the two cases. As a result, no generality is lost in the development of the theory of simple clustering by assuming that the cluster centers are distributed in space with statistical uniformity. This particular assumption will, then, be adopted in the following and the symbol λ will stand for the expected number of cluster centers per cubic parsec.

The following comments regarding the random variable ν may be useful. The probability generating function $G_\nu(t)$, which in the general theory is left unspecified, may in a particular case be $G_\nu(t) \equiv t$. This assumption is equivalent to the assertion that $P\{\nu = 1\} = 1$ and all "clusters" are composed of just one member galaxy. It follows that the galaxies are distributed in space uniformly. In the general case the probability that $\nu = 1$ is less than unity. If this probability is positive then a certain proportion, namely the proportion equal to $P\{\nu = 1\}$, of postulated clusters are composed of a single element each. These galaxies, each forming a separate "cluster" are uniformly distributed in space and will be referred to as "field galaxies". Their existence was visualized by HUBBLE as indicated in the quotation given in Sect. 1.

5. Observable random variables and the object of the theory of clustering of galaxies. The basic assumptions formulated in Sect. 4 refer to the hypothetical mechanism behind the distribution of galaxies in space. This distribution is

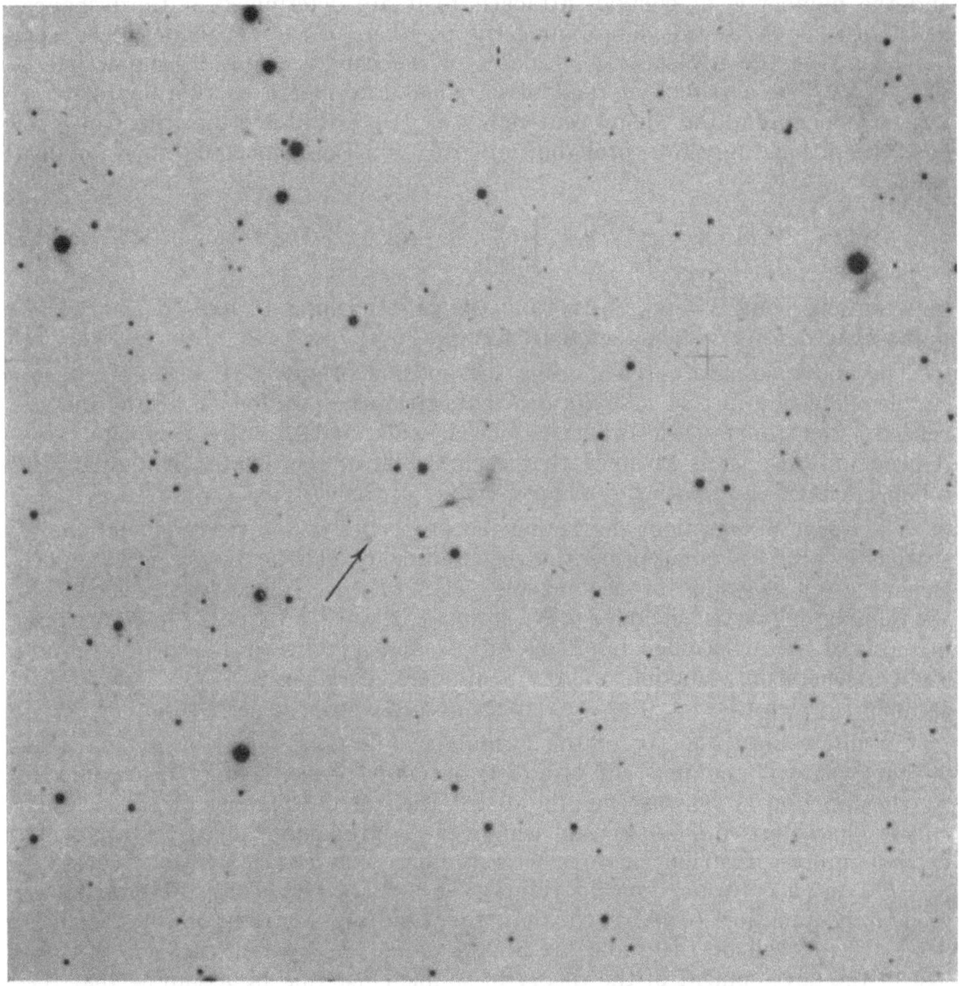

Fig. 1. Section of a plate taken by the 20″ astrographic telescope of the Lick Observatory.

not directly observable and the object of the theory of clustering consists in using the basic assumptions in order to deduce the distributions of the variables that are available for direct observation. There are three categories of such variables.

α) Counts of galaxies in regularly spaced squares. Systematic surveys of the sky, particularly over the two polar regions believed to be free, or relatively free, from interstellar obscuration yield photographs on which images of galaxies can be counted. Fig. 1 represents a section of a photograph of this kind with a number of images of galaxies. Several galaxies are quite distinct, but some others (see arrow) are faint and present difficulties of identification. This particular photograph is part of the material assembled by C. D. SHANE and C. A. WIRTANEN [3] at the Lick Observatory to whom the authors are deeply indebted for permission to reproduce Fig. 1 in the present article.

The usual method of summarizing the observations exemplified in Fig. 1 consists in dividing the photographic plate into relatively small squates, perhaps $10' \times 10'$ in angular dimensions, and in counting the identifiable images of galaxies. Table 1 reproduces a section of a plan of the photographic plate with figures in cells representing the numbers of images of galaxies counted by SHANE on a particular occasion on a particular plate. Fig. 2 shows a plot of the galaxies identified on this plate by SHANE.

Table 1. *Numbers of images of galaxies in $10' \times 10'$ squares counted by* C. D. SHANE *on Count made Nov. 15 to 20, 1948.*

Section of Plate 835, $\alpha = 14^{\mathrm{h}}\,40^{\mathrm{m}}$, $\delta = 5°$.

3	0	2	5	3	0	3	3	0	2	0	1	0	0	2	0	3	1
1	1	0	0	1	1	0	0	1	0	0	0	1	1	0	0	0	1
2	5	0	0	0	0	0	2	0	1	0	1	1	1	0	2	1	0
1	4	0	0	2	2	0	1	0	2	1	0	0	2	0	1	2	0
1	1	1	1	2	0	0	5	0	1	0	2	1	1	2	1	0	1
0	0	1	1	5	2	4	1	1	1	4	1	2	3	2	1	1	1
1	0	2	1	2	2	6	2	4	0	0	0	0	0	1	1	1	1
1	2	3	1	3	2	2	1	2	1	1	1	1	1	2	2	2	4
2	1	1	1	5	2	2	1	1	2	2	2	1	4	6	3	3	4
7	2	4	3	2	2	3	0	2	3	3	2	2	4	7	8	2	6
0	2	2	7	0	1	3	1	3	5	3	4	4	2	4	15	6	2
0	0	4	1	6	12	6	4	2	7	1	2	3	4	6	5	2	4
2	2	0	6	5	7	2	4	4	0	2	3	5	2	1	3	0	4
1	0	1	0	6	2	1	1	0	2	3	1	3	2	1	3	2	1
3	0	2	1	2	1	5	1	2	1	1	3	1	1	3	2	1	1
0	2	3	4	0	1	0	3	1	3	2	2	1	1	2	0	1	2
0	0	1	1	2	1	2	0	0	0	1	2	1	1	0	0	1	1
1	1	0	2	0	1	1	0	1	1	0	1	1	1	1	1	1	0

A glance at Fig. 1 will convince the reader that when counts on a particular plate are repeated, particularly if one takes care to insure that the second count is independent of the first, then, in general, the number of images of galaxies counted in a given square on the second count will not coincide with the number counted on the first. Of course, all the bright and distinct images of galaxies may reasonably be expected to be included in both counts. But, when one comes to very faint images, a subjective decision is necessary whether a particular condensation of silver grains in the emulsion is an image of a faint galaxy, or of a star or, simply, a local defect of the plate. And here variation in independently made decisions is unavoidable.

Up to recently, very little was known about the variability of counts, with most authors limiting themselves to expressions of confidence that the "error" does not exceed a few per cent. The word error is taken in quotation marks because, in the authors' opinion, it is not appropriate to describe the phenomenon discussed; a more appropriate term seems to be *variation*. In order to speak of an "error" in a count, one must first define the meaning of "correct count". Another glance at Fig. 1 should convince the reader that such a definition is not easy.

A systematic appraisal of the extent of variation in counts of images of galaxies, involving abour 2000 objects, is due to MAYALL[1]. A much more extensive

[1] N. U. MAYALL: Lick Obs. Bull. **458**, 177—198 (1934).

effort in the same direction is now in progress, by SHANE and WIRTANEN. The work involves duplicate independent counts on duplicate plates taken over 27 regions in the sky, each plate covering a square 6°×6° and containing between 2000 and 3000 countable images of galaxies. It is understood that similar work is being done by ZWICKY. When completed, this work will provide material for assignment of sources of variation. As of now it is quite clear that, for a fixed

Fig. 2. Reproduction made by C. D. SHANE of actual plate showing galaxies only. Plate centered at $\alpha = 14$ hour 40 min, $\delta = +5°$. Size of symbol indicates rough estimate of brightness on an arbitrary scale.

field in the sky, and for a fixed observational setup, including the telescope, the kind of emulsion, the exposure time, etc., the outcome of a count depends (a) upon the plate and (b) upon the threshold at which the observer can confidently distinguish between the images of galaxies and other objects on the plate. This threshold is subjective and varies from day to day.

β) Detailed observations of individual galaxies. Observations of varying extent and precision have been published from time to time and the Shapley-Ames catalogue[1] should be mentioned in the first place. The most recent relevant

[1] H. SHAPLEY and A. AMES: Ann. Harvard Obs. **88**, 43–75 (1932).

paper by HUMASON, MAYALL and SANDAGE [6], gives a very substantial set of accurate observations of the apparent magnitude and redshift of galaxies. Some of these galaxies are classified as field nebulae, others are identified as members of particular clusters.

γ) Statistics of clusters of galaxies. While detailed data on individual outstanding clusters of galaxies have been published for some time by a number of authors, a systematic accumulation of *statistics of clusters*, an accumulation based on clearly established definitions, is still a matter of the future. The first steps in this direction appear to be due to SHAPLEY and his collaborators[1]. Also, a considerable amount of work in the same direction has been done by ZWICKY. In particular, a recent paper [9] contains a table of estimated apparent diameters and of numbers of member galaxies of 64 clusters identified on a single 48-inch Schmidt plate taken at the Mt. Palomar Observatory. It is understood that similar data in considerably greater volume is already available. In fact, they were the subject of a paper by G. D. ABELL presented at the meeting of the American Astronomical Society held in Berkeley in August, 1956.

The above review of the available observational material indicates that the object of any theory of the spatial distribution of galaxies is to use the assumptions of this theory to deduce formulae characterizing.

(i) The joint distribution of counts of images of galaxies in regularly spaced squares on photographic plates.

(ii) The joint distribution of redshift and of apparent magnitude of both field galaxies and cluster member galaxies, for which measurements may be available, and

(iii) The joint distribution of the ascertainable characteristics of clusters.

It must be obvious that, in order to deduce any of the above distributions, the assumptions regarding the distribution of galaxies in space are not sufficient and that these assumptions must be supplemented by additional hypotheses relating the phenomena developing in space to what may be observed either on survey plates or in studies of individual galaxies. Perhaps unexpectedly, these assumptions must include one regarding the state of the universe, that is, whether it is static or expanding.

6. Assumptions relating the distribution of galaxies in space to the distribution of their counted images on a photographic plate. In conformity with well-established ideas, it is assumed that galaxies vary in brightness. More specifically, it is assumed that the absolute magnitude $\mathcal{M}$ of a galaxy is a random variable, independent of all other variables considered in Sect. 4, and having a probability density $p_{\mathcal{M}}(M)$, where the letter M stands for a particular value of $\mathcal{M}$. The absolute magnitude of a galaxy is, of course, not directly observable. Instead we observe the photographic apparent magnitude, say μ, which is connected with $\mathcal{M}$ by the equation

$$\left.\begin{aligned} \mu &= \mathcal{M} - 5 + 5\,\mathrm{Log}\,\xi + \chi(\xi) \\ &= \mathcal{M} - 5 + a\log\xi + \chi(\xi) \end{aligned}\right\} \tag{6.1}$$

where ξ stands for what we shall call the apparent distance of the galaxy, that is, the distance, measured in parsecs, of the position occupied by the galaxy at the time of emission of the light that is recorded by the instruments of the observer. The term $\chi(\xi)$ denotes a function of the apparent distance ξ and represents the effects of redshift, space-reddening, evolution and other factors causing deviations from the inverse square law of dimming. It is believed that for

[1] H. SHAPLEY: Proc. Nat. Acad. Sci. U.S.A. **19**, 591—596, 1001—1006 (1933) and several other papers.

moderately distant galaxies the dimming term $\chi(\xi)$ can be neglected but for faraway galaxies it is quite important.

A particular value of μ will be denoted by m. Formula (6.1) can be used to evaluate the conditional probability density of μ given the apparent distance ξ of a galaxy. This probability density will be denoted by

$$p_\mu(m \mid \xi) = p_{\mathcal{M}}[m + 5 - a \log \xi - \chi(\xi)]. \tag{6.2}$$

The early works on the subject, for example, those of HUBBLE [2], are characterized by the concept of a "limiting apparent magnitude" to which the counts on a given plate are made. This same concept was also used in a number of publications by the present authors [4] to [7]. The concept of limiting apparent magnitude presupposes the existence of a number m_1 such that, perhaps with minor mistakes, *every* galaxy in the region of the plate for which the photographic apparent magnitude is brighter than m_1 will be counted, and *none* fainter than m_1 will be counted. However, as illustrated by the variation between duplicate counts and between duplicate plates, any such presumption is highly unrealistic. Hope for a theory that is in close relationship with the facts observed must be based on a convincing and verifiable model of variability in counts of images of galaxies on a series of plates taken in conditions *intended* to be identical. The two alternative models considered in [6] do not appear to be satisfactory, but the following set of assumptions may have chances of survival.

To every observational setup, including the telescope in its location, the type of emulsion, the exposure and the development, and to every observer engaged in counting images of galaxies, there corresponds a function $\Phi(m)$, called the selection factor, defined over a range of apparent magnitudes and representing the probability that a galaxy with its photographic apparent magnitude equal to m will be identified and counted as such. For small values of m, that is, for bright images of galaxies, this function $\Phi(m)$ is close to unity. When m is increased, then $\Phi(m)$ is non-increasing and tends to zero. In order to conform with the fact that only finite numbers of galaxies are counted on photographic plates, it will be necessary to assume that the convergence to zero of $\Phi(m)$ is rapid enough for certain integrals to converge.

As mentioned above, the selection factor $\Phi(m)$ corresponds to the combination of the observational setup with the observer. However, the particular plates taken with the same observational setup differ among themselves and also the level of counts made by the same observer differs from one day to the next. In order to take this into account, the existence of two more random variables is postulated. These variables, Π and Ω, represent the *plate effect* and the *observer effect*, respectively. These variables are assumed to be mutually independent and independent of any other variable of the model.

Denoting by π and ω the particular values of Π and Ω respectively, the assumption is made that the conditional probability that the image of a galaxy of a photographic apparent magnitude m will be counted on a particular plate on a particular day is $\Phi(m+\pi+\omega)$. The final assumption is that the counting of one galaxy is independent of that of any other.

Using the concept of selection factor we can now write the expression for the conditional probability, given the apparent distance ξ of a galaxy and the values π and ω of the plate and of the observer factors, that a galaxy will be counted. In order to denote this probability we shall use the symbol $\Theta(\xi, \pi, \omega)$. Thus

$$\Theta(\xi, \pi, \omega) = \int_{-\infty}^{+\infty} \Phi(m + \pi + \omega)\, p_{\mathcal{M}}[m + 5 - a \log \xi - \chi(\xi)]\, dm. \tag{6.3}$$

Whenever an interpolatory formula for $\Phi(m)$ is needed, intuition suggests the possibility of using, say

$$\Phi^*(m, \gamma, \delta) = 1 - L\left(\frac{m-\gamma}{\delta}\right), \tag{6.4}$$

where γ and δ are two adjustable constants and

$$L(t) = \int_{-\infty}^{t} l(x)\,dx, \qquad l(x) = \frac{1}{\sqrt{2\pi}}\,e^{-\frac{1}{2}x^2}. \tag{6.5}$$

7. Hypotheses regarding the possible expansion of the universe. In the preceding section it was necessary to refer to the apparent distance of a galaxy. This term was used to describe the distance ξ to the position occupied by the galaxy at the moment of emission of the light signal registered by the photographic plate. In addition to the apparent distance it will be necessary for us to consider the similarly defined *apparent coordinates* of a galaxy and to establish their relation with the *present coordinates* X_1, X_2, X_3 to which reference was made in Sect. 5.

Visualizing the possibility of an expanding universe and taking into account the empirical evidence we assume that the velocity of recession of clusters of galaxies is a function of their distance. Further, with reference to the cosmological principle, we assume that this function is the same irrespective of the cluster center with respect to which the distances and the velocities of other clusters are measured. Easy reasoning shows then that for this invariance to persist is both necessary and sufficient that the clusters recede from each other with velocities proportional to distance, so that at any time t, the i-th coordinate of a cluster is represented by

$$u_i(t) = u_i(0)\,H_1(t) \tag{7.1}$$

with

$$H_1(t) = e^{\int_0^t h_1(t)\,dt} \tag{7.2}$$

for $i = 1, 2, 3$. If $h_1(t)$ is positive for all t then the universe is expanding. If $h_1(t)$ is negative for all t, then the universe is contracting. If $h_1(t)$ is identically zero then the universe is static. If $h(t)$ changes its sign then the universe may be termed pulsating.

In addition to the possible recession of cluster centers, it is appropriate to consider the possible expansion of clusters. If this expansion conforms with hypotheses similar to those adopted for the recession of clusters, then the coordinates of a galaxy at time t have the following form

$$x_i(t) - u_i(t) = [x_i(0) - u_i(0)]\,H_2(t) \tag{7.3}$$

where

$$H_2(t) = e^{\int_0^t h_2(t)\,dt}. \tag{7.4}$$

Details of the theory are given in [4]. Restricting ourselves to the simplest case, it will be assumed here that $h_1(t) = h_2(t) = h = gc = \text{const}$, where c stands for the velocity of light. The constant g will be called the expansion factor. The origin from which time t is measured is arbitrary. It will be convenient to assume that the origin is at the "present", that is to say, at the moment when a photograph of the sky is taken.

Taking into account that the velocity of light is independent of the velocity of the source of light with respect to the observer and denoting by $-\tau$ the moment in the past when the light now registered by the photographic plate was emitted by a galaxy, we find that this galaxy's apparent distance, say $\xi=\xi(-\tau)$, is connected with its present distance, say $\xi^*=\xi(0)$ by the equation

$$\xi = \xi^* \,\mathrm{e}^{-\tau h} = c\,\tau. \tag{7.5}$$

This establishes the following relation between the *apparent coordinates* of the galaxy, say $y=(y_1, y_2, y_3)$, on the one hand and its present coordinates $x=(x_1, x_2, x_3)$ on the other

$$x_i = y_i \,\mathrm{e}^{g\xi} \tag{7.6}$$

for $i=1, 2, 3$, where

$$\xi^2 = y_1^2 + y_2^2 + y_3^2. \tag{7.7}$$

Using formula (7.6) and referring to the assumptions regarding the conditional probability density $f^*(\eta)$ (see Sect. 4) of the present coordinates X_1, X_2, X_3 of a galaxy known to belong to a cluster that has $u=u_1, u_2, u_3$ as the present coordinates of its center, it is now possible to write an expression for the conditional probability density, say $f(y, u)$, of the apparent coordinates Y_1, Y_2, Y_3 of the same galaxy. We have

$$f(y, u) = f^*\left\{\left[\sum_{i=1}^{3} (y\,\mathrm{e}^{g\xi} - u_i)^2\right]^{\frac{1}{2}}\right\} J \tag{7.8}$$

where J represents the Jacobian of the transformation $X \to Y$, namely

$$J = \mathrm{e}^{3g\xi}(1 + g\,\xi). \tag{7.9}$$

8. Fundamental formula. A variety of seemingly disconnected results of the theory of simple clustering of galaxies is a consequence of a single formula which we label the fundamental formula [*4*]. In order to achieve this degree of generality, the fundamental formula has to be couched in terms that, at first sight, may appear artificial.

Let R_1 and R_2 be two arbitrary regions in space, disjoint or not. Consider a cluster C of galaxies and let $\nu(C)$ denote the number of its member galaxies. Further, let $\nu_i(C)$ be the number of these galaxies with their apparent position in R_i, for $i=1, 2$. Imagine now that for each galaxy apparently located in R_i a random experiment, say E_i, is performed with the probability of "success" represented by a measurable function of the galaxy's apparent coordinates $\Theta_i^*(y_1, y_2, y_3) \equiv \Theta_i^*$. If the experiment E_i yields success then the galaxy will be called "successful in R_i". Assume further that for every galaxy apparently located in the intersection $R_1 \cap R_2$ of the two regions there exists a measurable function $\varrho(y_1, y_2, y_3) \equiv \varrho$ of the galaxy's apparent coordinates such that the product $\Theta_1^* \Theta_2^* \varrho$ is equal to the probability that this galaxy will be simultaneously "successful" in both R_1 and R_2. Our last assumption will be that the experiments E_i performed for different galaxies are mutually independent.

Let $n_i(C)$ represent the number of galaxies that belong to the cluster C and are successful in the region R_i, for $i=1, 2$.

Now denote by $N(r_1, r_2)$ an arbitrary non-negative integer-valued function defined over all combinations of non-negative integer values of the two arguments $r_1 \leq r_2$. Finally, let $N_i(C) = N[n_i(C), \nu_i(C)]$ and

$$N_i = \sum_C N_i(C). \tag{8.1}$$

In other words, the random variable N_i is defined as the sum, extending over all clusters C in space, of the random variables $N_i(C)$ with $i=1, 2$.

With this notation, the fundamental formula gives the following expression for the joint probability generating function of N_1 and N_2

$$G_{N_1,N_2}(t_1, t_2) = e^{-\lambda \iiint \{1 - {}^*G_\nu(t_3[1-p_1(1-t_1)-p_2(1-t_2)-p_3(1-t_1t_2)])\} du_1 du_2 du_3}. \quad (8.2)$$

Formula (8.2) requires some explanation. The triple integral in the exponent extends over the range $-\infty < u_i < +\infty$ for $i = 1, 2, 3$. The symbol p_1 is defined as a function of $u = (u_1, u_2, u_3)$, representing the conditional probability that a galaxy belonging to a cluster, now centered at u, will be "successful" in R_1 but not in R_2. Similarly, for each $u \in E$, the symbol p_2 represents the conditional probability that a galaxy from a cluster now centered at u will be successful in R_2 but not in R_1. Finally, p_3 is the similarly defined probability that a galaxy from a cluster centered at u will be successful in the intersection of R_1 and R_2. Formulae for the probabilities p_i are easily written. Thus, for example,

$$p_1 = \iiint_{R_1 - R_1 \cap R_2} f(y, u)\, \Theta_1^*\, du_1\, du_2\, du_3 + \iiint_{R_1 \cap R_2} f(y, u)\, \Theta_1^* (1 - \varrho\, \Theta_2^*)\, du_1\, du_2\, du_3. \quad (8.3)$$

Our last comment refers to the operation denoted by the asterisk ${}^*G_\nu$, to be performed on the probability generating function G_ν evaluated at the point $t_3[1 - p_1(1-t_1) - p_2(1-t_2) - p_3(1-t_1t_2)]$. This operation is called the operation of reduction and consists in the following. First, the probability generating function in question is expanded in a triple power series,

$$G_\nu\{t_3[1 - p_1(1-t_1) - p_2(1-t_2) - p_3(1-t_1t_2)]\} = \sum_{u,v,w} a_{u,v,w}\, t_1^u t_2^v t_3^w. \quad (8.4)$$

Second, the product $t_1^u t_2^v t_3^w$ is replaced by $t_1^{N(u,w)}\, t_2^{N(v,w)}$. Thus, by definition,

$${}^*G_\nu\{t_3[1 - p_1(1-t_1) - p_2(1-t_2) - p_3(1-t_1t_2)]\} = \sum_{u,v,w} a_{u,v,w}\, t_1^{N(u,w)}\, t_2^{N(v,w)}. \quad (8.5)$$

The fundamental formula (8.2) refers to only two regions R_1 and R_2. However, generalization to any number of regions is immediate.

9. First specialization of the fundamental formula: joint distribution of counts of images of galaxies in two equal, regularly spaced squares on the photographic plate. Let R_1 and R_2 stand for two equal and similarly oriented disjoint solid angles in space, with their vertices at the observer. The intersection of these two solid angles by the plane of the photographic plate forms two disjoint squares which we shall denote by R_1 and R_2. We consider one particular plate with plate effect π, and one particular set of counts characterized by the observer effect ω. Our problem is to use the fundamental formula (8.2) in order to deduce the probability generating function of the two counts of images of galaxies in the squares R_1 and R_2. In order to achieve this it is sufficient to specialize appropriately the "experiments" E_i and the function $N(r_1, r_2)$ discussed in Sect. 8, as follows.

Let E_i denote all the circumstances leading to the determination of the apparent magnitude of a galaxy with its apparent position within R_i; and to its being counted or not by the observer. The galaxy in question will be called "successful" if it is counted by the observer. The probability of this happening is given by (6.3) and we write

$$\Theta_i^* = \Theta_i^*(y_1, y_2, y_3) = \Theta(\xi, \pi, \omega). \quad (9.1)$$

Since the two regions R_1 and R_2 are disjoint, it is not necessary to define ϱ and we have $p_3 \equiv 0$. Now, it is obvious that by defining

$$N(r_1, r_2) \equiv r_1, \quad (9.2)$$

the random variable N_i defined in (8.1) will coincide with the number of images of galaxies counted on the given occasion in the square R_i of the given plate. Referring to (8.2) we find that, with this particular specialization,

$$G_{N_1 N_2}(t_1, t_2) = e^{-\lambda \iiint \{1 - G_\nu [1 - p_1(1-t_1) - p_2(1-t_2)]\} d u_1 d u_2 d u_3} \tag{9.3}$$

with

$$p_i = \iiint_{R_i} f(y, u)\, \Theta(\xi, \pi, \omega)\, d y_1\, d y_2\, d y_3. \tag{9.4}$$

Naturally, in order to obtain numerical results that can be compared with empirical data, all the unknown functions involved in (9.3) and (9.4) have to be completely specified. With the present lack of astronomical information, we are forced to use appropriate interpolatory formulae, with a moderate number of adjustable parameters. Unfortunately, even with the simplest substitutions of this kind, the complexity of (9.3) does not make it probable that it can be used directly for numerical computations of the probability that, simultaneously, $N_1 = m$ and $N_2 = n$, with m and n representing arbitrary integers. On the other hand, formula (9.3) lends itself to easy computation of the moments of N_1 and N_2 and then, with appropriate substitutions of interpolatory formulae, these moments can be computed numerically and compared with their empirical counterparts.

Taking the logarithm of (9.3), differentiating with respect to t_1 and t_2, substituting $t_1 = t_2 = 1$ and using the familiar relations between the derivatives so obtained and the moments of the two variables N_1 and N_2, we have

$$E(N_i) = \lambda \nu_1 S_{10}, \tag{9.5}$$

$$\sigma^2_{N_i} = E\{[N_i - E(N_i)]^2\} = (\nu_2 - \nu_1) S_{20} + E(N_i), \tag{9.6}$$

$$\sigma_{N_1 N_2} = E\{[N_1 - E(N_1)][N_2 - E(N_2)]\} = (\nu_2 - \nu_1) S_{11} \tag{9.7}$$

where generally,

$$\nu_k = E(\nu^k) \tag{9.8}$$

and

$$S_{m,n} = \iiint_{-\infty}^{+\infty} p_1^m p_2^n\, d u_1\, d u_2\, d u_3. \tag{9.9}$$

The remarkable aspect of the above formulae is that the quotient

$$Q = \frac{\sigma_{N_1 N_2}}{\sigma^2_{N_1} - E(N_i)} = \frac{S_{11}}{S_{20}} \tag{9.10}$$

is independent of the density λ of cluster centers per unit volume of space and of the moments ν_1 and ν_2 of the distribution of the number ν of galaxies per cluster. As a result, the quotient Q, called the quasi-correlation of the counts of images of galaxies in two squares R_1 and R_2 on the photographic plate, depends on only three factors that are not under the observer's control: (i) on the function f^* which may be regarded as a characterization of the internal structure of the clusters, (ii) on the luminosity function $p_{\mathcal{M}}(M)$ of galaxies, and (iii) upon the value of the expansion factor g. The quasi-correlation also depends upon several factors that are, more or less, at the observer's disposal: (i) on the dimensions $2\alpha \times 2\alpha$ of the squares R_1 and R_2, (ii) on the distance, say $2k\alpha$ between these squares, and (iii) on the selection factor $\Phi(m)$. For any observational setup and regime of counting [which, together, fix $\Phi(m)$] and a fixed value of α, a sequence of empirical quasi-correlations can be computed corresponding to varying $k = 1, 2, 3, \ldots$. The resulting values can then be compared with the theoretical serial quasi-correlations computed by evaluating the right-hand side of (9.10) for selected combinations of the value of g and of functions to represent f^*, Ψ and $p_{\mathcal{M}}(M)$. The theoretical computations are continued until (if possible) a value of g and a set of functions are found that give a satisfactory fit to the empirical quasi-correlations for each k. The whole process may be repeated with different values of α, and, even, with a survey having different observational setup should this be available. If and when a combination of a value of g with a pair of functions f and $p_{\mathcal{M}}(M)$ are found that provide theoretical quasi-correlations always fitting the corresponding empirical quasi-correlations, for every k and every α and every observational setup, $\Phi(m)$ being adjusted appropriately, this might be treated as an indication that a system of simple clustering has been found that represents a reasonable approximation to the actual spatial distribution of galaxies. In addition, the value of g with which the satisfactory fit of the observations has been achieved would serve as an indication of the rate of expansion of the universe.

Unfortunately, the evaluation of the right-hand side of (9.10), though not prohibitive, is very laborious and the present state of knowledge in the above respects is limited to the following [6], [7]. The luminosity function $p_{\mathcal{M}}(M)$ was assumed to be either a normal density or a sum of normal densities with several alternative values of the parameters (the mean M_0 and the variances σ_M^2). Also assumed was the existence of a limiting magnitude m_1 of the plates of the Lick Observatory combined with a model of "errors in counting". The case of a static universe was considered, so that $g=0$, and the internal structure of a cluster was assumed to be of the form

$$f^*(\eta) = \frac{1}{\sigma\sqrt{2\pi}}\, e^{-\frac{1}{2\sigma^2}\eta^2}. \tag{9.11}$$

The computations now depend on $m_1 - M_0$, on σ_M and on σ, as far as unspecified parameters in space are concerned. Efforts were made to determine $m_1 - M_0$, σ_M and σ so as to fit empirical serial quasi-correlations based on counts[1] by SHANE and WIRTANEN made in two large areas of the sky near the north galactic pole for two sizes of squares, $2\alpha = 10'$ and $2\alpha = 1°$ (by combining adjacent smaller squares).

The results [6], [7] are exhibited in Fig. 3. Here, each of the four panels corresponds to a particular combination of values $m_1 - M_0$ and σ_M. On the horizontal axis is plotted the distance $2k\alpha$ between the centers of squares in which counts are made, while the quasi-correlations are measured on the ordinate axis. Open circles represent the empirical quasi-correlations for $2\alpha = 1°$ and crosses those for $2\alpha = 10'$. In each panel there are exhibited two pairs of curves, each pair corresponding to a fixed trial value of σ measured in parsecs. The continuous curve corresponds to $2\alpha = 1°$ and the dashed curve of each pair to $2\alpha = 10'$.

It will be seen that, whenever the continuous curve is adjusted so as to fit the open circles, the corresponding dashed curve passes above all the crosses and vice versa. Since the results reproduced in Fig. 3 are a fair sample of a number of other equally unsuccessful attempts to fit simultaneously the two sequences of empirical quasi-correlations, the conclusion is that, probably, there is something basically wrong among the hypotheses underlying the theoretical values of Q. Of course, any one of these hypotheses might be false. However, the authors are particularly uneasy about the following: (i) The error model used

[1] The counts in one of the areas are reported in C. D. SHANE and C. A. WIRTANEN: Astronom. J. **59**, 285–304 (1954); the others are to be published.

in [6], (ii) the hypothesis of *simple* rather than of higher order clustering, and (iii) the hypothesis that the universe is static so that $g=0$.

All the contents of the present section are based on a particular specialization of the fundamental formula, leading to the joint distribution of counts, made on one occasion, of the number of images of galaxies, in two squares on the same plate. Very similar specializations lead to the following joint distributions:

(i) Counts made on two different occasions in the same square of a given plate.

(ii) Counts made in homologous squares of two duplicate plates, taken over the same field in the sky and intended to be identical.

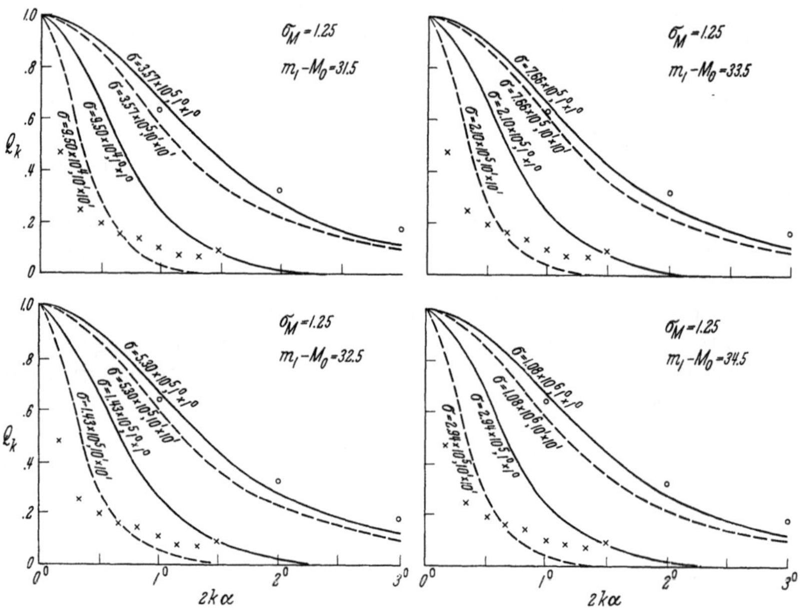

Fig. 3. Empirical quasi-correlations between counts of images of galaxies in $1^\circ \times 1^\circ$ and $10' \times 10'$ squares compared with theoretical values corresponding to $\sigma_M = 1.25$ and to varying trial values of the parameters $m_1 - M_0$ and σ.

(iii) Counts made in homologous squares of two plates, taken over the same field in the sky, one taken with a small and the other with a large telescope.

The framework of the present account prevents entering into details of the method of dealing with the variability of the random plate and observer effect Π and Ω; the following brief comments must suffice. It is assumed that, for a fixed observational setup and for a given observer, the variations in Π and Ω are small with expectations equal to zero. The method used to evaluate moments of the N_i consists then in first computing their conditional values, corresponding to fixed values π and ω of Π and Ω, respectively. Then these conditional moments are expanded in a Taylor series in powers of π and ω, keeping only a few terms. Finally, the expectations are taken with respect to variation of Π and Ω.

10. Problems of verification of the hypothesis regarding the expansion of the universe. In the preceding section reference was made to the possibility that the study of quasi-correlations combined with the theory of clustering of galaxies, may lead to the verification of the hypothesis that the observed shifts of spectral lines of faint galaxies are due to velocities of recession rather than to some other, thus far unidentified, factor. This problem is of sufficient importance to deserve a special section in the present paper even though, thus far, no definite results are reportable.

Currently, in a substantial section of astronomical literature the velocity interpretation of the redshift of galaxies is simply taken for granted and efforts are made to use the empirical magnitude-redshift relation to study the changes in these velocities that occur in time. Thus, for example [3], a study of this kind led to the tentative conclusion that the galaxies are decelerating in their recession.

Following HUBBLE [2] and in conformity with some ideas of ZWICKY [9], the authors are of the opinion that the velocity interpretation of the redshift requires verification. Also, to be convincing, this verification must be, as far as possible, independent of the spectroscope. HUBBLE's own method of verifying the hypothesis of the expansion of the universe consisted in plotting against the independent variable m the number of galaxies with their apparent photographic magnitude brighter than m. This number, denoted by $N(m)$ by HUBBLE, was estimated from counts of images of galaxies on the various survey plates. The authors agree with ZWICKY [9] that this particular method is not very promising because, although the functional relation between $N(m)$ and m does depend on whether or not the universe is expanding, there are considerable difficulties in securing observational data accurate enough to detect this difference. The reason is that, as discussed in Sects. 5 and 6, in relation to routine counts of galaxies on survey plates, the concept of the limiting apparent magnitude is a fiction. In fact, the authors are most doubtful as to whether, with present-day instrumentation, it is possible to establish for reasonably large areas of the sky, reliable values of $N(m)$ for $m = m_1, m_2, \ldots, m_s$.

In order to be promising, the method of verification of any hypothesis regarding the expansion of the universe must be insensitive to observational difficulties such as the unavoidable plate-to-plate variation in intensity and definition of images of the same galaxies as they appear on photographic plates taken with a given observational setup. This remark applies to the method of HUBBLE just discussed. Also it applies to the use of the magnitude-redshift relation discussed in Sect. 14.

In order to be reliable, the method must also be insensitive to any details of the spatial distribution of galaxies that are difficult to estimate with precision. This last remark also applies to the opinion of ZWICKY [9] that the distribution of the apparent diameters of clusters of galaxies identified on a collection of plates may be the key to the problem of expansion of the universe.

It is true that ZWICKY's own formulae for the distribution of diameters of clusters in a static and in an expanding universe are very simple and lead to sharply different numerical results. However, the authors are inclined to think that these formulae were deduced under too sweepingly simplifying assumptions. The clusters of galaxies are treated as discs, all of the same size, with the dimensions of their images on photographic plates inversely proportional to distance. In actual fact, of course, the variation in the apparent diameters of images of clusters is caused, in a complicated manner, by a number of factors. For example, as the distance of the cluster is increased, some of its fainter galaxies become invisible, which contributes to a decrease in the apparent dimensions. Also, the effect just described is likely to be felt more strongly with a cluster which has but few members than with a rich cluster. Thus, it must be obvious that the actual distribution of apparent diameters of clusters must depend on all elements of the theory of clustering, on the luminosity function $p_{\mathcal{M}}(M)$, on the structure of clusters f^*, on the distribution of the number ν of galaxies per cluster, as well as on the method of selection Φ. A formula taking into account all these elements is reproduced in Sect. 14.

The authors' own hopes for a verification of the hypothesis of expansion are connected with the study of serial quasi-correlations of counts of images of galaxies as discussed above. Here Fig. 3 is very relevant. The four panels of this figure illustrate the fact that the shape of the curve representing theoretical quasi-correlations, corresponding to a fixed size of squares in which the counts are made, is remarkably insensitive to simultaneous changes in the various constants involved. As a result, the empirical quasi-correlations for $1^\circ \times 1^\circ$ counts can be fitted very well starting with a great variety of the constants. To a lesser degree, this is also true for the empirical quasi-correlations corresponding to counts in $10' \times 10'$ squares. On the other hand, when a *simultaneous fit* of two series of quasi-correlations is attempted, one for $10' \times 10'$ and the other for $1^\circ \times 1^\circ$ squares, striking divergences develop and there appears to be no system of constants which fits both series to any reasonable degree of approximation. Thus, it appears that, while single curves of quasi-correlations corresponding to fixed size squares in which counts of images of galaxies are made, are insensitive to the details of the hypotheses regarding the clustering of galaxies, sets of such curves, corresponding to varying sizes of squares, are sensitive to the "structural" characteristics of these hypotheses. Also, the same is likely to be true for quasi-correlations corresponding to different observational setups. Thus, the authors attach considerable hopes to the results obtainable from survey plates to be taken with the new 120-inch telescope at the Lick Observatory. Similar hopes are connected with the gradually accumulating observations made with radio-telescopes.

While expressing these hopes, the authors are well aware of the existing limitations. It is obvious that the contemplated verification of the hypothesis of expansion amounts to a choice between the various cosmological theories. The authors are inclined to expect that the analysis of quasi-correlations will lead to a choice between two categories of these theories. On the one hand, we place the flat space theory of a stationary universe and the Bondi-Gold-Hoyle steady state theory [*1*]. On the other hand, there are all the other expansionist theories as described, for example, by ROBERTSON[1].

11. Index of clumpiness of the distribution of galaxies. Let $N_1, N_2, \ldots, N_r$ stand for the numbers of images of galaxies counted in r non-overlapping squares (with side 2α) on a survey plate. Claims have been made repeatedly that if the quotient

$$k_r^2(\alpha) = \frac{\frac{1}{r}\sum_{i=1}^{r}(N_i - \overline{N})^2}{\overline{N}} \tag{11.1}$$

with

$$\overline{N} = \frac{1}{r}\sum_{i=1}^{r} N_i \tag{11.2}$$

is plotted against $\overline{N}$, then the behavior of the resulting curve is indicative of the presence or absence of intergalactic extinction. Namely[2], the fact that $k_r(\alpha)$ appears to be a monotone increasing function of $\overline{N}$ was treated as evidence of the existence of intergalactic absorbing material.

[1] H. P. ROBERTSON: Publ. Astronom. Soc. Pacific **67**, 82–98 (1955).
[2] F. ZWICKY: Helvetica Physica Acta **26**, 241–254 (1953).

A closer study of the problem[1] revealed that, at least over a substantial range of observational setups, the monotone increasing character of the dependence of $k_r(\alpha)$ on $\bar{N}$ is a consequence of the phenomenon of clustering of galaxies alone and that, therefore, this particular behavior of the quotient $k_r^2(\alpha)$ provides no evidence of the existence of intergalactic clouds. Still, such clouds may well exist.

12. Second specialization of the fundamental formula: Problem of interlocking of clusters. The theoretical setup, described in the preceding sections, for studying the spatial distribution of galaxies is based on the presumption that this distribution is a conglomeration of randomly constructed and randomly distributed clusters. This is not, of course, the only approach to the problem. Another possibility is to consider that the actual distribution of galaxies in space is the realization of a (spatially) stationary stochastic process of unspecified properties.

In fact, a study of this kind, due to CHANDRASEKHAR and MUNCH[2], generated several papers by other authors. Essentially, the treatment of the problem consists in using the counts of images of galaxies in small squares on survey plates in order to estimate the first two moments and the correlation function of the stochastic process governing the distribution of galaxies in infinitesimal regions in space. An unfortunate oversight made the original formulae inapplicable to counts of galaxies. However, with the necessary modifications[3], the work may be continued. Nevertheless, the method does not appear specific enough to yield important information regarding the spatial distribution of galaxies. In fact, the same first two moments and the same correlation function can correspond to an immense variety of stochastic processes, including the process of simple and multiple clustering [*7*].

The advantage of the approach through the concept of clustering appears to be that, once the best fitting values of the constants involved are established, it is possible to obtain a variety of valuable information about the spatial relationships between the accumulations of galaxies that are interpreted as "clusters". One question that may be answered is whether clusters of galaxies are themselves uniformly distributed in space or, as visualized by CHARLIER (Sect. 1), are organized into clusters of higher order (Sect. 15). Another question, to be dealt with below, is whether clusters of galaxies interlock with one another or, somewhat like globular star clusters, are separated by considerable distances and thus offer the possibility of treatment as isolated dynamical systems.

The main difficulty in treating *phenomenal* problems of this kind consists in the formulation of an appropriate *mathematical* problem. Three different attempts[4] were made in this direction [*4*] but only one of them is reported in the present section.

Consider an arbitrary cluster C_0 of galaxies, to be labeled the selected cluster, and let $\nu_0 \geqq 1$ be its number of members. This number ν_0 will be treated as a random variable subject to the hypotheses explained in Sect. 4. Now we shall number the ν_0 galaxies of the selected cluster in the order of decreasing distance from the center of C_0. These distances, denoted by

$$\eta_1 \geqq \eta_2 \geqq \cdots \geqq \eta_{\nu_0}, \tag{12.1}$$

[1] J. NEYMAN, E. L. SCOTT and C. D. SHANE: Astrophys. Journ. Suppl. **1**, No. **8**, 269–294 (1954).

[2] S. CHANDRASEKHAR and G. MUNCH: Astrophys. Journ. **115**, 94–123 (1952). Further literature is listed in Ref. [*7*].

[3] J. NEYMAN and E. L. SCOTT: Proc. Nat. Acad. Sci. U.S.A. **40**, 873–881 (1954). — Astronom. J. **60**, 33–38 (1955).

[4] J. NEYMAN and E. L. SCOTT: Proc. Nat. Acad. Sci. U.S.A. **39**, 737–743 (1953).

will be treated as random variables satisfying the hypotheses of Sect. 4. For any positive number x let $S(x)$ denote a sphere of radius x centered at the center of the selected cluster C_0. Thus, $S(\eta_k)$ will stand for a sphere of the random radius η_k for $k=1, 2, \ldots, \nu_0$.

Now consider C, a cluster of galaxies different from the selected cluster C_0. If any one of the galaxies belonging to C is located *within* the sphere $S(\eta_k)$, we shall say that the cluster C *penetrates* the selected cluster *to the depth k*. Denote by τ_k the random variable defined as the number of clusters that penetrate the selected cluster to the depth k.

It is obvious that the distribution of τ_k characterizes the degree of mutual interlocking of clusters. Suppose, for example, that $P\{\tau_1=0\}$ is a number close to unity. The interpretation of this result would be that nearly all clusters are isolated clusters. Contrary to this, if $E(\tau_5)$ is, for example, equal to 12, this would mean that an isolated cluster is a fiction and that, as a general rule, each cluster is penetrated to the depth 5 by many others, on the average, twelve others.

Because of lack of space it is impractical to reproduce here the details [*4*] of the deduction of the probability generating function $G_{\tau_k}(t)$ of the random variable τ_k. However, the following brief remarks might be interesting because they illustrate the possibilities offered by the fundamental formula (8.2).

Our first remark is that the definition of τ_k requires that $\nu_0 \geqq k$. Thus, the function $G_{\tau_k}(t)$ must be calculated on the hypothesis $\nu_0 \geqq k$. Our second remark is that the four assumptions of Sect. 4 are sufficient for the evaluation of the conditional distribution of η_k, given that $\nu_0 \geqq k$. The details of this evaluation are given elsewhere and are not reproduced in the present account.

Proceeding further, we fix a positive number ζ and consider a cluster $C(\zeta)$ with center at a distance ζ from the center of the selected cluster C_0. For any positive number η, the assumptions of Sect. 4 determine the probability, say $P(\zeta, \eta)$, that a galaxy from the cluster $C(\zeta)$ will be located within the sphere $S(\eta)$.

Now let $\tau(\eta)$ denote the number of clusters, other than C_0, that possess at least one galaxy within the sphere $S(\eta)$. Also, let $G_{\tau(\eta)}(t)$ be the probability generating function of $\tau(\eta)$. It is obvious that the probability generating function of τ_k is the expectation of that of $\tau(\eta_k)$,

$$G_{\tau_k}(t) = E\{G_{\tau(\eta_k)}(t)\}, \tag{12.2}$$

so that the problem is reduced to the evaluation of $G_{\tau(\eta)}(t)$. This is done by a specialization of the fundamental formula (8.2) which makes $N_2 \equiv 0$ and $N_1 \equiv \tau(\eta)$, as follows. We identify R_2 with an empty set and R_1 with the sphere $S(\eta)$. Next we put $\Theta^* \equiv 1$. Finally, we put $N(r_1, r_2) \equiv 0$ for $r_1=0$ and $N(r_1, r_2) \equiv 1$, for $r_1 > 0$. With these specializations $p_1 = P(\zeta, \eta)$, and $p_2 = p_3 = 0$. Also the operation of reduction gives

$$^*G_\nu\{t_3[1-p_1(1-t_1)]\} = t_1 + (1-t_1)\, G_\nu[1-P(\zeta, \eta)]. \tag{12.3}$$

Now, easy algebra completes the solution of the problem. The numerical characteristics of the distribution of τ_k, and of two other similarly defined variables also describing the degree of interlocking of clusters, were calculated using the estimates of the constants of the theory of simple clustering [*6*], [*7*] obtained on the assumption that the universe is static. With the range of estimates of the parameters illustrated in Fig. 3, the characteristics of τ_k vary widely and their implications range from (a) clusters of galaxies are mutually interlocked with almost no isolated clusters to (b) most clusters are isolated not penetrated by their neighbors. Since none of the systems of parameters illustrated in Fig. 3

fits the observations, the assumptions on which they were obtained are now discredited and further preliminary work is required before embarking on computations regarding the degree of interlocking of clusters.

13. Apparent magnitude and redshift of single galaxies. As was noted in Sect. 5, there are three kinds of observable random variables available for verifying any theory of the spatial distribution of galaxies. Thus far, notably in Sects. 9 to 11, only the first kind of data was considered, namely counts of images of galaxies in small squares on photographic plates. In the present section, the second type of material will be considered. These are the gradually accumulated observations of apparent magnitude and of redshift of individual galaxies. In particular, we have in mind the collection of data on redshift and apparent magnitude recently published by HUMASON, MAYALL and SANDAGE [*3*].

The theoretical developments reported below form a, thus far, unpublished extension due to NEYMAN of the important paper by MALMQUIST[1]. The basic assumptions underlying the theory coincide with those explained in Sects. 4 and 6 with an adjusted interpretation of the selection factor $\Phi(m)$.

The point is that whatever the instrumentation available to an observer may be, it always implies some restrictions on the objects for which measurements of apparent magnitude and redshift can be made. Thus, the observer is bound to make some selection of these objects. The ultimate source of objects for which the measurements are actually performed must be, directly or indirectly, survey plates. In the case of the measurements of redshift of HUMASON and MAYALL, the reference to survey plates is often indirect since many of their objects stem from the Shapley-Ames catalogue.

In the present section, the selection factor $\Phi(m)$ is the conditional probability that a galaxy with apparent photographic magnitude m will be *selected for measurements of the apparent magnitude and redshift*; this is likely to be a different function of m from that serving as a selection factor in the preceding sections.

Because of the importance of the results of MALMQUIST, it may be useful to indicate briefly two main points at which the theory reported here deviates from MALMQUIST'S work. The first important point is that, whereas MALMQUIST was concerned with the classical integral equation of stellar statistics, which may be written as

$$p_\mu(m) = \int_0^\infty r^2 D(r)\, p_{\mathcal{M}}(m + 5 - 5 \operatorname{Log} r)\, dr, \tag{13.1}$$

where $D(r)$ represents the density of objects studied at distance r from the observer, in the present theory this equation is replaced by

$$p_\mu(m) = \Phi(m) \int_0^\infty r^2 D(r)\, p_{\mathcal{M}}(m + 5 - 5 \operatorname{Log} r)\, dr, \tag{13.2}$$

where $\Phi(m)$ is the selection factor discussed above. The second important point of difference is that, whereas MALMQUIST'S principal purpose was to estimate the spatial distribution $D(r)$ of the objects studied, starting with various *a priori* assumptions regarding the luminosity function $p_{\mathcal{M}}(M)$, in the theory here reported, the reverse problem is considered: starting with the assumptions of the theory of clustering of galaxies and using additional assumptions regarding the mechanism of selection of objects for measurements of apparent magnitude and redshift, establish formulae connecting the selection factor $\Phi(m)$ and the luminosity function $p_{\mathcal{M}}(M)$ with the distributions of random variables that are directly

[1] K. G. MALMQUIST: Ark. Mat., Astronomi Fys. **16**, No. 23, 1–52 (1922).

observable. Thus, for example, one of MALMQUIST'S *a priori* assumptions was that $p_{\mathcal{M}}(M)$ may be approximated by a normal probability density. His formulae for estimating the constants of this density were later used by HUBBLE and others. However, in the present work no specific assumptions are made regarding the luminosity function $p_{\mathcal{M}}(M)$ of galaxies, and methods are developed for estimating the shape of this function from the observations.

The results reported in the present section use the assumptions of Sects. 4 and 6 supplemented by the additional assumption that the redshift z is proportional to the apparent distance ξ of the galaxy, so that

$$c\,z = c\,\frac{d\lambda}{\lambda} = H\,\xi, \tag{13.3}$$

where c is the velocity of light and H stands for the HUBBLE constant. From these assumptions NEYMAN has deduced various characterizations of the conditional joint distribution of apparent magnitude and redshift, given that the galaxy to which these quantities refer has been *selected* for measurement.

When we are considering the apparent magnitude and the redshift of selected galaxies as random variables, we shall use the notation μ and Z. The letters m and z will be used to denote the particular values of μ and Z, respectively. Through a new specialization of the fundamental formula, the joint conditional distribution of μ and Z is obtained, given that the galaxies were selected on the basis of apparent magnitude alone, irrespective of the clusters to which they may happen to belong. This distribution depends, in a complicated way, on both the distribution of the number ν of galaxies per cluster and the internal structure of clusters characterized by the function f^*. Because of the complicated character of the general results only some easy specializations, involving certain additional hypotheses, will be reported. The principal additional restriction is that the observational material is composed of "field galaxies" only (that is, galaxies that are the sole member of a "cluster" of just one galaxy). Also, the theory would apply to observations of cluster galaxies, provided that the contribution to the data of each cluster is limited to a single randomly selected galaxy. Presumably, some relaxation of this condition would not affect the final results too seriously.

The other additional conditions assumed in this particular specialization are that the material is composed of not too distant galaxies for which the effects of redshift (or reddening, etc.) on the apparent magnitude and the effect of possible expansion of the universe can be neglected.

Proposition I. If the process of selection of field galaxies satisfies the conditions stated, then the distribution of apparent photographic magnitude μ among the selected galaxies is independent of the luminosity function $p_{\mathcal{M}}(M)$ and is given by

$$p_\mu(m) = C\,\Phi(m)\,10^{\frac{3m}{5}}, \tag{13.4}$$

where C is the norming factor such that the integral

$$C\int_{-\infty}^{+\infty}\Phi(m)\,10^{\frac{3m}{5}}\,dm = 1. \tag{13.5}$$

It follows from this proposition that the empirical distribution of apparent magnitude among the selected field nebulae may be used to estimate the selection factor $\Phi(m)$, and that this empirical distribution is entirely irrelevant to the problem of estimating the luminosity function $p_{\mathcal{M}}(M)$.

In order to formulate the next proposition we introduce the symbol $\mathfrak{M}$ to denote the "modified" absolute magnitude of a galaxy

$$\mathfrak{M} = \mu + 5 - 5 \operatorname{Log} c Z. \tag{13.6}$$

The modified absolute magnitude is computable for each galaxy for which measurements of the apparent magnitude μ and of the redshift Z are available. Under the present assumptions, the true absolute magnitude is

$$\left.\begin{aligned} \mathcal{M} &= \mu + 5 - 5 \operatorname{Log} \xi = \mu + 5 + 5 \operatorname{Log} H - 5 \operatorname{Log} c Z \\ &= \mathfrak{M} + 5 \operatorname{Log} H. \end{aligned}\right\} \tag{13.7}$$

Proposition 2. Under the assumptions specified, whatever be the selection factor $\Phi(m)$, the conditional probability density of the modified absolute magnitude of the selected galaxies is perfectly determined by the luminosity function $p_{\mathcal{M}}(M)$ of galaxies and is of the form

$$p_{\mathfrak{M}}(x) = C\, p_{\mathcal{M}}(x + 5 \operatorname{Log} H)\, e^{-\frac{3x}{a}}, \tag{13.8}$$

where C is the norming factor such that the integral of (13.8) taken from minus infinity to plus infinity equals unity with $a = 5 \operatorname{Log} e$.

As an interesting corollary to Proposition 2, notice that an estimate of the luminosity function $p_{\mathcal{M}}(M)$ can be obtained by three easy operations, as follows:

(i) The empirical values of the modified absolute magnitude $\mathfrak{M}$ of selected field galaxies are used to estimate their probability density $p_{\mathfrak{M}}(x)$, perhaps by fitting one of the Pearson curves.

(ii) The result of this estimation is multiplied by the factor $e^{3x/a}$.

(iii) The resulting function is normed by multiplying by a constant so that its integral for $-\infty < x < +\infty$ is unity.

These three steps may, of course, be conveniently combined into a single step. The result, in either case, is an estimate of the luminosity function $p_{\mathcal{M}}(x + 5 \operatorname{Log} H)$ evaluated at $M = x + 5 \operatorname{Log} H$. If H is presumed known then an easy change in the origin will yield $p_{\mathcal{M}}(x)$. Otherwise, our knowledge of $p_{\mathcal{M}}(x)$ lacks information about the so-called zero point.

As a further corollary of Proposition 2, we note that if, as assumed by HUBBLE, the luminosity function of galaxies is a normal probability density, then the probability density $p_{\mathfrak{M}}(x)$ of the modified absolute magnitude of the selected galaxies will also be normal and *vice versa*. Also if, as expected by ZWICKY[1], the luminosity function of the galaxies is exponential, then the probability density $p_{\mathfrak{M}}(x)$ will also be exponential and *vice versa*.

Since the above results are valid whatever be the selection factor $\Phi(m)$, *provided the apparent photographic magnitude is the sole basis for selection of objects for measurement,* Proposition 2 opens the possibility for an unambiguous estimation of the luminosity function of field galaxies.

The methods just described have been applied to several categories of galaxies selected from the data of HUMASON, MAYALL and SANDAGE [3] so as to satisfy the assumptions of the theory. This involved the elimination of objects believed to be cluster members and also of those that were included in the program because of certain special reasons, mentioned by the authors, rather than solely because of their apparent magnitude. The results obtained are exemplified in Figs. 4 and 5 and have the following general characteristics:

[1] F. ZWICKY: Helv. phys. Acta **26**, 241–254 (1953).

(i) Formula (6.4) appears to offer a satisfactory approximation to the actual selection factor (see Fig. 4).

(ii) The luminosity function $p_{\mathcal{M}}(M)$ of the field galaxies estimated by formula (13.8) appears to be represented by a unimodal curve, almost symmetrical, with a slight elongation toward fainter magnitudes. The approximation of this curve by a normal probability density (HUBBLE), though not exact, will probably not produce serious discrepancies in further applications.

(iii) The normal distributions fitted to observations of several categories of galaxies have their standard deviation σ_M approximately equal to unity.

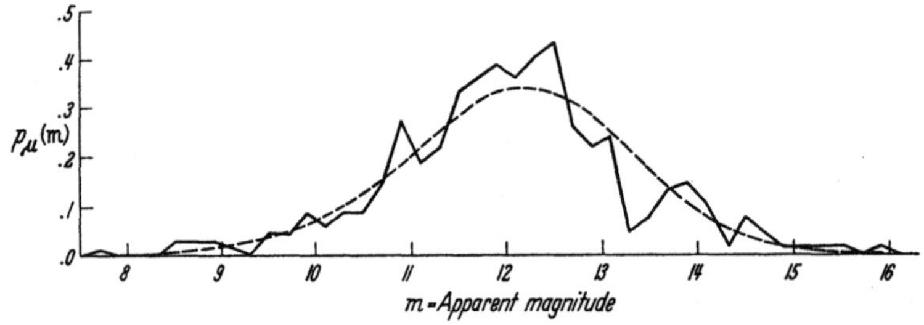

Fig. 4. Theoretical (dashed) and observed (solid) distribution of apparent magnitude for 349 field galaxies of type S. The theoretical curve corresponds to a selection function with estimated parameters, $\gamma = 11.4$ mag, $\delta = 1.0$ mag.

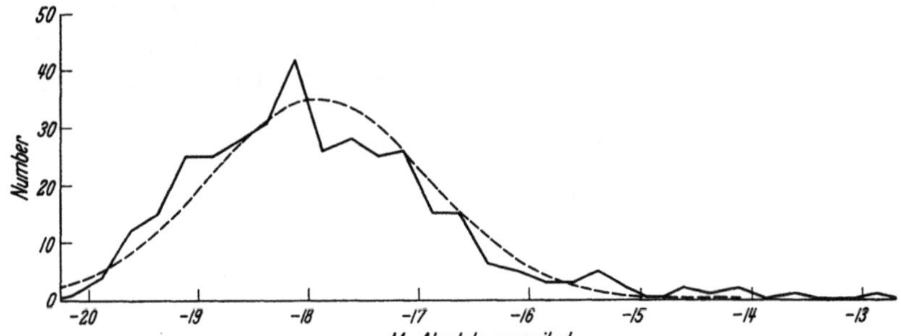

Fig. 5. Theoretical (dashed) and observed (solid) distribution of absolute magnitude (plus a constant) for 349 field galaxies of type S. The theoretical curve corresponds to a normal distribution with $\sigma_M = 1.0$ mag.

Further work is necessary to establish a closer relation between the theory and observations. For this reason the graphs and estimates given in Figs. 4 and 5 should be considered provisional.

The following proposition, together with its numerical applications, tends to illustrate that the adoption of the normal probability density to represent the luminosity function of field galaxies and of formula (6.4) to represent the selection factor $\Phi(m)$, are likely to produce a reasonable first approximation to the empirical data.

Proposition 3. If the process of selection of galaxies conforms with the basic assumptions of the present section, if the luminosity function $p_{\mathcal{M}}(M)$ of field galaxies is represented by the normal law and if the selection factor $\Phi(m)$ has the form (6.4), then the conditional probability density of the apparent photographic magnitude μ of the selected galaxies, given their redshift z, is represented by the formula, say

$$p_\mu(m \mid z) = \frac{e^{-\frac{1}{2\sigma_M^2}[m-q(z)]^2}}{\sigma_M\sqrt{2\pi}} \; \frac{1 - L\left(\frac{m-\gamma}{\delta}\right)}{1 - L\left(\frac{q(z)-\gamma}{\sqrt{\sigma_M^2+\delta^2}}\right)}, \tag{13.9}$$

where, for the sake of brevity,

$$q(z) = 5 \operatorname{Log} c z - 5 \operatorname{Log} H + M_0 - 5. \tag{13.10}$$

Also, the regression function of μ on z is given by, say,

$$\mu(z) = E(\mu \mid z) = q(z) - \frac{\sigma_M^2}{\sqrt{\sigma_M^2 + \delta^2}} \, \frac{l\left(\dfrac{q(z) - \gamma}{\sqrt{\sigma_M^2 + \delta^2}}\right)}{1 - L\left(\dfrac{q(z) - \gamma}{\sqrt{\sigma_M^2 + \delta^2}}\right)}. \tag{13.11}$$

Here the function $l(x)$ and $L(x)$ are those defined in (6.5) and (6.6).

Fig. 6 exhibits the scatter diagram of the apparent photographic magnitude and of the redshift of the 349 spiral field galaxies to which Figs. 4 and 5 also refer.

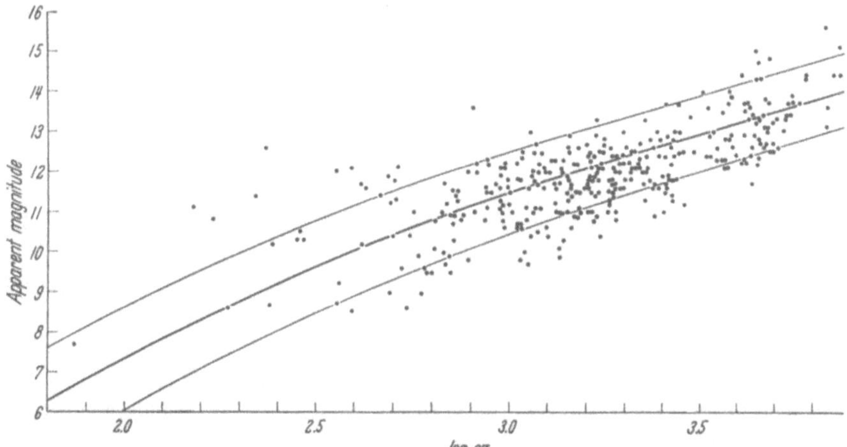

Fig. 6. Apparent photographic magnitude and redshift for 349 spiral field galaxies. The heavy curve represents the theoretical regression of magnitude on Log redshift using the estimated parameters from Figs. 4 and 5. The two lighter curves are expected to enclose about 80% of the points.

The heavy continuous line in Fig. 6 corresponds to Eq. (13.11) and was calculated using the following combination of constants. The parameters γ and δ were those obtained in the process of fitting the histogram in Fig. 4. It will be remembered that this process is entirely independent of any assumption regarding the luminosity function $p_{\mathcal{M}}(M)$. The constants σ_M and M_0 used in calculating (13.11) were those obtained in the process of fitting the histogram of the modified absolute magnitudes, exhibited in Fig. 5. This latter process is entirely independent of any assumption regarding the selection factor $\Phi(m)$. In other words, in calculating the heavy continuous curve in Fig. 6 two pairs of constants were used, each pair obtained independently from the other. The two light curves in Fig. 6 were computed using formula (13.9) with the same constants. These curves indicate the area that should contain 80% of the observations. The judgment as to whether a given fit is satisfactory or not is always somewhat subjective. In the present case, the authors are inclined to think that the observations tend to support both the basic assumptions of the theory and also the special assumptions that the selection factor $\Phi(m)$ has a form similar to (6.4) and that the luminosity function of spiral field galaxies does not differ very much from the normal law exhibited in Fig. 5. Similar results, with somewhat different numerical constants were obtained for elliptical field galaxies.

14. Statistics of clusters of galaxies. Under this heading it is appropriate to mention two kinds of data. First there are data collected by Shapley, Zwicky and their collaborators giving counts of visible members of identified clusters and the, appropriately defined, diameters of these clusters. Second, the already quoted memoir by HUMASON, MAYALL and SANDAGE [3] gives a substantial amount of measurements of apparent magnitude and of redshift of selected galaxies in individual clusters. Of the theory developed in relation to these observations very little has been published and only two results, both due to SCOTT, will be reported here.

The first result[1] to be described is concerned with the apparent radius of a cluster of galaxies [7]. Consider a cluster with its center at the "present" distance u from the observer. In accordance with the assumptions of the present theory of clustering, this cluster has a number ν of galaxies. This number ν is treated as a random variable with probability generating function denoted by $G_\nu(t)$. However, if the distance u is considerable, not all of the ν galaxies forming the cluster need have images that can be identified on a given photographic plate. We make the simplifying assumption that, for identification of the image of a galaxy, the apparent photographic magnitude of the galaxy must be brighter than a specified limit m_1. For the sake of brevity, such galaxies will be termed "visible" galaxies. Thus, in parallel with the random variable ν representing the total number of galaxies belonging to the given cluster, we have to consider another random variable, say ν^*, representing the number of galaxies in the cluster that are visible. Naturally, the distribution of ν^* depends upon the distribution of ν, upon the luminosity function of galaxies, upon the structure f^* of the clusters and on whether or not the universe is expanding.

We shall say that the cluster itself is visible if ν^* exceeds a specified limit, say n. This limit n is conventional. For example, in compiling the statistics of clusters, the observer may make a rule of counting only those clusters for which he is able to identify, say 25 member galaxies, etc.

Denote by $\Theta(\xi, m_1)$ the conditional probability that a galaxy with its apparent distance ξ will be visible and by $\Theta^\dagger(u, m_1)$ the analogous conditional probability that a galaxy, belonging to a cluster with its center located at its present distance u from the observer, will be visible. The probability $\Theta(\xi, m_1)$ can be obtained from an easy modification of formula (6.3), the probability $\Theta^\dagger(u, m_1)$ is given by

$$\Theta^\dagger(u, m_1) = \iiint_{-\infty}^{+\infty} \Theta(\xi, m_1)\, f(y, u)\, dy_1\, dy_2\, dy_3. \tag{14.1}$$

As may be expected, if the universe is static and if u is considerable, then $\Theta^\dagger(u, m_1)$ is indistinguishable from $\Theta(u, m_1)$.

One category of results can be exemplified by the formula giving the probability generating function of the variable ν^* which we quote for the particular case $n = 1$, as follows

$$G_{\nu^*}(t \mid \nu^* \geq 1) = 1 - \frac{\int_0^\infty u^2 \{1 - G_\nu[1 - \Theta^\dagger(u, m_1)(1 - t)]\}\, du}{\int_0^\infty u^2 \{1 - G_\nu[1 - \Theta^\dagger(u, m_1)]\}\, du}. \tag{14.2}$$

Now consider a cluster with ν^* visible galaxies and denote by Φ the angular distance between any one of these galaxies and the center of the cluster. Let ϱ be the greatest of the ν^* values of Φ. The random variable ϱ so defined will be described as the apparent radius of the cluster. For any positive number x

[1] An abstract appears in E. L. SCOTT: Proc. Internat. Congr. of Math., vol. 2, p. 303—304. Amsterdam: North-Holland Publ. Co. 1954,

denote by $F_\Phi(x|u)$ the conditional distribution function of Φ, given that the present distance of the cluster center is u, and by $F_\varrho(x|\nu^* \geqq n)$ the conditional distribution of the apparent radius ϱ, given that the cluster contains at least n visible galaxies. With this notation we have the following two formulae

$$F_\Phi(x \mid u) = \frac{1}{\Theta^\dagger(u, m_1)} \iiint\limits_{\omega(x)} \Theta(\xi, m_1)\, f(y, u)\, dy_1\, dy_2\, dy_3, \tag{14.3}$$

where $\omega(x)$ stands for a circular cone with axis passing through the center of the cluster considered, defined by the condition $0 \leqq \Phi \leqq x$.

The other formula is

$$F_\varrho(x \mid \nu^* \geqq 1) = 1 - \frac{\int\limits_0^\infty u^2 (1 - G_\nu\{1 - \Theta^\dagger(u, m_1)\,[1 - F_\Phi(x \mid u)]\})\, du}{\int\limits_0^\infty u^2 (1 - G_\nu\{1 - \Theta^\dagger(u, m_1)\})\, du}. \tag{14.4}$$

In the above formulae the value of n adopted is unity. It must be apparent that theoretical results corresponding to higher arbitrary values of n are much more useful. However, the corresponding formulae are substantially more complicated and, in order to save space, only their simplest versions corresponding to $n = 1$ are reproduced here.

It will be seen that formula (14.4) represents the solution of a problem analogous to that treated by ZWICKY and mentioned in Sect. 10. However, the methods of approach adopted in the two cases are very different.

The second category of SCOTT's results refers to a bias, first noticed by BEHR[1], in the estimated distance to clusters due to their selection, and affecting the magnitude-redshift relation [7]. The subject of study[2] includes the conditional regression function of the apparent magnitude, say μ_r, of the r-th brightest galaxy of a cluster on its redshift, given that this cluster has been identified as a cluster and given that the galaxy in question is bright enough for measurements of the redshift. The basic assumption in this problem is that this regression has some definite form (for example, that the regression of μ_r on Log z is linear) and the specific problem is to evaluate the difference between the regression for selected and for non-selected clusters. For illustration, the assumption will be adopted that among the non-selected clusters the regression of μ_r on Log z is strictly linear.

While the actual mechanism of selecting clusters is probably very complicated, it is assumed that the following two natural assumptions summarize the process of selection.

(a) In order that a cluster of galaxies be identifiable as a cluster, it must possess at least n visible galaxies, where the value of n is conventionally fixed, perhaps two dozen or so. Here, as above, the term "visible" is understood to mean that the photographic apparent magnitude of the galaxy is brighter than the limiting value m_1.

(b) In order that the redshift of the r-th brightest galaxy can be measured, it must be bright enough. More specifically it is assumed that the apparent photographic magnitude μ_r of the r-th brightest galaxy must be brighter than another fixed number, say $m_2 \leqq m_1$.

Obviously the two conditions (a) and (b) are necessary for the inclusion of a given cluster with its r-th brightest galaxy into a program for the study of the

[1] A. BEHR: Astronom. Nachr. **279**, 97–104 (1951).
[2] ELIZABETH L. SCOTT: Astronom. J. **62**, 248–265 (1957).

magnitude-redshift relation. The essential assumption of the theory developed is that these two conditions are also sufficient.

A basic result represents the conditional distribution function of the apparent photographic magnitude μ_r of the r-th brightest galaxy in a cluster satisfying the conditions (a) and (b), given that it has a fixed redshift z. The general formula for this distribution function is fairly complicated and must be given in terms of two auxiliary functions, as follows. Let

$$F(\xi, m_1, s, r, x) = \Theta^s(\xi, m_1) - \sum_{k=0}^{r-1} \binom{s}{k} \Theta^k(\xi, x)\,[\Theta(\xi, m_1) - \Theta(\xi, x)]^{s-k} \tag{14.5}$$

and also

$$F^*(\xi, m_1, n, r, x) = \sum_{1=n}^{\infty} \frac{1}{s!} F(\xi, m_1, s, r, x) \left.\frac{d^s G_\nu(t)}{dt^s}\right|_{t=1-\Theta(\xi, m_1)}. \tag{14.6}$$

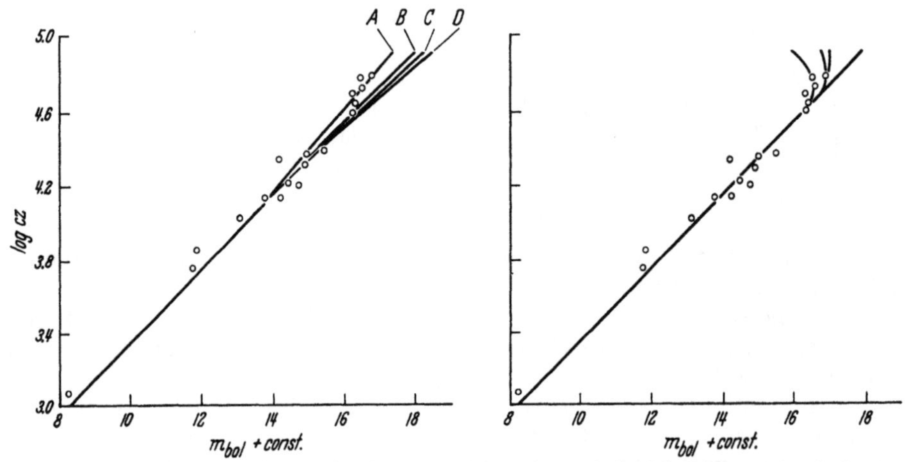

Fig. 7. The apparent velocity of 18 clusters with the apparent bolometric magnitude of the brightest galaxy in the cluster. The left panel, reproduced from HOYLE and SANDAGE shows four theoretical curves corresponding to four cosmological theories, all with no selection. The right panel shows four curves corresponding to one cosmological theory but four intensities of selection: (from left to right) $m_1 = 21.5$, $m_2 = 19.5$, $n = 25$; $m_1 = 21.5$, $m_2 = 20.5$, $n = 25$; $m_1 = m_2 = 21.5$, $n = 25$; no selection. All curves assume average number of galaxies per cluster is 100 and luminosity function is normal with mean -17 and $\sigma_M = 1$ mag.

Then, for $x \leq m_2$, the conditional distribution function of μ_r, among the clusters satisfying conditions (a) and (b), given that the redshift of the cluster is a known function of ξ, is represented by the formula

$$F\{\mu_r \leq x \,|\, (\nu^* \geq n), (\mu_r \leq m_2), \xi\} = \frac{F^*(\xi, m_1, n, r, x)}{F^*(\xi, m_1, n, r, m_2)}. \tag{14.7}$$

This formula can be used in order to compute the regression of μ_r on ξ.

In order to yield numerical results that can be compared with observations the general formula (14.7) must be specialized by substituting into it well-defined functions to represent $G_\nu(t)$ and the luminosity function of galaxies, and also some specific values for m_1 and $m_2 \leq m_1$. As already mentioned, thus far no reliable estimates of $G_\nu(t)$ are available. However, as a matter of interest, the vaguely determined estimates obtained in [6] may be tried. In particular this may be done to study the effects of selection on the regression, say $\mu_1(z)$, of μ_1 on z, recently considered by HOYLE and SANDAGE[1]. Interested in certain cosmological theories, these authors deduced the dependence of $\mu_1(z)$ on Log cz for

[1] F. HOYLE and A. R. SANDAGE: Publ. Astronom. Soc. Pacific **68**, 301–307 (1956).

unselected clusters and compared their deductions with the results of observations [3] that, however, *must have bren affected by selection.* The left-hand panel of Fig. 7 reproduces the graph of HOYLE and SANDAGE. Here each continuous curve represents the dependence of $\mu_1(z)$ on Log cz as determined by a particular cosmological theory. The dots represent the observations and it will be seen that they suggest a deviation from linearity. The right-hand panel of Fig. 7 contains the same observational points as the left. Also it contains four curves, all four computed using formula (14.7) on a single assumption regarding the state of the universe, namely that the universe is static or in the steady state. The difference in shape of the four curves depends solely upon the differences in assumptions regarding the details of selection of clusters for observation. The curves are not extended much beyond the presently available observations because intensive efforts, implying a different selection rule, would probably be required. Equally different curves are produced by a fixed rule of selection combined with different assumptions about the details of the clustering of galaxies.

With this amount of variation in the shape of the regression line, due to the effects of selection combined with different hypotheses about the clustering of galaxies, it must be clear that if one started with the hypothesis that among the totality of clusters not subjected to selection the regression $\mu_1(z)$ is not linear but has one of the shapes computed by HOYLE and SANDAGE, then, by adjusting appropriately the hypotheses about clustering, one could also attain a satisfactory fit to the data. The inescapable conclusion is that, in order to be able to make a choice among the competing cosmological theories on the basis of the magnitude-redshift relation, it is necessary to obtain more information on the spatial distribution of galaxies.

IV. Theory of multiple clustering of galaxies.

15. Basic ideas and some formu lae relating to second order clustering. Cluster ing of particles of an arbitrary order is defined recursively. Suppose that the clustering distribution of order $s \geq 1$ has been defined. In order to define clustering of order $s+1$, we begin by making hypotheses, exactly those enumerated in Sect. 4, concerned with the number ν of particles per cluster and about the internal structure f^* of clusters. However, instead of postulating that the cluster centers are distributed in space either uniformly or quasi-uniformly, we assume that the cluster centers are themselves clustered, following a distribution of order s. A formula, easy to deduce but somewhat complicated to write, connects the clustered distribution of $(s+1)$-th order with the clustered distribution of s-th order.

The general nature of results [7] may be examplified by a brief discussion of clustering of second order. Here we shall have to distinguish three kinds of "particles" distributed in space: (i) the "centers of superclusters", say C assumed to be uniformly distributed in space, (ii) the centers of clusters, say $C^{(1)}$, with a simply clustered distribution in space and with a random number ν_1 of them attached to a single supercluster, and (iii) galaxies, with a random number ν_2 of them attached to each cluster center $C^{(1)}$. Here both ν_1 and ν_2 are treated as random variables, all completely independent among themselves and independent from all other variables of the system, with probability generating functions $G_{\nu_1}(t)$ and $G_{\nu_2}(t)$, respectively.

In perfect analogy with the theory of simple clustering, we assume that, given the positions of centers of superclusters, the positions of centers of clusters are completely independent. Denoting by the single letter U the triplet of present coordinates of the center $C^{(1)}$ of a supercluster and by the single letter V the triplet of present coordinates of the cluster center C, and finally, by the single letter X the triplet of present coordinates of a galaxy, we postulate the existence of two functions $f_1^*(v, u)$ and $f_2^*(x, v)$, one to represent the conditional probability density of V given u and the other the conditional probability density of X given v.

With this notation, the joint probability generating function of the random variables N_1 and N_2 defined in Sect. 8 is given by the formula

$$G_{N_1, N_2}(t_1, t_2) = e^{-\lambda \int (1 - G_{v_1}\{\int f_1^*(v,u)\, G_{v_2}[1 - p_1(1-t_1) - p_2(1-t_2) - p_3(1-t_1 t_2)]\, dv\})\, du} \tag{15.1}$$

where p_1, p_2 and p_3 are functions of v defined in terms of f_2^* in a manner analogous to that explained in Sect. 8.

One of the interesting results of the theory of second order clustering is that [7], should this theory correspond to actual facts and, in ignorance of this, should the formulae of simple clustering be applied to estimate the structure f^* of clusters, there would result a substantial overestimate of the average volume occupied by a cluster. Indications are that the discrepancy exhibited in Fig. 3 between the empirical quasi-correlations on the one hand and the quasi-correlations deduced for the static universe from the theory of simple clustering, would diminish if the theoretical quasi-correlations were obtained from the theory of clustering of the second order.

As suggested in the introduction, the theory of clustering of galaxies is treated as a means of answering, approximately, the question: *How* are galaxies distributed in space? With reference to the possibility of considering clustering of higher order, it is believed that, eventually, a reasonably accurate answer to this question can be obtained. However, it must be clear that the theory outlined, or any theory of this kind, somewhat comparable in spirit to the Ptolemean attempts to use sequences of epicycles in order to represent the apparent motions of planets, cannot be expected to answer the more important question, *Why* are the galaxies distributed as they actually are? The answer to this "*Why*", even if this word is taken in (quite appropriately) quotes, may be forthcoming from studies of a novel kind, combining probabilistic and statistical considerations with those of dynamics. ULAM's attempts reported in Sect. 2 and a paper by McVITTIE[1] offering a relativistic treatment of clustering of galaxies, may be the forerunners of this new direction of research.

General references.

[1] BONDI, H.: Cosmology. Cambridge: University Press 1952.
[2] HUBBLE, E. P.: Astrophys. Journ. **84**, 517—554 (1936). See also a series of papers published by HUBBLE in this journal in the 1930's.
[3] HUMASON, M. L., N. U. MAYALL and A. R. SANDAGE: Astronom. J. **61**, 97—162 (1956).
[4] NEYMAN, J.: Ann. Inst. Henri Poincare **14**, 201—244 (1955).
[5] NEYMAN, J., and E. L. SCOTT: Astrophys. Journ. **116**, 144—163 (1952).
[6] NEYMAN, J., E. L. SCOTT and C. D. SHANE: Astrophys. Journ. **117**, 92—133 (1953).
[7] NEYMAN, J., E. L. SCOTT and C. D. SHANE: Proc. Third Berkeley Symposium on Math. Stat. and Prob., vol. 3, p. 75—111. Berkeley: Univ. of Calif. Press 1956.
[8] SHANE, C. D., and C. A. WIRTANEN: Astronom. J. **59**, 285—304 (1954).
[9] ZWICKY, F.: Proc. Third Berkeley Symposium on Math. Stat. and Prob., vol. 3, p. 113—144. Berkeley: Univ. of Calif. Press 1956.

[1] G. C. McVITTIE: Astronom. J. **60**, 105—115 (1955).

Distance and Time in Cosmology: The Observational Data.

By

G. C. McVittie.

With 9 Figures.

1. Introduction. The galaxies, or extra-galactic nebulae, are regarded in cosmology as the constituent units of the astronomical universe. The vast majority of them are so remote that all the ordinary methods of distance-determination used by astronomers fail and indirect methods of low precision are alone available. In discussing these questions, the operational point of view will be adopted: the distance of a galaxy must be found by an instrumental procedure which astronomers can carry out with the apparatus they in fact possess. The same criterion will apply when measurements of time are under consideration. For example, a distance derived from the apparent luminosity of a galaxy is acceptable because operations with the equipment available in observatories can be described for carrying out the required measurements. But a distance derived from the exchange of light-signals between galaxies, while it can be imagined, is not one describable in terms of operations accessible to terrestrial astronomers.

The problem of large-scale distance is further complicated by the theoretical conclusions of the theories of special and general relativity which will here be briefly summarized. In classical mechanics, the geometry of the universe is assumed to be Euclidean, the distance between two objects in it is absolute, and there is also an absolute time. All operations for measuring the distance from our Galaxy to some other, if properly carried out, should therefore give the same unique result. If a displacement $d\lambda$ is observed in the spectral line of wavelength λ in the spectrum of a galaxy, the classical theory of the Doppler effect shows that the radial velocity of the galaxy is

$$V = c\,\delta, \tag{1.1}$$

where c is the velocity of light and[1] $\delta = d\lambda/\lambda$. Here V is the rate of change of absolute distance with respect to absolute time of the galaxy at the instant when the light by which it is now seen left the galaxy. It also follows that, if $\delta = 1$, then the velocity of the galaxy is equal to c.

The special theory of relativity, while retaining the Euclidean hypothesis, asserts that time and distance measurements *do* depend on the circumstances of measurement, being affected by the relative velocities of the inertial coordinate frames with respect to which the measurements are made. When a spectral displacement δ is present, two different observational procedures, e.g. measurements of apparent luminosity, on the one hand, or of angular diameter, on the other, need no longer give the same result. Formula (1.1) also now becomes,

[1] The parameter δ denotes the same quantity as Neyman and Scott's z in the preceding article.

for a red-shift of the spectral lines[1],

$$1+\delta=\left(\frac{1+V/c}{1-V/c}\right)^{\frac{1}{2}}, \tag{1.2}$$

which shows that $V=0.6c$ if $\delta=1$ and $V=c$ only if $\delta=+\infty$. General relativity incorporates special relativity as a "local" theory and takes account of gravitational effects as well. The uniform model universes of this theory have metrics of the form [*11a*]

$$ds^2=dt^2-\frac{R^2(t)}{c^2}\,\frac{(dr^2+r^2\,d\vartheta^2+r^2\sin^2\vartheta\,d\varphi^2)}{(1+\frac{1}{4}k\,r^2)^2}. \tag{1.3}$$

where r is a dimensionless radial coordinate, $R(t)$, an undetermined function of t having the dimensions of length and k is the space-curvature constant ($k=-1, 0$, or $+1$). The observer O is presumed to be located at $r=0$ and a typical galaxy, P_i, at $(r_i, \vartheta_i, \varphi_i)$, these coordinates *remaining constant* and thus independent of t. If the galaxy emits light at time t_i that reaches O at time t_o, then

$$\delta=\frac{R(t_o)}{R(t_i)}-1=\frac{R_o}{R_i}-1, \tag{1.4}$$

and the following three operationally independent types of distance, amongst others, can be defined for P_i [*11b*]:

Luminosity-distance

$$D=\frac{R_o^2}{R_i}\,\frac{r_i}{1+\frac{1}{4}k\,r_i^2}. \tag{1.5}$$

Distance by apparent size

$$\xi=\frac{R_i\,r_i}{1+\frac{1}{4}k\,r_i^2}. \tag{1.6}$$

Distance by future apparent size

$$\bar{r}=\frac{R_o\,r_i}{1+\frac{1}{4}k\,r_i^2}. \tag{1.7}$$

Luminosity-distance is, by definition, that distance relative to which the intensity of a light-source falls of as D^{-2}; ξ is the distance obtained by measurements of the angular diameter of the source at the instant t_o; and $\bar{r}$, the distance by apparent size which will be found when light that is leaving P_i at the instant t_o eventually reaches O. We also have

$$\xi=D(1+\delta)^{-2}, \qquad \bar{r}=D(1+\delta)^{-1}, \tag{1.8}$$

and, each distance having its own rate of change with respect to t_o, the velocity corresponding to δ is not unique and only reduces to $c\delta$ if the square and higher powers of δ are negligible. The Hubble and acceleration parameters are, respectively,

$$h_1=\frac{R_o'}{R_o}, \qquad h_2=\frac{R_o''}{R_o}, \tag{1.9}$$

where a prime denotes a derivative of R with respect to t, and D can then be expanded as a power-series in δ of which the first two terms are [*11c*]

$$D=\frac{c\,\delta}{h_1}\left(1+\frac{h_1^2+h_2}{2h_1^2}\,\delta\right). \tag{1.10}$$

[1] H. Dingle: The Special Theory of Relativity, 3rd edn., p. 50. London: Methuen & Co., New York: J. Wiley & Sons 1950.

I. Observational methods of determining the distances of galaxies.

2. Local distance. The operational methods that astronomers employ to find the distances of galaxies will now be described. The process is a step by step one beginning with objects in our own Galaxy, working out to the galaxies in its immediate neighbourhood and thence to the remoter regions of space. Since the ultimate aim will be to interpret the entire system of galaxies through one of the uniform model universes (1.3), it is necessary to make a preliminary assumption. It will be supposed that, whatever the space-time may prove to be, it is of such a character that the region occupied by the Galaxy and the group of galaxies that are its immediate neighbours, is locally Euclidean. By this term we shall mean, firstly that the curvature of space is presumed to be so small that the region in question is "locally flat"; and, secondly, that the relative velocities of the objects in it are all small compared with the velocity of light, for example, do not exceed about 1000 km/sec. The first condition is not really a restriction on the nature of the space-time for it is a known result of Riemannian geometry that all space-times are locally flat: all that the condition does is to identify the objects that are found in the locally flat region round our Galaxy. The second condition is, of course, a restriction on the relative velocities for there is no reason why these should be small in the locally flat region of a Riemannian space-time.

These properties of the locally Euclidean region mean that, within it, the different types of distance reduce to one another and, in particular, to luminosity-distance. Until Sect. 7 is reached, it will be assumed that all distances are identical and equal to "local distance", a Euclidean distance that will simply be referred to as "distance" unless it is required to emphasize that it corresponds, in fact, to some particular type of distance. We begin with the operations that are used to find distances in the solar system because they are fundamental to the whole question. The first operation in this connection is to find the dimensions of the Earth and the distances between observatories on its surface. Surveyors begin by laying down a base-line which consists of measuring the spatial extension between two points on the Earth's surface with the aid of invar tapes stretched to a given tension. This step in itself would seem to dispose of the notion that the fundamental way of measuring distance is by means of a "rigid rod". The tapes are certainly not rigid in any sense of the word and, indeed, a little reflection will show that the concept of the rigid rod is but another way of introducing the idea that distance is an absolute quantity, i.e. of abandoning relativity and returning to classical mechanics. The surveyors are engaged in establishing a network of distance-measurements which will be *internally consistent*: whether the base-line is set up in Canada or in Peru, the consequential distance between New York and Boston must come out to have the same value, within pre-assigned errors. It is true that base-lines can also nowadays be established by sending light-signals between the ends of the base-line. But this is a derivative method which would never have come into use had it not been found to be consistent with the system based on stretched invar tapes. Working from the base line, and assuming that the Earth's surface is a mathematical surface in a 3-dimensional Euclidean space, distances between astronomical observatories are computed and these are taken as new base-lines for the extra-terrestrial measurements made by astronomers. It is well-known that the principle employed for the determination of distances in the solar system is to use Newtonian gravitational theory, that leads, through KEPLER'S Third Law, to a scale model of the solar system. One distance measured from the Earth to one planet, using Euclidean trigonometry, will then fix the scale-factor. The mean diameter of the Earth's orbit can then be found and it, in turn, becomes a new base-line for the distances of the nearer stars of our Galaxy. The fundamental distance-unit is the *parsec* one parsec corresponding to the distance in Euclidean geometry at which an object would subtend an angle of one second of arc when viewed from the two extremities of the mean radius of the Earth's orbit. In terms of the centimeters defined by the invar tape distances, one parsec (1 pc) is 3.0871×10^{18} cm and multiples of the parsec are the *kiloparsec* (1 kpc = 1000 pc) and the *megaparsec* (1 Mpc = 10^6 pc). A star, or other celestial object, distant one parsec from the Sun is said to have a *parallax*, p, of amount 1″; conversely, if the parallax is p'', the distance of the object is $1/p$ parsecs. Unfortunately, the mean radius of the Earth's orbit is so small, relatively speaking, that a direct measurement of parallax soon falls within the errors of measurement as we proceed outwards from the Sun into the stellar regions of the Galaxy. Indeed as soon as the parallax falls below 0″.01, the errors begin to vitiate the results very seriously. However, it is the case that the so-called "fixed" stars do, in fact, have measureable motions, at least in many cases. The displacement of the spectral lines in a star's spectrum gives the velocity-component in the line of sight (the radial velocity) through the theory of the Doppler effect; careful measurements of the position of a star, relatively to fainter and therefore presumably more distant stars, may reveal changes of position over the years. In this way the velocity-component perpendicular to the line of sight (the proper-motion) is found. The analysis of the proper-motions of a group of stars, assuming that their velocities are distributed at random relative to the Sun, will provide an average or *statistical parallax*, π, for the entire group. The radial velocities and the theory of galactic rotation will likewise establish a statistical parallax[1]. We shall see that these statistical parallaxes play an important role in cosmological distance determinations.

Another way of finding the distance (essentially, the luminosity-distance) of a luminous object depends on the manner in which the brightness of the object falls off with distance. It is always taken that this brightness decreases proportionately to the inverse square of the distance. The connection with measured quantities is made through the apparent magnitude, m, and the absolute magnitude, M, of the object. A detailed theory will be found in [*11d*]; for the moment it suffices to say that the magnitudes we shall consider throughout this article are photographic[2], that the absolute magnitude is the apparent magnitude the object would have if it were located at 10 pc from the Sun, and that the distance D is connected with m and M by the formula

$$\operatorname{Log} D = 0.2\,(m - M) + 1\,, \tag{2.1}$$

if space is assumed to be perfectly transparent to light. In terms of the parallax, (2.1) becomes

$$\operatorname{Log} p = 0.2\,(M - m) - 1\,, \tag{2.2}$$

and in either case the quantity $(m - M)$ is called the *distance-modulus* of the object which will be denoted by $\mathcal{M}$. Unfortunately the Sun lies fairly close to the central plane of the Galaxy, within the band of interstellar gas and dust which attenuates the light that must pass through it on its way to the observer.

[1] P. Couderc: The Expansion of the Universe, Chap. II. New York: Macmillan 1952. — R. J. Trumpler and H.F. Weaver: Statistical Astronomy, Chap. 3.5 and 3.7. Berkeley: University of California Press 1953.

[2] H.F. Weaver: Pop. Astron. **54**, 211, 287, 339, 389, 451, 504 (1946). — Astrophys. Journ. **106**, 306 (1947).

To allow for this, we shall introduce the *observed apparent magnitude*, m^*, which is related to m by $m^* = m + A$, where A is the total absorption[1] that the light has suffered during the course of its journey. For objects within the layer of absorbing material, $A = aD$, where a is the diminution in apparent magnitude per kiloparsec of D[2]. For objects lying altogether outside the obscuring band, such as other galaxies, the absorption depends not only on the material inside our Galaxy, but also on that in the galaxy in question and, if this exists, on the material in the space between the galaxies. The formula for the first of these effects, which is the only one known with any accuracy, given by OORT[3] and based on observational material due to HUBBLE, is

$$A = 0^{\mathrm{m}}.31 \operatorname{cosec} b, \tag{2.3}$$

where b is the galactic latitude of the object. A modified form of this *cosecant law* occurs in an investigation by MINEUR [*12*] who has

$$\left.\begin{aligned} A &= (0.22 \pm 0.02) \operatorname{cosec} b, \quad (b = \text{N.Gal.Lat.}) \\ &= (0.28 \pm 0.02) \operatorname{cosec} b. \quad (b = \text{S.Gal.Lat.}) \end{aligned}\right\} \tag{2.4}$$

It happens that the value adopted by SANDAGE [*9a*] is the average of these, viz.

$$A = 0.25 \operatorname{cosec} b. \tag{2.5}$$

In addition to the observed apparent magnitude, m^*, it is possible to define a fictitious absolute magnitude for an object by

$$M^* = M + A, \tag{2.6}$$

and an *apparent distance-modulus* by

$$\mathscr{M}^* = m^* - M = m - M + A = \mathscr{M} + A, \tag{2.7}$$

and therefore we also have

$$\mathscr{M} = m^* - M^*, \qquad M = m^* - \mathscr{M}^*,$$

the second formula showing that the absolute magnitude could be determined from a knowledge of m^* and $\mathscr{M}^*$ even if the absorption were unknown. Finally (2.1) may also be written

$$\operatorname{Log} D = 0.2\mathscr{M} + 1 = 0.2(\mathscr{M}^* - A) + 1. \tag{2.8}$$

Operationally, therefore the steps that are needed to determine D are, firstly, the measurement of m^*; secondly, the determination of A; and, thirdly, the estimation of M. Our concern will chiefly lie with the methods for finding M. For this purpose, variable stars will play an important role and the following terminology will be used: The *mean magnitude* of a variable star will be defined as the arithmetic average of its magnitudes at maximum and minimum light. Thus the mean observed apparent magnitude is

$$m^* = \tfrac{1}{2}(m^*_{\max} + m^*_{\min}), \tag{2.9}$$

---

[1] J. DUFAY: Nébuleuses Galactiques et Matière Interstellaire, Chap. VIII and IX Paris: Albin Michel 1954.

[2] As photometry improves, it should be possible to determine the absorption for individual objects by the formula $A = \chi E$, where E is the colour excess and χ, the ratio of total to selective absorption.

[3] J. H. OORT: Bull. Astr. Inst. Netherl. **8**, 233 (1938).

and the mean apparent magnitude, after allowance for absorption, is

$$m = \tfrac{1}{2}(m_{\max} + m_{\min}). \tag{2.10}$$

The mean absolute magnitude of the variable star is the absolute magnitude corresponding to the mean apparent magnitude given by (2.10). When we are dealing with a large number of variables forming a recognised group, for instance, those lying in a particular globular cluster, it will be convenient to speak of the *average magnitude* (observed apparent, apparent, absolute) when referring to the arithmetic average of the individual magnitudes of the stars. The *median magnitude* of the group is that magnitude which divides the individual magnitudes symmetrically, as many being less than the median as are greater than it. In contrast to the foregoing, it is also possible to define what we shall call the *mean-light magnitude* for an individual variable star. This kind of magnitude will be denoted by the subscript l and it is determined from the light-curve of the star by a process of averaging the energy emitted during each moment of a light-cycle. The mean-light observed apparent magnitude m_l^* is not necessarily identical with the m^* defined by (2.9).

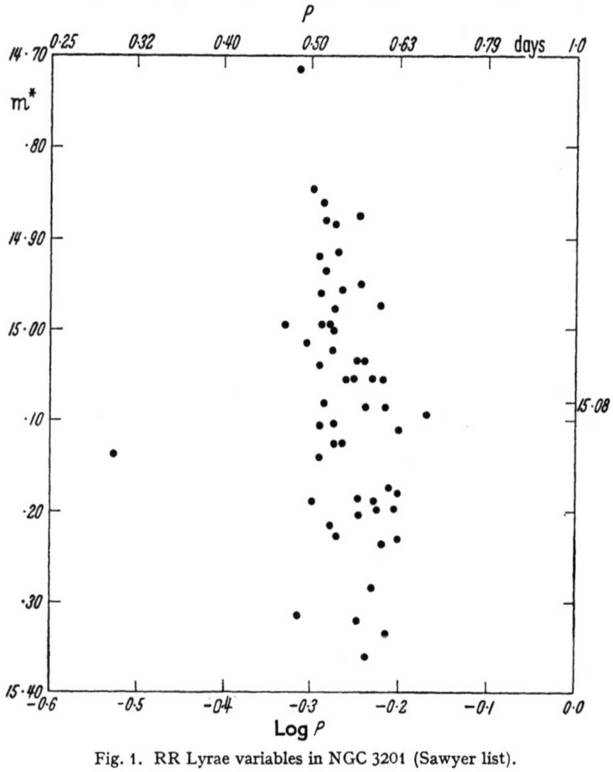

Fig. 1. RR Lyrae variables in NGC 3201 (Sawyer list).

This terminology is not universal: Shapley[1] calls our m^* the "median" presumably because it lies half-way between the maximum and minimum. This practice is followed by Arp [2] and by Roberts and Sandage, who also call our mean-light magnitude, the "mean" magnitude[2].

3. Variable stars of Population II. The importance of variable stars as distance indicators lies in the fact that they are easily recognizable and also because of the empirically established conclusion that the period of light-fluctuation of the star is an index of its absolute magnitude. Thus if a variable, or better still a group of variables, lying in some galaxy has had its period determined, it is possible to calculate the absolute magnitude. The mean apparent magnitude also being known, the distance of the star, and therefore of the galaxy in which it lies, can be computed from (2.8). The variables of Population II are the RR Lyrae stars, with periods of less than one day, and the Type II

[1] H. Shapley and Virginia McK. Nail: Proc. Amer. Philos. Soc. **92**, 310 (1948). Tab. 1. Harv. Reprints Ser. II, No. 25.

[2] M. Roberts and A. R. Sandage: Astronom. J. **60**, 185 (1955). See also P. Pismis: Publ. Astronom. Soc. Pacific **67**, 253 (1955).

Cepheids, of which the W Virginis stars, with periods of between 10 and 40 days, and the RV Tauri stars, whose periods range from 40 to 150 days, are the principal sub-types. These variables are recognisable, not only by their periods, but also by the forms of their light-curves and, for the W Virginis stars, by their spectral peculiarities. The RR Lyrae stars are numerous in the globular clusters of our Galaxy, constituting 93% of the variables there found. These clusters also contain W Virginis stars to the extent of 4% of the variables [*1a*]. Since all the stars in a globular cluster are at approximately the same distance from the Sun, differences in apparent magnitude are presumably due, in the main, to intrinsic differences in absolute magnitude. A list of known variables in the globular clusters has been compiled by HELEN SAWYER [*15a*], the maxima and minima being given amongst other data. In Figs. 1 to 5 are shown plots of mean observed apparent magnitudes against the logarithm of the period P in days, of RR Lyrae stars for five clusters each of which contains many such variables. The material for these clusters is described as "fairly" to "very" homogeneous except in the case of NGC 5272 [*15b*]. All five diagrams show the characteristic paucity of periods in the neighbourhood of 0.4 to 0.45 days. Figs. 1, 3 and 4 suggest that there is no change, on the average, of m^* with period and that RR Lyrae stars therefore have approximately the same mean absolute magnitude. But Fig. 2, for ω Centauri, and to a lesser extent, Fig. 5 for M 15 suggest that there is a *decrease* of mean absolute magnitude with period, i.e. that the stars are intrinsically brighter the greater their period. PARENAGO [*14*] from his study of the RR Lyrae stars that are not members of globular clusters and that are found in the Galaxy, also concludes in favour of this decrease of mean absolute magnitude and gives the formula $M = +0.44 - 0.20 \operatorname{Log} P$ as the connection between absolute magnitude and $\log P$. On the other hand, Fig. 6 shows ROBERTS and SANDAGE'S[1]

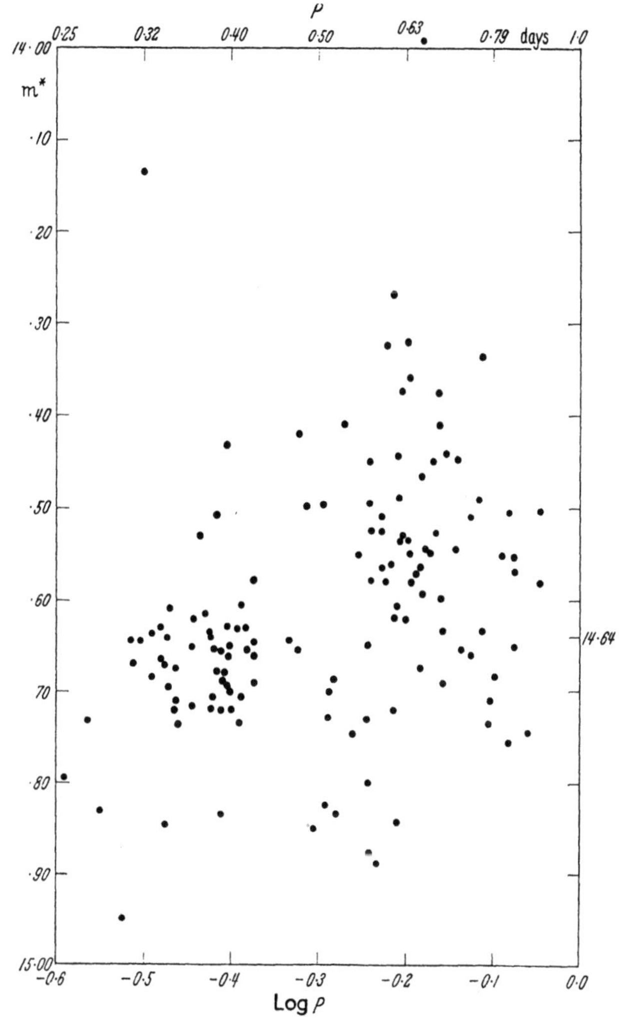

Fig. 2. RR Lyrae variables in NGC 5139 (ω Centauri) (Sawyer list).

[1] See footnote 2, p. 450.

results for 49 stars in M 3 in which observed apparent mean-light magnitudes, m_l^*, are plotted against Log P. There is a strong suggestion of an *increase* of absolute mean-light magnitude with period, estimated at 0.23 magnitude for the range of period 0.28 to 1.00 days. A similar result has been detected by Emilia Belserene[1] with respect to selected groups of RR Lyrae stars in ω Centauri and M 3. It may therefore be said that there is a strong presumption in favour of the conclusion that the RR Lyrae stars are, on the average, of the same mean absolute magnitude but that further detailed studies of individual globular clusters, or of other groups of RR Lyrae stars, may show that there is a small variation with period.

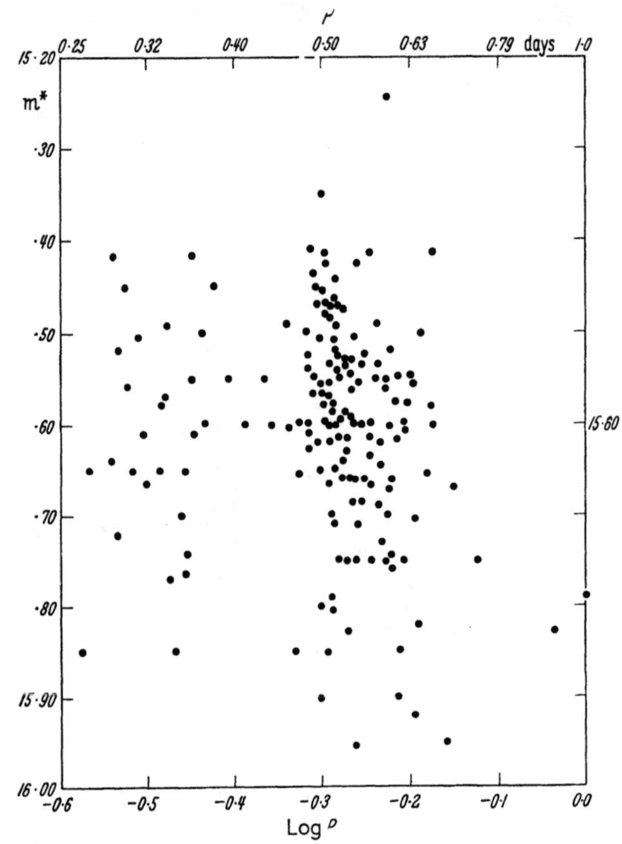

Fig. 3. RR Lyrae variables in NGC 5272 (M 3) (Sawyer list).

The next step is to decide for each cluster, what observed apparent magnitude is to correspond to this presumed unique mean absolute magnitude. The spread in magnitudes exhibited by Figs. 1 to 5 does not encourage the belief that this absolute magnitude can be fixed with an error of less than 0.1 or 0.2 of a magnitude. If there are a sufficient number of RR Lyrae stars identified in the cluster, it may be argued that the median of the mean observed apparent magnitudes should be chosen. Or the method of Schwarzschild[2], Arp, Baum and Sandage[3], and Sandage[4] may be employed. This consists in establishing the colour-magnitude array for the non-variable stars in the cluster, observing that the RR Lyrae stars fall in a compact group in the diagram, and choosing the observed apparent magnitude of the group as the appropriate one. Different techniques having been used, it is not surprising that the two methods give somewhat different results (Table 1, Cols. 4 and 5). The adopted mean observed apparent magnitude for the RR Lyrae stars of a cluster—by whichever method obtained—will be denoted by m_{RR}^* and the mean absolute magnitude by M_{RR}. The final and crucial step is the determination of M_{RR} without the use of luminosity criteria alone. The space motions of those RR Lyrae stars that are not members of globular clusters in our Galaxy can serve this purpose. In 1939, R. E. Wilson [*17*] used

[1] Emilia Belserene: Astronom. J. **59**, 406 (1954).
[2] M. Schwarzschild: Harv. Circ., No. 437 (1940).
[3] H.C. Arp, W.A. Baum and A.R. Sandage: Astronom. J. **58**, 4 (1953).
[4] A.R. Sandage: Astronom. J. **58**, 61 (1953).

the proper motions of 55 of them to derive the statistical parallax of the group, which he found to be 0''.00085. The stars were reckoned to be far enough, on the average, from the galactic plane not to suffer appreciable absorption, and their mean apparent magnitude was $m^*_{RR} = m_{RR} = 10.50$. Thus by (2.2),

$$M_{RR} = 10.50 + 5 + 5 \operatorname{Log}(0.00085) = +0.15.$$

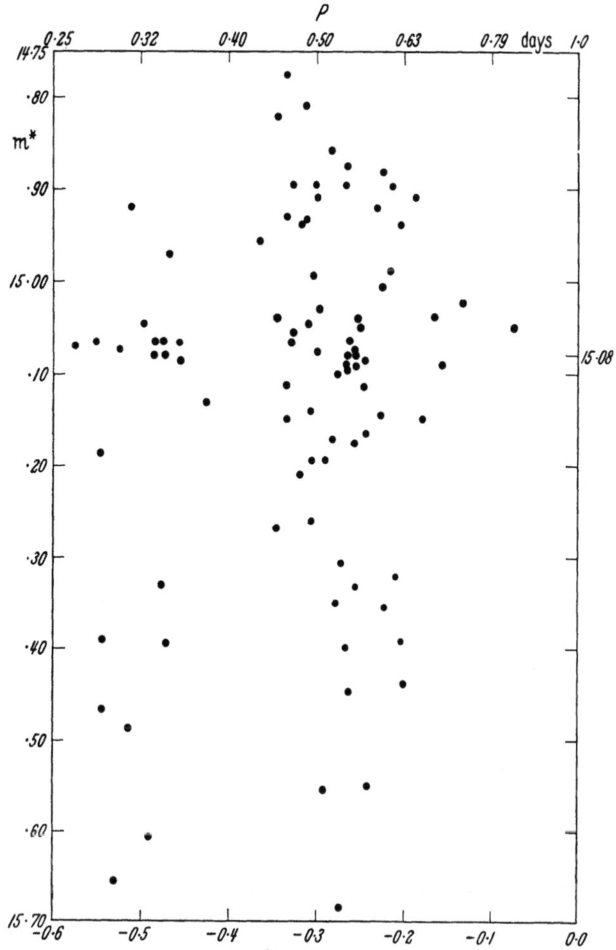

Fig. 4. RR Lyrae variables in NGC 5904 (M 5) (Sawyer list).

In 1944, MINEUR [12] took up the problem again basing his work on SCHNELLER'S catalogue[1], supplemented by the lists of JOY[2] and R. E. WILSON [17]. He started by assuming a provisional absolute magnitude of $+0.3$ for RR Lyrae stars of period 0.78 days, and the method of statistical parallaxes then gave the formula

$$M_{RR} = 0.32 - 5 \operatorname{Log} \lambda,$$

where

$$\lambda = 1.17 \pm 0.17$$

(from proper motions, solar-motion terms),

$$\lambda = 1.51 \pm 0.17$$

(from proper motions, residual velocities).

Thus M_{RR} is equal to either -0.02 or -0.58, whose mean is -0.30, which agrees with the statement made by MINEUR in 1952 [13] that, on revising his 1944 work, he had concluded that the "zero" of the RR Lyrae stars was -0.32. It is perhaps preferable to take the solar motion value of λ, which gives $M_{RR} = -0.02$, as the better of the two. Thus both WILSON'S and MINEUR'S work suggests that the mean absolute magnitude of the RR Lyrae stars is

$$M_{RR} = 0.0. \tag{3.1}$$

Still more recently, however, PARENAGO[3], using the velocity dispersion of the RR Lyrae stars at right angles to the galactic plane, finds

$$M_{RR} = +0.53 \pm 0.38,$$

[1] H. SCHNELLER: Kl. Ver. Univ. Sternwarte Berlin-Babelsberg, No. 20, 1938.
[2] A. JOY: Astrophys. Journ. **86**, 363 (1937); **89**, 356 (1939).
[3] P.P. PARENAGO: Astronom. J. USSR. **31**, 425 (1954).

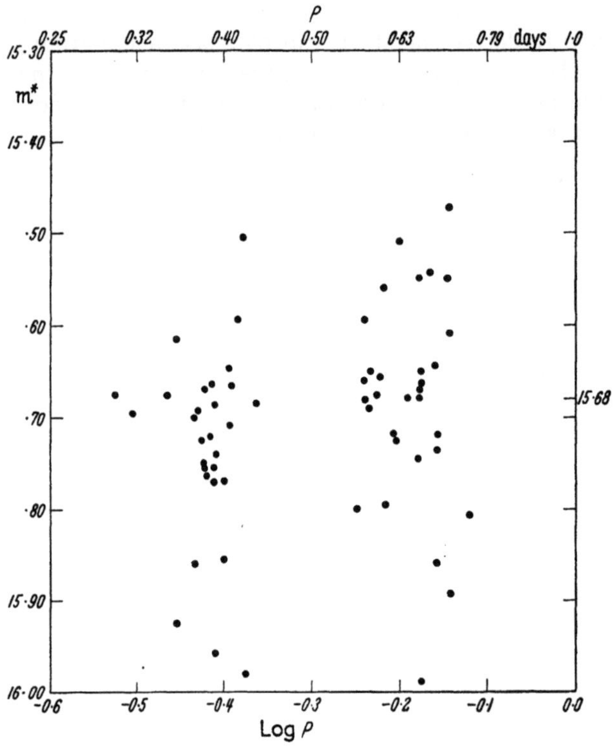

Fig. 5. RR Lyrae variables in NGC 7078 (M 15) (Sawyer list).

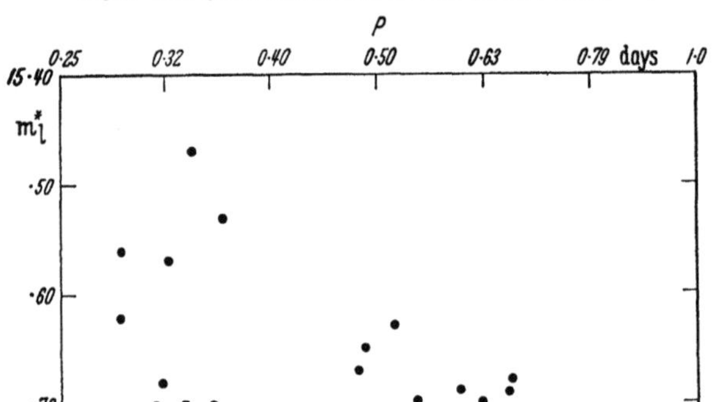

Fig. 6. Selected RR Lyrae variables in NGC 5272 (M 3) (Roberts and Sandage).

and Pavlovskaya[1], by considering the proper motions and radial velocities of the nearer galactic RR Lyrae stars of period less than 0.4 days obtains

$$M_{\mathrm{RR}} = +0.5 \pm 0.18 .$$

These two results would therefore indicate that the RR Lyrae stars are intrinsically fainter than Eq. (3.1) would make them.

In contrast to the foregoing methods, which are all based on statistical parallaxes, the colour-magnitude array method already referred to has also been used to find M_{RR}. The array (in terms of absolute magnitudes) for the main sequence stars in the globular cluster M 3 was fitted to the corresponding sequence for the stars in the solar neighbourhood. This showed that the RR Lyrae stars in the cluster fell in a position indicating that their (photographic) absolute magnitude[2] was $M_{\mathrm{RR}} = 0.2 \pm 0.2$. However when this method is applied to a few more globular clusters[3] it is found not to be free of difficulties and ambiguities; moreover, it does presuppose that the colour-magnitude array for the predominantly Population I stars near the Sun is identical with that of a Population II object such as a globular cluster. Whilst it serves to confirm in a general way the statistical parallax work of Mineur and Wilson,

[1] E. D. Pavlovskaya: Var. Stars **9**, 349 (1953).

[2] A. R. Sandage: Mimeographed Symposium on Astrophysics. Ann Arbor: University of Michigan Observatory 1953. — G. de Vaucouleurs: Publ. Astronom. Soc. Pacific **67**, 350 (1955).

[3] W.A. Baum: Astronom. J. **57**, 222 (1952); **59**, 422 (1954). — H.L. Johnson: Proc. 3rd Berkeley Symp. on Math. Stat. and Probability, Vol. III, p. 31. Berkeley: University of California Press 1956. — H.C. Arp: Astronom. J. **60**, 317 (1955), § 6.

it would appear that the present-day caution of astronomers, who tend to speak of "the assumption" that the RR Lyrae stars have absolute magnitude zero is well-founded; and the Russian astronomers appear to be convinced that $M_{RR} = +0.5$ [*14*].

The apparent distance-modulus of the RR Lyrae stars in a globular cluster is

$$\mathcal{M}^* = m^*_{RR} - M_{RR},$$

and therefore if m^* be the observed apparent magnitude of any other object in the cluster, or of the entire cluster itself, and if M be its absolute magnitude, then

$$\mathcal{M}^* = m^* - M = m^*_{RR} - M_{RR}, \tag{3.2}$$

and therefore

$$M = m^* - \mathcal{M}^* = m^* - m^*_{RR} + M_{RR}. \tag{3.3}$$

which determines M even if the precise value of the absorption for the cluster is unknown. This method will now be applied to the Type II Cepheids in clusters

Table 1. *Type II Cepheids in globular clusters.*

Cluster	Star	Log P	m^*_{RR}	m^*_{RR}	m^*	m^*	M	M
(1)	(2)	(3)	(4)	(5)	(6)	(7)	(8)	(9)
NGC 5272 M 3	154	1.184	15.72	15.60	12.68	13.45	−3.04	−2.15
NGC 5904	42	1.410	15.24	15.08	11.61	11.72	−3.63	−3.36
M 5	84	1.724			11.88	12.07	−3.35	−3.01
NGC 6205	1	0.164	14.83		13.94	13.94	−0.89	
M 13	6	0.325			14.31	14.31	−0.51	
	2	0.708			13.28	13.28	−1.54	
NGC 6254	3	0.896	15.22		13.46	13.46	−1.76	
M 10	2	1.574			12.62	12.68	−2.59	
NGC 7078 M 15	1	0.158	15.90	15.68	15.07	14.95	−0.83	−0.73
NGC 7089	1	1.192	16.19		14.04	14.04	−2.15	
M 2	5	1.545			13.88	13.88	−2.30	
	6	1.286			13.69	13.69	−2.50	
	11	1.827			12.68	12.68	−3.50	
NGC 5139	43	0.063	14.65	14.64	13.98	13.98	−0.67	−0.66
ω Cent.	92	0.129			14.34	14.34	−0.31	−0.30
	60	0.130			13.90	13.90	−0.75	−0.74
	61	0.356			14.10	14.10	−0.55	−0.54
	48	0.651			13.52	13.52	−1.13	−1.12
	29	1.198			12.97	12.97	−1.68	−1.67

and to the total absolute magnitudes of clusters as a whole. A study of 19 Type II Cepheids that have been found in 7 clusters is due to ARP [2] and the data are summarised in Table 1. Cols. 1, 2 and 3 of this Table contain, respectively, the name of the cluster, the number assigned to the star and the logarithm of its period. The periods range from 1.16 to 67 days. In Col. 4 is given ARP's observed apparent magnitude for RR Lyrae stars, determined by the colour-magnitude array method; Col. 5 contains the median of the mean observed apparent magnitudes of all RR Lyrae stars in the cluster, when they were sufficient in number in HELEN SAWYER's list to warrant such a determination (Figs. 1 to 5). In Cols. 6 and 7 are the mean observed apparent magnitudes of the stars of Col. 2, taken from ARP's paper and from HELEN SAWYER's list, respectively. Finally the corresponding absolute magnitudes for the stars of Col. 2, calculated from (3.3) with $M_{RR} = 0.0$ are given: in Col. 8 are ARP's values and in Col. 9 are those deduced from Cols. 7 and 5. If the absolute magnitudes of Col. 8 are plotted against Log P, Fig. 7 is obtained, which forms an example of a *period-luminosity diagram*. Unlike the RR Lyrae stars, the Type II Cepheids reveal a marked increase of luminosity with period. A curve of Log P as abscissa plotted against M as ordinate, is called a *period-luminosity curve* and it is possible that Fig. 7 contains a number of more or less parallel curves, as argued by Arp. Whether this is so or not, it is clear that a Type II Cepheid is intrinsically a much brighter object than is an RR Lyrae star and so can be detected at greater distances. In the next section we shall have occasion to discuss linear period-luminosity curves of the form

$$M = C - 1.74 \operatorname{Log} P;$$

taking each of Arp's stars and computing the value of the constant C for each one of them, a mean C can be deduced for the whole group. The result is the formula

$$M = -0.26 - 1.74 \operatorname{Log} P, \tag{3.4}$$

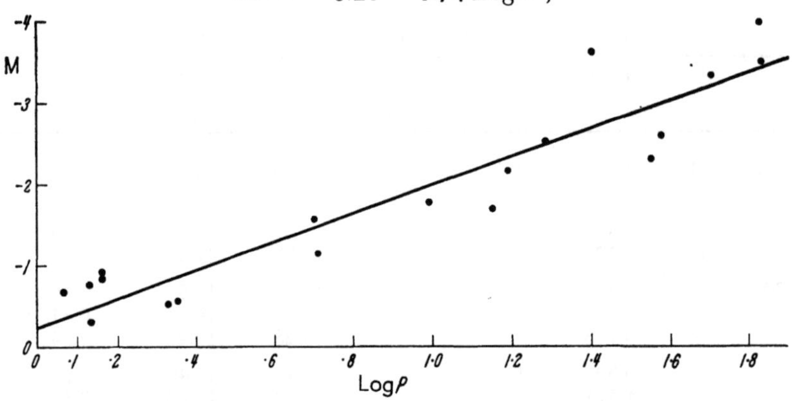

Fig. 7. Period-luminosity curve for Type II Cepheids.

which is the equation of the straight line shown in Fig. 7. The 23 Type II Cepheids in globular clusters discussed by Parenago [*14*], of which 15 also occur in Table 1, are evenly distributed, in a diagram corresponding to Fig. 7, about the line $M = +0.2 - 1.74 \operatorname{Log} P$. The difference of the constant term relative to (3.4) is accounted for by the fact that Parenago takes $M_{\mathrm{RR}} = +0.5$ in calculating the absolute magnitudes.

There are 94 globular clusters in our Galaxy that have been studied in some detail, the main part of the work being due to Shapley and his co-workers. The integrated magnitudes for entire clusters have been measured, and, for those clusters that contain a sufficient number of RR Lyrae stars, the absolute magnitudes are immediately determinable by (3.3) for any chosen value of M_{RR}. Details for the five clusters to which Figs. 1 to 5 refer are shown in Table 2 for $M_{\mathrm{RR}} = 0$ and it will be noticed that there are differences amounting to as much as 30% in the distances that can be assigned to a particular cluster. It is also clear from Col. 6 that globular clusters are intrinsically very bright objects. Lohmann [*10*] deduces from the work of Shapley[1] that the average absolute magnitudes for clusters range from -7.9 to -5.2, according to the "class", or degree of concentration, of the cluster [*16a*], with a median at -7.55; and from that of Mowbray[2], a range of -7.9 to -5.4 with median at -7.50; and Kron[3] makes the average for 16 least-reddened clusters -7.6.

[1] H. Shapley: Handbuch der Astrophysik, Bd. 5, Kap. 5. Berlin: Springer 1933.
[2] A. G. Mowbray: Astrophys. Journ. **104**, 47 (1946).
[3] G. E. Kron: Publ. Astronom. Soc. Pacific **68**, 230 (1956).

Table 2. *Distances of five globular clusters.*

Cluster	m^*	A	$\mathcal{M}^*$			M	D (kpc)		
	(1)	(2)	(3)	(4)	(5)	(6)	(7)	(8)	(9)
NGC 3201	8.8	2.2	14.52	15.08	—	−5.72	2.9	3.8	—
5139	5.34	1.3	14.62	14.64	14.65	−9.28	4.6	4.7	4.7
5272	7.23	0.5	15.44	15.60	15.72	−8.21	9.7	10.5	11.1
5904	6.94	0.5	15.26	15.08	15.24	−8.32	9.0	8.2	8.9
7078	7.35	0.2	15.63	15.68	15.90	−8.28	12.2	12.5	14.0

Col. 1 = observed app. mag. of cluster [*10*].
Col. 2 = absorption, in magnitudes [*10*].
Col. 3 = apparent distance-modulus [*10*].
Col. 4 = apparent distance-modulus (median of RR Lyrae [*15a*]).
Col. 5 = apparent distance-modulus [*2*].
Col. 6 = absolute magnitude of cluster, Col. 1 minus Col. 3.
Col. 7 = distance from $\text{Log}\, D = 0.2\,(\mathcal{M}^* - A) + 1$ and Cols. 1, 3.
Col. 8 = distance from $\text{Log}\, D = 0.2\,(\mathcal{M}^* - A) + 1$ and Cols. 1, 4.
Col. 9 = distance from $\text{Log}\, D = 0.2\,(\mathcal{M}^* - A) + 1$ and Cols. 1, 5.

The brightest non-variable stars in a cluster may also be used as a criterion of its absolute magnitude and therefore of its distance. SHAPLEY [*16b*] has found that if

m_{25}^* = average observed app. mag. of the 25 brightest stars in the cluster,
m_6^* = observed app. mag. of the 6th brightest star,
m_{30}^* = observed app. mag. of the 30th brightest star,

then $m_{RR}^* - m_{25}^*$, $m_{RR}^* - m_6^*$ and $m_{RR}^* - m_{30}^*$ are constants depending on the class of the cluster. Thus if a cluster contains few or no RR Lyrae stars, its m_{RR}^* can be estimated from the observations of m_{25}^*, m_6^* and m_{30}^*. By these methods the distances of the globular clusters in our Galaxy have been found to range from 1 or 2 kpc up to 70 kpc [*10*]. The actual value of m_{25}^* in a rich cluster is −1.3 according to BAADE[1] and more recent work on colour-magnitude diagrams of globular clusters indicates that it should be more nearly −1.5[2,3].

Summarizing the conclusions of this section, it may be said that: (a) the RR Lyrae stars are probably all of the same absolute magnitude, but more intensive study may well reveal a dependance of absolute magnitude on period of the order of 0.2 mag. over the period-range; (b) the statistical parallax and the colour-magnitude array methods indicate that, if there is a unique absolute magnitude, it lies close to zero. Assuming that this is so, then (c) the Type II Cepheids exhibit a relation between period and absolute magnitude represented by (3.4) and their absolute magnitudes for the longer periods reach to −3.5; (d) the absolute magnitudes of the globular clusters of our Galaxy have a median near to −7.5.

4. Variable stars of Population I. The variables that are now of paramount importance are the classical Cepheids, so called after the type star δ Cephei. Their periods range from just over one day to about 100 days; apart from their periods, the classical Cepheids can be recognised by their light-curves, the nature of the rise and fall of their surface velocities during an oscillation, their spectra, and so on. We shall be much concerned with the period-luminosity curves exhibited by definite groups of classical Cepheids, for example, those

[1] W. BAADE: Astrophys. Journ. **100**, 137 (1944).
[2] See footnote 3, p. 452.
[3] See footnote 4, p. 452.

found in one or other of the two Magellanic Clouds, and a matter of terminology must first be settled. Let it be supposed that it has proved possible to find the mean absolute magnitudes of a group of classical Cepheids and that the period, P, of each star in the group has also been determined. It is in fact the case that the periods can be found with a greater degree of certainty than can the absolute magnitudes. Suppose then that M is plotted as ordinate against log P as abscissa for the stars in the group, and a period-luminosity curve drawn through the resulting points. Let P_z be the value of P corresponding to $M=0$; then this point is called the *zero-point* of the period-luminosity curve. Now suppose that, for some reason, each absolute magnitude is revised by the same amount, the periods remaining unaltered. As a result the absolute magnitude corresponding to P_z now becomes ΔM, say; then ΔM is called the *correction to the zero-point* of the original period-luminosity curve. Clearly this correction is entirely meaningless unless the original curve is specified without possibility of doubt. It happens in practice that period-luminosity curves are often straight lines, as we have seen in the case of the Type II Cepheids; when this happens the original curve is

$$M = C - s \operatorname{Log} P,$$

where C and s are constants. The revised curve is

$$M = C' - s \operatorname{Log} P,$$

and therefore

$$0 = C - s \operatorname{Log} P_z,$$
$$\Delta M = C' - s \operatorname{Log} P_z,$$

and so the correction to the zero-point of the original curve is

$$\Delta M = C' - C, \tag{4.1}$$

i.e. it is simply equal to the difference of the constant terms in the original and the new linear period-luminosity curves.

Classical Cepheids are found in abundance in the Large Magellanic Cloud (LMC) and in the Small Magellanic Cloud (SMC), two galaxies that lie fairly close to our own. By studying the classical Cepheids in these two galaxies, the Harvard astronomers, notably Shapley and his co-workers, have established and refined the period-luminosity curve exhibited by variables of this kind. As long ago as 1930, Shapley [*16c*] had given a curve based on the then-known material from the Magellanic Clouds and this curve is listed in numerical form in Col. 1, Table 3. The method used for establishing this original curve need not be described, as better material was available for Shapley's later determination of 1940. Since all the stars in the LMC and the SMC are at approximately the same distance from the Sun, a correlation of mean observed apparent magnitude with period may be interpreted as one between mean absolute magnitude with period. Fig. 8 exhibits the period-luminosity diagram for 40 classical Cepheids (shown by points) in the LMC[1] and Fig. 9 for 49 in the SMC. In each Cloud they form a representative selection, with best determined periods and magnitudes, out of a far larger number. Periods have in fact been determined for 654 stars in the SMC[2] and for 339 in the LMC[3]. The diagrams illustrate the follow-

[1] See footnote 1, p. 450.

[2] Harv. Circ., No. 439 (1940); No. 444 (1942). — Proc. Nat. Acad. Sci. U.S.A. **26**, 541 (1940). — Harv. Bull. No. 920, 13 (1949), and No. 919, 4, 6 (1949).

[3] Harv. Circ. No. 439 (1940). — Harv. Bull. No. 919, 4, 6 (1949); No. 921, 1 (1952). — Astronom. J. **55**, 249 (1951).

ing points: (a) there is a scatter of mean observed apparent magnitude at each period, which is believed to reflect a real variation of absolute magnitude for a given period and which amounts to almost one magnitude; (b) there is a great lack of periods of less than 2.5 days ($\text{Log}\,P = 0.40$) in the LMC, whereas they are abundant in the SMC. Thus only three variables with periods less than 2.5 days, and greater than one day, have been detected in the LMC (shown by crosses in Fig. 8), whereas 186 occur in the SMC. (c) If we imagine a straight line drawn through the points representing the variables with $P > 2.5$ in either Fig. 8 or Fig. 9, inspection

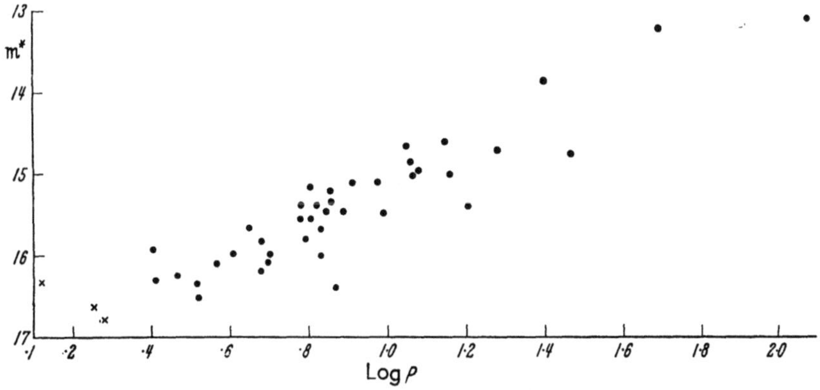

Fig. 8. Period-luminosity diagram for the Large Magellanic Cloud.

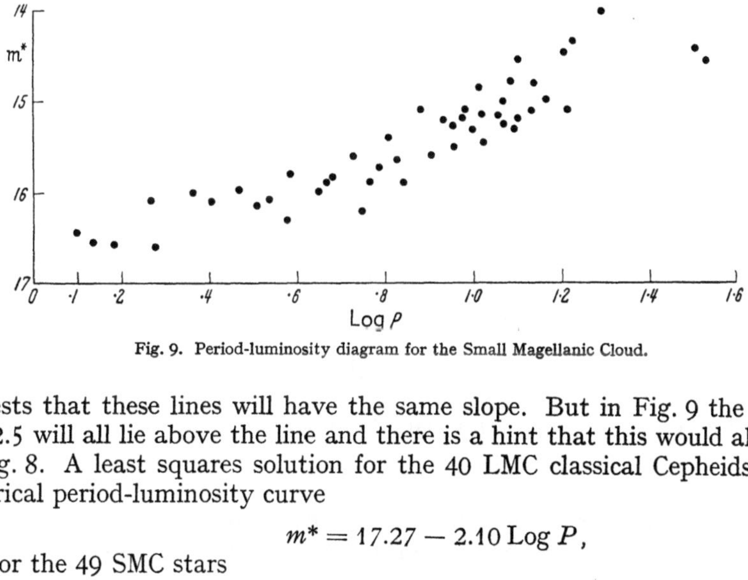

Fig. 9. Period-luminosity diagram for the Small Magellanic Cloud.

suggests that these lines will have the same slope. But in Fig. 9 the points for $P < 2.5$ will all lie above the line and there is a hint that this would also happen in Fig. 8. A least squares solution for the 40 LMC classical Cepheids gives the empirical period-luminosity curve

$$m^* = 17.27 - 2.10\,\text{Log}\,P, \tag{4.2}$$

and for the 49 SMC stars

$$m^* = 16.92 - 1.73\,\text{Log}\,P, \tag{4.3}$$

the slope of the second straight line being less than that of the first because of the effect pointed out in (c) above. A more complete analysis by SHAPLEY[1] gives the following formulae:

For 137 variables in the LMC $m^* = 17.14 \pm 0.08 - (2.08 \pm 0.09)\,\text{Log}\,P$, (4.4)

and for 307 in the SMC $m^* = 17.04 \pm 0.05 - (1.74 \pm 0.06)\,\text{Log}\,P$. (4.5)

[1] See footnote 2, p. 458.

Following Shapley, we take

$$m^* = 17.04 - 1.74 \operatorname{Log} P \tag{4.6}$$

as the empirical relationship between mean observed apparent magnitude and $\operatorname{Log} P$. It is established for the SMC and for the range of P from 1.2 to 40 days ($\operatorname{Log} P = 0.08$ to 1.60). Now the period-luminosity curve corresponding to (4.6) will have the same slope and will therefore be

$$M = C - 1.74 \operatorname{Log} P, \tag{4.7}$$

where C is a constant to be determined by criteria that do not involve luminosity considerations. Obviously if one absolute magnitude, M_0, and a corresponding period, P_0, can thus be determined, the value of C is

$$C = M_0 + 1.74 \operatorname{Log} P_0. \tag{4.8}$$

Up to the year 1952 it was assumed (i) that the absolute magnitude of a Type II Cepheid was the same as that of a classical Cepheid of the same period [*16d*]; and (ii) that the RR Lyrae stars formed a continuous sequence with the classical Cepheids. The determination of C made by Shapley in 1940 was based on these two assumptions and on Wilson's [*17*] work on the statistical parallaxes of classical Cepheids in our Galaxy. The method implies that the classical Cepheids of the SMC are identical in kind with those of our Galaxy, a conclusion which some astronomers have come to question[1] [*5b*]. Wilson used the proper motions of 86 Cepheids, including however the Type II Cepheid W Virginis in his list, and also the radial velocities of certain of these stars for an independent statistical parallax determination. Taking first the proper motions, it was found that the average of the mean observed apparent magnitudes for the group of 86 stars was $m^* = 7.96$, that their average $\operatorname{Log} P$ was $\operatorname{Log} P_0 = 0.89$, and that the statistical parallax was $\pi = 0''.00135$. Thus the fictitious average absolute magnitude for the group was

$$M_0^* = 7.96 + 5 + 5 \operatorname{Log} (0.00135) = -1.39.$$

To this value, Wilson made two corrections: he allowed -0.07 for the dispersion in magnitudes and, since the stars he was using were near the galactic plane and thus affected by absorption, he also allowed -0.63 (equivalent to an absorption of 0.85 mag. per kpc) for absorption. Thus the true average absolute magnitude was

$$M_0 = -(1.39 + 0.07 + 0.63) = -2.09,$$

at the period given by $\operatorname{Log} P_0 = 0.89$. Hence substituting into (4.8) we find

$$C_1 = -0.543. \tag{4.9}$$

By similar arguments, for those of his stars that had measured radial velocities, Wilson found that $M_0 = -1.47$ for $\operatorname{Log} P_0 = 0.97$ and therefore now

$$C_2 = +0.218. \tag{4.10}$$

To results obtained from radial velocities, Wilson assigned half-weight as compared with proper motion determinations. Thus the final value of C is

$$C = \tfrac{2}{3}(C_1 + \tfrac{1}{2} C_2) = -0.29, \tag{4.11}$$

a result derived directly and without any knowledge of some previous period-luminosity curve for the SMC. Shapley himself did, however, use an indirect

[1] J. Schilt, I. Epstein and S. J. Hill: Astronom. J. **60**, 317 (1955), § 6.

method that brought in his original curve, Col. 1, Table 3. Through a procedure equivalent to the calculation of C in (4.11), WILSON had corrected the zero-point of the original curve by $\Delta M = -0.14 \pm 0.02$. Now from the original curve, by linear interpolation, the value of M for $\operatorname{Log} P_0 = 0.89$ turns out to be -1.69 and therefore the corrected absolute magnitude is $M_0 = -(1.69 + 0.14) = -1.83$, substitution of which into (4.8) for $\operatorname{Log} P_0 = 0.89$ gives $C = -0.28$. This value of C differs from that in (4.11) in no important respect and therefore SHAPLEY'S 1940 period-luminosity curve for the SMC is

$$M = -0.28 - 1.74 \operatorname{Log} P, \tag{4.12}$$

which is listed numerically in Col. 2, Table 3. There is no reason to suppose that a period-luminosity curve must necessarily be linear and SHAPLEY[1] has adopted, as

Table 3. *Period-luminosity curves for classical Cepheids.*

P (days)	Log P	M (1)	M (2)	M (3)	M (4)	M_p (5)	M (6)
1.58	0.2	−0.61	−0.63	−0.68	−0.50	−0.29	−2.15
2.51	0.4	0.93	0.98	1.01	0.81	0.61	2.47
3.98	0.6	1.22	1.32	1.33	1.23	0.90	2.76
6.31	0.8	1.53	1.67	1.66	1.64	1.21	3.07
10.0	1.0	1.89	2.02	2.02	2.06	1.57	3.43
15.8	1.2	2.26	2.37	2.39	2.48	1.94	3.80
25.1	1.4	2.68	2.72	2.80	2.89	2.36	4.22
39.8	1.6	3.19	3.06	3.25	3.31	2.87	4.73
63.1	1.8	3.81	3.41	3.73	3.72	3.49	5.35

Col. 1 = SHAPLEY'S original curve [*163*].
Col. 2 = SHAPLEY'S 1940 curve (SMC data), $M = -0.28 - 1.74 \operatorname{Log} P$.
Col. 3 = SHAPLEY'S non-linear curve (SMC data), 1940.
Col. 4 = SHAPLEY'S 1940 curve (LMC data), $M = +0.02 - 2.08 \operatorname{Log} P$.
Col. 5 = MINEUR'S provisional curve, 1944: Col. 1 plus 0.32.
Col. 6 = MINEUR'S final curve, 1944..

a better representation of the SMC data, the non-linear curve given in numerical form in Col. 3, Table 3. It will be noticed that the entries of Cols. 1 and 3 differ by about -0.14 only in the neighbourhood of $\operatorname{Log} P = 0.8$. The equations (4.12) for classical Cepheids, and (3.4) for Type II Cepheids are essentially identical but this depends on the calculation of C through (4.11). If C_2 is rejected on the ground that radial velocities of classical Cepheids are not good enough for the determination of a statistical parallax, Eq. (4.12) becomes with the aid of (4.9), $M = -0.54 - 1.74 \operatorname{Log} P$, and there is now a correction to the zero-point of (3.4) of $\Delta M = -0.28$.

If we take a linear period-luminosity curve with the slope appropriate to the LMC we have $M = C - 2.08 \operatorname{Log} P$ which, with $M_0 = -1.83$ at $\operatorname{Log} P_0 = 0.89$, becomes

$$M = +0.02 - 2.08 \operatorname{Log} P, \tag{4.13}$$

and is tabulated in Col. 4, Table 3. These details are given in order to show that up to 1940 SHAPLEY had arrived at several slightly different period-luminosity curves for the Magellanic Clouds and to emphasize how necessary it is, when speaking of a "correction to the zero-point of SHAPLEY'S curve", to indicate exactly which curve is intended.

[1] H. SHAPLEY: Harv. Reprint 207 (1940).

Other investigations in the 1930's, for example, that of KUKARKIN[1], had shown that the original period-luminosity curve of Col. 1, Table 3, did not require revision except in detail. Nevertheless inconsistencies began to appear: for example, classical Cepheids had been found by HUBBLE [8] in the galaxy M 31. Their absolute magnitudes could be calculated, knowing their periods, and their observed apparent magnitudes then gave the distance-modulus of the galaxy. This in turn determined the absolute magnitudes of the globular clusters of M 31[2] and it turned out that the average absolute magnitude of 212 clusters fell at -5.0. In our Galaxy, the median is more nearly -7.5, a discrepancy that was puzzling. In 1944 came BAADE'S[3] important conclusion that the stars in galaxies were divisible into two broad classes, Population I to which the classical Cepheids belong, and Population II that embrases the RR Lyrae stars and the Type II Cepheids. By 1952, BAADE[4] had concluded through his inability to detect RR Lyrae stars in M 31 with the 200-inch telescope that the assumptions (i) and (ii) of p. 460 were not tenable, that a classical Cepheid was about 1.5 mags. brighter than a Type II Cepheid of the same period and that the RR Lyrae stars did not in fact constitute a continuation of a classical Cepheid period-luminosity curve[5]. After the event, MINEUR [13] pointed out that a very similar conclusion was implied in his 1944 investigation [12] though it had passed unnoticed at the time. MINEUR had used as provisional absolute magnitudes, M_p, for classical Cepheids those listed in Col. 5, Table 3, which were obtained by adding $+0.32$ to SHAPLEY'S original values, a modification that MINEUR attributes to EMDEN. MINEUR'S determination of the statistical parallax of the classical Cepheids of our Galaxy had included an improved mathematical technique for dealing with the absorption. He found for the true absolute magnitude

$$M = M_p - 5 \operatorname{Log} \lambda,$$

where

$$\left.\begin{aligned} \lambda &= 1.90 \pm 0.20 \quad \text{(from radial velocities, galactic rotation terms)},\\ &= 2.57 \pm 0.40 \quad \text{(from proper motions, solar motion terms)},\\ &= 2.58 \pm 0.40 \quad \text{(from proper motions, residual velocities)}. \end{aligned}\right\} \tag{4.14}$$

It may be argued that the first value of λ is too uncertain to use; be this as it may, MINEUR takes the average of the three, which is 2.35, and so

$$M = M_p - 5\,(0.3711) = M_p - 1.86. \tag{4.15}$$

He describes the value $M_p = +0.32$ as his "provisional zero" and therefore his new "zero" is $M = -1.54$ as stated by him in 1952 [13]. Instead of conducting the further discussion solely in terms of "corrections to a zero-point", we have tabulated in Col. 6, Table 3, the values of M obtained from the entries in Col. 5 and the formula (4.15). Suppose then that we seek a linear representation of Col. 6 by the formula

$$M = C - 1.74 \operatorname{Log} P,$$

---

[1] B. V. KUKARKIN: Astronom. J. USSR. **14**, 317 (1937).

[2] C. K. SEYFERT and J. J. NASSAU: Astrophys. Journ. **102**, 377 (1945).

[3] See footnote 1, p. 451.

[4] W. BAADE: Trans. Internat. Astronom. Union **8**, 397; Cambridge: Cambridge University Press 1954.

[5] CECILIA PAYNE-GAPOSCHKIN: Variable Stars and Galactic Structure Chap. IV. London: Athlone Press 1954. But see W. IWANOVSKA: Bull. Astronom. Cop. Univ. Torun **3**, 15 (1953), for an interesting suggestion that there are variables with periods less than one day that do lie on the extension of the classical Cepheid curve.

having the slope appropriate to the SMC. It is found that the entries from $\operatorname{Log} P = 0.2$ to 1.4 are represented to within ± 0.06 in M by

$$M = -1.74 - 1.74 \operatorname{Log} P, \tag{4.16}$$

though the entries for $\operatorname{Log} P = 1.6$ and 1.8 do not fit. Adopting this as the period-luminosity curve for the classical Cepheids of the SMC and comparing it with (3.4), it follows that the zero-point of the latter has been corrected by an amount

$$\Delta M = -1.74 + 0.26 = -1.48,$$

which is in good agreement with BAADE's suggestion that a classical Cepheid is about 1.5 mags. brighter than a Type II Cepheid of the same period. However, BAADE[1] believes that the results (4.14) and (4.15) are accidental, because MINEUR had found by his method a solar motion of 30.2 km/sec relative to the classical Cepheids of our Galaxy. BAADE takes 20 km/sec as being the correct value and says that he then obtains a "zero" in MINEUR's sense of -0.92 instead of -1.54. This is equivalent to taking, in place of (4.15),

$$M = M_p - 1.24.$$

The present writer has verified that this formula is obtainable by retaining the first value of λ in (4.14) and modifying the other two by putting $S_0 = 20$ km/sec in the relevant formula, $S_0/\lambda = 11.7$, given by MINEUR. In place of (4.16), we should then have

$$M = -1.12 - 1.74 \operatorname{Log} P, \tag{4.17}$$

and this gives a zero-point correction to (3.4) of only

$$\Delta M = -1.12 + 0.26 = -0.86. \tag{4.18}$$

The justification of BAADE's procedure is perhaps open to question, because MINEUR did not *assume* his value for the solar motion but *deduced* it from his data. Thus it is hardly legitimate to substitute some other value for the solar motion without also modifying MINEUR's data and/or his theory. BAADE's criticism seems to amount to questioning the whole of MINEUR's procedure rather than to finding, in (4.17) and (4.18), the results he should have obtained.

A recent attack on the determination of the statistical parallax of a group of classical Cepheids in our Galaxy is due to BLAAUW and H. R. MORGAN[2]. Their paper is very brief, but the present writer has had the opportunity of discussing the work with Dr. BLAAUW. It appears that the aim of the investigation was to determine a zero-point correction to SHAPLEY's original curve, Col. 1, Table 3, by using improved proper motions for 18 classical Cepheids and better information on the absorption correction. Since the periods of the stars ranged from $\operatorname{Log} P = 0.4$ to 1.0, it is only in this range that the correction was intended to apply. The value of ΔM obtained was -1.4 and its probable error was estimated at ± 0.3. In view of the importance that has been attached to this result, it is worth while to examine it in some detail. The method consists of comparing an "old" statistical parallax for the group of 18 stars with a "new" parallax. Firstly, taking the data of JOY[3] as starting point, the "old" parallax is found to be $0''.00468$ and the "new" is $0''.00307$. Suppose that m is the average of the mean apparent magnitudes of the 18 stars; then the "old" parallax is associated with an absolute

[1] W. BAADE: Publ. Astronom. Soc. Pacific **68**, 5 (1956).
[2] A. BLAAUW and H. R. MORGAN: Bull. Astr. Inst. Netherl. **12**, 95 (1954).
[3] A. JOY: Astrophys. Journ. **89**, 356 (1939).

magnitude M_1 for the group, this absolute magnitude referring to Col. 1, Table 3, and

$$M_1 = m + 5 + 5 \operatorname{Log}(0.00468).$$

Let M_o^* be a fictitious absolute magnitude for the group corresponding to the "new" parallax so that

$$M_o^* = m + 5 + 5 \operatorname{Log}(0.00307).$$

Hence eliminating m,

$$M_o^* = M_1 - 5 \operatorname{Log}(468/307) = M_1 - 0.92.$$

But the average absorption for the group of stars, implicit in Joy's work, is found to be 0.25 and Blaauw and Morgan consider that 0.7 is a better value. Thus the true absolute magnitude for the group is M_o where

$$M_o^* = M_o + (0.7 - 0.25) = M_o + 0.45$$

and the correction to the zero-point of the curve given by Col. 1, Table 2, is

$$\Delta M = M_o - M_1 = -1.37,$$

which agrees with the correction -1.35 given in the last entry of the column headed "Joy" in Blaauw and Morgan's Table 2. Secondly, starting again with the data of Parenago[1], "old" and "new" parallaxes of 0".00495 and 0".00272, respectively, were computed and the reduction of the Parenago to the Shapley zero-point produced a correction of $\Delta M = -1.44$. Combining these results, the zero-point correction to Col. 1, Table 3 is taken to be -1.4 ± 0.3.

The remarkable feature of this procedure is that numerical values of the symbols are not needed, except for the two parallaxes. The question arises whether the same result would be attained if more empirical information were introduced into the computations. To investigate this point, the average of the mean observed apparent magnitudes for the 18 stars has been calculated from the data in Joy's paper and it proves to be $m_o^* = 5.68$, the average $\operatorname{Log} P$ being $\operatorname{Log} P_o = 0.768$. Since the new parallax for the group of stars is 0".00307, we have

$$M_o^* = 5.68 + 5 + 5 \operatorname{Log}(0.00307) = -1.88.$$

But the absorption is 0.7 and therefore $M_o = -2.58$. If the Parenago data are used the average of the mean observed apparent magnitudes for the group is obtained from the individual magnitudes listed by Blaauw and Morgan in their Table 1, because these magnitudes are very nearly identical with those in the Kukarkin and Parenago Catalogue[2]. We thus obtain $m_o^* = 5.82$ and the new parallax is now 0".00272, so that $M_o^* = -2.01$; and allowing 0.7 for the absorption, we have $M_o = -2.71$. The average of these two values of M_o is -2.65 and, taking this to correspond to $\operatorname{Log} P_o = 0.768$, it follows from (4.8) that $C = -1.31$ and therefore that the linear period-luminosity curve for classical Cepheids in the SMC is

$$M = -1.31 - 1.74 \operatorname{Log} P.$$

But now comparing this formula with (3.4), the correction to the zero-point of the latter is

$$\Delta M = -1.31 + 0.26 = -1.05,$$

[1] P.P. Parenago: Publ. Sternberg State Astr. Inst. **16**, 71 (1949).

[2] B.V. Kukarkin and P.P. Parenago: General Catalogue of Variable Stars. Moscow: Akademia Nauk U.S.S.R. 1948.

which falls considerably short of the correction suggested by BLAAUW and MORGAN's ratio of old and new distances method. The correction is certainly an improvement on the one deduced by BAADE from MINEUR's work [Eq. (4.18)] but it differs substantially from the -1.5 that is being sought for.

Another kind of investigation is due to WEAVER[1], who has kindly supplied the following summary of his work: he used radial velocities of classical Cepheids in our Galaxy together with the linear theory of circular galactic rotation to derive a zero-point correction of -1.56 mags. to SHAPLEY's "adopted curve" listed in Col. 3, Table 3. WEAVER's method involves two steps, firstly, radial velocities arising from motion in circular orbits round the centre of the Galaxy of critically placed classical Cepheids serve to establish the distance to the galactic centre in terms of the distances provided by the adopted period-luminosity curve. Secondly, the zero-point of this curve is adjusted to make the derived distance to the galactic centre agree with the one found, for example, from the RR Lyrae variables. The correction of -1.56 mags. is based upon only 21 Cepheids, data for which were available in the literature. The estimated uncertainty of the correction is 0.2 to 0.3 magnitudes; this uncertainty could be reduced by observing radial velocities, magnitudes and colours of additional known, critically placed, classical Cepheids. The method is capable of yielding a zero-point correction with as much accuracy as (i) the distance to the galactic centre is known; (ii) the motion of the galactic classical Cepheids is circular and a function of distance from the galactic centre alone; and (iii) the effects of interstellar absorption can be eliminated from the distances of the Cepheids used in the determination.

PARENAGO [*14*] approaches the period-luminosity curve for classical Cepheids by placing less reliance on the SMC value of -1.74 for the coefficient of Log P in (4.7). From the space motions of the galactic classical Cepheids, he finds

$$M = -1.24 - 1.67 \operatorname{Log} P \quad (P < 9),$$
$$M = -0.85 - 2.08 \operatorname{Log} P \quad (P > 9).$$

There are also in the recent literature, period-luminosity curves for classical Cepheids that are quoted without an explanation of their modes of derivation. Such is the formula

$$M = -1.60 - 1.74 \operatorname{Log} P,$$

given by INGRID TORGARD[2], which would give a zero-point correction relative to (3.4) of $\Delta M = -1.34$, C. W. ALLEN [*1b*] also tabulates a period-luminosity curve that agrees very roughly with our Col. 6, Table 3.

When classical Cepheids are observed in remote galaxies it may be relatively easier to determine the observed apparent magnitude at maximum light than at minimum. In such cases it would be useful to deduce the mean observed apparent magnitude from that at maximum. To this end, attempts have been made to relate Log P to the amplitude of the light-variation

$$\mathscr{A} = m^*_{\max} - m^*_{\min}$$

for classical Cepheids in the LMC, the SMC and the Galaxy. Thus for 40 stars in the LMC, SHAPLEY, MCKIBBEN and MOHR[3] have found that

$$\operatorname{Log} P = 1.32\,\mathscr{A} - 0.68,$$

[1] H. F. WEAVER: Astronom. J. **59**, 375 (1954).

[2] INGRID TORGÅRD: Astrophys. Journ. **120**, 370 (1954).

[3] H. SHAPLEY, VIRGINIA MCKIBBEN and JENKA MOHR: Proc. Nat. Acad. Sci. U.S.A. **26**, 326 (1940). Harv. Reprint 202.

a relation valid for P between 2.5 and 20 days. But Shapley[1] later showed that, for 365 stars in the SMC, the relation between average amplitude $\bar{\mathscr{A}}$ and Log P was not linear. For short periods $\bar{\mathscr{A}}$ was about 0.8 mags. and it increased to 1.3 mags. at Log $P=1.5$, *i.e.* at a period of 31.65 days. On the other hand, for 232 classical Cepheids in the Galaxy, $\bar{\mathscr{A}}$ at Log $P=1.5$ was more nearly 1.75. The overall $\bar{\mathscr{A}}$ for the SMC was 0.94 compared with 1.22 for the Galaxy, a difference attributed to systematic error. In Table 4, we reproduce the part of Shapley's table of period-range against average amplitude[2] that refers to longer periods. It is based on the fully explored regions of the SMC. In view of the size of the mean errors, it is not to be expected that the mean observed apparent magnitude of a classical Cepheid can be calculated from its observed apparent magnitude at maximum with any great accuracy. But if the latter can be measured for a classical Cepheid in some distant galaxy, and the period is also known, one-half the corresponding entry under $\bar{\mathscr{A}}$ added to the observed apparent magnitude at maximum should give an estimate of the mean observed apparent magnitude of the star. It is also interesting to note that Baade has urged[3] the desirability of using the relation between Log P and the absolute magnitude of a variable star at *minimum* light as the fundamental equation.

Table 4. *Amplitudes of the classical Cepheids.*

Interval of Log P	Mean Log P	Mean Amp. $\bar{\mathscr{A}}$ (mags.)	Mean error (mags.)
0.8—0.9	0.845	0.89	±0.06
0.9—1.0	0.949	0.92	0.09
1.0—1.1	1.055	0.97	0.08
1.1—1.2	1.164	1.01	0.08
1.2—1.3	1.233	1.23	0.11
1.3—1.5	1.386	1.25	0.13
>1.5	1.712	1.32	0.09

The contents of this section make it abundantly clear that the period-luminosity curve for classical Cepheids is still very inaccurately known. It is not that empirical relationships between m^* and Log P, such as (4.4) and (4.5) cannot be found, but rather that their conversion into relations between M and Log P is uncertain. Statistical parallaxes for groups of classical Cepheids in our Galaxy do not, it would seem, fix a point on the (M, Log P) curve with the desired accuracy. At the present time therefore, the adoption of a particular period-luminosity curve is to some extent a matter of subjective estimation. We shall choose the curve given by Col. 6, Table 3 on the grounds that (a) Mineur believed that he had correctly deduced it from his data; (b) it gives a zero point correction of -1.54 as compared with Shapley's original curve, Col. 1, Table 3; and (c) a linear representation of Col. 6, Table 3, shows that a classical Cepheid is brighter by -1.48 in absolute magnitude than a Type II Cepheid of the same period.

Table 5. *Absolute magnitudes of galactic novae.*

Method	M	Weight
Direct trigonometric parallax	-6.7	1
Statistical parallax	-7.6	$\frac{1}{2}$
Rate of nebular expansion. .	-7.9	1
Interstellar line intensities .	-7.6	1
Galactic rotation effect . . .	-6.5	$\frac{1}{2}$
Weighted average	-7.3 ± 0.2	

5. Other distance indicators. In addition to the foregoing types of objects, there are in our Galaxy various other classes of luminous objects whose absolute magnitudes can be determined with some degree of accuracy and which, if similar

[1] H. Shapley: Proc. Nat. Acad. Sci. U.S.A. **28**, 501 (1942). Harv. Reprint 249.
[2] H. Shapley: Loc. cit., Tab. 3.
[3] B. J. Bok: Sky and Telescope **15**, 348 (1956).

objects can be identified in other galaxies, can be used for rough distance-determinations. Amongst such are the following:

α) Novae. A study of novae in the Galaxy has been made by McLaughlin[1] from whose work Table 5 is taken. The weighted average agrees as well as can be expected with the average of −7.5 due to Lundmark[2] and with an average of −7.6 for seven galactic novae given by McLaughlin in 1948[3].

β) Supergiant stars. The brightest stars in our Galaxy are, according to Keenan and Morgan[4], classifiable within each spectral type into the classes Ia, the most luminous, and Ib, the less luminous. In Table 6, their photovisual magnitudes have been converted to photographic. The statistical parallax method of distance determination is again used to compute absolute magnitudes. van Rhijn[5] gives similar values for the brightest *O* and *B* stars of our Galaxy. The P Cygni stars, which, like novae, are surrounded by an expanding shell of gas, can also be very brilliant. Beals[6] has found that some 60 stars of this type range from absolute magnitude −8.7 (HD 223385), through −6.3 (P Cygni) to as low as +6.0, with a mean at −3.8. It might appear that, if the brightest stars in a galaxy were to be observed, we should have an easily-applied criterion of distance. This is not so, however, because there are groups of *O* stars that are surrounded by regions of interstellar gas which they ionize. The total light of the gas and stars is greater than that of individual bright stars. Thus there is a danger that these bright H II regions will be mistaken for single stars and far too low an intrinsic luminosity will be assigned to them [9b].

Table 6. *Absolute magnitudes of bright stars.*

Spectral type	M (Ib)	M (Ia)
B 0	−6.33	−7.03
B 5	−5.88	−7.18
A 0	−4.8	−7.0
A 5	−4.3	−6.8

γ) Planetary nebulae. These objects, recognisable by their emission-line spectra and disc-like structure, can also serve as distance-indicators. An analysis of some 136 planetary nebulae of our own Galaxy has been made by Berman[7] who computed absolute magnitudes by finding distances through (a) the method of statistical parallaxes, (b) the angular sizes of the planetary nebulae, and (c) the effect of galactic rotation. He concluded that, if m_s was the apparent magnitude of the central star and m_n that of the nebula surrounding it, then, on the average, the absolute magnitude of the planetary nebula was

$$M = -0.88 - 0.32\,(m_s - m_n)\,. \tag{5.1}$$

In Berman's list there are 80 planetaries for which m_s and m_n have been independently determined; a calculation shows that the average $(m_s - m_n)$ for these stars is +2.75, forty-three of them having values between 3 and 0 and the extremes ranging from 7.3 to −1.4. With this average, it follows from (5.1) that the average absolute magnitude is −1.76, but individual planetaries may attain −3.22.

---

[1] D. B. McLaughlin: Pop. Astron. **50**, 233 (1942). — Publ. Astronom. Soc. Pacific **57**, 69 (1945).

[2] K. Lundmark: Vistas in Astronomy, Vol. II, § 15. London and New York: Pergamon Press 1957.

[3] D. B. McLaughlin: Pop. Astron. **56**, 459 (1948). Tab. 1.

[4] P. C. Keenan and W. W. Morgan: Astrophysics (Ed.: J. Hynek), p. 23. New York: McGraw-Hill 1951.

[5] P. J. van Rhijn: Publ. Astronom. Lab. Groningen No. 51 (1946).

[6] C. S. Beals: Publ. Dominion Astrophys. Obs. Victoria, B.C. **9**, 1 (1951).

[7] L. Berman: Lick Obs. Bull. **18**, 57 (1937).

6. Distances and absolute magnitudes of some nearby galaxies. The methods described in the previous sections will now be applied to the study of some of the galaxies nearest to our own. The general hypothesis will be made that similar objects in all galaxies have the same absolute magnitudes as corresponding objects in our own Galaxy and in the Magellanic Clouds. It must however be remembered that an observer on Earth is at an advantage when he is observing an external galaxy, as compared with his situation when studying our own. He can usually view the external galaxy almost in its entirety, whereas he observes our own Galaxy in a restricted region near the Sun, because of the presence of obscuring clouds. Thus the proportions of bright objects identified in the external galaxy may be different from those he notices in the Galaxy.

α) The Small Magellanic Cloud. Two methods that have been used for finding the distance of this galaxy depend on the assumption that the RR Lyrae stars have absolute magnitude 0.0. Firstly, Thackeray and Wesselink[1] have identified three RR Lyrae stars whose average mean observed apparent magnitude is $m^* = 19.1$, in the globular cluster NGC 121 lying near the edge of the SMC. By (2.7) this would give an apparent distance-modulus of $\mathcal{M}^* = 19.1$. Secondly, Shapley and Nail[2] have used the method of brightest stars in globular clusters on the three SMC clusters NGC 121, 416 and 419, to determine that $\mathcal{M}^*$ lies between 19.13 and 19.20, though they regard the determination as of somewhat low weight. A method independent of these, is to use the classical Cepheids: if the linear representation (4.16) is adopted, we have, by elimination of $\operatorname{Log} P$ from (4.6), that $\mathcal{M}^* = 18.78$. Contributory evidence comes from objects of other types: for example, Henize, Hoffleit and Nail[3] have concluded that the probable maxima of four novae in the SMC correspond to values of m^* ranging from 11.5 to 11.0 with a mean at 11.22. Hence if these novae have average absolute magnitude -7.5, as in our Galaxy, $\mathcal{M}^* = 18.72$. Buscombe and de Vaucouleurs[4] arrive at similar conclusions from their study of the SMC novae, giving $\mathcal{M}^* = 18.6 \pm 0.2$. But if the absolute magnitudes of novae at 15 days after maximum are considered, de Vaucouleurs [*6*] finds $\mathcal{M}^* = 18.8$. Again Lindsay[5] reports the discovery of 17 objects in the SMC that he identifies with planetary nebulae and he deduces that, if $\mathcal{M}^* = 18.9$, their absolute magnitude is -3.6 ± 0.6. This is rather high when compared with the -1.76 for our Galaxy (see Sect. 5γ) but is perhaps explicable by the fact that only the brightest planetaries of the SMC are likely to be observed.

It seems possible that the apparent distance-modulus of the SMC may be 19 or even somewhat higher, though Buscombe, Gascoigne and de Vaucouleurs [*5*] gave $\mathcal{M}^* = 18.6$ in 1954 and de Vaucouleurs in 1955 [*6*] revised this to $\mathcal{M}^* = 18.75 \pm 0.1$. To find the distance of the galaxy, it is still necessary to know the appropriate absorption. Up to about 1940 Shapley had adopted values near to 0.25 for this quantity and had stressed the transparency of space in the direction of the SMC, a point of view strongly supported by the measurements of the colours of classical Cepheids in this galaxy due to Gascoigne and Kron[6] and by the work on early type stars of Code and Houck[7]. Nevertheless

[1] A.D. Thackeray and A. J. Wesselink: Observatory **75**, 33 (1955). — Nature, Lond. **171**, 693 (1953).
[2] H. Shapley and Virginia McK. Nail: Proc. Nat. Acad. Sci. U.S.A. **40**, 1 (1954).
[3] K.G. Henize, Dorrit Hoffleit and Virginia McK. Nail: Proc. Nat. Acad. Sci. U.S.A. **40**, 365 (1954).
[4] W. Buscombe and G. de Vaucouleurs: Astronom. J. **60**, 155 (1955).
[5] E.M. Lindsay: Monthly Notices Roy. Astronom. Soc. London **115**, 248 (1955).
[6] S.C.B. Gascoigne and G.E. Kron: Publ. Astronom. Soc. Pacific **65**, 32 (1953).
[7] A.D. Code and T.E. Houck: Astronom. J. **61**, 173 (1956).

many astronomers use the absorption $A_s = 0.44$ given by (2.3) for S. Gal. Lat. 45°; BAADE [3] adopts 0.3 and DE VAUCOULEURS, 0.45 ± 0.1 [6]. It has therefore seemed best to give in Table 7 the distance D of the SMC for various values of $\mathscr{M}^*$ and of A_s, showing that the largest D differs from the smallest by 43% of the latter. At the time of writing (July 1956), favourite values of D range from 45 to 55 kpc. To find the absolute magnitude of the SMC, its observed apparent magnitude is required, and this curiously enough, is still very poorly known. BUSCOMBE, GASCOIGNE and DE VAUCOULEURS have discussed this question [5] and they adopt $m^* = 2.0 \pm 0.5$. The absolute magnitudes in Table 7 have been calculated for $m^* = 2.0$, by (2.7).

β) The Large Magellanic Cloud. Here again the assumption that the RR Lyrae stars have absolute magnitude zero yields distance-moduli. THACKERAY and WESSELINK[1] have found 21 RR Lyrae stars in NGC 1466, a globular cluster on the edge of the Cloud, and two in NGC 1978, a cluster within the Cloud. The average of their mean observed apparent magnitudes leads to $\mathscr{M}^* = 19.2$. SHAPLEY and NAIL[2] have used the method of brightest stars for the LMC globular clusters NGC 1783, 1806, 1835, 1846, 1856, 1978 and 2056 to find $\mathscr{M}^* = 19.01 \pm 0.06$, which they regard as having more weight than the corresponding SMC determination. Turning next to the classical Cepheids, it is possible to obtain a linear representation of Col. 6, Table 3 for the range $\text{Log}\,P = 0.6$ to 1.6 to within ± 0.1 in M, by $M = -1.42 - 2.08\,\text{Log}\,P$. Eliminating $\text{Log}\,P$ from this and the empirical formula $m^* = 17.14 - 2.08\,\text{Log}\,P$, we find the low value of $\mathscr{M}^* = 18.56$. Again six novae in the LMC have probable maxima that range from 13 to 10.5 with a mean at 11.5[3] [5a]. If this is taken to correspond to absolute magnitude -7.5, then $\mathscr{M}^* = 19.00$, but if the absolute magnitudes of novae at 15 days after maximum are used instead [6], then $\mathscr{M}^* = 18.4$.

Table 7. *Distances for the Small Magellanic Cloud.*

$\mathscr{M}^*$	D (kpc) $A_s = 0$	0.25	0.4	M
19.0	63	56	52	−17.0
18.8	58	51	48	−16.8
18.6	52	47	44	−16.6

It therefore seems to be again probable that the apparent distance-modulus of the LMC is 19 though BUSCOMBE, GASCOIGNE and DE VAUCOULEURS [5] adopt 18.7, which is revised to 18.75 ± 0.1 by DE VAUCOULEURS [6]. To calculate the distance, the absorption is required and again there does not seem to be agreement. For S. Gal. Lat. 33°, the cosecant law gives $A_L = 0.57$; but counts of external galaxies suggest that $A_L = 0.3$ and BAADE [3] adopts $A_L = 0.4$. In Table 8, the same method has been adopted for the distance as in Table 7. The observed apparent magnitude of the LMC is also doubtful, being perhaps $m^* = 0.5 \pm 0.5$ [5], and the absolute magnitudes of Table 8 have been calculated for $m^* = 0.5$.

The apparent distance-moduli of Table 8 will now be used to compute the absolute magnitudes of highly luminous stars in the LMC, which are numerous. FEAST, THACKERAY and WESSELINK[4] have found 28 stars brighter than $m^* = 11$, which are believed to be supergiants of type A and B, from their spectra. The LMC also contains the P Cygni star S Doradus. Table 9 contains the absolute magnitudes of the two brightest of the 28 stars, of S Doradus and also the absolute

[1] See footnote 1, p. 468.
[2] See footnote 2, p. 468.
[3] See footnote 3, p. 468.
[4] M.W. FEAST, A.D. THACKERAY and A.J. WESSELINK: Observatory 75, 216 (1955).

magnitude corresponding to $m^* = 11$. The brightest of the Hα emission objects associated with *Be* stars in the LMC also have m^* of 9.2 and 9.8[1] and ten P Cygni stars—excluding S Doradus—have m^* ranging from 10.5 to 13.0 with a mean at 11.7[2]. This mean corresponds to a range of -7.5 to -7.1 in absolute magnitude according to the $\mathcal{M}^*$ which is chosen from Table 8, in rough agreement with the results for our Galaxy (see Sect. 5β). Thus a consideration of the brightest stars of the LMC does not preclude the possibility of its apparent distance-modulus being 19.0 ± 0.2, but it does not permit us to fix it with any precision. The galaxy evidently contains stars as luminous as, if not more luminous than, those of the Galaxy.

Table 8. *Distances for the Large Magellanic Cloud.*

$\mathcal{M}^*$	D (kpc)			M
	$A_L = 0$	0.3	0.6	
19.2	69	60	52	−18.7
19.0	63	55	49	−18.5
18.8	58	50	44	−18.3

In contrast to the foregoing methods, all of which determine luminosity-distance, Gum and de Vaucouleurs[3] have suggested that for the LMC—and for some other galaxies—a determination of distance by apparent size is possible[4]. They base this on the empirical conclusion that certain ringe-like H II regions in our Galaxy and in certain others have an average maximum diameter of 85 pc. In the LMC, two rings lead to distances of 41 and 43 kpc respectively [*5*],

Table 9. *Brightest stars of the Large Magellanic Cloud.*

Star	m^*	M		
		$\mathcal{M}^* = 19.2$	19.0	18.8
HD 33579 (type *A* 2)	9.1	−10.1	−9.9	−9.7
HDE 269781 (type *B* 9)	9.7	−9.5	−9.3	−9.1
S Doradus (P Cygni)	10.1	−9.1	−8.9	−8.7
	11.0	−8.2	−8.0	−7.8

which would be roughly in accord with the entry in Table 8 for $\mathcal{M}^* = 18.8$ and $A_L = 0.6$, giving $M = -18.3$ for the LMC.

γ) The Andromeda galaxy M 31 (NGC 224). This galaxy is also of the greatest importance for cosmological distances. Taking first the methods that depend on the assumption that RR Lyrae stars have absolute magnitude zero, there is the list of globular clusters due to Seyfert and Nassau[5], though it is true that 10 of these objects have been identified as Orion Nebula Hα emission regions by Haro[6]. The median of the 212 observed apparent magnitudes measured by Seyfert and Nassau is $m^* = 17.32$ and therefore, if this corresponds to the median absolute magnitude -7.5 of the globular clusters in the Galaxy, then $\mathcal{M}^* = 24.8$ for M31. But unpublished observed apparent magnitudes, determined photoelectrically by Mayall and Kron, for 68 of the brighter globular clusters, when converted to photographic magnitudes, have a median at 16.36 which makes $\mathcal{M}^* = 23.9$. Again Baade and Swope[7] have found that the average observed

[1] K.G. Henize and F.D. Miller: Publ. Obs. Univ. Michigan **10**, 75 (1950).
[2] See footnote 6, p. 467.
[3] C.S. Gum and G. de Vaucouleurs: Observatory **73**, 152 (1953).
[4] See G. de Vaucouleurs: Ann. d'Astrophys. **11**, 247 (1948) for a definition of the "effective dimension" of a galaxy.
[5] See footnote 2, p. 462.
[6] G. Haro: Astronom. J. **55** 66 (1950).
[7] W. Baade and Henrietta H. Swope: Astronom. J. **60**, 151 (1955).

apparent magnitude of the brightest stars of Population II in M31 is 22.7; if this corresponds to −1.5, as suggested by the globular clusters of our Galaxy, then $\mathcal{M}^* = 24.2$. Turning now to the classical Cepheids, BAADE and SWOPE[1] report that they have found 38 in one field of M31, whose average period is 20.0 days, and 237 in another, of average period 10.9 days. But observed apparent magnitudes and period-luminosity diagrams have not been published; it is merely stated that "by fitting the period-luminosity relation with revised zero-point to the upper boundary of the scatter diagram of the (second) field, we obtain (an apparent distance-modulus) of 24.25". We are therefore thrown back on the observations of HUBBLE [*8b*] who gives periods and observed apparent magnitudes at maximum of 39 classical Cepheids in M31. BAADE has shown that these magnitudes are systematically too bright and has given a table of the required corrections[2]. By linear interpolation in this table, the correction to the observed apparent magnitude at maximum of each of HUBBLE'S stars has been evaluated. The mean observed apparent magnitude for each star was then calculated from its period with the aid of Table 4 and it was then found that these magnitudes could be represented by the formula

$$m^* = 22.19 - 1.74 \operatorname{Log} P.$$

Then using (4.16) we have

$$\mathcal{M}^* = 22.19 + 1.74 = 23.9. \quad (6.1)$$

Table 10. *Distances for M 31.*

$\mathcal{M}^*$	D (kpc)			M
	$A_{31} = 0.2$	0.6	1.0	
24.0	575	479	398	−19.7
24.2	631	525	436	−19.9
24.4	692	575	479	−20.1

The study of novae in M31 also yields distance-moduli: in 1929 HUBBLE [*8a*] found that 85 novae had an average observed apparent magnitude at maximum light of $m^* = 16.43$; MAYALL[3] showed that 8 novae had $m^* = 16.3$; and ARP[4] lists 30 novae whose average $m^* = 16.9$. If these magnitudes correspond to −7.5, then $23.8 \leq \mathcal{M}^* \leq 24.4$. Finally BAADE[5] has observed 6 objects that he identifies as planetary nebulae whose m^* range from 21.7 to 22.2, with a mean at 22.04; if this corresponds to −1.76, then $\mathcal{M}^* = 23.8$, if to −3.12 then $\mathcal{M}^* = 25.16$, and if we follow BAADE and use (5.1) with $m_s - m_n = 4$ or 5, then $24.2 \leq \mathcal{M}^* \leq 24.5$. The eleven possible values of $\mathcal{M}^*$ that have been quoted are not all of equal weight; nevertheless their average falls at 24.3 which agrees with the often quoted 24.2 or 24.25 for the $\mathcal{M}^*$ of M31. However, it is difficult to believe that this quantity is known to the first, let alone the second, place of decimals. We shall therefore employ the same procedure as for the Magellanic Clouds and give distances for the galaxy for three values of $\mathcal{M}^*$, as shown in Table 10. To find the distance, the absorption is again required and the cosecant law (2.3) gives the high value $A_{31} = 0.87$ for S. Gal. Lat. 21°. The value of A_{31} calculated from SHANE'S determination[6] of the galactic absorption in the longitude range 80 to 100° is 1.0. BAADE[7] however adopts 0.4, and MAYALL and KRON'S programme[8] of photoelectric measurements on some 140 globular clusters in

[1] See footnote 7, p. 470.

[2] W. BAADE: Astrophys. Journ. **100**, 137 (1944). Tab. 1.

[3] N. U. MAYALL: Publ. Astronom. Soc. Pacific **43**, 217 (1931).

[4] H. C. ARP: Astronom. J. **61**, 15 (1956).

[5] W. BAADE: Astronom. J. **60**, 151 (1955).

[6] C. D. SHANE and C. A. WIRTANEN: Astronom. J. **59**, 285 (1954). — C. D. SHANE: Astronom. J. **61**, 292 (1956).

[7] W. BAADE: Mimeographed Symposium on Astrophysics, p. 17. Ann Arbor: University of Michigan Observatory 1953.

[8] Ann. Report Lick Obs.: Astronom. J. **60**, 285 (1955).

M31, the Galaxy and the SMC have led Kron to estimate provisionally that A_{31} lies between 0.3 and 0.5. The absolute magnitudes of the galaxy have been calculated from the observed apparent magnitude, 4.33 ± 0.03 [7]; they show that M31 is a giant system much more intrinsically luminous than is the LMC, or, indeed, the Galaxy whose M is estimated to lie between -18.6 and -19.7[1] [1c].

Hubble and Sandage[2] have detected in M31 a very luminous new type of irregular variable for which $m^* = 15.6$; its absolute magnitude is therefore -8.6 ± 0.2 according to the value of $\mathscr{M}^*$ chosen from Table 10 (Sandage[3] takes -8.4). The star is thus comparable in intrinsic luminosity with the brightest of the P Cygni stars of our Galaxy and with the giants of the LMC (see Table 9).

δ) The Triangulum galaxy, M 33 (NGC 598). As far as the present writer has been able to find out, there are no observations that permit of a determination of the apparent distance-modulus of this galaxy through the assumption that the RR Lyrae stars have absolute magnitude zero. Nor are there recent published lists of periods and observed apparent magnitudes of classical Cepheids. However in 1926, Hubble[4] gave periods and observed apparent magnitudes at maximum of 35 classical Cepheids in M33. Applying the same technique as was done for Hubble's classical Cepheids in M31, we find

$$m^* = 22.04 - 1.74 \operatorname{Log} P,$$

and therefore using (4.16) as the period-luminosity curve the apparent distance-modulus of M33 is

$$\mathscr{M}^* = 22.04 + 1.74 = 23.8.$$

In 1926 Hubble had concluded on the basis of an early form of Shapley's period-luminosity curve for the SMC that the apparent distance-moduli of M33 and the SMC differed by 4.55. If therefore $\mathscr{M}^*_{\text{SMC}} = 19$, there is rough agreement between Hubble's 23.55 and the one we have just found. In 1936 Hubble[5] had concluded that the brightest stars of M33 have observed apparent magnitude of 15.6 and if these are taken to have absolute magnitudes lying between -8 and -9, as suggested by the results for our Galaxy, the LMC and M31, then $23.6 \leq \mathscr{M}^* \leq 24.6$. Again Hubble and Sandage[2] have found four irregular variables in M33 of the same type as the one they discovered in M31. The average m^* for these four stars in 15.38 (15.7 to 15.1) and therefore, if the value -8.4 is accepted for the absolute magnitude, $\mathscr{M}^* = 23.8$. It therefore seems likely that, for M33, $\mathscr{M}^*$ is about 24.0 [9c], but clearly much more work is needed before certainty is approached.

In S. Gal. Lat. 31° [1c], the cosecant law gives $A_{33} = 0.6$, Baade [3] adopts 0.4 and Shane's formula[6] gives 0.43. With these values and $\mathscr{M}^* = 24$, the formula (2.8) shows that the distance of M33 lies in the range 525 to 479 kpc. Since the total observed apparent magnitude of the galaxy is $+6.19 \pm 0.03$ [7], its absolute magnitude, with the same value of $\mathscr{M}^*$, is -17.8.

ε) The Local Group. These four galaxies and our own are the five intrinsically brightest members of the small cluster of about 24 galaxies called the Local Group. The distances and absolute magnitudes of the others are not known

[1] Cecilia Payne-Gaposchkin: Introduction to Astronomy, p. 457. New York: Prentice Hall 1954.
[2] E. Hubble and A.R. Sandage: Astrophys. Journ. **118**, 353 (1953).
[3] A.R. Sandage: Astronom. J. **59**, 180 (1954).
[4] E. Hubble: Astrophys. Journ. **63**, 236 (1926). Tab. 2.
[5] E. Hubble: Astrophys. Journ. **84**, 157 (1936). Tab. 2.
[6] See footnote 6, p. 471.

with any greater accuracy than for the larger members of the group; they are all intrinsically faint objects, their absolute magnitudes ranging from about -15 to -8.5. Lists of these galaxies will be found in the literature[1] [*1c*], [*7*].

ζ) M81 (NGC 3031); NGC 4321; M87 (NGC 4486). Recent data for the determination of the distances and absolute magnitudes of other bright galaxies lying within the (presumably) locally Euclidean region around the Galaxy, become increasingly scanty once the Local Group is over-passed. Take, for example, the spiral galaxy M81, reported on by SANDAGE[2]. The data used are: one classical Cepheid of observed apparent magnitude at maximum light of 21.4 and of period 30.65 days; blue irregular variables of which the two brightest have m^* of 18.4 and 18.7, respectively; and 20 novae. For the Cepheid, $\operatorname{Log} P = 1.49$, which by Table 4 implies an amplitude of $1^{\mathrm{m}}.25$. Thus the mean observed apparent magnitude is $21.4 + 0.62 = 22.02$. Linear interpolation in Col. 6, Table 3, shows that $\operatorname{Log} P = 1.49$ corresponds to $M = -4.45$ and thus $\mathcal{M}^* = 26.47$ for M81. SANDAGE obtains 26.8 but assumes that the amplitude of the Cepheid is $2^{\mathrm{m}}.0$. If the two blue variables are of absolute magnitude -8.4, then $\mathcal{M}^* = 26.8$ or 27.1. Again HUBBLE had found that the average magnitude of the 20 novae of M81 was 3.8 mags. fainter at 14 days after maximum light than for the novae of M31. In his report, SANDAGE uses the apparent distance-modulus for M31 given by (6.1) and thus obtains $\mathcal{M}^* = 23.9 + 3.8 = 27.7$; but if the apparent distance-modulus of M31 is raised to 24.2, then $\mathcal{M}^* = 28.0$. From these meagre data SANDAGE's conclusion that $\mathcal{M}^* = 27.1 \pm 0.3$ would seem to be a trifle optimistic. At any rate, the total observed apparent magnitude of the galaxy being $+7.9$ according to BIGAY [*4*] and 7.85 ± 0.3 according to HOLMBERG [*7*], its absolute magnitude may be computed to lie between -18.6 ($\mathcal{M}^* = 26.5$) and -20.1 ($\mathcal{M}^* = 28$); it is -19.2 if $\mathcal{M}^* = 27.1$. Thus M81 appears to be a galaxy comparable in intrinsic luminosity with M31. It lies in N. Gal. Lat. 42° [*1c*] and the cosecant law would give an absorption of 0.46. Hence if $\mathcal{M}^* = 27.1$, the distance is 2.13 Mpc.

NGC 4321 is a spiral belonging to the Virgo Cluster of galaxies and is discussed in HMS [*9a*]. Its red-shift corresponds to a velocity of recession of about 1500 km per sec and therefore the galaxy does not quite satisfy the conditions for inclusion in the locally Euclidean region. Its brightest stars are first observed at about $m^* = 20.8$, assuming that their colour-index is about zero. Thus, if they are taken to have absolute magnitudes between -8 and -9, then $28.8 \leq \mathcal{M}^* \leq 29.8$. The N. Gal. Lat. of the galaxy is 77° and the cosecant law would give $A = 0.32$. Thus by (2.7) the distance-modulus is $\mathcal{M} = 28.5$ to 29.5, and the distance ranges from 5.01 to 7.94 Mpc. In HMS the distance-modulus is given as 29.05, by taking the brightest stars to have $M = -8.5$ and $A = 0.25$. BIGAY [*4*] gives $+9.89$ as the m^* for the galaxy and therefore its absolute magnitude lies in the range -18.9 to -19.9. With $\mathcal{M} = 29.05$ the distance of the galaxy would be 6.46 Mpc and its absolute magnitude -19.4. In the Magellanic Clouds where there are many kinds of objects that can serve as distance-indicators, it still seems likely that the apparent distance-moduli are subject to an error of ± 0.2. That the apparent distance-modulus of NGC 4321 can be known to the second place of decimals from the apparent magnitude at which its brightest stars first appear is hardly credible.

The third brightest galaxy of the Virgo cluster is M87 (NGC 4486); by comparing photometrically its globular clusters with those of M31, BAUM[3] has derived

---

[1] A. G. WILSON: Publ. Astronom. Soc. Pacific **67**, 27 (1955).

[2] See footnote 3, p. 472.

[3] W. A. BAUM: Publ. Astronom. Soc. Pacific **67**, 329 (1955).

a distance-modulus of 30.2. With Bigay's [4] $m^* = 9.84$ and an absorption of 0.26, the absolute magnitude of M87 has the high value of -20.6 with a correspondingly great distance of 11 Mpc. These results on NGC 4321 and 4486 emphasize the uncertainty in the distance of the Virgo cluster as a whole (see below, Table 14).

η) Conclusions. Perhaps the most important conclusion that emerges from this discussion of the nearby galaxies is that a great deal of observational work still remains to be done before their distances and absolute magnitudes can be known with precision. There does not seem to be any short-cut to the solution of these problems, as the monumental work of Shapley and his co-workers on the Magellanic Clouds has demonstrated. Many different kinds of distance-indicators in a galaxy will have to be identified and studied before its distance and absolute magnitude can be found and the probable errors in the determinations stated. Of the two quantities, the absolute magnitude should be the easier to obtain, for it does not involve a precise knowledge of the absorption. The assumption that the absorption applicable to the observed apparent magnitude of the entire galaxy is the same as that for the individual distance-indicators it contains, should suffice in the first instance. On the other hand, the distance is uncertain, not only because of the uncertainty in the absolute magnitude, but also because of the correction for absorption. For instance, in each line of Table 8, the greatest distance is about 1.3 times the least, due to this effect. For this reason alone, it is illusory to speak, as has been done since 1952, of a "distance-scale of the universe" which is to be "revised" by multiplication by 2, or 3 or 4. The "distance-scale of the universe" really means the distances of individual galaxies, or of clusters of galaxies, and the "revision" means a change of opinion regarding the absolute magnitudes assigned to galaxies consequent on the alteration of the zero-point of the period-luminosity curve for classical Cepheids. In the text-books and the literature, distance-moduli for galaxies are given to the first, or even the second, decimal place, and a unique distance for each galaxy is listed. Such lists have the appearance of tables of distances from the Sun of planets in the solar system and the unwary reader may well be excused if he jumps to the conclusion that the two sets of distances are known with comparable accuracies. This practice will not be followed here, the reader being asked to consider Tables 7, 8, 10 and the various other distance-determinations contained in this section, before he accepts any one distance for a particular galaxy.

The conclusion that does emerge with some degree of certainty is that the intrinsically brightest galaxies have absolute magnitudes lying between about -18.9 and -19.9. This is an unfortunately wide range, but it is better to face the fact than to persuade oneself that some one figure, -19.4 or -19.82, or whatever it may be, is known with certainty. In Sect. 7, we shall try to use this conclusion when we come to employ the red-shift as a distance-indicator.

7. The red-shift as distance-indicator. The vast majority of galaxies are so faint in apparent luminosity—presumably because of their great distances—that individual objects can no longer be identified in them and therefore all the preceding methods of distance-determination fail. Nor is it now still possible to assume that they lie in the locally Euclidean region round our Galaxy. The spectra however reveal the displacement of spectral lines toward the red, compared with laboratory sources, which, as a very rough approximation, increases the fainter the galaxy under observation. Thus the conclusion may be drawn that, on the average, the displacement is proportional to the distance of the galaxy. The red-shift δ is constant for all the lines in the spectrum of any particular galaxy[1] and therefore it has the characteristic property of a Doppler displacement. This interpretation is accepted in general relativity; it is also assumed that the relative motions, spatial distribution and other large-scale properties of the whole system of galaxies can be described by identifying each galaxy with a point in a space-time (1.3) whose coordinates (r, ϑ, φ) are constants [*11b*]. Under these conditions, the luminosity-distance of a galaxy whose red-shift is δ is given by (1.10), h_1 and h_2 being defined by (1.9). Thus, if the numerical values of h_1 and h_2 can be found from observation, the luminosity-distance D can be calculated from a measurement of δ and then, if desired, the distances ξ and $\bar{r}$ can also be obtained from Eqs. (1.6) and (1.7). We now proceed to show how the numerical values of the parameters h_1 and h_2 can be found.

The most recent evaluation is to found in the contribution by SANDAGE to HMS [*9a*], a paper that was completed in July 1955, and which also gives the data on red-shifts and apparent magnitudes accumulated in the past twenty years at the Mount Wilson and Palomar and at the Lick Observatories. A year earlier, preliminary data on clusters of galaxies were supplied to the present author and were used as an illustrative example in the calculation of h_1 and h_2 [*11c*]. The methods used by SANDAGE and McVITTIE differ in certain respects, though they are theoretically equivalent; both will be employed in the account that follows. A glossary of notations will be found in Sect. 7ε (p. 485) below.

α) *Luminosity-distance and apparent photographic magnitude.* The first step in both treatments is to relate the theoretical formula (1.10) for D with apparent photographic magnitude and with absolute magnitude. The simple formula (2.1) will not now do for the reason that, each laboratory wavelength λ being increased by an amount $\delta\, d\lambda$, the energy-distribution curve in the spectrum of a galaxy, as received by the observer, is distorted compared with its shape at emission. The allowance that must be made for this effect is called the K-correction and the formula that replaces (2.1) is [*11d*]

$$\operatorname{Log} D = 0.2\,(m - K - M) + 1\,, \tag{7.1}$$

where

$$\left.\begin{aligned} K &= -\frac{5}{2}\log\left\{\frac{1}{(1+\delta)\, I_0}\int_0^\infty \sigma(\lambda)\, B\left(\frac{\lambda}{1+\delta}\right) d\lambda\right\},\\ I_0 &= \int_0^\infty \sigma(\lambda)\, B(\lambda)\, d\lambda, \end{aligned}\right\} \tag{7.2}$$

λ is the wavelength as received by the observer, $\sigma(\lambda)$ is the extinction function and $B\left(\frac{\lambda}{1+\delta}\right)$ is the energy-distribution function of the spectrum of the galaxy. In addition to the wavelength and δ, B depends on such physical parameters as the temperature of the galaxy emitting the radiation. In both McVITTIE's and SANDAGE's treatments the development is complicated by the assumption that these physical parameters are undergoing secular changes, a hypothesis that was mainly based on the Stebbins-Whitford effect, an excess reddening observed in the spectra of elliptical galaxies[2]. However WHITFORD has now shown that this effect is almost entirely due to errors in the method of observation[3] and that, if it exists at all, the effect amounts to less than 10% of what it was originally

[1] R. MINKOWSKI and O.C. WILSON: Astrophys. Journ. **123**, 373 (1956). — A.E. LILLEY and E.F. McCLAIN: Astrophys. Journ. **123**, 172 (1956).

[2] J. STEBBINS and A.E. WHITFORD: Astrophys. Journ. **108**, 413 (1948). — J. STEBBINS: Monthly Notices Roy. Astronom. Soc. London **110**, 416 (1950).

[3] A.E. WHITFORD: Astronom. J. **61**, 352 (1956).

thought to be. In view of this conclusion, and also because of the many other uncertainties in the data which will be described as we proceed, an allowance for secular changes in the physical parameters appears to be an unwarranted refinement and it has been omitted in (7.2). If the function $B\left(\frac{\lambda}{1+\delta}\right)$ in the latter formula is expanded by Taylor's theorem as a power-series in δ, it can be proved [*11d*] that

$$K = K_1\,\delta + K_2\,\delta^2 + \cdots, \tag{7.3}$$

where K_1 and K_2 are constants that could be evaluated if the functions σ and B were known. Sandage [in HMS, formula (B7)] arrives at (7.2) by a slightly different path because of his method of reducing photographic to bolometric magnitudes. He calculates numerically the K-correction for various values of the red-shift δ, using the function σ due to Stebbins and Whitford. His results are shown in Table 11, according as the galaxy is identified with an E galaxy (Col. 1), an Sb galaxy (Col. 2) or an Sc galaxy (Col. 3). For the given range of δ, the K-correction can be represented by a quadratic function of δ (first two terms of the series in (7.3)) to within 2% by the following formulae:

Table 11. *Photographic K-correction.*

δ	K (ph. mag.)		
	(1)	(2)	(3)
0.00	0.00	0.00	0.00
0.05	0.22	0.25	0.14
0.10	0.44	0.50	0.30
0.15	0.66	0.75	0.47
0.20	0.89	0.99	0.62
0.25	1.10	1.24	0.81

$$\left.\begin{aligned} &\text{Table 11, Col. 1: } K = 4.38\,\delta + 0.34\,\delta^2,\\ &\text{Table 11, Col. 2: } K = 5.02\,\delta - 0.34\,\delta^2,\\ &\text{Table 11, Col. 3: } K = 2.70\,\delta + 2.00\,\delta^2.\end{aligned}\right\} \quad (7.4$$

The important point for our purposes is that the value of K_1 is *positive* and not less than 2.7.

There are now two formulae for D, namely, Eqs. (1.10) and (7.1) through which, by eliminating D, a relation between m and δ results. Sandage's and McVittie's treatments differ in that this elimination is performed in different ways. The Sandage method is to eliminate Log$D$ and therefore (1.10) is written

$$\operatorname{Log} D = \operatorname{Log}(c\,\delta) + \frac{1}{E}\log_e\left\{1 + \frac{1}{2}\left(1 + \frac{h_2}{h_1^2}\right)\delta\right\} - \operatorname{Log} h_1,$$

where $E = \log_e 10 = 2.303$. Thus if the second term is expanded to the first power of δ, we have

$$\operatorname{Log} D = \operatorname{Log}(c\,\delta) + \frac{1}{2E}\left(1 + \frac{h_2}{h_1^2}\right)\delta - \operatorname{Log} h_1.$$

Then using (7.1) we obtain[1]

$$m - K = 5\operatorname{Log}(c\,\delta) + 1.086\left(1 + \frac{h_2}{h_1^2}\right)\delta + (M - 5 - 5\operatorname{Log} h_1). \tag{7.5}$$

In McVittie's method, the aim is to find m directly as a function of δ, using the formula (7.3) for K. To do this (7.1) is written as

$$D = 10^{1-0.2M}\,x\exp(-b_1\,\delta - b_2\,\delta^2),$$

where

$$x = 10^{0.2m}, \quad b_1 = 0.2\,E\,K_1, \quad b_2 = 0.2\,E\,K_2, \tag{7.6}$$

[1] This is identical with HMS, Eq. (3), remembering that Sandage writes m_{bol} for $m - K$, and that we are omitting his μ, which depends on secular changes in the physical parameters contained in the function B, and on the highly uncertain absorption between galaxies.

or, expanding to the order δ^2,

$$D = 10^{1-0.2M} x \{1 - b_1 \delta - (b_2 - b_1^2/2)\, \delta^2\}.$$

Thus eliminating D with the aid of (1.10) and expanding to the order δ^2, we find

$$x = A_1 \delta + A_2 \delta^2, \tag{7.7}$$

where the constants A_1 and A_2 are defined by

$$A_1 = \frac{c}{h_1} 10^{0.2M-1}, \quad A_2 = \frac{c}{h_1}\left(b_1 + \frac{h_1^2 + h_2}{2h_1^2}\right) 10^{0.2M-1}, \tag{7.8}$$

so that

$$h_1 = \frac{c}{A_1} 10^{0.2M-1}, \tag{7.9}$$

$$h_2 = -\left(1 + 2b_1 - \frac{2A_2}{A_1}\right) h_1^2, \tag{7.10}$$

$$\frac{h_2}{h_1^2} = -\left(1 + 2b_1 - \frac{2A_2}{A_1}\right) = -q_0. \tag{7.11}$$

The formulae (7.5) and (7.7) are theoretically equivalent to one another; but they are not quite equivalent when numerical data come to be fitted to them by the method of least squares. By (7.6), x depends only on m and it is expressed in (7.7) as a polynomial in the other variable δ. But (7.5) is really a formula of the form

$$m - K - 5\,\mathrm{Log}\,(c\,\delta) = S_1 + S_2\,\delta, \tag{7.12}$$

in which S_1 and S_2 are the constants to be determined. Thus it is really the composite variable $m - (K_1\delta + K_2\delta^2) - 5\,\mathrm{Log}\,(c\delta)$ that is being expressed as a polynomial in the second variable δ. The effect of this compositeness is avoided by using (7.7). Moreover it follows from (7.6) and (7.10) that the acceleration parameter h_2 depends on K_1, and not at all on K_2, a fact obscured when (7.12) is employed.

β) The observational data on clusters of galaxies. The importance of a cluster of galaxies lies in the fact that we can assume with some degree of certainty that the brightest members of a cluster have absolute magnitudes lying within the range -18.9 to -19.9 that was found by a study of the galaxies near to our own. The definition of a cluster is not a very precise one: a large cluster contains 500 or more members all of which are seen in approximately the same direction in the sky. For example, the 400 or so members of the Corona Borealis cluster (1520+2754) cover an area of the celestial sphere about equal to that of the full Moon. The spread in apparent magnitude is about 5 mags. in most of the large clusters[1]. Such general qualitative definitions may be good enough to identify the members of relatively nearby clusters, but it can be appreciated that, for very remote clusters, difficulties in identification increase with the number of intrinsically faint galaxies of the general field that happen to lie between the observer and the cluster. Moreover the matter is complicated by the fact that clusters are of different types[2].

The data on red-shifts, for both individual galaxies in the general field and for cluster-members, are given in HMS in terms of $c\delta$ rather than in terms of the observed quantity δ. The accuracy in $c\delta$ is estimated to be such that the values listed are within 175 km/sec of their true values with more than half within

[1] E. HUBBLE: The Realm of the Nebulae. Oxford: University Press 1936.
[2] C.D. SHANE and C.A. WIRTANEN: Astronom. J. 59, 285 (1954). § XI.

35 km/sec. These numbers would correspond to accuracies of 0.0006 and 0.0001 in δ and therefore we shall assume with Sandage that the red-shifts for clusters are known to the third decimal place in δ. The red-shifts are given both as observed and also as corrected by taking the Local Group of galaxies as the standard of rest[1]. We shall employ throughout these corrected red-shifts because in this way the solar motion is eliminated. It will be necessary to determine the red-shift for a cluster as a whole and here a difficulty presents itself in the data given in HMS. The Virgo cluster and the Coma cluster are two clusters, the members of which have observed apparent magnitudes that are, relatively speaking, bright. They are therefore presumably amongst the nearer of the large clusters and it has proved possible to measure the red-shifts of comparatively large numbers of the individual members, 73 in Virgo and 23 in Coma. As is seen in Col. 4 of Table 12, the range in δ is large, amounting to the equivalent of 2944 km/sec for Virgo and to 4198 km/sec for Coma [*9d*]. The second figure is rivalled by that for the Corona Borealis cluster, for only eight measurements. It will also be noticed that for 7 clusters only two measurements are given and for 3, only one. Now the red-shift for a cluster as a whole is taken to be the average of the individual values; this amounts to a $c\delta$ of 1136 km/sec for Virgo (range: 2944 km/sec) and to 6657 km/sec for Coma (range: 4198 km/sec). It is also a remarkable fact that, when only two or three measurements are made for a cluster, the range of δ is much smaller than it is in the Virgo, Coma or Corona Borealis clusters, and, of course, when there is only one measurement, there is no range at all. The question therefore arises: Is the average δ for a cluster in which only a few measurements have been made really comparable with the average for a cluster with numerous measurements? It is hard to avoid the conclusion that there may be a strong effect of selection in the data for the last seven clusters of Table 12. Clearly,

Table 12. *Data on clusters of galaxies.*

(1)	(2)	(3)	(4)	(5)				(6)	(7)	(8)	(9)
Cluster	δ	No. of δ	Range of δ	m^* for 1st	3rd	5th	10th	$\overline{m}^*$	A	$m_1 = \overline{m}^* - A$	$x = 10^{0.2\,m_1}$
Virgo*	0.004	73	0.010	9.10	9.80	9.90	10.50	9.18	0.26	8.92	60.814
Perseus*	0.018	5	0.004	13.02	14.47	14.49	14.75	13.54	1.20	12.34	293.77
Coma*	0.022	23	0.014	12.90	13.31	13.60	14.52	12.94	0.25	12.69	345.15
Hercules	0.035	7	0.007	—	14.81	—	15.77	14.40	0.37	14.03	639.74
2308 + 0720* (Peg. II)	0.043	3	0.004	14.85	15.47	16.13	—	15.06	0.33	14.73	883.08
2322 + 1425	0.044	2	0.003	15.34	15.87	16.22	16.60	15.36	0.36	15.00	1000.0
1145 + 5559	0.052	4	0.003	15.88	—	16.89	—	15.99	0.29	15.70	1380.4
0106 − 1536	0.053	2	0.002	15.20	15.74	16.70	16.80	15.47	0.26	15.21	1101.6
1024 + 1039	0.065	1	—	16.25	—	—	—	16.25	0.31	15.94	1541.7
1239 + 1852* (Virgo 3)	0.072	2	0.003	15.41	16.14	16.32	16.89	15.55	0.25	15.30	1148.2
1520 + 2754* (Cor. Bor.)	0.072	8	0.014	16.57	16.67	16.96	—	16.31	0.30	16.01	1592.2
0705 + 3506	0.078	2	0.002	17.11	—	—	—	17.11	0.73	16.38	1888.0
1431 + 3146* (Boötes)	0.131	2	0.001	17.93	18.36	18.78	19.26	17.94	0.27	17.67	3419.8
1055 + 5702	0.134	2	0.006	18.22	18.33	18.73	19.25	17.99	0.30	17.69	3451.5
0025 + 2223*	0.159	2	0.001	18.59	18.80	18.88	19.38	18.27	0.39	17.88	3767.1
0138 + 1840*	0.173	1	—	18.40	18.55	18.84	19.14	18.09	0.37	17.72	3499.5
0925 + 2044*	0.192	1	—	18.58	19.15	19.30	19.72	18.54	0.35	18.19	4345.2
0855 + 0321* (Hydra)	0.202	3	0.001	19.26	19.56	19.66	20.16	19.02	0.49	18.53	5081.6

[1] M.L. Humason and H.D. Wahlquist: Astronom. J. **60**, 254 (1955).

there is a crying need for more individual measurements in most, and particularly in the faint, clusters, great as are the difficulties in obtaining the spectra of such faint objects. In any case, as SANDAGE points out, the average red-shift is a combination of (1) the systematic distance effect, (2) that part of the internal velocity dispersion remaining in the average of the red-shifts of the cluster-members, (3) the peculiar motion of the cluster itself.

Turning next to the apparent magnitudes, we have listed in Col. 5, Table 12, the data given in HMS [*9e*] for the observed apparent magnitudes, m^*, of the 1st, 3rd, 5th and 10th brightest galaxies in the clusters. It will be seen that all four galaxies have not, in fact, been measured for m^* in each case and therefore again a lack of comparability between clusters is to be feared. At any rate, SANDAGE computes a "synthetic" brightest galaxy for each cluster by subtracting from the m^* of the 3rd, 5th and 10th, respectively, average differences of 0.48, 0.80 and 1.29, *i.e.* the $\overline{m}^*$ of Col. 6 is obtained by the formula,

$$\overline{m}^* = \tfrac{1}{4}\{m_1^* + (m_3^* - 0.48) + (m_5^* - 0.80) + (m_{10}^* - 1.29)\},$$

with omission of the appropriate terms (and corresponding change of the factor $\tfrac{1}{4}$) when one or more of the entries in Col. 5 is a blank. In allowing for the effect of absorption in our Galaxy, SANDAGE subtracts the differential absorption 0.25 (cosec $b-1$) from each $\overline{m}^*$ and then uses these partially corrected apparent magnitudes until he comes to the final calculation of the Hubble parameter h_1. At this stage, he allows for the extra 0.25 of absorption by adding it to the M that occurs in the constant term in (7.5). There seems to be no advantage in this roundabout procedure; therefore in Col. 7 the absorption $A=0.25$ cosec b has been listed, and in Col. 8, the consequential apparent magnitude m_1. Lastly, the values of x are given in Col. 9. A least squares solution for A_1 and A_2 in (7.7), using the data of Cols. 2 and 9, Table 12, gives

$$A_1 = 21.91\times 10^3, \qquad A_2 = 9.168\times 10^3. \tag{7.13}$$

Applying the familiar technique of testing linear hypotheses, we compute the F-ratio and it turns out that, if the hypothesis $A_2=0$ were to be rejected, the risk of error is doing so would be 0.45. Thus there appears to be little danger in assuming that $A_2=0$ and that the *relation between x and δ is a linear one.* This, as we shall see below, does not imply that the relation between luminosity-distance D and δ is also linear; linearity between observed quantities is indeed a matter of no importance.

In SANDAGE's method the entries of Col. 6, Table 12, being corrected for differential absorption only, the m used in (7.12) are $\overline{m}^*-0.25$ (cosec $b-1$). For each cluster a mean $K=4.7\delta$ is calculated from the entries of Cols. 1 and 2 of Table 11, and thus the numerical value of $m-K$ for each cluster is obtained. SANDAGE's least squares solution for S_1 and S_2 in (7.12), and the probable errors, are

$$S_1 = -5.81 \pm 0.092, \qquad S_2 = -1.180 \pm 0.875. \tag{7.14}$$

However, these are not the values that SANDAGE in fact uses, at least when he comes to calculate h_2. He discusses with great care the question of the determination of the observed apparent magnitudes of the galaxies of the last six clusters of Col. 1, Table 12, concludes that they are systematically too faint, and that progressive corrections of from 0.0 to -0.2 mags. should be applied to them. Unfortunately he does not list these corrected magnitudes, but implies that when they are used in a least squares solution, the value of S_2 turns out to be -2.2, the corresponding S_1 not being stated. Two conclusions may be drawn from this, the first being that decreasing the apparent magnitudes for the last six clusters, leaving the rest of the data unchanged, preserves the negative character of S_2 and makes it smaller. And, secondly, that these comparatively minor changes in the data alter S_2 from -1.18 to -2.2, equivalent to multiplication by a factor of 1.9. Sandage also considers the S_1 and S_2 that arise when *photovisual*, instead of photographic, magnitudes are employed. The results are

$$S_1 = -6.71 \pm 0.094, \qquad S_2 = -1.976 \pm 0.895,$$

with S_2 decreased to -3.0 if the correction mentioned above is made to the observed apparent magnitudes of the last six clusters. He seeks to interpret the difference between this value of S_2 and that of (7.14) as a consequence of the Stebbins-Whitford effect, the existence of which is now highly problematical. A simpler explanation appears to be possible: the $m-K$ for photovisual magnitudes are very different from those for photographic but the δ for each cluster is not changed [9f]. Now Sandage has already shown that S_2 is multiplied by 1.9 if the *photographic* magnitudes of the last six clusters are slightly altered. A fortiori, modifications to *all* the apparent magnitudes, consequent on substituting photovisual for photographic, would be expected to alter the S_2 obtained by a least squares solution, and this indeed appears to have happened. That this is the most likely explanation also follows from the fact that it is not only S_2, but also the first-order constant S_1, which has been altered by the change of magnitude-system.

In addition to the data on clusters of galaxies, Sandage discusses those for galaxies in the general field, the majority of which have relatively smaller red-shifts. It is therefore not now necessary to include the second order term $S_2\delta$ in (7.12); S_1 alone can be evaluated from the data. For 474 field galaxies of all types, Sandage finds [9h]

$$S_1 = -4.235 \pm 0.128. \tag{7.15}$$

γ) The Hubble and acceleration parameters. The last step is to find the numerical values of h_1 and h_2 and we take McVittie's method first. If we work back from (7.9) through (7.3), (7.1) and (1.10), it is clear that the distance-unit involved in c is the parsec, and that the factor 10^{-1} in the right-hand side of (7.9) also represents the reciprocal of a distance in parsec. Thus h_1 is the reciprocal of a time and, taking the time-unit to be the second and the velocity of light to be 2.9978×10^{10} cm/sec, we have $c = 9.711 \times 10^{-9}$ pc/sec. Thus (7.9) and the A_1 of (7.13) yield

$$\operatorname{Log} h_1 = -14 + 0.647 + M/5. \tag{7.16}$$

But at the Mount Wilson and Palomar Observatories it is customary to give h_1, not as the reciprocal of a time, but as so many km/sec/Mpc. Following Sandage, h_1 expressed in these units will be denoted by H where

$$H = (3.087 \times 10^{13}) \times 10^6 \times h_1,$$

and therefore (7.16) becomes

$$\operatorname{Log} H = +6.136 + M/5. \tag{7.17}$$

The reciprocal of h_1 can also be expressed in years (1 year $= 3.156 \times 10^7$ sec) and so by (7.16)

$$\operatorname{Log}\{(h_1^{-1})_{\mathrm{yr}}\} = +5.854 - M/5. \tag{7.18}$$

It should be emphasized that the red-shift data of Table 12 determine, through A_1, the constant terms on the right-hand sides of (7.16) to (7.18). The numerical values of h_1, H or h_1^{-1} are obtained only after an estimate of M is available. Such values will be found in Cols. 1, 2, 3 of Table 13 for three values of M in the range -18.9 to -19.9 in which the absolute magnitude of the synthetic brightest galaxy in a cluster is presumed to lie. The assumption is here made, of course, that the same M is applicable to each cluster. It will be observed that H, for example, ranges in value from 227 to 143 km/sec/Mpc. Before the revision of the absolute magnitudes of the nearer galaxies, consequent on the alteration of the period-luminosity curve for the classical Cepheids, H was assigned the value 540 km/sec/Mpc [1]. To obtain this number the three values of H in Col. 2, Table 13, must be multiplied by 2.4, 3 and 3.75, respectively. That the old value of H had itself been uniquely determined is more than doubtful: there is therefore here an additional reason to that given in Sect. 6η why it is illusory to speak of a single factor by which the old "distance-scale of the universe" can be multiplied to obtain the "new".

Turning next to h_2, we have by (7.11) and (7.6),

$$\frac{h_2}{h_1^2} = -(1 + 0.92 K_1 - 2A_2/A_1), \tag{7.19}$$

and the Eqs. (7.4) indicate that K_1 ranges from 2.7 to 5.02, according to the type that is assigned to the synthetic brightest galaxy in a cluster. The ratio A_2/A_1 is, as has been indicated, probably equal to zero; but if the A_2 of (7.13) is accepted, it is equal to $+0.4$. In either case, h_2/h_1^2 is *negative* and this negativeness is revealed to be a consequence, in the main, of the fact that K_1 is a positive number whose value is at least 2.7. The sign of h_2 means that, if the data of Table 12 and the values of K_1 in (7.4), or any larger ones, are accepted, the expansion of the system of galaxies out to the distance of the Hydra cluster is being retarded at the present time [*11e*]. The F-ratio test was used, with $K_1 = 4.7$, to test the hypothesis that $h_2 = 0$, and it turned out that, if this hypothesis were rejected, the risk of error in doing so was 0.013. Thus it may be concluded that the right-hand side of (7.19) does give a significant negative value of h_2. It does not, of course follow that, for more distant galaxies with red-shifts greater than 0.202, this conclusion would necessarily still hold good. Nor need it continue to be true if other kinds of information are introduced into the computations. For instance, suppose that it is assumed that, as fainter clusters are measured for red-shift, the observer progressively selects clusters that are richer and richer in intrinsically brighter and brighter galaxies. This can be allowed for by replacing M in (7.1) by $M_0 + W_1\delta$, where M_0 is the same for all clusters and W_1 is a constant to be found. The only effect on h_2 of this alteration [*11f*] would be to change (7.19) to

$$\frac{h_2}{h_1^2} = -[1 + 0.92(K_1 + W_1) - 2A_2/A_1],$$

K_1 still being given by (7.4) and A_2, A_1 by (7.13). Assuming that $A_2 = 0$ the condition that $h_2 \geq 0$ is $W_1 \leq -(1.09 + K_1)$ which, with $K_1 = 4.7$, yields $M = M_0 - 5.8\delta$. This shows that, if M_0 refers to the brightest galaxies of the Virgo cluster ($\delta = 0$), then the Hydra cluster ($\delta = 0.2$) has been selected for red-shift measurements because it contains super-giant galaxies of absolute magnitudes -1.2 brighter than those of Virgo. Whether this is plausible or not is another question;

[1] W. Baade and R. Minkowski: Astrophys. Journn. **119**, 206 (1954).

but it has been argued by Elizabeth Scott[1] that selection effects of this kind can vitiate the conclusions derived from the data of Table 12.

In Sandage's method, the algebraic computation of h_1 from S_1 is a little more complicated. As has already been pointed out, the M in (7.5) must be replaced by $M+0.25$ to allow completely for the effect of absorption. In the equation referred to, the distance-unit is again the parsec and so, taking the time-unit to be the second, $c\delta$ should be in parsec/sec. But km/sec are in fact employed which means that the $c\delta$ are multiplied by $3\,087\times 10^{13}$. This can be compensated for by multiplying h_1 by the same factor in (7.5). Thus we have

$$M+0.25-5-5\,\mathrm{Log}\,(3.087\times 10^{13} h_1)=S_1,$$

which, with $S_1=-5.81$, leads to

$$\mathrm{Log}\,h_1=-14+0.722+M/5, \tag{7.20}$$

and hence

$$\mathrm{Log}\,H=+6.212+M/5, \tag{7.21}$$

$$\mathrm{Log}\,\{(h_1^{-1})_{\mathrm{yr}}\}=+5.779-M/5. \tag{7.22}$$

Comparing with (7.16), (7.17) and (7.18), it will be seen that the differences between the two sets of formulae are not large, though McVittie's values of H

Table 13. *Values of the Hubble parameter.*

M	h_1 (sec^{-1}) (1)	H (km/sec/Mpc) (2)	h_1^{-1} (yr.) (3)	h_1 (4)	H (5)	h_1^{-1} (6)
-18.9	7.36×10^{-18}	227	4.30×10^{9}	8.75×10^{-18}	270	3.62×10^{9}
-19.4	5.85×10^{-18}	180	5.42×10^{9}	6.95×10^{-18}	215	4.56×10^{9}
-19.9	4.64×10^{-18}	143	6.82×10^{9}	5.52×10^{-18}	171	5.74×10^{9}

turn out to be 84% of Sandage's (Table 13). The numerical values derived from (7.20) to (7.22) are shown in Cols. 4 to 6 of Table 13. Sandage himself uses a graphical method, together with the unique choice $M=-19.82$, to obtain $H=180$ km/sec/Mpc. An essentially identical result, $H=177$ km/sec/Mpc, follows from (7.21) with $M=-19.82$. But if, in the above calculations, the value -4.24, appropriate to the field galaxies, had been used for S_1 [see Eq. (7.15)], H would have been reduced to lie between 131 and 83 km/sec/Mpc. To recover the H of Col. 5, Table 13, one must take the average absolute magnitude for the field galaxies to lie between -17.33 and -18.33, which is reasonable since the field galaxies are of all kinds and are not chosen to be those of the greatest intrinsic luminosity. The uncertainty in H, which Sandage expresses by writing $H=180f$ km/sec/Mpc, where f is a "correction factor at present unknown", is evidently due, as Table 13 indicates, in the main to the uncertainty in the absolute magnitude that is to be selected as that of the synthetic brightest galaxy in a cluster. An alternative way of finding H employed by Sandage [*9g*] is to use the first approximation to (1.10), viz.

$$H=(c\,\delta)/D, \tag{7.23}$$

where $c\delta$ is in km/sec and D in Mpc. Taking $c\delta=1136$ km/sec for the Virgo cluster and calculating its distance from a true distance-modulus of 29.05 for NGC 4321, there comes $H=176$ km/sec/Mpc. This determination however

[1] Elizabeth L. Scott: Astronom. J. **61**, 190 (1956).

depends on the amount of faith that is attached to the number 29.05. If instead we take $28.5 \leq \mathcal{M} \leq 29.5$ for NGC 4321 (see Sect. 6ζ), the consequential values of H range from 227 to 143 km/sec/Mpc and agree with Col. 2, rather than with Col. 5, of Table 13. The selection of some "best value" for H within these limits or those indicated in Cols. 2 and 5 of Table 13, must therefore await a more certain determination of M—or its equivalent—than is at present available.

SANDAGE's value for h_2 with $S_2 = -2.2$ is

$$\frac{h_2}{h_1^2} = -(3.0 \pm 0.8), \tag{7.24}$$

a result that also follows from (7.19) if either $A_2/A_1 = 0$ and $K_1 = 2.17$ or $A_2/A_1 = +0.4$ and $K_1 = 3.04$. These values of K_1 are rather low and therefore it seems probable that (7.24) gives an upper limit for h_2/h_1^2. But the important point is that McVITTIE's and SANDAGE's methods give values of h_1 that do not differ too widely and also show that h_2 is negative.

Finally, a brief reference to the estimates of h_1 and h_2 made by McVITTIE in "*General Relativity and Cosmology*" [*11g*] may be made. Twelve clusters were employed, of which eleven are indicated by asterisks in Col. 1, Table 12, the twelfth being Pegasus 1 (Group G 7619 of Table 11, HMS) that is not listed in Table 12. The observed apparent magnitudes were provisional ones and the least squares solutions were made using the, comparatively faint, apparent magnitudes of the third brightest galaxy in each cluster corrected for differential absorption only. This had the effect of decreasing the values of h_1 as compared with those in Table 13, but nevertheless h_2 was found to be a negative number.

δ) Distances computed from the red-shift. Assuming that the Hubble and acceleration parameters are known, the luminosity-distance of a galaxy can be computed from its red-shift from Eq. (1.10), which, if D is in Mpc and c and $c\delta$ in km/sec can also be written

$$D = \frac{c\delta}{H}\,[1 + (1 + h_2/h_1^2)\,(c\delta)/(2c)]. \tag{7.25}$$

The method, of course, implies that the formula established from the cluster data is applicable to all galaxies, whether they are members of clusters or not. From (7.19) we have

$$(1 + h_2/h_1^2)/2 = -(0.46\,K_1 - A_2/A_1), \tag{7.26}$$

and therefore (1.10) becomes

$$D = \frac{c\delta}{h_1}\,[1 - (0.46\,K_1 - A_2/A_1)\,\delta]. \tag{7.27}$$

In the right-hand side of (7.26), let $K_1 = 4.7$, because from (7.4) this is the mean of the values for E and Sb galaxies and also because $K = 4.7\,\delta$ is the K-correction adopted by SANDAGE [*9d*]. Let us also take $A_2 = 0$. Then (7.27) becomes

$$D = \frac{c\delta}{h_1}\,(1 - 2.16\,\delta), \tag{7.28}$$

where D is in parsec if we take $c = 9.711 \times 10^{-9}$ pc/sec and h_1 is in (seconds)$^{-1}$. For each object that will be considered, a D_{min} and a D_{max} will be computed corresponding to the extreme values of h_1 of Col. 1, Table 13. This means that

$$\mathrm{Log}\, D_{min} = +9.120 + \mathrm{Log}(\delta\,(1 - 2.16\,\delta)), \quad (M = -18.9), \tag{7.29}$$

$$\mathrm{Log}\, D_{max} = +9.321 + \mathrm{Log}(\delta\,(1 - 2.16\,\delta)), \quad (M = -19.9). \tag{7.30}$$

The luminosity-distances of seven of the clusters in Col. 1, Table 12, are listed in Table 14. The first-order approximation to (1.10) is

$$D = (c\,\delta)/h_1, \tag{7.31}$$

a formula differing from (7.23) only in the units employed. The effect of the second-order term, 2.16δ, is already sensible for the Hercules cluster and is considerable by the time the distance of the Hydra cluster is reached. For this cluster, the $D_{\min}$ and $D_{\max}$ calculated from (7.31) are 266 and 423 Mpc, respectively, It is also important to notice that, for the distant clusters, it is essential to specify which kind of distance is meant. To take the Hydra cluster as an example: from the luminosity-distances of Table 14 and Eq. (1.8), the distance by future apparent size ranges from $\bar{r}_{\min} = 125$ Mpc to $\bar{r}_{\max} = 199$ Mpc; and if distance

Table 14. *Luminosity-distances of clusters and radio-sources.*

	δ	$D_{\min}$ (Mpc)	$D_{\max}$ (Mpc)
Cluster			
Virgo	0.004	5.27	8.00
Perseus	0.018	22.4	35.6
Hercules	0.035	42.2	70.0
0106 − 1536	0.053	62.9	98.4
1431 + 3146 (Boötes)	0.131	123	197
0138 + 2044	0.173	142	226
0855 + 0321 (Hydra)	0.202	150	239
Radio-source			
14 N 5 A (NGC 5457)	0.0013	1.71	2.72
12 N 4 A (NGC 4258)	0.0016	2.11	3.35
13 N 4 A (NGC 5194/95)	0.002	2.64	4.19
19 N 4 A (Cygnus A)	0.056	64.6	103

by apparent size is employed, the range is $\xi_{\min} = 104$ Mpc to $\xi_{\max} = 166$ Mpc. It is also evident that formula (7.28) cannot be extrapolated much beyond red-shifts equal to 0.202, if indeed it can be extrapolated at all. For when $\delta = 0.46$, the value of D is apparently zero, which means, of course, that the coefficient of the second-order term is no longer reliably taken as 2.16 and that higher order terms have also come into play.

It is now well-known that certain sources of radio-waves are galaxies of certain special kinds[1] whose distances can be computed from the red-shift observed in their optical spectra. Two such radio-sources are 03 N 4 A (NGC 1275) and 12 N 1 A (NGC 4486), that are member-galaxies of the Perseus and Virgo clusters, respectively. Their distances may therefore be taken as those of the clusters in which they lie. The case of the pair of colliding galaxies 19 N 4 A (Cygnus A) is an interesting one. Baade and Minkowski[2] in 1954 gave 16830 km/sec for its red-shift. With $H = 540$ km/sec/Mpc. they quoted 33 Mpc for the distance, but there is apparently an arithmetical error here, because, by (7.23), $D = 16830/540 = 31$ Mpc. Shakeshaft[3] allows for the "revision of the distance-scale of the universe" by doubling the Baade and Minkowski distance, and thus obtains 66 Mpc, instead of 62 Mpc. More recently, Minkowski and Wilson[4] have revised the red-shift to 16812 ± 9 km/sec, equivalent to $\delta = 0.056$, and

[1] J.L. Pawsey: Astrophys. Journ. **121**, 1 (1955).
[2] W. Baade and R. Minkowski: Astrophys. Journ. **119**, 206 (1954).
[3] J.R. Shakeshaft: Phil. Mag. **45**, 1136 (1954).
[4] R. Minkowski and O. C. Wilson: Astrophys. Journ. **123**, 373 (1956).

the resulting range of luminosity-distance is shown in Table 14, taking (7.29) and (7.30) as the formulae for D. In this Table, the luminosity-distances of three other radio-sources, whose identifications are however still somewhat uncertain, are also listed, using the same formulae.

ε) Glossary of notations.

	HMS	McVittie
Red-shift	z	δ
Luminosity-distance	r	D
Ratio of parameters	$R_0 \ddot{R}_0 / \dot{R}_0^2$	h_2/h_1^2
Obs. app. photogr. mag.	P	m^*
Diff. absorption	$\Delta P(b)$	$A - 0.25$
K-correction	$K(z, t_0)$, K_p	K
Corrected app. mags.	$\overline{P}_C$, m_{bol}, m_C	$m_1 - K + 0.25$
Least squares constants	B and D	S_1
	B	S_2
Extinction function	$S(\lambda)$	$\sigma(\lambda)$
Spectral energy-distribution	$I_0(\lambda)$	$B(\lambda)$
functions	$I_z(\lambda)$	$B\left(\frac{\lambda}{1+\delta}\right)$

II. Time and the age of the universe.

8. An operational method for measuring time implies the existence of some phenomenon which, in the experience of a human observer, possesses the characteristic of periodicity or of regular repetitiveness. The diurnal rotation of the heavens, the oscillations of a pendulum are examples of such periodic phenomena. The measurement of the lapse of time between two events consists essentially of counting the number of repetitions of the phenomenon that have taken place between the two occurrences. It is true that there are derivative methods of time-measurement depending on atomic theory, a specific instance being the quartz crystal whose internal vibrations, though directly inaccessible to observation, can be translated into a periodic phenomenon experienced as such by the scientist. The human imagination is however capable of conceiving of "clocks" which, from the operational point of view, are not time-measuring devices at all. The so-called "atomic clock" is a conceptual device of this kind. By means of a theory, each line in the spectrum of a given atom can be associated with a time-interval. But a spectral line, viewed in a spectroscope, is a purely statical phenomenon without periodicity; as such it provides no means of measuring time. The atom therefore by itself is not a clock that is accessible to astronomers.

In the preceding sections, the second and its multiple the year have been employed as units of measured time because operational methods have been developed by astronomers for counting the number of such units between the occurrence of two events. The identification of this time with the t-coordinate of (1.3) is purely a matter of definition. If, by using this definition, a contradiction with observation ever results—a contingency which has not yet arisen—a reappraisal of the definition would have to be made. This point of view however differs from that of some cosmologists who hold that an observer can define the t-coordinate in (1.3) by studying the motions of the galaxies in his immediate neighbourhood, or their spatial configurations at different values of t[1].

[1] H. Bondi: Cosmology, pp. 70, 101. Cambridge: University Press 1952. — H.P. Robertson: Publ. Astronom. Soc. Pacific 67, 82 (1955).

This is a picturesque description of a mathematical property of t, namely, that at successive instants $t = constant$, the 3-dimensional spatial manifold of points in (1.3) remains similar to itself but on a larger (or smaller) scale. But the definition has no operational content for a human astronomer, since the galaxies are not observed to be "in motion". Their relative angular separations do not vary (no proper motion) and the displacement of their spectral lines is a purely statical phenomenon providing no periodic repetition of an event whose successive occurrences can be counted. Indeed, the interpretation of the red-shift as evidence of motion (a radial velocity) is itself a matter of definition and is not an inherent property of the observed phenomenon.

The study of the system of galaxies lying in a spherical volume centered at our Galaxy and whose outer limit is at the distance of the Hydra cluster, has thrown up a parameter (h_1^{-1}) having the dimensions of a time. This reciprocal of the Hubble parameter is evidently a time-variable since by (1.9) and (7.26)

$$\frac{d}{dt}(h_1^{-1}) = \left\{\left(\frac{d}{dt}\frac{R}{R'}\right)\right\}_{t=t_0} = \left\{1 - \frac{R R''}{R'^2}\right\}_{t=t_0} = 1 - \frac{h_2}{h_1^2} = 2\left(1 + 0.46\, K_1 - \frac{A_2}{A_1}\right). \quad (8.1)$$

It is therefore increasing at the present time at a rate whose exact amount depends on the adopted values of K_1 and of the ratio A_2/A_1. Many astronomers have in the past jumped to the conclusion that h_1^{-1} is the "age of the universe", i.e. the time since the expansion of the system of galaxies commenced. We proceed to analyze this notion using general relativity as the interpretative theory. One definition of the start of the expansion is to make it coincide with the instant $t = t_a$, where t_a is the real solution of

$$R(t) = 0 \quad (8.2)$$

that is first attained on working backwards from $t = t_0$. The age of the universe is then

$$T = t_0 - t_a, \quad (8.3)$$

t_0, of course, being the present moment. This definition of the start of the expansion implies that initially all the galaxies were in an infinitesimally small volume [*11i*]; it is an entirely arbitrary one since neither theory nor observation forces us to make it[1]. But since it is a popular definition with some astronomers, it will be adopted here. Now to find t_a through (8.2) it is necessary to know the functional form of $R(t)$, i.e. to know this function for all values of t. But numerical values of h_1 and of h_2 merely determine, by (1.9), the values of R'/R and of R''/R at the particular instant $t = t_0$. Thus they neither give the functional form of R nor do they prove that R''/R is a constant and therefore independent of t_0. The use of Einstein's gravitational equations for the determination of $R(t)$ [*11j*] does not solve the problem because extrapolatory hypotheses regarding the pressure and density of matter in the universe for the unobservably remote past have then to be made[2]. One or two examples will show how specific numerical values of h_1 and h_2 can agree with "ages" that are as long or as short as we please.

Let h_1^{-1} now stand for any number in the range (Col. 3, Tabelle 13)

$$4.3 \times 10^9 \leq h_1^{-1} \leq 6.8 \times 10^9 \text{ (years)}, \quad (8.4)$$

and let h_2, in (years)$^{-2}$, be the consequentially deduced number

$$h_2 = -(1 + 0.92\, K_1 - 2 A_2/A_1)\, h_1^2 = -q_0 h_1^2, \quad (8.5)$$

[1] G.C. McVittie: Astronom. J. **58**, 129 (1953). § 5.

[2] For an example of this method see F. Hoyle and A.R. Sandage: Publ. Astronom. Soc. Pacific **68**, 301 (1956).

where, if $2.7 \leqq K_1 \leqq 5$ and $0 \leqq A_2/A_1 \leqq 0.4$, the pure number q_0 satisfies

$$2.7 \leqq q_0 \leqq 5.6. \tag{8.6}$$

Firstly, then, let us make the arbitrary assumption that

$$R(t) = q\,t, \tag{8.7}$$

where q is a constant with the dimensions of velocity. Hence by (1.9), (8.2) and (8.3)

$$T = t_0 - 0 = h_1^{-1}, \quad h_2 = 0.$$

The first equation shows that h_1^{-1} is now the age of the universe but the second contradicts (8.5) and (8.6). It is also clear that (8.7) is the only possible form for $R(t)$ which will make h_1^{-1} itself equal to T. Therefore this interpretation of the reciprocal of the Hubble parameter must be rejected if the data of Table 12 are accepted. Secondly, let us choose a form for $R(t)$ that will give us a small value of T; this will happen if we take

$$R(t) = \frac{q}{h_1}\left\{\frac{2}{q_0}\,h_1 t - (h_1 t - 1)^2\right\}, \tag{8.8}$$

which makes $t_0 = h_1^{-1}$ and satisfies (1.9) and (8.5), in contrast to (8.7). Hence, by (8.2) and (8.3)

$$T = t_0 - t_a = \left\{-\frac{1}{q_0} + \left(\frac{1}{q_0}\left(2 + \frac{1}{q_0}\right)\right)^{\frac{1}{2}}\right\} h_1^{-1},$$

and so, using also (8.4) and (8.6), we find

$$2.5 \times 10^9 \leqq T \leqq 3.9 \times 10^9 \quad \text{(years)}. \tag{8.9}$$

Thirdly, let us go to other extreme and deduce a very large value of T which we can do by assuming that

$$R(t) = \frac{q}{h_1}\exp^1\{h_1 t - \tfrac{1}{2}(1 + q_0)(h_1 t - 1)^2\}, \tag{8.10}$$

which again makes $t_0 = h_1^{-1}$ and satisfies (1.9) and (8.5). But now obviously $t_a = -\infty$ and therefore the age of the universe, T, becomes infinite. Incidentally, in the de Sitter universe of general relativity and also in the creation of matter theory of BONDI and GOLD[1], it is assumed that $R(t) = R_0\,e^{h_1(t-t_0)}$, where R_0 is a constant with the dimensions of length. This hypothesis makes the age of the universe also infinitely long, but it implies that $h_2 = +h_1^2$, which, as in the case (8.7), contradicts (8.5) and (8.6). Thus the de Sitter Universe and the Bondi and Gold theory are not in accord with the data of Table 12. These examples, which could be multiplied according to the mathematical ingenuity of the reader, reveal in detail that numerical values for R'/R and R''/R for some particular instant t, do not establish uniquely the age of the universe, defined as in (8.2).

There are other ways of extrapolating to the initial instant of the astronomical universe: the age of the Earth, for instance, is given as 3.4 to 5×10^9 years[2], the ages of the globular clusters NGC 5272 and 6205 are estimated at 5×10^9 [3] and 2×10^9 years[4], respectively, and the ages of meteorites at 4.5 to 5×10^9 years[5].

[1] H. BONDI: Cosmology. Cambridge: University Press 1952.
[2] This Encyclopedia, Vol. XLVII, p. 335. Berlin-Göttingen-Heidelberg: Springer 1956.
[3] A. R. SANDAGE: Mem. Soc. Roy. Sci. Liege **14**, 254 (1953).
[4] W. A. BAUM: Astronom. J. **59**, 422 (1954).
[5] H. C. UREY: Nature, Lond. **175**, 321 (1955).

These time-intervals are of value to the cosmologist because they rule out certain functions $R(t)$; indeed the only admissible functions are those giving ranges of T which include the ages estimated by such alternative methods.

Acknowledgments.

The author would like to express his gratitude to Drs. A. Blaauw, F. K. Edmondson, S. C. B. Gascoigne, H.L. Johnson, G.E. Kron, N.U. Mayall, C.D. Shane, H.F. Weaver and S.P. Wyatt jr., and to Miss Mary Hanania, for their help, advice or criticism concerning various points that arose during the preparation of this article.

General references.

[1] Allen, C.W.: Astrophysical Quantities. London: Athlone Press 1955. — [1a] p. 236. — [1b] p. 190, column headed $\overline{M}_{pg}$. — [1c] p. 241.

[2] Arp, H.C.: Astronom. J. **60**, 1 (1955).

[3] Baade, W.: Astrophys. Journ. **100**, 147 (1944), Table 1.

[4] Bigay, J.H.: Ann. Astrophys. **14**, 319 (1951).

[5] Buscombe, W., S.C.B. Gascoigne and G. de Vaucouleurs: Austral. J. Sci., Suppl. **17**, 3 (1954). — [5a] Tab. X. — [5b] Sect. C 5.

[6] de Vaucouleurs, G.: Publ. Astronom. Soc. Pacific **67**, 350 (1955).

[7] Holmberg, E.: Medd. Lund. Astron. Obs., Ser. II, No. 128 (1950).

[8] Hubble, E.: Astrophys. Journ. **69**, 103 (1929). — [8b] Table II.

[9] Humason, M.L., N.U. Mayall and A.R. Sandage: Astronom. J. **61**, 97 (1956). Referred to in text as HMS. — [9a] p. 136. — [9b] p. 160. — [9c] At p. 161, $\mathcal{M}^*$ is given as (24.15). — [9d] Tables II and XIII. — [9e] Table XII under $\boldsymbol{P}$. Table A 1 for 1st and 10th of Virgo. — [9f] Table XIII: compare $\overline{\boldsymbol{P}}_C$ and $\overline{\boldsymbol{V}}_C$. — [9g] p. 161. — [9h] Table IX.

[10] Lohmann, W.: Z. Astrophys. **30**, 234 (1952).

[11] McVittie, G.C.: General Relativity and Cosmology. London: Chapman & Hall 1956. — [11a] § 8.2. — [11b] § 8.5 and p. 192, Note 6. — [11c] § 9.1 and § 9.2. — [11d] § 8.6. — [11e] Eq. (9.215). — [11f] Eqs. (9.208) and (9.214). — [11g] § 9.2. — [11h] § 8.4. — [11i] § 8.3. — [11j] § 9.7, where the consequence of using Einstein's equations are also discussed.

[12] Mineur, H.: Ann. Astrophys. **7**, 160 (1944).

[13] Mineur, H.: C. R. Acad. Sci. Paris **235**, 1607 (1952).

[14] Pareńago, P.P.: Var. Stars **10**, 193 (1954).

[15a] Sawyer, Helen B.: Publ. Dav. Dunlap Obs. **2**, No. 2 (1955). — [15b] Private communication.

[16] Shapley, H.: Star Clusters. New York: McGraw-Hill 1930. — [16a] p. 11. — [16b] pp. 159—161. — [16c] p. 135. — [16d] p. 129, Fig. X, 3.

[17] Wilson, R.E.: Astrophys. Journ. **89**, 218 (1939).

Newtonsche und Einsteinsche Kosmologie.

Von

O. Heckmann und E. Schücking.

Mit 1 Figur.

Bezeichnungen. Griechische Indices laufen von 0 bis 3, lateinische von 1 bis 3. Über doppelt auftretende Indices ist stets zu summieren. Partielle Ableitungen nach t sind durch einen an das Symbol unten rechts angefügten senkrechten Strich mit dem nachfolgenden Index 0 abgekürzt. Es ist also $v_{i|0} = \frac{\partial v_i}{\partial t}$. Entsprechend ist $v_{i|k} = \frac{\partial v_i}{\partial x_k}$. Kovariante Ableitungen sind dadurch kenntlich gemacht, daß ihr Index durch einen voraufgehenden senkrechten Doppelstrich bezeichnet wurde. Gewöhnliche Ableitungen nach der Zeit sind durch einen Punkt bezeichnet. Das Kronecker-Symbol δ_{ik} ist gleich 1 für $i = k$ und sonst Null.

I. Einleitung.

1. Allgemeines und Historisches zum kosmologischen Problem. Die weitverbreitete Tendenz, der Kosmologie radikale Skepsis entgegenzubringen und sie anzusehen als fruchtlose Spekulation, die keine Konsequenzen für die normale Physik und Astronomie habe, übersieht ein Hauptproblem der exakten Naturwissenschaften, das verknüpft ist mit der Weltstruktur im Großen, also mit der Kosmologie: Jede Voraussage eines Ereignisses ist — mindestens im Rahmen der heutigen Feldphysik — darauf angewiesen, daß man den Anfangszustand des Feldes auf irgendeiner raumartigen Hyperfläche so genau wie möglich kennt. Voraussagen sind also nur dann möglich, wenn — irgendwie zu begründende — Annahmen über den Zustand des Feldes auf jener Hyperfläche *auch in großer Entfernung von unserem Standort* eingeführt werden. Denn es ist ja prinzipiell denkbar, daß Ereignisse, die vor sehr langer Zeit jenseits des Horizontes unserer heutigen Teleskope stattfanden, schon im nächsten Augenblick eine merkliche Wirkung ins Sonnensystem einstrahlen werden. Natürlich kann man die Möglichkeit solcher Wirkungen leugnen, indem man die Hypothese einführt, daß die fraglichen Felder auf der anfänglichen Hyperfläche in großer Entfernung verschwanden. Aber diese Annahme ist eine Hypothese über die möglichen *Anfangs-Randbedingungen*, also eine spezielle kosmologische Hypothese, die wahr oder falsch sein kann. Kosmologische Hypothesen über das Verhalten der Welt im Großen sind demnach enthalten in *jeder* lokalen Voraussage auf Grund der bekannten Naturgesetze, solange diese die Form der heute üblichen Differentialgleichungen haben.

Mit diesem Typus von Vorkommnissen ist die Beziehung der Kosmologie zur „lokalen" Physik aber nicht erschöpft. Räumliche Endlichkeit des Universums bedingt zum Beispiel, daß ein Teilchen nur noch diskrete Impulswerte besitzen kann[1]. Überdies bestimmen die Invarianzeigenschaften der Welt im Großen weitgehend die Grundgesetze der Feldtheorien von Elementarprozessen. Sowohl Erhaltungssätze wie die Vertauschungsrelationen hängen ab von den Bewegungsgruppen, die in der Welt zulässig sind. E. Wigner und seine Mitarbeiter haben

[1] Vgl. den Artikel von L. Infeld in dem Band Albert Einstein, Philosopher and Scientist. New York 1949.

in diesem Zusammenhang die Frage untersucht, welche Konsequenzen sich aus der Hypothese ergeben, daß die Welt die Metrik des de Sitter-Kosmos besäße[1].

Die moderne Kosmologie begann im Jahre 1917 als A. EINSTEIN[2] zeigte, daß gemäß seinen um das Λ-Glied erweiterten neuen Feldgleichungen der Gravitation eine homogene, statische, druckfreie Materieverteilung in einem Raume konstanter, positiver Krümmung möglich sei.

Etwas später, aber noch im gleichen Jahre 1917, fand W. DE SITTER[3], daß die von A. EINSTEIN aufgestellten Grundgleichungen noch eine ganz andere homogene Lösung zuließen, die einer materiefreien Welt entsprach. Wie wir heute wissen, war seine Darstellung der Lösung nicht sehr zweckmäßig und waren seine Folgerungen unvollständig. Doch veranlaßte er durch seine theoretische Entdeckung die Vermehrung der Radialgeschwindigkeitsmessungen außergalaktischer Nebel, da er glaubte, sein Modell an den Bewegungen möglichst ferner Objekte prüfen zu sollen.

Bevor aber noch die ersten Anzeichen einer Korrelation von Entfernung und Rotverschiebung in den Spektren solcher Nebel gefunden wurden[4], hatte A. FRIEDMANN schon 1922 und 1924[5] die ganze Reihe der expandierenden Weltmodelle von positiver und negativer Raumkrümmung entdeckt, in welchen die von A. EINSTEIN und W. DE SITTER gefundenen als Grenzfälle enthalten waren.

Nachdem E. HUBBLE 1929[6] die empirische Beziehung zwischen der Fluchtgeschwindigkeit und der Entfernung der Nebel sichergestellt hatte, verbreitete sich die Vorstellung von der Expansion der Welt sehr schnell. Zur Geschichte der theoretischen Entwicklungen bis 1933 vergleiche man den Bericht von H. ROBERTSON: Relativistic Cosmology. Rev. Mod. Phys. **5**, 62—90 (1933). Von 1932 bis 1936 erfolgte der Ausbau der theoretischen Vorstellungen auf der einen, die Vermehrung des empirischen Materials an immer ferneren Nebeln auf der anderen Seite. Das Buch von E. HUBBLE „The Realm of the Nebulae" und verschiedene zusammenfassende theoretische Werke gaben in den folgenden Jahren einen Überblick über die Einsteinsche Kosmologie[7].

Nach dem letzten Kriege setzte eine neue Entwicklung ein: die bisherigen empirischen Daten wurden als unzureichend empfunden. Mit Hilfe neuer Instrumente wurden sie teils auf sichere Fundamente gestellt, teils erweitert zu noch schwächeren und ferneren Nebeln hin. Auf der Seite der reinen Theorie dürfte das wichtigste Ergebnis im Rahmen der Einsteinschen Gravitationstheorie die Konstruktion eines Weltmodells durch K. GÖDEL sein[8]. Es ist homogen mit Materie erfüllt, die aber an jeder Stelle „absolut" rotiert, d.h. in bezug auf die Koordinaten eines lokal-geodätischen Koordinatensystems. Ferner mußte sich die relativistische Theorie mehr und mehr auseinandersetzen mit inzwischen

---

[1] E. WIGNER: Helv. phys. Acta, Suppl. **4**, 210 (1956).

[2] A. EINSTEIN: Sitzgsber. preuß. Akad. Wiss. **1917**, 142. Abgedruckt in: Das Relativitätsprinzip, herausgeg. von O. BLUMENTHAL, mehrere Auflagen z.B. 3. Aufl., Leipzig u. Berlin 1920. In englischer Übersetzung bei Dover Publications, Inc.

[3] W. DE SITTER: Proc. Akad. Wet. Amsterd. **19**, 1217 (1917). — Monthly Notices Roy. Astronom. Soc. London **78**, 3 (1917).

[4] C. WIRTZ: Astronom. Nachr. **222**, 21 (1924).

[5] A. FRIEDMANN: Z. Physik **10**, 377 (1922); **21**, 326 (1924).

[6] E. HUBBLE: Proc. Nat. Acad. Amer. **15**, 168 (1929).

[7] E. HUBBLE: The Realm of the Nebulae. Oxford 1936. Übersetzt ins Deutsche von K. KIEPENHEUER, Das Reich der Nebel. Braunschweig 1938. — E. HUBBLE: The observational Approach to Cosmology. Oxford 1937. — R. TOLMAN: Relativity, Thermodynamics, and Cosmology. Oxford 1934. — G. MCVITTIE: Cosmological Theory. London 1937. — O. HECKMANN: Theorien der Kosmologie. Berlin 1942.

[8] K. GÖDEL: Rev. Mod. Phys. **21**, 447 (1949).

herangewachsenen rivalisierenden Deutungen der Beobachtungen, die teilweise von ganz neuen Voraussetzungen ausgehen[1].

Es verdient hervorgehoben zu werden, daß eine schon 1934 von E. MILNE und W. McCREA[2] geschaffene Kosmologie auf der Grundlage der Newtonschen Mechanik sich seitdem als folgenreich erwiesen hat: Sie gab einmal Anlaß zu einer Neuinterpretation der Newtonschen Mechanik. Sodann lieferte sie mit der Einsteinschen Kosmologie so viele übereinstimmende Resultate, daß man zur Behandlung kosmisch-*lokaler* Probleme, etwa der Dynamik eines Spiralnebels oder eines Nebelhaufens mit der mathematisch einfacheren Methode der Newtonschen Theorie vielfach auskommen wird.

Eine ziemlich vollständige Bibliographie zur Kosmologie findet man für die Zeit bis 1932 bei H. P. ROBERTSON: Rev. Mod. Phys. **5**, 62—90 (1933). Für Literatur bis 1940 vgl. H. ROBERTSON, Science in Progress, 2. Ser. 1940. Siehe auch O. HECKMANN: Theorien der Komologie. Berlin 1942. Für eine Gesamtübersicht des Problemkreises und seiner Literatur vergleiche man G. C. McVITTIE: General Relativity and Cosmology. London 1956.

II. Newtonsche Kosmologie.

2. Grundlagen der Newtonschen Gravitationstheorie. *α) Lokale Inertialsysteme.* Die Anwendbarkeit der Newtonschen Theorie auf unendliche Materieverteilungen konstanter Dichte ist häufig bezweifelt worden[3], weil man für die Berechnung des Potentials die Formel

$$\Phi = \iiint \frac{\varrho}{r}\, dx_1\, dx_2\, dx_3$$

festhielt. Als dann 1934 E. MILNE und W. McCREA entdeckten[4], daß die bis dahin nur in der Einsteinschen Kosmologie abgeleitete Friedmannsche Differentialgleichung, die den zeitlichen Ablauf der Expansion eines homogen-isotropen Weltmodells mit inkohärenter Materie regelt, sehr einfach aus der Newtonschen Theorie abgeleitet werden kann, wurde diesem Ergebnis wenig Beachtung geschenkt. Die 1954/55 von D. LAYZER und W. McCREA[5] geäußerte Kritik der Newtonschen Theorie in ihrer Anwendung auf unendliche Systeme schien dieser Einstellung recht zu geben. Es läßt sich indessen zeigen, daß eine widerspruchsfreie Newtonsche Theorie auch für unendliche Systeme begründet werden kann, deren Dichte im Unendlichen nicht verschwindet. Sie ist eine gute Näherung der Einsteinschen Theorie und eignet sich zur Behandlung lokaler Probleme, wenn keine Ausartungen (extreme Dichten oder Geschwindigkeiten) vorliegen.

Man erhält eine einwandfreie und auch für kosmologische Anwendungen brauchbare Fassung der Newtonschen Theorie, wenn man in ihr eine neue Definition des Inertialsystems einführt[6]. Die herkömmliche Definition leidet bekanntlich

[1] E. MILNE: Relativity, Gravitation, and World-Structure. Oxford 1935. — H. BONDI: Cosmology. Cambridge 1952. — P. JORDAN: Schwerkraft und Weltall, 2. Aufl. Braunschweig 1955.

[2] Vgl. die in Anm. 4 zitierten Arbeiten.

[3] C. NEUMANN: Über das Newtonsche Prinzip der Fernwirkung, S. 1 u. 2. Leipzig 1896. — H. SEELIGER: Astronom. Nachr. **137**, 129 (1895).

[4] E. MILNE: A Newtonian expanding universe. Quart. J. Math., Oxford Ser. **5**, 64—72 (1934). — W. McCREA u. E. MILNE: Quart. J. Math., Oxford Ser. **5**, 73—80 (1934). — E. MILNE: Relativity, gravitation and world-structure. Oxford 1935. — O. HECKMANN: Nachr. Ges. Wiss. Göttingen, N.F. **3**, 169—181 (1940). — O. HECKMANN: Theorien der Kosmologie. Berlin 1942. — E. MILNE: Nature, Lond. **150**, 489 (1942). — H. BONDI: Cosmology. Cambridge 1952.

[5] D. LAYZER: Astronom. J. **59**, 268 (1954). — W. McCREA: Math. Gazette **39**, No. 330 (1955). — Nature, Lond. **175**, 466 (1955).

[6] O. HECKMANN u. E. SCHÜCKING: Z. Astrophys. **38**, 95 (1955). — Z. Astrophys., **40**, 81 (1956).

an dem Übelstand, daß gleichförmig, geradlinig bewegte Körper nur in einer leeren, gravitationsfreien Welt existieren könnten. Tatsächlich realisierbar sind aber nur *lokale* Inertialsysteme, in deren Ursprung ein „kräftefreier" Körper ruht und ein beliebiger linearer Oszillator seine Schwingungsrichtung beibehält. Solche Systeme sind durch ein Bezugssystem realisiert, das mit einem frei fallenden Fahrstuhl, der nicht rotiert, fest verknüpft ist. In solchen frei fallenden Systemen erklären wir lokal (d.h. für $x_i = 0$) die Newtonschen Bewegungsgleichungen

$$\ddot{x}_i = \frac{1}{m} K_i \tag{2.1}$$

für gültig. x_i sind die kartesischen Koordinaten eines euklidischen Raumes, t ist die absolute Zeitkoordinate, m ist die Masse eines Massenpunktes im Ursprung, und K_i ist der Vektor der auf ihn wirkenden Kraft, die keinen Gravitationsanteil mehr enthält. Das Fehlen von Coriolis-Termen drückt die Rotationsfreiheit des Koordinatensystems aus. Infolge der Wirkung der Gravitation gehen die ∞^{10} verschiedenen frei fallenden Systeme *nicht* durch Galilei-Transformationen auseinander hervor. In jedem der ∞^4 Weltpunkte existieren ∞^6 frei fallende Systeme, die durch Drehung (drei Parameter) und Translation mit konstanter Geschwindigkeit (drei Parameter) miteinander zusammenhängen. Denkt man sich die Gravitationsfelder gegen Null konvergierend, so werden die *lokalen* Inertialsysteme zu den üblicherweise betrachteten Inertialsystemen, die durch Galilei-Transformationen auseinander hervorgehen. Behandelt man in einem frei fallenden System die Bewegung eines Massenpunktes, der sich *nicht* im Ursprung befindet, so haben wir das Potential Φ des Gravitationsfeldes in (2.1) einzuführen. Dann lauten die Bewegungsgleichungen

$$\ddot{x}_i = \frac{1}{m} K_i - \Phi_{|i}, \tag{2.2}$$

wo $\Phi_{|i} = 0$ für $x_i = 0$ gelten soll. Φ soll eine reine Funktion der Ortskoordinaten x_i und der Zeit t sein. Eine Geschwindigkeitsabhängigkeit von Φ ist wiederum wegen der Rotationsfreiheit des Koordinatensystems auszuschließen. Die Aufspaltung der rechten Seite von (2.2) in zwei Anteile ist stets zweifelsfrei möglich, da der „Gravitationsanteil" $\Phi_{|i}$ von den individuellen Eigenschaften des Probekörpers unabhängig ist, also auf *jeden* Körper wirkt, der sich zu der gegebenen Zeit t an der Stelle x_i befindet. K_i hingegen hängt von den individuellen Eigenschaften des Probekörpers ab, ist also beispielsweise der Schub einer Rakete oder das Produkt aus Ladung und elektrischer Feldstärke, die auf einen Probekörper wirkt. $\frac{1}{m} K_i$ ist also die Beschleunigungsdifferenz, die irgendein beliebigen Kräften K_i ausgesetzter Probekörper relativ zu irgendeinem an seinem Ort frei fallenden Probekörper zeigt. Um das auszudrücken schreibt man (2.2) vorteilhaft

$$\ddot{x}_i + \Phi_{|i}(x_j) = \frac{1}{m} K_i(x_j). \tag{2.2'}$$

Man *könnte*, muß aber nicht, die Sprechweise einführen: $\frac{1}{m} K_i$ ist die Beschleunigung relativ zum absoluten Raum.

Wir beschreiben nun die Bewegung des Probekörpers von einem anderen lokalen Inertialsystem x'_i aus, dessen Achsen zu dem der x_i parallel seien. Dann gilt

$$x'_i = x_i - b_i(t) \tag{2.3}$$

mit beliebigen Zeitfunktionen b_i.

In dem System der x_i' erhält man an Stelle von (2.2') für die Bewegung des Probekörpers

$$\ddot{x}_i' + \Phi'_{|i} = \frac{1}{m} K_i(x_j' + b_j). \tag{2.2''}$$

Durch zweimalige Differentiation von (2.3) folgt mit (2.2') und (2.2'') wegen $K_i(x_j) = K_i(x_j' + b_j)$

$$\Phi'_{|i} - \ddot{b}_i = \Phi_{|i}. \tag{2.4}$$

Folglich gilt für die Transformation des Potentials auf das neue frei fallende System

$$\Phi' = \Phi + x_i \ddot{b}_i + f(t) \tag{2.5}$$

mit einer beliebigen Funktion $f(t)$:

Die obige Verbindung von Kraftbegriff und lokalem Inertialsystem wird ermöglicht durch die Proportionalität von träger und schwerer Masse, auf deren Bedeutung von A. EINSTEIN so nachdrücklich hingewiesen wurde[1]. Durch die obige Interpretation kann man bereits in der Newtonschen Theorie ausdrücken, was von H. WEYL als die „Grundannahme" der Einsteinschen Gravitationstheorie formuliert wurde, nämlich, „daß die Gravitation nicht eine *Kraft* ist, welche die Körper aus der ihnen durch das Führungsfeld vorgezeichneten Bahn ablenkt, sondern mit der Trägheit unlöslich verbunden, im Führungsfelde enthalten ist"[2].

Die Äquivalenz der frei fallenden Systeme ist jedoch im allgemeinen nicht vollständig. Denn obwohl in ihrem Ursprung der Gradient, d.h. der Vektor der ersten Ableitungen, des Gravitationspotentials verschwindet, bilden z.B. die zweiten Ableitungen des Gravitationspotentials einen symmetrischen Tensor zweiter Stufe, der im allgemeinen von Ort zu Ort verschieden ist. Eine wichtige Ausnahme liegt dann vor, wenn das Potential Φ gegen eine endliche kontinuierliche Gruppe invariant ist, d.h. wenn das Potential Φ' aus (2.5) als Funktion der x_i' die gleiche Funktion ist wie das Potential Φ als Funktion der x_i. Das ist für die Kosmologie von Bedeutung [vgl. z.B. (3.25)].

Da sich das Potential Φ gegenüber (2.3) gemäß (2.5) transformiert, bleibt die Randbedingung $\Phi = 0$ beim Übergang auf ein anderes frei fallendes System nicht erhalten. Sie ist für Systeme endlicher Gesamtmasse zu ersetzen durch

$$\Phi_{|i|k} - \tfrac{1}{3}\delta_{ik}\Phi_{|j|j} = 0 \quad \text{im Unendlichen,} \tag{2.6}$$

die Φ im Verein mit der Poissonschen Gleichung festzulegen gestattet bis auf eine in den Koordinaten x_i lineare Funktion mit zeitabhängigen Koeffizienten[3]. Der raumabhängige Anteil dieser Funktion läßt sich dadurch festlegen, daß man den Ursprung des Koordinatensystems mit einem frei fallenden Probekörper vereinigt.

β) Die Grundgleichungen. Wir behandeln im weiteren die das Weltall erfüllende Materie als ein beliebig kompressibles Gas, dessen Moleküle die außergalaktischen Nebel sind. Volumelemente sind entsprechend groß zu wählen.

In den frei fallenden Systemen postulieren wir nun in bekannter Weise die Gültigkeit der Grundgleichungen der Newtonschen Theorie kontinuierlicher

---

[1] A. EINSTEIN: Jahrbuch der Radioaktivität und Elektronik **4**, 411 (1907); **5**, 98 (1907).
[2] H. WEYL: Die mathematische Analyse des Raumproblems, S. 44. Berlin 1923.
[3] Siehe Fußnote 6, S. 491.

Materieverteilungen

$$\varrho_{|0} + (\varrho\, v_i)_{|i} = 0, \tag{2.7}$$

$$v_{i|0} + v_{i|k}\, v_k = -\,\Phi_{|i} - \frac{1}{\varrho}\, p_{|i}, \tag{2.8}$$

$$\Phi_{|i|i} + \Lambda_0 = 4\pi G\,\varrho. \tag{2.9}$$

In ihnen sind ϱ, v_i, p Funktionen von t und x_i, die nacheinander bedeuten: Materiedichte, Geschwindigkeit und Druck. G ist die Gravitationskonstante. Gl. (2.7) ist die Kontinuitätsgleichung, (2.8) enthält die drei Bewegungsgleichungen und (2.9) ist die Poissonsche Gleichung, die um das kosmologische Glied Λ_0 erweitert wurde in Analogie zu der Erweiterung der Einsteinschen Gravitationstheorie um das Λ-Glied[1]. Durch Anbringen von weiteren Zusatzgliedern ließe sich auch schon im Rahmen dieser Gleichungen Materie- und Impulserzeugung oder Veränderlichkeit von G berücksichtigen, falls man solche Verallgemeinerungen zu betrachten wünscht. Die Gln. (2.7) bis (2.9) sind durch Angabe einer Zustandsgleichung zu ergänzen, die p und ϱ verknüpft. Davon soll hier abgesehen werden, da die weiteren Überlegungen davon unabhängig sind. Wir setzen in der Folge einfach $p=0$, d.h. wir vernachlässigen die Pekuliarbewegungen der außergalaktischen Nebel.

Die Gln. (2.6) bis (2.9) genügen dem Kausalprinzip. Das heißt, sie erlauben, aus einer zur Zeit $t=t_0$ vorgegebenen Materieverteilung mit vorgegebenem Geschwindigkeitsfeld $v_i(t_0, x_j)$ Materieverteilung und Geschwindigkeitsfeld zu jedem früheren oder späteren Zeitpunkt eindeutig zu ermitteln. Für diese Festlegung ist die Wahl der Randbedingungen von entscheidender Bedeutung[2].

Da alle Differentialgleichungen (2.7) bis (2.9) streng lokale Natur besitzen, darf man ihre Gültigkeit auch für unendliche Systeme postulieren. Die Kugelsymmetrie des Tensorellipsoids der zweiten räumlichen Ableitungen des Potentials Φ im Unendlichen, die sich in (2.6) ausspricht, ist von A. STÖHR und H. RÜSSMANN[3] für verschiedene kosmologische Modelle ausgenutzt worden. Die Randbedingungen (2.6) liefern jedoch nicht das Analogon des von K. GÖDEL entdeckten Weltmodells. Um derartige Analogien nicht durch zu enge Wahl der Randbedingungen in der Newtonschen Theorie von vornherein auszuschließen, muß man im Falle beliebiger unendlicher Materieverteilungen Gl. (2.6) ersetzen durch

$$\Phi_{|i|k} - \tfrac{1}{3}\,\delta_{ik}\,\Phi_{|j|j} = A_{ik}(t), \quad A_{ii} = 0, \quad \text{im Unendlichen.} \tag{2.10}$$

Dabei repräsentieren die A_{ik} wegen der Nebenbedingung $A_{ii}=0$ fünf unabhängige, vorläufig noch unbekannte, Funktionen der Zeit. Um den Zusammenhang mit der Einsteinschen Theorie noch enger zu machen, könnte man daran denken, diese Funktionen festzulegen, nachdem man alle räumlich homogenen Weltmodelle in der Einsteinschen Gravitationstheorie untersucht hat. Das ist jedoch noch nicht in der nötigen Ausführlichkeit geschehen. Zunächst sichern jedenfalls die neuen Randbedingungen ebenso wie die Bedingungen (2.6) die Erfüllung der Kausalitätsforderungen, wenn man annimmt, daß die Materiedichte für den Grenzübergang ins Unendliche einem endlichen Grenzwert zustrebt[2].

---

[1] A. EINSTEIN glaubte irrtümlich, das kosmologische Glied in seinen Feldgleichungen stände in Analogie zu der C. Neumannschen Gleichung

$$\Phi_{|i|i} + \Lambda_0 \Phi = 4\pi G\,\varrho.$$

Das ist jedoch nicht der Fall.

[2] Siehe Fußnote 6, S. 491.

[3] A. STÖHR: Math. Nachr. **6**, 71 (1951). — H. RÜSSMANN: Math. Diplomarbeit Göttingen.

3. Weltmodelle. Die Behandlung bestimmter homogener Materieverteilungen im Rahmen der Newtonschen Gravitationstheorie besitzt ein zweifaches Interesse. Einmal kann man nämlich die Dynamik unbegrenzter Materieverteilungen an einfachen Beispielen studieren und zum anderen versuchen, diese Theorie auf die Dynamik des Universums anzuwenden. Besonderes Interesse besitzen die druckfreien Weltmodelle, deren Geschwindigkeitsfeld räumlich homogen ist. Die mathematische Forderung der räumlichen Homogenität des Geschwindigkeitsfeldes, das sog. Weltpostulat, läßt sich in der folgenden Weise formulieren: Die räumliche Translation von dem mitschwimmenden, frei fallenden System der x_i auf das entsprechende System der x_i'

$$x_i = x_i' + a_i(t) \tag{3.1}$$

liefert durch Differentiation nach t die Gleichungen

$$v_i(x_j, t) = v_i'(x_j', t) + \frac{d a_i}{d t}. \tag{3.2}$$

Das Weltpostulat verlangt dann

$$v_i'(x_j', t) = v_i(x_j', t). \tag{3.3}$$

Aus (3.1) bis (3.3) folgt sogleich durch Einsetzen

$$v_i(x_j, t) = v_i(x_j - a_j, t) + \frac{d a_i}{d t}. \tag{3.4}$$

Durch Differentiation nach x_k erhält man

$$v_{i|k}(x_j, t) = v_{i|k}(x_j - a_j, t) = a_{ik}(t), \tag{3.5}$$

weil die erste Gl. (3.5) zeigt, daß die $v_{i|k}$ nicht mehr von x_j abhängen. Integration von (3.5) liefert

$$v_i = a_{ik}(t)\, x_k + b_i(t). \tag{3.6}$$

Die Bedingung des Mitschwimmens der Beobachter verlangt, daß die $b_i(t) = 0$ gesetzt werden müssen. Die $a_i(t)$ bestimmen sich durch Einsetzen von (3.6) in (3.4). Das Weltpostulat für das Geschwindigkeitsfeld bedingt also, daß

$$v_i = a_{ik}(t)\, x_k \tag{3.7}$$

ist. Die Linearität des Strömungsfeldes sichert die in (3.3) zum Ausdruck kommende Forderung der „Mittelpunktslosigkeit" der Welt. Für jeden Beobachter einer dreiparametrigen Schar bietet das Strömungsfeld der Materie den gleichen Anblick.

Ein durch (3.7) beschriebenes Strömungsfeld läßt sich in einer gegen Translationen und zeitunabhängige Rotation invarianten Weise in drei Anteile zerlegen. Man definiert

$$a_{ik} = a_{\underline{ik}} + a_{\underset{\smile}{ik}} \tag{3.8}$$

mit

$$a_{\underline{ik}} = \tfrac{1}{2}(a_{ik} + a_{ki}), \qquad a_{\underset{\smile}{ik}} = \tfrac{1}{2}(a_{ik} - a_{ki}), \tag{3.9}$$

wo also $a_{\underline{ik}}$ und $a_{\underset{\smile}{ik}}$ die symmetrischen und schiefsymmetrischen Anteile bilden. Den symmetrischen Teil zerlegt man zweckmäßig weiter in

$$a_{\underline{ik}} = q_{ik} + \frac{\dot R}{R}\,\delta_{ik} \tag{3.10}$$

mit

$$\frac{\dot R(t)}{R(t)} = \frac{1}{3}\, a_{\underline{ii}} = \frac{1}{3}\, a_{ii}, \tag{3.11}$$

so daß

$$q_{ik} = a_{\underline{ik}} - \tfrac{1}{3}\,\delta_{ik}\,a_{\underline{jj}}, \quad \text{also} \quad q_{ii} = 0 \tag{3.12}$$

ist. Den schiefsymmetrischen Teil kann man auch durch einen Vektor w_i ausdrücken gemäß

$$w_1 = a_{\underset{\smile}{23}}, \quad w_2 = a_{\underset{\smile}{31}}, \quad w_3 = a_{\underset{\smile}{12}}, \tag{3.13}$$

also

$$a_{\underset{\smile}{jk}} = \varepsilon_{jkl}\,w_l, \tag{3.14}$$

wo der Tensor ε_{jkl} in allen Indices schiefsymmetrisch ist mit $\varepsilon_{123} = 1$. Zusammenfassend schreibt man also

$$a_{ik} = \frac{\dot R}{R}\,\delta_{ik} + q_{ik} + \varepsilon_{ikl}\,w_l. \tag{3.15}$$

Das erste Glied auf der rechten Seite von (3.15) beschreibt die Expansion (Kontraktion), das zweite die Scherung und das dritte die Rotation des Geschwindigkeitsfeldes. w_l ist der Vektor der Winkelgeschwindigkeit.

Führt man den Ansatz (3.15) in die Gln. (2.7) bis (2.9) ein, so liefert die Elimination des Potentials aus (2.8) und (2.9) die Aussage, daß ϱ nur noch von t abhängt. Wenn man (2.8) nach x_k differenziert und die Vertauschbarkeit der zweiten Ableitungen des Potentials beachtet, erhält man jetzt insgesamt die Gleichungen

$$\left.\begin{aligned} \frac{4\pi}{3}\,\varrho\,R^3 &= \mathfrak{M} = \text{const} > 0,\\ (R^2 w_i)^{\bullet} &= q_{ij}(R^2 w_j),\\ \ddot R - \frac{R}{3}(\Lambda_0 - q_{ij}q_{ij}) - \frac{2}{3}R\,w_i w_i + \frac{G\mathfrak{M}}{R^2} &= 0. \end{aligned}\right\} \tag{3.16}$$

Die Gln. (3.16) sind fünf Differentialgleichungen für die zehn Funktionen ϱ, R, w_i, q_{ik}. Ihre Lösungen werden erst bestimmt durch Hinzunahme der fünf Randbedingungen (2.10), die wegen (2.8) und (3.15) lauten

$$\left.\begin{aligned} -\frac{1}{R^2}(R^2 q_{ik})^{\bullet} - q_{ij}q_{jk} + \frac{1}{3}\,\delta_{ik}q_{jr}q_{jr} - w_i w_k + \frac{1}{3}\,\delta_{ik}w_j w_j &= A_{ik}(t),\\ A_{ii} &= 0. \end{aligned}\right\} \tag{3.17}$$

Diese Randbedingungen kann man ersetzen durch äquivalente Forderungen an die fünf Funktionen q_{ik} solange man über die A_{ik} noch nichts näheres weiß. Eine Vorgabe der q_{ik} ist natürlich weitaus bequemer, weil man in diesem Fall die Gln. (3.17) nicht zu lösen braucht. Aus dem Strömungsfeld (3.7) berechnet sich das Potential Φ gemäß (2.8) zu

$$\Phi = -\tfrac{1}{2}(\dot a_{ik} + a_{ij}a_{jk})\,x_i x_k, \tag{3.18}$$

wo das Potential so normiert wurde, daß es im Ursprung mitsamt seinen ersten räumlichen Ableitungen verschwindet.

Im folgenden werden zunächst kosmologische Modelle mit verschwindender Scherung des Weltsubstrates untersucht. Es gilt also

$$q_{ij} = 0. \tag{3.19}$$

α) Starre Rotation, $R = 1$.

Ohne Einschränkung der Allgemeinheit kann man dann aus der zweiten Gl. (3.16) entnehmen

$$w_1 = w_2 = 0, \quad w_3 = \omega.$$

Die letzte Gl. (3.16) liefert dann die folgende Beziehung zwischen Winkelgeschwindigkeit und Materiedichte

$$\Lambda_0 + 2\omega^2 = 4\pi G \varrho. \tag{3.20}$$

In diesem Fall rotiert das gesamte Weltsubstrat also überall mit der gleichen Winkelgeschwindigkeit ω im lokalen Inertialsystem.

Für $\omega = 0$ erhält man das Analogon des statischen Einstein-Kosmos und für $\Lambda_0 = -4\pi G\varrho$ ein Analogon des Gödelschen Modells.

β) Isotrope Expansion. Man erhält diese Modelle, wenn man in (3.16) $w_i = 0$ setzt. Mit der Integrationskonstanten h gewinnt man dann

$$\frac{4\pi}{3}\varrho R^3 = \mathfrak{M} = \text{const} > 0, \tag{3.21}$$

$$\frac{1}{2}\dot{R}^2 = \frac{\Lambda_0}{6} R^2 + h + \frac{G\mathfrak{M}}{R}. \tag{3.22}$$

(3.22) ist die Friedmannsche Differentialgleichung[1]. Die Integration der Gln. (3.17) des Strömungsfeldes liefert im vorliegenden Falle mit den Konstanten x_i^0

$$v_i = \frac{dx_i}{dt} = \frac{\dot{R}}{R} x_i, \quad x_i = R(t)\, x_i^0. \tag{3.23}$$

Ein wesentliches Charakteristikum der Gl. (3.22) ist die Singularität für $R = 0$, bei der die Lösungen vertikale Tangenten haben und die Dichte unendlich wird. Für $\Lambda_0 = 0$ lassen sich die Friedmannschen Gleichungen leicht integrieren. Man erhält mit der reellen Hilfsgröße τ, die durch die Gleichung

$$d\tau = \sqrt{2|h|}\, R^{-1} dt$$

definiert ist

$$\left.\begin{aligned}
h > 0:&\quad R = \frac{G\mathfrak{M}}{2h}(\operatorname{Cos}\tau - 1); &&\pm(t - t_0)\sqrt{2h} = \frac{G\mathfrak{M}}{2h}(\operatorname{Sin}\tau - \tau),\\
h = 0:&\quad R = \left(\frac{9G\mathfrak{M}}{2}\right)^{\frac{1}{3}} |t - t_0|^{\frac{2}{3}},\\
h < 0:&\quad R = \frac{G\mathfrak{M}}{2|h|}(1 - \cos\tau); &&\pm(t - t_0)\sqrt{2|h|} = \frac{G\mathfrak{M}}{2|h|}(\tau - \sin\tau).
\end{aligned}\right\} \tag{3.24}$$

Durch zweckmäßige Normierung von R kann man setzen $h = 1, 0, -1$. Dann liefern die Formeln (3.24) die entsprechenden Formeln der Einsteinschen Gravitationstheorie für die expandierenden Räume konstanter negativer, verschwindender, positiver Krümmung. Gl. (3.18) für das Potential liefert mit (3.23)

$$\Phi = \frac{2\pi}{3} G\varrho\, x_i x_i. \tag{3.25}$$

Transformiert man das Potential gemäß (2.5), so besitzt Φ für jeden anderen mitschwimmenden Beobachter die gleiche Form wie (3.25), wenn man von einer additiven Konstanten absieht.

γ) Isotrope Expansion und starre Rotation. Diese Modelle erhält man mit der x_3-Achse als Rotationsachse und konstantem k durch

$$w_1 = 0, \quad w_2 = 0, \quad w_3 = \frac{k}{R^2}.$$

[1] A. Friedmann: Z. Physik **10**, 377 (1922).

Als Verallgemeinerung der Friedmannschen Differentialgleichung kommt jetzt

$$\frac{1}{2}\dot{R}^2 = \frac{\Lambda_0}{6}R^2 + h + \frac{G\mathfrak{M}}{R} - \frac{k^2}{3R^2}. \tag{3.26}$$

Der wesentliche Unterschied dieser Modelle gegenüber (3.22) ist, daß jetzt die Singularität der isotrop expandierenden Modelle infolge der Rotation des Weltsubstrates fortfällt.

Diese Lösungen γ) enthalten die beiden Fälle α) und β) als Spezialfälle und stellen die allgemeinste Lösung ohne Scherung dar. In der Einsteinschen Gravitationstheorie wird jedoch die zu γ) analoge Lösung durch weitere Feldgleichungen ausgeschlossen.

δ) Lösungen mit Scherung. Einige solcher Lösungen ohne Spezifizierung der Randbedingungen wurden von F. ZAGAR angegeben[1]. Die stationären Fälle wurden von O. HECKMANN und E. SCHÜCKING behandelt[2].

In der Hydrodynamik wird gezeigt, daß das Auftreten von Scherungen sich bemerkbar macht in der inneren Reibung des Substrates. Ist η der Koeffizient der inneren Reibung, so wird die pro Zeit- und Raumeinheit irreversibel erzeugte Wärmemenge W[3]

$$W = \eta\, q_{ik} q_{ik}. \tag{3.27}$$

Die statistische Mechanik ergänzt dieses Ergebnis durch die Feststellung, daß das Auftreten von Scherungen, also $q_{ik} \neq 0$, notwendig zu dauernder Erhöhung der spezifischen Entropie führt. Lediglich die Modelle ohne Scherung besitzen konstante spezifische Entropie[4].

Es ist hierzu bemerkenswert, daß die Kinematik des Substrates von den Reibungstermen nicht beeinflußt wird. Denn wenn man die Bewegungsgleichungen (2.8) zu den Navier-Stokesschen Gleichungen erweitert, liefern die Reibungsterme im Falle linearer Strömungsfelder keinen Beitrag[5].

ε) Vacuolenmodell. Ein sehr einfaches *inhomogenes* Weltmodell, das zur Beschreibung von Nebelhaufen möglicherweise Verwendung finden könnte, ist das Vacuolenmodell. Aus einem der Modelle β) denke man sich beliebig viele mitexpandierende Kugeln von Materie entleert (Vacuolen), in deren Mittelpunkt man einen Massenpunkt hineinsetzt. Dann folgt sogleich aus der Forderung der Stetigkeit des Potentials und seiner ersten Ableitungen, daß die Massen der einzelnen Massenpunkte gerade gleich sein müssen den Massen, die sonst in den Vacuolen enthalten wären. Es ist lediglich zu fordern, daß die Vacuolen einander nicht durchdringen dürfen.

Eine mit dem isotrop expandierenden Weltsubstrat sich ausdehnende Vacuole habe den Radius

$$r_0 = R(t)\,a, \qquad a = \text{const}. \tag{3.28}$$

Das Potential eines in ihrem Zentrum stehenden Massenpunktes der Masse M beträgt

$$\Phi_{\text{innen}} = -\frac{MG}{r} + A; \qquad r \leqq r_0. \tag{3.29}$$

Das Potential des Substrates ist gemäß (3.25)

$$\Phi_{\text{außen}} = \frac{2\pi}{3} G\varrho r^2; \qquad r \geqq r_0. \tag{3.30}$$

---

[1] F. ZAGAR: R. C. Acad. Lincei (8) **18**, 452 (1955).
[2] Siehe Fußnote 6, S. 491.
[3] S. ROSSELAND: Astrophysik, S. 63. Berlin 1931.
[4] Siehe Fußnote 6, S. 491.
[5] Vgl. die Lehrbücher der Hydrodynamik.

Die Forderung

$$\left(\frac{\partial \Phi_{\text{innen}}}{\partial r}\right)_{r=r_0} = \left(\frac{\partial \Phi_{\text{außen}}}{\partial r}\right)_{r=r_0} \tag{3.31}$$

liefert

$$M = \frac{4\pi}{3} \varrho r_0^3. \tag{3.32}$$

Dieses M ist zeitunabhängig wegen der ersten Gl. (3.16). M ist genau die Masse innerbalb einer Kugel vom Radius r_0 im homogenen Substrat. Aus (3.32) folgt wegen

$$(\Phi_{\text{innen}})_{r=r_0} = (\Phi_{\text{außen}})_{r=r_0} \tag{3.33}$$

mit (3.29) und (3.30)

$$A = 2\pi G \varrho r_0^2. \tag{3.34}$$

Die hier nur für eine Vacuole angeführten Bedingungen (3.31) und (3.33) lassen sich sofort auf beliebige Anzahlen einander nicht durchdringender Vacuolen ausdehnen.

Auch das analoge Modell in der Einsteinschen Gravitationstheorie läßt sich leicht angeben (vgl. Ziff. 11).

Anmerkung bei der Korrektur: Im Rahmen der Newtonschen Kosmologie behandelt Inhomogenitäten W. BONNOR: Monthly Notices Roy. Astronom. Soc. London **117**, 104 (1957).

III. Einsteinsche Kosmologie.

4. Vorbemerkungen zur Einsteinschen Gravitationstheorie. Im voranstehenden Abschnitt II über Newtonsche Kosmologie war eingangs gezeigt worden, daß die Newtonsche Theorie bereits in eine Form gebracht werden kann, die der von A. EINSTEIN erkannten Bedeutung der Gleichheit von schwerer und träger Masse gerecht wird[1]. Fordert man nun die Gültigkeit der speziellen Relativitätstheorie in den „Fahrstühlen" der frei fallenden aber relativ zueinander beschleunigten Systeme, so fordert man damit den Riemannschen Charakter des Weltkontinuums. Dann muß einerseits die Newtonsche Theorie schnell bewegter Körper nach den Erfordernissen der speziellen Relativitätstheorie modifiziert werden. Andererseits bedingt aber bereits die lokale Gültigkeit der Newtonschen Theorie für beliebig langsame Bewegungen, daß die Vektorübertragung nicht integrabel wird. Ebenso wie man auf der Erdoberfläche durch Herumführen eines Vektors um eine geschlossene Kurve feststellen kann, daß der Riemann-Tensor dieser zweidimensionalen Mannigfaltigkeit nicht verschwindet, erkennt man aus höchst einfachen Experimenten, daß auch der Riemann-Tensor der vierdimensionalen Raumzeit nicht verschwinden kann (vgl. Fig. 1, und ihre Erläuterung). Die Gravitation ist mit dem Trägheitsfeld zum übergeordneten Begriff des Führungsfeldes[2] vereinigt. Sie ist — wie man sagt — geometrisiert und äußert sich im Nichtverschwinden des Riemann-Tensors. Diese Tatsache schließt alle Gravitationstheorien aus, die mit einem streng pseudoeuklidischen Raumzeitkontinuum auszukommen suchen, weil in ihnen eine empirisch brauchbare Definition des Inertialsystems fehlen *muß*.

Der angedeutete Weg zur Einsteinschen Gravitationstheorie soll hier nicht ausgebaut werden. Wir setzen vielmehr die Einsteinsche Gravitationstheorie als

---

[1] O. HECKMANN u. E. SCHÜCKING: Z. Astrophys. **38**, 95 (1955).

[2] Diese Interpretation der Einsteinschen Gravitationstheorie geht auf die Analyse von H. WEYL zurück, die zum ersten Male in der 5. Auflage seines grundlegenden Lehrbuches „Raum. Zeit. Materie", Berlin 1923, ausführlich dargelegt worden ist.

bekannt voraus[1] und erläutern kurz, welche Stellung die in den vorhergehenden Ziffern diskutierte Newtonsche Theorie kontinuierlicher, inkohärenter Materie in ihrem Rahmen einnimmt.

In der Einsteinschen Gravitationstheorie ist die Bahn eines Massenpunktes (Probekörpers), der nur den Kräften von Trägheit und Schwere unterworfen ist, eine zeitartige geodätische Linie. Es ist stets möglich, ein Koordinatensystem zu wählen, in dem diese Linie die Gleichung $x^i = \text{const}$ hat und in dem zugleich die ersten Ableitungen der $g_{\mu\nu}$ längs dieser Linie verschwinden (Fermi-Koordinaten[2]) und die pseudoeuklidischen Werte annehmen. Da die Christoffel-Symbole in diesen Koordinaten längs der Geodätischen verschwinden, ändert ein Vektor ξ^μ bei Parallelverschiebung längs dieser Linie seine Komponenten nicht, wie die Formel

$$\delta \xi^\mu = -\Gamma^\mu_{\nu\lambda} \xi^\nu \, dx^\lambda$$

zeigt. Deutet man beispielsweise ξ^μ als Drehimpuls-Vektor eines kräftefreien Kreisels, so entspricht diese Formel dem Sachverhalt, daß im lokalen Inertialsystem (oder, in EINSTEINs Sprache, in einem frei fallenden „Fahrstuhl") ein reibungsfrei rotierender Kreisel, dessen Schwerpunkt im Koordinatenursprung ruht, stets in die gleiche Richtung weist. Man hat also das lokale Inertialsystem der Newtonschen Theorie mit dem Fermi-Koordinatensystem der Einsteinschen Theorie zu identifizieren. Formuliert man die Einsteinschen Feldgleichungen in diesem speziellen Fermi-Koordinatensystem, so erhält man bei Vernachlässigung der Krümmung des dreidimensionalen Raumes lokal exakt die Poissonsche Gleichung und die Newtonschen Bewegungsgleichungen des Substrates. Diese Tatsache zeigt, daß die Newtonsche Theorie eine zweckmäßige und legitime Approximation der Einsteinschen Gravitationstheorie ist, insbesondere, daß sie die Krümmungen in den zweidimensionalen Flächen, die eine Zeitrichtung enthalten, korrekt liefern kann. Vielleicht wird das einfachste Beispiel dafür geliefert durch die Vorstellung eines langen, geradlinigen Schachtes vom Nord- zum Südpol durch den Erdmittelpunkt, innerhalb dessen ein frei fallender Probekörper um den Erdmittelpunkt oszilliert. Dabei ist das mit dem Erdmittelpunkt verbundene Koordinatensystem, sofern es nicht rotiert, ebenso ein Inertialsystem wie das mit dem frei fallenden Probekörper verbundene. Die Weltlinie des Probekörpers ist eine Sinuskurve, die in Fig. 1 wiedergegeben ist. Die Newtonsche Interpretation sagt, daß der Probekörper dem Kraftfeld der Erde folge. Die Einsteinsche Theorie sagt, daß er auf einer Geodätischen in einem gekrümmten Weltgebiet laufe. Die weitgehende Übereinstimmung der Newtonschen Theorie mit der Erfahrung erlaubt die nicht-ver-

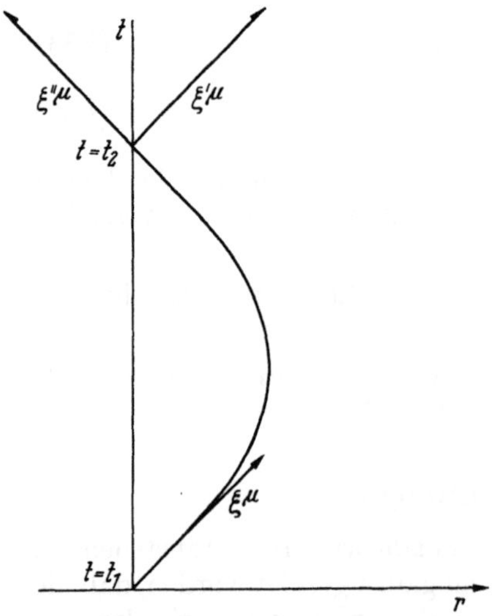

Fig. 1. Weltlinie eines Probekörpers, der um den Erdmittelpunkt oszilliert. ξ^μ ist der Tangentenvektor an die Weltlinie des Probekörpers beim Durchgang durch den Erdmittelpunkt. Die Parallelverschiebung dieses Vektors in dem lokalen Inertialsystem, das mit dem Erdmittelpunkt fest verknüpft ist, liefert den Vektor ξ'^μ. Parallelverschiebung in dem lokalen Inertialsystem, das mit dem frei fallenden Probekörper fest verknüpft ist, führt zu dem Vektor ξ''^μ. Die Nichtintegrabilität der Parallelverschiebung äußert sich darin, daß ξ'^μ und ξ''^μ verschieden sind.

[1] Vgl. die ausführliche Darstellung im Rahmen des Artikels von P. G. BERGMANN über allgemeine Relativitätstheorie in Bd. IV dieses Handbuches.

[2] E. FERMI: Rendiconti dei Lincei **31**, 21—23, 51—52 (1922).

schwindenden Komponenten des Riemann-Tensors im Erdmittelpunkt zu berechnen aus der Nichtintegrabilität der Vektorübertragung von t_1 nach t_2.

Man wird also nicht überrascht sein, im weiteren wesentliche Resultate der Newtonschen Kosmologie im Rahmen der Einsteinschen wiederzufinden, dort allerdings in höheren Zusammenhängen.

5. Grundgleichungen der Einsteinschen Gravitationstheorie. Für den Energieimpulstensor einer vollkommenen Flüssigkeit setzt man in der Relativitätstheorie

$$T^{\mu\nu} = (\varrho + p)\, u^\mu u^\nu - p\, g^{\mu\nu}. \tag{5.1}$$

u^μ ist dabei die Vierergeschwindigkeit

$$u^\mu = \frac{dx^\mu}{ds} \quad \text{mit} \quad u^\mu u_\mu = 1. \tag{5.2}$$

p ist der isotrope Druck und ϱ die Materiedichte. Für $p = 0$ erhalten wir den für die Kosmologie besonders interessanten Grenzfall der inkohärenten Materie. Für $p = \frac{1}{3}\varrho$ definiert (5.1) ein isotropes Strahlungsfeld. Der Vektor u^μ ist in diesem Falle jedoch nicht mehr der Vektor einer Materieströmung, sondern legt nur die Zeitachsen der Ruhsysteme fest, in denen das Strahlungsfeld lokal isotrop erscheint.

Das Führungsfeld der $g_{\mu\nu}$ ist Feldgleichungen unterworfen, die in der Newtonschen Theorie ihr Analogon in der Poissonschen Gleichung haben. Sie lauten

$$G_{\mu\nu} = (\Lambda + \tfrac{1}{2}\varkappa T)\, g_{\mu\nu} - \varkappa T_{\mu\nu}. \tag{5.3}$$

Dabei bedeutet $\varkappa = \frac{8\pi G}{c^2}$ die relativistische Gravitationskonstante. Ferner ist

$$T = T_{\mu\nu} g^{\mu\nu}, \quad G_{\mu\nu} = \Gamma^\lambda_{\lambda\mu|\nu} - \Gamma^\lambda_{\mu\nu|\lambda} + \Gamma^\varrho_{\mu\lambda}\Gamma^\lambda_{\nu\varrho} - \Gamma^\lambda_{\mu\nu}\Gamma^\varrho_{\lambda\varrho}. \tag{5.4}$$

$\Gamma^\lambda_{\mu\nu}$ sind die Christoffel-Symbole zweiter Art. Sie sind definiert durch

$$\Gamma^\lambda_{\mu\nu} = g^{\lambda\varrho}\,\Gamma_{\mu\nu,\varrho} = g^{\lambda\varrho}\cdot\tfrac{1}{2}\,(g_{\mu\varrho|\nu} + g_{\nu\varrho|\mu} - g_{\mu\nu|\varrho}). \tag{5.5}$$

Λ ist das kosmologische Glied. Es wurde 1917 von A. EINSTEIN eingeführt als Erweiterung seiner im Jahr vorher mitgeteilten Feldgleichungen. H. WEYL[1] hat jedoch gezeigt, daß die folgenden Forderungen auf die Feldgleichungen *mit* Λ-Glied führen:

Die linken Seiten von

$$G_{\mu\nu} - \tfrac{1}{2} g_{\mu\nu} G + \Lambda g_{\mu\nu} = -\varkappa T_{\mu\nu} \tag{5.3'}$$

(G ist hier ausnahmsweise nicht die Newtonsche Gravitationskonstante sondern der verjüngte Ricci-Tensor!) sollen erstens divergenzfrei sein, sie sollen zweitens nur vom metrischen Fundamentaltensor $g_{\mu\nu}$ und seinen ersten und zweiten Ableitungen abhängen, sie sollen drittens in diesen zweiten Ableitungen linear sein.

Die Weylschen Forderungen sind der einzige zwingende Weg zu den Feldgleichungen. Man könnte das Λ-Glied nur dann aus ihnen ausschließen, wenn man eine zusätzliche Forderung einführt.

6. Felder mit inkohärenter Materie. Als Folge von (5.3) ergibt sich bekanntlich die lokale Energieimpulserhaltung in der Form

$$T^{\mu\nu}{}_{\|\nu} = 0. \tag{6.1}$$

[1] H. WEYL: Raum. Zeit. Materie, 5. Aufl. Berlin 1923. Anhang 1.

Für inkohärente Materie erhält man daraus

$$(\varrho u^\mu u^\nu)_{||\nu} = (\varrho u^\nu)_{||\nu} u^\mu + \varrho u^\mu_{||\nu} u^\nu = 0. \tag{6.2}$$

Überschieben dieser Gleichung mit u_μ liefert wegen (5.2) die Kontinuitätsgleichung, weil der letzte Term verschwindet, wie man durch kovariante Ableitung von (5.2) bestätigt:

$$(\varrho u^\nu)_{||\nu} = \frac{1}{\sqrt{-g}} (\varrho u^\nu \sqrt{-g})_{|\nu} = 0. \tag{6.3}$$

Setzt man (6.3) in (6.2) ein, so folgt, falls ϱ nicht verschwindet, für das Vektorfeld u^μ die Differentialgleichung der geodätischen Linien

$$u^\mu_{||\nu} u^\nu = 0. \tag{6.4}$$

Für die Untersuchung von Materiefeldern mit inkohärenter Materie empfiehlt sich die Benutzung spezieller, dem Strömungsfeld besonders angepaßter, Koordinaten. Das wichtigste unter den in Frage kommenden Koordinatensystemen ist dasjenige, in dem die Materie „auf Ruhe" transformiert ist. Löst man in diesem Koordinatensystem die Feldgleichungen, so hat man das Bewegungsproblem der Materie, also die Integration der Differentialgleichungen ihrer geodätischen Linien zugleich mitgelöst. Diese mit der Materie mitschwimmenden Koordinaten sind in der Hydrodynamik als Lagrangesche Koordinaten bekannt. In einem endlichen Gebiet lassen sich die mitschwimmenden Koordinaten stets einführen, indem man das Vektorfeld u^μ in die Form transformiert

$$u^\mu = \delta^\mu_0. \tag{6.5}$$

Dann gilt

$$u_\nu = g_{\nu\mu} u^\mu = g_{\nu 0}. \tag{6.6}$$

Die Bewegungsgleichungen der Materie sind dann gegeben durch $x^i = \text{const}$. Aus (6.5) und (6.6) folgt mit (5.2)

$$g_{00} = 1. \tag{6.7}$$

Setzt man (6.5) in (6.4) ein, so kommt

$$u^\mu_{||\nu} u^\nu = \delta^\mu_{0|\nu} \delta^\nu_0 + \Gamma^\mu_{\nu\lambda} \delta^\nu_0 \delta^\lambda_0 = \Gamma^\mu_{00} = 0. \tag{6.8}$$

Mit (5.5) erhält man daraus

$$\Gamma^\mu_{00} = \tfrac{1}{2} g^{\mu\lambda} (2 g_{0\lambda|0} - g_{00|\lambda}) = 0. \tag{6.9}$$

Wegen (6.7) gilt dann

$$g_{0i|0} = 0; \tag{6.10}$$

die g_{0i} hängen also nicht von der Zeit ab. Für die Metrik folgt

$$ds^2 = (dx^0)^2 + 2 g_{0i}(x^j)\, dx^0\, dx^i + g_{ij}(x^\mu)\, dx^i\, dx^j. \tag{6.11}$$

α) Die Einführung einer universellen Zeit. Da für die Materieweltlinien $dx^i = 0$ ist, gilt auf ihnen $ds = dx^0$. x^0 mißt also die Eigenzeit jeder im räumlichen Koordinatensystem der x^j ruhenden Uhr. Alle diese Uhren laufen gleich schnell, können aber beliebige Nullpunktsunterschiede aufweisen. Denn man kann durch die Transformationen

$$x'^0 = x^0 + \varphi(x^j), \quad x'^i = x^i \tag{6.12}$$

mit beliebigem $\varphi(x^j)$ zu einer neuen Zeitkoordinate x'^0 übergehen, derart, daß die Nullpunkte der Uhren um den ortsabhängigen Betrag $\varphi(x^j)$ geändert werden.

Erweitert man (6.12) zu

$$x'^0 = x^0 + \varphi(x^j), \quad x'^i = x'^i(x^j), \tag{6.13}$$

so hat man damit die allgemeinste Transformation, die die Gültigkeit von (6.7) und (6.10) nicht berührt. Zu einem universellen Gleichzeitigkeitsbegriff führt (6.11) erst dann, wenn man eine *Konvention* über die g_{0i} einführt. Die g_{0i} transformieren sich bei Anwendung von (6.12) nach dem Gesetz

$$g'_{0i} = g_{0i} - \varphi_{|i}. \tag{6.14}$$

Kommt man also überein, daß ein bestimmtes einmal vorgegebenes Funktionentripel g_{0i} ungeändert bleiben soll, so erreicht man, daß φ konstant wird, macht also per conventionem die durch die Festlegung der Funktionen g_{0i} ausgezeichnete Zeit t zur „kosmischen" Zeit[1].

Durch die Transformation (6.14) lassen sich die g_{0i} nur dann wegtransformieren, wenn sie Gradient eines Skalars $\psi(x^j)$ sind. Falls aber die Größe $g_{0i|k} - g_{0k|i}$ nicht identisch verschwindet, kann man die g_{0i} nicht durch eine Transformation (6.13) beseitigen. Die eben erwähnte Rotation des Vektors g_{0i} bestimmt die Rotation des Weltsubstrates relativ zu einem beispielsweise durch Kreisel ausgerichteten mitschwimmenden lokalen Inertialsystem. Dort und nur dort, wo $g_{0i|k} - g_{0k|i} = 0$ ist, verschwindet die Rotation des Weltsubstrates. Dann und nur dann also, wenn die Rotation des Weltsubstrates verschwindet, kann man durch eine Transformation (6.13) g_{0i} zum Verschwinden bringen. Die oben erwähnte Bedingung der Rotationsfreiheit des Weltsubstrates kann man auch bei Berücksichtigung von (6.6), (6.7) und (6.10) kovariant formulieren durch

$$w_{\mu|\nu} = \tfrac{1}{2}(u_{\mu|\nu} - u_{\nu|\mu}) = 0. \tag{6.15}$$

Damit hat man eine vom Koordinatensystem unabhängige Formulierung für das Verbot absoluter Rotation gefunden.

β) Kinematische Aufspaltung des Strömungsfeldes. Es empfiehlt sich, neben den $w_{\mu\nu} = -w_{\nu\mu}$ den Scherungstensor $q_{\mu\nu}$ einzuführen, der definiert ist durch

$$q_{\mu\nu} = \tfrac{1}{2}(u_{\mu\|\nu} + u_{\nu\|\mu}) - \tfrac{1}{3}(g_{\mu\nu} - u_\mu u_\nu)\, u^\lambda_{\|\lambda}. \tag{6.16}$$

Gemäß (6.4) und (5.2) genügen die $q_{\mu\nu}$ und $w_{\mu\nu}$ den Gleichungen

$$q_{\mu\nu} = q_{\nu\mu}, \quad q^\mu_\mu = 0, \quad q_{\mu\nu} u^\nu = 0, \quad w_{\mu\nu} u^\nu = 0. \tag{6.17}$$

Schließlich kann man noch einen Skalar einführen, welcher die Expansion des Substrates beschreibt. Man setzt

$$u^\lambda_{\|\lambda} = \frac{3R_{|\mu}}{R}\, u^\mu \tag{6.18}$$

und betrachtet diese Gleichung als partielle Differentialgleichung für R. Jede Lösung von (6.18) liefert eine mögliche Funktion R. Aus (6.15) und (6.16), (6.18) folgt nun sogleich

$$u_{\mu\|\nu} = (g_{\mu\nu} - u_\mu u_\nu)\, \frac{R_{|\lambda}}{R}\, u^\lambda + q_{\mu\nu} + w_{\mu\nu}. \tag{6.19}$$

Die analoge Zerlegung von $v_{i|k}$ in drei Anteile war schon im Rahmen der Newtonschen Kosmologie in (3.15) durchgeführt worden. Die jetzt kovariant definierten $w_{\mu\nu}$, $q_{\mu\nu}$, R sind gerade so gewählt, daß sie in einem mitschwimmenden lokalen Inertialsystem lokal übereinstimmen mit den entsprechenden Größen

[1] Näheres hierzu in O. HECKMANN: Theorien der Kosmologie, 2. Aufl. (in Vorbereitung).

aus Ziff. 3. In diesem mitschwimmenden lokalen Inertialsystem verschwinden nämlich im Ursprung alle ersten Ableitungen der $g_{\mu\nu}$, die $g_{\mu\nu}$ selbst nehmen dort die pseudoeuklidischen Werte an und der Vektor des Strömungsfeldes reduziert sich im Ursprung auf $u^\mu = \delta_0^\mu$. Alle Komponenten der Tensoren $w_{\mu\nu}$ und $q_{\mu\nu}$, die den Index „0" enthalten, verschwinden gemäß (6.17) im Ursprung, und weil kovariante Ableitungen in (6.19) im Ursprung in gewöhnliche übergehen, fällt man auf die Formel der Ziff. 3 zurück. Analog wie im Newtonschen Fall kann man auch die Rotation des Substrates durch einen Vektor der Winkelgeschwindigkeit beschreiben. Mit seiner Hilfe kann man dann aus Erhaltungssätzen und Feldgleichungen kovariante Gleichungen gewinnen, die in einem mitschwimmenden lokalen Inertialsystem exakt auf die Gln. (3.16) führen[1]. In dieser Tatsache drückt sich wieder die enge Analogie zwischen Newtonscher Theorie und Einsteinscher Gravitationstheorie aus.

7. Das Weltpostulat. Die Mehrzahl der heutigen Forscher neigt dazu, die Forderung der Mittelpunktslosigkeit der Welt, also ein Homogenitätspostulat, zur Grundlage der Kosmologie zu machen. Die Beobachtungen aus der weiteren Umgebung (Radius einige hundert Millionen Lichtjahre) des Milchstraßensystems legen tatsächlich die Vermutung nahe, daß die Homogenität annähernd verwirklicht sei. Aber schon in entfernteren Teilen des Weltalls, wo die speziellen Eigenschaften eines kosmologischen Modells merklich in die Reduktion der Beobachtungen eingehen, ist der Grad der vorausgesetzten Homogenität schwer zu prüfen. Es dürfte schwierig sein, die empirisch gefundenen stochastischen Nebelverteilungen in eine kosmologische Dynamik einzubauen. Vielmehr wird man vorläufig durch Verwischung von Fluktuationen stetige Modelle konstruieren, bei welchen eine strenge Homogenität in abstracto der empirisch nur bedingt feststellbaren Homogenität in concreto entspricht. Die mathematische Formulierung dieser strengen Homogenität im Rahmen der Riemannschen Geometrie benutzt Hilfsmittel aus der Theorie der kontinuierlichen Gruppen. Ein erster Schritt zur Einführung eines Homogenitätspostulates ist die Einführung einer universellen Zeit: Nachdem man die Materie auf Ruhe transformiert und das vierdimensionale Linienelement (6.11) gewonnen hat, soll eine Funktion $\varphi(x^j)$ existieren, derart daß die nach einer Transformation des Typs (6.12) entstehende Materiedichte nur noch von der Zeit abhängt. Das ist eine Konvention über die g_{0i}. Durch

$$\varrho = \varrho(x^0) \tag{7.1}$$

hat man also eine kosmische Zeitkoordinate x^0 festgelegt. Alle weiteren noch erlaubten Transformationen, die (7.1) unverändert lassen, sind die Transformationen

$$x'^i = x'^i(x^j), \tag{7.2}$$

also die Transformationen der drei Raumkoordinaten unter sich. Dabei wurde abgesehen von den trivialen Transformationen $x'^0 = x^0 + \text{const}$, die auf allen Weltlinien die Zeigerstellungen aller Uhren um den gleichen Betrag ändern. Bei dieser Definition einer kosmischen Zeitkoordinate sind zwei Ereignisse gleichzeitig, wenn in diesen Raumzeitpunkten die mittlere Materiedichte des Weltalls gleich groß ist. Man benutzt also die Expansion des hinsichtlich der Dichte homogenen Universums, um eine kosmische Zeitkoordinate zu definieren. Nur in dem physikalisch weniger interessanten Spezialfall, wo die Materiedichte ϱ nicht von der Zeit abhängt, bei auf Ruhe transformierter Materie, wo also in kovarianter Schreibweise $\varrho_{|\mu} u^\mu = 0$ ist, also bei den sog. stationären Modellen,

[1] A. RAYCHAUDHURI: Z. Astrophys. **43**, 161 (1957). In einem speziellen Koordinatensystem wurde diese Frage von A. RAYCHAUDHURI, Phys. Rev. **98**, 1123 (1955) behandelt.

läßt sich die Funktion $\varphi(x^j)$ in (6.12) nicht konstant machen. Tatsächlich führt in dem stationären Modell von GÖDEL die Definition einer kosmischen Zeit auf fundamentale Schwierigkeiten[1].

Die Forderung der Homogenität der Dichte erschöpft aber die strenge Homogenitätsforderung noch keineswegs. Die dreidimensionalen Unterräume $\varrho = \text{const}$ sollen auch in jeder anderen Hinsicht homogen sein. Insbesondere müssen diese Räume notwendig eine (negativ definite) homogene Metrik g_{ik} besitzen. Da die Einsteinschen Feldgleichungen die Entwicklung des metrischen Feldes für alle Zeiten zu berechnen erlauben, wenn die $g_{\mu\nu}$ und ihre ersten zeitlichen Ableitungen gegeben sind, genügt es, die Homogenität zu irgendeiner Zeit für die $g_{\mu\nu}$ und ihre ersten zeitlichen Ableitungen zu postulieren. Anderenfalls würde man Postulate aufstellen, die möglicherweise den Feldgleichungen widersprechen könnten. Wir fordern also: Zur kosmischen Zeit $t = t_0$ soll der dreidimensionale Raumschnitt metrisch homogen sein; genauer, es soll eine dreiparametrige transitive kontinuierliche Gruppe existieren, gegen die die Metrik $g_{\mu\nu}$ invariant ist. Da jetzt nur noch die dreidimensionalen Koordinatentransformationen (7.2) frei sind, verhalten sich die Größen $g_{ik|0}$ und g_{0i} gegenüber diesen Transformationen wie ein dreidimensionaler symmetrischer Tensor und ein Vektor in dieser dreidimensionalen Mannigfaltigkeit. Wir fordern weiter, daß auch diese Tensoren gegen die dreiparametrige Gruppe invariant sind. Denkt man sich die g_{ik} (damit auch $g_{ik|l}$ usw.), sowie $g_{ik|0}$ und g_{0i} invariant auf dem dreidimensionalen Raum $t = t_0$ vorgegeben, so erlauben die Einsteinschen Feldgleichungen, aus diesen Größen die zweiten Ableitungen der g_{ik} nach der Zeit zu berechnen und das Cauchysche Anfangswertproblem zu lösen. Dazu benötigt man jedoch nur sechs von den zehn Feldgleichungen. Die übrigen vier enthalten nicht die zweiten Ableitungen der g_{ik} nach der Zeit. Sie stellen Bedingungen zwischen den g_{ik} und deren ersten zeitlichen Ableitungen und den g_{0i} sowie ihren räumlichen Ableitungen und der Materiedichte ϱ dar. Diese vier Bedingungen schränken also die Wahl der Anfangsbedingungen ein[2].

Alle möglichen homogenen Weltmodelle erhält man durch die Aufstellung aller dreidimensionalen Riemannschen Räume, die eine dreiparametrige transitive Bewegungsgruppe zulassen. Diese Aufgabe wurde bereits im Jahre 1897 von L. BIANCHI gelöst[3] durch die Auffindung von neun verschiedenen Raumtypen, die alle im genannten Sinne homogen sind. Dabei sind die dreidimensionalen Räume vom Bianchischen Typ I und V notwendig Räume konstanter verschwindender und negativer Krümmung. Dreidimensionale Räume vom Typ IX enthalten unter anderen Räume konstanter positiver Krümmung. In diesen drei Fällen lassen die Räume nicht nur eine drei- sondern sogar eine sechsparametrige Bewegungsgruppe zu. Das bedeutet jedoch noch keineswegs, daß der dreidimensionale Tensor $g_{ik|0}$ der ersten zeitlichen Ableitungen der g_{ik} und der Vektor g_{0i} ebenfalls gegen diese sechsparametrige Gruppe invariant sind. Den höchsten Grad räumlicher Homogenität besitzt das vierdimensionale Linienelement

$$ds^2 = (d x^0)^2 - R^2(x^0)\, d\sigma^2, \tag{7.3}$$

wo $d\sigma^2$ das Linienelement eines dreidimensionalen Raumes konstanter Krümmung ist. Weil die Zeitabhängigkeit als gemeinsamer Faktor herausgezogen ist,

[1] K. GÖDEL: Rev. Mod. Phys. **21**, 447 (1949). Vgl. auch einen Beitrag desselben Verfassers in dem Sammelband ALBERT EINSTEIN, Philosopher and Scientist. New York 1949. K. GÖDEL diskutiert die hier nicht berührten topologischen Probleme, die mit der Einführung einer kosmischen Zeit verknüpft sind.

[2] K. STELLMACHER: Math. Ann. **115**, 136 (1937). Y. FOURÈS-BRUHAT: Acta Math. **88**, 141 (1952).

[3] L. BIANCHI: Mem. Soc. Ital. Sci., Ser. IIIa **11** (1897).

sind auch die ersten zeitlichen Ableitungen auf der Anfangshyperfläche gegen eine sechsparametrige Gruppe invariant. Auch die g_{0i} zeigen wegen ihres Verschwindens den Invarianzcharakter. Diese Räume konstanter Krümmung besitzen die Eigenschaft, durch eine beliebige Drehung um einen beliebigen ihrer Punkte wieder in sich überzugehen. Das bedeutet: sie haben neben der Eigenschaft der Homogenität auch die Eigenschaft der Isotropie. Jeder Beobachter hat also auch in jeder Richtung die gleiche Ansicht von der Welt. Es sind diese speziellen homogenen Modelle, die man bisher mathematisch untersucht und vor anderen zur Beschreibung des Kosmos herangezogen hat[1]. Der näheren Darstellung dieser Modelle ist der Rest dieses Artikels gewidmet.

Das Homogenitätspostulat ist, wie anfangs bemerkt, sicherlich im Universum nicht streng erfüllt. Wenn man deshalb daran denkt, es abzuschwächen, so könnte man es beispielsweise auf der Anfangshyperfläche auf das Unendliche beschränken, sofern die Anfangshyperfläche nicht geschlossen ist. In diesem Fall hätte das abgeschwächte Homogenitätspostulat dann die Form einer „Anfangs-Randbedingung" für das metrische Feld.

8. Isotrope Modelle. Die in der vorigen Ziffer erwähnten homogenen Modelle mit isotroper Expansion sind bei ruhender Materie durch das Linienelement (7.3) gekennzeichnet. Für die Ableitung dieses Linienelementes ist es nicht erforderlich, gruppentheoretische Betrachtungen heranzuziehen. Man kann vielmehr, wie z.B. von A. RAYCHAUDHURI[2] gezeigt wurde, diese Modelle gewinnen aus dem *speziellen* Homogenitätspostulat: Rotation und Scherung des Weltsubstrates sollen überall verschwinden. Diese Forderung bedeutet: die Tensoren (6.15) und (6.16) sollen *überall* verschwinden.

Bei auf Ruhe transformierter Materie bedeutet das Verschwinden von (6.15), daß das Linienelement auf die Form

$$ds^2 = (dx^0)^2 + g_{ik}(x^\mu)\, dx^i\, dx^k \tag{8.1}$$

gebracht werden kann. Da nun $u_\mu = g_{0\mu} = \delta_{0\mu}$ ist, lautet (6.16)

$$q_{\mu\nu} = -\Gamma_{\mu\nu,0} - \frac{1}{3}(g_{\mu\nu} - u_\mu u_\nu)\frac{1}{\sqrt{-g}}\left(\sqrt{-g}\right)_{|0} = 0. \tag{8.2}$$

Daraus folgt weiter

$$g_{\mu\nu|0} = (g_{\mu\nu} - u_\mu u_\nu)\left(\log\sqrt[3]{-g}\right)_{|0}. \tag{8.3}$$

Oder, da die Gleichungen für $\mu = 0$ oder $\nu = 0$ identisch erfüllt sind,

$$g_{jk|0} = g_{jk}\left(\log\sqrt[3]{-g}\right)_{|0}. \tag{8.4}$$

Setzt man

$$\sqrt[3]{-g} = S^2(x^\mu), \tag{8.5}$$

so folgt aus (8.4) durch Integration

$$g_{jk} = -S^2(x^\mu)\,\bar{\gamma}_{jk}(x^l), \quad \text{Det} \quad \|\bar{\gamma}_{jk}\| = 1. \tag{8.6}$$

Damit von (8.6) auf (7.3) geschlossen werden kann, muß zweierlei gezeigt werden: S^2 darf die Zeitabhängigkeit nur in einem Faktor enthalten, d.h. es soll gelten

$$g_{jk} = -R^2(x^0)\,\gamma_{jk}(x^l) = -S^2(x^\mu)\,\bar{\gamma}_{jk}(x^l), \tag{8.7}$$

und die γ_{jk} müssen den metrischen Fundamentaltensor eines Raumes konstanter Krümmung darstellen. Beide Ergebnisse werden von den Feldgleichungen geliefert, in die man jetzt mit dem Linienelement (8.7) eingeht:

[1] H. ROBERTSON: Rev. Mod. Phys. 5, 62 (1933).

[2] A. RAYCHAUDHURI: Phys. Rev. 98, 1123 (1955).

Die Feldgleichungen (5.3) für inkohörente Materie kann man auch schreiben

$$G_{\mu\nu} = \left(\Lambda + \frac{\varkappa\varrho}{2}\right) g_{\mu\nu} - \varkappa\varrho\, u_\mu u_\nu. \tag{8.8}$$

Aus (8.8) folgt speziell

$$G_{0k} = 0. \tag{8.9}$$

Diese drei Gleichungen garantieren bereits, daß $S^2(x^\mu)$ die Zeitabhängigkeit abspaltet (vgl. Anhang).

Das Linienelement lautet also mit $x^0 = t$

$$ds^2 = dt^2 - R^2(t)\,\gamma_{jk}(x^l)\,dx^j dx^k. \tag{8.10}$$

Mit diesem Ansatz sind die „$0k$"-Gleichungen bereits erfüllt. Die „00"- und „ik"-Gleichungen liefern (vgl. Anhang)

$$G_{00} = \frac{3}{R}\ddot{R} = \Lambda - \frac{\varkappa\varrho}{2}, \tag{8.11}$$

$$G_{ik} = G^*_{ik} - \gamma_{ik}[\ddot{R}R + 2\dot{R}^2] = -\left(\Lambda + \frac{\varkappa\varrho}{2}\right) R^2\gamma_{ik}. \tag{8.12}$$

Dabei ist G^*_{ik} in (8.12) als Ricci-Tensor des dreidimensionalen Raumes mit der Metrik γ_{ik} definiert. Aus Gl. (8.11) folgt, daß die Materiedichte ϱ nur eine Funktion von t sein kann, da auch R nur von t abhängt und Λ sowie $\varkappa$ Konstanten sind. Die durch t definierte Zeitkoordinate ist also nach der Terminologie der vorigen Ziffer die kosmische Zeit. (8.12) läßt sich in die Form setzen

$$G^*_{ik} = -\left[\left(\Lambda + \frac{\varkappa\varrho}{2}\right) R^2 - R\ddot{R} - 2\dot{R}^2\right]\gamma_{ik}. \tag{8.13}$$

Da G^*_{ik} und γ_{ik} Funktionen der Raumkoordinaten x^j sind, die eckige Klammer aber nur die Zeit t enthält, so muß die Klammer konstant sein. Die Konstante setzen wir gleich $2\varepsilon K$. Dabei soll K positiv sein, während ε nur der Werte $1, 0, -1$ fähig ist.

Die Gl. (8.13) spaltet folglich auf in die Gleichungen

$$\left(\Lambda + \frac{\varkappa\varrho}{2}\right) R^2 - R\ddot{R} - 2\dot{R}^2 = +2\varepsilon K, \tag{8.14}$$

$$G^*_{ik} = -2\varepsilon K\gamma_{jk}. \tag{8.15}$$

(8.15) ist nun die notwendige und hinreichende Bedingung dafür, daß der dreidimensionale Raum mit der Metrik γ_{ik} ein Raum konstanter Krümmung ist[1].

Der zeitliche Ablauf der Expansion wird durch die beiden Gln. (8.11) und (8.14) beschrieben. Aus ihnen oder auch unmittelbar aus (6.3) erhält man die Kontinuitätsgleichung

$$(R^3\varrho)^{\boldsymbol{\cdot}} = 0. \tag{8.16}$$

Die Integration dieser Gleichung liefert

$$\frac{4\pi}{3} R^3\varrho = \mathfrak{M} = \text{const} > 0. \tag{8.17}$$

Aus (8.11) und (8.14) erhält man sodann mit (8.17)

$$\dot{R}^2 = \frac{\varkappa\mathfrak{M}}{4\pi R} + \frac{\Lambda}{3} R^2 - \varepsilon K. \tag{8.18}$$

---

[1] L. Eisenhart: Riemannian Geometry. Princeton 1926 § 26.

Das ist die sog. Friedmannsche Differentialgleichung, die von A. FRIEDMANN für $\varepsilon = \pm 1$ gewonnen wurde[1]. Die Lösungen dieser Gleichung befriedigen im Verein mit (8.17) auch wieder die Gl. (8.11). Das Problem der Integration der Feldgleichungen reduziert sich also darauf, die Lösungen der Friedmannschen Differentialgleichung anzugeben. Da die Funktion $R(t)$ nur bis auf einen beliebigen Faktor festgelegt ist, bedeutet es keine Einschränkung der Allgemeinheit, die Konstante $K = 1$ zu setzen. Die dann entstehende Gleichung lautet

$$\dot{R}^2 = \frac{\varkappa \mathfrak{M}}{4\pi R} + \frac{1}{3} \Lambda R^2 - \varepsilon. \tag{8.19}$$

Die Lösungen von (8.19) hängen von zwei Parametern $\mathfrak{M}$ und Λ ab. Durch die Transformation

$$x = \sqrt{\frac{1}{3}|\Lambda|}\, t, \quad y = \sqrt{\frac{1}{3}|\Lambda|}\, R, \quad \mu = \frac{\varkappa \mathfrak{M}}{4\pi} \sqrt{\frac{1}{3}|\Lambda|}, \quad \Lambda \neq 0, \tag{8.20}$$

kommt man auf die Gleichung

$$\left(\frac{dy}{dx}\right)^2 = \frac{\mu}{y} + \eta y^2 - \varepsilon, \qquad \eta, \varepsilon = +1, 0, -1, \tag{8.21}$$

die zwar zunächst nur für $\eta = \pm 1$ gilt, aber auch für $\eta = 0$ den Fall $\Lambda = 0$ umfaßt, wenn man $\mu = \frac{\varkappa \mathfrak{M}}{4\pi}$ setzt. Insgesamt erhält man also neun Lösungstypen, in denen der Parameter μ beliebige positive Werte annehmen kann[2].

Führt man in die Friedmannsche Differentialgleichung (8.19) statt der Variablen t die Variable ct ein, so lautet wegen $\varkappa = \frac{8\pi G}{c^2}$ die Gleichung

$$\frac{1}{2}\dot{R}^2 = \frac{G\mathfrak{M}}{R} + \frac{\Lambda_0}{6} R^2 + h. \tag{8.22}$$

Dabei wurde abkürzend gesetzt $\frac{-\varepsilon c^2}{2} = h$ und $\Lambda c^2 = \Lambda_0$. Diese Gleichung ist mit der entsprechenden Gl. (3.22) der Newtonschen Theorie identisch. Die in (3.24) angegebenen Lösungen der Friedmannschen Gleichung für $\Lambda = 0$ können sofort im relativistischen Fall übernommen werden. Die allgemeine Lösung für $\Lambda \neq 0$ ist von G. LEMAÎTRE[3] durch elliptische Funktionen angegeben worden. Den Fällen der elliptischen, parabolischen und hyperbolischen Bewegung in der Newtonschen Theorie entsprechen in der relativistischen Theorie die Fälle positiver, verschwindender und negativer Krümmung der Räume $x^0 = \text{const}$.

Über den allgemeinen Verlauf der Lösungen $R(t)$ vergleiche man die Literatur[4], die auch zahlreiche Abbildungen enthält.

Für das Linienelement der Räume konstanter Krümmung wird im folgenden der Ausdruck

$$d\sigma^2 = \gamma_{ik}\, dx^i\, dx^k = d\chi^2 + S^2(\chi)\,(d\vartheta^2 + \sin^2\vartheta\, d\varphi^2) \tag{8.23}$$

benutzt, wo

$$S(\chi) = \frac{1}{\sqrt{\varepsilon}} \sin\sqrt{\varepsilon}\,\chi = \left\{\begin{array}{lll} \sin\chi & \text{für} & \varepsilon = 1, \\ \chi & \text{für} & \varepsilon = 0, \\ \operatorname{Sin}\chi & \text{für} & \varepsilon = -1, \end{array}\right\} \tag{8.24}$$

---

[1] A. FRIEDMANN: Z. Physik **10**, 377 (1922); **21**, 326 (1924).

[2] Vgl. z.B. W. DE SITTER: The astronomical aspect of the theory of relativity, Univ. Calif. Publ. Math. **2**, No. 8, 143–196 (1933) oder auch R. TOLMAN: Relativity, Thermodynamics, and Cosmology. Oxford 1934.

[3] G. LEMAÎTRE: Ann. Soc. Sci. Bruxelles, Ser. A **53**, 51 (1933).

[4] Vgl. z.B. O. HECKMANN: Göttinger Nachr. **1932**, 97. — R. TOLMAN: Relativity, Thermodynamics, and Cosmology. Oxford 1934. — H. BONDI: Cosmology. Cambridge 1952.

ist. Auch die Form

$$d\sigma^2 = (1 + \tfrac{1}{4}\varepsilon r^2)^{-2} dx^i dx^i; \quad r^2 = x^i x^i, \tag{8.25}$$

wird oft verwendet (vgl. z.B. den vorhergehenden Artikel von McVittie in diesem Bande).

9. Beobachtbare Beziehungen. α) *Rotverschiebung.* In dieser Ziffer sollen diejenigen Formeln abgeleitet werden, die es erlauben, die isotrop expandierenden homogenen Modelle mit der Erfahrung zu vergleichen[1]. Es ist hierzu erforderlich, das Verhalten von Lichtstrahlen in dem metrischen Felde zu untersuchen, welches durch das Linienelement (8.10) mit (8.23)

$$ds^2 = dt^2 - R^2(t)\left[d\chi^2 + S^2(\chi)\,(d\vartheta^2 + \sin^2\vartheta\, d\varphi^2)\right] \tag{9.1}$$

gegeben ist. Infolge der Homogenität des Modells ist es keine Einschränkung der Allgemeinheit, den Ort des Beobachters in den räumlichen Nullpunkt des Koordinatensystems, also nach $\chi = 0$ zu legen. Für einen Lichtstrahl, der zur Zeit t_1 an einem Ort der räumlichen radialen Koordinate χ_1 ausgestrahlt wird und zur Zeit t_2 vom Beobachter bei $\chi = 0$ empfangen wird, besitzt die Lichtbahn aus Symmetriegründen konstante Winkelkoordinaten ϑ und φ. Für die Bahn des Lichtstrahls gilt also $d\vartheta = d\varphi = 0$. Setzt man das in (9.1) ein, so erhält man, da $ds = 0$ für jede Lichtbahn charakteristisch ist,

$$dt = -R(t)\, d\chi \tag{9.2}$$

(mit dem Minuszeichen, da wir Einstrahlung betrachten). Die Integration dieser Gleichung ergibt

$$\chi_1 = -\int_{\chi_1}^{0} d\chi = \int_{t_1}^{t_2} R^{-1}(t)\, dt. \tag{9.3}$$

Ein Lichtsignal, das zu einer späteren Zeit $t_1 + dt_1$ von der Lichtquelle ausgesandt wird, trifft beim Beobachter zur Zeit $t_2 + dt_2$ ein. Wenn Lichtquelle und Beobachter nun relativ zum expandierenden Substrat ruhen, also keine Peculiargeschwindigkeit gegenüber dem expandierenden Feld zeigen, tragen sie für alle Zeiten die festen Koordinaten χ_1 und 0. Wenn man die Integration von (9.2) also für das zur Zeit $t_1 + dt_1$ von der Quelle ausgesandte Lichtsignal durchführt, erhält man analog zu (9.3)

$$\chi_1 = -\int_{\chi_1}^{0} d\chi = \int_{t_1+dt_1}^{t_2+dt_2} R^{-1}(t)\, dt. \tag{9.4}$$

Subtrahiert man (9.3) von (9.4), so wird mit infinitesimalem dt_1

$$\frac{dt_2}{R(t_2)} = \frac{dt_1}{R(t_1)}. \tag{9.5}$$

Wie man aus (9.1) erkennt, sind wegen $d\chi = d\vartheta = d\varphi = 0$ die Zeitintervalle dt_1 und dt_2 unmittelbar als Eigenzeitintervalle aufzufassen, die angezeigt werden von Uhren, die in (9.1) räumlich ruhen. Wir bezeichnen sie mit ds_1 bzw. ds_2. Wenn man also ds_1 mit einer bestimmten Wellenlänge identifiziert, die ein bestimmtes Atom in einem ruhenden irdischen Laboratorium aussendet, hat man ds_2 mit der entsprechend geänderten Wellenlänge zu identifizieren, die vom

[1] H. Robertson: Publ. Astronom. Soc. Pacific **67**, 82 (1955). — Helv. phys. Acta, Suppl. **4**, 128 (1956).

Beobachter empfangen wird. Das Verhältnis dieser beiden Wellenlängen wird gemäß (9.5)

$$\frac{d s_2}{d s_1} = \frac{\lambda_2}{\lambda_1} = \frac{R(t_2)}{R(t_1)} = 1 + z = \frac{R_2}{R_1}. \tag{9.6}$$

Wenn man diese Formel zur Deutung der Rotverschiebung des Lichtes extragalaktischer Nebel benutzt, geht natürlich darin die Annahme ein, daß man die Eigenzeit $d s_1$ zur Zeit der Emission im irdischen Laboratorium zur Empfangszeit t_2 messen kann, also die Annahme, daß die Eigenfrequenzen der Atome im Laufe der Zeit keine Veränderung erfahren. In die Deutung der Beobachtungen geht also die Annahme ein, daß die Naturkonstanten — soweit sie die Eigenfrequenzen der Atome festlegen — von der Zeit unabhängig sind.

Wenn die Lichtquelle zur Zeit der Emission relativ zum Substrat die rein radiale Peculiargeschwindigkeit $\beta_1 = v_1/c$ hatte und der Beobachter zur Zeit des Empfanges relativ zum Weltsubstrat die rein radiale Geschwindigkeit $\beta_2 = v_2/c$ hatte, lautet die entsprechende Formel für die Linienverschiebung

$$\frac{\lambda_2}{\lambda_1} = (1 + z)\,\frac{(1 + \beta_2)(1 + \beta_1)}{\sqrt{(1 - \beta_2^2)(1 - \beta_1^2)}}. \tag{9.7}$$

Dabei bedeutet wieder wie vorher $1 + z = R_2/R_1$. In dieser Formel drückt sich das relativistische Additionstheorem der Geschwindigkeiten aus. Zwei Anwendungen dieser Formel sind bemerkenswert: Setzt man einmal $\beta_1 = 0$, so sieht man, daß die von der Bewegung der Erde um die Sonne herrührende Dopplerverschiebung mit dem Faktor $1 + z$ multipliziert ist. Das ist ein Effekt, der prinzipiell beobachtbar wäre bei sehr entfernten Nebeln, die nahe der Ekliptik stehen. Er dürfte jedoch zur Zeit noch unter der Fehlergrenze heutiger Meßgenauigkeit liegen. Wenn man andererseits $\beta_2 = 0$ setzt, sieht man, daß die Peculiargeschwindigkeiten von Nebeln in einem Nebelhaufen aus den üblicherweise direkt in Geschwindigkeitsmaß ausgedrückten Dopplerverschiebungen erst nach Division durch $(1 + z)$ gewonnen werden können. Wie die Behandlung der Bewegung von Probekörpern zeigt, ist diese Geschwindigkeit noch einmal durch $(1 + z)$ zu dividieren, wenn man sie für den Zeitpunkt der Ankunft des Lichtsignals auf der Erde errechnen will.

Prinzipiell wäre es möglich, das Verhältnis (9.6) der beiden Eigenzeiten nicht nur durch die Verschiebung von Spektrallinien zu ermitteln. Die Beobachtung irgendeines, möglicherweise bekannten zeitlichen Ablaufs, wie z. B. des An- oder Abklingens der Helligkeit einer Supernova könnte prinzipiell ebensogut verwendet werden.

Die Gln. (9.6) und (9.7) zeigen, daß die von ihnen beschriebenen relativen Wellenlängenänderungen über das ganze Spektrum hinweg die gleichen sind. Diese Folgerung ist in einem Spezialfall mit außerordentlich hoher Genauigkeit bestätigt worden. Die Verschiebungen der 21 cm-Linie des atomaren Wasserstoffs zweier extragalaktischer Radioquellen führten auf einen im Rahmen der Fehlergrenzen gleichen Wert der „Fluchtgeschwindigkeit" wie die Messungen der Wellenlängenverschiebungen von Spektrallinien im Bereich des sichtbaren Lichtes[1].

Ehe jetzt die Relationen zwischen Helligkeit und Entfernung extragalaktischer Nebel abgeleitet werden, kann man schon darauf hinweisen, daß die Messung von Radialgeschwindigkeiten und scheinbaren Helligkeiten von extragalaktischen Nebeln (Messungen an fast 1000 Nebeln liegen bisher vor[2]) bereits

[1] R. MINKOWSKI u. O. WILSON: Astrophys. Journ. **123**, 373 (1956).
[2] M. HUMASON, N. MAYALL u. A. SANDAGE: Astronom. J. **61**, 97 (1956).

dazu benutzt werden können, die Isotropie der Nebelverteilung in bezug auf unseren Standort zu prüfen. Nebel gleicher scheinbarer Helligkeit sollten im Durchschnitt die gleiche Rotverschiebung zeigen. Dabei ist natürlich vorausgesetzt, daß es sich im wesentlichen um ähnliche Nebel handelt, die man in verschiedenen Richtungen beobachtet. Näherungsweise ist diese Isotropie der Nebelflucht bestätigt, obwohl sich anzudeuten scheint, daß die Nebel südlich der Milchstraßenebene im Durchschnitt etwas heller sind als die nördlichen Nebel gleicher Rotverschiebung[1]. Die wenigen bisher untersuchten fernen Nebelhaufen zeigen keine Anzeichen einer Anisotropie der Nebelflucht und bestätigen insofern teilweise die vorausgesetzten Eigenschaften des hier benutzten Modells.

β) Durchmesser, Helligkeit und *„Entfernung"*. Die Beziehung zwischen scheinbarem und linearem Durchmesser eines Nebels folgt aus (9.1). Wenn $\varDelta$ der lineare Durchmesser ist, gilt zur Zeit t_1 der Emission des Lichtes

$$\varDelta = R(t_1)\, S(\chi)\, d\vartheta. \tag{9.8}$$

Dieser Winkel $d\vartheta$ ist auch der Winkel, unter dem einem irdischen Beobachter der Nebel zur Zeit t_2 erscheint, nämlich als derjenige Teil des Vollkreises, den er bei der Emission des Lichtes am Himmel ausmachte. Das spätere Schicksal des Nebels nach Aussendung des Lichtes spielt für die Berechnung keine Rolle.

Die Beziehung zwischen scheinbarer und absoluter Helligkeit eines Nebels läßt sich vorteilhaft mit Hilfe der Lichtquantenvorstellung ableiten. Ein Nebel der absoluten Helligkeit L möge während des Eigenzeitintervalles der Länge 1 n_1 Lichtquanten der Frequenz ν_1 in den Raumwinkel 1 ausstrahlen. Dann gilt

$$L = n_1 h \nu_1. \tag{9.9}$$

Zur Zeit t_2 hat sich die emittierte Lichtenergie $4\pi L$ auf eine Kugeloberfläche der Größe $4\pi S^2(\chi) R^2(t_2)$ ausgebreitet und besitzt auf dieser die Gesamtenergie $4\pi n_2 h \nu_2$. Der auf die Flächeneinheit (Öffnung eines Fernrohrobjektivs) auftreffende Betrag l ist ein Maß der scheinbaren Helligkeit des Nebels. Es gilt

$$l = \frac{4\pi n_2 h \nu_2}{4\pi S^2(\chi) R^2(t_2)}. \tag{9.10}$$

Da keine Lichtquanten verloren gehen, wenn keine Absorption des Lichtes unterwegs stattfindet, verhalten sich die Zahlen der emittierten zu den empfangenen Quanten genau so wie die ausgesandten zu den empfangenen Frequenzen, also umgekehrt proportional zu den Eigenzeiten. Es gilt mithin nach (9.5)

$$\frac{n_2}{n_1} = \frac{\nu_2}{\nu_1} = \frac{R(t_1)}{R(t_2)}. \tag{9.11}$$

Setzt man das in (9.10) ein, so kommt mit (9.6)

$$l = \frac{L}{S^2(\chi) R^2(t_2)\, (1+z)^2}. \tag{9.12}$$

Dies ist die gesuchte Formel für den Zusammenhang zwischen absoluter und scheinbarer Helligkeit eines Nebels. Führt man zur Abkürzung den Begriff der photometrischen Distanz ein, definiert durch

$$D = R(t_2)\, S(\chi)\, (1+z), \tag{9.13}$$

so läßt sich (9.12) auch in der gewohnten Form schreiben

$$l = \frac{L}{D^2}. \tag{9.14}$$

[1] Vgl. die vorige Arbeit, S. 138.

Aus (9.8) und (9.12) kann man auch sogleich die Formel für die scheinbare Flächenhelligkeit extragalaktischer Objekte ableiten. Es wird

$$\frac{l}{(d\vartheta)^2} = \frac{L\,R^2(t_1)}{\Delta^2(1+z)^2\,R^2(t_2)}. \tag{9.15}$$

Aus (9.15) folgt, daß die Flächenhelligkeit extragalaktischer Objekte jetzt stark von der Rotverschiebung z abhängt. Mit Hilfe von (9.6) erhält man aus (9.15)

$$\frac{l}{(d\vartheta)^2} = \frac{L}{\Delta^2(1+z)^4}. \tag{9.16}$$

Die Messung von Totalhelligkeiten bei extragalaktischen Nebeln stößt auf Schwierigkeiten, weil ihre Konturen schwer definierbar sind und ihre Flächenhelligkeit sehr langsam nach außen abnimmt. Deshalb spielt (9.16) für die Photometrie entfernter Nebel eine Rolle[1], weil man an Stelle von Totalhelligkeiten von Nebeln die Helligkeiten der Nebelteile benutzt, die innerhalb einer gegebenen Kontur konstanter Flächenhelligkeit (Isophoten) liegen. Diese Konturen liegen gemäß (9.16) bei absolut gleichen Nebeln von verschiedenen Entfernungen, also von verschiedenem z in verschiedenen linearen Abständen vom jeweiligen Nebel-Zentrum. Die schwächsten Außenpartien der Nebel gelangen also mit zunehmender Entfernung nicht mehr zur Abbildung. Schon deshalb wird das Licht immer fernerer Nebel immer mehr von den spektralen Eigenschaften der Kerne bestimmt.

Die linearen Durchmesser, absoluten Helligkeiten und absoluten Flächenhelligkeiten der Nebel, die in die obigen Formeln eingehen, beziehen sich sämtlich auf den Zeitpunkt der Emission des Lichtes. Will man also ferne Nebel mit nahen vergleichen, so hat man zeitliche Änderungen des Durchmessers und der absoluten Helligkeit des Nebels in Betracht zu ziehen, ebenso wie Änderungen der spektralen Energieverteilung. Darüber ist jedoch sehr wenig bekannt. Neben diesen Korrekturen hat man bei der Ermittlung der photometrischen Distanz eines fernen Nebels die Möglichkeit einer allgemeinen internebularen Absorption ins Auge zu fassen. Deutliche Anzeichen für ihr Vorhandensein sind bisher noch nicht gefunden worden. Eine weitere wichtige Korrektur, die schließlich bei der Bestimmung von photometrischen Distanzen entfernter Nebel ins Spiel kommt, ist der sog. K-Term. Die selektive Empfindlichkeit der Aufnahmeapparatur (Atmosphäre, Spiegel, Photoplatte usw.) bedingt, daß die Änderung der spektralen Energieverteilung infolge der Rotverschiebung des Lichtes über den bolometrischen Effekt hinaus noch berücksichtigt werden muß[2]. Zur Berechnung dieser Korrektur ist die Kenntnis der spektralen Energieverteilung des Nebels Voraussetzung. Vergleicht man insbesondere noch einen sehr fernen Nebel mit einem nahen Nebel, so ist es erforderlich, die Energieverteilung des nahen Nebels im Ultravioletten zu kennen, weil dieser Spektralbereich bei dem fernen Nebel infolge der Rotverschiebung des Nebellichtes in den Empfindlichkeitsbereich der Apparatur gerückt ist. Der K-Term ist genähert proportional zu z, gibt also bei fernen Nebeln einen beträchtlichen Beitrag zur oben erwähnten Flächenhelligkeitskorrektur. Aus allem dürfte deutlich werden, daß die Bestimmung der photometrischen Distanz eines entfernten Nebels mit beträchtlicher Rotverschiebung auf der experimentellen Seite eine schwierige Aufgabe ist, die zu ihrer Lösung umfangreiche Beobachtungen an extragalaktischen Nebeln zur Voraussetzung hat, die in dem erforderlichen Umfang bisher noch nicht vorliegen.

---

[1] J. STOCK u. E. SCHUECKING: Astronom. J. **62**, 98 (1957).

[2] Siehe Fußnote 2, S. 510.

In diesem Zusammenhang ist noch auf eine erst kürzlich erschlossene Methode zur Bestimmung von Rotverschiebungen hinzuweisen. Jenseits von $z=0{,}2$ werden die Nebel so lichtschwach, daß das Spektrum des Nachthimmellichtes das Nebelspektrum stark überdeckt, so daß die Messung von Linienverschiebungen im Nebelspektrum ein außerordentlich schwieriges Problem darstellt. W. BAUM[1] gelang es, die Rotverschiebung durch photoelektrische Messung der spektralen Intensitätsverteilung zu ermitteln. Aus der verschobenen Verteilung berechnete BAUM Rotverschiebungen bis zu $z=0{,}4$. Das Baumsche Verfahren fußt auf der Annahme, daß die spektrale Energieverteilung ferner Nebel sich im Verlauf der letzten ein bis zwei Milliarden Jahre, der vermutlichen Laufzeit des Lichtes, nicht geändert hat. Entgegen früheren Messungen von STEBBINS und WHITFORD an elliptischen Nebeln glaubte BAUM festzustellen, daß bis $z=0{,}2$ keine solchen Änderungen der spektralen Energieverteilung des vom Nebel ausgesandten Lichtes festzustellen waren.

γ) Theorie und Erfahrung. Für die Prüfung kosmologischer Modelle kommen im Prinzip alle Beziehungen zwischen der photometrischen Distanz und anderen beobachtbaren Daten in Frage. Außer der Rotverschiebung sind besonders zu nennen Nebelanzahlen und Durchmesser. Auch die Hintergrundhelligkeit des Himmels liefert einen Anhaltspunkt. Das Verfahren der Nebelzählungen, das der große Pionier moderner kosmologischer Forschungen, E. HUBBLE[2], für die Kosmologie nutzbar gemacht hatte, hat bisher nicht Ergebnisse der Präzision geliefert, die man erhofft hatte. Eine Reduktion des Hubbleschen Materials nach modernen Gesichtspunkten scheint nicht sehr lohnend zu sein. Neuere Zählungen bis zu Nebeln der 23. Größenklasse herunter sind nicht begonnen worden, so daß vermutlich im Laufe der nächsten Jahre keine für die Kosmologie entscheidenden Ergebnisse von diesem an sich sehr wichtigen Verfahren erwartet werden dürfen. Die Zählung spezieller Objekte insbesondere zusammenstoßender Spiralnebel, die Radioquellen sind, könnte möglicherweise aus noch größeren Tiefen des Weltraumes entscheidende kosmologische Daten liefern. Es herrscht aber noch keine Einigkeit darüber, ob die Radioquellen mit fehlender Konzentration zur Milchstraßenebene hin tatsächlich als entfernteste extragalaktische Objekte anzusprechen sind[3].

Die Grundlage für die kosmologische Auswertung der Zählungen ist die Integralgleichung

$$N = \frac{4\pi}{Q} R_2^3 \int\limits_0^{\chi} n[\chi(t)]\, S^2(\chi)\, d\chi. \tag{9.17}$$

Dabei ist die Streuung der absoluten Helligkeiten der zu zählenden Objekte vernachlässigt. Mit $N(l)$ ist die Anzahl der Objekte pro Quadratgrad bis zur scheinbaren Helligkeit l bezeichnet[4]. Q ist die Zahl der Quadratgrade auf der Sphäre und $n[\chi(t)]$ die Anzahl der Objekte je Volumeneinheit. Die weitere Durchführung der Rechnungen soll hier nicht besprochen werden.

Wegen der Helligkeit des Nachthimmellichtes läßt sich die Hintergrundhelligkeit des Himmels im optischen Bereich vom Erdboden aus nicht ermitteln. Radioastronomisch könnte diese Größe aber möglicherweise meßbar sein, wenn

[1] Vgl. Sky a. Telescope, Dec. **1956**, 60.

[2] E. HUBBLE: The Realm of the Nebulae. Oxford 1936. The observational Approach to Cosmology. Oxford 1937.

[3] J. BALDWIN: Nature, Lond. **174**, 320 (1954).

[4] W. PRIESTER: Z. Astrophys. **34**, 283 (1954); **46**, 179 (1958). — J. PAWSEY in Int.-Astr. Union, Draft Reports 1958, S. 371.

es gelänge, den Beitrag der Quellen innerhalb der lokalen Gruppe abzuziehen. Für die nähere Diskussion dieses Problemes verweisen wir auf die Literatur[1].

Für die weitere Diskussion beobachtbarer Beziehungen erweist es sich als zweckmäßig, $T = t_2 - t_1$, die zwischen Emission und Empfang des Lichtsignals verstrichene kosmische Zeit, als Variable einzuführen. Dann wird aus (9.3) und (9.6)

$$\chi(T) = \int_{t_2 - T}^{t_2} R^{-1}(t)\, dt, \qquad z = \frac{R(t_2)}{R(t_2 - T)} - 1. \tag{9.18}$$

Aus der ersten Gleichung erhält man nun leicht durch Reihenentwicklung und Einsetzen $r = S(\chi)$ als Potenzreihe in T. Entsprechend gewinnt man aus der zweiten Gl. (9.18) die Rotverschiebung z als Potenzreihe in T. Kehrt man die erste Reihe um und setzt dann $T(r)$ in die Reihe $z(T)$ ein, so kommt als Beziehung zwischen r und z bis auf Glieder höherer Ordnung

$$z = \dot{R}\, r - (\ddot{R}\, R - \dot{R}^2)\, \frac{r^2}{2!} + \cdots. \tag{9.19}$$

Vernachlässigt man das Glied mit r^2, so erhält man daraus

$$z \approx \frac{\dot{R}}{R}(r R) \approx \frac{\dot{R}}{R} D. \tag{9.20}$$

Es folgt also in erster Näherung die Proportionalität von Rotverschiebung und Entfernung, die von HUBBLE entdeckt wurde. $\dot{R}/R$ ist dabei die sog. Hubble-Konstante. Ebenso wie die anderen Koeffizienten der Reihenentwicklungen bezieht sie sich auf die Zeit t_2 des Lichtempfanges.

Logarithmiert man die Formeln (9.13) und (9.14) und führt vermittels

$$m = -2{,}5\, \mathrm{Log}\, l, \qquad M = -2{,}5\, \mathrm{Log}\, L \tag{9.21}$$

scheinbare und absolute Größenklassen ein, so wird

$$0{,}2\, m = 0{,}2\, M + \mathrm{Log}\, R + \mathrm{Log}\, r(1 + z). \tag{9.22}$$

Eliminiert man r durch Umkehrung von (9.19), erhält man

$$\left.\begin{aligned} \mathrm{Log}\, z &= 0{,}2\,(m_z - \Delta m_z) - 0{,}2\, M + \mathrm{Log}\, \frac{\dot{R}}{R}, \\ \Delta m_z &= 1{,}086 \left[\left(1 + \frac{\ddot{R} R}{\dot{R}^2}\right) z + \cdots\right]. \end{aligned}\right\} \tag{9.23}$$

Bei der Berechnung von m hat man insbesondere K-Term, Leuchtkraftänderung des Nebels usw. zu berücksichtigen[2]. Als neuer dimensionsloser Koeffizient tritt hier die Größe

$$\frac{\ddot{R} R}{\dot{R}^2} \tag{9.24}$$

auf. Gemäß (9.23) pflegt man Log z gegen m aufzutragen, um auf diese Weise die Hubble-Konstante und (9.24) zu bestimmen.

Es ist jedoch zu betonen, daß ein Arbeiten mit abgebrochenen Potenzreihen nur von zweifelhaftem Wert ist. Durch einen Ausgleich nach der Methode der

[1] Siehe Fußnote 3, S. 513.
[2] Siehe Fußnote 2, S. 510.

kleinsten Quadrate werden nämlich die ersten Glieder einfach zur Interpolation der mit zufälligen Fehlern behafteten Beobachtungen benutzt. Es wird kaum möglich sein, zwei Terme verschiedener Ordnung durch Ausgleichung richtig zu ermitteln, wenn sie die Interpolation gleich gut leisten, weil sie dann durch Beobachtungsfehler statistisch gekoppelt werden. Wenn auch die ersten Glieder mit kleinen mittleren Fehlern herauskommen, sind ihre Werte deswegen noch nicht verbürgt, weil alle höheren Glieder der Entwicklung formal zu Null angenommen wurden[1].

Aus der Hubble-Konstanten in Verbindung mit (9.24) und der mittleren Materiedichte im Weltall kann man im Prinzip über die Friedmannsche Differentialgleichung das Vorzeichen der Krümmung, den Krümmungsradius R und das kosmologische Glied bestimmen. Angesichts der großen Ungenauigkeit der Ausgangsdaten wäre es jedoch verfrüht, den erhaltenen Werten allzuviel Bedeutung beizulegen[2]. Durch Integration der Friedmannschen Differentialgleichung läßt sich nach Bestimmung der Anfangsbedingungen und der Konstanten, die in diese Gleichung eingehen, das „Weltalter" ableiten.

10. Topologische Fragen. Als A. EINSTEIN im Jahre 1917 sein statisches Weltmodell vorschlug, deutete er die räumliche dreidimensionale Metrik als die eines sphärischen Raumes. Der Mathematiker F. KLEIN wies damals sogleich darauf hin, daß die Metrik des dreidimensionalen Raumes ebensowohl aufgefaßt werden dürfe, als die eines elliptischen Raumes. Der dreidimensionale elliptische Raum läßt sich darstellen als eine dreidimensionale Kugel, die in einen vierdimensionalen euklidischen Raum eingebettet ist, und in der die Punkte, die durch Spiegelung am Mittelpunkt der Kugel auseinander hervorgehen, als identisch angesehen werden. Allgemeiner gilt aber, daß jede eigentlich diskontinuierliche, fixpunktfreie Symmetriegruppe zur Identifizierung von Punkten herangezogen werden kann, wenn die Metrik gegen die Operationen der Gruppen invariant ist. Im Fall der dreidimensionalen Räume konstanter Krümmung gewinnt man auf diese Weise eine Fülle von neuen Raumtypen, die lokal alle die Eigenschaften von Räumen konstanter Krümmung haben, jedoch topologisch nicht äquivalent sind. Dreidimensionale Räume konstanter positiver Krümmung sind jedoch notwendig geschlossen (sie besitzen ein endliches Volumen). Von ihnen gibt es unendlich viele topologisch verschiedene Raumformen. Bei den dreidimensionalen Räumen konstanter verschwindender Krümmung gibt es 18 Typen, darunter insbesondere auch geschlossene Raumformen wie z.B. den Raum, der durch Identifizierung der gegenüberliegenden Flächen eines Würfels entsteht. Bei den Räumen konstanter negativer Krümmung gibt es wieder unendlich viele Typen, insbesondere auch geschlossene Raumformen[3]. Entsprechende Aussagen für die anderen homogenen dreidimensionalen Räume sind uns nicht bekannt geworden. Die Lösung des Raumtypenproblems in der vierdimensionalen Raumzeit scheint bisher noch nicht die gebührende Aufmerksamkeit der Mathematiker gefunden zu haben. Insbesondere sind die topologischen Probleme von Mannigfaltigkeiten mit indefiniter Metrik noch wenig beachtet worden. Lediglich E. SCHRÖDINGER[4] hat versucht, im Falle des de Sitter-Kosmos (vierdimensionale Kugel im pseudoeuklidischen fünfdimensionalen Raum) gegenüberliegende *Raumzeit*punkte miteinander zu identifizieren. Die wichtige Frage, wie die topologischen Eigenschaften der Raumzeit prinzipiell beobachtet werden könnten, ist noch ungelöst.

---

[1] Siehe Fußnote 1, S. 503.
[2] F. HOYLE u. A. SANDAGE: Publ. Astronom. Soc. Pacific **68**, 301 (1956).
[3] E. PESCHL: 8. Semesterber. Math. Seminar Universität Münster 1936.
[4] E. SCHRÖDINGER: Expanding Universes. Cambridge 1956.

Die Untersuchung der Eigenschaften eines Weltmodells im Großen gewinnt an Übersichtlichkeit, wenn man sich des Hilfsmittels der „Einbettung" bedient. Wie H. ROBERTSON[1] gezeigt hat, lassen sich die homogen-isotropen Weltmodelle in einen fünfdimensionalen pseudoeuklidischen Raum einbetten. Da es für viele Untersuchungen ausreichend ist, zwei Dimensionen des Raumes zu vernachlässigen, lassen sich viele Fragen anschaulich an dem Modell einer zweidimensionalen Mannigfaltigkeit klären, die in den dreidimensionalen pseudoeuklidischen Raum eingebettet ist. Diese Flächen stehen in voller Analogie zu den in der Physik häufig benutzten ebenen Raumzeitdiagrammen der Minkowski-Welt.

Aus den Robertsonschen Diagrammen lassen sich sogleich einige wichtige Tatsachen ablesen: In vielen Fällen existieren „Koordinatenhorizonte" in den einzelnen kosmologischen Modellen; die Koordinaten, die zur Beschreibung der Mannigfaltigkeit gewählt wurden, überdecken dann nur ein Teilgebiet der ganzen, durch das Linienelement gegebenen Mannigfaltigkeit (dabei werden Identifizierungen von Weltpunkten nicht berücksichtigt). Die an den Koordinatenhorizonten auftretenden Singularitäten des Linienelementes sind in solchen Fällen nur durch die Koordinatenwahl bedingt. Bei den homogen isotropen Modellen ist die gern mit dem „Urknall" identifizierte Singularität hingegen reell. Nur das linear expandierende Modell mit konstanter negativer Krümmung bildet eine Ausnahme. Dieses Modell, in dem die Materiedichte verschwinden muß, ist nur ein Ausschnitt der vierdimensionalen pseudoeuklidischen Welt, der all jene Weltpunkte umfaßt, die zeitartig zu einem ausgezeichneten Ereignis liegen.

Auch die Existenz von „optischen Horizonten" läßt sich anschaulich in den Robertsonschen Diagrammen erkennen. Es handelt sich hier um die Tatsache, daß in einigen Modellen der Nachkegel eines Substratbeobachters im Laufe der Zeit nicht die ganze vierdimensionale Mannigfaltigkeit überstreicht. Es gibt dann Ereignisse, von denen dieser Beobachter niemals etwas erfahren kann. Das sind genau die Ereignisse, die jenseits seines Horizontes stattfinden[2].

11. Verschiedene Verallgemeinerungen. Außer den homogen-isotropen Modellen mit inkohärenter Materie spielen gelegentlich noch andere kosmologische Modelle wichtige Rollen. Obwohl jetzt die Energiedichte der Lichtstrahlung (vermutlich auch der Neutrinostrahlung und der kosmischen Strahlung) im Weltall außerordentlich klein ist gegenüber der mit c^2 multiplizierten Materiedichte der gleichmäßig über das All verteilt gedachten extragalaktischen Nebel, so folgt daraus noch nicht, daß man die Strahlung im Energieimpulstensor völlig vernachlässigen darf. Da die Materiedichte proportional zu R^{-3}, die Energiedichte der Strahlung jedoch proportional zu R^{-4} abnimmt in den homogen-isotropen Modellen, so war in den Frühstadien des Universums die Energiedichte der Strahlung größer als die mit c^2 multiplizierte mittlere Materiedichte. Ein Weltmodell mit inkohärenter Materie stellt also für die Frühstadien des Universums eine schlechte Näherung dar. Für diese Zeiträume empfiehlt es sich dann, Modelle eines sog. Lichtkosmos zu verwenden.

Eine der bedeutendsten Entdeckungen innerhalb der theoretischen Kosmologie ist das rotierende Weltmodell von K. GÖDEL[3]. Da im Gödelschen Weltmodell kein Rotverschiebungseffekt zu beobachten ist, kommt es zwar sicherlich zur Beschreibung der Wirklichkeit nicht in Betracht. Trotzdem hat dieses Modell der Theorie wichtige Hinweise gegeben. Einmal lieferte es ein drastisches Beispiel

[1] H. ROBERTSON: Rev. Mod. Phys. **5**, 62 (1933).
[2] W. RINDLER: Monthly Notices Roy. Astronom. Soc. London **116**, 662 (1956).
[3] Siehe Fußnote 8, S. 490.

dafür, daß das Machsche Prinzip sicherlich keine Folge der Einsteinschen Gravitationstheorie ist. Die homogen verteilte Materie dieses statischen Weltmodells zeigt hier nämlich absolute Rotation gegenüber dem Trägheitskompaß. Außerdem machte K. GÖDEL mit diesem Modell, in dem es geschlossene zeitartige Linien gibt, auf tiefliegende Schwierigkeiten aufmerksam, die sich an den Begriff einer absoluten Zeitkoordinate knüpfen. In einer wegweisenden späteren Arbeit[1] hat K. GÖDEL dann viele Sätze, zumeist ohne Beweise, für allgemeinere räumliche homogene Modelle, die Expansion, Rotation und Scherung des druckfreien Weltsubstrates zeigen, angegeben.

Die homogen-isotropen Weltmodelle stellen sicherlich eine weitgehende Idealisierung der wirklichen Weltmetrik dar. Es taucht deshalb die Frage auf, ob die Abweichungen von der Homogenität und Isotropie einen merklichen Einfluß auf die zeitliche Entwicklung des Weltmodells haben, ob also die kosmologischen Lösungen der Feldgleichungen stabil sind oder nicht. Bisher sind nur sehr spezelle und tastende Versuche zur Lösung dieses wichtigen Problems bekannt geworden[2].

Die allgemeine Untersuchung inhomogener Modelle, die der wirklichen Materieverteilung im Kosmos sehr viel eher entsprechen dürften, ist durch A. RAYCHAUDHURI[3] bedeutend gefördert worden. Besonderes astronomisches Interesse beanspruchen dabei die Ansätze, die bisher zur einer Theorie der Nebelhaufen vorliegen. Da nämlich das Milchstraßensystem einem solchen kleinen Nebelhaufen, der lokalen Gruppe, angehört, taucht die Frage auf, wieweit die lokalen Abweichungen von der Homogenität bei der Reduktion kosmologischer Beobachtungsdaten berücksichtigt werden müssen und ob andererseits die Expansion des Weltalls Rückwirkungen auf die Vorgänge innerhalb der lokalen Gruppe oder des Milchstraßensystems nach sich zieht. Dieses schwierige Problem, dessen Lösung auch durch die Unkenntnis zahlreicher wichtiger Beobachtungsdaten behindert ist, läßt sich qualitativ zunächst durch die Behandlung inhomogener kugelsymmetrischer Materieverteilungen studieren, die in großer Entfernung vom Symmetriezentrum in das expandierende homogen-isotrope Modell übergehen. Besonders einfach ist in dieser Hinsicht das von A. EINSTEIN und E. STRAUS behandelte Vacuolenmodell, dessen Newtonsche Version wir schon früher diskutiert haben[4]. An Hand dieses Modells läßt sich erkennen, daß die Nebelhaufen und Nebel selbst an der allgemeinen Expansion nicht teilzunehmen brauchen. Aus der näheren Diskussion dieses Modells ergibt sich noch, daß die Zeitskala des Schwarzschild-Feldes in der Vacuole nicht mit der kosmischen Zeitskala der Substratbeobachter übereinstimmt. Das bedeutet genauer: Die Uhr, die durch den Umlauf eines Planeten um die Sonne geliefert wird, mißt nicht die kosmische Zeit t, die in das kosmologische Linienelement eingeht. Der Umrechnungsfaktor von Planetenzeit auf kosmische Zeit ist zeitlich langsam veränderlich. Dem endlichen Weltalter entsprechen mehr Planetenumläufe (also Erdjahre) als der jetzige Umrechnungsfaktor von Planetenzeit auf kosmische Zeit liefert. Beobachtbare Effekte scheinen jedoch aus diesem Sachverhalt vorläufig nicht hervorzugehen. Die exzentrische Lage der Milchstraße in der lokalen Gruppe könnte vielleicht eine leichte scheinbare Anisotropie der Nebelflucht erzeugen. Aber wahrscheinlich wird dieser Befund von anderen größeren Effekten überdeckt.

[1] K. GÖDEL: Proc. Internat. Congr. Math. (Cambridge, Mass., 30. Aug. bis 5. Sept. 1950), 1. p. 175. 1952.
[2] W. BONNOR: Z. Astrophys. **35**, 10 (1954).
[3] Siehe Fußnote 2, S. 506; s. a. A. SELMANOW, Dokl. Akad. Nauk SSSR. **107**, 815 (1956).
[4] A. EINSTEIN und E. STRAUS: Rev. Mod. Phys. **17**, 120 (1945). Für weitere Literatur s. E. SCHÜCKING: Z. Phys. **137**, 595 (1954). Siehe auch C. GILBERT: Monthly Notices Roy. Astronom. Soc. London **116**, 678 (1956).

Recht ausgiebig untersucht worden ist das Verhalten von Probekörpern in kosmologischen Modellen. Hier handelt es sich offenbar darum, die Bahnen der geodätischen Linien zu integrieren. Im Falle, wo die Metrik gegen eine Bewegungsgruppe invariant ist, lassen sich sofort so viel erste Integrale der Bewegungsgleichungen angeben wie die kontinuierliche Gruppe Parameter hat. Sei $u^\mu = dx^\mu/ds$ der Tangentenvektor der geodätischen Linie und ξ_μ der zu einer infinitesimalen Transformation der Gruppe gehörende Killing-Vektor, so gelten die Gleichungen

$$u^\mu_{\|\nu} u^\nu = 0, \qquad \xi_{\mu\|\nu} + \xi_{\nu\|\mu} = 0. \tag{11.1}$$

Wenn D/Ds die Ableitung längs der geodätischen Linie bezeichnet, folgt gemäß (11.1)

$$\frac{D}{Ds}(u^\mu \xi_\mu) = u^\mu_{\|\nu} u^\nu \xi_\mu + u^\mu \xi_{\mu\|\nu} u^\nu = 0. \tag{11.2}$$

Als erstes Integral erhält man also

$$u^\mu \xi_\mu = \text{const.} \tag{11.3}$$

Im Falle der homogen-isotropen Modelle erhält man insbesondere das wichtige Ergebnis: der lokale Relativimpuls eines Probekörpers ist dem Weltradius $R(t)$ umgekehrt proportional. Wenn man den Probekörper als de Broglie-Welle auffaßt, bedeutet dieses Ergebnis, daß sich Materiewellen in bezug auf ihre Wellenlängenänderung bei der Expansion des Weltalls ebenso verhalten wie die Lichtquanten.

Der Aufbau einer statistischen Mechanik, die auf dem Verhalten von Probekörpern zu fußen hätte, ist im Rahmen der Einsteinschen Gravitationstheorie bisher noch nicht weit gediehen. Lediglich die Grundzüge einer phänomenologischen Thermodynamik sind bisher in diese Theorie eingebaut worden[1]. Immerhin ist diese Theorie wichtig, wenn man Fragen wie die nach der Gültigkeit des zweiten Hauptsatzes für das Weltall als Ganzes zu diskutieren wünscht. Auf die Vorstellungen über die Entstehung der Elemente, soweit sie mit der Kosmologie verknüpft sind, gehen wir in diesem Artikel nicht ein. Wir verweisen hierfür auf die Literatur[2].

12. Mathematischer Anhang. Die in Ziff. 8 benötigten Formeln für den Ricci-Tensor des Linienelementes der isotrop expandierenden homogenen kosmologischen Modelle können folgendermaßen gewonnen werden. Für das Linienelement

$$(ds)^2 = (dx^0)^2 - S^2(x^\mu)\,\bar{\gamma}_{ik}(x^l)\,dx^i\,dx^k \tag{12.1}$$

muß man zunächst die Christoffel-Symbole zweiter Art berechnen. Wegen $g_{0\mu} = \delta^\mu_0 = g^{\mu 0}$ wird

$$\Gamma^0_{0\mu} = g^{0\nu}\Gamma_{0\mu,\nu} = \Gamma_{0\mu,0} = 0, \qquad \Gamma^l_{00} = g^{l\nu}\Gamma_{00,\nu} = 0. \tag{12.2}$$

Das bedeutet, daß alle Symbole, in denen der Index „0“ zwei- oder dreimal vorkommt, verschwinden. Weiter gewinnt man

$$\Gamma^j_{0k} = \Gamma_{0k,r}\,g^{jr} = \frac{1}{2}\,g_{kr|0}\,g^{jr} = \frac{S_{|0}}{S}\,g_{kr}\,g^{jr} = (\log S)_{|0}\,\delta^j_k. \tag{12.3}$$

Gemäß (5.4) wird

$$G_{0k} = \Gamma^\mu_{\mu 0|k} - \Gamma^\mu_{0k|\mu} + \Gamma^\nu_{0\mu}\Gamma^\mu_{k\nu} - \Gamma^\mu_{0k}\Gamma^\nu_{\nu\mu}. \tag{12.4}$$

---

[1] R. TOLMAN: Relativity, Thermodynamics, and Cosmology. Oxford 1934.

[2] R. ALPHER u. R. HERMAN: Rev. Mod. Phys. **22**, 153, 406 (1950). — R. ALPHER, J. FOLLIN u. R. HERMAN: Phys. Rev. **92**, 1347 (1953), dort auch weitere Literatur. Ferner G. GAMOW: Mat.-fys. Medd. Akad. Kopenh. **1953**, Nr. 10. Für andere Vorstellungen zur Elemententstehung vergleiche man den Bericht von A. CAMERON: AECL. No. 454 Scientific Document Distribution Office Atomic Energy of Canada Ltd., Chalk River, Ontario, Canada 1957. Dort finden sich zahlreiche Literaturangaben. Siehe auch E. BURBIDGE, G. BURBIDGE, W. FOWLER u. F. HOYLE: Rev. Mod. Phys. **29**, 548 (1957).

Zerlegt man die von 0 bis 3 laufenden Summationen in solche, die nur den Index 0 enthalten und die übrigbleibenden, in denen der (latainische) Index von 1 bis 3 läuft, so erhält man aus (12.4) bei Fortlassung aller Terme, die wegen (12.2) verschwinden

$$G_{0k} = \Gamma^j_{j0|k} - \Gamma^j_{0k|j} + \Gamma^l_{0j}\Gamma^j_{kl} - \Gamma^l_{0k}\Gamma^j_{jl}. \tag{12.5}$$

Daraus wird mit (12.3)

$$G_{0k} = 3\,(\log S)_{|0|k} - (\log S)_{|0|k} = 2\,(\log S)_{|0|k}. \tag{12.6}$$

Aus der Bedingung $G_{0k} = 0$ folgt, daß

$$S(x^\mu) = R(x^0)\,\overline{S}(x^i) \tag{12.7}$$

ist. Damit ist gezeigt, daß (12.1) die Form (8.10) haben muß. Für das in (12.1) als Spezialfall enthaltene Linienelement

$$(d s)^2 = (d x^0)^2 - R^2(x^0)\,\gamma_{ik}(x^l)\,d x^i\,d x^k \tag{12.8}$$

sind nun die 00- und *ik*-Komponenten des Ricci-Tensors zu berechnen. Die Christoffel-Symbole, die mehr als einmal den Index 0 enthalten, verschwinden gemäß (12.2). Für die Christoffel-Symbole mit einem Index 0 gilt wegen (12.3)

$$\Gamma^j_{0k} = (\log R)_{|0}\,\delta^j_k \tag{12.9}$$

und

$$\Gamma^0_{jk} = \Gamma_{jk,0} = -\tfrac{1}{2} g_{jk|0} = R\,R_{|0}\,\gamma_{jk}. \tag{12.10}$$

Führt man weiter durch die Definition

$$\Gamma^{*l}_{jk} = \tfrac{1}{2}\gamma^{lr}\,(\gamma_{jr|k} + \gamma_{kr|j} - \gamma_{jk|r}) \tag{12.11}$$

mit

$$\gamma_{ik}\gamma^{kl} = \delta^l_j \tag{12.12}$$

die Christoffel-Symbole des dreidimensionalen Raumes mit der Metrik γ_{jk} ein, so wird

$$\Gamma^{*k}_{ij} = \Gamma^k_{ij}. \tag{12.13}$$

Gemäß (5.4) wird

$$G_{00} = \Gamma^\mu_{\mu 0|0} - \Gamma^\mu_{00|\mu} + \Gamma^\nu_{0\mu}\Gamma^\mu_{0\nu} - \Gamma^\mu_{00}\Gamma^\nu_{\mu\nu}. \tag{12.14}$$

Wenn man die nicht verschwindenden Glieder sammelt, so kommt

$$G_{00} = 3\,(\log R)_{|0|0} + \Gamma^i_{0k}\Gamma^k_{0i}. \tag{12.15}$$

Wegen (8.8) erhält man als „00"-Feldgleichung bei inkohärenter Materie

$$G_{00} = 3\,\frac{R_{|0|0}}{R} = \Lambda - \frac{\varkappa\varrho}{2}. \tag{12.16}$$

Aus (5.4) folgt weiter

$$G_{ik} = \Gamma^\mu_{\mu i|k} - \Gamma^\mu_{ik|\mu} + \Gamma^\nu_{i\mu}\Gamma^\mu_{k\nu} - \Gamma^\mu_{ik}\Gamma^\nu_{\mu\nu}. \tag{12.17}$$

Bezeichnet man mit G^*_{ik} den mit den Γ^{*k}_{ij} berechneten Ricci-Tensor des dreidimensionalen Raumes mit der Metrik γ_{ij}, so kommt

$$G_{ik} = G^*_{ik} - \Gamma^0_{ik|0} + \Gamma^l_{i0}\Gamma^0_{kl} + \Gamma^0_{il}\Gamma^l_{k0} - \Gamma^0_{ik}\Gamma^i_{0i}. \tag{12.18}$$

Einsetzen der Christoffel-Symbole führt zu

$$G_{ik} = G^*_{ik} - \gamma_{ik}\,[R\,R_{|0|0} + 2\,(R_{|0})^2]. \tag{12.19}$$

Wegen (8.8) erhält man also als *ik*-Komponenten der Feldgleichungen für inkohärente Materie

$$G^*_{ik} = -\left[\left(\Lambda + \frac{\varkappa\varrho}{2}\right)R^2 - R\,R_{|0|0} - 2\,(R_{|0})^2\right]\gamma_{ik}. \tag{12.20}$$

Andere kosmologische Theorien.

Von

O. Heckmann und E. Schücking.

1. Übersicht. Kurz nach Aufstellung der Einsteinschen Gravitationstheorie wurde von verschiedenen Autoren versucht, mit Hilfe dieser Theorie Stabilität und Eigenschaften des Elektrons und Protons zu verstehen. Mehrere Forscher vermuteten damals, daß ein inniger Zusammenhang zwischen den Problemen der Mikrophysik und der Struktur der Welt im Großen bestehen könne. Einen umfassenden Ausdruck fand diese Vermutung dann in den dreißiger Jahren in der „fundamental theory" von A. Eddington [1]. Sie entsprang dem Versuch, die reziproke Feinstrukturkonstante als ganze Zahl zu interpretieren und große dimensionslose Zahlen, wie das Verhältnis der elektrischen zur Schwere-Anziehung im Wasserstoffatom, mit kosmologischen Daten zu verknüpfen. Im weiteren Verfolg seiner Arbeit gelangte A. Eddington schließlich zu einer Theorie der Elementarteilchen und zahlreicher Naturkonstanten, die eng mit einem bestimmten Weltmodell der Einsteinschen Kosmologie verknüpft war. Eine Diskussion dieser Theorie gehört im wesentlichen in die Mikrophysik und wird von uns nicht versucht.

Ein anderer Erklärungsversuch für die großen dimensionslosen Naturkonstanten, die im Mittelpunkt der Eddingtonschen Theorie standen, wurde 1937 von P. Dirac vorgeschlagen. Dirac vermutete, daß sie in Wahrheit Funktionen des Weltalters seien [2]. Dieser von P. Dirac skizzierte Gedanke wurde von P. Jordan weiter entwickelt [3]. Mit Hilfe einer Erweiterung der projektiven Relativitätstheorie [4] versuchte P. Jordan dann seit 1944, den Diracschen Gedanken im Rahmen einer solchen Feldtheorie durchzuführen. Er entwarf dazu ein linear mit der Zeit expandierendes sphärisches Weltmodell, in dem fortlaufend Materie entsteht und die Gravitations„konstante" umgekehrt proportional zum Weltalter abnimmt. Diese Jordansche Theorie ist eine der zahlreichen Erweiterungen der Einsteinschen Gravitationstheorie. Jeder dieser Theorien kann man eine eigene Kosmologie zuordnen. Daß hier die Jordansche Theorie ausführlicher diskutiert wird, hat zwei Gründe: Einmal ist diese Theorie von vornherein im Hinblick auf die zu ihr gehörige Kosmologie entwickelt und zum anderen sind ihre Konsequenzen für die Kosmologie eingehend untersucht worden [5].

In die Klasse der für die Kosmologie bedeutsam gewordenen Erweiterungen der Einsteinschen Gravitationstheorie gehört ebenfalls die 1948/49 von F. Hoyle geschaffene Theorie des stationären Universums. Die Einsteinschen Gravitationsgleichungen sind in ihr derart abgeändert worden, daß ein homogen-isotropes stationäres Modell des Kosmos mit dauernder Neuerzeugung von Materie die Feldgleichungen löst [6].

Das gleiche Modell eines stationären Universums wurde 1948 von H. Bondi und T. Gold postuliert [7]. Die von diesen Autoren skizzierte Theorie eines stationären Universums stützt sich jedoch im Gegensatz zu F. Hoyle nicht auf ein System von Feldgleichungen einer erweiterten Feldtheorie. Das Homogenitätspostulat spielt hier nicht mehr die Rolle einer Anfangsbedingung, die weitgehend unabhängig von den Feldgleichungen eingeführt werden kann, sondern wird zum zentralen Prinzip, das auch die lokal gültigen Naturgesetze mitbestimmt.

Diese neue Einstellung zum kosmologischen Problem war schon seit 1933 von E. MILNE bis in viele Einzelheiten hinein durchdacht worden. Als Homogenitätspostulat hatte MILNE jedoch die Invarianz der Weltmetrik und der Materieströmung gegenüber der homogenen Lorentz-Gruppe postuliert. Ausgehend von seinem Homogenitätspostulat hat E. MILNE unter Ablehnung der Einsteinschen Gravitationstheorie große Teile der Physik in andersartiger Weise zu begründen versucht [*8*].

Mit diesen Theorien sind die Hauptlinien der kosmologischen Forschung gekennzeichnet, die nicht zugleich auf der Einsteinschen Gravitationstheorie und der heutigen Physik fußen. Da ernstliche Widersprüche der Einsteinschen Gravitationstheorie und der aus ihr abgeleiteten Kosmologie mit der Erfahrung nicht zu bestehen scheinen, liegt vorerst kein hinreichender Grund vor, an der Einsteinschen Kosmologie zu zweifeln. Die erwähnten „anderen kosmologischen Theorien" zeigen also höchstens neue Denkmöglichkeiten auf.

Die vielfach unternommenen Versuche, die Rotverschiebung des Lichtes extragalaktischer Nebel durch Einführung von ad hoc postulierten Effekten zu verstehen, die ein statisches Universum ermöglichen sollen, werden hier nicht diskutiert. Bei solchen Versuchen wird im allgemeinen nicht beachtet, daß die Stabilität eines statischen Weltmodells höchst fraglich sein könnte, so daß man auf diese Weise der Vorstellung einer Expansion der Welt — jedenfalls aber einer Kosmologie — nicht ausweicht. Die Frage, wie der Doppler-Effekt von dem neuen Effekt zu trennen sei, würde zudem neue Probleme aufwerfen.

2. Kosmologie und Mikrophysik. Die in den dreißiger Jahren dieses Jahrhunderts einsetzenden Spekulationen über die Existenz einer elementaren Längeneinheit von der Größenordnung 10^{-13} cm regten verschiedene Forscher dazu an, wichtige kosmologische Daten wie die mittlere Materiedichte des Weltalls oder das Weltalter in „natürlichen" Maßeinheiten auszudrücken. Man wählte z.B. als „Elementarzeit" die Zeit, die das Licht benötigt, um die Strecke der Elementarlänge zurückzulegen, also etwa 10^{-23} sec. Als Masseneinheit wählte man z B. die Masse eines Nukleons, also etwa 10^{-24} g, und erhielt somit als Einheit der Materiedichte etwa die Dichte der Atomkerne, also rund 10^{15} g cm^{-3}. In diesen Maßeinheiten ergab sich die mittlere Materiedichte der Welt zu etwa 10^{-43} elementaren Dichteeinheiten und das Weltalter zu etwa 10^{40} Elementarzeiten. Größenordnungsmäßig entsprach die letzte Zahl dem Verhältnis der elektrischen Anziehung zur Gravitationskraft zwischen einem Elektron und einem Proton. Eine weitere „kosmische" Zahl, welche die Größenordnung des Quadrats von 10^{40} haben sollte, war die Zahl der Elementarteilchen in der Welt, die ebenfalls in die Spekulationen über die kosmischen Zahlen miteinbezogen wurde [*1*].

Eine Deutung der kosmischen Zahlen der Größenordnung 10^{40} und 10^{80} bzw. der Zahl 10^{-40} wurde von P. DIRAC diskutiert [*2*]. DIRACs Gedanke war, die großen Zahlenwerte von dimensionslosen Konstanten, wie z.B. des Verhältnisses von elektrischer Anziehung zur Gravitationsanziehung in einem Wasserstoffatom, als instantane Werte aufzufassen, die dem in Elementarzeiten gemessenen Weltalter direkt proportional seien mit einem Proportionalitätsfaktor der Größenordnung 1. Unter der Voraussetzung, daß Ladung und Masse von Elementarteilchen sowie die Dielektrizitätskonstante des Vakuums zeitlich unveränderlich seien, wäre aus DIRACs Vermutung gefolgt, daß die Gravitationskonstante in

Wahrheit zeitlich veränderlich sein müsse und mit zunehmendem Weltalter abnehme.

Gegen Vorstellungen dieser Art läßt sich folgendes einwenden:

1. Dimensionslose Zahlen der Größenordnung 10^{40} brauchen keineswegs in physikalischem Zusammenhang mit dem in Elementarzeiten gemessenen Weltalter zu stehen. Es wäre durchaus denkbar — um nur eine von vielen Möglichkeiten herauszugreifen — sie mit dem Ausdruck $e^{\frac{2}{3}\alpha}$ (α = reziproke Feinstrukturkonstante) in Beziehung zu setzen, der ebenfalls eine dimensionslose Zahl der Größenordnung 10^{40} ist.

2. Die in Zusammenhang mit den kosmischen Zahlen häufig zitierte Zahl der Elementarteilchen in der Welt ist zur Zeit auf Grund der Ungenauigkeit der Beobachtungen nicht verifizierbar. Sie ist selbst der Größenordnung nach unbekannt.

3. Die Hypothese einer elementaren Länge, auf die sich das System der natürlichen Maßeinheiten stützt, ist zur Zeit wenig mehr als der von einigen Forschern vermutete Bestandteil einer noch nicht existierenden korrekten Theorie der Elementarteilchen.

3. Jordansche Kosmologie. Der Diracsche Gedanke einer Veränderlichkeit der Gravitations„konstanten" wurde von P. JORDAN erstmalig feldtheoretisch formuliert. Eine Erweiterung der projektiven Relativitätstheorie, die von P. JORDAN und anderen Verfassern in den letzten 13 Jahren entwickelt wurde, bot sich dazu als Hilfsmittel an [*4*]. Diese Theorie liefert neben den Komponenten des metrischen Feldes und den vier elektromagnetischen Potentialen noch eine weitere skalare Feldgröße als Funktion der vier Raumzeitvariablen, die von P. JORDAN als „Gravitationsinvariante", d.h. als veränderliche Gravitations-„konstante" gedeutet wurde.

Zum Verständnis der Jordanschen Theorie ist es nicht erforderlich, den Formalismus der erweiterten projektiven Relativitätstheorie zu benutzen. Es genügt, die vierdimensionale Fassung der Theorie zu verwenden. P. JORDAN gewinnt seine Feldgleichungen aus dem invarianten Variationsprinzip

$$\delta \int \varkappa^{\eta} \left[G - \zeta \frac{\varkappa_{|\mu} \varkappa^{|\mu}}{\varkappa^2} - \varkappa L \right] \sqrt{-g}\, dx^0\, dx^1\, dx^2\, dx^3 = 0 \tag{3.1}$$

als Lagrange-Gleichungen. η und ζ sind dabei neue dimensionslose Naturkonstanten, G ist die Verjüngung des Ricci-Tensors, $\varkappa$ ist die Gravitationsinvariante und L ist die Lagrange-Funktion der Materie und des elektromagnetischen Feldes [*5*]. Bei der Auswahl dieses Variationsprinzips wurde unter anderem die wesentliche Forderung berücksichtigt, daß das Variationsprinzip für das Vakuum keine dimensionsbehaftete Naturkonstante enthalten solle. Auf diese Weise wurde ein kosmologisches Glied ausgeschlossen. Zur weiteren Spezifizierung von (3.1) benutzt P. JORDAN die Bedingung, daß die Lagrange-Gleichungen von (3.1) ein homogen-isotropes Weltmodell mit

$$R \sim t, \quad \varkappa \sim t^{-1}, \quad \varrho \sim t^{-1}, \quad \varepsilon = 1, \quad p = 0 \tag{3.2}$$

(Bezeichnungen wie im vorigen Artikel) als partikuläre Lösung besitzen sollen. Dabei wird noch angenommen, daß die Lagrange-Funktion L im Zeitmittel verschwindet[1]. Diese Bedingungen führen auf $\eta = 1$[2]. Auf diese Weise läßt sich die Diracsche Überlegung in den Rahmen der Jordanschen Theorie ein-

[1] Zur Begründung vergleiche man [*5*] § 19.

[2] G. LUDWIG: Fortschritte der projektiven Relativitätstheorie. Braunschweig 1951.

bauen. Allerdings ist dabei zu beachten, daß die Jordansche Theorie neben der von vornherein geforderten partikulären Lösung der kosmologischen Differentialgleichungen noch eine mehrparametrige Mannigfaltigkeit anderer Lösungen für die Funktionen $R(t)$, $\varkappa(t)$ und $\varrho(t)$ eines homogen-isotropen Modells mit verschwindendem Druck und positiver Raumkrümmung enthält. Wenn man dementsprechend fordern wollte, daß die Postulate (3.2) nur noch asymptotisch für große t jedoch für alle Lösungen des Modells gelten sollen, hätte man das asymptotische Verhalten der Lösungen der kosmologischen Differentialgleichungen für beliebiges η zu untersuchen, um die Vermutung zu bestätigen, daß auch in diesem Falle $\eta = 1$ gesetzt werden muß, um die Forderungen (3.2) asymptotisch zu erfüllen. Eine solche Untersuchung ist jedoch bisher noch nicht bekannt geworden.

Wenn man sich für die Wahl $\eta = 1$ entschieden hat, muß man versuchen, Aussagen über die dimensionslose Naturkonstante ζ zu gewinnen. Ganz allgemein ergibt sich aus den Jordanschen Feldgleichungen [*3*], daß für sehr große Werte von ζ bei konstantem $\varkappa$ asymptotisch die Einsteinschen Feldgleichungen herauskommen. Dieses Ergebnis allein berechtigt jedoch noch nicht zu dem Schluß, daß ζ in der Jordanschen Theorie bei $\eta = 1$ notwendig sehr groß sein muß. Zur Klärung dieser Frage wurden die kugelsymmetrischen Lösungen der Jordanschen Vakuumfeldgleichungen recht eingehend untersucht. Dabei ergab sich, daß die *statischen* Lösungen nur für große ζ mit den astronomischen Beobachtungen im Planetensystem verträglich sind[1]. Dieses Ergebnis rechtfertigt jedoch noch nicht die Annahme, daß ζ groß sein muß, weil in der Jordanschen Theorie — im Gegensatz zur Einsteinschen — eine Fülle von nichtäquivalenten, nichtstatischen, kugelsymmetrischen Vakuumlösungen existiert, die auf ihre Brauchbarkeit hin untersucht werden müßten zur Darstellung der astronomischen Beobachtungen im Planetensystem[2]. Dabei wäre besonders darauf zu achten, daß die Lösungen in großer Entfernung vom Zentrum in das metrische Feld des expandierenden Weltalls übergehen müßten.

JORDAN setzt die Lagrange-Funktion des Materiefeldes im Mittel gleich Null und erhält durch Variation von (3.1) die Feldgleichungen für $\varkappa$ und die $g_{\mu\nu}$[3]

$$\left.\begin{aligned} G + \zeta\left(2\,\frac{\varkappa_{\|\mu}^{|\mu}}{\varkappa} - \frac{\varkappa^{|\mu}\,\varkappa_{|\mu}}{\varkappa^2}\right) &= 0,\\ G_{\mu\nu} - \frac{1}{2}\,g_{\mu\nu}\,G + \frac{\varkappa_{|\mu\|\nu}}{\varkappa} - g_{\mu\nu}\,\frac{\varkappa_{\|\lambda}^{|\lambda}}{\varkappa} - \zeta\left(\frac{\varkappa_{|\mu}\,\varkappa_{|\nu}}{\varkappa^2} - \frac{1}{2}\,g_{\mu\nu}\,\frac{\varkappa^{|\lambda}\,\varkappa_{|\lambda}}{\varkappa^2}\right) &= -\varkappa\,T_{\mu\nu}. \end{aligned}\right\} \qquad (3.3)$$

Dabei ist $T_{\mu\nu}$ der Energie-Impulstensor. Für $\zeta = \frac{3}{2}$ sind diese Gleichungen nur miteinander verträglich, wenn $T = 0$ ist. T ist hier die Verjüngung des Energie-Impulstensors. Aus den Gln. (3.3) läßt sich der Erhaltungssatz

$$(\varkappa^2\,T^{\mu\nu})_{\|\nu} = 0 \qquad (3.4)$$

ableiten. Die Berechnung kosmologischer Modelle verläuft ähnlich wie in der Einsteinschen Gravitationstheorie. Für homogen-isotrope Modelle mit inkohärenter Materie setzt man

$$d s^2 = d t^2 - R^2(t)\,d\sigma^2, \qquad T_{\mu\nu} = \varrho(t)\,\delta_{\mu 0}\,\delta_{\nu 0}, \qquad (3.5)$$

wo $d\sigma^2$ das Quadrat des Linienelementes eines dreidimensionalen Raumes konstanter Krümmung ist. Dem Homogenitätspostulat entsprechend wird ange-

[1] O. HECKMANN, P. JORDAN u. W. FRICKE: Z. Astrophys. **28**, 113 (1951). Vgl. auch [*5*] S. 184.

[2] Vgl. [*5*] § 33 und E. SCHÜCKING: Z. Physik **148**, 72 (1957).

[3] Vgl. [*5*] § 27.

nommen, daß die Gravitationsinvariante $\varkappa$ nur von t abhängt. Der Erhaltungssatz (3.4) liefert jetzt die Gleichung[1]

$$\varkappa^2 \varrho R^3 = B = \text{const} > 0. \tag{3.6}$$

Für positive Krümmung und $\zeta > 2$ erhält man als partikuläre Lösung der Feldgleichungen[1]

$$\left.\begin{aligned} R &= \left(\frac{2}{\zeta-2}\right)^{\frac{1}{2}} t, \quad \varkappa = \varkappa_0 t^{-1}, \quad \varrho = \varrho_0 t^{-1}, \\ \text{mit} \quad \varkappa_0 \varrho_0 &= 2\zeta - 3, \quad \varkappa_0 = \text{const}, \quad \varrho_0 = \text{const}. \end{aligned}\right\} \tag{3.7}$$

Für die Funktion $R(t)$ ergibt sich bei positiver Krümmung des Raumes die Differentialgleichung[1]

$$\left.\begin{aligned} &3\ddot{R}R\left[\dot{R} + (\zeta-1)\frac{R}{t}\right]^2 - \zeta\left[2 + \ddot{R}R + 2\dot{R}^2\right]^2 + \\ &\quad + \left[3\dot{R} + (\zeta-3)\frac{R}{t}\right]\left[\dot{R} + (\zeta-1)\frac{R}{t}\right]\left[2 + \ddot{R}R + 2\dot{R}^2\right] = 0. \end{aligned}\right\} \tag{3.8}$$

Diese Differentialgleichung vertritt in der Jordanschen Theorie die Stelle der Friedmannschen Differentialgleichung der Einsteinschen Kosmologie. Jede Lösung von (3.8) liefert mit Ausnahme der singulären Lösungen[2] auch ein Weltmodell der Jordanschen Kosmologie. $\dot{\varkappa}$ und ϱ errechnen sich dabei aus den Gleichungen

$$\left.\begin{aligned} \varrho &= (3-2\zeta)\frac{\dot{\varkappa}}{\varkappa}\cdot\frac{1}{\varkappa t}, \\ \frac{3}{R^2}(\dot{R}^2+1) + 3&\left(\frac{\dot{R}}{R} + \frac{2\zeta-3}{t}\right)\frac{\dot{\varkappa}}{\varkappa} + \frac{\zeta}{2}\frac{\dot{\varkappa}}{\varkappa^2} = 0 \end{aligned}\right\} \tag{3.9}$$

für gegebenes R, $\dot{R}$, t und ζ mit $\varkappa = 1{,}86 \cdot 10^{-27}\,\text{cm}\,\text{g}^{-1}$. Die mathematische Theorie der Differentialgleichung (3.8) ist in der zweiten Auflage von „Schwerkraft und Weltall“ [5] ausführlich dargestellt worden[2]. Aus den dort geschilderten Untersuchungen folgt: Für $\zeta > \frac{9}{4}$ oszillieren die Lösungen von (3.8) um die linear expandierende Lösung (3.7) herum und gehen für große t asymptotisch in sie über. Näherungsweise gilt nach Jordan[3]

$$\left.\begin{aligned} R(t) &= \left(\frac{2}{\zeta-2}\right)^{\frac{1}{2}} t + P_0 t^{\frac{1}{2}} \cdot \cos\left\{\left(\zeta - \frac{9}{4}\right)^{\frac{1}{2}} \ln\frac{t}{t_0}\right\}, \\ P_0 &= \text{const}, \quad t_0 = \text{const}. \end{aligned}\right\} \tag{3.10}$$

In der Jordanschen Kosmologie ist der Vergleich mit der Erfahrung schwieriger als in der Einsteinschen Kosmologie mit Λ-Glied. Bei Jordan taucht zwar kein kosmologisches Glied auf, dafür aber die dimensionslose Konstante ζ. Insgesamt braucht man also zur Festlegung eines kosmologischen Modells in der Jordanschen Theorie vier verschiedene Daten (abgesehen von $\varkappa$), nämlich R, $\dot{R}$, t und ζ, während in der Einsteinschen Kosmologie drei, etwa R, $\dot{R}$ und Λ, genügen.

Immerhin ließe sich zumindest die lineare Lösung (3.7) mit der Erfahrung vergleichen und über die allgemeineren Lösungen könnte man mit Hilfe von Ungleichungen einige allgemeinere Aussagen ableiten. Daß dies hier nicht versucht wird, hat seinen Grund darin, daß die physikalische Deutung der Jordanschen Theorie und damit auch die Interpretation der Beobachtungsdaten noch

[1] Vgl. [5] § 30.
[2] Vgl. [5] § 31.
[3] P. Jordan: Z. Physik **132**, 655 (1952).

nicht abschließend geklärt werden konnte. Wie man aus (3.7) erkennt, wird die Gesamtmasse des Weltalls nach P. JORDAN proportional zum Quadrat des Weltalters. Es ist im sphärischen Fall

$$M = 2\pi^2 R^3 \varrho = 2\pi^2 \varrho_0 \left(\frac{2}{\zeta - 2}\right)^{\frac{3}{2}} t^2. \tag{3.11}$$

Damit wäre eine weitere der Diracschen Koinzidenzen großer Zahlen begründet. Die dadurch geforderte Massenzunahme im Kosmos (bei abnehmendem $\varkappa$) soll nach P. JORDAN so zustande kommen, daß laufend Sterne oder gar Sternsysteme neu entstehen. Bisher ist es jedoch noch nicht gelungen, diese Vorstellung in mathematisch befriedigender Weise aus der Theorie abzuleiten. Wendet man den Erhaltungssatz (3.4) jedoch auf die Bewegung eines Planeten der Masse m an, so folgt, daß sich der Planet auf einer geodätischen Linie bewegt und daß $\varkappa^2 m$ längs der Bahn konstant sein muß. W. PAULI hat dementsprechend gefordert, $\varkappa^2 m$ mit der Masse zu identifizieren, also die Jordansche Theorie neu zu interpretieren[1]. Die daraufhin von M. FIERZ durchgeführte Analyse der Jordanschen Theorie zeigt, daß diese Theorie auf eine Veränderlichkeit der Dielektrizitätskonstanten des Vakuums hinausläuft, wenn die Konstante η im Variationsprinzip (3.1) von -1 verschieden ist[2]. Eine Veränderlichkeit der Dielektrizitätskonstanten des Vakuums würde sich wiederum in einer Veränderlichkeit der Feinstrukturkonstanten ausdrücken. Diese eventuelle Veränderlichkeit ist jedoch in hohem Maße der Prüfung zugänglich. Wie R. MINKOWSKI bemerkt hat[3], beweisen die Messungen der Rotverschiebung im Spektrum der Radioquelle Cygnus A, daß sich die Feinstrukturkonstante in den letzten 700 Jahrmillionen nicht nennenswert geändert haben kann. In das Verhältnis der Rotverschiebungen im optischen Bereich und im Bereich der 21 cm-Linie geht nämlich im wesentlichen das Quadrat der Feinstrukturkonstanten ein. Aus diesem Ergebnis läßt sich schließen, daß tatsächlich $\eta = -1$ sein muß. Eine vollständige Interpretation der Jordanschen Theorie und ihrer Anwendungen auf die Kosmologie unter diesem Aspekt liegt jedoch noch nicht vor.

Auch die kosmogonischen Konsequenzen der $\varkappa$-Theorie in Jordanscher Interpretation sind in großer Ausführlichkeit untersucht worden[4]. Da insbesondere viele Probleme der Erdgeschichte zur Zeit noch einer exakten Behandlung nicht zugänglich sind, wäre es verfrüht, ihnen zur Bestätigung oder Widerlegung der Jordanschen Theorie eine entscheidende Bedeutung beizumessen.

Da es nicht unbedingt erforderlich ist, die Jordansche Erweiterung der projektiven Relativitätstheorie mit den Diracschen Überlegungen zu verknüpfen, kann man auch nach anderen Prinzipien Umschau halten, die zur Auswahl geeigneter Feldgleichungen dienen könnten. Fragen dieser Art und ihre Konsequenzen für die Kosmologie sind speziell von K. JUST untersucht worden[5]. Es ist übrigens nicht schwer auch für die Jordansche Theorie eine Newtonsche Fassung zu formulieren, die ihren physikalischen Gehalt mit Begriffen der klassischen Mechanik zu interpretieren erlaubt.

4. Die Theorie des stationären Universums. Die Einsteinschen Feldgleichungen der Gravitation wurden im Jahre 1917 noch einmal wesentlich erweitert durch Einführung des sog. kosmologischen Λ-Gliedes. Unter den vielen denkbaren kovarianten Gleichungen für das Gravitationsfeld, die im Rahmen der Riemannschen

---

[1] W. PAULI: Briefliche Mitteilung an P. JORDAN.
[2] M. FIERZ: Helv. phys. Acta **29**, 128 (1956).
[3] R. MINKOWSKI u. O. WILSON: Astrophys. Journ. **123**, 373 (1956).
[4] Vgl. [5] § 34.
[5] K. JUST: Z. Physik **140**, 485, 542, 648 (1955); **143**, 472 (1955).

Geometrie zulässig sind, besitzen die durch das kosmologische Glied erweiterten Feldgleichungen eine einzigartige Stellung: Sie folgen als Lagrange-Gleichungen eines invarianten Variationsprinzips (geben also gemäß den Untersuchungen E. NOETHERs[1] zu Erhaltungssätzen Anlaß), dessen Lagrange-Funktion der Forderung genügte, im Vakuumfall nur die $g_{\mu\nu}$ und ihre ersten und zweiten Ableitungen, die letzteren jedoch linear, zu enthalten. Wer also diesen Weg zur Gewinnung der Feldgleichungen anerkennt (ein anderer ist nicht bekannt), der ist zur Einführung eines kosmologischen Gliedes in die Feldgleichungen *gezwungen*.

Der subjektive Grund zur Einführung des kosmologischen Gliedes in die Feldgleichungen war für EINSTEIN jedoch der Wunsch, das Modell eines statischen Kosmos als Lösung der Feldgleichungen zu erhalten. Als aus E. HUBBLEs Beobachtungen 1929 hervorging, daß dieses Modell empirisch unbrauchbar war, schlug A. EINSTEIN vor, das kosmologische Glied aus Gründen „der logischen Ökonomie" gleich Null zu setzen. Gegen diesen Schritt sprachen jedoch nicht nur theoretische sondern auch wichtige empirische Gründe. Die Beobachtungen an extragalaktischen Nebeln in Verbindung mit den Altersbestimmungen von Gesteinen verlangten nämlich die Einführung eines von Null verschiedenen kosmologischen Gliedes, wenn man das Weltalter aus der Expansion größer halten wollte als das Alter der Erde. Modelle mit verschwindendem Λ lieferten nämlich ein Weltalter, das kleiner war als das Alter der Erde.

Wer EINSTEINs Gründe der „logischen Ökonomie" glaubte billigen zu sollen, der konnte sich bis 1952, als die kosmische Distanzskala durch BAADE[2] verdoppelt wurde, wegen der Altersdiskrepanz auf den Standpunkt stellen, daß EINSTEINs Feldgleichungen ohne kosmologisches Glied die Wirklichkeit nicht darstellten. Deshalb schlug F. HOYLE 1948 und 1949 zwei verschiedene Erweiterungen der Einsteinschen Feldgleichungen ohne kosmologisches Glied vor, die es erlauben sollten, als Lösung der Gleichungen ein Modell zu finden, in dem die Altersdiskrepanz nicht mehr auftreten konnte[3]. Nur die zweite Fassung der Hoyleschen Theorie soll hier kurz betrachtet werden, da F. HOYLE selbst sie als der ersten überlegen ansieht[4].

Ähnlich wie EINSTEIN bei Einführung des kosmologischen Gliedes und P. JORDAN bei der Auswahl der Feldgleichungen seiner $\varkappa$-Theorie ging HOYLE bei der Auswahl seiner Feldgleichungen vor. Auch er hatte das Wunschbild eines bestimmten kosmologischen Modells, das die Feldgleichungen lösen sollte. Zur Gewinnung dieses Modells schlug er vor, an Stelle des üblichen kosmologischen Prinzips, das Homogenität und zumeist auch Isotropie des dreidimensionalen Weltraumes fordert, das *Perfect Cosmologic Principle* (PCP) einzuführen, das darüber hinaus noch die Stationarität der Expansion behauptet. Während bisher viele Forscher einen wichtigen Fingerzeig in der näherungsweisen Übereinstimmung der reziproken Hubble-Konstanten mit dem Alter des Milchstraßensystems zu erblicken glaubten, wurde jetzt die Vorstellung eines endlichen Weltalters von der Steady State (stationären) Theorie der Expansion abgelehnt. Das PCP behauptet genauer, daß die Weltmetrik mitsamt dem Strömungsfeld der Materie gegen eine siebenparametrige kontinuierliche Bewegungsgruppe invariant sein soll. Jeder im Weltsubstrat mitschwimmende Beobachter soll zu *jeder* Zeit den gleichen Anblick vom Weltall haben. Mit dieser Forderung des PCP sind nach den Untersuchungen von H. ROBERTSON[5] nur zwei kosmologische Modelle

[1] E. NOETHER: Invariante Variationsprobleme. Nachr. Ges. Wiss. Göttingen **1918**.
[2] W. BAADE: Trans. Internat. Astronom. Union **8**, 397 (1952).
[3] Vgl. [*6*] sowie Nature, Lond. **163**, 196 (1949) und H. BONDI: Cosmology. Cambridge 1952.
[4] F. HOYLE: Monthly Notices Roy. Astronom. Soc. London **109**, 365 (1949).
[5] H. ROBERTSON: Rev. Mod. Phys. **5**, 62 (1933), Anhang.

verträglich: Der statische Einstein-Kosmos und die DeSitter-Welt, wenn sie als exponentiell expandierender Raum verschwindender Krümmung aufgefaßt wird. Da der statische Einstein-Kosmos mit dem empirisch gesicherten Expansionsphänomen nicht in Einklang ist, bleibt nur die zweite Möglichkeit. Das Linienelement lautet dann

$$d s^2 = d t^2 - e^{2t/T}\left[(d x^1)^2 + (d x^2)^2 + (d x^3)^2\right], \qquad T = \text{const}. \tag{4.1}$$

Dabei ist die Materie in diesem Koordinatensystem auf Ruhe transformiert zu denken. Es soll also gelten

$$u^\mu = \frac{d x^\mu}{d s} = \delta_0^\mu. \tag{4.2}$$

Da $R(t) = \exp(t/T)$ ist, wird die Hubble-Zahl $\dot{R}/R = 1/T$, also konstant. Die Einsteinschen Feldgleichungen mit kosmologischem Glied für inkohärente Materie liefern mit (4.1) und (4.2) aber, daß die Materiedichte ϱ verschwinden muß. Das stationäre Modell kann also unter diesen Bedingungen nur dann eine Lösung der Feldgleichungen sein, wenn die Feldgleichungen in geeigneter Weise abgeändert werden. Von der Einführung eines negativen Druckes nach dem Vorschlag von W. McCrea[1] sehen wir hierbei ab.

Das PCP angewandt auf die Materiedichte bedingt deren Konstanz. Infolge des Expansionsphänomens ist das nur möglich, falls fortlaufend Materie erzeugt wird, um die durch die Expansion bewirkte Verdünnung auszugleichen.

Hier soll skizziert werden, wie eine Theorie der Erzeugung inkohärenter Materie phänomenologisch formuliert werden kann. Wenn λ die Dichte der pro Zeiteinheit entstehenden Materie ist, tritt an die Stelle der Kontinuitätsgleichung der Newtonschen Theorie

$$\varrho_{|0} + (\varrho v_i)_{|i} = \lambda(x_j, t). \tag{4.3}$$

In einem lokalen Inertialsystem gilt weiter

$$(\varrho v_i)_{|0} + (\varrho v_i v_k)_{|k} = K_i + \lambda V_i. \tag{4.4}$$

Dabei ist K_i die Volumkraft und λV_i der Vektor der pro Zeiteinheit entstehenden Impulsdichte. Es wurde angenommen, daß dieser Impuls ausschließlich gegeben wird dadurch, daß die entstehende Materie die Geschwindigkeit V_i besitzt. Die relativistische Fassung von (4.3) und (4.4) für inkohärente Materie lautet

$$T^{\mu\nu}_{\ \ \|\nu} = S^\mu, \qquad T^{\mu\nu} = \varrho u^\mu u^\nu \tag{4.5}$$

mit

$$u^\mu = \frac{d x^\mu}{d s}, \qquad S^\mu = \lambda V^\mu, \qquad V^\mu V_\mu = 1. \tag{4.6}$$

Dabei ist jetzt S^μ der Vierervektor der pro Zeiteinheit entstehenden Energie-Impulsdichte und V^μ der Vektor der Vierergeschwindigkeit der entstehenden Materie.

F. Hoyle faßt nun den Vierervektor S^μ als Divergenz eines symmetrischen Tensors zweiter Stufe auf[2]:

$$S^\mu = -\frac{1}{\varkappa} C^{\mu\nu}_{\ \ \|\nu}. \tag{4.7}$$

[1] W. McCrea: Proc. Roy. Soc. Lond., Ser. A **206**, 562 (1951).
[2] Siehe Fußnote 4, S. 526.

$\varkappa$ ist dabei wie üblich die relativistische Gravitationskonstante. Mit Hilfe von (4.7) kann nun die erste Gl. (4.5) geschrieben werden

$$\left(T^{\mu\nu} + \frac{1}{\varkappa} C^{\mu\nu}\right)_{\|\nu} = 0. \tag{4.8}$$

In dieser Weise konstruiert F. HOYLE einen divergenzfreien symmetrischen Tensor zweiter Stufe, den er nach Multiplikation mit $-\varkappa$ der linken Seite der Einsteinschen Feldgleichungen gleichsetzt. So erhält er dann unter Fortlassung des kosmologischen Gliedes die Feldgleichungen

$$R_{\mu\nu} - \tfrac{1}{2} g_{\mu\nu} R = -\varkappa T_{\mu\nu} - C_{\mu\nu}. \tag{4.9}$$

Es ist indessen zu betonen, daß im Falle von Materieerzeugung, also bei Ungültigkeit der Erhaltungssätze, die Wahl divergenzfreier geometrischer Größen als linke Seite der Feldgleichungen ihre Berechtigung verliert. Denn das Aufsuchen eines symmetrischen Tensors zweiter Stufe verschwindender Divergenz, der aus den $g_{\mu\nu}$ und ihren Ableitungen gebildet wird, war gerade *durch die Gültigkeit der Erhaltungssätze motiviert.*

Läßt man jedoch diese Einwände außer acht und hält sich an (4.9), so verbleibt die Aufgabe der Konstruktion des Tensors $C^{\mu\nu}$ aus den S^{μ} gemäß (4.7). Da $C^{\mu\nu}$ zehn Komponenten besitzt, sind vier Gleichungen für zehn Unbekannte zu lösen. Das Problem ist also weitgehend unbestimmt. Die Tatsache, daß empirisch nichts über S^{μ} bekannt ist, macht das ganze Problem noch unbestimmter. Ein sicher gangbarer Weg um dennoch zu einer Theorie der Materieerzeugung zu gelangen, bestände darin, in das Variationsproblem für die Feldgleichungen einen Zusatzterm aufzunehmen, der in invarianter Weise aus den noch unbekannten materieerzeugenden Feldgrößen und ihren Ableitungen aufgebaut wäre. In diesem Falle erhielte man sogleich nach bekannten Methoden den Tensor $C^{\mu\nu}$. Überdies wäre weiter garantiert, daß die zusätzlichen Feldgleichungen kovariant wären, daß man ebensoviel Gleichungen wie Unbekannte erhielte und daß die Gleichungen miteinander kompatibel wären.

Dieser geradlinige Weg zur Gewinnung der zusätzlichen Feldgleichungen ist von F. HOYLE jedoch nicht beschritten worden. Er verwendet die Feldgleichungen (4.9) und führt ein Vierervektorfeld $*C^{\mu}$ ein, für das er ohne nähere Begründung die weiteren Feldgleichungen

$$\varepsilon^{\mu\nu\lambda\sigma} S_{\nu} S_{\lambda|\sigma} = 0 \quad \text{mit} \quad S_{\nu} = -\frac{1}{2\varkappa}\left(*C^{\mu}{}_{\|\nu\|\mu} + *C_{\nu}{}^{\|\mu}{}_{\|\mu}\right), \quad *C^{\mu}_{\|\mu} = \text{const} \tag{4.10}$$

postuliert. Die erste Gleichung stellt den hier kovariant formulierten Ausdruck für die Hoylesche Forderung dar, daß das Vektorfeld der S_{ν} ein Feld von Normalenvektoren einer einparametrigen Hyperflächenschar in der vierdimensionalen Raumzeit bildet. Den Tensor $C_{\mu\nu}$ definiert er durch

$$C_{\mu\nu} = \tfrac{1}{2}\left(*C_{\mu\|\nu} + *C_{\nu\|\mu}\right). \tag{4.11}$$

Eine Deutung des Vektors $*C_{\mu}$ wird von F. HOYLE nicht versucht. Die in die Gln. (4.10) eingehende Konstante wird von F. HOYLE so festgelegt, daß sein PCP-Modell als Lösung der Gleichungen auftritt. Geht man nämlich in die Hoyleschen Feldgleichungen mit dem Ansatz

$$\left.\begin{aligned} ds^2 = dt^2 - R^2(t)\, d\sigma^2, \quad *C_{\mu} = \alpha(t)\,\delta_{\mu 0}, \quad T_{\mu\nu} = \varrho\, u_{\mu} u_{\nu}, \\ u_{\mu} = \delta_{\mu 0}, \quad d\sigma^2 = \gamma_{ik}(x^j)\, dx^i dx^k \end{aligned}\right\} \tag{4.12}$$

ein, wo $d\sigma^2$ das Quadrat des Linienelementes eines dreidimensionalen Raumes konstanter Krümmung ist, so sind die ersten Gln. (4.10) identisch erfüllt. Die letzte Gl. (4.10) mit $*C^{\mu}_{\|\mu} = A = \text{const.}$, und die Feldgleichungen (4.9) lauten dann[1]

$$\left.\begin{aligned} (\alpha R^3)^{\cdot} &= A R^3, \\ 3\frac{\ddot{R}}{R} &= -\dot{\alpha} + \frac{A}{2} - \frac{\varkappa\varrho}{2}, \\ \frac{\ddot{R}}{R} &= -2\frac{\dot{R}^2}{R^2} + \frac{A}{2} + \frac{\varkappa\varrho}{2} - \frac{\dot{R}\alpha}{R} - \frac{2\varepsilon}{R^2}. \end{aligned}\right\} \tag{4.13}$$

Für

$$\alpha = \frac{3}{T}, \quad A = \frac{9}{T^2}, \quad \varepsilon = 0, \quad R = e^{t/T}, \quad \varkappa\varrho T^2 = 3 \tag{4.14}$$

erhält man das durch (4.1) und (4.2) charakterisierte Modell eines stationären Universums als partikuläre Lösung. Aus (4.5) gewinnt man

$$\lambda = \frac{3\varrho}{T}, \quad V_\mu = \delta_{\mu 0}. \tag{4.15}$$

Für inkohärente Materie folgt aus (4.8) in bekannter Weise[1]

$$\varkappa\varrho\, u^{\mu}_{\|\nu} u^{\nu} + D^{\mu} = 0, \quad D^{\mu} = C^{\mu\nu}_{\|\nu} - u^{\mu}_{\|\lambda} C^{\lambda\nu}_{\|\nu}. \tag{4.16}$$

Da D^μ im allgemeinen nicht verschwinden wird, bedeutet dies, daß sich die Materie nur in Ausnahmefällen — wie z. B. im stationären Hoyleschen Weltmodell — auf geodätischen Linien bewegen wird. Von F. HOYLE ist jedoch eingewendet worden, daß dieser Schluß nicht zwingend sei, da die neu entstehende Materie unter der Wirkung von Kernkräften und elektromagnetischen Kräften durch Stoßprozesse auf geodätische Bahnen gebracht werde[2]. Im Rahmen der phänomenologischen Gleichungen ist jedoch für solche Vorstellungen kein Platz solange diese Kräfte und Felder nicht explizit in die Feldgleichungen hereingenommen werden und die behaupteten Wechselwirkungen tatsächlich angegeben werden.

Obwohl F. HOYLE die Einführung eines kosmologischen Gliedes ablehnt, laufen seine Feldgleichungen doch *im wesentlichen* darauf hinaus, daß das Gravitationsgesetz durch Einführung einer kosmischen Abstoßungskraft abgeändert wird. Die Hoyleschen Feldgleichungen lauten nämlich gemäß (4.9)

$$R_{\mu\nu} + C_{\mu\nu} - \tfrac{1}{2} g_{\mu\nu} C^{\lambda}_{\lambda} = -\varkappa\left(T_{\mu\nu} - \tfrac{1}{2} g_{\mu\nu} T\right). \tag{4.17}$$

Die "00"-Gleichung, die im Newtonschen Fall der Poissonschen Gleichung entspricht und jedenfalls den Hauptteil des Gravitationsphänomens beschreibt, lautet also

$$R_{00} + C_{00} - \tfrac{1}{2} g_{00} C^{\lambda}_{\lambda} = -\varkappa\left(T_{00} - \tfrac{1}{2} g_{00} T\right). \tag{4.18}$$

An Stelle der Poissonschen Gleichung hat man zugleich in der Newtonschen Näherung der Hoyleschen Theorie zu setzen

$$\Phi_{|i|i} + C_{00} - \tfrac{1}{2} C^{\lambda}_{\lambda} = 4\pi G\varrho. \tag{4.19}$$

Der zweite und dritte Term der linken Seite entsprechen aber genau der Einführung eines kosmologischen Gliedes. Man kann das bestätigen, wenn man das PCP-Modell im Rahmen der Newtonschen Näherung ableitet mit $v_i = \frac{1}{T} x_i$. Dann ergibt die Berücksichtigung von (4.3) zunächst, daß

$$\frac{3}{T}\varrho = \lambda \tag{4.20}$$

---

[1] Vgl. den Artikel über Einsteinsche Kosmologie.
[2] Siehe Fußnote 4, S. 526.

ist, wo jetzt ϱ die mittlere Materiedichte im PCP-Modell ist. Aus (4.4) erhält man sodann weiter unter Benutzung von (4.3) mit $V_i = v_i$

$$\frac{1}{T^2} x_i = -\Phi_{|i}. \tag{4.21}$$

Es ergibt sich also durch Einsetzen in die Poissonsche Gleichung mit Λ-Glied

$$\Lambda - \frac{3}{T^2} = 4\pi G \varrho. \tag{4.22}$$

Also ist die Theorie des stationären Weltalls mit der Poissonschen Gleichung ohne Λ-Glied nicht verträglich. Qualitativ war dieses Ergebnis aber von vornherein klar. Wegen $v_i = \frac{1}{T} x_i$ lauten die Bewegungsgleichungen eines festen Substratelementes

$$x_i = e^{t/T} x_i^0, \quad x_i^0 = \text{const}. \tag{4.23}$$

Die Substratelemente erfahren also eine dauernde Abstoßung während nach der Newtonschen Theorie ohne Λ-Glied eine Anziehung vorliegen müßte, die durch die Wirkung neu entstehender Materie noch verstärkt würde. Betrachtungen zur Kondensation von Nebeln im Rahmen der Hoyleschen Theorie, die sich auf die Poissonsche Gleichung ohne Λ-Glied stützen, sind wegen (4.22) von vornherein widerspruchsvoll.

Der Vergleich der Hoyleschen Kosmologie mit den Beobachtungen an extragalaktischen Nebeln geschieht mit Hilfe der gleichen Formeln, die im vorhergehenden Artikel über Einsteinsche Kosmologie dargestellt sind. Aus dem Linienelement (4.1) folgt, daß im stationären Universum $\ddot{R}$ ein positives Vorzeichen besitzen muß. Die neuesten Beobachtungen von A. SANDAGE und W. BAUM scheinen nicht dafür zu sprechen, daß dieses Ergebnis der Hoyleschen Theorie mit den Beobachtungen in Einklang steht.

Eine empirische Widerlegung des PCP wäre sofort gegeben, falls der Nachweis gelänge, daß sehr entfernte Spiralnebel im Durschschnitt andere scheinbare Eigenschaften besitzen als nähere Nebel, wie das in den Modellen der Einsteinschen Kosmologie der Fall sein muß. Bei HOYLE sollte nämlich wegen der Stationarität des Universums der zeitliche Unterschied zwischen der Emission des Lichtes ferner Nebel und seiner Ankunft auf der Erde nichts ausmachen, weil das Weltall zu jeder Zeit den gleichen Anblick bieten soll. Vielleicht ließe sich diese Frage durch die Untersuchung ferner Radioquellen entscheiden. Eine weitere Prüfungsmöglichkeit der Hoyleschen Theorie soll hier ebenfalls nur angedeutet werden: Da dauernd Nebel neu entstehen sollen, müßten sich schon in der näheren Umgebung des galaktischen Systems Nebel sehr hohen Alters finden. In ihnen sollte das Verhältnis von Masse zur Leuchtkraft Werte annehmen, die weit über dem Durchschnitt liegen.

Von H. BONDI und T. GOLD [7] wurde 1948 ebenfalls eine Theorie des stationären Universums entwickelt. Im Gegensatz zu F. HOYLE stützt sich die Theorie dieser Autoren nicht auf Feldgleichungen, sondern ausschließlich auf das PCP. Bei der Betrachtung des Weltmodells eines stationären Universums führt die Bondi-Gold-Theorie zu den gleichen Ergebnissen wie die Hoylesche Lehre, postuliert aber — da sie über keine Feldgleichungen verfügt — keinen Zusammenhang zwischen Materiedichte und Hubble-Zahl. Für ihre nähere Besprechung vgl. man H. BONDI, Cosmology. Cambridge 1952.

5. Milnesche Kosmologie. *α) Einleitung.* E. A. MILNE hatte, teilweise unterstützt von einigen Mitarbeitern, seine Kosmologie im wesentlichen um 1935 ausgebildet [*8*]. In der Folge aber hat er eine vollständig neue Mechanik, Gravitationstheorie und Elektrodynamik entwickelt. .Seine zahlreichen Abhandlungen nach 1935 hat er zusammengefaßt in dem Werk „Kinematic Relativity" [9], das auch ein vollständiges Literaturverzeichnis bis 1948 enthält. Wesentliche Arbeiten nach dem Tode von MILNE (1950) sind uns nicht bekannt geworden. Im folgenden interessieren uns nur die eigentlich kosmologischen Teile der Milneschen Theorie; die neue Dynamik und Gravitationstheorie sowie die Elektrodynamik behandeln wir nicht.

Das Charakteristikum der Theorie ist, daß sie kein empirisches Element in Form von lokal prüfbaren und gültigen „Naturgesetzen", etwa Feldgleichungen irgendwelcher Art, einführt, sondern nur aus Definitionen und Axiomen ein widerspruchsfreies System von Sätzen aufzubauen sucht. Äußerlich ist die Theorie nicht in die in der Mathematik üblichen Form einer strengen Axiomatik gekleidet. Doch erhebt sie den Anspruch, eine weitgehend apriorische Arithmetisierung der Physik zu liefern. Eine teilweise strenge Fassung der Axiome der Theorie stammt von G. C. MCVITTIE[1].

Obwohl MILNE stets mit polemischem Nachdruck den radikalen Unterschied seines Weltmodells gegenüber den Modellen der Einsteinschen Gravitationstheorie betont hat, soll im folgenden mehr Wert gelegt werden auf die Herausarbeitung der Zusammenhänge beider Gebiete.

β) Die Strömung des homogenen Substrates. Wir denken uns den Raum wie in anderen Kosmologien mit einem kontinuierlichen Substrat erfüllt; jedem Element des Substrates ordnen wir einen von ∞^3 mitschwimmenden Beobachter zu. Die Beobachter werden mit „kongruenten" Uhren ausgerüstet ([9], § 8). Sie können einander Lichtsignale zusenden und zurückreflektieren. Die Sende-, Empfangs- und Reflexionszeiten können beobachtet werden an den Uhren, die ein Lichtsignal passiert. Jeder Beobachter soll die Uhr eines jeden anderen Beobachters ablesen können. MILNE fordert, daß diese Beobachter eine „Äquivalenz" bilden sollen ([9], Teil I, Kap. II und III) und zeigt, daß diese Forderung sich immer in die folgende Form bringen läßt: Die Fundamental-Beobachter sollen sich alle relativ zueinander geradlinig und gleichförmig bewegen; die Koordinaten, die irgend zwei von ihnen dem gleichen Ereignis zuschreiben, stehen miteinander in der Beziehung

$$x' = \frac{x - U\,t}{\sqrt{1 - U^2}}; \qquad y' = y; \qquad z' = z; \qquad t' = \frac{t - U\,x}{\sqrt{1 - U^2}}. \tag{5.1}$$

Hier ist die x-Richtung in die Verbindungslinie beider Beobachter gelegt, U ist ihre Relativgeschwindigkeit. Die Lichtgeschwindigkeit wurde der Einfachheit halber gleich 1 gesetzt.

MILNE gibt die Lorentz-Transformation nicht einfach vor, sondern folgert sie aus dem Begriffskomplex der Äquivalenz und den in ihm enthaltenen Operationen, die wir im vorigen Absatz andeuteten. In der Sprache der Gruppentheorie müßte man sagen, daß die Forderung der Äquivalenz identisch ist mit der Forderung, daß zwischen den Fundamentalbeobachtern die abstrakte Gruppe der speziellen, homogenen Lorentz-Transformationen gilt.

An der Spitze der Milneschen Kosmologie steht das Postulat, daß die makroskopische Struktur des Weltmodells für jeden Beobachter isotrop und invariant sei gegenüber den Transformationen (5.1). Es ist zweckmäßig, zunächst die Konsequenzen dieses Homogenitäts-Postulates für das Strömungsfeld zu untersuchen. Ein Fundamentalbeobachter P beschreibe in seinem Koordinatensystem die

[1] G. C. MCVITTIE: Axiomatic treatment of kinematical relativity. Proc. Roy. Soc. Edinburgh, Sect. A **61**, 210 (1942). Siehe auch die Erwiderung von G. J. WHITROW, a. a. O. **61**, 298 (1943).

Strömung des Weltsubstrates mit Hilfe des Vektorfeldes

$$\bar{u}(x, y, z, t), \quad \bar{v}(x, y, z, t), \quad \bar{w}(x, y, z, t). \tag{5.2}$$

Ein zweiter Beobachter P', der mit P durch (5.1) verknüpft ist, schreibt dann dem Substratelement an der Stelle x, y, z, t den Vektor

$$\bar{u}'(x', y', z', t'), \quad \bar{v}'(x', y', z', t'), \quad \bar{w}'(x', y', z', t') \tag{5.2'}$$

zu. Nach dem Additionstheorem der Geschwindigkeiten ist

$$\bar{u}' = \frac{\bar{u} - U}{1 - \bar{u}\,U}; \quad \bar{v}' = \frac{\bar{v}\sqrt{1 - U^2}}{1 - \bar{u}\,U}; \quad \bar{w}' = \frac{\bar{w}\sqrt{1 - U^2}}{1 - \bar{u}\,U} \tag{5.3}$$

gleichzeitig mit sechs analogen Gleichungen für die Relativbewegung in der y- und z-Richtung, die bzw. die Geschwindigkeitskomponenten V und W enhalten.

Die Forderung der Homogenität des Strömungsfeldes besagt jetzt, daß $\bar{u}', \bar{v}', \bar{w}'$ von x', y', z', t' ebenso abhängen wie die $\bar{u}, \bar{v}, \bar{w}$ von x, y, z, t, man hat also, um die Abhängigkeit der $\bar{u}, \bar{v}, \bar{w}$ von ihren Argumenten zu bestimmen, einfach in (5.2′) die Akzente fortzulassen. Dadurch erhält man z.B. für $\bar{u}$ die Funktionalgleichungen

$$\left.\begin{aligned}
\bar{u}\left(\frac{x - U t}{\sqrt{1 - U^2}}, y, z, \frac{t - U x}{\sqrt{1 - U^2}}\right) &= \frac{\bar{u}(x, y, z, t) - U}{1 - \bar{u}(x, y, z, t)\cdot U},\\
\bar{u}\left(x, \frac{y - V t}{\sqrt{1 - V^2}}, z, \frac{t - V x}{\sqrt{1 - V^2}}\right) &= \frac{\bar{u}(x, y, z, t)\sqrt{1 - V^2}}{1 - \bar{u}(x, y, z, t)\cdot V},\\
\bar{u}\left(x, y, \frac{z - W t}{\sqrt{1 - W^2}}, \frac{t - W x}{\sqrt{1 - W^2}}\right) &= \frac{\bar{u}(x, y, z, t)\sqrt{1 - W^2}}{1 - \bar{u}(x, y, z, t)\cdot W}.
\end{aligned}\right\} \tag{5.4}$$

Sechs analoge Gleichungen gelten für $\bar{v}$ und $\bar{w}$. Die Parameter U, V, W können beliebige Werte annehmen, sofern $U^2 + V^2 + W^2 < 1$ bleibt. Die einzige Lösung dieses Systems von neun Funktionalgleichungen lautet, wenn man mit Milne die Isotropie der Strömung fordert,

$$\bar{u} = \frac{x}{t}, \quad \bar{v} = \frac{y}{t}, \quad \bar{w} = \frac{z}{t}. \tag{5.5}$$

Dieses Strömungsfeld hat also die gewünschten Homogenitätseigenschaften; es bleibt invariant gegenüber den Transformationen (5.1) und (5.3); es hat universellen Charakter. Der Beweis ist im Anhang gegeben.

γ) Die Dichte des homogenen Substrates. Nachdem das Strömungsfeld des Weltsubstrates festliegt, ist es leicht, die Folgerungen aus dem Homogenitäts-Postulat für die Dichte des Substrates zu ziehen. Wir fordern zunächst, daß

$$n(x, y, z, t)\,dx\,dy\,dz \tag{5.6}$$

die Anzahl von Substratpartikeln im Volumenelement $dx\,dy\,dz$ bei x, y, z zur Zeit t, invariant sei gegenüber (5.1). Da nun

$$dV = dx\,dy\,dz\,\frac{dt}{ds} \tag{5.7}$$

mit

$$ds^2 = dt^2 - dx^2 - dy^2 - dz^2 \tag{5.8}$$

der invariante Ausdruck für das dreidimensionale euklidische Volumenelement, da ferner

$$X = t^2 - x^2 - y^2 - z^2 \tag{5.9}$$

die einzige Lorentz-invariante Funktion der Koordinaten ist, so hat man in

$$n(x, y, z, t)\,dx\,dy\,dz = \varrho(X)\,dV \tag{5.10}$$

mit einer vorläufig noch unbestimmten Funktion $\varrho(X)$ den geforderten invarianten Ausdruck für die Dichte. Will man ihn näher festlegen, so ist es erstens notwendig, die Bewegung des Substrates zu berücksichtigen, um dt/ds zu erhalten; zweitens muß man eine Aussage machen über das Neuentstehen oder Vergehen von Substratpartikeln. MILNE entschließt sich, die Erhaltung der Partikelzahl zu fordern. Damit muß gelten

$$\frac{\partial n}{\partial t} + \frac{\partial(n\bar{u})}{\partial x} + \frac{\partial(n\bar{v})}{\partial y} + \frac{\partial(n\bar{w})}{\partial z} = 0, \tag{5.11}$$

wo $\bar{u}, \bar{v}, \bar{w}$ aus (5.5) zu entnehmen sind. Setzt man in (5.11) nach (5.10) ein [mit Rücksicht auf (5.5), (5.7), (5.8), (5.9)]

$$n = t\,\varrho(X)\,X^{-\frac{1}{2}}, \tag{5.12}$$

so wird

$$\varrho = B\,X^{-\frac{3}{2}}, \tag{5.13}$$

wo B eine dimensionslose Integrationskonstante ist. Ferner wird

$$n = B\,t\,X^{-2}. \tag{5.14}$$

Die Ausdrücke (5.5) und (5.14) beschreiben die invariante Strömung und Dichte des Weltsubstrates: Jeder Fundamentalbeobachter sieht das Substrat radial strömen mit zeitlich konstanter Geschwindigkeit jedes einzelnen Substratelementes. Das instantane Strömungsfeld aber ist das einer überall isotropen Expansion. Jeder Fundamentalbeobachter selbst ruht in dem Substratelement im Ursprung seines Koordinatensystems, das mit dem Substrat schwimmt. Die Dichte nimmt an einer festen Stelle monoton ab mit der Zeit, sie verschwindet für $t \to \infty$, sie war unendlich für $t = 0$. In einem bestimmten Zeitpunkt aber ist sie für kleine x, y, z konstant, wächst nach „außen" monoton an und wird unendlich auf dem „Horizont" $X = 0$, der seinerseits mit Lichtgeschwindigkeit forteilt.

δ) Die Neueinteilung von Uhren und die Zeitskala. In der Milneschen Theorie werden kosmische Distanzen definiert mit Hilfe von gedachten Lichtsignalen, die zwischen Fundamentalbeobachtern hin- und herreflektiert werden. Die Vorschrift, nach der verfahren wird, ist diese: Geht zur Zeit t_1 ein Signal vom Koordinatenursprung O ab, kommt bei einem Fundamentalbeobachter P an, wird dort reflektiert und kommt im Augenblick t_2 wieder in O an, so soll der Beobachter O dem Ereignis des Lichtreflexes in P die Entfernung

$$l = \tfrac{1}{2}(t_2 - t_1) \tag{5.15}$$

und die Epoche

$$t = \tfrac{1}{2}(t_2 + t_1) \tag{5.16}$$

zuschreiben. Wenn aber O das Zifferblatt seiner Uhr irgendwie neu einteilt, so daß sie nun die Zeit τ anzeigt, — wobei $\tau(t)$ eine monoton wachsende und stetige Funktion sein soll — so führt MILNE konsequenterweise nun auch neue Distanzen λ ein, die nach der alten Vorschrift gewonnen werden:

$$\lambda = \tfrac{1}{2}(\tau_2 - \tau_1)\,; \qquad \tau = \tfrac{1}{2}(\tau_2 + \tau_1)\,. \tag{5.17}$$

Von besonderem Interesse ist für MILNE die Transformation, welche man durch Integration aus

$$\frac{d\tau}{t_0} = \frac{dt}{t} \tag{5.18}$$

erhält. Die Konstante t_0 liefert sozusagen die Zeiteinheit. Man hat dann die neue Zeit

$$\tau = t_0 + t_0 \log \frac{t}{t_0}. \tag{5.19}$$

Dem Zeitpunkt $t = 0$ in der alten Zeitrechnung entspricht der Zeitpunkt $\tau = -\infty$ in der neuen. Schreibt man (5.19) für Zeitpunkte $\tau_1(t_1)$ und $\tau_2(t_2)$ hin und eliminiert dann zwischen diesen Gleichungen sowie (5.15) bis (5.17) die Konstanten t_1, t_2, τ_1, τ_2, so folgt

$$t = t_0 \, e^{\frac{\tau - t_0}{t_0}} \operatorname{Cos} \frac{\lambda}{t_0}; \qquad l = t_0 \, e^{\frac{\tau - t_0}{t_0}} \operatorname{Sin} \frac{\lambda}{t_0}. \tag{5.20}$$

An die Zeit τ und die Transformation (5.20) knüpfen sich bei MILNE weitgehende Folgerungen und Ansprüche.

Bildet man die Geschwindigkeiten $d\lambda/d\tau$ und dl/dt, so hängen sie nach (5.20) zusammen gemäß

$$\frac{d\lambda}{d\tau} = \frac{\frac{dl}{dt} - \operatorname{Tan} \frac{\lambda}{t_0}}{1 - \frac{dl}{dt} \cdot \operatorname{Tan} \frac{\lambda}{t_0}} = \frac{\frac{dl}{dt} - \frac{l}{t}}{1 - \frac{l}{t} \frac{dl}{dt}}. \tag{5.21}$$

Diese Formel leistet die Umrechnung irgendeiner Geschwindigkeit im (t, l)-System in das neue (τ, λ)-System. Wir haben gesehen, daß aber

$$\frac{dl}{dt} = \frac{l}{t} \tag{5.22}$$

die Geschwindigkeit einer Fundamentalpartikel ist relativ zum Ursprung des Systems (t, l). Dann folgt aus (5.21), daß die Geschwindigkeit der gleichen Partikel im System (τ, λ) verschwindet. Das System (τ, λ) ist also so gewählt, daß in ihm alle Fundamentalbeobachter ruhen. Man bringt den Sachverhalt in eine etwas allgemeinere Form. wenn man an Stelle von (l, t) die Koordinaten x, y, z, t benutzt und sie transformiert in $\lambda, \varphi, \psi, \tau$ gemäß den Gleichungen

$$\left.\begin{aligned} x &= t_0 \, e^{\frac{\tau - t_0}{t_0}} \operatorname{Sin} \frac{\lambda}{t_0} \sin\varphi \sin\psi, & y &= t_0 \, e^{\frac{\tau - t_0}{t_0}} \operatorname{Sin} \frac{\lambda}{t_0} \sin\varphi \cos\psi, \\ z &= t_0 \, e^{\frac{\tau - t_0}{t_0}} \operatorname{Sin} \frac{\lambda}{t_0} \cos\varphi, & t &= t_0 \, e^{\frac{\tau - t_0}{t_0}} \operatorname{Cos} \frac{\lambda}{t_0}. \end{aligned}\right\} \tag{5.23}$$

Damit wird

$$X = t^2 - x^2 - y^2 - z^2 = t_0^2 \, e^{\frac{(\tau - t_0)}{t_0}}, \tag{5.24}$$

und die invariante Dichte $\varrho(X)$ hängt nicht mehr vom Ort, sondern nur noch von der Zeit τ ab. Sie ist in den neuen Koordinaten stets räumlich homogen. Aus der Tatsache, daß das Produkt $\varrho(X)\, dV$ überhaupt zeitunabhängig ist, wird die Berechtigung hergeleitet zu sagen „die" Dichte im $(\lambda, \varphi, \psi, \tau)$-System sei stationär.

MILNE nennt das für jeden Fundamentalbeobachter andere System der x, y, z, t ein „privates", das der $\lambda, \varphi, \psi, \tau$ ein „öffentliches" Koordinatensystem. Er behauptet (vgl. [9], Kap. VIII, wo viele Einzelheiten diskutiert werden), in dem dargestellten Sachverhalt eine zweifache Möglichkeit zur Interpretation der Rotverschiebung der Nebel zu haben: In der Sprache der privaten Koordinaten handele es sich um eine echte Expansion in Verbindung mit dem gewöhnlichen Doppler-Effekt. Ein Photon habe nach dieser Interpretation eine Frequenzänderung erlitten infolge der Relativbewegung von Lichtquelle und Beobachter.

In der Sprache der öffentlichen Koordinaten dagegen handele es sich um ein stationäres Weltmodell, die Expansion sei verschwunden; die Atome aber liefen nach der t-Zeit und nicht nach der τ-Zeit; ein ankommendes Photon berichte uns von der geringeren Frequenz der Atomuhren zur Zeit der Emission.

ε) Beziehungen zur Einsteinschen Kosmologie. Das ganze Werk von MILNE ist durchzogen von der Tendenz, sich scharf abzugrenzen gegen die Kosmologie der Einsteinschen Gravitationstheorie. Doch ist es zweckmäßig, Verbindungen, die sich immer noch ziehen lassen, aufzuzeigen. Wir greifen hier nur die Frage der zwei Zeitskalen auf und prüfen den Unterschied gegen die gewöhnliche Kosmologie. In der Sprache der Einsteinschen Gravitationstheorie ist es üblich, eine Weltmetrik

$$ds^2 = g_{\mu\nu}\, dx^\mu\, dx^\nu, \qquad \mu, \nu = 0, 1, 2, 3$$

dann und nur dann stationär zu nennen, wenn die $g_{\mu\nu}$ nicht von der Zeitkoordinate x^0 abhängen. Wenn MILNE daher die Transformation (5.23) verwendet, um die private Metrik (5.8) umzurechnen in

$$ds^2 = e^{\frac{2(\tau - t_0)}{t_0}} \left[d\tau^2 - d\lambda^2 - t_0^2 \operatorname{Sin}^2 \frac{\lambda}{t_0} (d\varphi^2 + \sin^2\varphi\, d\psi^2)\right], \tag{5.25}$$

so würde man diese als nichtstationär bezeichnen müssen wegen des Exponentialfaktors, der von τ abhängt. Besteht man aber auf der Milneschen Terminologie und nennt die Metrik (5.25) statisch, weil die eckige Klammer zeitunabhängig ist, so kann die gewöhnliche Kosmologie mit der Metrik

$$ds^2 = dt^2 - R^2(t)\, d\sigma^2; \qquad d\sigma^2 = \gamma_{ik}(x^1, x^2, x^3)\, dx^i\, dx^k, \qquad i, k = 1, 2, 3, \tag{5.26}$$

durch die Transformation

$$\tau = \int \frac{dt}{R(t)} \tag{5.27}$$

in die Form

$$ds^2 = R^2(\tau)\, [d\tau^2 - d\sigma^2] \tag{5.28}$$

gebracht werden, die nun konsequenterweise wiederum (nach MILNE) „statisch" genannt werden müßte. Es ist klar, daß also in der Frage der zwei Zeitskalen die Milnesche Theorie nicht mehr und nicht weniger leistet als die Kosmologie der Einsteinschen Gravitationstheorie.

Sollte irgendwann einmal gefunden werden, daß die Perioden atomarer und planetarischer Uhren nicht in einem konstantem Verhältnis stehen, so ist damit gewiß eine wichtige Erkenntnis über die physikalische Natur beider Arten von Systemen verbunden. Die Milnesche Kosmologie wird dadurch aber ebensowenig bestätigt oder widerlegt wie die Einsteinsche Gravitationstheorie.

Die Frage der Lichtfortpflanzung findet man bei MILNE ([*9*], Kap. VIII) ausführlich behandelt. Doch ist die Milnesche Theorie den neueren Beobachtungsdaten an extragalaktischen Nebeln (vgl. den Artikel von MCVITTIE) nicht gegenübergestellt worden.

6. Mathematischer Anhang. Die Strömung des Weltsubstrates werde in der Milneschen Kosmologie beschrieben durch einen Vierervektor u_μ (μ läuft von 1 bis 4 mit $x_4 = it$, also $c = 1$). Abstatt der Gültigkeit der Gl. (5.4) etc. des Textes mit der nachträglichen Forderung der Isotropie setzen wir hier fest, daß die Äquivalenz der mit dem Substrat schwimmenden Beobachter sich darin äußern soll, daß der Vektor u_μ gegen die Transformationen der homogenen Lorentz-Gruppe

$$x'_\mu = a_{\mu\nu} x_\nu, \qquad a_{\mu\nu} a_{\lambda\nu} = \delta_{\mu\lambda}, \qquad \operatorname{Det} \|a_{\mu\nu}\| = +1, \qquad a_{44} > 0 \tag{6.1}$$

invariant sei. Diese Forderung läßt sich mathematisch formulieren durch

$$u'_\mu(x_\lambda) = u_\mu(x_\lambda). \tag{6.2}$$

Diese Gleichung sagt aus, daß der transformierte Vektor u'_μ als Funktion seiner Koordinaten betrachtet die gleiche funktionale Abhängigkeit besitzt wie der ursprüngliche Vektor. Für den Beweis genügt es, sich auf infinitesimale Koordinatentransformationen zu beschränken. Wegen der Orthogonalität der Transformationsmatrix in (6.1) lauten die sechs verschiedenen durch das schiefsymmetrische Indicespaar $\underset{\smile}{\alpha\beta}$ gekennzeichneten infinitesimalen homogenen Lorentz-Transformationen

$$\delta x_\mu = x'_\mu - x_\mu = \varepsilon\, \xi_{\mu \underset{\smile}{\alpha\beta}} = \varepsilon\,(\delta_{\mu\alpha}\,\delta_{\nu\beta} - \delta_{\mu\beta}\,\delta_{\nu\alpha})\, x_\nu. \tag{6.3}$$

In dieser Gleichung ist ε eine infinitesimale Größe. Wenn wir für den Augenblick die Indices α und β an den sog. Killing-Vektoren $\xi_{\mu\underset{\smile}{\alpha\beta}}$ fortlassen, besitzen die infinitesimalen Transformationen die Form

$$\delta x_\mu = \varepsilon\, \xi_\mu. \tag{6.4}$$

Wenn man Glieder in ε von höherer als erster Ordnung vernachlässigt, gilt für eine infinitesimale Transformation

$$u_\mu(x_\lambda) = \frac{\partial x'_\nu}{\partial x_\mu}\, u'_\nu\,[x'_\varrho(x_\lambda)] = (\delta_{\nu\mu} + \varepsilon\, \xi_{\nu|\mu})\, u'_\nu(x_\lambda + \varepsilon\, \xi_\lambda). \tag{6.5}$$

Daraus folgt mit (6.2)

$$u_{\mu|\lambda}\, \xi_\lambda + u_\lambda\, \xi_{\lambda|\mu} = 0. \tag{6.6}$$

Wenn wir in diese Gleichung für die sechs verschiedenen Killing-Vektoren $\xi_{\lambda\underset{\smile}{\alpha\beta}}$ die Ausdrücke aus (6.3) einführen, erhalten wir das folgende System von 24 partiellen Differentialgleichungen

$$u_{\mu|\alpha}\, x_\beta - u_{\mu|\beta}\, x_\alpha = u_\beta\, \delta_{\alpha\mu} - u_\alpha\, \delta_{\beta\mu} \tag{6.7}$$

für die vier Funktionen u_μ.

Wenn wir jetzt die Summationsvorschrift außer Kraft setzen und in diese Gleichungen ohne Einschränkung der Allgemeinheit den Ansatz

$$u_\mu = x_\mu\, \Phi_\mu\,(x_1^2, x_2^2, x_3^2, x_4^2) \tag{6.8}$$

einführen, erhalten wir

$$2\, x_\mu\, x_\alpha\, x_\beta \left(\frac{\partial \Phi_\mu}{\partial x_\alpha^2} - \frac{\partial \Phi_\mu}{\partial x_\beta^2}\right) = \delta_{\mu\alpha}\, x_\beta (\Phi_\beta - \Phi_\mu) - \delta_{\mu\beta}\, x_\alpha (\Phi_\alpha - \Phi_\mu). \tag{6.9}$$

Falls α und β von μ verschieden sind, folgt

$$\frac{\partial \Phi_\mu}{\partial x_\alpha^2} = \frac{\partial \Phi_\mu}{\partial x_\beta^2}. \tag{6.10}$$

Falls sodann $\alpha = \mu$ und $\beta \neq \mu$ ist, lauten die Gln. (6.9)

$$2\, x_\mu\, x_\mu \left(\frac{\partial \Phi_\mu}{\partial x_\mu^2} - \frac{\partial \Phi_\mu}{\partial x_\beta^2}\right) = \Phi_\beta - \Phi_\mu. \tag{6.11}$$

in Verbindung mit (6.10) folgt aus diesen Gleichungen, daß die Funktionen Φ_μ gleich einer Funktion Φ sind. Demgemäß gilt

$$\frac{u_k}{u_4} = \frac{x_k}{i t} = -i\, v_k\,; \qquad v_k = \frac{x_k}{t}. \tag{6.12}$$

Andererseits kann man leicht bestätigen, daß jeder Vektor der Form

$$u_\mu = x_\mu \Phi\left(\sum_{\lambda=1}^{4} x_\lambda^2\right) \tag{6.13}$$

auch gegenüber den Transformationen der homogenen Lorentz-Gruppe invariant ist. Damit ist der im Text behauptete Satz von E. MILNE bewiesen.

Literatur.

[1] EDDINGTON, A.: Fundamental Theory. Cambridge 1945. — Relativity Theory of Electrons and Protons. Cambridge 1936. — SLATER, N.: The Development and Meaning of EDDINGTON's Fundamental Theory. Cambridge 1956.

[2] DIRAC, P.: Nature, Lond. **139**, 323 (1937). — Proc. Roy. Soc. Lond., Ser. A **165**, 199 (1938).

[3] JORDAN, P.: Naturwiss. **25**, 513 (1937); **26**, 417 (1938).

[4] BERGMANN, P.: Artikel über allgemeine Relativitätstheorie in Band IV dieses Handbuches.

[5] JORDAN, P.: Schwerkraft und Weltall, 2. Aufl. Braunschweig 1955.

[6] HOYLE, F.: Monthly Notices Roy. Astronom. Soc. London **108**, 372 (1948); **109**, 365 (1949).

[7] BONDI, H., and T. GOLD: Monthly Notices Roy. Astronom. Soc. London **108**, 252 (1948). BONDI, H.: Cosmology. Cambridge 1952.

[8] MILNE, E.: Relativity, Gravitation and World-Structure. Oxford 1935.

[9] MILNE, E.: Kinematic Relativity. Oxford 1948.

Zum ganzen Problemkreis der Kosmologie vergleiche man auch den Bericht der Solvay-Konferenz 1958.

Sachverzeichnis.

(Deutsch-Englisch.)

Bei gleicher Schreibweise in beiden Sprachen sind die Stichwörter nur einmal aufgeführt.

Subject Index.

(English-German.)

Where English and German spelling of a word is identical the German version is omitted.

MIX
Papier aus verantwortungsvollen Quellen
Paper from responsible sources
FSC® C105338

If you have any concerns about our products,
you can contact us on
ProductSafety@springernature.com

In case Publisher is established outside the EU,
the EU authorized representative is:
Springer Nature Customer Service Center GmbH
Europaplatz 3, 69115 Heidelberg, Germany

Printed by Libri Plureos GmbH
in Hamburg, Germany